Joint International Conference of The 2nd International Conference on Industrial and System Engineering (IConISE) and The 7th Annual Conference on Industrial and System Engineering (ACISE)

IOP Conference Series: Materials Science and Engineering
Volume 1072

Surabaya, Indonesia
22-23 July 2020

ISBN: 978-1-7138-4579-9
ISSN: 1757-8981

Printed from e-media with permission by:

Curran Associates, Inc.
57 Morehouse Lane
Red Hook, NY 12571

Some format issues inherent in the e-media version may also appear in this print version.

Institute of Physics
Dirac House, Temple Back
Bristol BS1 6BE UK

Phone: 44 1 17 929 7481
Fax: 44 1 17 920 0979

techtracking@iop.org

Additional copies of this publication are available from:

Curran Associates, Inc.
57 Morehouse Lane
Red Hook, NY 12571 USA
Phone: 845-758-0400
Fax: 845-758-2633
Email: curran@proceedings.com
Web: www.proceedings.com

TABLE OF CONTENTS

IOP Conf. Series: Materials Science and Engineering 1072 (2021) 011001 doi:10.1088/1757-899X/1072/1/011001

Joint International Conference of
The 2nd International Conference on Industrial and System Engineering (IConISE)
and
The 7th Annual Conference on Industrial and System Engineering (ACISE)

Basic Information

IConISE is a biannual international conference organized by the Industrial and Systems Engineering Department of Institut Teknologi Sepuluh Nopember, Surabaya. At the same time, ACISE is an international conference which has been conducted annually by the Industrial Engineering Department of Diponegoro University.

The theme of this joint conference is "Big Data Analytics: A competitive Tools toward Smart Industrial System". As may all of us realize, in the last decade, industries experienced a massive shift towards digitalization. As the benefit of digitalization, a vast amount of data could be collected almost effortlessly throughout the operational activities within companies and organizations. Due to the advancing information technologies, a company or organization also can access a huge amount of external data. The internal and external big data can be capitalized further to improve the effectiveness and efficiency of the industrial systems or in another word, make the industrial systems smarter.

This fascinating forum is dedicated to discussing the role and issues of big data analytics in every significant aspect of the industrial system. The joint organizing committee has received over 200 submissions from Indonesia, Malaysia, Italy, Philippines, Taiwan, United Kingdom and Japan. After the double-blind review process, 149 papers were accepted, then presented and discussed in the parallel session. Selected papers were then published in IOP Conference Series: Materials Science and Engineering, a Scopus-indexed proceeding.

IOP Conf. Series: Materials Science and Engineering 1072 (2021) 011001 doi:10.1088/1757-899X/1072/1/011001

Committee Information

Advisory Committee:
- Nurhadi Siswanto, S.T., MSIE, Ph.D, Head of Industrial and System Engineering Dept. ITS
- Dr. Ratna Purwaningsih, S.T., M.T., Head of Industrial Engineering Dept. Diponegoro University

IConISE 2020 Chair: Ratna Sari Dewi, S.T., M.T., Ph.D
ACISE 2020 Chair: Dr. Manik Mahachandra, S.T., M.Sc
Secretary:
- Dr.-Ing Novie Susanto, S.T., M.Eng
- Retno Widyaningrum, S.T., MBA., Ph.D

Finance & Registration Team:
- Anny Maryani, S.T., M.T.
- Diesta Iva Maftuhah, S.T., M.T
- Suko Endri Royaningtyas, S.E.

Program Team:
- Dr. Sri Hartini, S.T., M.T.
- Dewanti Anggrahini, S.T., M.T.
- Yudha Prasetyawan, S.T., M.Eng.

Paper Team:
- Prof. Dr. Aries Susanty, S.T., M.T.
- Dr. Ir. Heru Prastawa, DEA
- Dr. Naniek Utami Handayani, S.Si
- Dr.Eng. Erwin Widodo
- Dyah Santhi Dewi, S.T., M.Eng.Sc, Ph.D
- Nani Kurniati, Ph.D
- Niniet Indah Arvitrida, S.T., M.T., Ph.D
- Putu Dana Karningsih S.T., M.Eng.Sc, Ph.D

Publication Team:
- Dr. Denny Nurkertamanda, S.T., M.T.
- Arief Rahman, S.T., M.Sc
- Atikah Aghdhi Pratiwi, S.T., M.T.

Sponsorship Team:
- Dr. Adithya Sudiarno, S.T., M.T.
- Dr. Purnawan Adi Wicaksono, S.T., M.T.
- Dody Hartanto, S.T., M.T.
- Niken Anggraini Savitri, S.T., M.T.
- Rindi Kusumawardani, S.Si., M.Sc.

IOP Conf. Series: Materials Science and Engineering 1072 (2021) 011001 doi:10.1088/1757-899X/1072/1/011001

ICONISE-ACISE 2020 IOP Publishing
IOP Conf. Series: Materials Science and Engineering 1072 (2021) 011002 doi:10.1088/1757-899X/1072/1/011002

Joint International Conference of

The 2nd International Conference on Industrial and System Engineering (IConISE)

and

The 7th Annual Conference on Industrial and System Engineering (ACISE)

Peer review declaration

All conference organisers/editors are required to declare details about their peer review. Therefore, please provide the following information:

- Type of peer review: Double-blind
  The author(s) were not informed who their reviewers are, as well as the two reviewers were also not known about the author(s).

- Conference submission management system: Easychair (https://easychair.org/my/conference?conf=iconiseacise2020#)

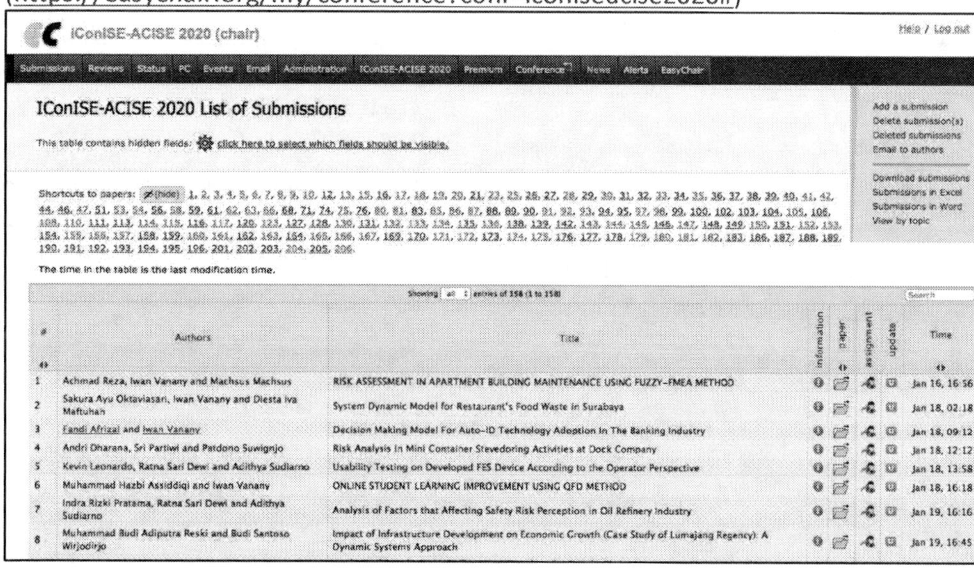

- Number of submissions received: 206 (two hundred and six)

- Number of submissions sent for review: 160 (one hundred and sixty)
  The rest of 46 submissions were abstracts, therefore were not continued to the reviewers.

- Number of submissions accepted: 150 (one hundred and fifty)

- Acceptance Rate (Number of Submissions Accepted / Number of Submissions Received X 100): 150 / 206 x 100 = 72.81%

IOP Conf. Series: Materials Science and Engineering 1072 (2021) 011002 doi:10.1088/1757-899X/1072/1/011002

- Average number of reviews per paper: 2 (two)

- Total number of reviewers involved: 20 (twenty)

- Any additional info on review process:
    - The paper committee assigned a submission to any reviewer outside the author's own institution.
    - Some of the submissions with significantly different results in review (for example got accept and reject decision at the same time), were then reviewed by the third additional reviewer.
    - Attached below the review evaluation example.

- Contact person for queries: Dr. Manik Mahachandra (manik.mahachandra@ft.undip.ac.id, +62-878-8191-8305)

IConISE-ACISE 2020 (chair)

Help / Log out

Submissions | Reviews | Status | PC | Events | Email | Administration | IConISE-ACISE 2020 | Premium | Conference | News | Alerts | EasyChair

Add Review on IConISE-ACISE 2020 Submission 206

View submission
Add associate

To ensure independent and unbiased reviews, EasyChair does not show reviews of other PC members to a PC member who is assigned this submission but did not enter her or his review yet. As a consequence of this policy, if you enter a review and then revise it so that the score changes, EasyChair will show both the original and the revised reviews.

For this reason, please **make sure that the first review you enter for the submission is a valid review and not a draft** and that the scores you give to the submission are the intended scores. If you revise a review and change the scores, we recommend to add a note for PC members explaining the reason for changing the scores.

Submission and Reviewer Information

Title: Understanding the Effects of Satisfaction and Trust on E-Repeat Purchase Intention: An Ergonomic Cognitive Approach

Authors: Dyah Santhi Dewi, Adithya Sudiarno, Ratna Sari Dewi and Joshua Aritonang

PC member: Manik Mahachandra

Enter subreviewer information. Leave these fields empty if reviewed by yourself

Click on "Add associate" in the context menu to add your associate as a reviewer.

First name: *

Last name: *

Email: *

Evaluation

TECHNICAL CRITERIA

Scientific merit. * Notably scientific rigor, accuracy, and correctness.

- 5: excellent
- 4: good
- 3: fair
- 2: poor
- 1: very poor

Communication. * Clarity of expression, communication of ideas, readability, and discussion of concepts.

- 5: excellent
- 4: good
- 3: fair
- 2: poor
- 1: very poor

Discussion. * Sufficient discussion of the context of the work and suitable referencing.

- 5: excellent
- 4: good
- 3: fair
- 2: poor
- 1: very poor

QUALITY CRITERIA

Motivation. * Clearly problem statement and motivation; Scientific interest of the results.

- 5: excellent
- 4: good
- 3: fair
- 2: poor
- 1: very poor

Originality. * Manuscript has minimum repetition; Significant parts of the manuscripts has not already been published elsewhere.

- 5: excellent
- 4: good
- 3: fair
- 2: poor
- 1: very poor

Length. * Content of the work has sufficient length to express the scientific interest.

- 5: excellent
- 4: good
- 3: fair
- 2: poor
- 1: very poor

PRESENTATION CRITERIA

Title. * Adequacy and appropriate content of the article.

- 5: excellent
- 4: good
- 3: fair
- 2: poor
- 1: very poor

Abstract. * Contain essential information; Complete (contain background, objective, method, result).

- 5: excellent
- 4: good
- 3: fair
- 2: poor
- 1: very poor

Figure and table. * Diagrams, figures, tables, and captions are essential and clear.

- 5: excellent
- 4: good
- 3: fair
- 2: poor
- 1: very poor

Text and mathematics. * All text are brief but still clear.

- 5: excellent
- 4: good
- 3: fair
- 2: poor
- 1: very poor

Conclusion. * Manuscript contain a carefully written conclusion, summarizing what has been learned and why it is interesting and useful.

- 5: excellent
- 4: good
- 3: fair
- 2: poor
- 1: very poor

REVIEW RESULT

Overall evaluation. * You may provide a detailed review, including a justification for your scores.

- 5: Accept with no revisions
- 4: Accept with minor revisions
- 3: Accept with major revisions
- 2: Borderline paper
- 1: Strong reject

Comment for author. * Please give short explanation of your review judgement here.

IOP Conf. Series: Materials Science and Engineering 1072 (2021) 011002 doi:10.1088/1757-899X/1072/1/011002

Reviewer's confidence. * Mark your confidence level in reviewing the manuscript, based on your expertise.

- 5: Expert
- 4: High
- 3: Medium
- 2: Low
- 1: None

Confidential remarks for the program committee. If you wish to add any remarks intended only for PC members please write them below. These remarks will only be seen by the PC members having access to reviews for this submission. They will not be sent to the authors. This field is optional.

Submit review

IOP Conf. Series: Materials Science and Engineering 1072 (2021) 012001 doi:10.1088/1757-899X/1072/1/012001

Risk assessment for apartment building maintenance using Fuzzy-FMEA methods

A Reza[1*], I Vanany[2] and Machsus[3]

[1]Department of Technology Management, Institut Teknologi Sepuluh Nopember, Surabaya, Indonesia.
[2]Department of Industrial and System Engineering, Institut Teknologi Sepuluh Nopember, Surabaya, Indonesia.
[3]Department of Civil Infrastructure Engineering, Institut Teknologi Sepuluh Nopember, Surabaya, Indonesia.

Email: *Achmadreza1995@gmail.com

Abstract. In apartment buildings, damage often occurs during maintenance. This study aims to identify risks, estimate risks, and provide mitigation actions for risks in apartment building maintenance. Types of risks that often occur, namely for the kind of work (1) civil and (2) MEP (Mechanical Electrical and Plumbing). Risk assessment in this research uses the Fuzzy-FMEA method (Failure mode and effect analysis). The results of risk assessment, the most critical risk is the risk of damage to marble floors with an FRPN value of 64.7. By using the FMEA method, as we know that there are seven causes of marble floor damage, namely (1) no standard work procedures, (2) no employee training, (3) unhealthy workers, (4) drugs influence, (5) no there is an inspection at the time of handover, (6) no routine checklist, and (7) strong winds. Some mitigation actions are also describes in this paper.

1. Introduction

In building management, risk management is one of the important issues to the risk event can be controlled, accepted/paid for, or transferred [1]. The managing risk in building management will be effective if it implies the risk assessment method including determining faults of reach risk events. Risk assessment to evaluate risk events in order to determine the mitigation plans is important for company managers. Building management that applies risk assessment activity will have a level of awareness and readiness in dealing with the potential of risks. [1] suggested that risk could be accepted with proper building maintenance.

FMEA (Failure mode and effect analysis) has become a popular method and extensively used in many areas such as nuclear, maintenance waste, engine, aerospace, and others [2, 3, 4]. Authors [4] believed that the results of FMEA analysis could be used to provide information for decision making in risk management. Fuzzy in the FMEA approach has the advantage in conducting risk evaluation and prioritization based on the preferences of the experts [5]. Fuzzy-FMEA has also become extensively used in several industries. Authors [6] use Fuzzy-FMEA to analyse volume and chemical control systems in a nuclear power plant. Authors [7] developed the fuzzy-FMEA technique for estimating risk in turbo stems for a diesel engine. Authors [8] also use fuzzy-FMEA to rank RPN values from the results of evaluating risk factors for severity, occurrence, and detection as input variables of the fuzzy system.

In an apartment building, there are many uncertainties in the occurrence of a risk, according to authors [1,9], uncertainties are a state of mind filled with doubt. In reality, the building management has made various efforts such as doing routine maintenance for each building part. This does not provide a significant impact on minimizing the risks that occurred. Building management must carry out risk mitigation efforts to focus on minimizing risks. Risk identification also purposes to find risks that have not arisen but have the potential to become risks. The risk that arises also affects the increasing complaints from residents and guests. This study aims to (a) identify the risk event in an apartment building maintenance, (b) asses the all-risk events with the Fuzzy-FMEA method, and (c) determine risk mitigation action. The problem in this research is how a company can conduct a risk assessment to prevent risks from happening. This research is significant to be doing to provide recommendations for companies not to let a risk occur repeatedly and cause significant losses for the company.

2. Research Method

The application of the Fuzzy-FMEA approach for risk assessment to reduce risk in an apartment building department consists of several research stages. A previous research related with research stages and case study was used as base to determine research stages [1,10,11,12, 13]. A case study in apartment "A" was conducted to understand the applicability of the fuzzy-FMEA approach for risk assessment. This research was conducted with three steps, namely (1) data collection, (2) risks identification, (3) risk assessment, and (4) risk mitigation action. Several steps are needed to complete each phase. Fuzzy method is used because there is a number of uncertainties or doubts of the severity. The severity is different for each incident. Fuzzy can analyse the uncertainty and then produce output in the form of an FRPN values.

2.1. Data Collection

In this research, data collection was carried out in six months of 2019 years. Data collection method used by taking primary data from the daily operational report of the apartment building and interview with engineering supervisor to determining the score of detection parameter. The Figure 1 showed the flow of data collection.

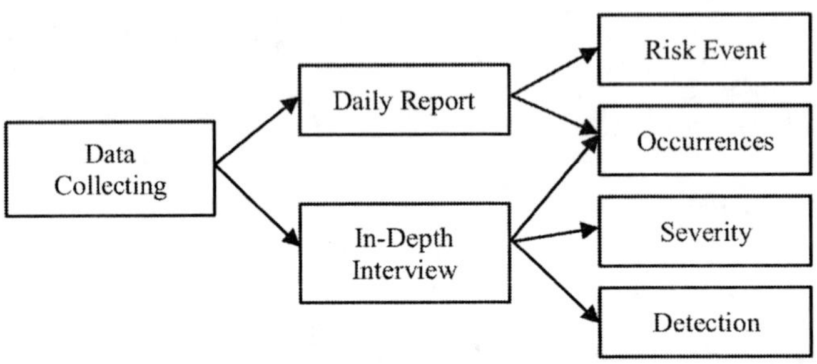

Figure 1. Research data collecting flowchart.

IOP Conf. Series: Materials Science and Engineering 1072 (2021) 012001 doi:10.1088/1757-899X/1072/1/012001

2.2. Risk Identification

Risk identification was carried out by the daily operational reports of an apartment buildings. Risk identification results are dividing into two types, which are civil and MEP. The Table 1 shows the failure risk and observation results in case study.

Table 1. Result of risk identification in an apartment building.

Risk Code		Risk Event	Observations Results		
			Occurrence	**Severity (IDR)**	**Detection**
Civil	C1	Cracked Ceiling	58.00	500,000.00	460.00
	C2	Broken Main Door	57.00	110,000.00	115.00
	C3	Loose Parquet	22.00	700,000.00	605.00
	C4	Cracked Wall	69.00	700,000.00	845.00
	C5	Torn Wallpaper	12.00	300,000.00	340.00
	C6	Marble Damage	10.00	2,000,000.00	1110.00
	C7	Broken Ceramic	17.00	300,000.00	555.00
	C8	Broken Glass	16.00	1,200,000.00	630.00
	C9	Balcony frame broken	6.00	450,000.00	390.00
	C10	Cracked Natural Stone	32.00	450,000.00	220.00
MEP	M1	AC temperature is not appropriate	77.00	300,000.00	220.00
	M2	Electrical installation damage	82.00	300,000.00	255.00
	M3	Lights Out	98.00	130,000.00	95.00
	M4	Clogged Floor Drain	53.00	100,000.00	165.00
	M5	Broken at Building automation system	16.00	150,000.00	185.00
	M6	Sink damage	40.00	200,000.00	305.00
	M7	Broken Urinal/Closet	40.00	500,000.00	245.00
	M8	Cleaning Air Conditioner	84.00	100,000.00	135.00
	M9	Lack of Electrical power	10.00	1,200,000.00	165.00
	M10	Damage at Sky pool area	19.00	150,000.00	160.00
	M11	Clogged Grease trap	15.00	400,000.00	245.00
	M12	AHU/FCU damage	15.00	500,000.00	185.00
	M13	Gas installation leakage	16.00	100,000.00	105.00
	M14	Broken Exhaust fan/blower	11.00	500,000.00	185.00
	M15	Sprinklers/Smoke detector/hydrant	22.00	500,000.00	330.00

The next step for each risk event is measured S, O, and D values from the daily report and also interviews with the engineer.

2.3. Measurement and Data Processing

Data collected from several sources, such as daily reports and interview results, are then measured, and is the values of each parameter. The next step in data processing is changing the value of each parameter into a value on a scale that has been determined for the input parameter on the step calculating RPN and FRPN. Change the score with interpolation formula and Calculation RPN with FMEA classical method. The Table 2 shows the score of FMEA parameter for each risk event. Calculation RPN with FMEA classical method is multiplying severity, occurrence, and detection.

$$RPN = S \times O \times D \quad (1)$$

Calculation of the scale value by using the interpolation formula to change the value of the input parameter from the actual value to the scale value of area of the fuzzy input one to five for the input. The input detection value has a scale value of one to four. The following is an example of calculating the scale value for the ceiling damage occurrence input parameter.

Table 2. The score of FMEA parameter for each risk event.

Risk Code		**Parameter**			**RPN**
		Occurrences	**Severity**	**Detection**	
Civil	C1	4.3	1.83	2.02	15.90
	C2	4.24	1	1	4.24
	C3	2.02	2.25	2.72	12.36
	C4	5	2.25	3.11	34.99
	C5	1.38	1.4	1.83	3.54
	C6	1.25	5	4	25.00
	C7	1.7	1.4	1.72	4.09
	C8	1.63	3.31	3.22	17.37
	C9	1	1.72	2.83	4.87
	C10	2.65	1.72	1.63	7.43
MEP	M1	4.05	1.73	2.44	17.10
	M2	4.27	1.73	2.5	18.47
	M3	5	1.11	1	5.55
	M4	2.95	1	1.66	4.90
	M5	1.27	1.18	2.38	3.57
	M6	2.36	1.36	3.5	11.23
	M7	2.36	2.45	2.94	17.00
	M8	4.36	1	1.38	6.02
	M9	1	5	1.66	8.30
	M10	1.41	1.18	1.88	3.13
	M11	1.23	2.09	2.94	7.56
	M12	1.23	2.45	2.38	7.17
	M13	1.27	1	1.09	1.38
	M14	1.05	2.45	2.38	6.12
	M15	1.55	2.45	4	15.19

2.4. Fuzzy-FMEA design

Fuzzy-FMEA design using MATLAB Software. The system has three input parameters like severity, occurrence, and detection, and one output is fuzzy risk priority number. After designing inputs and outputs, fuzzy rules are making from values determined by researchers with a total are a hundred rules. The Table 3 shows the fuzzy rules in risk assessment.

Table 3. Fuzzy rules.

D=VL						
RPN		**F**				
		VL	**L**	**M**	**H**	**VH**
S	VL	VL	VL	VL	VL	VL
	L	VL	VL	L	L	L
	M	VL	L	M	M	M
	H	L	M	M	M	H
	VH	L	M	M	H	H
D=L						
RPN		**F**				
		VL	**L**	**M**	**H**	**VH**
S	**VL**	VL	VL	L	L	L
	L	L	L	M	M	M
	M	L	M	M	M	H
	H	M	M	H	H	H
	VH	M	H	H	H	H
D=M						
RPN		**F**				
		VL	**L**	**M**	**H**	**VH**
S	**VL**	L	L	L	M	M
	L	L	L	M	M	M
	M	L	M	M	H	H
	H	M	M	H	H	H
	VH	M	H	H	H	VH
D=H						
RPN		**F**				
		VL	**L**	**M**	**H**	**VH**
S	**VL**	L	L	M	M	M
	L	L	M	M	M	M
	M	M	M	H	H	H
	H	M	H	H	VH	VH
	VH	H	H	VH	VH	VH

FRPN value calculation manually by inputting the value of each parameter into the fuzzy rules that have been made.

Table 4. FRPN score for each risk event in an apartment building.

Risk Code		Severity	Occurrence	Detection	FRPN
Civil	C1	1.83	4.3	2.02	26.20
	C2	1.00	4.24	1	7.52
	C3	2.25	2.02	2.72	26.60
	C4	2.25	5	3.11	45.00
	C5	1.40	1.38	1.83	8.20
	C6	5.00	1.25	4	64.70
	C7	1.40	1.7	1.72	8.41
	C8	3.31	1.63	3.22	45.00
	C9	1.72	1	2.83	14.10
	C10	1.72	2.65	1.63	14.10
MEP	M1	1.73	4.05	2.44	41.60
	M2	1.73	4.27	2.5	45.00
	M3	1.11	5	1	7.52
	M4	1.00	2.95	1.66	15.00
	M5	1.18	1.27	2.38	21.70
	M6	1.36	2.36	3.5	25.00
	M7	2.45	2.36	2.94	38.40
	M8	1.00	4.36	1.38	7.52
	M9	5.00	1	1.66	32.30
	M10	1.18	1.41	1.88	7.99
	M11	2.09	1.23	2.94	25.00
	M12	2.45	1.23	2.38	23.50
	M13	1.00	1.27	1.09	7.52
	M14	2.45	1.05	2.38	23.50
	M15	2.45	1.55	4	25.00

The table 4 describes the highest RPN for civil and MEP risk event are risk code C2 and M2. Risk code C2 is for risk event Marble damage, and M2 is for risk event in electrical installation damage.

3. Results and Discussions

In this study, the result of the calculation of the value FRPN with the fuzzy-FMEA method can be used as a reference value, which states that it can be the most critical risk if it has the highest FRPN value. Following the result analysis of all fields both Civil and MEP to obtain the most critical risks that must be immediately handled by Building management in case study. The average scores for work categories each element and the ranking of risk event FPRN value in the civil field is shown in Table 5 and Table 6.

Table 5. The average score of FMEA parameter for each risk types.

Element	Average score	
	Civil	MEP
Severity	2.19	1.88
Occurrence	2.52	2.36
Detection	2.41	2.28
RPN	12.97	8.85
FRPN	25.98	23.10

Table 6. Ranking of risk event FRPN value in the civil field.

Code of Risk Event		FRPN	Ranking
Civil	C1	26.20	5
	C2	7.52	10
	C3	26.60	4
	C4	45.00	2
	C5	8.20	9
	C6	64.70	1
	C7	8.41	8
	C8	45.00	3
	C9	14.10	6
	C10	14.10	7

In the civil risk-ranking (see Table 6), the highest rating is obtained on the C6 risk code, namely broken or cracked marble with an FRPN value of 64.70. This risk is state as a critical risk in case study. For this reason, risk mitigation is needed to reduce the level of occurrence or severity. The Figure 2 showed the fault three analysis in a marble damage.

Making fault tree analysis is based on the causes of each top event, and the causes varied as technician errors and lack of experience from cleaning service. There are also causes of natural factors, like strong winds that make the marble fall and break. following the relationship between causes and top events.

In marble damage, there are six cut sets and seven basic events, where the damage is caused by collision or release. Top event collision has two cut sets, which are less experienced technicians because there are no SOPs and no training, or human error due to lack of health and emotional factors. Whereas the top event apart has two cut sets, namely strong winds and initial installation, because there is no checking during the handover or there is no routine checklist.

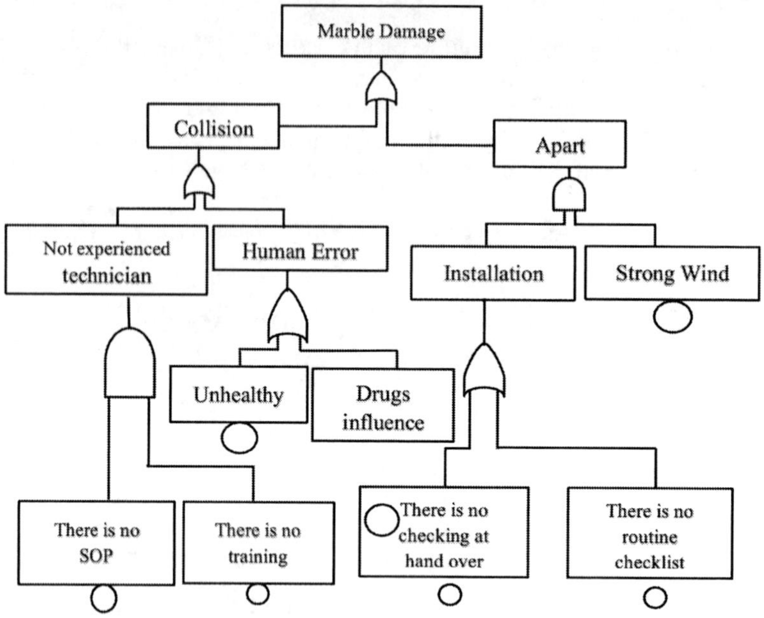

Figure 2. Fault tree analysis of a marble damage

From the data above, the writer can create a basic event table and recommendations for each of the basic events. This is necessary for the company as a preventative measure to prevent such damage. Table 7 shows the basic events and recommendations.

Table 7. Recommendations for each basic event in marble damage.

No	Basic Event	Recommendation
1	There is no procedure standard for work	Made procedure standard for work in an area containing marble in detail.
2	There is no training	Given training or upgrading knowledge about marble, how it is treated for the technicians and cleaning services. Supervision during the field.
3	Unhealthy	Make a form that states the technician's condition is healthy, and that form signed for accountability.
4	Drugs influence	Make a form that states the technician is not on drugs influence, and that form signed for accountability.
5	There is no checking at hand over	Layered area checklist (more than one person) when there is marble installation in the new unit area.
6	There is no routine checklist	Making a checklist form and schedule based on area and for a certain time.

These recommendations can be made to reduce the risk of marble damage in case study. After the above recommendations are executed, it is analyzed again how often the risk event occurs. After that, it is re-analyzed how the recommendation is right on target or not.

4. Conclusions

Based on the fuzzy-FMEA method, the most critical risk out of 25 identified risks is the C6 risk code. There is marble damage which has an FRPN value of 64.7. Based on the FTA method of the most critical risks obtained seven basic events: There is no SOP for work, no training, not feeling well, mental disorders, no checking during handover, no routine checklist, and strong winds. This research was only carried out at the analysis stage and the improvement plan and had not yet reached the implementation and Plan-Do-Check-Action stages. Many other factors, such as safety risk and fire risk that exist in an apartment building that can be assessment and improved to reduce the risk. Basic events that fondly through the FTA method can be recalculated using the FMEA table. Suggestions for further research from this research are that actual calculations can be applied to other apartment building case studies that have similar characteristics. The value of existing risks can be compared between one building and another.

5. References

[1] Talamo C, Bonanomi M. Knowledge Management and Information Tools for Building Maintenance and Facility Management. Springer International Publishing; 2015; doi.org/10.1007/978-3-319-23959-0.

[2] Kutlu AC, Ekmekçioğlu M. Fuzzy failure modes and effects analysis by using fuzzy TOPSIS-based fuzzy AHP. *Expert Syst. Appl.* Elsevier BV; 2012 Jan;39(1):61–7. doi.org/10.1016/j.eswa.2011.06.044.

[3] Sutrisno A, Gunawan I, Vanany I, Khorshidhi HA. A maintenance waste risk appraisal model based on modified failure mode and effect analysis (FMEA). *2016 IEEE International Conference on Industrial Engineering and Engineering Management (IEEM).* IEEE; 2016 Dec; doi.org/10.1109/ieem.2016.7798112.

[4] Pillay A, Wang J. Modified failure mode and effects analysis using approximate reasoning. *Reliab. Eng. Syst. Saf.* Elsevier BV; 2003 Jan;79(1):69–85. doi.org/10.1016/s0951-8320(02)00179-5.

[5] K Meng Tay K, Peng Lim C. Fuzzy FMEA with a guided rules reduction system for prioritization of failures. *Int. J. Qual. Reliab. Manag.* Emerald; 2006 Oct;23(8):1047–66. doi.org/10.1108/02656710610688202.

[6] Guimarães ACF, Lapa CMF. Fuzzy FMEA applied to PWR chemical and volume control system. *Prog. Nucl. Energy.* Elsevier BV; 2004 Jan;44(3):191–213. doi.org/10.1016/s0149-1970(04)90010-8.

[7] Xu K, Tang L., Xie M, Ho S., Zhu M. Fuzzy assessment of FMEA for engine systems. *Reliab. Eng. Syst. Saf.* Elsevier BV; 2002 Jan;75(1):17–29. doi.org/10.1016/s0951-8320(01)00101-6.

[8] Wang Y-M, Chin K-S, Poon GKK, Yang J-B. Risk evaluation in failure mode and effects analysis using fuzzy weighted geometric mean. *Expert Syst. Appl.* Elsevier BV; 2009 Mar;36(2):1195–207. doi.org/10.1016/j.eswa.2007.11.028.

[9] Chanter B, Swallow P, editors. Building Maintenance Management. Blackwell Publishing Ltd; 2007 Jan 1; doi.org/10.1002/9780470692011.

[10] Vanany I, Zailani S. Urgency in Managing the Risk in Supply Chain Amongst Indonesian Manufacturing Companies. *Int. Bus. Manag.* Medwell Publications; 2010 Apr 1;4(4):199–208. doi.org/10.3923/ibm.2010.199.208.

[11] Wahyuni HC, Vanany I, Ciptomulyono U. Identifying risk event in Indonesian fresh meat supply chain. *IOP Conference Series: Materials Science and Engineering.* IOP Publishing; 2018 Apr;337:012031. doi.org/10.1088/1757-899x/337/1/012031.

[12] Sutrisno A, Gunawan I, Vanany I, Asjad M, Caesarendra W. An improved modified FMEA model for prioritization of lean waste risk. *Int. J. Lean Six Sigma.* Emerald; 2018 Oct 26;11(2):233–53. doi.org/10.1108/ijlss-11-2017-0125

[13] Dağsuyu C, Göçmen E, Narlı M, Kokangül A. Classical and fuzzy FMEA risk analysis in a sterilization unit. *Comput Ind Eng*. Elsevier BV; 2016 Nov;101:286–94. doi.org/10.1016/j.cie.2016.09.015.

IOP Conf. Series: Materials Science and Engineering 1072 (2021) 012002 doi:10.1088/1757-899X/1072/1/012002

Decision-making model for Auto-ID technology adoption in the banking industry

F Afrizal[1]* and I Vanany[1]

[1]Department of Industrial Engineering, Sepuluh Nopember Institute of Technology, Surabaya, Indonesia

Email: *fandiafrizal177@gmail.com

Abstract. The justification of investment technology Auto-ID (automatic identification) in banks aims to assist practitioners in solving banking problems. A study is needed to find out the importance of investing in Auto-ID technology in a company. This technology will be made in an integrated system that can be used to monitor important documents and assets in the bank with two alternatives that are suitable for this case namely barcode and RFID. In the process of technology adoption, there are various forms of approaches ranging from strategic, economic, analytical and comprehensive where in the case of IT justification is more suitable to use analytic and comprehensive namely DEMATEL and ANP because it needs to capture intangible factors. In the DEMATEL calculation, the operator criteria group with the highest weight is eternity, followed by IT and economic optimization. In addition, it was also found that the most influential recipient groups were satisfaction, cost savings, and expansion. Then in the ANP method, the calculation results are obtained that the global weighting of the criteria with the highest value is an opportunity, followed by benefits, costs, and risks. So that the alternative decision obtained is RFID is the technology chosen for adoption in the banking industry.

1. Introduction

In a country, banks have an important role in the economy. They are intermediaries among people with a lack of capital and a surplus [1]. Since the digital revolution, information technology has been widely applied to the financial sector and provides impetus to traditional banks to carry out digital transformation. The emergence of the internet and digitalization has caused disruption to the banking industry in the world in general and Indonesia in particular [2]. Research has shown that service quality is increasingly recognized as being critical in the success of any business [3]. Service quality has been widely used to evaluate the performance of banking services [4]. Banks understand that customers will be loyal if they receive greater value than competitors banks can get high profits if they are able to position themselves better than their competitors in certain markets [5].

In the opinion of some experts, needs such as tracking inventory such as laptops, cars, and important customer documents such as principal documents (land certificates, book vehicle owners), agreement documents, and the like that often move between branches become an important focus [6]. Customer documents that are usually in the bank are often processed either for the credit application process or processes that require external parties such as a notary public or the National Land Agency so that this process often creates the risk of slipping documents [7]. According to Ambira & Kemoni [8] the need to strengthen records management as a critical success factor in risk mitigation within banks. So, it is proposed an effort that can be done by banks to use standardized Auto-ID (Automatic Identification) so that the company can monitor the movement of assets and records carried out by employees or couriers. Other factors that cause the use of Auto-ID are due to the speed of reading data, the lack of errors in reading data, and flexibility [9].

Wireless identification technologies such as Auto-ID make it easier to develop finer environments for doing business [10]. This is supported from previous research on smart blood tracking systems in the health care industry with a type of system that has a web-based platform (online) and is supported by automatic identification technology (Auto-ID) such as barcodes and radio frequency identification (RFID) believed to help companies in developing its business [11]. It was mentioned that there were various kinds of automatic identification systems (Auto-ID) including (1) Barcodes, (2) Optical Character Recognition, (3) Radio Frequency ID, (4) Magnetic Stripe, (5) Voice Recognition, (6) Fingerprint. Of the many Auto-ID technologies suitable for completing document traceability systems are barcode and RFID [12]. The process of document mobilization at banks in Indonesia especially in Surabaya is still classified as using a manual. The process of taking and tracking documents is done and carried out by humans so that the potential for human error is high. In a survey said in 2018, 43% of bankers surveyed were likely to invest in transformation technology in the next 18 months (see Table 1). This is known to get the highest percentage when compared to other factors, namely changing customer needs, competition, operational excellence, risk management, regulation so that the need for technological transformation is in great demand in the banking world.

Table 1. Technology business transformation.

Technology Business Transformation	Scoring scale
Technology	43 %
Changing Costumer Needs	34 %
Competition	14 %
Operational Excelence	6%
Risk Management	2%
Regulation	1%

In each document, a tag will be installed which will be detected by the reader, so that through the tag the document can be monitored through the system [13]. To use this technology an investment cost is needed for the procurement of equipment at the outset. There are various price variations that can be taken into consideration for initial investment [13]. Besides the reliability of the reading level, level of organizational readiness, and the technological risks that may be caused are also a consideration related to the negative impacts in the adoption process [14].

To evaluate the adoption of new technology, a decision analysis process is needed to find out which criteria are most influential. The decision analysis process requires several supporting criteria to determine the alternatives to be chosen. In the process of technology adoption, there are various forms of approaches ranging from strategic, economic, analytical and comprehensive where in the case of IT justification will be more suitable to use analytical and comprehensive, namely MCDM (multi-criteria decision making) because it needs to capture intangible factors [15]. Many researchers believe that a comprehensive approach developed with some integrated techniques is more appropriate when used in the justification of IT investment compared to economic, strategy and analytic approaches [16].

In the case of MCDM, many methods that can be used including the popular one is AHP. The AHP process is a multi-criteria decision-making process used to choose an alternative. AHP has been widely used to reflect the importance of criteria. many studies of AHP have been carried out, including the application of AHP in various fields of planning, choosing the best alternative, and resolving conflicts [14]. In this research, there is a decision-making problem that cannot be structured in a hierarchical manner because it involves interaction and dependency between its elements. In addition, there is also the possibility of interdependence and feedback on the problem. Therefore, the ANP (Analytical Network Process) method was chosen because it is a method for considering this. ANP can also illustrate the strength of interconnectedness so that it can understand complex relationships in the adoption system and find the best alternative [17]. To improve the quality of decisions in determining alternatives, the DEMATEL method was chosen to complete the ANP method as a basis for describing the structure of the ANP network. Although the use of the ANP method has been able to capture the influence and dependence between the criteria, adopting the DEMATEL method can optimize the causal relationship. The merging of the DEMATEL and ANP (DANP) methods is considered to be complementary to complete complex problems with more accurate results and is able to describe the conditions of the scope under real study [18].

2. Methodology

The object of this study is a decision analysis of the implementation of new Auto-ID technology in the banking industry in Surabaya. While the subjects of this study are experts who work at banks and already have work experience in filing important documents. Interviews will be conducted with the position of the head, deputy head of the Surabaya branch or head of the credit department. In the process of sampling and population, this method needs to be done so that the results obtained can be optimal. In this study, the Delphi method is used for the sampling process. The method used in collecting data is by observation, interview, questionnaire and literature study. The questionnaire used is a semi-open questionnaire. While the data source used is primary data because the data taken is interview data collected directly through experts based on their experience.

2.1 Preparation phase

In this preparation, the stage is preparing everything related to research. The first thing to do is study literature. In this literature, the study is used to study methods and theories as to the initial foundation in thinking. This literature study helps researchers determine the initial dimensions and criteria before submitting them to the expert. This was taken from previous similar studies as well as several journals and supporting books. The second thing is to do a field study. At this stage, the researchers collected supporting data to support subsequent data processing. It also obtains preliminary information from the surrounding environment as a basis for starting a study.

2.2 Data Collection Phase

The criteria were adopted from several previous studies obtained from national and international journals. With this adoption, the researcher finally has a reference to several criteria. Then from the aforementioned criteria, the researcher proposes them to the head of the credit sub-unit in the bank in order to get the criteria that are really needed by the company with the current conditions. So, at this stage, the final criteria will be obtained which will be used at a later stage which of course has been approved by the company. Then the researcher arranged the questionnaire design in 3 (three) types. Questionnaire 1 was used to validate the criteria to the expert, questionnaire 2 was to identify the relationship between the sub-criteria using the DEMATEL method, questionnaire 3 was a pairwise comparison to get the criteria weights. Questionnaires 1, 2 and 3 will be given to the same expert. The interview process will be conducted with experts in the company. The experts will be asked to answer questions relating to the relationship between the influence of criteria. From the results of interviews that have been carried out obtained data which is then processed to see the effect of the relationship between the sub-criteria using the DEMATEL method. In the DEMATEL calculation is only done to see the pairwise comparisons on sub-criteria only. In addition, this method is also to find out which criteria are included as influencers. The details of this process are as follows [19]:

1) Creating a direct relationship matrix
   In this step, a direct relationship matrix will be made using four levels of comparison scale, namely 0 (no influence), 1 (low influence), 2 (moderate effect), 3 (high effect), 4 (very high effect). Previously it is necessary to calculate the number of each column and row of matrix A, then find the maximum value to enter into the formula.

2) Normalization of the direct relationship matrix
Using matrix A, normalizing the relationship with matrix D directly can be obtained using the following formulation.

$$D = z \times A \quad (1)$$

$$Z = \min \{\frac{1}{\max i \sum_{j=1}^{n} aij}, \frac{1}{\max j \sum_{i=1}^{n} aij}\} \quad (2)$$

3) Calculate the Total Matrix T
The total T matrix relationship symbolized as an identity matrix can be obtained through the following equation:

$$T = D + D2 + D3 + ... + Dh$$
$$= D(I + D + D2 + ... + Dh\text{-}1)[(I\text{-}D)(I\text{-}D)\text{-}1]$$
$$= D (I - Dh)(I\text{-}D)\text{-}1$$

Where I can be defined as an identity matrix

$$I = (I - D)(I - D)\text{-}1 \quad (3)$$

$$T = D (I - D)^{-1}, h \rightarrow \infty, D^h = [0]_{n \times n} \quad (4)$$

4) Compute vector D (dispatcher) and R vector (receiver)
The number of rows and columns is separately represented as vector D and vector R. Then the vector (D + R) is referred to as "prominence" horizontally. This vector was created to add D to R which shows how important a criterion is. Besides that, there is also a vector (D-R) called "relation" into a group of causes and effects. The above can then be formulated as follows:

$$T = [tij]n \times n' \qquad i,j = 1,2,3, ..., n \quad (5)$$

$$D = [ri]_{nx1} = [\sum_{j=1}^{n} tij]_{n \times 1} \qquad = [tj]_{n \times 1} \quad (6)$$

$$R = [sj]_{nx1} = [\sum_{i=1}^{n} tij]_{1 \times m} \qquad = [tj]_{n \times 1} \quad (7)$$

5) Get an impact-diagram map
From the values obtained from the total relationship matrix, each value provides information on how much influence the group of criteria i with criteria j. So that if you only rely on the results of the total relationship matrix, there will be difficulties in determining decision making. For this reason, a threshold value for the level of influence needs to be given so that only a few elements have values above the threshold in the T matrix that can be entered on the impact-diagram map. This threshold value is usually determined by the expert. Impact-diagram maps can be obtained by mapping values (D + R) and (D-R) so that information can be obtained to make decisions.

2.3 Data Processing Stage

Please The first step in processing data is to design the ANP structure based on the calculation results obtained using DEMATEL. The design of the ANP model includes the creation of dimensional structures that show relationships between dimensions. The design of this model will later be used to determine what criteria are compared in the comparison matrix at a later stage. Having made the ANP network structure and knowing what criteria should be compared to the interview again to be asked to compare in pairs each of the criteria using the ANP scale. This second interview aims to get the priority weights of each criterion. Then the paired comparison results are performed for each dimension using the ANP principle where the matrix contains the inverse value of the paired comparison results according to expert judgment. After obtaining information about the comparison results of the criteria obtained from the expert, the researcher will then calculate the weight of each dimension. afterward, a consistency test is performed to see the accuracy of the judgment value given by the expert. The consistency test is carried out twice by looking at the consistency ratio (CR) and the consistency index (CI). A matrix is said to be consistent if the CR value is not more than 10%. If the consistency ratio is getting closer to zero, then the better the value and shows that the matrix is said to be consistent. If the results of the consistency test indicate that the weighting is not consistent, then the number that is rated the most outlier will be eliminated and then will be re-weighted.

2.4 Data Analysis and Interpretation Phase
From the evaluation results, it is known which criteria are good enough and which need to be improved. So with this information, the researcher can analyse the implications or effects of these criteria on the successful implementation of Auto-ID, and the impact that might occur if the criteria are not implemented. Then BOCR analysis is carried out to determine the factors of benefit, opportunity, cost, risk. This analysis is done by looking at the weighting results obtained from comparing the criteria derived from the BOCR

3. Result and Discussion
3.1 The banking world
The business process of a credit application that occurs within a bank is generally handled by 4 internal parties, namely the credit unit, management, document units, and operational units. The process starts by receiving the debtor who will apply for credit by the document unit in the bank. The debtor conveys the purpose of applying for credit along with collateral that can be provided as a condition. The process continues until the file is stored in a safe. After this process is completed, a memo is made to the operational unit that the documents provided are complete. The operational unit then processes the loan funds submitted by the debtor along with the loan agreement regarding the instalment and repayment process. Business process for submitting credit and business process credit repayment are shown in Figure 1 and Figure 2.

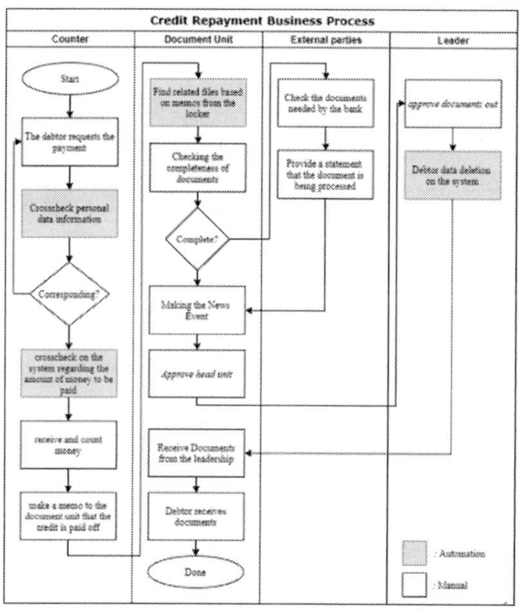

Figure 1. Business process for submitting credit.

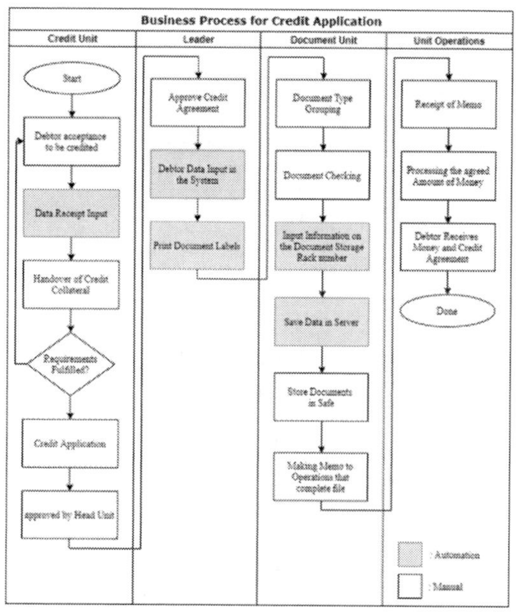

Figure 2. Business process credit repayment.

On repayment of the credit, the process starts from the debtor applying for repayment at the counter. then the counter will crosscheck the customer's personal information and so on until the process ends with the debtor receiving the main documents back from the leader after all administration has been fulfilled. The cloud service architecture system is described in Figure 3.

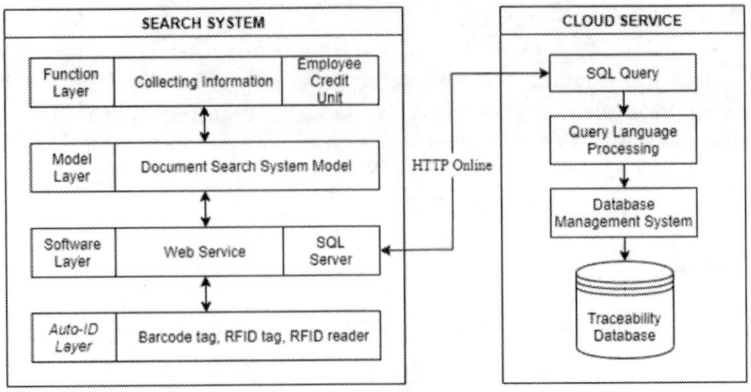

Figure 3. Cloud service architecture system.

3.2 Selection Criteria

Criteria consist of (4) four types consisting of benefits, opportunities, costs, and risks. Each criterion will then be reduced to sub-criteria and alternatives. The benefit criterion relates to the benefits to be gained by the company when adopting information technology (IT). While the opportunity criteria relate to opportunities that can be taken in the future. Then the cost criteria are related to the costs that will arise when adopting new information technology (IT) for the company. Furthermore, risk associated with risks that arise with the existence of this technology for internal companies and customers. Table 2 shows the value of prominence and relations.

Table 2. The value of prominence and relation.

Sub-Criteria	D	R	D+R	D-R
Economic	3.848	3.764	7.612	0.084
Satisfaction	3.227	4.208	7.434	-0.981
Effort	3.006	3.631	6.637	-0.625
Time	3.566	3.650	7.216	-0.084
Functionality	3.929	3.856	7.786	0.073
Cost Saving	3.711	4.023	7.734	-0.313
Expansion	3.029	3.901	6.930	-0.872
Innovation	3.733	3.655	7.389	0.078
Image	2.916	3.486	6.403	-0.570
Optimization of IT	3.951	3.698	7.648	0.253
Infrastructure	3.358	3.258	6.616	0.101
Maintenance	3.458	3.194	6.652	0.263
Switching	3.657	3.076	6.732	0.581
Service	3.016	3.556	6.572	-0.540
Human Resource	2.650	3.285	5.935	-0.636
Data Access & Integrity	3.543	3.312	6.855	0.231
Lack of Control	3.368	3.158	6.526	0.209
Security	3.808	3.729	7.537	0.079
Internet Latency	3.785	3.071	6.856	0.713
Permanence	4.500	3.000	7.500	1.500
Vendor Lock-in	3.503	3.050	6.553	0.454

This method is done with the aim to find out how much influence between sub-criteria one with other sub-criteria that exist in a criterion. Respondents (experts) were asked to rate the direct effects of factors using integer scales from 0 to 4 to represent from "absolutely no influence (0)" to "very high influence (4)". The results of the questionnaire then calculated the average value among the three respondents to form the initial relationship matrix. Then the data is processed using Microsoft Excel with five calculation stages. After obtaining a direct relationship matrix, the normalization matrix is calculated. After normalizing the direct relationship matrix, then the total relationship matrix is performed using equations (2.3) and (2.4) in chapter 2 about DEMATEL. After getting the total relationship matrix, the next step is to calculate the vectors D and R. The results of prominence (D + R) indicate the importance of the criteria for the traceability system. While the results of relations show a causal relationship on the criteria. The following is shown in table 3 calculation of vector D, vector R, vector D-R, and vector D + R. The boundary value is obtained by finding the average of the total relationship matrix. In this study, the value of the threshold value is 0.1668. Figure 4 shows hierarchy model of auto-ID technology decision making.

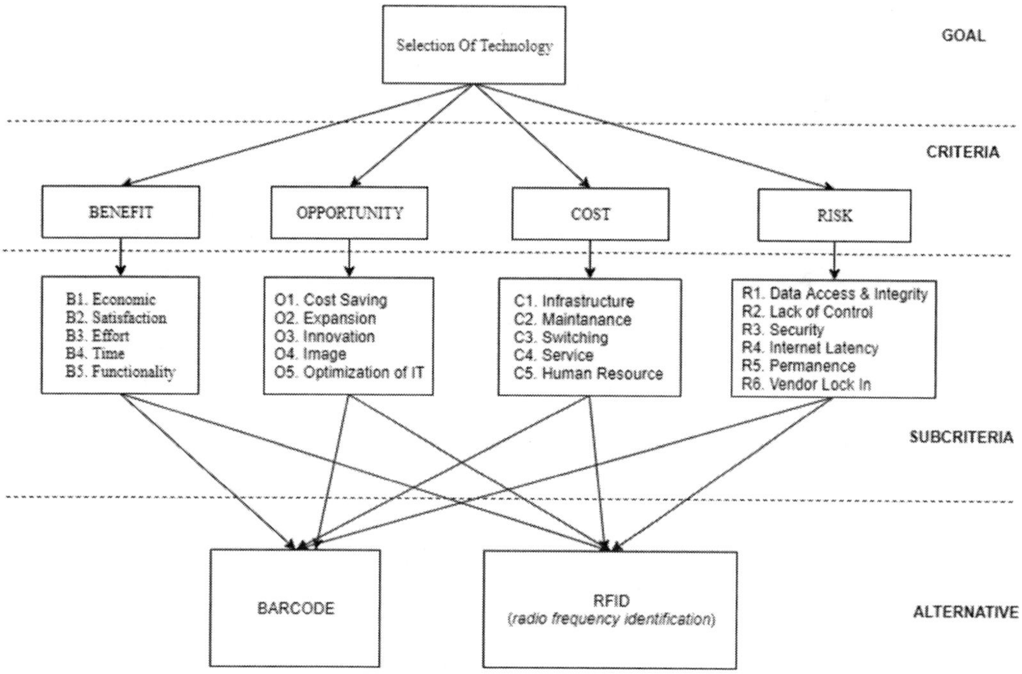

Figure 4. Hierarchy model of Auto-ID technology decision making.

In the DEMATEL calculation is only done to see the pairwise comparisons on sub-criteria only. But by knowing the relationship between the sub-criteria later it will also be known the relationship between the criteria because in the criteria there are sub-criteria which then there are alternatives as well as the decision of this study. So, all three can be described in an ANP network model.

IOP Conf. Series: Materials Science and Engineering 1072 (2021) 012002 doi:10.1088/1757-899X/1072/1/012002

Table 3. Sub-criteria weights.

Name	Normalized by cluster	Limiting
Economic	0.0975	0.0325
Effort	0.0953	0.0318
Functionality	0.2346	0.0783
Satisfaction	0.4006	0.1337
Time	0.1719	0.0574
Human Resource	0.1112	0.0120
Infrastructure	0.3267	0.0351
Maintenances	0.0921	0.0099
Service	0.4037	0.0434
Switching	0.0662	0.0071
Cost Saving	0.3246	0.1607
Expansion	0.2943	0.1458
Image	0.1127	0.0558
Innovation	0.1564	0.0775
Optimization of IT	0.1120	0.0555
Data Access & Integrity	0.2039	0.0130
Internet Latency	0.0400	0.0025
Lack of Control	0.0514	0.0033
Permanence	0.0225	0.0014
Security	0.6791	0.0432
Vendor Lock in	0.0031	0.0002

Table 3 shows the local weights of the sub-criteria found in the Normalized by Cluster column. While the global weights between the criteria and sub-criteria are shown in the Limiting column. Then to get the local weight from the criteria, the weight in the Limiting column is divided by the weight in the Normalized by Cluster column. From the results of the division, the calculation of the average for each grouping is then carried out to get the local criteria weights. Local weight criteria can be seen in Table 4.

Table 4. Criteria weights.

Name	Weights
Benefit	0.3337
Opportunity	0.4952
Cost	0.1075
Risk	0.0636

At this stage, the sub-criteria network model calculation is calculated after obtaining the criteria and sub-criteria local weights. Alternative pair comparisons are obtained through network diagrams between sub-criteria. Furthermore, the calculation of the sub-criteria network model is done by filling out a pairwise comparison questionnaire on node elements with the aim of comparing all related alternatives. The pairwise comparison results generated local weights in all alternatives shown in Table 5.

Table 5. Alternative weights.

Name	Ideals	Normals	Raw
Barcode	0.3593	0.2643	0.0000
RFID UHF	1.0000	0.7357	0.0000

4. Conclusion

Research on the Auto-ID adoption decision model in the banking industry has several conclusions. First, in the process of adopting Auto-ID technology in the banking industry, there are several criteria that can serve as benchmarks in determining decisions. There is 1 sub-criterion which is rejected, that is agility and 1 suggested sub-criterion, image. There are 14 cause sub-criteria (permanence, optimization of IT, & functionality) and 7 sub-criteria due (human resource, image, service). Second, from the comparison of the criteria that have been obtained the relationship between the criteria using the DEMATEL method obtained ANP network relationship that has an inner dependency and outer dependency relationship on each criterion. Then from each comparison of the criteria of the ANP network model that has been made it is known that a reasonable decision taken in adopting technology at the bank is to use RFID technology. This is based on the 3 criteria of benefit, opportunity, and risk being the criteria that affect RFID alternatives as viable alternatives for adoption. for future research, it is better to use more experts who can represent answers from several banking industries in Indonesia. Furthermore, the authors hope this research can be developed into a study to make a decision analysis in the adoption of technology in the banking industry in Indonesia, with the criteria and sub-criteria derived according to this research.

References

[1] Jeucken MHA, Bouma JJ. The Changing Environment of Banks. *Greener Management International*. GSE Research Limited; 1999 Sep 1;1999(27):20–35. doi.org/10.9774/gleaf.3062.1999.au.00005

[2] F. Khanboubi and A. Boulmakoul. A roadmap to lead risk management in the digital era. Hassan II University of Casablanca, B.P. 146 Mohammedia, Morocco, 2018. Retrieved from: researchgate.net/publication/325056409_A_roadmap_to_lead_risk_management_in_the_digital_era

[3] A. Parasuraman, V. A. Zeithaml and L. L. Berry. SERVQUAL – a multiple-item scale for measuring consumer perceptions of service quality. *Journal of Retailing*, pp. 12-40, 1988

[4] Cowling A, Newman K. Banking on people. *Pers. Rev*. Emerald; 1995 Nov;24(7):25–40. doi.org/10.1108/00483489510095789

[5] Davies F, Moutinho L, Curry B. Construction and testing of a knowledge-based system in retail bank marketing. *Int. J. Bank Mark*. Emerald; 1995 Mar;13(2):4–14. doi.org/10.1108/02652329510078613

[6] M. C. O'Connor. Banking Group to Set RFID Roadmap. *RFID Journal*, p. 2, 2008

[7] M. Ismail, Perbankan Syariah, Jakarta: Prenamedia Group, 2011.

[8] Ambira C, Kemoni H. Records management and risk management at Kenya Commercial Bank Limited, Nairobi. *SA Journal of Information Management*. AOSIS; 2011 Oct 6;13(1). doi.org/10.4102/sajim.v13i1.475

[9] Liawatimena S. Analisis dan Perancangan Sistem Karcis Elektronik Pada Gerbang Masuk Busway Dengan Menggunakan RFID. *ComTech: Computer, Mathematics and Engineering Applications*. Universitas Bina Nusantara; 2010 Dec 1;1(2):942. doi.org/10.21512/comtech.v1i2.2650

[10] B. Violino. RFID May Boost Service at Banks. *RFID Journal*, pp. 1-2, 2003

[11] Vanany I, Shaharoun AM. The Comprehensive Framework for RFID Justification in Healthcare. *Int. Bus. Manag*. Medwell Publications; 2011 Feb 1;5(2):76–84. doi.org/10.3923/ibm.2011.76.84
[12] H. Finkenzeller, RFID Handbook: Fundamentals and Application in Contactless Smartcard, Radio Frequency Identification and Near Field Communication, United Kingdom: Wiley, 2003
[13] F. Ropraz. Using RFID for Supply Chain Management. Electronic Business Course, pp. 1-36, 2008
[14] I. Vanany. An AHP Based Method to Prioritize the Barriers and Critical Success Factors of RFID Adoption in Healthcare. *International Business Management*; 2011:vol. 5, pp. 427-35
[15] B. Farbey, F. F. Land and D. Targett. How to evaluate your IT investment. Oxford: Butterworth Heinemann. 1993
[16] Irani Z, Love PED. Developing a frame of reference for ex-ante IT/IS investment evaluation. *Eur J Inf Syst*. Informa UK Limited; 2002 Mar;11(1):74–82. doi.org/10.1057/palgrave.ejis.3000411
[17] T. L. Saaty. Decision Making with the Analytic Network Process. 2006
[18] Chen S, Lee H, Wu Y. Applying ANP approach to partner selection for strategic alliance. *Manag. Decis*. Emerald; 2008 Apr 4;46(3):449–65. doi.org/10.1108/00251740810863889
[19] Uygun Ö, Kaçamak H, Kahraman ÜA. An integrated DEMATEL and Fuzzy ANP techniques for evaluation and selection of outsourcing provider for a telecommunication company. *Comput Ind Eng*. Elsevier BV; 2015 Aug;86:137–46. doi.org/10.1016/j.cie.2014.09.014

IOP Conf. Series: Materials Science and Engineering 1072 (2021) 012003 doi:10.1088/1757-899X/1072/1/012003

Usability testing on developed FES device according to the operator perspective

K Leonardo[1,2], R S Dewi[1,3] and A Sudiarno[1,4]

[1]Department of Industrial Engineering, Institut Teknologi Sepuluh Nopember, Surabaya, Indonesia

Email: [2]kevleo1994@gmail.com, [3]ratna@ie.its.ac.id, [4]adithya.sudiarno@gmail.com

Abstract. The post-paralysis due to stroke, spinal cord injury or musculoskeletal disorder can be restored by rehabilitation. A popular rehabilitation method used today are physiotherapy and therapy using electrical stimulation. Functional Electrical Stimulation (FES) is one of the methods of rehabilitation using electrical stimuli, and has been proven to be effective enough to restore muscle function that is weakened due to paralysis. Nowadays, FES devices on the market are designed to rehabilitate only one type of movement and not yet applied wearable and ergonomics design. Meanwhile, the FES devices developed by the laboratory has applied a wearable design and has been able to rehabilitate several movements at once, but not yet for ergonomics design. This study developed a usability instrument and do the usability evaluation for the FES medical devices which is rehabilitate multiple movements on the upper limb. The usability variables chosen for the usability evaluation are simplicity, learnability, memorability, informativeness, and trust ability. Usability evaluation is carried out by distributing questionnaires. Data obtained from the propagation of the questionnaire are processed by statistic method such as average value and standard deviation. The results indicate that the simplicity, learnability, and memorability variables were well applied to the developed FES devices. While the variables informativeness and trust ability dimension are still not properly applied to developed FES devices.

1. Introduction

Stroke is a disease that affects many people in the world, especially in large urban areas. This disease is characterized by brain tissue death caused by reduced blood flow and oxygen to the brain due to blockages, narrowing, or rupture of blood vessels [1]. After the patient recovered from stroke, the joints in the paralyzed limbs become more rigid. Spinal cord injury or spinal cord damage can cause sufferers to experience stiffness in some parts of the body due to pain caused by spinal cord damage. Through research that has been done, one way to rehabilitate patients after stroke and spinal cord injury is the Physiotherapy method. Nowadays, technology has developed so rapidly that physiotherapy no longer has to be done with the help of other humans, but instead uses a method called Functional Electrical Stimulation (FES). The FES rehabilitation method is carried out by providing electrical stimulation to the muscles through the skin, so that the stimulated muscles can contract and cause movement of the joints.

Some of the consequences of a stroke that attacks the upper limb of the human body cause weakness or paralysis that occurs in wrist movements such as twisting and swinging, finger movements, elbow joint movements, and also in the shoulder joints [2]. Rehabilitation using the FES method in several previous studies has been proven to reduce joint stiffness and increase muscle strength of limbs in post stroke patients [2].

There are several researches about FES devices that are conducted by researcher from Institut Teknologi Sepuluh Nopember Surabaya, Indonesia. First research started by Arifin which make a low cost joint angle sensor for measurement movement using gyroscope sensor, accelerometer sensor, and Kalman filter. [3] At 2015, that research developed by Fauzan to redesign the device produce from the Arifin research [3] using wearable design [4]. At 2016, The research was continues by Erwin, which did an experimental study to use the wearable design to rehabilitate elbow joint, and wrist joint using PID controller [5], and Stanley, which did an experimental study to use the wearable design to rehabilitate knee joint using fuzzy logic with cycle to cycle controller [6]. Latest the research developed by Kevin by redesign of the wearable device by applying the ergonomics design [7].

Usability is a science related to human interaction with technology, both computers and machines, or it can be said that knowledge is related to the user interface and user experience. Usability also is an important factor to make a successful product [8]. Evaluation of usability in the field of medical equipment itself has not been done much as a result, there are many of the medical equipment that can give bad perception in patients who want to use it and also caused unconfident on operators while operating the medical device, especially FES device.

2. Methodology

This research aims to design a usability instrument and make a measurement tools that can be used to evaluate the FES medical devices and also conduct a usability evaluation for the FES device that used to rehabilitate multiple movement on the upper limb that were designed by Kevin [7] as shown on Figure 1 until Figure 6. Figure 1 and Figure 2 shown the Position Module (PM) which is place on both left and right side on palm, lower arm, and upper arm as shown in Figure 5. This module has a function to read the movement of the arm. Figure 3 shown the Central Processing Module (CPM) main unit of this device, function to process the data read by PM and combine with the parameters input from the Remote Module (RM) to calculate the right coefficient of the controller used. Figure 4 shown the RM which is the operator use to input the parameter for the rehabilitation session. Both CPM and RM place on the back of the jacket shown in Figure 6.

Figure 1. Position module upper side.

Figure 2. Position module lower side.

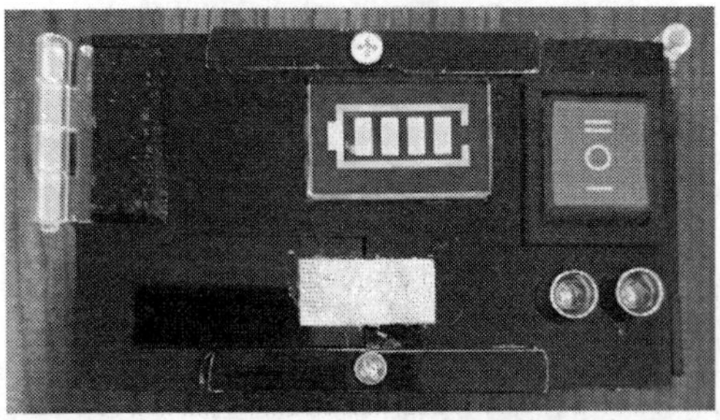

Figure 3. Central processing module place on the back of the jacket.

Figure 4. Remote module for input the parameters of the rehabilitation.

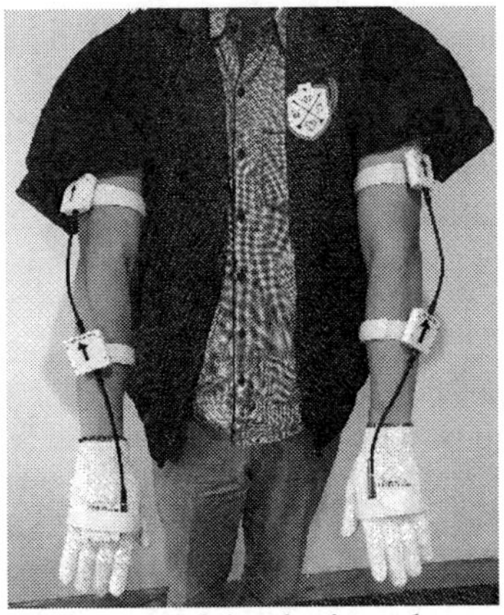

Figure 5. Front of the device when put on.

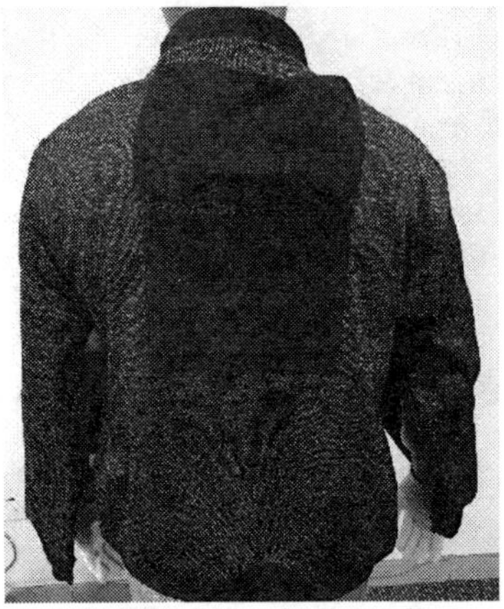

Figure 6. Back of the device when put on.

2.1. Literature Study
Literature study conducted by find out what is usability, what is the function, how to conduct the research, how to do data retrieval, and also how to process the data and make the conclusion. This phase help to make a way of thinking to make the initial dimension and criteria. The subject was the student of the biomedical engineering from Institut Teknologi Sepuluh Nopember Surabaya, which concentration is the rehabilitation, so the student has the basic to become an engineer to design a medical device, and also to operate it.

2.2. Design of Usability Measurement
The design of the usability measurement tool is done by sampling the example of the usability dimension used in the previous journal which performs the usability evaluation on electronic equipment, and medical equipment to determine the dimension of suitable usability to evaluate the FES medical device.

2.3 Design a Questionnaire and Data Retrieval
Questionnaire question design was conducted by taking references from several previous journals on questionnaire questions that could be proposed for medical devices in both hospital, operator, and patient perspectives. The questionnaire on this study was intended for medical equipment operators, so that it could support the purpose of this research to evaluate the usability of the medical equipment made against the operator's perspective.

Data retrieval for the evaluation of usability is done by requesting the respondent to operate the created FES device. Once the respondent has finished operating the created FES device, the respondent will be asked to fill out the questionnaire sheet that was created in the previous step.

2.4 Data Analysis
The data obtained is then analysis to be processed and evaluated with statistical tools on every aspect of the usability used. The processing of data used to process the data of the questionnaire obtained is the average and standard deviation on the respective variable of the usability dimension that is used to produce an evaluation of the usability that can be used to assess the FES device developed by Kevin [7].

3. Result
3.1 Usability Tool Measurement
Usability dimension in this study, determined by consideration of several aspects, including aspects of medical devices, medical, and electronic equipment. On a journal written by Marko Ylikulju [9], the standard of medical devices is for the evaluation of the ability, requiring at least 5 major dimensions.

According to specialist Jeff Rubin and the Chisnell Fund on the journal, a product that can be used has a dimension of efficient (efficiency), effective (effectiveness), satisfaction (satisfaction), can be learned (learnability), and accessible (accessibility). While according to the specialist Jakob NIELSESN, also on the same source of journal, defines five dimensions of the most important usability, among them can be learned (learnability), efficient (efficiency), can remember (memorability), errors (and satisfaction). On other journal written by Whitney Quesenberry [10], determination of the dimension of usability for a medical product, also used the dimension mentioned by Jeff Rubin, and the Chisnell fund. Meanwhile, on the journal written by Jongseo Kim, and Sung H. Han [8], a good electronic product, has a dimension of simplicity, consistency, accessibility, error prevention, learnability, and efficiency. In a book titled "Usability Testing of Medical devices" written by Michael Wiklund, Jonathan Kendler, and Allison Strochlic [11], it is mentioned that the most important criterion of a medical device is informative and trustworthy (trustable).

Based on the reference, the selected dimension of the usability in this study are simplicity, learnability, memorability, informativeness, and trust ability. Determining the dimension of this usability is based on a number of considerations, such as an efficient and effective product is a simple product, so the chosen dimension of simplicity as an alternative, then dimension of learnability and memorability in the select as it reflects the resulting device. Because the FES device is an electronic device and also a medical device, it is mandatory to be trusted and has clear and easy to understand information, so that the dimension informativeness in select. While the dimension of confidence in the evaluation of usability is still not found, so this research proposes the name trust ability as a new dimension of usability reflecting an evaluation for medical devices that is very important. In the Table 1 are outlined the five dimensions of the usability used along with their respective definitions.

Table 1. Usability dimension used with its definition.

No	Usability Dimension	Description
1	Simplicity	User interface and method of interaction of a product should be simple, clear, and identifiable intuitively
2	Learnability	The effort required to learn the user interface and interaction methods should be small
3	Memorability	User interface and interaction methods should be easy to remember
4	Informativeness	User interface presented to the user should be easy and clear to understand
5	Trust ability	User's ability to trust and want to use products that have been made

3.2 Usability Dimension Variable

Usability dimension variables are a development of the usability dimension used so that a bullet point can be used as an indicator to determine the questionnaire question. The main purpose of variable dimension is to find out if each of the selected dimension that is selected has an attachment to the usability evaluation.

The first dimension is simplicity, with three pieces of variable of usability dimension, i.e. the number of cables, the number of buttons, and the ease to wear. The simplicity dimension variable can be seen in Table 2. In the number of cables variable, it is selected to explain whether the number of cable connections used complicate the operator in the device installation. In the variable number of buttons, explain whether the number of buttons used already represents all the functions contained in the device and has made it easier for operators to operate. While the variables are easy to apply, explain the level of ease of device to wear to the patient by the operator.

Table 2. Usability dimension of simplicity variable.

Usability Dimension	Code	Usability Dimension Variable	Description
Simplicity	B1	Number of cables	The number of cable connections used on the device
	B2	Number of buttons	Number of buttons already represent function
	B3	Ease to use	Easy device to be worn by the operator to the patient

The second dimension of usability is learnability, with variables of usability dimension are two pieces, i.e. operator duration to recognize the function and operator duration to recognize the placement. Both variables are selected to determine how easily the FES device was studied by the operator, both from the function side and on the laying side. This dimension represents the convenience of the device to learn. The learnability dimension usability can be seen in Table 3.

Table 3. Usability dimension of the learnability variable.

Usability Dimension	Code	Usability Dimension Variable	Description
Learnability	C1	Operator duration to recognize the function	Duration required by the operator to recognize the function and button of each module
	C2	Operator duration to recognize placements	Duration required by the operator to recognize the placement of each module

The third variable of usability dimension is memorability, with two pieces variables, i.e. how often the operator sees the manual book to find an explanation of the function, and how often the operator sees the manual book to find an explanation of the placement and installation steps of the device to the patient's body. This dimension represents that the resulting device is easy to remember, so it does not make it difficult for the operator to operate and pair the device. Usability dimension variable memorability can be seen in Table 4.

The fourth variable of usability dimension is informativeness with variables of the ease of UI to understand, and the feedback from the UI. This variable is more directed towards the UI, where the UI is focused to facilitate the operator while operating the device. Informativeness dimension variables can be seen in Table 5.

The fifth variable of usability dimension is trust ability with variable of two pieces, i.e. the operator’s trust to try the device for first time, and the operator's desire to operate the device again. These two variables are chosen to meet the trust ability dimension criteria, where trust is the main benchmark in the use of medical devices. Both variables also represent the research objectives at the patient's confidence level of the FES device. Trust ability dimension variables can be seen in Table 6.

Table 4. Usability dimension of memorability variable.

Usability Dimension	Code	Usability Dimension Variable	Description
Memorability	D1	How often the operator sees the manual book to find an explanation of the function	How often the operator opens the manual book to learn the function of buttons and function modules
	D2	How often the operator sees the manual book to find an explanation of the placement and installation steps of the device to the patient's body	How often an operator opens a manual book to learn placements and how to install devices

Table 5. Usability dimension of informativeness variable.

Usability Dimension	Code	Usability Dimension Variable	Description
Informativeness	E1	Ease of UI to understand	Ease of use of UI to be understood by the operator
	E2	Feedback from the UI	The UI used has good feedback, making it easier for operators to operate the device

Table 6. Usability dimension of trust ability variable.

Usability Dimension	Code	Usability Dimension Variable	Description
Trustability	F1	Operator's trust to try the device for first time	Operator's trust to operate the FES device on first-time trial
	F2	Operator's desire to operate the device again	Operator's desire to operate the FES device again in the future

3.3 Questionnaire Design

The question drafting for the evaluation questionnaire is done by developing a variable of the usability dimension that has been determined, so that it becomes a form of question with a scaling method of 1 to 5 that is presented with a highly disagreed, disagreed, normal, agreed, and highly agreed.

3.4 Usability Analysis

3.4.1 Validity and Reliability Test Methods

Validity is a measure that indicates the level or validity of an instrument. Validity in data collection testing can be differentiated into two, i.e. the validity of the factor and the validity of the item. The validity factor is measured when an item is used using more than one factor. The validity measurement of this factor is performed by correlate between the factor values with the total factor value. Item validity measurements are performed by correlate between item values with the total value of the item. The testing used in this research is the validity of the item. The validity test result for all query items can be seen in Table 7. In the table there is a variable B1, B2, and B3 representing each variable of usability dimension of the simplicity, the C1 and C2 variables represent each variable of usability dimension of the learnability, variables D1 and D2 represent each variable of usability dimension of memorability, the E1 and E2 variables represent each variable of usability dimension of the informativeness, and the F1 and F2 variables represent each of usability dimension of the trust ability.

Table 7. Validity test results.

Variable	Correlation Coefficient	Significance	Decision
B1	0.698	0.004	valid
B2	0.784	0.001	valid
B3	0.737	0.002	valid
C1	0.776	0.001	valid
C2	0.867	0.000	valid
D1	0.959	0.000	valid
D2	0.917	0.000	valid
E1	0.740	0.002	valid
E2	0.780	0.001	valid
F1	0.835	0.000	valid
F2	0.702	0.004	valid

Based on the correlation calculation result will be obtained a correlation coefficient. These coefficients are used to measure the validity level of an item and determine if an item is worth using. The significance limit used in this validity test is $\alpha = 0.05$. These significance limits determine whether an item deserves to be used. If the value of significance is less than the limit, it can be concluded that the question item can be used to measure a predefined variable (valid).

Reliability testing is a test to determine whether a questionnaire is used to collect reliable research data. The reliability in question is that when the measurement tool is reused to make the measurement, the results will not be much different from the results that have been obtained before. One of the commonly used methods in research is Cronbach's Alpha. The reliability level determination criteria can be seen in the Table 8. As for the calculation result of Cronbach's Alpha value for each dimension can be seen in Table 9. According to Table 9, it can be concluded that the questions in the B, C, and D dimensions have been able to provide precise and consistent results on every measurement performed (reliable). While the question on the dimensions of E and F are less capable of delivering reliable results.

Table 8. Criteria for determining reliability levels.

Criteria	Interval
Very low	0.00 - 0.19
Low	0.20 - 0.39
Middle	0.40 - 0.59
High	0.60 - 0.79
Very High	0.80 - 1.00

Table 9. Reliability test results.

Dimension	Cronbach's Alpha	Numbers of N	Conclusion
Simplicity (B)	0.580	3	Middle
Learnability (C)	0.518	2	Middle
Memorability (D)	0.839	2	Very high
Informativeness (E)	0.268	2	Low
Trust ability (F)	0.315	2	Low

3.4.2 Data Processing Questionnaire using Descriptive Statistical Methods

The results of the questionnaire obtained were processed using the descriptive statistic method. The simplicity, learnability, and memorability usability dimension have a good value, meaning that the dimension had been well applied in the device development. Thus, the informativeness and trust ability dimension has a fair value, meaning that the dimension was not well applied. Table 10 shows the result for the calculation result of the data obtained.

Table 10. Calculated result obtained from the questionnaire.

No	Dimension	Variable	Code	Mean	Total Mean	Standard Deviation	Total Standard Deviation
1	Simplicity	Cable amount	B1	2.32	2.30	1.54	1.55
2		Button amount	B2	2.25		1.48	
3		Easy to use	B3	2.31		1.63	
4	Learnability	Operator's duration recognizes the function	C1	2.18	2.18	1.43	1.39
5		Operator's duration recognizes the placement	C2	2.18		1.29	
6	Memorability	Frequency operators see manual book related functions	D1	2.25	2.27	1.34	1.44
7		Operator frequency see manual book related installation	D2	2.31		1.63	
8	Informativeness	UI easy to understand	E1	1.81	1.81	1.17	1.16
9		UI gives a good feedback	E2	1.81		1.13	
10	Trust ability	Operator confidence for the first time using the device	F1	1.86	1.78	1.11	1.09
11		Operator willing to operates the device in the future	F2	1.63		1.04	

4. Conclusion

This research has been developed a measurement evaluation measuring instrument that consists of the dimension of simplicity, learnability, memorability, informativeness, and trust ability. FES devices have been tested using the usability measuring instrument developed in this research. The result of the validity test on a shared probability questionnaire indicates that the proposed question has been valid to meet the chosen usability variable. While the reliability test results show that the simplicity and learnability dimension are at moderate reliability, the memorability dimension is at a very reliable level, and the informativeness and trust ability dimension are at a low reliability level. From the results of data processing by looking for average and standard deviation, the results of the variables B1, B2, B3, C1, C2, D1, and D2 were well applied to the FES devices designed on this research. While the variables E1, E2, F1, and F2 are still not properly applied to FES devices designed on this research.

References

[1] M. Brainin and W. D. Heiss, Textbook of Stroke Medicine, New York: Cambridge University Press, 2010.

[2] D. D. A. D. Silva, D. N. Venketasubramanian, D. A. A. Roxas, Jr., D. L. P. Kee and D. Y. Lampl, Understanding Stroke a Guide for Stroke Survivors and Their Families, 2014. Retrieved from moleac.com/ebook/Understanding_Stroke_-_Guide_for_Stroke_Survivors.pdf

[3] A. Arifin, H. Saito and T. Watanabe, "An Error Reduction Method of Portable, Low-Cost Joint Angle Sensor System," *IEICE Technical Report*, 2008. Retrieved from https://www.researchgate.net/publication/231316061_An_Error_Reduction_Method_of_Po

rtable_Low-Cost_Joint_Angle_Sensor_System_for_Human_Movement_Measurement_and_Control

[4] F. Arrofiqi, "Pengembangan Perangkat Wearable untuk Pengukuran Gerakan Manusia dan Rehabilitasi Kemampuan Berjalan Menggunakan Functional Electrical Stimulation System," *Jurusan Teknik Elektro*, Fakultas Teknologi Industri, Institut Teknologi Sepuluh Nopember, Surabaya, 2015.

[5] Widjaja ES, Arifin A, Arrofiqi F, Nuh M. Sistem Restorasi Gerak Sendi Siku Menggunakan Functional Electrical Stimulation. *Jurnal Teknik ITS*. Lembaga Penelitian dan Pengabdian kepada Masyarakat ITS; 2016 Sep 30;5(2). doi.org/10.12962/j23373539.v5i2.16067.

[6] S. Setiawan, "Studi Eksperimental Wearable Functional Electrical Stimulation pada Pergerakan Knee Joint dengan Menggunakan Sistem Kendali Cycle-to-Cycle Control," Jurusan Teknik Elektro, Fakultas Teknologi Industri, ITS, Surabaya, 2016.

[7] K. Leonardo, "Evaluasi Usabilitas pada Rancang Bangun Sistem Restorasi Gerak Pergelangan Tangan dan Sendi Siku menggunakan Functional Electrical Stimulation," Surabaya, 2020.

[8] Kim J, Han SH. A methodology for developing a usability index of consumer electronic products. *Int. J. Ind. Ergon.* Elsevier BV; 2008 Mar;38(3-4):333–45. doi.org/10.1016/j.ergon.2007.12.002.

[9] M. Ylikulju, "PLANNING A SUMMATIVE USABILITY TEST FOR A MEDICAL," 2018.

[10] W. Quesenbery, "Dimensions of Usability: Defining the Conversation, Driving the Process," *Proceedings of the UPA 2003 Conference*, 2003.

[11] M. Wklund, J. Kendler and A. Strochlic, "Usability Testing of Medical Devices," Taylor & Francis Group, Boca Raton, 2016.

[12] M. a. H. p. R. Agency, Human Factors and Usability Engineering – Guidance for Medical Devices Including Drug-device Combination Products, 2017.

IOP Conf. Series: Materials Science and Engineering 1072 (2021) 012004 doi:10.1088/1757-899X/1072/1/012004

QFD applications for quality improvements online student learning

M H Assiddiqi[1*] and I Vanany[2]

[1]Department of Technology Management, Institut Teknologi Sepuluh Nopember, Surabaya, Indonesia.
[2]Department of Industrial and System Engineering, Insitut Teknologi Sepuluh Nopember, Surabaya, Indonesia

Email: *hasbijobs@gmail.com

Abstract. Online student learning is in great demand by students both junior and senior high school, but there are still dissatisfaction and complaints about online student learning, such as software factors, teaching factors, customer services, and learning materials. This study aims to identify the quality of online student learning based on the dissatisfaction of junior and senior high school students and to give improvement priority to improve the quality of online student learning using the QFD (Quality Function Development) method. QFD, which is arranged with 2 phases and each phase is arranged based on VOC (Voice of Customer) of junior and senior high school students. Phase 1 has two dimensions. The First dimension is the process dimension, and the second dimension is the factor dimension. Process dimension phase 1 was designed using VOC and online student learning process to get the rank of dissatisfaction of junior and senior high school students in the online student learning process. Factor dimension phase 1 was designed using VOC and online student learning critical factor to get the rank of junior and senior high school student dissatisfaction from online student learning factor while in phase 2 QFD was designed using online student learning critical factor and improvement programs to get priority ranking and effectiveness of these improvement activities. So that the two phases of QFD that have been designed can identify problems in the processes that occur in the online student learning and the factors that influence them and make effective improvements for junior and senior high school education.

1. Introduction

Technological progress in the current era of globalization is so fast and has greatly influenced life in all fields, including education. Offline student learning is one of the institutions that focus on the field of education, such as Primagama, Ganesha Operation. The offline student learning is increasingly in demand by students and parents. This is due to the establishment of a minimum passing grade standard set by the government, the level of competition in schools, or competition in university entrance exams. Offline student learning has also begun to utilize technology to achieve progress. One of the technological advances in student learning is the change of learning methods that were previously offline to online. Online student learning is a study guide that uses learning methods that can be done without having to face to face and do not depend on time, distance, and transportation costs. Online student learning provides learning via the internet with a view to saving time, resources in implementing learning activities. The use of online student learning methods has now been used by several online tutors, such as Ruangguru, Zenius.

Based on Table 1, it can be concluded that has many users, and features are more attractive compared to offline student learning, but the quality for the online student learning process still has some dissatisfaction. There are several things that influence the dissatisfaction in the learning process of online student learning, namely the factor of the software used, the ease of users in using online student learning, the quality of the instructor who is lacking in providing lessons to features in online student learning and incomplete material.

Table 1. Student learning comparison.

Student Learning	Years	Student	Time Learning	Place	Features
Primagama (Offline)	1982	90,000/years	Determined by the institution	Primagama	Book, Try-out Offline, Classroom
Ruangguru (Online)	2014	6 million users	Determined by the student	Everywhere	Learning Video, Try-out Online, Application

Online student learning has students from different levels of education, and the majority of users are junior and senior high school students. Junior and senior high school students have different characteristics, such as the ability to master technology and an understanding of learning material. In addition, senior high school students have more material and exams needed than junior high school students, such as TPA lesson/test and university entrance examination exams, which can cause the level of need or dissatisfaction from each level of education to be different from online student learning. From this description, it can be seen that the organizers of online student learning need to improve their quality at every level of education, such as junior high and senior high school.

The application of QFD has been implemented in many research areas. Authors [1] uses QFD to identify key process and critical halal factors and propose priority improvement programs for halal meat factories. Author [2] uses the QFD to uses QFD to identify quality procedures and propose improvements in service quality at ATPM authorized workshops. Author [3] applied the QFD to select process of sustainable production indicators. Process dimension phase 1 QFD is designed using VOC and online student learning process to get the junior and senior high school student dissatisfaction ranking in the online student learning process. Factor dimension Phase 1 QFD was designed using VOC and online critical factor lessons (critical factors in online lessons) to get the junior and senior high school student dissatisfaction ranking from online student learning factors, Phase 2 in proposed model was designed using online problems critical factors and improvement action to get the ranking of activities priority or improvements priority. Two phases of QFD that have been designed can identify problems in the processes that occur in online student learning and the factors that influence them to make effective improvements for junior and senior high school education.

2. Research Design

The research method was carried out by distributing questionnaires, then weighting the questionnaire and from the weighting values in designed process dimension phase 1 QFD, dimension factor phase 2 QFD, and QFD phase 2 to determine the ranking of dissatisfaction and the effectiveness ranking of improvement programs.

2.1 Characteristics of Online Students Learning

The characteristics of online student learning are carried out from the demographic aspect of the level of education so that the questionnaire data will be grouped based on the level of education, namely junior and senior high school. Characteristics are done in order to identify the needs and dissatisfaction of junior and senior high school students for the quality of online student learning.

2.2 Phase 1: Process dimension

Process Dimension phase is designed using the VOC of junior and senior high school students with the online student learning process. The making of the questionnaire will be classified according to the dimensions of e-service quality, according to [4] and the attributes are adjusted to online student learning. Attributes obtained from complaints contained in the online student learning comments column on google play and then checking the suitability between complaints on google play with complaints known by online student learning management. The attributes are described in Table 2.

Table 2. Quality attributes dimensions online student learning.

Dimensions	Attributes
Efficiency	1. Ease to install, register, and log in 2. Ease of running online student learning applications 3. Ease of finding the desired material and information
Fulfilment	1. Completeness of learning material 2. Completeness of the test and discussion 3. The ability of online tutors instructors in explaining the material
System availability	1. The ability to have no problems (errors) during install, registration, login 2. The ability of online student learning applications to have no problems (errors) when running. 3. Update online student learning application
Responsiveness	1. Responsiveness and handling of customer service in answering questions and complaints of students

The VOC obtained from the student questionnaire is then entered into the matrix, and the relative dissatisfaction calculation is used to determine the priority of dissatisfaction with the process in online student learning such as (1) install, registration and login, (2) learning text, (3) test and test answer, and (4) review. A Likert scale value of the VOC from 1 to 5. The VOC will be weighted relations with the online student learning process based on the perception of online student learning management with the [5]. Factor Dimension phase is designed using the voice of customers of junior and senior high school students with online student learning critical factors is shown in Table 3. Factor dimension phase 1QFD will be generated online student learning critical factor priorities. Adding a score will determine the priority of things that must be improved.

Table 3. Online student learning critical factor references.

Process	Online Student Learning Critical Factor	References
Install, Registration and Login Learning Text /Video Material Test and Test Answer	Technology	[6,7,8]
Install, Registration and Login Learning Text Video Material Test and Test Answer Review	Student	[6,7,8]
Learning Text /Video Materials Test and Test Answer	Instructors	[6,7,8]
Review	Support	[9,10]
Learning Text /Video Materials Test and Test Answer	Learning content	[10,11]

3. Data Analysis

3.1 Voice of Customer

The Voice of a customer is obtained through a questionnaire. The questionnaire is intended for respondents who use online student learning. The majority of online student learning users are junior and senior high school students, so the respondent data used are students from junior and senior high schools. Respondents filled out questions regarding the quality attributes of online student learning dimensions (see Table 4).

Table 4. Respondents data.

Level Education	Number of Respondents			Total
	Online Student Learning A	Online Student Learning B	Online Student Learning C	
Junior High School	81	1	0	82
Senior High School	138	4	2	144
Total	219	5	2	226
Percentage	96.90%	2.21%	0.8%	100%

Respondent data that can be used only one online student learning is online student learning A. The respondent data from online student learning has a percentage of 96.90%. Respondent data from another online student learning are not used because it does not meet the required amount of data. Phase 1 QFD is based on the voice of customer from online student learning A and improvement programs designed to be used for online student learning online A.

Junior high school student dissatisfaction higher than senior high school students is described, such as(1) the ability of online student learning applications to have no problems (errors) when running, (2) ease of running online student learning applications, (3) the ability to have no problems (errors) during install, registration, login, (4) ease to install, register, log in, (5) ease of finding the desired material and information, (6) responsiveness of customer service in answering questions and complaints of students, (7) the ability of customer service in answering questions and complaints of students, and (8) update the online student learning application

From the dissatisfaction, it can be concluded that junior high school students tend to have higher dissatisfaction with the use of application systems compared to senior high school students. This can happen because junior high school students do not have much experience using smartphones or know online student learning than senior high school students so that in junior high school students feel more difficult and ask more about customer service and because a large number of questions allows customer service to not respond properly.

The dissatisfaction of senior high school students that is higher than junior high school students is described, such as (1) completeness of the test and discussion, (2) completeness of learning material, (3) the ability of online tutors instructors in explaining the material. From the dissatisfaction, it can be concluded that senior high school students tend to have high dissatisfaction with the content of learning materials ranging from examinations to instructors. This can happen because the subjects of learning and variations of senior high school student exams are more than junior high school students, such as TPA, UNAS, SBMPTN, and UM. Dissatisfaction comparisons are described in the Table 5.

Table 5. Junior high school and senior high school dissatisfaction comparison.

Dissatisfaction Ranking	Attributes of Junior High School Quality Dimensions		Attributes of Senior High School Quality Dimensions	
	Score	Attributes	Score	Attributes
1	2.82	The ability of online student learning applications to have no problems (errors) when running.	2.68	The ability of online student learning applications to have no problems (errors) when running
2	2.75	The ease of running online student learning applications	2.60	Complete exam and discussion
3	2.70	The ability to have no problems (errors) during install, registration, login	2.50	The ease of running online student learning applications
4	2.60	Ease to install, register and log in	2.34	Completeness of learning material
5	2.58	Ease of finding the desired material and information	2.32	Ease of finding the desired material and information
6	2.51	The responsiveness of customer service in answering questions and complaints of students	2.27	The responsiveness of customer service in answering questions and complaints of students
7	2.50	The ability of customer service in answering questions and complaints of students	2.25	The ability of customer service in answering questions and complaints of students
8	2.32	Completeness of learning material	2.23	The ability of online tutors in explaining the material
9	2.21	Complete exam and discussion	2.03	Ease to install, register and log in
10	2.06	Update online student learning applications	1.86	The ability to have no problems (errors) during install, registration, login
11	1.84	The ability of online tutors in explaining the material	1.67	Update online student learning applications

3.2 Process Dimension Phase 1 QFD Analysis

In the process dimension phase 1 QFD was designed using the results of dissatisfaction questionnaires for online lessons and online process lessons. (see Table 6). The value of online student learning process dissatisfaction from junior high school students which is higher than senior high school students is described, such as: (1) Install, Registration and Login, (2) Review. The dissatisfaction of the online student learning process is connected to the dissatisfaction of the dimensions of the online student learning quality dimension so that it can be concluded that junior high school students tend to have a higher dissatisfaction with the use of the application system compared to senior high school students. This can happen because junior high school students do not have much experience using smartphones or know online student learning than senior high school students so that in junior high school students feel more difficult and ask more about customer service. Because a large number of questions allow customer service not to respond properly.

Table 6. Online student learning process dissatisfaction comparison.

Dissatisfaction Ranking	JHS Online Student Learning Process		SHS Online Student Learning Process	
	Score	Details	Score	Details
1.	4.33	Learning Text/Video Materials	4.57	Learning Text/Video Materials
2	3.69	Test and Test Answer	3.85	Test and Test Answer
3	1.84	Install, Registration and Login	1.70	Review
4	1.74	Review	1.46	Install, Registration and Login

The dissatisfaction of senior high school students which is higher than junior high school students is described, such as: (1) Learning Text / Video Materials, and (2) Test and Test Answer. From the dissatisfaction, it can be concluded that high school students tend to have high dissatisfaction with the content of learning materials ranging from exam variations to instructors. This can happen because the subjects of learning and variations of senior high school student exams are more than junior high school students, such as TPA, UNAS, SBMPTN, and UM.

3.3 Factor Dimension Phase 1 QFD Analysis

Factor dimension phase 1QFD was designed using the results of the dissatisfaction questionnaire on online student learning and online student learning critical factors. The comparison of the critical factors dissatisfaction with online student learning is described in the Table 7.

Table 7. Online student learning critical factor dissatisfaction comparison.

Dissatisfaction Ranking	SHS Online Student Learning Critical Factor		SHS Online Student Learning Critical Factor	
	Score	Factor	Score	Factor
1.	4.77	Technology	4.4	Technology
2	2.16	Student	2.28	Learning Content
3	1.93	Learning Content	2.05	Student
4	1.67	Support/ Customer Services	1.97	Instructor
5	1.62	Instructor	1.63	Support/ Customer Services

The online student learning critical factor dissatisfaction score for junior high school students which is higher than senior high school students are described, such as: (1) Technology, (2) Student, (3) and support / customer Services. The dissatisfaction of online student learning critical factors is connected to the dissatisfaction of the dimensions of online student learning quality dimensions. It can be concluded that junior high school students tend to have a higher dissatisfaction with the use of the application system than senior high school students and is evidenced by the dissatisfaction of student factors that are ranked 2. This can occur because junior high school students have not had much experience using a smartphone or getting to know online student learning than senior high school students so that in their use junior high school students feel more difficult and ask more about customer service and because a large number of questions allows customer service to not respond properly.

The dissatisfaction of senior high school students that is higher than junior high school students is described, such as: (1) learning content, and (2) instructor. From the dissatisfaction, it can be concluded that senior high school students tend to have high dissatisfaction with the content of learning materials ranging from examinations to instructors. This can happen because the subjects and variations of senior high school student exams are more numerous than junior high school students, and the difficulty level of senior high school learning materials is more difficult than junior high school.

3.4 Phase 2QFD Analysis

In phase 2QFD, it was designed to use online student learning critical factors problem and improvement programs. Online student learning critical factor problems are based on existing problems of junior and senior high school dissatisfaction students. The QFD phase 2 aims to determine the priority of improvements that must be done first.

The highest improvement programs are found in IT staff training programs on mastering safe, efficient, and efficient programs so that improvement is a priority for improvement programs that must be carried out. There are differences in the ranking of effective improvement programs between junior and senior high school students. These differences cause the priorities of improvements that can be done will be different. There are several effective improvement programs for junior high school students higher than senior high school students, and it can be concluded that the level of education has an effect on effective improvement programs. The value of effective improvement dissatisfaction programs for junior high school students is higher than senior high school students described, such as: (1) training IT staff on mastering an efficient, safe and attractive program, (2) provide clear information regarding instructions on each feature or application usage, (3) check the program and test the program before uploading the program on various types of android and smartphone versions, (4) create a place or feature so that students can ask questions and report complaints about the application system, and (5) training on customer service regarding online student learning application knowledge collaborating with IT experts and experts who can overcome these problems

Effective improvement programs are connected with dissatisfaction with the dimensions of online student learning quality dimensions and senior high student factors so that effective improvement programs that are carried out tend to improve the use of application systems so that junior high school students are much easier to use online student learning. The dissatisfaction of senior high school students that is higher than junior high school students is described, such as: (1) gather material that often appears on exams and collect various types of exams, (2) providing questions and learning discussions for students and instructors, (3) training on online student learning instructors on how to teach and deepen related knowledge, (4) discussion or ask the opinions of others about the location of errors how to teach course.

Improvement programs effectiveness is connected with dissatisfaction with the dimensions of online student learning quality dimensions and senior high student factors so that effective improvement programs undertaken tend to improve the quality of learning material, examinations, and instructors in order to meet the needs of senior high school students. The improvement programmes comparison is described in the Table 8.

Table 8. Improvement programmes comparison.

Ranking	Improvement Programmes (JHS)		Improvement Programmes (SHS)	
	Score	Details	Score	Details
1.	5.17	Training of IT staff on mastery of efficient, safe and attractive programs	4.71	Training of IT staff on mastery of efficient, safe and attractive programs
2	3.85	Provides clear information about the instructions on each feature or application usage	3.57	Provides clear information about the instructions on each feature or application usage
3	3.40	Check programs and program tests before uploading programs on various types of android and smartphone versions	3.06	Check programs and program tests before uploading programs on various types of android and smartphone versions
4	2.82	Create a separate place or feature so students can ask questions and report complaints about the application system	2.66	Gather material that often appears on exams and collect various types of exams
5	2.39	Gather material that often appears on exams and collect various types of exams	2.65	Create a separate place or feature so students can ask questions and report complaints about the application system
6	1.67	Training on customer service regarding online student learning application knowledge Collaborating with IT experts and experts who can overcome these problems	1.69	Provides a means of asking questions and learning discussions for students and instructors
7	1.42	Provides a means of asking questions and learning discussions for students and instructors	1.63	Training on customer service regarding online student learning application knowledge Collaborating with IT experts and experts who can overcome these problems
8	1.1	Training on online student learning instructors on how to teach and deepen related knowledge Discussion or ask for opinions of others about the location of errors how to teach it	1.4	Training on online student learning instructors on how to teach and deepen related knowledge Discussion or ask for opinions of others about the location of errors how to teach it

4. Conclusions

The highest dissatisfaction ranking of junior and senior high school students on the online student learning process is found in the learning process text /video materials and the lowest in the review process. The difference in the level of dissatisfaction between junior and senior high school students is in the process of install, registration, and login, where junior high school students have higher dissatisfaction while in the test and test answer process where senior high school students have higher dissatisfaction scores. That is because junior high school students do not yet have enough knowledge and experience about smartphones and online student learning compared to senior high school students, while senior high school students have more types of examinations and subjects than junior high school students.

The highest dissatisfaction ranks of junior and senior high school students on the online student learning critical factor are found in the technology factor. The difference in the level of dissatisfaction between junior and senior high school students is on the student factor where junior high school students are dissatisfied with technology while in learning content factors where senior high school students have a higher dissatisfaction value. That is because the factors of junior high school students who do not yet have enough knowledge and experience about smartphones and online student learning thus affect the various dimensions of online student learning quality attributes while senior high school students require more variations in subjects and examinations, such as TPA, UM (Independent Test).

The highest-ranking effectiveness of the improvement program for junior and senior high school students is found in the training program for IT staff on mastering an efficient, stable, and attractive program. The difference in the ranking effectiveness of improvement programs between junior and senior high school students. In the improvement program, development new feature to students can ask questions and report complaints about the application system where junior high school students score higher effectiveness while the program collects material that often appears on examinations and collects various types of examinations where senior high school students value higher effectiveness.

This is due to the needs and dissatisfaction at each different level of education. Junior high school students are beginners in the use of smartphones and online student learning more in need of direction or answers to questions or complaints on the online student learning application, while senior high school students have enough experience of online student learning more in need of a variety of material and examinations. Firstly, online student learning can carry out improvement actions in accordance with effective improvement programs ranking priority that researchers have made to improve their quality. Secondly, Future research can focus on specific or more specific online student learning and add elementary education level segmentation or regional segmentation.

References

[1] Vanany I, Maarif GA, Soon JM. Application of multi-based quality function deployment (QFD) model to improve halal meat industry. *J. Islam. Mark.* Emerald; 2019 Mar 4;10(1):97–124. doi.org/10.1108/jima-10-2017-0119.

[2] Desiawan VA. Penerapan quality function deployment dengan mengadopsi penggabungan metode service quality dalam upaya meningkatkan kualitas layanan pada bengkel resmi atpm. Universitas Indonesia; 2010. Retrieved from: http://lib.ui.ac.id/file?file=digital/20297291-T29699-Victor%20Assani%20Desiawan.pdf

[3] Pusporini P, Vanany I. Selection process of sustainable production indicators using eco-quality function deployment. *IOP Conference Series: Materials Science and Engineering*. IOP Publishing; 2019 Jun 12;528:012019. doi.org/10.1088/1757-899x/528/1/012019.

[4] Parasuraman A, Zeithaml VA, Malhotra A. E-S-QUAL. *J Serv Res*. SAGE Publications; 2005 Feb;7(3):213–33.doi.org/10.1177/1094670504271156.

[5] Akao Y. Quality function deployment: integrating customer requirements into product. *Productivity Press*. Cambridge; 1990.

[6] Dillon and Guawardena. A Framework For The Evaluation Of Telecommunications-Based Distance Education, Milton Keynes: Open University, 1995. Retrieved from http://www.ndsl.kr/ndsl/search/detail/article/articleSearchResultDetail.do?cn=NPAP0256956

[7] Leidner DE, Jarvenpaa SL. The Information Age Confronts Education: Case Studies on Electronic Classrooms. *Inf. Syst. Res.* Institute for Operations Research and the Management Sciences (INFORMS); 1993 Mar;4(1):24–54. doi.org/10.1287/isre.4.1.24.

[8] Volery T, Lord D. Critical success factors in online education. *Int. J. Educ. Manag*. Emerald; 2000 Sep;14(5):216–23. doi.org/10.1108/09513540010344731.

[9] Benigno V, Trentin G. The evaluation of online courses. *J Comput Assist Learn*. Wiley; 2008 Oct 9;16(3):259–70. doi.org/10.1046/j.1365-2729.2000.00137.x.
[10] Govindasamy T. Successful implementation of e-Learning. *Internet High Educ*. Elsevier BV; 2001 Jan;4(3-4):287–99. doi.org/10.1016/s1096-7516(01)00071-9.
[11] Papp R. Critical success factors for distance learning. *Americas Conference on Information Systems*. Long Beach, California; 2000.

IOP Conf. Series: Materials Science and Engineering 1072 (2021) 012005 doi:10.1088/1757-899X/1072/1/012005

Inventory model for multi-perishable goods with limited storage capacity

F F Afiffi[1]* and N I Arvitrida[1]

[1]Department of Industrial Engineering, Sepuluh Nopember Institute of Technology, Surabaya, Indonesia

Email: *azrifziaf@gmail.com

Abstract. The warehouse has an important role, namely storing finished products as well as semi-finished, especially in the Agrifood Supply Chain (ASC) which has special characteristics such as limited shelf life, changing variations in demand and prices that can be adjusted to make ASC look more complex and more difficult than other supply chains. In this study, various thoughts are offered to fit the real conditions. Stochastic demand distribution, the presence of constant lead time on each product, exponential decay rates, to limited storage space capacity. Related to this assumption, the ongoing review (Q, r) is aimed at supply policies that are relevant to the agricultural-food supply chain problem. The development of a mathematical model with decision variables Q and r to minimize costs on warehouse procurement is then analyzed using a comparison using parameter ordering costs, storage costs, and shortage costs that have been determined in this study. on items with limited warehouse capacity. In previous research, the process of storing and procuring multi-products did not contain short-lived products. Also, other studies also explain the storage of products in a short shelf life does not contain a shortage of limited storage capacity. It can be done on research that will be carried out adjusted to conditions that occur in the real world It is estimated that it will make a practical contribution and be a reference model of inventory policies for warehouses that have relevant requirements.

1. Introduction

Inventory control is described as the design and management of storage policies for raw materials, semi-finished goods, and finished goods and one of them is the perishable product. Currently, perishable products have deep challenges regarding the process [1] of processing to product storage process in various commodities e.g. medicines, health products, blood storage, agricultural products, Livestock products that have an expiry date. [2] mentions that any perishable items must have shelf life and damage inherent to the product itself. So according to [3], there is increasing complexity to determine the optimal booking in the inventory as it is added perishable into research objects that have different properties compared to non-perishable products. However, when demand is assumed to be random and perishable in accordance with the situation will show results of unsatisfactory performance when resolved by a classic approach without entering the perishable variable.

According to [4], commodity products perishable has a variety of difficult constrains one of which is random lifetimes that have high flexibility to the product life and are affected by some aspects surrounding items such as, temperature, product handling, etc. Coupled with the demand that is always changing in each period will complicate the inventory policy used. By classifying product Inventory system perishable into two categories namely, random lifetime and fixed lifetime.

In general, perishable products have a limited lifespan, such as garment products, blood, medicines, and agricultural products will be depleted if not used before the expiry date, resulting in the cost of the result of damaged products and the wasted [3]. In fact, according to [5], products such as medicines and food require special attention because it has a shelf life restriction that cannot be simply ignored. Which will directly impact the performance of inventory management as described by [6], which records when at least 10% of products-perishable damaged and become garbage before consumers buy. [7] mentions in the United States in 2006 at least 10.9% of blood platelets become damaged because it has expired the life of the bait as a result of not being done.

In previous research, the first category of research studies was conducted by [5], [8] to [10] implemented the model inventory Replenishment continuous review (Q, R) policy, single item. Among these three studies, [5], [8] which included a constant shelf life of the product with a deterministic lead times distribution. Whereas [9] did not incorporate a lifetime variable or zero lifetime in its research with a constant lead times distribution, likewise with [10] with a stochastic lead times distribution. The second category of previous research is about perishable items that have been done by [2], [5], [11], [12] mentioning when their models were either used on product storage models using various approaches and constraints and assumptions. However, these studies have not found any studies considering the limitations of warehouse and multi-item capacity therein. Similarly, research conducted by [9], [10] which entered the warehouse capacity limit with a single item, also with [13] with multi-item, but no one considering shelf life period or decaying on the stored product.

Preceded by research conducted by [2], which describes the lot size and reorder point models are appropriate used in inventory that has a gradual change in shelf life due to product damage or decaying using Approach method and analysis of Hadley – Within and use the simulation to develop the model (Q, R). Assuming the inventory review is done routinely or continuously. [2] also explained the relationship of the decaying model and perishable inventory and using the parameters cost, there are setup or ordering cost, replacement cost, holding cost, and atom cost.

Plus, the research has been done by [10] which uses mathematical models and focuses on the problem of limited storage capacity in the warehouse. Unlike [2], research by [10] using four components cost, among others, ordering cost, inventory holding cost on internal storage, inventory holding cost on external storage, and atom cost. Assuming demand and lead time are stochastic, so the demand for lead times is used to use empirical distributions. The development of model conducted by [14], is known to have several assumptions one of them is to use (Q, R) inventory policy on a single item that carries out the procurement continuously.

The purpose of this research is to develop a mathematical model that is appropriate and in accordance with continuous review details (Q, r) that consider the product's shelf life and storage space capacity in the form of simple simulations to determine the flexibility (Q, r) policy of multi-item products. In addition, also to find out the influence of shelf life (lifetime) on the product cost incurred by the warehouse, so as to minimize the costs that must be incurred by the warehouse to store the product in accordance with the demand as a consideration of the warehouse in determining the capacity needed.

2. Literature Review

In previous studies, the first category of research studies was conducted by [8], [5], [9], [10] applied a continuous review inventory replenishment model (Q, r) policy, single item. Among the three studies [8] and [5] included a constant product shelf life with deterministic lead time distribution. Whereas A. [9] did not include the variable lifetime or zero lifetime in his study with a constant lead time distribution, as well as [10] with a stochastic lead time distribution.

The second category of previous research is about perishable items that have been conducted by [11], [12], [5], and [2], said that if their model is well used in the product storage model by using various approaches and limitations and assumptions. However, none of these studies have found any research that considers the limitations of warehouse and multi-item capacity. Same with research conducted by [9] and [10] who put a limit on warehouse capacity with single items, also with [13] with multi-items, but none considered the shelf life or decaying of stored products.

Preceded by research conducted by [2], which explains the lot size and reorder point models that are appropriate for use in inventories that have gradual changes in shelf life due to product damage or decaying using the Hadley - Within method and analysis approach and using simulations for develop a model (Q, r). Assuming inventory review is done regularly or continuously. [2] also explain the relationship of decaying models and perishable inventory and using cost parameters among others, setup or ordering costs, replacement costs, holding costs, and shortage costs.

Added research has been conducted by [10], which uses a mathematical model and focuses on the problem of limited storage capacity in the warehouse. In contrast to [2]'s research, [10] used four cost components including ordering costs, inventory holding costs in internal storage, inventory holding costs in external storage, and shortage costs. Assuming demand and lead time are stochastic, so the demand for lead time just happens using an empirical distribution. The model development carried out by [14], is known to have several assumptions, one of which is to use (Q, r) inventory policies on single items that carry out continuous procurement. According to [14], the fundamental difference between a continuous review and periodic review is at the time of different replenishment orders. In Continuous review, an order is carried out if the inventory position on-hand stock has reached point R or lower than R. While the periodic review depends on the specified review time interval. To support multi-item research, so based on previous research conducted by [15], which explains the heuristic model used in the problem of multiple product procurement with dynamic demand conditions coupled with limited warehouse capacity. By adopting a previous study conducted by [16] for single-item lot sizing with no well-proven capacity, it was able to increase lot-for-lot schedules by combining procurement based on cost-saving priority rules on multi-item capacity problems. Perishable product is an additional problem for the model that will be developed based on previous research conducted by [2], which explains and considers perishability into two general classifications based on product age, namely, fixed lifetime and random lifetime. By adding a multi-product model to be more difficult, this is explained by [4] with a multi-product perishable inventory model where when there are two products, one product has a limited product life of m periods, and one product has a product life infinite. Product units that have a limited product life in the period have a holding cost that is greater or equal to the holding cost of products that have an unlimited shelf life. Same with the procurement cost or procurement costs of products with a limited shelf life will certainly have a greater frequency of ordering compared to products that have an unlimited shelf life.

The formulations listed above are used assuming zero inventory at the beginning and end of the period. Problems using the assumption of initial and ending inventory above zero or positive can be easily solved by allocating initial inventory to meet the demand for the first period plus ending inventory needed to meet demand in the last period. For the record, multi-item problems with limited warehouse capacity result in many problems that arise suddenly and unexpectedly, thus requiring a good initial inventory. Because zero inventory levels at the beginning of the period for all items will have a bad impact going forward.

3. Optimization Model

3.1 Overview

This study uses various parameters that affect the decision variables in this study. The parameters used in this study include, CP = Replenishment Cost (IDR / order), CH = Holding Cost (IDR / unit), CS = Shortage Cost (IDR / unit). Whereas in this study there are 2 decision variables, among others, Q = Replenishment order quantity (unit) and r = Reorder point (unit).

Likewise, the variables that support this research. Among them have a value that has been given previously as variable warehouse capacity which is denoted W, the existence of warehouse capacity in this study is a limitation needed so that inventory levels in the warehouse do not exceed available storage capacity. Then D or d represents the demand in units of units and μD or μd which represents the average demand in units of units within a certain period. The existence of τ and Lt as the Leadtime and Lifetime owned by each commodity product is certainly different so that different reorder points for each commodity are also different.

Every time replenishment is done, several units are damaged within a certain period, ψ in units. So, the expectation of a damaged product at the lead time, $\psi\tau$, due to the influence of Lt on perishable products that limits the shelf life of a product. This making product damage. Same with the average unit demand which may occur at a lead time, $\mu d\ \mu\tau$.

Parameter

C_P	Replenishment Cost (IDR/order)
C_H	Holding Cost (IDR/unit)
C_S	Shortage Cost (IDR/unit)

Variable

W	Warehouse capacity (unit)
D	demand (unit)
μD	Demand rate (unit)
τ	Lead time (hari)
Lt	Life time (hari)
ψ	The unit that was decaying in the period (unit)
$\mu d\ \mu\tau$	Lead time demand average (unit)
$\psi\tau$	Product expectations are decaying at the lead time (unit)
PS	The probability of a unit shortage that is assumed to be lost sales (unit)
ES	Expect the number of items that are shortage (unit)
IP	Random level of inventory position when ordering (unit)
Wpr	Warehouse priority (Ton)
Pi	Price product-*i* *(IDR)*
Tp	Percentage of price weighted *(%)*
Td	Percentage of demand weighted *(%)*

Decision Variable

Q	Replenishment order quantity (unit)
r	Reorder point (unit)

Assumption

(1) The product under study is part of food products that are included in the list of necessities that have a finite lifetime.
(2) If the inventory position reaches or below r point, it will be ordered again.
(3) Data demand or demand is stochastic.
(4) Leadtime is constant or fixed in each commodity.
(5) One particular commodity has a fixed shelf life and lead time at every replenishment.
(6) Do not consider the existence of joint replenishment between products.
(7) Storage warehouse capacity is assumed not to change.
(8) The warehouse also functions as a market assuming demand of 5% of the total needs of East Java.
(9) If there is a shortage, it will be assumed as lost sales.
(10) Data were available in the form of historical demand data, lead time of each product, warehouse capacity, and costs that become calculation parameters.
(11) The cost of money on holding costs is assumed to be the same for each product.
(12) The reduced lifetime of the product does not affect the price reduction of the product, assuming the price of the product is considered to be fixed as long as the product's shelf life is still thereby eliminating the cost of quality.

4. Inventory Model

All items carried out in a certain period do not exceed the available warehouse capacity. This is done to increase the reserves issued and reduce the risks arising from excess product stock, $W \geq \sum_{i=1}^{di} \sum_{t=1}^{dit} dit$, where W as fixed warehouse capacity is the limit in ordering products $i = 1,2, \ldots, N$ to meet demand d in periods $t = 1,2, \ldots, T$. So, the number of products owned in the warehouse is not permitted to exceed the available warehouse capacity. To find out the average demand for each product i is $\mu Di = Di / T$.

For perishable products that have a limited lifetime so it is necessary to find out the number of products that are damaged in each order made,

$$\psi = (Q - \mu_{Di} T) / T \tag{1}$$

and we got,

$$\psi = Q / T - \mu_{Di} \tag{2}$$

at the inventory level during the cycle that is affected by damage and product demand. The existence of this linkage can be used to determine the safety stock that can be obtained through reorder points minus demand at the time of lead time and damage at the time of lead time.

$$s = r - \mu_{Di}\tau - \psi\tau = r - Q\tau / T \tag{3}$$

4.1 Inventory reviews and demand during stockouts.
Having an item with a limited shelf life, coupled with lead time and random requests, making perishable products requires continuous review of storage to anticipate the occurrence of stockout, with the probability of the occurrence of demand with a quantity higher than the stock of goods available at the warehouse, $PS = P\ (D > EOH_{ordering})$. This can be said if the stock of products available at the warehouse has reached the reorder point and makes the incoming demand is higher than the reorder point, $PS = P\ (D > r)$ and means that through the probability of a stockout demand,

$$PS = \sum_{d=r+1}^{Dmax} f(d) \tag{4}$$

we can get an estimate of the number of stockout units that cannot be met,

$$ES = \sum_{d=r+1}^{Dmax} (D - r) f(d) \tag{5}$$

4.2 Average inventory management costs per cycle.
Assuming that on-hand inventory at the beginning of the cycle is equal to the number of order quantity, $OH_{begin} = Q$, then on-hand inventory in the middle of the cycle is the same as the reorder point, $OH = r$, the amount of on-hand inventory at the end of the cycle is equal to the number of items at the reorder point is added to the estimated product that is not met and then reduced by the average demand for the cycle, $OH_{end} = r - D + ES$, so that it is obtained,

$$OH_{begin} = Q + r - D + ES = Q + OH_{end} \tag{6}$$

In this model it is assumed to only use internal warehouse capacity without involving external parties by entering the external warehouse capacity so that expected over-ordering at warehouse capacity is zero, EO = 0, Over-ordering is not allowed to occur so that IP - W $\leq$ 0, this is intended to limit the number of product orders so as not to exceed the warehouse capacity limit previously set.

$$IP = Q + r + ES \tag{7}$$

From equation (7) above we can find out the inventory position, *IP*, by knowing the number of product orders of *Q*, plus the number of products at r, reorder points, plus the number of items estimated to have a shortage of *ES*, plus ψ which is an estimated number of products damaged during lead time. Inventory level in the warehouse can be determined. $IL = IP - D$, i.e. inventory position is reduced by the amount of demand for several products in one cycle. So, it is known that the total cost in the warehouse is,

$$TC = \sum \frac{\mu Di\ (CPi + CSi\ ESi + (CHi\ Qi/\ \mu Di)EOH)}{Qi + ESi} \tag{8}$$

IOP Conf. Series: Materials Science and Engineering 1072 (2021) 012005 doi:10.1088/1757-899X/1072/1/012005

With,

$$EOH = Qi/2 + Ri - \mu di\, \mu\tau i + ESi \tag{9}$$

4.3 Determine r and Q of each item.
The warehouse can determine the number of items to be ordered, Q, through the equation below,

$$Q = Lti\text{-}\mu Li\left(\sqrt{\frac{2\mu Di}{CHi}\left[CPi + CSi\,ES + \frac{CHi\,ESi}{\mu Di}\left(\mu i - ri - ES/2\right)\right.} - ES\right) \tag{10}$$

Equation (10) above, is obtained based on the basic formula looking for EOQ as initial Q i.e,

$$Q = \sqrt{\frac{2\mu Di\, CPi}{CHi}} \tag{11}$$

Then determine the reorder point with the equation,

$$r = \mu d\, \mu\tau + s \tag{12}$$

With restrictions, the number of items at the reorder point plus the order quantity of a product must not be more than equal to the warehouse capacity owned, r + $Q \leq W$.

4.4 Determine the capacity allocation of each item.
Considering the function of the warehouse is as a buffer stock of the needs and the limited capacity of the warehouse that is owned so it needs a factor as a difference for each product commodity based on product prices and demand on the product. Assuming priority over the capacity of the warehouse to meet demand and priority on product prices to control prices so that an equation is obtained,

$$WPr = \left(\left(\frac{Pi}{Pmin} \times \%Tp\right) + \left(\frac{\mu Di}{\mu Dmin} \times \%Td\right)\right) \times W \tag{13}$$

The allocation of warehouse capacity for each product is determined as a limitation on the value of Q not to order more than the predetermined WPr.

5. Numerical Example.
5.1 Determine r and Q.
Each of the commodities above has different cost parameters except for replenishment cost (*CP)* which has the same cost on each object commodity studied. *CP* of onion, *CP* of garlic, *CP* of red chili, and *CP* of cayenne have a value of IDR 94,687 each time a replenishment is performed. As shown in Table 1 above, each type of product has a holding cost (CH) and a shortage cost (CS) that varies depending on the supporting costs contained therein.

Table 1. Data input of each commodity.

	Commodity (IDR)			
	Onion	**Garlic**	**Red Chili**	**Cayenne**
Replenishment Cost (C_P)	Rp 94,687	Rp 94,687	Rp 94,687	Rp 94,687
Holding Cost (C_H)	Rp 266,554,475	Rp 268,754,975	Rp 281,152,475	Rp 292,923,975
Shortage Cost (C_S)	Rp 22,292,000	Rp 26,693,000	Rp 51,488,000	Rp 75,031,000
μD (ton)	4.3639	3.3836	4.6823	9.9807
μL (day)	1	1	1	1
Lt	4	4	5	6
Warehouse Capacity (ton)	70			

Generally, demand and prices on products will affect variable costs as an important part of finding the order quantity of a product, but demand and prices on products can also be used as a benchmark for determining the capacity of warehouses for each product commodity. As shown in Table 2 which explains if the distribution of warehouse capacity in each commodity by involving product price factors in a certain time period. By dividing the priority scale by 40% on the price of the product and 60% on the average demand for the product, then each product capacity available at the warehouse can be determined. Onion commodities that have a product sale price of IDR 22,292 per kg or IDR 22,292,000 per ton occupies the fourth position of the four commodities so that it becomes the last priority in determining warehouse capacity.

Table 2. Price priority to warehouse capacity.

Komoditas	Price (IDR)	Price Priority (40%)	X	Commodity Prio	WH Capacity (Ton)
Onion	22,292	4	1	1.1738	13.6307
Garlic	26,693	3	1.197425085	1.0790	10.5687
Red Chili	51,488	2	2.309707518	1.7542	14.6253
Cayenne	75,031	1	3.365826305	3.1162	31.1751
Total				**7.1232**	**70**

However, in the aspect of demand, shallots have a fairly high average demand of 6.9738 tons per day and occupy the second position of the four commodities, with the highest demand being cayenne pepper of 10.4998 tons per day. So in determining the capacity of the warehouse, it is necessary to have a simple calculation to get the warehouse capacity of each product where shallots get a portion of 16.48% of the total available capacity and cayenne pepper with the highest priority capacity of 43.75% of the total available capacity as shown in Table 3.

Table 3. Demand priority to warehouse capacity.

Komoditas	Ave. demand ton per day	Demand Priority (60%)	X	Demand %	WH Capacity (Ton)
Onion	4.3639	3	1.2897	16.42%	11.4968
Garlic	3.3836	4	1	15.10%	10.5677
Red Chili	4.8171	2	1.423	24.88%	17.4150
Cayenne	9.9807	1	2,949	43.60%	30.5206
Total	**22.5453**			**100%**	**70**

The existence of a warehouse capacity limit on each commodity has an impact on the $IL \leq W$ constraint, namely over-ordering of related products. This certainly will be very important for companies that have limited space in storing products so that it becomes difficult to get access to additional storage space or external warehouses. The distribution of warehouse capacity allocation for each product commodity can be seen in Figure 1 assuming the conditions that have been attached to the previous table. Then after the storage space capacity of each product is determined, the storage space capacity becomes a limit for each product in determining the number of products to be ordered for Q with no more than the warehouse capacity. That way each product has different W, Q, and r.

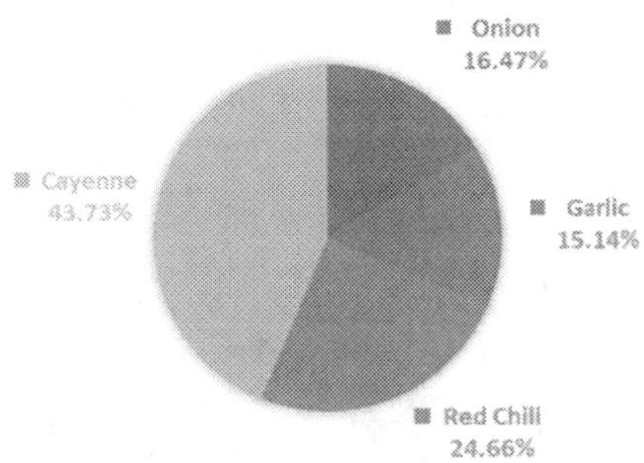

Figure 1. Warehouse capacity allocation of each commodity.

The model in this study can find out the value of each Q and r of each product. These results are listed in Table 4 which explains based on calculations made using certain assumptions and restrictions. For example, in this study using a fixed warehouse capacity of 70 tons. With constant lead time and exponential lifetime, and demand has taken in normal distribution refers to primary data obtained through the East Java food security service.

Table 4. Q and r value each commodity.

	Commodity			
Warehouse	**Onion**	**Garlic**	**Red Chili**	**Cayenne**
Total Warehouse Capacity (Ton)		70		
Warehouse Capacity (Ton)	11.5299	10.5981	17.2637	30.6084
Safety Stock (Ton)	6.6418	4.7571	6.6421	10.5402
Remaining Capacity (Ton)	4.8881	5.8410	10.6216	20.0682
Order Quantity (Q)	10.8704	8.9744	13.2310	25.9809
Order Quantity after constraint (Q)	4.2287	4.2174	6.5890	15.4407
Reorder point (r)	11.0056	8.1406	11.3435	20.5209

5.2 Cost and Revenue Analysis.

Based on the above model, the total cost in the warehouse can be known through equation (8), so that it can also be seen the optimal capacity limit needed for each commodity. This refers to the cost, revenue, and profit that can potentially be obtained by the warehouse. This certainly becomes a reference and input for each business model warehouse. However, there is a significant impact on the growing capacity of the warehouse to store these perishable products. As shown in Figure 2, which explains the costs that may be incurred by the warehouse for all related commodities.

At a warehouse capacity of 3 tons to 10 tons, the cost that may be incurred by the warehouse is very high or the highest among others. With a total average demand of 22.54 tons, the capacity of 3 tons certainly does not meet demand, so that the biggest cost burden lies in the shortage cost *(CS)*, which is the cost of not meeting the demand for each commodity. With a total cost of IDR 34,128,169,305, and revenue of only IDR. 4,789,759,823 makes 3 tons capacity not an option for warehouses that have high demand.

The high cost occurs to a certain capacity limit, so the difference from the revenue obtained by the warehouse becomes profit. Warehouse issued the lowest cost at the point of 27 tons warehouse capacity. Figure 3 and the detail in Table 5, it shows that the revenue obtained by the warehouse reaches the maximum point and cannot increase again at the point of 29.7 tons with IDR 35,355,860,515. Increased warehouse capacity, at this point, does not affect the revenue obtained by the warehouse. On the contrary, the costs that must be incurred by the warehouse actually increase, because the holding cost *(CH)* in the warehouse which continues to increase is not offset by increased demand.

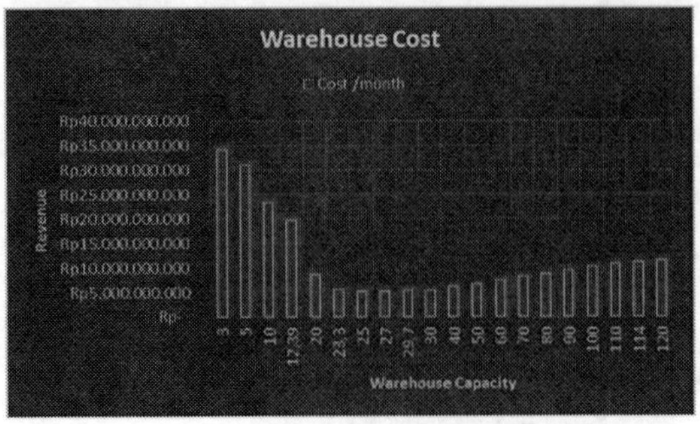

Figure 2. Cost per warehouse capacity graph.

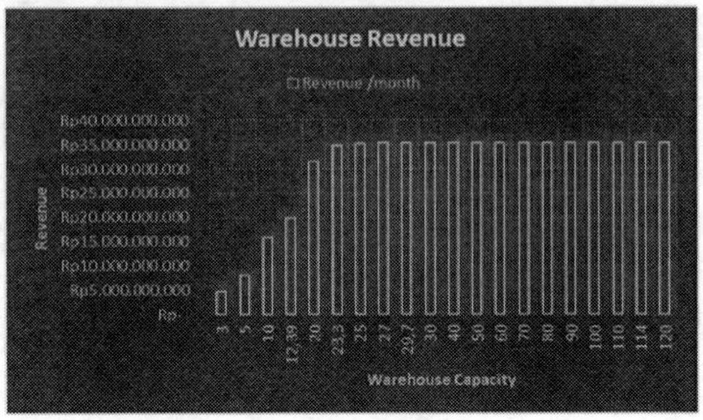

Figure 3. Revenue per warehouse capacity graph.

After reaching the highest point with the same demand and lifetime conditions and lead time, an increase in warehouse capacity results in a decrease in total profits for all commodities. This is due to the increasing burden of warehouse maintenance which is not offset by increased demand for these commodities.

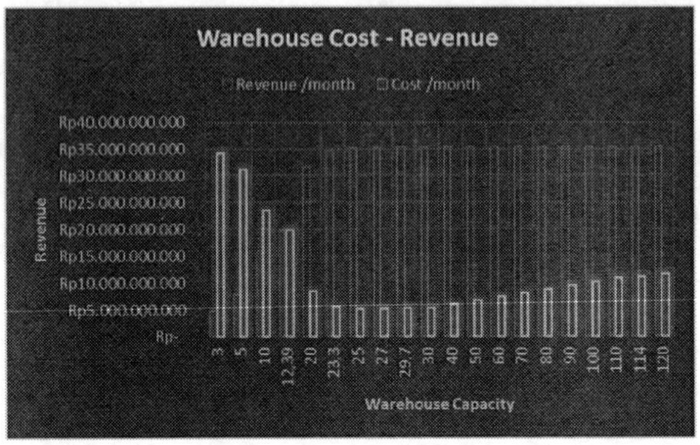

Figure 4. Revenue and cost pairing per capacity graph.

As shown in Figure 4 which compares the amount of warehouse revenue and expenditure to products by taking into account the available warehouse capacity, where the higher the warehouse capacity, the higher the expenditure charged to the storage warehouse. So that it affects the profit per month earned. So, to increase the benefits with a high warehouse capacity, it is necessary to increase the demand accordingly so that it does not burden the storage warehouse. The difference between revenue and cost decreases and becomes increasingly unprofitable to invest in warehouses with a large capacity if it is not matched by large demand.

Table 5. Cost, revenue, & profit per warehouse capacity.

Total WH Capacity	Cost /month (IDR)	Cost (%)	Revenue /month (IDR)	Revenue (%)	Profit per month (IDR)	Status
0	-		-	0.00%		
3	34,128,169,305	551.93%	4,789,759,823	13.55%	-29,338,409,482	Newsboy 4 products
5	31,068,320,232	493.48%	7,982,933,039	22.58%	-23,085,387,194	Newsboy 4 products
10	23,418,697,550	347.35%	15,965,866,077	45.16%	-7,452,831,472	Newsboy 4 products
20	8,514,669,269	62.65%	31,550,515,071	89.24%	23,035,845,801	Newsboy 4 products
23.3	5,484,749,397	4.77%	34,814,419,778	98.47%	29,329,670,380	Newsboy 4 products
25	5,236,392,851	0.03%	35,176,101,845	99.49%	29,939,708,993	Newsboy 4 products
27	5,234,953,494	0.00%	35,310,865,344	99.87%	30,075,911,851	Newsboy 4 products
29,7	5,369,945,916	2.58%	35,355,860,515	100.00%	29,985,914,599	Newsboy 4 products
30	5,389,944,537	2.96%	35,355,860,515	100.00%	29,965,915,978	Newsboy 4 products
40	6,069,523,696	15.94%	35,355,860,515	100.00%	29,286,336,819	Newsboy 4 products
50	6,746,167,568	28.87%	35,355,860,515	100.00%	28,609,692,947	Newsboy 2 products
60	7,424,989,234	41.83%	35,355,860,515	100.00%	27,930,871,281	Newsboy 1 product
70	8,103,621,526	54.80%	35,355,860,515	100.00%	27,252,238,989	Newsboy 0 product
80	8,785,189,105	67.82%	35,355,860,515	100.00%	26,570,671,410	Newsboy 0 product
90	9,469,407,911	80.89%	35,355,860,515	100.00%	25,886,452,604	Newsboy 0 product
100	10,152,395,790	93.93%	35,355,860,515	100.00%	25,203,464,725	Newsboy 0 product
110	10,834,815,549	106.97%	35,355,860,515	100.00%	24,521,044,966	Newsboy 0 product
114	11,115,463,834	112.33%	35,355,860,515	100.00%	24,240,396,681	Newsboy 0 product
120	11,515,436,261	119.97%	35,355,860,515	100.00%	23,840,424,254	Newsboy 0 product

References

[1] Beamon BM. Supply chain design and analysis: *Int J Prod Econ*. Elsevier BV; 1998 Aug;55(3):281–94. doi.org/10.1016/s0925-5273(98)00079-6.

[2] Nahmias S, Wang SS. A Heuristic Lot Size Reorder Point Model for Decaying Inventories. *Manage Sci*. Institute for Operations Research and the Management Sciences (INFORMS); 1979 Jan;25(1):90–7. doi.org/10.1287/mnsc.25.1.90.

[3] Siriruk P. The optimal ordering policy for a perishable inventory system. *Eng. Comput. Sci.* Proceedings of the World Congress; 2012:vol. 2, pp. 1453–55. Retrieved from: http://www.iaeng.org/publication/WCECS2012/WCECS2012_pp1453-1455.pdf

[4] Nahmias S. Perishable Inventory Theory: A Review. *Oper Res.* Institute for Operations Research and the Management Sciences (INFORMS); 1982 Aug;30(4):680–708. doi.org/10.1287/opre.30.4.680.
[5] Kouki C, Jemaï Z, Minner S. A lost sales (r, Q) inventory control model for perishables with fixed lifetime and lead time. *Int J Prod Econ.* Elsevier BV; 2015 Oct;168:143–57. doi.org/10.1016/j.ijpe.2015.06.010.
[6] B M Roberti *RFID Will Help Keep Perishables Fresh* Rfid J., pp. **1–2**, 2005.
[7] Fontaine MJ, Chung YT, Rogers WM, Sussmann HD, Quach P, Galel SA, et al. Improving platelet supply chains through collaborations between blood centers and transfusion services. *Transfus Med.* Wiley; 2009 Oct;49(10):2040–7. doi.org/10.1111/j.1537-2995.2009.02236.x.
[8] Berk E, Gürler Ü. Analysis of the (Q,r) Inventory Model for Perishables with Positive Lead Times and Lost Sales. *Oper Res.* Institute for Operations Research and the Management Sciences (INFORMS); 2008 Oct;56(5):1238–46. doi.org/10.1287/opre.1080.0582.
[9] Pan A, Hui C-L, Ng F. An Optimization of(Q,r)Inventory Policy Based on Health Care Apparel Products with Compound Poisson Demands. *Math. Probl. Eng.* Hindawi Limited; 2014;2014:1–9. doi.org/10.1155/2014/986498.
[10] Singha K, Buddhakulsomsiri J, Parthanadee P. Mathematical Model of (R,Q) Inventory Policy under Limited Storage Space for Continuous and Periodic Review Policies with Backlog and Lost Sales. *Math. Probl. Eng.* Hindawi Limited; 2017;2017:1–9. doi.org/10.1155/2017/4391970.
[11] Baron O, Berman O, Perry D. CONTINUOUS REVIEW INVENTORY MODELS FOR PERISHABLE ITEMS WITH LEADTIMES. *Probab Eng Inf Sci.* Cambridge University Press (CUP); 2017 Jun 5;1–26. doi.org/10.1017/s0269964817000225.
[12] Dobson G, Pinker EJ, Yildiz O. An EOQ model for perishable goods with age-dependent demand rate. *Eur. J. Oper. Res.* Elsevier BV; 2017 Feb;257(1):84–8. doi.org/10.1016/j.ejor.2016.06.073.
[13] Zhao X, Fan F, Liu X, Xie J. Storage-Space Capacitated Inventory System with (r, Q) Policies. *Oper Res.* Institute for Operations Research and the Management Sciences (INFORMS); 2007 Oct;55(5):854–65. doi.org/10.1287/opre.1070.0394.
[14] Singha K, Buddhakulsomsiri J, Parthanadee P. Optimal Cycle Service Level for Continuous Stocked Items with Limited Storage Capacity. *KnE Life Sciences.* Knowledge E; 2018 Mar 1;4(2):82. doi.org/10.18502/kls.v4i2.1659.
[15] Minner S. A comparison of simple heuristics for multi-product dynamic demand lot-sizing with limited warehouse capacity. *Int J Prod Econ.* Elsevier BV; 2009 Mar;118(1):305–10. doi.org/10.1016/j.ijpe.2008.08.034.
[16] Wagner HM, Whitin TM. Dynamic Version of the Economic Lot Size Model. *Manage Sci.* Institute for Operations Research and the Management Sciences (INFORMS); 1958 Oct;5(1):89–96. doi.org/10.1287/mnsc.5.1.89.

IOP Conf. Series: Materials Science and Engineering 1072 (2021) 012006 doi:10.1088/1757-899X/1072/1/012006

Modelling tank truck routing for city park watering problems in Surabaya

N I Arini[1], N Siswanto[1]* and A Rusdiansyah[1]

[1]Department of Industrial and Systems Engineering, Institut Teknologi Sepuluh Nopember, Surabaya, Indonesia

E-mail: *siswanto@ie.its.ac.id

Abstract. Surabaya City Government tries to have 20% public and 10% private green open space to obey regulation of Indonesian Law No. 26/2007, so that the City Government is actively increasing the number of public green space by creating city parks. With the ever-increasing amount of green open space, maintenance activities are needed, one of which is to do routine watering every day. The problem often faced by the City Government is determining the truck routes to optimize traveling time. Many variables can affect this traveling time, among others, demand, number of vehicles, vehicle capacity, time windows, operating hours and vehicle routes. In practice, a large number of watering locations, irregular routes, and uneven service area assignments often result in overtime which impacts the next shift working hours. The water truck routing problem can be categorized as NP Hard so that the process of scheduling and determining routes use Ant Colony. The objective of this study is to develop a model that can minimize the total routing time and the number of vehicles in the process of filling and watering from the depot to the park. The scenarios developed provides savings in terms of time minimization and distribution of assignments.

1. Introduction

According to Indonesian Law No. 26/2007, the city area should have minimum 30% of green open space, consisting 20% of public and 10% of private. To carry out this regulation, the Surabaya City Government is actively focusing on the addition number of parks, while keeping maintaining the existing parks by watering them every day. One of the problems faced by the city government in watering activities is determining the number of vehicles used to fulfill the needs of watering plants at each point, where each park has a different demand. In addition to demand at several points, there are still other variables which effects this activity such as vehicle capacity, shift time limits, operational working hours and vehicle routes. The variation and irregular values of these variables often results in employee overtimes or vehicles returning late to the rayon office. These overtime and lateness result in disruption of the next work shift (afternoon shift) because the allocation of available vehicles is only 1 vehicle per district.

The watering truck routing problem is the extention of Vehicle Routing Problem (VRP). While the VRP is a problem of which the vehilce visiting each point only once, the problem needs to be split the service so that the division of tasks between vehicles is evenly distributed. So, it needs additional constraints, especially the time windows constraints. Therefore, the problems in this study can be classified as a models of Vehicle Routing Problem with Time Windows (VRPTW) by adding split service.

The approach used to solve the problem is an Ant Colony Optimization (ACO), a meta-heuristic method, by considering a number of parks that must be visited by vehicles before they are returning to the depot within a time limit shift. The use of ACO algorithm is expected to help find the best value with faster computing time than the one of exact algorithm.

2. Literature Review

In previous studies, routing research related to picking up and deliveries were carried out by [1] regarding the customer requests fullfilment for delivery and pickup of goods in one trip. Latiffianti et. al [2] develops exact algorithm for solving pickups and deliveries of workers from several points of origin and several destinations with time windows called as split delivery vehicle routing problem (SDVRP). Dayanara et. al [3] also develops an ACO model for waste collection in the city of Surabaya. In Dayanara's problem, the waste collection can be pick up splitly within the time window by using an intermediate facility.

Another variant, called as VRP Delivery and Pickup (VRPDP), was conducted by [4] solved by using a Tabu Search approach based on variations in single demand. The problem of VRPDP basically models a set of vehicles which visit to each node are done exactly once. However, in some cases, nodes can be visited more than once (split service) if there are significant savings, distance, time, cost, and the number of routes compared to without the split service option. The extention of this problem was conducted by [5] concerning multiple node visits which can be called as Split Delivery Vehicle Routing Problems with Time Windows (SDVRPTW).

Similar research was conducted by [6] using the Branch and Cut algorithm to solve SDVRPTW problems which produce an optimal output of 5% compared to SDVRPTW with exact calculations. Other research on SDVRPTW was also conducted by [7] which compared the method of solving the problem using Ant Colony Optimization (ACO) and Hybrid Meta-heuristic Algorithm which is a combination of ACO, Genetic Algorithm (GA), and heuristics. This study tested the three meta-heuristic algorithms on several datasets used in several previous studies [8]. The test results obtained promising results for ACO, where 14 of the 21 datasets used in the study of [8] yield better values than the other two algorithms. Another meta-heuristic methods called Crow Search Algorithm is also implementated in solving complex scheduling problem as discussed in [9]. This research concluded that the Crow Search Algorithm overtake other algorithms such as the combination of Particle Swarm Optimization and Simulated Annealing (PSO + SA), hybrid Genetic algorithm (hGA) in term of computational time and optimum solution.

3. Methodology

3.1 Data Collecting

This study utilises data from direct observation, Google Maps application as well as the historical data. The detailed data as follow:

1. The location and the length of 46 parks in east part of Surabaya.
2. The location of 40 water filling locations (rivers).
3. The number and capacity of tank trucks.
4. The operating hours when vehicles will be departed from the depot at 06.00 AM and must return back before 03.00 PM.
5. Various travel time data taken from Google Maps, which includes data from depot to park, between parks, park to each river, depot to river and vice versa.
6. Loading and unloading service time are assumed to be constant for 15 minutes and 20 minutes, respectively for each 5000 Liter.
7. Route and assignment of tank trucks

3.2 Data Processing

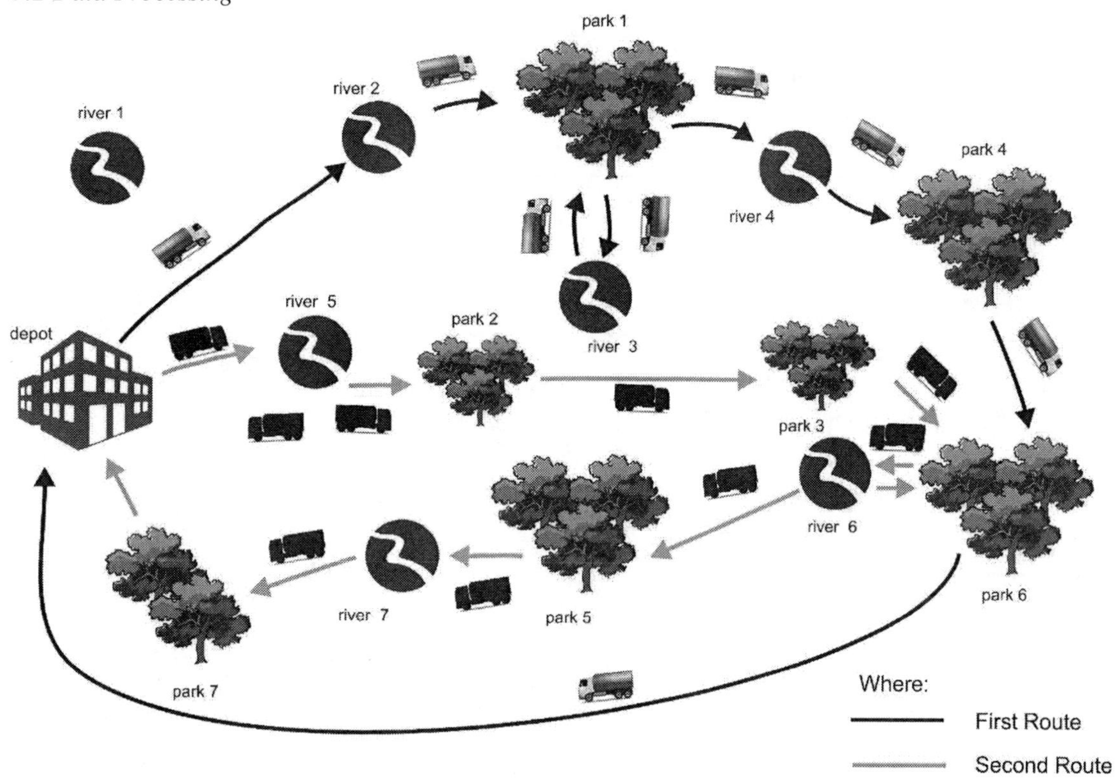

Figure 1. Illustration model.

As seen in Figure 1, the truck departs from the depot with empty tank and fills the water in the nearest river (pickup node). Then, the vehicle goes to the nearest park for watering. For park that have demand less than or equal to truck capacity, trucks only need once to do watering (unloading). However, when the demand for a park is more than the truck capacity, the park can be visited several times until all demands are met, either with the same or different trucks. For example, as seen in Figure 1, vehicle 1 initially fills with water in river 2 and unload it in park 1. Because the demand for park 1 exceeds the capacity of truck 1, then truck 1 returns to refill water in the nearest river, in this case at river 3. After all, 1 park demand is fulfilled, then truck 1 continues the journey to the next watering location. This means that in this case, park 1 is served twice by truck 1.

3.2.1 Ant Colony Optimization for Watering Routes for City Parks

There are two main stages for developing ACO model to solve this problem, namely the construction of Traveling Salesman Problem (TSP) routes and improvement routes by dividing into several sub-routes.

1. Construction of a TSP route using the Ant Colony Optimization approach

The process of TSP construction routes with the Ant Colony Optimization algorithm is carried out in several stages as seen in Figure 2.

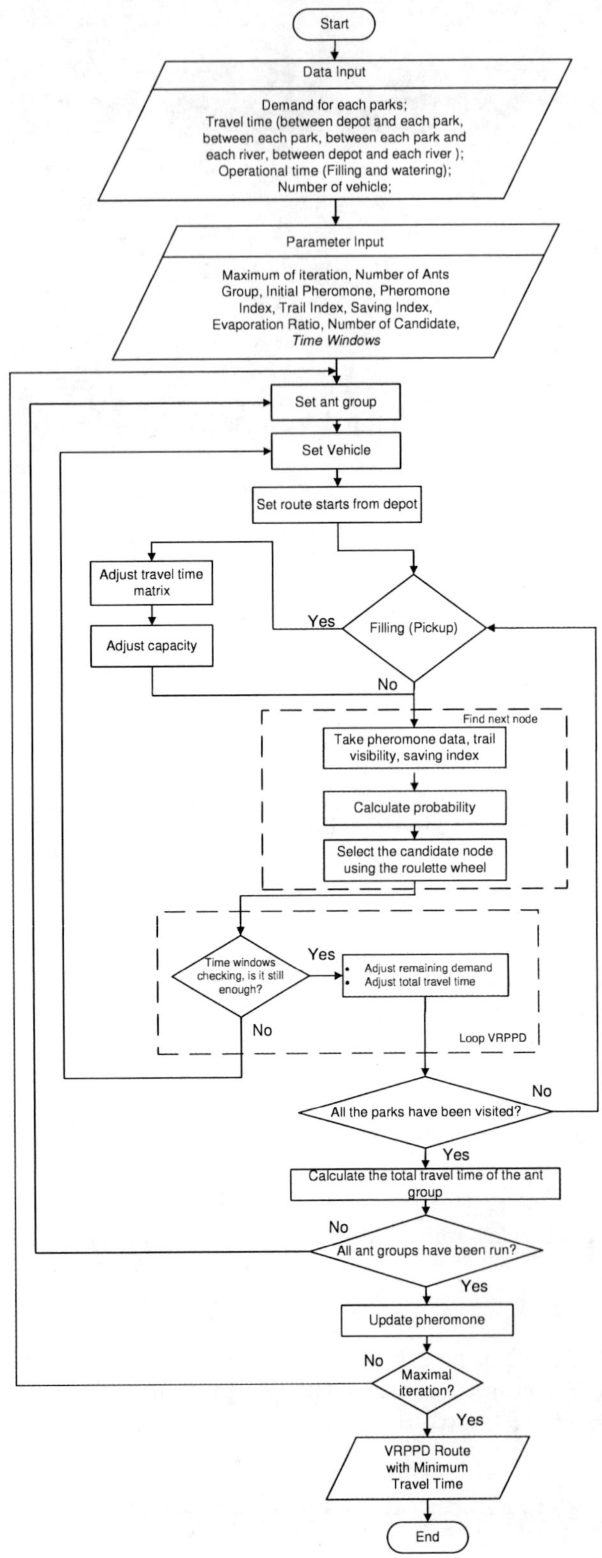

Figure 2. Flowchart of developing ant colony optimization algorithm.

a. Initialization

- Data:
  a) Travel time from depot to the park.
  b) Travel time from depot or park to the river.
  c) Travel time between two parks.
  d) Loading and unloading time.

- Parameter:
  a) Early pheromones (τ_{ij}^{1})
  The pheromone value will always be updated in every algorithm iteration, starting from the first iteration until the maximum iteration is determined or has reached optimal results.
  b) Visibility Index *(h)*
  The visibility value between nodes is the inverse of the distance ($\frac{1}{distance}$) of each node.
  c) Evaporation ratio (ρ)
  The evaporation ratio is the rate of pheromone evaporation which has value $0 < \rho < 1$.
  d) Number of ant groups (*N*)
  The ACO algorithm is a population-based metaheuristic, so the search for solutions directly based on certain groups. The specified number of ant groups will be run in one iteration. There is no definite provision regarding the number of ant groups to produce an optimal value. So, it takes several trials to get the best results.
  e) Pheromone level index (α) and visibility index (β)
  The α value is the pheromone weight τ and β is the weight that controls the visibility of the pheromone level. To simplify the calculation, the values of the two indices are equalized to 1.
  f) Number of candidates
  To facilitate the selection of the next node for each vehicle, selected candidate nodes are created. For candidate nodes that have been selected by one vehicle, they will not be selected by another vehicle.
  g) Maximum iteration
  The maximum number of iterations is fixed as long as the algorithm runs. For the optimal value of a maximum iteration, several trials need to be done to get the best results.

b. Probability and selection of the next node

Calculation of the probability to select a segment using the $P = \frac{\tau_{ij}^{\alpha} \times h_{ij}^{\beta}}{\sum \tau_{ij} \times h_{ij}}$. For each segment, a cumulative probability range is related to the segment selection. Then certain values are chosen based on random numbers in the range (0.1). Specific segments will be selected based on random numbers generated. After each ant occupies each of the specified points, the ant will start to travel from the first point of each as a point of origin and head for one of the other points as a destination. Then from the second point of each, the ant will continue its journey by selecting one of the points that have not been visited as the next destination point. Points that have been served will be stored in a taboo list so that they are not visited again. The ant journey continues until all points are visited one by one.

c. Calculation of travel time

The total travel time is calculated after an ant has completed the entire route and returned to the depot, starting from the depot to the river, river to park, park to park, and park to depot. The total travel time value is stored in memory as optimal local (lk).

d. Adjust pheromone
After all groups are run, adjust the pheromone values $\tau_{n+1} = (1-\rho)\tau_n + \rho\tau_0$

e. Iterate
For each iteration that is done, the taboo list is first emptied to be refilled in the order of new points in the next iteration. Then iterates for stages (b) to (d) until it reaches a maximum iteration or has reached convergence. From the overall iteration will be obtained the optimal global value of the minimum total travel time.

2. Dividing VRP Sub-Routes
The following is an algorithm carried out in finding route recommendations for watering city parks in Surabaya. The results of the Ant Colony Optimization TSP will be used as a basis for making VRP sub-routes that have been adapted to existing constraints.

a. Initialization
- Data used as input on this VRP sub-route are:
  1. The optimum TSP route from the previous ACO TSP
  2. Travel time from the depot to the park
  3. Travel time from park to park
  4. Travel time from the park to the river
  5. Travel time from the depot to the river
  6. Water withdrawal operational time
  7. Watering operational time
  8. Number of trucks
  9. Demand of park
  10. Truck capacity
  11. End time shift
- Initialize the truck to be used and the park visited

b. Truck Selection
In this case, there are nine units of trucks with 5000-liter tank capacity. From a total of nine trucks, there will be a selection of candidates for the number of parks that can be visited first. For each park that has been selected as a candidate by one truck, it cannot be chosen by another truck. The process of selecting park candidates is to divide the park into four sections. Out of 46 parks, there are 12 park candidates. The purpose of selecting candidates is so that the directions of the vehicles are spread.

c. Selection of starting point
The choice of a starting point for each scenario always starts from the depot. For every truck at the starting point, it has a capacity = 0. Each truck will return to the starting point after all the parks have been visited and the deadline for operating hours has finished.

d. Park selection
In this study, 46 parks have to be visited. Each park has different demands depending on the size of the park. There is no time limit for a visit to the park or in other words the park is free to visit at any time. Park selection relates to the smallest travel time from the last position of the truck by considering the capacity and the time remaining.

e. Selection of watering point
In the selection of watering points, considering the travel time from the starting point with the watering point and saving weight to determine which point will be visited next. For the next selection of visit points, initially, the closest potential points are determined based on travel time, then a list of saving values for each potential point candidate is made, from the largest to the smallest. The point that has the shortest time and the highest saving value will be chosen for

the next visit. This point selection follows the route produced by TSP ACO. Then see the travel time and capacity, is it possible to visit the point. If the truck's capacity does not meet the demand at the watering point, a visit to the nearest water filling point is made. But if the time does not meet, the truck will return to the depot. In this process, it is possible to have a split service.

f. Selection of water uptake
A visit to the water collection point is a soft constraint, so this visit is carried out if the truck capacity = 0. There is no specific process in selecting the water filling point because it only considers the smallest travel time from the last truck location and the next watering point.

4. Result and Discussion

The algorithm is tested using Intel® core ™ i5-2450m CPU @ 2.50GHZ 4 GB RAM. The parameters used in this testing as follow: the evaporation value ρ=(0.3; 0.6; 0.9), the number of ants=(5; 10) and the maximum iteration of (10; 100; 500). Running models are carried out by 10 replications and the best objective function value is selected from these ten replications. Table 1 shows that the minimum, average, and standard deviation values of ten replications for these various parameters.

Table 1. The results of ACO Model for various parameters.

ρ	M	I	Min	Mean	SD	ρ	m	i	Min	Mean	SD
0.3	5	10	3890.0	3911.0	9.5	0.6	5	10	3727.1	3740.0	11.5
			9	9	0				8	8	0
			404	471.5	66.3				358	421.3	37.4
		50	3634.9	3647.6	12.2			50	3640.0	3647.4	6.6
			8	8	0				8	8	0
			1552.0	1661.6	120.8				1843	1889.6	49.3
		100	3639.6	3644.6	3.8			100	3595.7	3626.5	14.3
			8	8	0				8	8	0
			4080.0	4255.6	110.6				3200.0	3362.7	146
0.3	10	10	3657.8	3688.2	16.8	0.6	10	10	3681.1	3696.7	10.8
			8	8	0				8	8	0
			1034.0	1039.3	5.2				625.0	676.4	66.4
		50	3611.9	3630.4	14.5			50	3611.5	3621.3	6.9
			8	8	0				8	8	0
			3251.0	3469.2	123.1				3060.0	3400.6	236
		100	3618.1	3621.5	2.5			100	3623.5	3624.6	1.0
			8	8	0				8	8	0
			5353.0	5852.1	369.4				6405.0	6466.4	45.4
0.9	5	10	3635.5	3659.4	12.8	0.9	10	10	3727.1	3740.0	11.5
			8	8	0				8	8	0
			1553,0	2045,4	696.6				358	421.3	37.4
		50	3635.5	3650.5	13.4			50	3640.0	3647.4	6.6
			8	8	0				8	8	0
			1469.0	1492.5	17.3				1843	1889.6	49.3
		100	3620.3	3640.0	16.0			100	3595.7	3626.5	14.3
			8	8	0				8	8	0
			2981.0	3206.4	177.1				3200.0	3362.7	146

Based on the results on Table 1, it can be concluded that the greater the number of ant groups and the number of iterations, the higher the required average constant computing time. While the greater the value of evaporation does not constantly affect the required computing time. The objective function of the problem, total travel time, it can be achieved by the parameter ρ = 0.6 and the number of ant groups m = 5 and the iteration is 50, with a minimum travel time of 3611.5 minutes and an average travel time of 3621.3 minutes.

From above trials and considerations, a new experiment is conducted which uses the evaporation coefficient parameter $\rho = 0.6$, the number of ant groups m = 5 and the number iteration i = 50 with ten replications. The result of this experiment can be seen in Table 2.

Table 2. Result ten replications of second round experiment.

No	Objective Function	Number of Vehicle	Time	No	Objective Function	Number of Vehicle	Time
1	3576.1	8	3851	6	3566.9	8	3835
2	3564.2	8	3802	7	3560.7	8	3839
3	3560.5	8	3822	8	3559.1	8	3860
4	3558.6	8	3840	9	3559.5	8	3857
5	3562.3	8	3848	10	3558.9	8	3844

Objective function (in minutes), number of vehicle (in units), and computational time (in seconds).

The assignment of the test results using the ACO model does not violate existing constraints in terms of capacity or time window with all demands in each park can be served all. The total number of vehicles available at the city is currently 9 units, while the recommendation of this study only requires eight units of vehicles. So, the demands of each park will be able to be served all of them. The city also still have a spare tank truck that will stay at the depot as a backup vehicle in the event of an increase in the number of demands or when there is a vehicle which has break down condition. The back up vehicle is very important due to every day the park must always be watered. Otherwise, the survival of plants as a green space in the city of Surabaya will be disrupted. This optimization result with the 3558.6 minutes total traveling time is better than the one of existing condition, with the 3866 minutes.

5. Conclusion

The development of ACO model for been applied for finding the truck routing in watering parks in the city of Surabaya. The model tries to determine the routes of the vehicles and the number of vehicles needed. The parameters used in the ACO model test for watering parks in the city of Surabaya were obtained from a trial of several parameters with 10 replications in each combination. The results of the selected parameters are parameters that have a small standard deviation can provide the minimum objective function value. The recommended route generated by the ACO algorithm model with the parameter evaporation value 0.6, the number of ants 5 and a maximum number iteration of 50 for watering parks in the city of Surabaya produces a goal function value of 3558.6 minutes by using 8 units of vehicles. This assignment does not violate existing contraints in terms of vehicle's capacity or time window with all demands in each park can be served all.

References

[1] Desaulniers G, Desrosiers J, Erdmann A, Solomon MM, Soumis F. VRP with Pickup and Delivery. *The vehicle routing problem.* 2002 Jan 1;9:225-42. doi.org/10.1137/1.9780898718515.ch9.

[2] Latiffianti E, Siswanto N, Firmandani RA. Split delivery vehicle routing problem with time windows: a case study. *IOP Conference Series: Materials Science and Engineering*. IOP Publishing; 2018 Apr;337:012012. doi.org/10.1088/1757-899x/337/1/012012.

[3] Dayanara DH, Arvitrida NI, Siswanto N. Vehicle Routing Problem with Split Service, Time Window and Intermediate Facility for Municipal Solid Waste Collection in Surabaya City with Ant Colony Optimization Algorithm. *IOP Conference Series: Materials Science and Engineering*. IOP Publishing; 2019 Sep 6;598:012020. doi.org/10.1088/1757-899x/598/1/012020.

[4] Wassan N, Nagy G. Vehicle Routing Problem with Deliveries and Pickups: Modelling Issues and Meta-heuristics Solution Approaches. *International Journal of Transportation*. NADIA; 2014 Apr 30;2(1):95–110. doi.org/10.14257/ijt.2014.2.1.06

[5] Toth P, Vigo D, editors. Problems, Methods, and Applications, Second Edition. *Vehicle Routing*. Society for Industrial and Applied Mathematics; MOS-SIAM Series on Optimization; 18. 2014. 463 p. ISBN 9781611973587.

[6] Bianchessi N, Irnich S. Branch-and-Cut for the Split Delivery Vehicle Routing Problem with Time Windows. *Transp. Sci*. Institute for Operations Research and the Management Sciences (INFORMS); 2019 Mar;53(2):442–62. doi.org/10.1287/trsc.2018.0825.

[7] Rajappa GP, Wilck JH, Bell JE. An Ant Colony Optimization and Hybrid Metaheuristics Algorithm to Solve the Split Delivery Vehicle Routing Problem. *International Journal of Applied Industrial Engineering*. IGI Global; 2016 Jan;3(1):55–73. doi.org/10.4018/ijaie.2016010104.

[8] Derigs U, Li B, Vogel U. Local search-based metaheuristics for the split delivery vehicle routing problem. *J Oper Res Soc*. Informa UK Limited; 2010 Sep;61(9):1356–64. doi.org/10.1057/jors.2009.100.

[9] Adhi A, Santosa B, Siswanto N. A meta-heuristic method for solving scheduling problem: crow search algorithm. *IOP Conference Series: Materials Science and Engineering*. IOP Publishing; 2018 Apr;337:012003. doi.org/10.1088/1757-899x/337/1/012003

IOP Conf. Series: Materials Science and Engineering 1072 (2021) 012007 doi:10.1088/1757-899X/1072/1/012007

Design and implementation for performance measurement system in the power plant operation and maintenance company

P M Andrias[1*] and I Vanany[2]

[1]Department of Technology Management, Institut Teknologi Sepuluh Nopember, Surabaya, Indonesia.
[2]Department of Industrial and Systems Engineering, Institut Teknologi Sepuluh Nopember, Surabaya, Indonesia.

Email: *marga.andryazh@gmail.com

Abstract. The purpose of this paper is to re-design and implementation performance measurement system using Balanced Scorecards (BSc) in power plant operations and maintenances companies. Three main lack of existing performance measurement system that are reasoning to conduct the re-design performance measurement system in case study such as lack in strategy map, weights on perspectives, strategy objectives, and Key Performance Indicators (KPI's), and some KPI's do not represent relevant performance and are not measurable in quantitative data. Determining 16 strategy objectives and 43 KPI's based on strategy company and weights and target each KPI's was been conducted. Based measurement using scoring system were calculated and analyzed. The poor performance (red color) was analyzed to be improvement in next year. The recommendations given are increasing human resources competencies and capabilities as electricity regulation and electricity market requirement, as well as enhance information technology system to support data base that aim completion of work more effectively.

1. Introduction

Balanced scorecard (BSc) model was developed by Kaplan and Norton [1] and many companies in Forbes Company listing [2] including in Indonesia [3] was implemented BSc model as company performance system. Vision, mission and strategy objectives of company are needed as basic to determine the objectives strategy each perspective (financial, customers, internal business process, and learning and growth) [1]. Several previous research in BSc design and implementation were conducted such as manufacturing industry [4,5] and healthcare industry [6,7]. For power plant companies, company performance system is significant management system to manage and control performance and achieving the objectives of company strategy.

PT. X as case study in this paper, it has had a performance measurement system which is balanced scorecard adopted. The intensive discussions result with some managerial company's case study, three main lack of existing performance measurement system that are reasoning to conduct the re-design performance measurement system in case study. Firstly, a strategy map to present the interrelationships between strategy objectives in four (4) perspectives has not yet been made. Secondly, weights on perspectives, strategy objectives, and KPIs that represent the level of importance and which are also dominant in influencing company performance are also not yet determined based on management opinions. Thirdly, Some KPIs do not represent relevant performance and are not measurable in quantitative data. For example, for SCM KPIs that only use assessment preferences of related managers without using data such as inventory turnover ratio, days of supply and others.

The purpose of this study is to re-design and implementation the company performance measurement system using Balanced Scorecard (BSc). Analytical Hierarchy Process (AHP) methods is used to determine th weight of perspectives, strategy objectives and Key Performance Indicators

(KPI). Scoring system and traffic light system also used to determine the results performance each KPI and leveling of performance.

2. Research Method

The research processes are described systematically ally through the stages of data collection and evaluation of performance measurement system, data processing, analysis and interpretation of data, and the stages of conclusions and suggestion. Several stages were conducted to complete this research that are described. In the stage 1 (problem identification), the problem identification is an initial stage of research as a determinant. At this stage is carried out field studies and identification of problem as well as a preliminary study both in direct observation and literature studies that relevant to this research. At this stage a general framework of research has been formed as a starting point and a reference for the next stage. In stage 2 (observation), study and direct observation of managerial and operational activities as well as the review of existing documents in the generating business unit of the research object company that aims to obtain a real condition in the company compared to the theoretical basis obtained from the study of literature.

In stage 3(literature study), after conducting a direct observation, literature study was conducted to study various theories directly related to support research. The reference used in this research is to conduct a literature study in the form of books, manual book/manual manufacture, journals, proceeding and papers that support problem solving in this research related to the concept of Performance measurement, Balance Scorecard, KPI, Analytical Hierarchy Process, Scoring System and traffic light system (TLS). In stage 4 (data gathering), this stage is carried out by brainstorming with the Heads of Corporate Planning and senior leaders that supported by documents and literature. Furthermore, identification of the organization, generating business units and business processes that using existing data either in the form of quality guidelines, procedures, maps of the company's business processes, generating unit business processes, company profile, annual reports and data obtained from direct observation. The result is primary data and secondary data. Primary data is obtained in fact and factually in the field with interviews, questionnaires, and direct observations in the field to obtain opinions from decision makers and experts in the company regarding the KPI of the generating business unit. Secondary data is data that has been previously recorded by PT. X, such as Company Profile, procedures, work instructions, unit performance reports, O&M Services Contracts, Services Level Agreement of the O&M Services Contract, generating business unit performance indicators, management reports, strategic reports and other data needed.

3. Results and Discussions

The initial process of performance measurement system design included the identification of Key Performance Indicator (KPI) and the performance measurement system evaluation of generating business unit of PT. X in measuring its performance in order to achieve company's vision and mission. The next step is designing and mapping of the Strategic Theme, Strategic Objective and KPI design referring to the Balanced Scorecard Perspective then distributing the questionnaire and interviewing the Head of Corporate Planning Unit and Senior Leader to determine and validate the Strategic Objective and KPI. Designing of the Strategic Objective and KPI of the generating business unit refers to the company's Strategic Objective and KPI, holding company regulations, SLA of the O&M Services Contract, Asset Management guidelines, which are subsequently adjusted according to Generating business unit needs. The Table 1 shows strategy objectives (SO) and KPI existing in case study

AHP approach that was invented by Saaty was used to determine weighting each perspectives, objective strategy and KPI [8] method The next activity is processing data that has been collected previously through questionnaires to determine the Perspective, Strategic objective and KPI weighting of the respondents, as a Kasat Renkor and Senior Leaders who are parties who are directly involved in the company's business activities and performance measurement of generating business units. The results will be processed using the AHP method to get the consistency of the data obtained and determine the overall weight of each KPI Unit. The Performance Measurement in case study was developed using the Scoring System method. Strategic Objectives and KPIs that have been reorganized in the form of performance measurement matrices based on the level of importance will be given a score for the performance in case study. Then the Traffic Light System will be used according to the range score. Appendix A shows KPI measurement results in case study.

Table 1. The strategy objectives (SO) and KPI existing in case study.

Perspective	Strategic Theme	Code	Strategic Objectives	KPI Code	Key Performance Indicators
Financial Perspective	Maximize Share Holder Value	SOF1	Profitability Increase	F11	Gross Profit Margin
		SOF2	Cashflow Optimation	F21	Weekly Cash Usage Report (Current Ratio)
		SOF3	Sales Increase	F31	Reimbuse + VO Monitoring
				F32	Monthly Report
				F33	BA Job Inspection
				F34	Total Income
		SOF4	Cost Optimization	F41	% Actual RKAU
				F42	% Actual Monthly Cashy Budget
Customer Perspective	Enhance Market Share	SOC1	Services Providing whose exceed customer satisfaction	C11	Customer Satisfaction Index
Internal Process Bussiness Perspective	Cost Optimization and Operational Excellence	SOI1	25 NERC Operation Excellence	I11	Sales
				I12	EAF
				I13	EFOR
				I14	SOF
				I15	SdOF
				I16	NPHR
				I17	Afpm/NCF
				I18	PS (Own use)
		SOI2	Work Planning & Control Optimization	I21	RAV
				I22	PM Compliance
				I23	WO Planned Bcklog
				I24	Wrech Time
				I25	Rework Ratio
				I26	Contractor Cost Ratio
				I27	Reactive Work
		SOI3	Operation Management Optimization	I31	FLM to SR FLM Ratio
				I32	Start Up Duration
		SOI4	Efficiency Management Optimization	I41	Number of Efficiency Improvement Program
				I42	Ratio Completion Efficiency Improvement Program
		SOI5	Management Optimation Outage	I51	Performance Rating Score
		SOI6	HSE Compliance Enhancement	I61	Zero Accident
				I62	Emission
				I63	Proper
		SOI7	Material Management Optimization	I71	Inventory Turn Over
				I72	Service Level Material
				I73	Purchase Effectiveness
				I74	Supplier Satisfaction Level
		SOI8	Realiability Management Optimization	I81	% Life Cycle Cost Analysis to RKAP
				I82	% Asset Health
		SOI9	Risk Management Optimization	I91	Number of Risk Register
				I92	Number of Risk Mitigation
Learning & Growth Perspective	Agile and Competence Organization and Work Force	SOL1	Ensuring of the capability, capacity, and readiness of HR system	L12	Certified Employees Number
		SOL2	Continous Innovation Program	L21	Total Number of Innovations Applied
				L22	Number of Innovations that Commercialized

In this study, the result of the calculation KPI in case study based on Appendix A, there are 33 KPIs categorized as green, 6 KPIs categorized as yellow and 4 KPIs categorized as red. The main score of all KPIs is 99,4%. There are 4 KPIs that do not reach the target and the KPI performance is still far from the maximum. For this reason, it is necessary to conduct an analysis in order to find out the cause of the KPI not being achieved.

3.1 Analysis of % Actual Monthly Cash Budget Achivement (F42)

This KPI is the lowest achievement in case study which is 53.19% obtained from comparison of the cumulative budget actual in January till December 2019 amount 147% against 100% targeted, following its distribution in Figure 1.

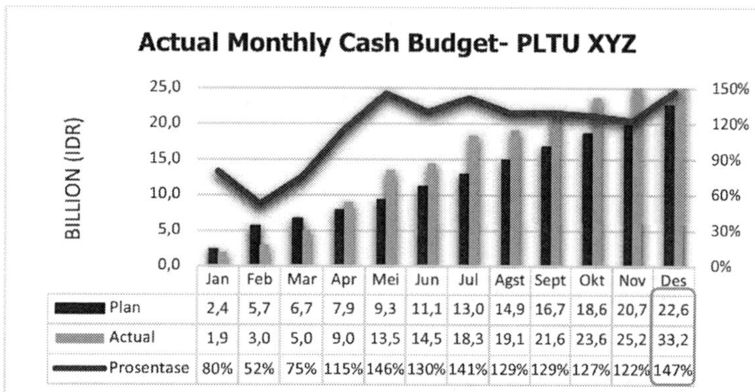

Figure 1. Distributing of % monthly cash budget achievement.

Based on Figure 1 can be concluded that in general the KPI since April 2019 has been achieved in red. The worst performance value was in December, amounting to 147% of the target. If you look at the graph of the development of performance above, there are a number of conditions from this KPI that need to be underlined and paid close attention to, as shown in Figure 2.

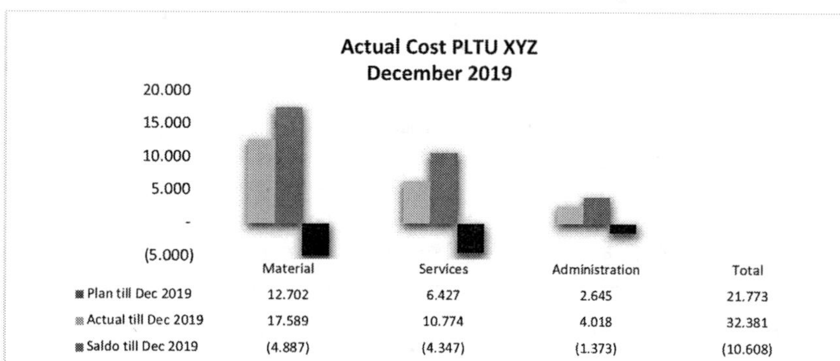

Figure 2. Monthly cash budget breakdown.

Based on Figure 2, the actual of the material, service and administrative cost items exceeds the provisions, so the need for controlling the execution of maintenance and procurement processes is based on top priority for unit reliability compared to other programs. The KPI applies to other generating business units in order to prioritize the work to be executed. The KPI needs serious attention, how this condition can occur and what corrective steps can be taken. Because it will affect the company's cash flow as well as company revenue.

3.2 Analysis of Gross Profit Margin Achievement (F11)

Gross profit accounting is the difference between sales and COGS (Cost of Goods Sold) not including overhead, salaries, taxes, interest payments. Average gross profit for O&M Services determined by PT. X in amount of 15.5%. The KPI of Gross Profit Margin (GPM) will experience a domino effect on the increase in KPI performance if the KPI% achievement of IMR, the% achievement of the RKAU also increases its performance, so that the purpose of the KPI as a form of control and awareness ranging from Power Plant business units to the head office in order to maintain corporate GPM.

3.3 Analysis of Certified Employees Number (L12)

The number of actual certified employees to the contract conditions and according to the energy and mineral resources ministry regulation at 3rd level of each competency. From a total of 48 organic personnel, still 66.7% have been certified up to 3rd level, can be seen in table 2. Fulfilment of HR competency requirements can support the internal implementation of business processes and the achievement of KPI generating business units that have an impact on the company's performance targets, to be needed integrated control and awareness starting from the HR of generating business units and head office through monitoring and evaluating the development of employee competency profiles.

Table 2. Number of certified employees.

Position	Amount	1st Level Certification	2nd Level Certification	3rd Level Certification	Percentage of 3rd Level Certification
Management / Structure	16	16	16	12	75.00%
Technical					
Operation	17	17	17	13	76.47%
Maintenance	5	5	5	4	80.00%
Planner	4	4	4	1	25.00%
HSE & Chemical	3	3	3	2	66.67%
Non-Technical					
HR, Finance & Administration	3	3	0	0	0%

4. Improvement Analysis of KPI

In case study, a performance measurement system that is designed to be applicable to all generating business units. It can be interpreted that the KPI does not have significant points to be called a problem. For the four KPIs that have a low score of KPI can affect the generating business unit performance. Priority can be applied to KPIs that have a high level of importance or influence on the overall performance of the company. The biggest weight of KPI in sequence is the number of Certified Employees in accordance with the requirements (0.2446), Gross Profit Margin (1.00), the number of unit risk mitigation (0.5) and% MCB (AKB) Achievement (0.305).

Gross Profit Margin is affected by the % actual of Monthly Cash Budget (AKB) and % actual RKAU achievement. The greater the percentage of achievement of MCB and RKAU, the smaller percentage of GPM produced. Then improvements can then be prioritized in the % actual MCB achievement. From the results of the improvement analysis above, two KPI can be obtained with the highest priority to be carried out such as (1) number of Certified Employees that required and (2) % Actual Monthly Cash Budget Achievement.

The following are some suggestions for improvement that are two prioritized KPI in case study, which are:

4.1 Number of certified employees according to conditions.
The management of PT X needs to develop manpower planning to meet customer requirements and in accordance with the company's grand strategic namely Enlarge Market through Product Customization & New Technology Expertise Development. Some suggestions given are in Table 3.

Table 3. Recommendations for each basic event in number of certified employees.

No	Basic Event	Recommendation
1	There is no optimization Manpower Planning	Made procedure standard and guidelines for competencies and career path manpower planning that needed generating business unit. find out the number of buffered human resources and then to collaborate with educational institutions which are not only in the scope of capacity fulfillment, but also in the scope of capabilities. Generating business units are also challenged to create experts in their fields to enlarge the market from the differentiation of O&M service products
2	There is no optimization training planning	Given training or upgrading knowledge about power plant (technical and non-technical).

4.2 Percentage of actual of monthly cash budgeting
The accuracy of the results of achieving MCB depends on the pattern of existing historical data and the planning techniques used. Some suggestions given are Table 4.

Table 4. Recommendations for each basic event in % actual of monhtly cash budgeting.

No	Basic Event	Recommendation
1	There is no historical data pattern used in the MCB planning process	Made data base system to support accuration Monthly Cost Budget planning
2	There is no refresh training	Given training or upgrading knowledge regarding RKAU planning technique show to choose the right forecasting techniques for existing data patterns, such as internal and external training processes. With ability of proper planning and control patterns, it is expected that the right level of achievement according to plan.
3	There is no historical breakdown equipment	Made data base system emergency work and implementation procedures for predict the age or breakdown each equipment. to support well-planned of the costs needed.

5. Conclusion and Recommendation

The design of generating business unit performance measurement system at PT. X based on four perspectives of the Balanced Scorecard. Based on 3 respondence combined result using AHP, the biggest weighting from the BSC perspective is on the Learning and Growth perspective and in order of weights Learning & Growth (0.354), Finance (0.338), Internal Business Process (0.17), and Customer (0.137). And the biggest weighting KPI on GPM (1,00), Curent Ratio (1.00) and CSI (1.00). Each perspective in the BSC has a strategic theme developed into several Strategic Objectives (SO) and there are a total of 16 SO generated from all BSC perspectives and 43 KPIs were obtained consisting of 8 KPIs from a Financial perspective, 1 KPI from a Customer perspective, 31 KPIs from an Internal Business Process perspective, and 3 KPIs from a Learning and Growth perspective.

Based on the AHP, Scoring System and Traffic Light System that implemented in case study, there are 33 KPIs categorized as green, 6 KPIs categorized as yellow and 4 KPIs categorized as red. The main score of all KPIs is 99.4%. The two KPI can be obtained with the highest priority to be carried out is the L12 and F42 KPI code. There is F42 which has a lowest achievement value of 53, 19% and the seconds L21 66, 7%. Based on the analysis method of the most critical effort obtained 2 basic events: (1) need to be enhance the managing develop manpower planning capability and capacity fulfilment, which meet customer requirements and in accordance with the company's grand strategic likes Enlarge Market through Product Customization & New Technology Expertise Development, also challenged to create experts in their fields to enlarge the market from the differentiation of O&M service products. (2) need to be enhance integrated information system to aim completion work more effectively.

This newly created performance measurement model as one of the reference models that needs to be considered by the field of the office of strategic management division to plan, evaluate and monitor the performance of Generating business units and improve their performance. There was a redesign of the previous KPI calculation such as WPC or maintenance management, material management, operations management, outage management, and reliability management were not SMART (Specific, Measureable, Achievable, Reliability, and Time) contained caused the KPI calculation used a Likert scale so there was no accurate achievement. Suggestions for further research from this research are that actual calculations can be applied to other generating business unit case studies that have similar characteristics but used other performance measurement method to be compared between difference method.

References

[1] Kaplan. R. S., Norton, D. P. Translating Strategy into Action: The Balanced Scorecard, Harvard Bussiness School Press, Boston; 1996

[2] Behn RD. Why Measure Performance? Different Purposes Require Different Measures. *Public Adm Rev*. Wiley; 2003 Sep;63(5):586–606. doi.org/10.1111/1540-6210.00322.

[3] Vanany I. Performance Measurement: Model & Aplikasi. ITS Press. Surabaya; 2009

[4] Lee AHI, Chen W-C, Chang C-J. A fuzzy AHP and BSC approach for evaluating performance of IT department in the manufacturing industry in Taiwan. *Expert Syst. Appl*. Elsevier BV; 2008 Jan;34(1):96–107. doi.org/10.1016/j.eswa.2006.08.022.

[5] Vanany I, Suwignjo P, Yulianto Design of supply chain performance measurement system for lamp industry. 1st International Conference on Operations and Supply Chain Management; 2005.

[6] Leksono EB, Suparno, Vanany I. Development of Performance Indicators Relationships on Sustainable Healthcare Supply Chain Performance Measurement Using Balanced Scorecard and DEMATEL. *Int J Adv Sci Eng Inf Technol*. Insight Society; 2018 Feb 27;8(1):115. doi.org/10.18517/ijaseit.8.1.3852.

[7] Leksono EB, Suparno S, Vanany I. Integration of a Balanced Scorecard, DEMATEL, and ANP for Measuring the Performance of a Sustainable Healthcare Supply Chain. *Sustainability*. MDPI AG; 2019 Jul 2;11(13):3626. doi.org/10.3390/su11133626.

[8] Saaty, T. L. Decision Making for Leaders: The Analytical Hierarchy Process for Decisions in Complex World. Pittsburgh: University of Pittsburgh; 1993

Appendix A

Table A1. Score each KPI in case study

Code	KPI Code	Key Performance Indicators	Target	Unit	Scoring System	Score	% Score	KPI Weight	KPI Score	SO Weight	SO Score	Perspective Weight	Perspective Score	Total Score
SOF1	F11	Gross Profit Margin	15.5	%	Higher is better	8.78	56.67%	1.000	0.567	0.312	0.177			
SOF2	F21	Weekly Cash Usage Report (Current Ratio)	100	%	Must be one	100	100.00%	1.000	1.000	0.373	0.373			
SOF3	F31	Reimbuse + VO Monitoring	15	year/month	Lower is better	15	100.00%	0.244	0.244	0.200	0.187	0.137	0.113	
	F32	Monthly Report	10	year/month	Must be one	10	100.00%	0.197	0.197					
	F33	BA Job Inspection	15	year/month	Lower is better	15	100.00%	0.165	0.165					
	F34	Total Income	100	%	Higher is better	83.06	83.06%	0.395	0.328					
SOF4	F41	% Actual RKAU	100	%	Lower is better	110.46	89.54%	0.695	0.622	0.116	0.091			
	F42	% Actual Monthly Cashy Budget	100	%	Lower is better	146.81	53.19%	0.305	0.162					
SOC1	C11	Customer Satisfaction Index	83	Index	Higher is better	84.86	102.24%	1.000	1.022	1.000	1.022	0.338	0.346	
SOI1	I11	Sales	196.579	MWh	Higher is better	195,029.36	99.21%	0.175	0.174	0.035	0.043	0.170	0.193	0.994
	I12	EAF	79.92	%	Higher is better	90.45	113.18%	0.144	0.163					
	I13	EFOR	5.93	%	Lower is better	1.46	175.41%	0.130	0.288					
	I14	SOF	13.42	%	Lower is better	8.13	139.42%	0.105	0.146					
	I15	SdOF	8.5	frequency / unit	Lower is better	3.50	158.82%	0.105	0.167					
	I16	NPHR	4168.6	kCal/kWh	Lower is better	4,358.63	95.44%	0.105	0.100					
	I17	Afpm/NCF	72.05	%	Higher is better	77.11	107.02%	0.105	0.112					
	I18	PS (Pemakaian Sendiri)	13.42	%	Lower is better	9.23	131.23%	0.105	0.138					
SOI2	I21	RAV	3 – 5	%	Lower is better	3.35	88.19%	0.141	0.124	0. 115	0.151			
	I22	PM Compliance	280	%	Higher is better	92.40	115.50	0.169	0.195					
	I23	WO Planned Bcklog	4 – 6	Week	Lower is better	2.43	139.25%	0.141	0.196					
	I24	Wrech Time	55	%	Lower is better	81.96	149.02%	0.120	0.179					
	I25	Rework Ratio	< 5	%	Lower is better	1	180.00%	0.119	0.214					
	I26	Contractor Cost Ratio	20 – 40	%	Lower is better	23.19	84.05%	0.167	0.140					
	I27	Reactive Work	5 – 10	%	Lower is better	0.62	187.60%	0.141	0.265					

Table A1. Score each KPI in case study (Cont.)

Code	KPI Code	Key Performance Indicators	Target	Unit	Scoring System	Score	% Score	KPI Weight	KPI Score	SO Weight	SO Score	Perspective Weight	Perspective Score	Total Score
SOI3	I31	FLM to SR FLM Ratio	≥ 85	%	Higher is better	100	117.65%	0.568	0.668	0.117	0.129			
	I32	Start Up Duration	100 – 150% KOK	%	Lower is better	100	100.00%	0.432	0.432					
SOI4	I41	Number of Efficiency Improvement Program	20	Poin	Higher is better	28	140.00%	0.432	0.605	0.094	0.151			
	I42	Ratio Completion Efficiency Improvement Program	≥ 50	%	Higher is better	87.82	175.64%	0.568	0.998					
SOI5	I51	Performance Rating Score	≥ 95	%	Higher is better	96.65	101.73%	1.000	1.017	0.107	0.109			
SOI6	I61	Zero Accident	0	frequency	Must be zero	0	100.00%	0.517	0.517	0.249	0.249			
	I62	Emission	100	%	Must be one	100	100.00%	0.304	0.304					
	I63	Proper	Biru		Must be one	Biru	100.00%	0.179	0.179					
SOI7	I71	Inventory Turn Over	≥ 95		Higher is better	126.87	133.55%	0.247	0.330	0.098	0.103			
	I72	Service Level Material	88	%	Higher is better	87.00	98.86%	0.304	0.301					
	I73	Purchase Effectiveness	88	%	Higher is better	70.00	79.55%	0.247	0.196					
	I74	Supplier Satisfaction Level	90	%	Higher is better	98.10	111.48%	0.202	0.225					
SOI8	I81	% Life Cycle Cost Analysis to RKAP	75	%	Higher is better	100	133.33%	0.432	0.576	0.098	0.132			
	I82	% Asset Health	61 – 85	%	Higher is better	93.50	135.00%	0.568	0.769					
SOI9	I91	Number of Risk Register	100	%	Higher is better	100	100.00%	0.500	0.500	0.088	0.069			
	I92	Number of Risk Mitigation	50	%	Higher is better	28.00	56.00%	0.500	0.280					
SOL1	L12	Certified Employees Number	100	%	Higher is better	66.67%	66.67%	0.245	1.000	0.691	0.691			

Table A1. Score each KPI in case study (Cont.)

Code	KPI Code	Key Performance Indicators	Target	Unit	Scoring System	Score	% Score	KPI Weight	KPI Score	SO Weight	SO Score	Perspective Weight	Perspective Score	Total Score
SOL2	L21	Total Number of Innovations Applied	>5	Poin	Higher is better	4	80.00%	0.529	0.423	0.309	0.276	0.354	0.342	
	L22	Number of Innovations that Commercialized	>2	Poin	Higher is better	2	100.00%	0.471	0.471					

An optimization model for course scheduling in Undergraduate Industrial Engineering Program of Universitas Sebelas Maret

M Pratiwi[1,2]**, C N Rosyidi**[1,3] **and Yuniaristanto**[1,4]

[1]Industrial Engineering Department of Universitas Sebelas Maret, Jl. Ir. Sutami, 36 A, Surakarta, Indonesia

Email: [2]mega.pratiwimp01@gmail.com, [3]cucuknur@gmail.com, [4]yuniaristanto@gmail.com

Abstract. Course timetabling is a common problem in many universities. With different number of students enrolled in every course, the number of lecturers, and various size of room's capacity, then it would be difficult for scheduler to assign courses manually. The assigning process must meet all of educational rules in order to satisfy students and lecturers for a better educational quality. There are sets of time slot, courses, and rooms. The room efficiency and the effectiveness needed due to limited available room. The problem is formulated in the form of binary integer linear programming. The objective function of the model is to maximise the room utility. The model also considers probability of a course taken by different student groups in order to avoid clash. The results show a significant difference between manual scheduling and optimization results can satifsy more constraints than manual scheduling.

1. Introduction

The problem of courses scheduling at universities is a combinatorial problem which consists of scheduling a set of courses with a number of rooms and time slots. Industrial Engineering Department of Universitas Sebelas Maret faces a complicated scheduling problem. There are four level of students: first year, second year, third year, and fourth year. They enrolled to different courses including the ones that must be retaken to remake the mark of certain courses. Hence the scheduler must separate them into several groups. The student groups separation is made based on the student-level, class, and the courses. Besides that, with different number of students, the scheduler has other difficulties regarding to the room availability as it is used together with other study programs. There are also elective courses which must be taken by the third and fourth-year students provided by six laboratories. The delivery time of each course depends on credit of each course whether two or three credits. The data used in this study which consist of the room available, the students, the courses, the time slots, and the lecturers are retrieved from a database provided by the industrial engineering department of Universitas Sebelas Maret.

Every course is taught by different lecturers based on the competency and teaching load they have. The lecturers must also be assigned to certain time slot where their mandatory and elective courses are not held simultaneously. Several research in course timetabling has been conducted. For example, [1] developed graph coloring, network flow techniques and artificial intelligence (simulated annealing, taboo search, genetic algorithms, constraint satisfaction). [1] also categorized scheduling problems in educational institutions into 3 groups as follows: scheduling courses at universities, scheduling examinations at universities, and scheduling subjects at school. The problem of scheduling in educational institutions (educational timetabling) involves teachers or lecturers, students, and courses or subjects in a certain time and space. The problem can be formulated by finding the best allocation for both student and lecturer into the right time slot and limited rooms regarding to some educational rules or constraints. There are two constraints in timetabling, namely hard constraints and soft constraints. According to [2] hard constraint is a constraint in scheduling that must be met in any way so that the schedule is feasible. While soft constraints are constraint that must be met as much as possible, as long as they do not violate the hard constraints. Sometimes due to the complexity of scheduling problems at the university, not all hard constraints can be met.

This study develops course scheduling based on the model of [3] to maximize educational quality. Hence each course should be assigned to the most qualified lecturer. The educational quality can be maximized by assigning courses to the lecturer at the right time slot of a day for specific student group in different major. The study conducted by Komijan develop a model that assigning courses for the right time slot and room regarding to the student and lecturer. There are some things affecting the courses that should be considered before the assignment like student groups who take the courses, number of students, courses load, and the lecturer who taught the course. The student groups will later be an effective way to solve problem where student can't attend more than one course at a time. Then the number of students is used to assign courses to appropriate room. It also shows that the room is used efficiently. Next the courses load will be used to put courses to the appropriate time slot which has different duration. Last, the lecturer that taught the course must be considered carefully as the lecturer have their unavailable time, and different courses to teach

2. Methods

Several authors have solved scheduling problems using integer linear programming techniques such as [4], [5], [6], [7] and [8]. Specifically, [7] used the Langrangian Relaxation technique and in [8] expanded his technique to complete subproblem grouping. Meanwhile, Ferland and Roy (1985) formulated the problem as an assignment problem, and solved the model with quadratic programming. This study use integer linear programming for the model. Linear programming allows us to formulate mathematical model based on real world problem. According to [9] One of the most popular methods for linear programming is simplex method. Simplex method tends to find the most feasible solution by modifying the value of basic variable, so that maximum value for objective function is obtained. And another method in linear programming specifically to solve transportation problem is northwest corner and least cost method. Both of these methods' objective is to find the cheapest or smallest unit cost for demand and supply. The northwest corner method commonly used to find and calculate feasible solutions which include demand and supply problems involving many sources. The number of total demands should meet the number of total supplies is required for this method to work. And for the least cost method, based on [10] could find the better solution with the cheapest routes by assigning as much as possible cells with smallest unit cost. This method is conducted by crossing out the row or column and adjusting the amounts of demand of supply, until there's only one row or column left uncrossed. Thus, the smallest unit cost can be obtained. The linear programming in this study use binary logic because according to [11], binary logic could indicate whether particular alternative should be performed or used and to force logical relations among decisions. Other than that, because of different constraints exist in this study also become one of the factors why binary logic is used. By implementing binary logic which consist of zero-one decisions for yes/no answer, then the suitable solution will show up when there are two options is given.

2.1Notations

Parameter

i = time slot

j = courses

k = rooms

K_r= sets of regular rooms
I_t = sets of time slots with load 2 credits
I_d= sets of time slots each day
J_m= sets of mandatory courses
J_e= sets of elective courses
J_c= sets of courses with load 3 credits
J_r= sets of courses thought by lecturer
J_s = sets of student groups that take the courses
T_r = sets of time slot where the lecturer is unavailable
e_j= number of students enrolled in a course
f_k= room capacity

Decision Variable

$$x_{i,j,k} = \begin{cases} 1, & \textit{if the courses j is assigned to time slot i and room k} \\ 0, & \textit{otherwise} \end{cases}$$

2.2 Model Formulation

Objective function of the model is to maximize the number of course assignment as shown in Equation (1).

$$Max = \sum_i \sum_j \sum_k x_{ijk} \tag{1}$$

The model consists of some constraints. The first constraint is to ensure that every course is assigned to only one room. It prevents one course getting held in different room. Equation (2) show the constraint.

$$\sum_i^I \sum_k^K x_{i,j,k} = 1 \qquad ;\forall j \tag{2}$$

The second constraint is used to ensure that each course held in one room and one time slot. So, there will be no other courses carried out in one room and one specific time slot.

$$\sum_j^J x_{i,j,k} \leq 1 \qquad ;\forall i,k \tag{3}$$

The third constraint is needed to ensure the match the credit course with the appropriate time slot. Thus, courses with a load of 3 credits cannot be held in a time slot for courses with a load of 2 credits. However, courses with a load of 2 credits can be held in a time slot for courses with a load of 3 credits.

$$\sum_i^{I_t} \sum_k^{K_r} \sum_j^{J_c} x_{i,j,k} = 0 \tag{4}$$

Equation (5) is used to ensure the lecturer assignment to exactly one time slot and one course. So, the lecturer can't teach more than one course in one time slot.

$$\sum_{j \in Jr} \sum_k^{K_r} x_{i,j,k} \leq 1 \qquad \forall i \tag{5}$$

Equation (6) is used specifically if there's any other agenda the lecturer must attend. So, then the teacher can't be assigned to time slot where he's unavailable, and there will be no course held at that time.

$$\sum_{j \in Jr} \sum_{k}^{K_r} \sum_{i \in T_r} x_{i,j,k} = 0 \tag{6}$$

The sixth constraint in Equation (7) ensures the student group as the main focus. This constraint prevent student who probably re-take the course from previous year to sit in two courses or more at the same slot time. This also prevent student from different student group to have a clashed schedule.

$$\sum_{j \in Js} \sum_{k}^{K_r} x_{i,j,k} \leq 1 \qquad \forall i \tag{7}$$

The seventh constraint consists of two constraint: the first is to limit the use of regular classroom and second the use of laboratory room. Due to the room capacity, mandatory course is not possible to be held in laboratory room due to its capacity. Instead, laboratory room is used for elective course. The constraint to limit the use of regular room for elective course is shown in Equation (8).

$$\sum_{k}^{K_r} \sum_{i}^{I} \sum_{j}^{J_e} x_{i,j,k} = 0 \tag{8}$$

Equation (9) is used to match the number of students in a course with the room capacity.

$$e_j x_{i,j,k} \leq f_k \qquad \forall i,j,k \tag{9}$$

The last constraint is used to restrict the same course for different student group held in the same day. This constraint prohibit lecturer to teach the same course simultaneously. The lecturer will teach the same course for different student group in the next day. Equation (10) shows the constraints.

$$\sum_{i}^{I_d} \sum_{k}^{K_r} \sum_{j}^{J_m} x_{i,j,k} \leq 1 \tag{10}$$

3. Result and discussion

The model is solved using Lingo 18 which provides report with all possibility combinations from every variable. The result from the Lingo report should be translated first into Microsoft Excel table because the notation is coded in number. The combinations of decisions variable contained in Lingo as binary code zero-one then listed to know which combination is possible for yes/no answer. Later, with the translated result in Microsoft Excel table, the schedule is formed. Table 1 shows the listed result of Lingo report with one/yes decision.

Table 1. Lingo report.

No	Variable	No	Variable	No	Variable	No	Variable	No	Variable
1	X(I1, J59, K7)	14	X(I6, J25, K3)	27	X(I8, J62, K8)	40	X(I14, J29, K3)	53	X(I17, J50, K3)
2	X(I2, J27, K3)	15	X(I6, J28, K4)	28	X(I9, J16, K4)	41	X(I14, J40, K4)	54	X(I18, J33, K3)
3	X(I2, J30, K4)	16	X(I6, J56, K5)	29	X(I9, J20, K2)	42	X(I14, J53, K1)	55	X(I18, J38, K4)
4	X(I3, J1, K1)	17	X(I6, J60, K7)	30	X(I9, J43, K1)	43	X(I15, J6, K1)	56	X(I18, J57, K5)
5	X(I3, J5, K2)	18	X(I6, J61, K8)	31	X(I9, J52, K3)	44	X(I15, J8, K2)	57	X(I18, J65, K9)
6	X(I3, J24, K3)	19	X(I7, J3, K1)	32	X(I11, J26, K4)	45	X(I15, J41, K4)	58	X(I19, J2, K1)
7	X(I3, J64, K9)	20	X(I7, J7, K3)	33	X(I11, J39, K1)	46	X(I15, J44, K3)	59	X(I19, J4, K2)
8	X(I4, J32, K3)	21	X(I7, J42, K4)	34	X(I12, J9, K2)	47	X(I16, J31, K3)	60	X(I19, J55, K5)
9	X(I4, J37, K4)	22	X(I7, J45, K2)	35	X(I12, J11, K3)	48	X(I16, J36, K4)	61	X(I20, J14, K4)
10	X(I4, J54, K1)	23	X(I8, J15, K1)	36	X(I12, J13, K4)	49	X(I16, J63, K9)	62	X(I20, J21, K1)
11	X(I5, J12, K1)	24	X(I8, J19, K2)	37	X(I12, J49, K1)	50	X(I17, J10, K1)	63	X(I20, J22, K3)
12	X(I5, J17, K2	25	X(I8, J23, K3)	38	X(I13, J46, K4)	51	X(I17, J18, K2)	64	X(I1, J35, K4)
13	X(I5, J48, K4)	26	X(I8, J58, K6)	39	X(I13, J51, K1)	52	X(I17, J47, K4)	65	X(I1, J34, K3)

The report listed on above then one by one is translated based on the number code. For example X (I1, J59, K7) means the notation for i which means timeslot for timeslot number 1, and j which means courses for course number 59, and k which means rooms for room number 7, have value 1. It means that courses number 59 can be conducted in timeslot number 1, at the room number 7. Later, after all the report is translated into what timeslot, courses, and room, we can build an appropriate schedule form. And since every course is taught by different lecturer, we can also put the lecturer into suitable timeslot and room. Table 2 shows the results.

Based on Table 2, every course is well scheduled into appropriate timeslot and rooms. There are no clashes found between courses. Each course is taught by specific lecturer and taken by specific student group. As we can see above, different courses are assigned to different timeslot and rooms to prevent the lecturer teaching different student group by the same timeslot. For example, mandatory courses 'Optimization' taught by lecturer with initial SS is assigned to timeslot 2, room k=3. Later, lecturer SS has another assigned elective course of 'Human Factor Based Product Development' by timeslot 3, room k=9 which is suitable laboratory for that specific elective courses. By that, lecturer SS has 2 assigning courses for different student group 4A and group that take elective courses. We can also conclude that during assigning for elective courses 'Human Factor Based Product Development' or 'Supply Chain Management', no student group from third and fourth year is assigned to mandatory courses. The same condition is obtained in the following schedule.

In Table 3, some elective courses are held simultaneously. That make the student group taking elective course in specific field of interest has no option but to choose one elective course at one time. Even though, they can only choose one, it doesn't limit the student group for having only 1 elective course in a day within the same laboratory. For example, student group taking elective course 'Multi Variable Analysis' can still take 'Standardization' in one day. This also work for student group taking different elective course from different laboratory.

From the Table 4, there's a vacant timeslot from 09.20 WIB to 12.00 WIB where no courses are allowed to be held. This due to the unavailable time of lecturer to teach because of special agenda including department meeting routine and *competitive improvement program* held by each laboratory. This means that the constraint where the lecturer can't teach on their unavailable time is achieved.

Table 2. The schedule optimization result.

Timeslot	Day	Hour	*k=1*	*k=2*	*k=3*	*k=4*	*k=5*	*k=6*	*k=7*	*k=8*	*k=9*	*k=10*
1	MONDAY	07.30-08.20			Manufacturing Process II 4B (LH)	Work System Analysis and Design 4A (RDA,BS)			Supply Chain Management (WS)			
		08.25-09.15			Manufacturing Process II 4B (LH)	Work System Analysis and Design 4A (RDA,BS)			Supply Chain Management (WS)			
2		09.20-10.10			**Optimization 4A (SS)**	**Industrial Statistics 4B (FF,EP)**						
		10.15-11.05			**Optimization 4A (SS)**	**Industrial Statistics 4B (FF,EP)**						
		11.10-12.00			**Optimization 4A (SS)**	**Industrial Statistics 4B (FF,EP)**						
3		13.00-13.50	**Calculus II 2A (IWS)**	**Introduction to Engineering and Design 2B (MNF,TR)**	Bahasa 2C (AYW)						Human Factor Based Product Development (SS)	
		13.55-14.45	**Calculus II 2A (IWS)**	**Introduction to Engineering and Design 2B (MNF,TR)**	Bahasa 2C (AYW)						Human Factor Based Product Development (SS)	
		14.50-15.40	**Calculus II 2A (IWS)**	**Introduction to Engineering and Design 2B (MNF,TR)**								
4		15.45-16.35	Entrepreneurship 8B (YA,MH)		Database 4B (YP,EL,MNF)	Engineering Economics 4A (WS,FF)						
		16.40-17.30	Entrepreneurship 8B (YA,MH)		Database 4B (YP,EL,MNF)	Engineering Economics 4A (WS,FF)						

Table 3. The schedule optimization result.

Timeslot	Day	Hour	*k=1*	*k=2*	*k=3*	*k=4*	*k=5*	*k=6*	*k=7*	*k=8*	*k=9*	*k=10*
5	TUESDAY	07.30-08.20	Computer Programming 2C (YP,EL)	Materials Engineering 2B (TR,RHS)		Occupational Health and Safety 6B (IRF,RDA)						
		08.25-09.15	Computer Programming 2C (YP,EL)	Materials Engineering 2B (TR,RHS)		Occupational Health and Safety 6B (IRF,RDA)						
6		09.20-10.10			**Multiple Variable Calculus 4A (SBW,VK)**	**Optimization 4B (SS)**	Industrial Ecology (BS,RDA)		Performance Measurement (YA)	Multi Variable Analysis (EP)		
		10.15-11.05			**Multiple Variable Calculus 4A (SBW,VK)**	**Optimization 4B (SS)**	Industrial Ecology (BS,RDA)		Performance Measurement (YA)	Multi Variable Analysis (EP)		
		11.10-12.00			**Multiple Variable Calculus 4A (SBW,VK)**	**Optimization 4B (SS)**						
7		13.00-13.50	**Calculus II 2C (IWS)**	Design of Production Aids 6A (LH)	Basic Physics II 2A (MA,IRF)	**System Modeling 6B (YP,CNR)**						
		13.55-14.45	**Calculus II 2C (IWS)**	Design of Production Aids 6A (LH)	Basic Physics II 2A (MA,IRF)	**System Modeling 6B (YP,CNR)**						
		14.50-15.40	**Calculus II 2C (IWS)**			**System Modeling 6B (YP,CNR)**						
8		15.45-16.35	Engineering Mechanics 2C (RHS,TR)	Pancasila 2A (IRF)	Bahasa 2B (AP)			Tolerance Design (CNR)		Standardization (FF)		
		16.40-17.30	Engineering Mechanics 2C (RHS,TR)	Pancasila 2A (IRF)	Bahasa 2B (AP)			Tolerance Design (CNR)		Standardization (FF)		

Table 4. The schedule optimization result.

Timeslot	Day	Hour	*k=1*	*k=2*	*k=3*	*k=4*	*k=5*	*k=6*	*k=7*	*k=8*	*k=9*	*k=10*
9	WEDNESDAY	07.30-08.20	Facility Layout Design 6A (BS,RDA)	Pancasila 2B (IRF)	Marketing 6B (RZ,YA)	Materials Engineering 2A (TR,RHS)						
		08.25-09.15	Facility Layout Design 6A (BS,RDA)	Pancasila 2B (IRF)	Marketing 6B (RZ,YA)	Materials Engineering 2A (TR,RHS)						
10		09.20-10.10										
		10.15-11.05										
		11.10-12.00										
11		13.00-13.50	**Management information System 6A (YP,MNF)**			**Multiple Variable Calculus 4B (SBW,VK)**						
		13.55-14.45	**Management information System 6A (YP,MNF)**			**Multiple Variable Calculus 4B (SBW,VK)**						
		14.50-15.40	**Management information System 6A (YP,MNF)**			**Multiple Variable Calculus 4B (SBW,VK)**						
12		15.45-16.35	Organizational Behavior and Design 6A (RZ)	Basic Physics II 2C (MA,IRF)	Computer Programming 2B (YP,EL)	Engineering Mechanics 2A (RHS,TR)						
		16.40-17.30	Organizational Behavior and Design 6A (RZ)	Basic Physics II 2C (MA,IRF)	Computer Programming 2B (YP,EL)	Engineering Mechanics 2A (RHS,TR)						

Table 5. The schedule optimization result.

Timeslot	Day	Hour	*k=1*	*k=2*	*k=3*	*k=4*	*k=5*	*k=6*	*k=7*	*k=8*	*k=9*	*k=10*
13	THURSDAY	07.30-08.20	Marketing 6A (RZ,YA)			Design of Production Aids 6B (LH)						
		08.25-09.15	Marketing 6A (RZ,YA)			Design of Production Aids 6B (LH)						
14		09.20-10.10	Entrepreneurship 8A (YA,MH)		**Industrial Statistics 4A (FF,EP)**	**Management information System 6B (YP,MNF)**						
		10.15-11.05	Entrepreneurship 8A (YA,MH)		**Industrial Statistics 4A (FF,EP)**	**Management information System 6B (YP,MNF)**						
		11.10-12.00			**Industrial Statistics 4A (FF,EP)**	**Management information System 6B (YP,MNF)**						
15		13.00-13.50	**Introduction to Engineering and Design 2C (MNF,TR)**	Basic Physics II 2B (MA,IRF)	Facility Layout Design 6B (BS,RDA)	**System Modeling 6A (YP,CNR)**						
		13.55-14.45	**Introduction to Engineering and Design 2C (MNF,TR)**	Basic Physics II 2B (MA,IRF)	Facility Layout Design 6B (BS,RDA)	**System Modeling 6A (YP,CNR)**						
		14.50-15.40	**Introduction to Engineering and Design 2C (MNF,TR)**			**System Modeling 6A (YP,CNR)**						
16		15.45-16.35			Database 4A (YP,EL,MNF)	Work System Analysis and Design 4B (RDA,BS)					Project Management (SS)	
		16.40-17.30			Database 4A (YP,EL,MNF)	Work System Analysis and Design 4B (RDA,BS)					Project Management (SS)	

From Table 5, the course with 3 credits load (bold letters) is perfectly held in the time slot dedicated for that courses. The average number of mandatory courses assignment for student groups are no more than 3. For example, in table 5, the student group 6B has three different courses in a day as following courses : 'Design of Production Aids' from 07.30 WIB until 09.15 WIB, 'Information System Management' from 09.20 WIB until 12.00 WIB', and last 'Facility Layout Design' from 13.00 WIB until 14.45 WIB. That means the student at least have a break between or after courses they take in a day.

Table 6. The schedule optimization result.

Timeslot	Day	Hour	*k=1*	*k=2*	*k=3*	*k=4*	*k=5*	*k=6*	*k=7*	*k=8*	*k=9*	*k=10*
17	FRIDAY	07.30-08.20	Computer Programming 2A (YP,EL)	Materials Engineering 2C (TR,RHS)	Organizational Behavior and Design 6B (RZ)	Occupational Health and Safety 6A (IRF,RDA)						
		08.25-09.15	Computer Programming 2A (YP,EL)	Materials Engineering 2C (TR,RHS)	Organizational Behavior and Design 6B (RZ)	Occupational Health and Safety 6A (IRF,RDA)						
18		09.20-10.10			Manufacturing Process II 4A (LH)	Engineering Economics 4B (WS,FF)	Environmental Ergonomics (BS,RDA)				Product Creativity and Innovation (RHS)	
		10.15-11.05			Manufacturing Process II 4A (LH)	Engineering Economics 4B (WS,FF)	Environmental Ergonomics (BS,RDA)				Product Creativity and Innovation (RHS)	
		11.10-12.00										
19		13.00-13.50	**Calculus II 2B (IWS)**	**Introduction to Engineering and Design 2A (MNF,TR)**			Work Improvement Methods (RDA,BS)					
		13.55-14.45	**Calculus II 2B (IWS)**	**Introduction to Engineering and Design 2A (MNF,TR)**			Work Improvement Methods (RDA,BS)					
		14.50-15.40	**Calculus II 2B (IWS)**	**Introduction to Engineering and Design 2A (MNF,TR)**								
20		15.45-16.35	Pancasila 2C (IRF)		Bahasa 2A (FS)	Engineering Mechanics 2B (RHS,TR)						
		16.40-17.30	Pancasila 2C (IRF)		Bahasa 2A (FS)	Engineering Mechanics 2B (RHS,TR)						

Another example can be found in Table 6 for student group 2A. They have three courses in a day, but there's a long break between courses due to religious event held every Friday. Because of this event, the timeslot number 18, which supposed for 3 credits load is decreased for 2 credits load only. The duration is shortened from 09.20 WIB until 12.00 WIB to 09.20 WIB until 11.05 WIB to provide the time needed for religious event. And so, for other day, the number of assigned courses for every student group doesn't exceed three. However, this doesn't apply to lecturer. Some lecturer has a system team teaching which allows them to share teaching load/time based on material taught. The number of courses the lecturer has to teach in one day depends on their material sharing deals. It can be mid semester period teaching, or based on the number of the exam given.

4. Conclusion

This paper provides solution for course scheduling. The courses are well assigned to appropriate rooms and to specific time slot as expected. The previous work without optimization clearly has some difficulties in course-room assignment due to limited room available. But with optimization, it was obvious that with the limited number of rooms, the course can still be scheduled appropriately. The result shows that all constraint is fully satisfied even. There might be new constraint needed to limit the average number of course taken by student in a day. The limitation of courses will help student to have a break depending on credit load they have in a day, and also to prevent lecturer for having a nonstop teaching. Other than that, further research also needed to consider the agenda every lecturer has, because it can be vary from one lecturer to another lecturer. This mean to prevent lecturer absence due to sudden events or meetings.

References

[1] Schaerf A, Schaerf M. (1995). Local search techniques for high school timetabling. *In Proc. of the 1st International Conference on the Practice and Theory of Automated Timetabling, pp. 313-323*

[2] Mohmad KMN, Kendall G. Universiti Malaysia Pahang examination timetabling problem: scheduling invigilators. *J Oper Res Soc*. Informa UK Limited; 2014 Feb;**65**(2):214–26. doi.org/10.1057/jors.2012.6

[3] Komijan A, Koupaei MN. A Mathematical Model for University Course Scheduling: A Case Study. *Int. J. Tech. Res. Appl.* 2015;**19**:2320-8163

[4] Breslaw JA. A linear programming solution to the faculty assignment problem. *Socio-Econ. Plan. Sci*. Elsevier BV; 1976 Jan;**10**(6):227–30. doi.org/10.1016/0038-0121(76)90008-2

[5] McClure RH, Wells CE. A mathematical programming model for faculty course assignments. *Decis. Sci. J. Innov. Educ*. Wiley; 1984 Jul;**15**(3):409–20. doi.org/10.1111/j.1540-5915.1984.tb01226.x

[6] McClure RH, Wells CE. A mathematical programming model for faculty course assignments. *Decis. Sci. J. Innov. Educ*. Wiley; 1984 Jul;**15**(3):409–20. doi.org/10.1111/j.1540-5915.1984.tb01226.x

[7] Tripathy A. School Timetabling—A Case in Large Binary Integer Linear Programming. *Management Science*. Institute for Operations Research and the Management Sciences (INFORMS); 1984 Dec;**30**(12):1473–89. doi.org/10.1287/mnsc.30.12.1473

[8] Tripathy A. Computerised decision aid for timetabling — a case analysis. *Discret. Appl. Math.* Elsevier BV; 1992 Mar;**35**(3):313–23. doi.org/10.1016/0166-218x(92)90253-7

[9] Sra S, Nowozin S, Wright SJ. *Optimization for Machine Learning*. London: The MIT Press Cambridge. Massachusetts Institute of Technology. 2012.

[10] Mishra S. Solving Transportation Problem by Various Methods and Their Comparison. *International Journal of Mathematics Trends and Technology*. Seventh Sense Research Group Journals; 2017 Apr 25;**44**(4):270–5. doi.org/10.14445/22315373/ijmtt-v44p538

[11] Stevens SP, Palocsay SW. Teaching Use of Binary Variables in Integer Linear Programs: Formulating Logical Conditions. *INFORMS Transactions on Education*. Institute for Operations Research and the Management Sciences (INFORMS); 2017 Sep;**18**(1):28–36. doi.org/10.1287/ited.2017.0177

Tensile properties of epoxy/1 wt% graphene nanocomposites prepared with ethanol

M S Saharudin[1,5], S Hasbi[2,6], C Okolo[3,7] and F Inam[4,8]

[1]Universiti Kuala Lumpur Malaysia Italy Design Institute (UniKL MIDI), 119 Jalan 7/91, Taman Shamelin Perkasa, 56100, Kuala Lumpur, Malaysia
[2]Department of Mechanical Engineering, Faculty of Engineering, National Defence, University of Malaysia (NDUM), Kem Sungai Besi, 57000, Kuala Lumpur, Malaysia
[3]Department of Mechanical and Construction Engineering, Northumbria University, Newcastle upon Tyne NE1 8ST, UK
[4]University of East London, Docklands Campus, 4-6 University Way, London E16 2RD, United Kingdom

E-mail: [5]mshahneel@unikl.edu.my, [6]syafawati@upnm.edu.my, [7]chinyere.okolo@northumbria.ac.uk, [8]f.Inam@uel.ac.uk

Abstract. In this research, solution casting technique was applied to produce four types of nanocomposites. Different ethanol dosages; 0g, 1g, 3g, and 5g were used to disperse graphene in the epoxy matrix. It was observed that 1g dosage of ethanol was the most effective concentration to disperse 1wt% graphene in the epoxy matrix. At 1 g dosage of ethanol used, the Young's modulus, tensile strength, and toughness were increased by 130% ,76%, and 187% respectively. SEM images illustrated that the graphene was able to inhibit the advancing cracks and detoured cracks propagation. It is observed that the ethanol needs to be removed completely during processing to ensure its effectiveness, otherwise, the remaining ethanol can cause porosity which is undesirable to the tensile properties of the nanocomposites.

1. Introduction

Composites are materials which are comprising at least two or more phases that are chemically and physically bonded. Composite materials have a number of advantages, such as good resistance to corrosion, high fatigue strength and very low weight. In industry, composite materials have been widely used in aerospace, automotive, military, biomedical and sports leisure goods [1].

Because of their excellent thermal, chemical stability and mechanical properties when combined with fillers. Epoxy resin are used in numerous industrial applications particularly in aerospace, automotive and construction industries [2][3][4]. In recent years, epoxy resins have gained research interest in engineering applications due to their unique balance of chemical and mechanical and advantages of ease of processing [4][5]. However, due to the crosslinking structures which make the epoxy prone to crack and brittleness, the applications of epoxy resins remain limited.

Various studies have been carried out to increase the epoxy toughness. One of the popular methods is by adding Nano-filler such as graphene in the epoxy. Graphene-based polymer composite has attained great attention since the explosion of graphene research in 2004 due to its excellent performance in terms of thermal, mechanical and gas barrier [6]. Graphene-based materials have been extensively used in various fields such as composites, coatings, electronic devices, energy storage, sensors and biomedical [6]. Asif et al in their previous research have established that the Young's modulus and micro hardness of multi-layer graphene/epoxy have improved by 26% and 18% respectively [7]. In addition, the glass transition temperature (T_g) and storage modulus also improved in comparison to unreinforced epoxy.

Addition of Nano-filler such as graphene in epoxy can significantly enhance the physical and chemical properties of the composite at extremely low loadings provided that the filler is homogenously dispersed in the matrix [8][9]. The graphene; when dispersed homogenously acts as a barrier to hinder crack propagation and share the external stress in order to evade stress concentration which in turn improves the mechanical properties of the composite. In contrast, poorly dispersed graphene can act as a stress raiser and increase stress concentration, resulting the reduction of the mechanical properties [10][11]. Graphene also tends to re-aggregate in the matrix due to the strong van der Waal forces between discretely graphene particles, thus manual mixing is impractical to disperse graphene in epoxy [3][12][13].

Solvents are usually used to ensure uniform dispersion in matrices [12] by manipulating the presence of function group attached on the graphene surface. This method allows direct dispersion of graphene in organic solvents. This method is widely used for processing epoxy/graphene nanocomposites. By using mild sonication method for instance, physical and chemical interaction between the functionalised graphene and polymeric matrices can be enhanced [14]. Likewise, graphene can be easily dispersed in a suitable solvent using bath sonication technique, mixed with epoxy resin and the solvent is removed in a high temperature [15].

Solvents such as ethanol showed good dispensability and steadiness to disperse graphene. Besides that, several other solvents such as tetrahydrofuran, acetone, dichloromethane, isopropyl alcohol and water also used to produce epoxy/graphene composites. Nonetheless, not all are efficient dispersant agents for graphene [10]. Using less toxic and volatile dispersing solvents such as alcohols and water, dispersion stability and scalable routes to dispersion still remain as challenges [16]. In addition, accurate characterisation, safe handling and post-processing poses greater challenges from the manufacturing point of view. Another vital challenge to overcome is achieving uniform dispersion at an attractive cost.

Ethanol has been widely used to homogenously spread graphene in epoxy matrix. The usage of ethanol at low content (1 wt%) to disperse graphene has not been reported elsewhere. Several types of epoxy/1 wt%-graphene nanocomposites were fabricated and their tensile properties were studied. In our previous work, flexural strength and flexural modulus increased by 62% and 61% for samples toughened with 1wt% graphene [13]. Likewise, the highest impact strength was recorded for 1 wt% graphene dispersed with 1g ethanol, where an improvement of 9.5% was observed [13]. To the best of author's knowledge, very few studies have been reported on the tensile properties of epoxy/1 wt% graphene nanocomposites prepared with small percentage of ethanol. None of the previous studies, nevertheless, have given due attention to the influence of adding different concentration of ethanol to a fixed concentration of graphene.

2. Materials and Methods

Graphene Nano platelets with particle size of 2 μm and surface area of 300 m^2/g was acquired from Sigma Aldrich, UK. Ethanol concentrations of 0 g, 1 g, 3 g and 5 g was used to disperse 1wt% of graphene in epoxy. Epoxy (Miracast 1517A, density 1.13 g/cm3) and hardener (Miracast 1517B, density 1.1 g/cm3) was purchased from Miracon Sdn. Bhd, Seri Kembangan Malaysia. It has been demonstrated that low viscosity of hardener inhibit an underpinning of agglomeration by refining the dispersion as well as responsible for fast curing [17]. The time taken for the resin to reach gelation was 40 minutes, and it was cured at room temperature.

Bath sonicate was used to disperse graphene in ethanol and this procedure was carried out for 5 minutes. Then the graphene was thoroughly mixed with epoxy. The temperature of the ethanol and epoxy mixture then was set to 60°C for about 10 minutes to fully remove ethanol through evaporation. Then, hardener was added in the epoxy-graphene mixture. The mixture was hand mixing and stirred manually for another 10 minutes before it was poured into silicone mould for 12 hours curing process at room temperature. Finally, all samples were cured for 6 hours at 150^0C to achieve complete crosslink [18].

The procedures of making the silicone mould were rather simple and the gelation time was only about 25 minutes. The silicone mould materials which consists silicone rubber and curing agent were purchased from a local company namely Portal Trading Sdn. Bhd, Penang, Malaysia. The mixture of ratio 98:2 (silicone rubber: hardener) was poured into mould made from acrylic and left for curing around 24 hours. The schematic illustration of the samples fabrication is shown in Figure 1. The tensile samples were produced as per ASTM D638 standard and the schematic drawing is shown in Figure 2. Minimum five tensile samples were tested for each nanocomposite system.

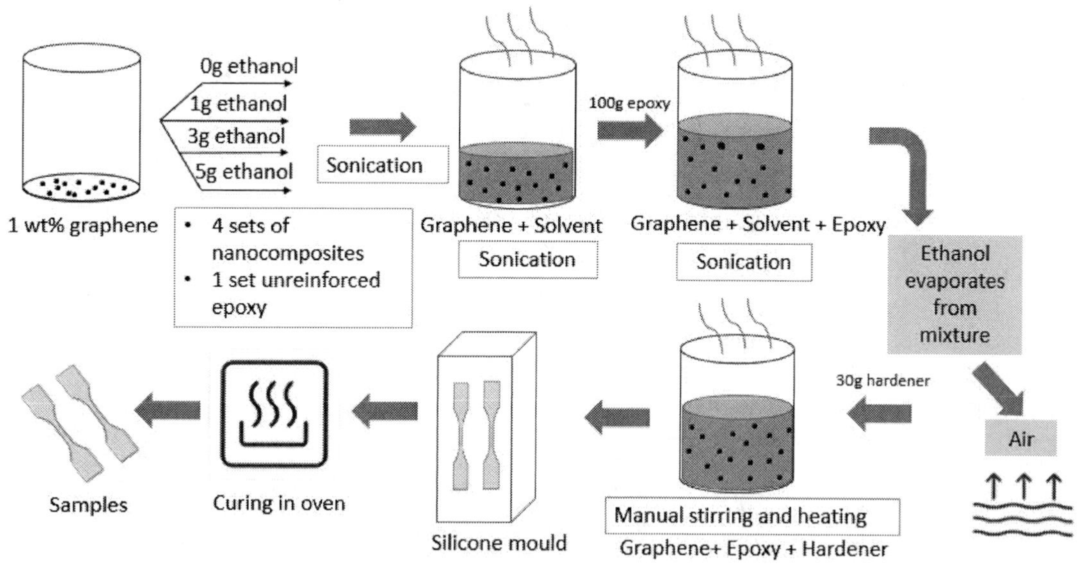

Figure 1. Preparation of nanocomposite samples.

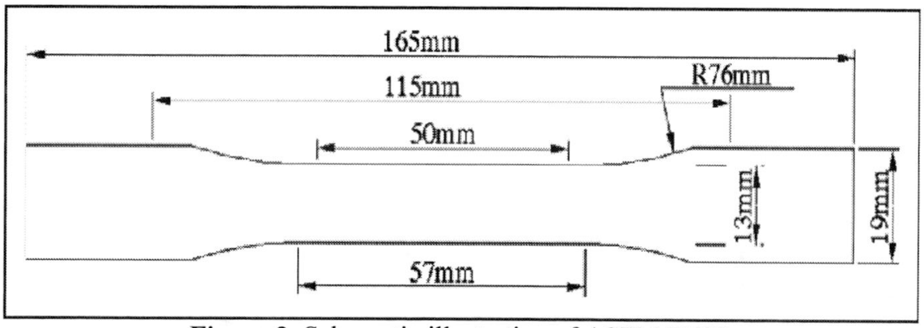

Figure 2. Schematic illustration of ASTM D638.

Five samples from each composition were prepared and tested by using Victor Universal Testing Machine (Victor VE 2302). The tensile crosshead speed was maintained 2 mm/min for all samples. Young's modulus (E) was obtained by the slope of elastic of a stress and strain graph, the value was calculated by the ratio of stress to its corresponding strain value.

$$E = \frac{Stress}{Strain} \quad (1)[19]$$

Where stress $(\sigma) = \frac{F}{A}$ and strain $(\epsilon) = \frac{\Delta l}{l_0}$ (2)[19]

Tensile strength (MPa) taken in this research was the maximum stress that a composite material can withstand while being stretched or pulled before breaking. As for the tensile strain, it was taken from the maximum point before failure. Finally, for toughness, ($J.mm^{-3}$) the value was obtained from the area under the stress-strain curve within the elastic limit as shown in Figure 3.

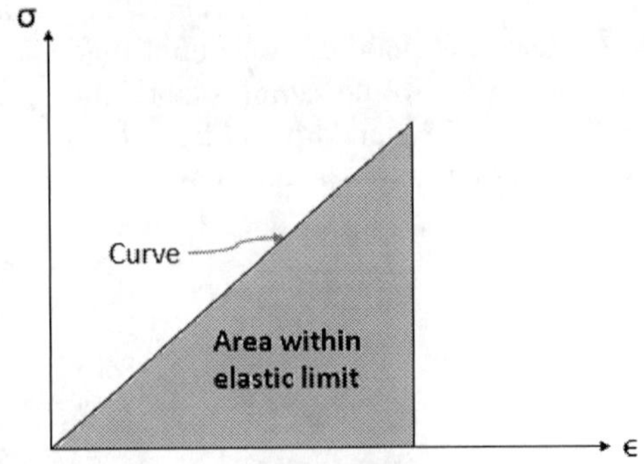

Figure 3. Schematic of area below stress-strain curve.

3. Results and Discussion

Figure 4 shows the stress versus strain curve for all nanocomposite samples prepared with different content of ethanol. It is observed that there is a variation in the stress-strain curve for samples prepared with low ethanol content. In general, all nanocomposite samples prepared with ethanol show good improvement compared with unfilled epoxy (ME). However, it is interesting to point out, without ethanol presence, the tensile stress value is somehow significantly high, however it is slightly lower than epoxy-1wt% graphene prepared with 1g ethanol. It is clearly seen that this sample showed the highest maximum value of stress compared to unfilled epoxy (ME). This can be related to the high aspect ratio of graphene and good bond between graphene and epoxy [20][1].

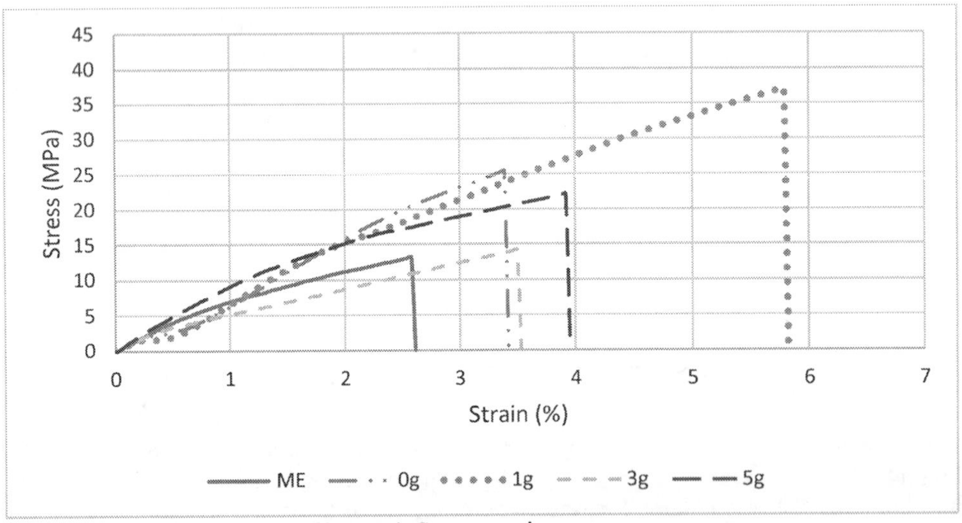

Figure 4. Stress-strain curve.

Figure 5 shows Young's modulus for unfilled epoxy and epoxy/1 wt% graphene samples. Unfilled epoxy recorded the lowest Young's modulus value of 4.1 GPa. The maximum improve in Young's modulus was observed in the sample of epoxy/1 wt% graphene prepared with 1g of ethanol, as 119% of enhancement was achieved. At higher ethanol content, the Young's modulus was improved 36% and 73% for samples prepared with 3g and 5g ethanol content. It is worth to mention, in the case of 3g ethanol content, the Young's modulus was slightly lower than expected value. This can be associated with several factors such as the porosity of samples caused by retained ethanol that acted as stress raiser and could be due to the presence of aggregates in epoxy matrix.

The variation of tensile strength for unfilled epoxy and its nanocomposite systems is shown in Figure 6. The lowest tensile strength recorded was 15.9 MPa in the case of unfilled epoxy. The highest tensile strength was in the case of epoxy/1 wt% graphene prepared with 1g ethanol, where an improvement of 76% was recorded. The lowest increase of 14% was obtained for epoxy/1 wt% graphene dispersed in 3g ethanol.

Figure 7 shows the graph of tensile strain for unfilled epoxy (ME) and its nanocomposites. In general, samples reinforced with graphene showed lower tensile strain compared to unreinforced epoxy. Unfilled epoxy recorded the highest flexural strain, this can be associated to lower strength value. In contrast, the minimum flexural strain was observed in the case of epoxy/1 wt% graphene prepared with 1g ethanol. Minimum flexural strain can be attributed to high stiffness.

The variation of toughness for all samples is presented in Figure 8. This toughness was obtained from area under stress-strain curve which represents energy required before the composites failure. Samples prepared with and without ethanol show good toughness. Epoxy/1 wt% graphene prepared without ethanol showed an increase of 152% compared with unreinforced epoxy. The maximum increase in toughness was observed in the case of epoxy/1 wt% graphene prepared with 1g ethanol, where a maximum increase of 187% was obtained. For 1 wt% graphene samples prepared with 3g and 5g of ethanol, the toughness improved 62% and 115 respectively.

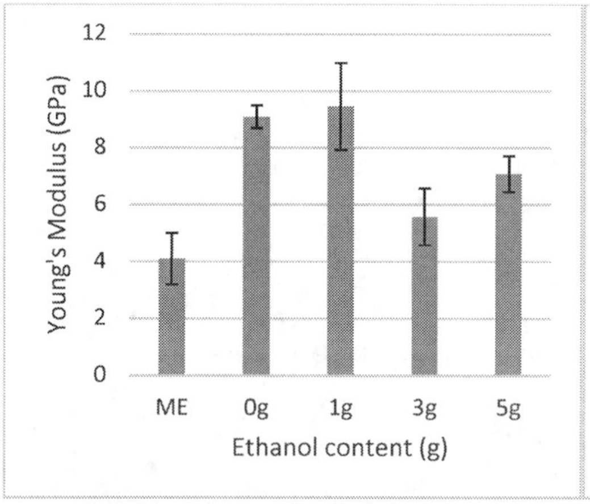

Figure 5. Young's modulus.

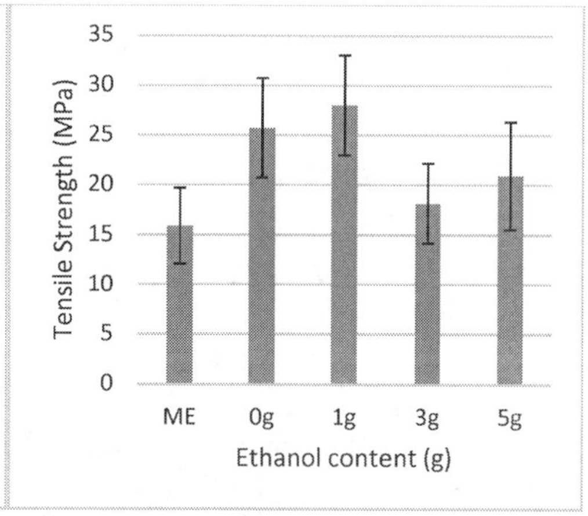

Figure 6. Tensile strength.

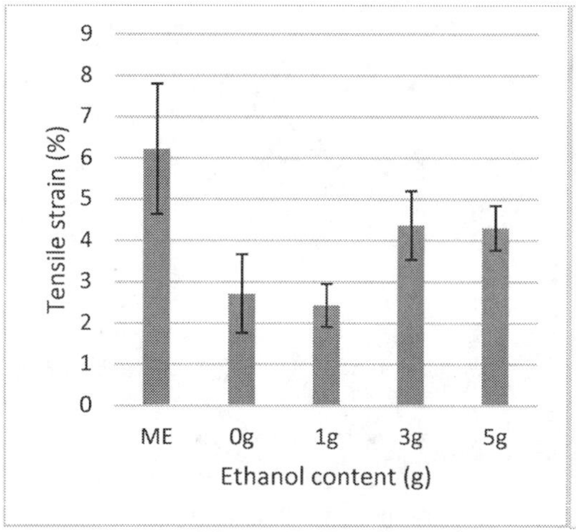

Figure 7. Tensile strain.

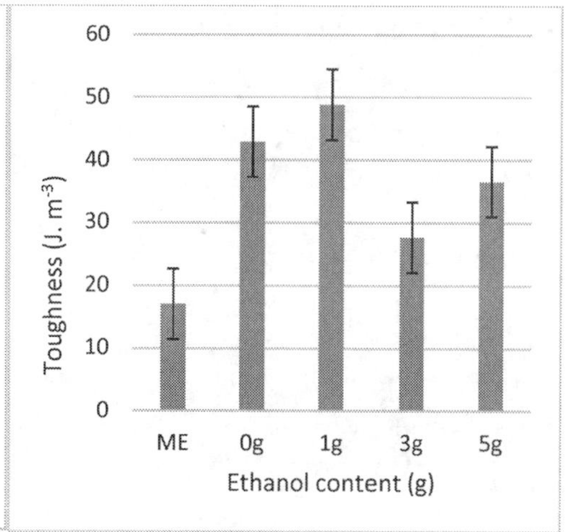

Figure 8. Tensile toughness.

4. SEM Images

Figure 9(a) shows a fractured surface of unfilled epoxy. The image shows a that a brittle fracture has taken place. In unfilled epoxy sample, there was no crack bridging pattern which can be observed. When the crack started, it travels without resistance resulting in straight fracture path. When graphene was introduced, the fracture mechanisms changed due to resistance by reinforcement as shown in Figure 9(b). The result indicates that the dispersal of graphene in the polymer matrix has lower efficiency compared with samples prepared with 1g ethanol [17]. Since this sample was prepared with mild sonication and without graphene being dispersed in ethanol agglomerates of 5 μm can be observed. When the resin cures, the epoxy chain slowly becomes fixed and impedes graphene from homogenously dispersed. Therefore, dispersion of graphene is not easy because of high viscosity and sticky nature of epoxy matrix. Graphene tends to form a cluster because of their weak intermolecular Van der Waals forces [6]. In Figure 9(c) the sample was prepared with 1g of ethanol, the size of agglomerates was significantly reduced to smaller size, hence produce better tensile properties. Figure 9(d) shows sample prepared with 3g ethanol where there is no specific crack pattern can be seen. However, during the processing, the ethanol was not completely removed and porosity was likely to occur. The porosity caused by retained ethanol can be observed for the samples prepared with 3g and 5g of ethanol. The porosity arises from the remained ethanol created stress raiser effect [6]. Pronounced effect in a brittle material is very likely compared to ductile material. When yield stress is surpassed at the point of maximum stress, plastic deformation take place particularly for ductile material. If the crack propagates and face the agglomerate, it will either diminished by agglomerate or diverted if loads are higher. Nevertheless, most part of the epoxy matrix is not completely toughened since graphene is only present in the form of clusters. This can be attributed to sample prepared with 3g of ethanol where the crack escalate via the brittle epoxy and fracture occurs due to poorly dispersed graphene. Hence the graphene reinforcement was not able to improve the tensile properties of epoxy.

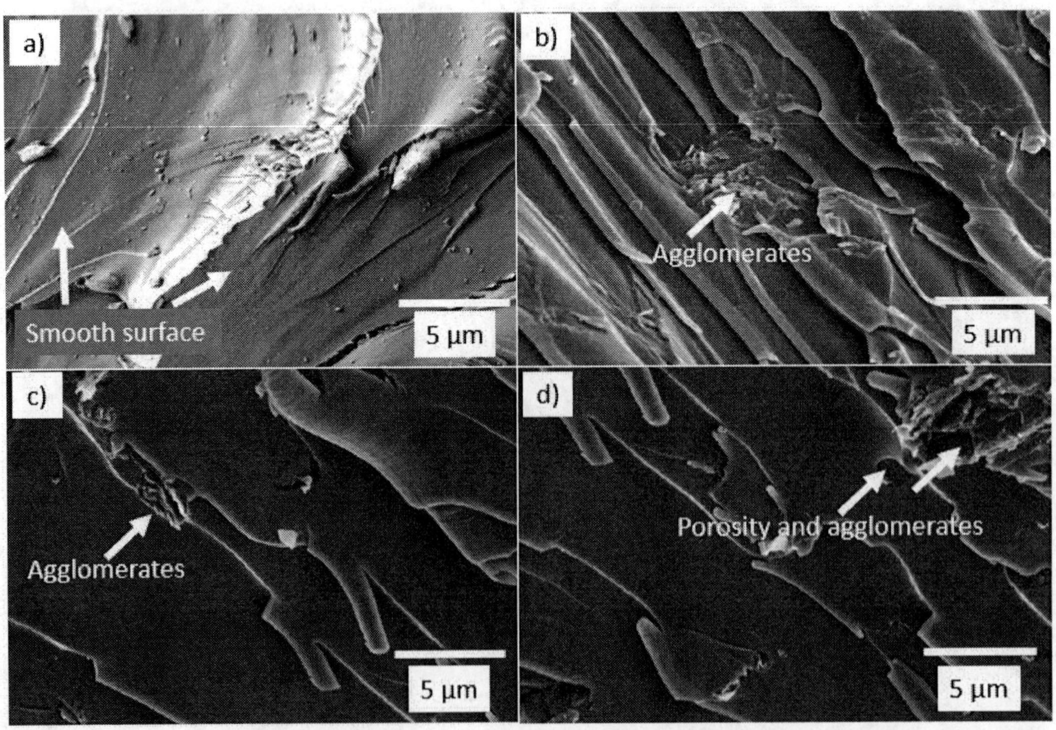

Figure 9. SEM Images of unfilled epoxy (a), Epoxy/1 wt% graphene prepared without ethanol (b), Epoxy/1 wt% graphene prepared with 1g ethanol (c), and Epoxy/1 wt% graphene prepared with 3g ethanol (d).

5. Conclusion

In this research unfilled epoxy (ME) and four different type of epoxy/1 wt% graphene composites dispersed with several ethanol concentrations were successfully produced. All composite samples were fabricated using solution casting technique. Overall, the results indicate that ethanol is an excellent processing solvent for 1 wt% graphene dispersed in the epoxy matrix. The highest increase in Young's modulus and tensile strength were recorded for sample 1 wt% graphene dispersed in 1g ethanol, with an improvement of 130% and 76% respectively. About 187% improvement in toughness was recorded, also dispersed with 1g of ethanol. SEM images revealed that graphene was able to inhibit the propagation of cracks and significantly increased the tensile properties by detouring the advancement of cracks. Inefficient evaporation of ethanol during processing led to the presence of tiny holes or also known as porosity, which of course very undesirable to the mechanical properties of the nanocomposites. For future research, it is recommended that the temperature of manual stirring to be increased from 60°C to 80°C to completely remove the remaining solvent.

References

[1] Kumar K, Davim JP, editors. Composites and Advanced Materials for Industrial Applications. *Adv. Chem. Mater. Eng.* IGI Global; 2018. doi.org/10.4018/978-1-5225-5216-1.

[2] Shyha I, Fu GY, Huo DH, Le B, Inam F, Saharudin MS, et al. Micro-Machining of Nano-Polymer Composites Reinforced with Graphene and Nano-Clay Fillers. *Key Eng. Mater.* Trans Tech Publications, Ltd.; 2018 Oct;**786**:197–205. doi.org/10.4028/www.scientific.net/kem.786.197.

[3] Saharudin M S, Shyha I, Inam F. Viscoelastic and mechanical properties of multi-layered-graphene polyester composites. *2nd International Conference on Advances in Mechanical Engineering.* 2016: 41–5.

[4] Shahneel Saharudin M, Atif R, Hasbi S, Naguib Ahmad Nazri M, Ubaidah Saidin N, et al. Synergistic effects of halloysite and carbon nanotubes (HNTs + CNTs) on the mechanical properties of epoxy nanocomposites. *AIMS. Mater. Sci.* American Institute of Mathematical Sciences (AIMS); 2019;**6**(6):900–10. doi.org/10.3934/matersci.2019.6.900.

[5] Saharudin MS, Jumahat A, Kahar AZA, Ahmad S. The Influence of Alumina Filler on Impact Properties of Short Glass Fiber Reinforced Epoxy. *Appl. Mech. Mater.* Trans Tech Publications, Ltd.; 2013 Sep;**393**:88–93. doi.org/10.4028/www.scientific.net/amm.393.88.

[6] Atif R, Shyha I, Inam F. Mechanical, Thermal, and Electrical Properties of Graphene-Epoxy Nanocomposites—A Review. *Polym.* MDPI AG; 2016 Aug 4;**8**(8):281. doi.org/10.3390/polym8080281.

[7] Atif R, Wei J, Shyha I, Inam F. Use of morphological features of carbonaceous materials for improved mechanical properties of epoxy nanocomposites. *RSC Adv.* Royal Society of Chemistry (RSC); 2016;**6**(2):1351–9. doi.org/10.1039/c5ra24039e.

[8] N,N-Dimethylformamide (DMF) Usage in Epoxy/Graphene Nanocomposites: Problems Associated with Reaggregation. *Polym.* MDPI AG; 2017 May 27;**9**(12):193. doi.org/10.3390/polym9060193.

[9] Wei J, Saharudin MS, Vo T, Inam F. Effects of surfactants on the properties of epoxy/graphene nanocomposites. *J. Reinf. Plast. Compos.* SAGE Publications; 2018 Apr 10;**37**(14):960–7. doi.org/10.1177/0731684418765369.

[10] Atif R, Shyha I, Inam F. The degradation of mechanical properties due to stress concentration caused by retained acetone in epoxy nanocomposites. *RSC Adv.* Royal Society of Chemistry (RSC); 2016;**6**(41):34188–97. doi.org/10.1039/c6ra00739b.

[11] Saharudin MS, Atif R, Shyha I, Inam F. The degradation of mechanical properties in polymer nano-composites exposed to liquid media – a review. *RSC Adv.* Royal Society of Chemistry (RSC); 2016;**6**(2):1076–89. doi.org/10.1039/c5ra22620a.

[12] Wei J, Saharudin MS, Vo T, Inam F. Dichlorobenzene: an effective solvent for epoxy/graphene nanocomposites preparation. *Royal Soc. Open Sci.* The Royal Society; 2017 Oct;**4**(10):170778. doi.org/10.1098/rsos.170778.

[13] The Processing of Epoxy/1 wt%-graphene Nanocomposites: Effects of Ethanol on Flexural Properties. *Int. J. Innov. Technol. Explor.* Eng. Blue Eyes Intelligence Engineering and Sciences Engineering and Sciences Publication - BEIESP; 2019 Nov 10;**9**(1):5440–4. doi.org/10.35940/ijitee.a8105.119119.

[14] Wei J, Vo T, Inam F. Epoxy/graphene nanocomposites – processing and properties: a review. *RSC Adv.* Royal Society of Chemistry (RSC); 2015;**5**(90):73510–24. doi.org/10.1039/c5ra13897c.

[15] Wajid AS, Das S, Irin F, Tanvir Ahmed HS, Shelburne JL, Parviz D, et al. Erratum to "Polymer-stabilized graphene dispersions at high concentrations in organic solvents for composite production" [Carbon 50 (2012) 526–534]. *Carbon.* Elsevier BV; 2012 Apr;**50**(5):2065. doi.org/10.1016/j.carbon.2012.01.002.

[16] Johnson DW, Dobson BP, Coleman KS. A manufacturing perspective on graphene dispersions. *J. Colloid Interface Sci.* Elsevier BV; 2015 Oct;**20**(5-6):367–82. doi.org/10.1016/j.cocis.2015.11.004.

[17] Wei J, Atif R, Vo T, Inam F. Graphene Nanoplatelets in Epoxy System: Dispersion, Reaggregation, and Mechanical Properties of Nanocomposites. *J. Nanomater.* Hindawi Limited; 2015:1–12. doi.org/10.1155/2015/561742.

[18] Saharudin M S, Hasbi S, Mohd Rashidi N, Jamal Nordin M S. Effect of short-term water exposure on mechanical properties of multi-layer graphene and multi-walled carbon nanotubes-reinforced epoxy nanocomposites. *J. Eng. Sci. Technol.* 2018;**13**(42):26–39.

[19] Zherebtsov S, Semenova IP, Garbacz H, Motyka M. Advanced mechanical properties. Nanocrystalline Titanium. Elsevier; 2019;103–21. doi.org/10.1016/b978-0-12-814599-9.00006-7.

[20] Zaman I, Phan TT, Kuan H-C, Meng Q, Bao La LT, Luong L, et al. Epoxy/graphene platelets nanocomposites with two levels of interface strength. *Polym.* Elsevier BV; 2011 Mar;**52**(7):1603–11. doi.org/10.1016/j.polymer.2011.02.003.

[21] Guan L-Z, Wan Y-J, Gong L-X, Yan D, Tang L-C, Wu L-B, et al. Toward effective and tunable interphases in graphene oxide/epoxy composites by grafting different chain lengths of polyetheramine onto graphene oxide. *J. Mater. Chem. A.* Royal Society of Chemistry (RSC); 2014 Jul 22;**2**(36):15058. doi.org/10.1039/c4ta02429j.

IOP Conf. Series: Materials Science and Engineering 1072 (2021) 012010 doi:10.1088/1757-899X/1072/1/012010

Investment recommendation system on agriculture peer-to-peer lending websites

A S Ahmadiyah[1,2], T I I Apriwibowo[1,3], A E Raharja[1,4], S A Amosea[1,5], and H A Radriyantami[1,6]

[1]Informatics Department, Institut Teknologi Sepuluh Nopember, Jl. Raya ITS Kampus ITS Sukolilo, Surabaya, 60111, Indonesia

E-mail: [2]adhatus@if.its.ac.id, [3]tubagusirkham@gmail.com, [4]adnaner0608@gmail.com, [5]salsha.amosea239@yahoo.com, [6]hayuajeng@gmail.com

Abstract. With the growth of peer-to-peer investment websites, selecting the precise investment is tedious for amateur investors. This study aims to help the no-to-low experienced investors to identify which investments are prospects for them to invest their money. We have collected data from four trusted agriculture peer-to-peer lending websites then used a Decision Tree technique to train and test the data then classify them into a feasible to be invested or not. The result shows that both return and tenor parameters are the top two criteria to be considered. Moreover, the experiment result shows that our recommendation system has successfully performed at 98.29%.

1. Introduction

Peer-to-Peer (P2P) landings are foreign to most people. P2P lending itself is the practice of giving money to individuals or businesses where borrowers are matched directly with investors through an online lending platform [1]. Recently, P2P lending platforms are growing fast because it offers relatively more comfortable requirements and a lower interest rate for the borrower to apply for a loan than most banks do. Meanwhile, it proposes a swiftly emerging alternative financial market for investors with potentially higher investment returns. In the agricultural field in Indonesia, the P2P lending platform contributes to the growth of the local farming by meeting the farmers with the investors directly through the platform. In the end, it helps the growth of the economy as well. However, some no-to-low experienced investors still unsure of which to invest their money. The researchers have addressed this problem to succeed in the P2P lending process.

Previous researches have discussed the P2P lending mechanism. In [2], U Yunus conducted qualitative research on the different characteristics of the P2P lending platform in Indonesia and Singapore. From this research, one knows the type of P2P lending platform growing in Indonesia, but it does not cover the more specific P2P platform on an agricultural field. While in [3], the author studied on how to build trust between borrowers and investors in this platform. Analysis of mobile application for P2P lending has been conducted by [4] and [5] has conducted an empirical study on the predicted risk of P2P lending platform. In fact, there has not been a practical solution addressing which loan is a prospect to be invested.

Therefore, we propose an investment recommendation system to help the no-to-low experienced investor to pick the best options when investing in agricultural P2P lending platforms. We operate machine learning to investigate lists of loan proposals from the farmers and narrow them down to the prospect ones to invest. From our experiment, we find two crucial parameters to consider a potential loan. Overall, our proposed system performs at a high confidence level.

2. Methodology

This research consists of two main phases, i.e., the training and testing phases, as shown in Figure 1.

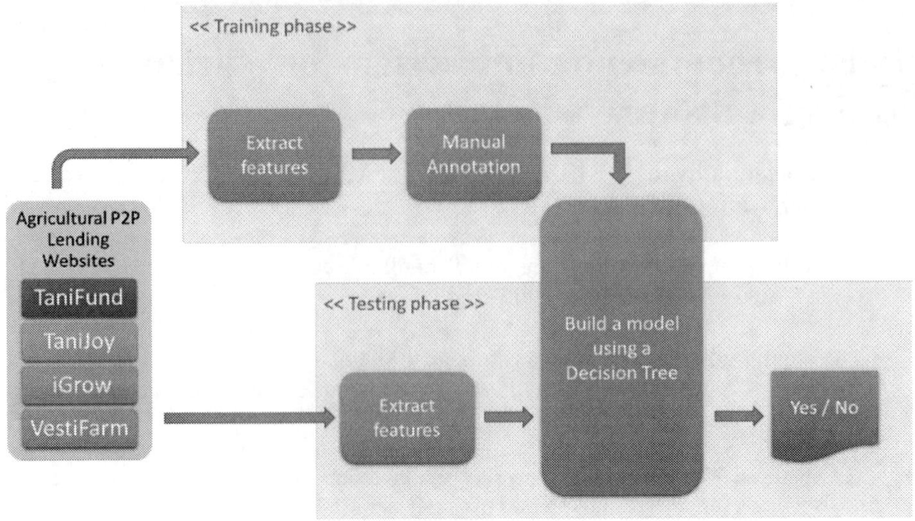

Figure 1. Proposed investment recommendation system.

We start by collecting investment datasets from four trusted agricultural P2P lending sites based in Indonesia: Tanifund [6], iGrow [7], Vestifarm [8], and Tanijoy [9]. Each dataset has several features, specifically the partner name, product name, yield, tenor, the value per unit, and profit per month (based on value per unit, tenor, and yield). Those features are treated as the inputs of our proposed system to determine the investment decision. Table 1 shows an example of the training dataset.

The training phase starts by labelling the dataset. Annotation by an investment expert is first employed to mark each loan as a prospect or not. The annotation step is essential since the newbie investors do not have any information about how to determine the feasibility investment of the proposed loans. Having known the judgment, we then build a model to draw the pattern.

Table 1. Training dataset example.

P2P lending Site	Partner name	Product name	Value/unit (IDR)	Return (p.a)	Tenor (mo)	Profit / mo
TaniFund	Berkah Pisang Mas	Banana (Pisang Mas)	100,000	0.18	24	1,500
TaniFund	Berkah Tani	Papaya (Calina)	100,000	0.18	24	1,500
TaniJoy	Agrindo Karya Persada	Corn seeds B5	5,575,000	0.15	12	69,688
TaniJoy	Agrindo Karya Persada	Corn seeds B4	5,575,000	0.15	12	69,688
TaniJoy	Agrindo Karya Persada	Corn seeds B3	5,575,000	0.15	12	69,688
iGrow	CV. Damarayu	Oyster	500,000	0.45	13	18,750
iGrow	PT. Ihsan Agritama Teknologi	Fruit trading	500,000	0.26	3	10,833
VestiFarm	Bambang Purwanto	Potatoes (Granola)	2,918,000	0.21	6	51,065
VestiFarm	Dahlia Girsang, dkk	Potatoes (Granola)	2,535,000	0.11	6	23,238
VestiFarm	Daniel P.K Sihombing	Chili	2,236,000	0.12	8	22,360

We utilize the J48 classifier in WEKA, which implements of C.45 Decision Tree technique [10]. J48 is employed due to its ability to simplify decision making. In addition, it is capable of pruning insignificant input information.

In the second phase, new data without prior information of either candidate or not a chance for loans are examined using the J48 classifier model built in the first phase. In the end, the system labelled those data into yes or no for investment.

3. Result and Discussion

The following are the results of our implementation using the J48 (C4.5 Decision Tree algorithm). We use the J48 because of its effectiveness in solving the problem, which has characteristics like the one we are currently used, in terms of analysis all input features and removing insignificant features from the consideration.

We treat 85% of the collected dataset as the training data and the rest as testing data. After manually label the training data, we utilize the J48 Decision Tree algorithm to train our data. From the training data visualization, as shown in Figure 2, we noticed that J48 prunes value per unit and profit per month features and relies on return per annum and tenor features for the decision making. In other words, the pattern used by the expert to classify a loan into feasible to invest or not, mainly return per annum is more than 14.85% and tenor is not more than 13 months.

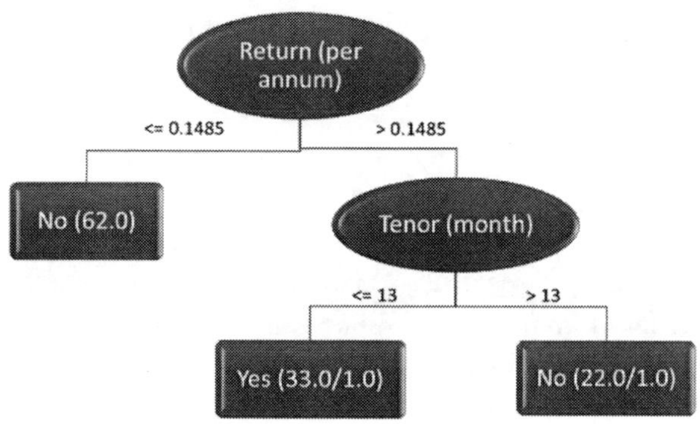

Figure 2. Tree structure of the trained data.

Next, we apply the previous Decision Tree Model to test the unlabelled dataset and classify them into achievable to be invested or not. Table 2 displayed some of our tested datasets.

Table 2. Testing dataset example.

P2P lending Site	Partner name	Product	Value/unit (IDR)	Return (p.a)	Tenor (mo)	Profit / mo	Invest
TaniFund	TF Tani Makmur	Melon (Sakata Glamour)	100,000	0.18	3	1,500	?
TaniJoy	Tanjung Lesung Agropolis	Avocado	2,500,000	0.15	180	31,250	?
iGrow	Komunitas Petani	cinnamon (Kerinci)	500,000	0.03	1	1,250	?
VestiFarm	Wildan Yuda Asmara	Golden Melon	2,190,000	0.01	6	18,250	?
VestiFarm	Wildan Yuda Asmara	Rock Melon	2,814,000	0.01	6	23,450	?

We have measured the accuracy of our investment recommendation model, and it obtains a 98.29% accuracy rate. To test the performance of our method, we have compared our J48 Decision Tree-based model against one of the Artificial Neural Network algorithms, specifically Multi-Layer Perceptron (MLP). It turns out; our proposed method outperforms MLP by 3% accuracy.

4. Conclusion

Applying the J48 Decision Tree algorithm is a straightforward solution for conducting investment recommendations. J48 helps to remove insignificant input criteria from consideration. Besides, it is proven to be efficient since the depth of our tree structure only two levels. Furthermore, having a 98.29% accuracy rate, this system may help a newbie investor who does not have prior knowledge of the investment analysis. Lastly, tenor and return value are the top two criteria to be considered for this type of investment out of five attributes input to the system.

References

[1] A EO. Peer-to-Peer Lending: Business Model Analysis and the Platform Dilemma. IJFET SciDoc Publishers LLC; 2018 Sep 29;31–41. doi.org/10.19070/2643-038x-180005.

[2] Yunus U. A Comparison Peer to Peer Lending Platforms in Singapore and Indonesia. *JPCS*. IOP Publishing; 2019 Jun;**1235**:012008. doi.org/10.1088/1742-6596/1235/1/012008.

[3] Sukmaningsih DW. A Model for Lender-Borrower Trust in Peer-To-Peer Lending. *ComTech*. Universitas Bina Nusantara; 2018 Jul 1;**9**(1):15. doi.org/10.21512/comtech.v9i1.4287.

[4] Wang G, Richad, Octavius Ong YB. Analysis the use of P2P lending mobile applications in Indonesia. *JPCS*. IOP Publishing; 2019 Nov;**1367**:012006. doi.org/10.1088/1742-6596/1367/1/012006.

[5] Qian M, Hu F. An Empirical Study on Prediction of the Default Risk on P2P Lending Platform. *IOP Mater. Sci. Eng*. IOP Publishing; 2019 Apr 10;**490**:062048. doi.org/10.1088/1757-899x/490/6/062048.

[6] TaniFund. Pembiayaan berdampak sosial untuk pertanian Indonesia. Jakarta: PT Tani Fund Madani Indonesia. 2020. Retrieved from https://tanifund.com/.

[7] iGrow. Pembiayaan pertanian berdampak sosial, aman, dan terpercaya. Depok: PT iGrow Resources Indonesia. 2020. Retrieved from https://igrow.asia/.

[8] Vestifarm. Jakarta: PT. Vestifarm Agro Indonesia. 2020. Retrieved from https://vestifarm.com/.

[9] Tanijoy. Your Trusted Agriculture Partner. Jakarta: PT. Tanijoy Agriteknologi Nusantara. 2020. Retrieved from https://www.tanijoy.id/.

[10] Salzberg SL. C4.5: Programs for Machine Learning by J. Ross Quinlan. Morgan Kaufmann Publishers, Inc., 1993. *Mach Learn*. Springer Science and Business Media LLC; 1994 Sep;**16**(3):235–40. doi.org/10.1007/bf00993309.

IOP Conf. Series: Materials Science and Engineering 1072 (2021) 012011 doi:10.1088/1757-899X/1072/1/012011

Cluster analysis to determine business strategy for MSMEs in Yogyakarta

R S Nugraha[1,2], W S Jatiningrum[1,3] and R D Astuti[1,4]

[1]Department of Industrial Engineering, Ringroad Selatan, Bantul, Special Region of Yogyakarta, Indonesia

E-mail: [2]ranggasatyanugraha@gmail.com, [3]wandhansari.sekar@ie.uad.ac.id, [4]reni_dwiastuti@ie.uad.ac.id

Abstract. Micro, Small, and Medium Enterprises (MSMEs) in Indonesia need to be encouraged to develop themselves. This is due to their contribution in the national economies. Yogyakarta, known as a creative city, has various kinds of micro enterprises, particularly food business. The appropriate strategies need to be applied to develop their business. Nevertheless, many food businesses in Yogyakarta city has different characteristics. The purpose of this study is to cluster food businesses as micro enterprises in Yogyakarta. K-means method is used to analyse 38 food businesses based on 13 variables related to characteristics of food businesses. 2 clusters obtained were named 'prepared to level up cluster' and 'developing cluster'. 'Prepared to level up cluster' was considered to be more potential than 'developing cluster'. SWOT analysis was then used to formulate the appropriate strategy for each cluster. By calculating total score of internal and external factor evaluation, obtained appropriate strategies for each cluster. Growth strategy through vertical integration was suitable for 'prepared to level up cluster'. Meanwhile, 'developing cluster' need to adopt growth strategy through horizontal integration to develop its business.

1. Introduction

Micro, Small, and Medium Enterprises (MSMEs) has an important role to contribute national economy improvements in Indonesia. This was marked by the success of MSMEs to survive in the period 1997-1998 national crisis. Positive growth of MSMEs in each year also show that MSMEs has contributed significantly [1]. The Ministry of Cooperatives and Small and Medium Enterprises (SMEs) for the Republic of Indonesia stated that total number of MSMEs in Indonesia reached 62.9 million. Therefore MSMEs able to absorb 120 million workers or 3 million more compared to 2016. However, MSMEs deal with several problems, such as limited capital, low quality of human resources, and lack of understanding and using for technology [2][3][4].

Yogyakarta, known as a creative city, also has a variety of MSMEs with positive growth in each year. According to interview with Cooperative, SMEs, Manpower and Transmigration Office of Yogyakarta city, as many as 23.604 MSMEs scattered in Yogyakarta city with various types of business sectors, namely food, handicraft, metal, and service businesses. The most developed is food business, with total 6.373 MSMEs. This is caused by Yogyakarta as a student and tourist destination from various regions [5] [6]. Therefore, various types of food are provided to accommodate various tastes [7] [8]. Cooperative, SMEs, Manpower and Transmigration Office of Yogyakarta city was asked to coordinate MSMEs to be fostered so that they could develop properly. This office appointed a Management and Business Consultant to assist MSMEs in Yogyakarta city. Coaching for MSMEs has been carried out for micro business that are known to be active. These micro businesses have also succeeded through curation process by Cooperative, SMEs, Manpower and Transmigration Office of Yogyakarta city. Coaching programmes conducted consist of training, assistance, consultation, monitoring, collaboration with business partners, and capital assistance.

Each of micro business has their own characteristic. Therefore, each of them needs different business strategy to be applied. The absence of clustering or grouping for micro businesses in Yogyakarta city, has made difficulties for related office to determine an appropriate strategy in each micro businesses, particularly in food businesses. Fathania [9] has conducted a cluster analysis for MSMEs in Pasuruan district using k-means. Similar method was also used by Setyaningsih [10] to conduct SMEs cluster analysis in Indonesia. Mustaniroh et. al. [11] was using AHP to determine strategic priorities for each cluster formed. Another method can be used to analyse business strategies is Strength Weakness Opportunity Threat (SWOT) analysis. This method is able to identify various factors systematically to formulate company's strategy [12].

According to Management and Business Consultants for MSMEs, clustering needs to be conducted to cluster micro businesses for food category. This must be carried to facilitate properly coaching process and to determine appropriate business strategies to MSMEs. Thus, food businesses that have similar characteristics will gather in same cluster. Cluster analysis needed to facilitate MSMEs for determining business strategy for each cluster of food business. Therefore, this study aimed to conduct cluster analysis for micro businesses in food category or called food business. Prior similar studies usually determine the business strategies directly for each cluster formed. In this study, business strategies were developed for each cluster using SWOT analysis.

2. Method

The object of this research was 61 food businesses in Yogyakarta city. These micro businesses have been curated by Cooperative, SMEs, Manpower and Transmigration Office of Yogyakarta city. Based on literature review and discussion with Management and Business Consultant for MSMEs in Yogyakarta city, 13 variables related to MSMEs were obtained. They were composed of capital (X1), average gross profit per month (X2), average net profit per month (X3), number of workers (X4), business life (X5), certificate and legality (X6), production capacity per month in Rupiah (X7), increasing for production capacity (X8), management and organization (X9), packaging (X10), branding (X11), modern market (X12), and financial inclusion and literacy (X13). Analysis were conducted through two stages, cluster analysis and SWOT analysis.

2.1. Cluster Analysis

K-means clustering was used to cluster food businesses in several cluster. Micro business with the same characteristics would be included in one cluster. K-means clustering is a non-hierarchical clustering method that needs to partition data into one or more clusters [13]. The initial stage from this method is choosing K (initial from centroid). Random point can be taken to determine the centroid. Each point around centroid will establish a grouping called cluster. Calculation of distance between data and centroid is conducted by the concept of Euclidean Distance. Questionnaires containing questions from 13 variables related to characteristics of MSMEs were arranged. Respondents were owners of food businesses in Yogyakarta city.

2.2. SWOT Analysis

The result of cluster analysis was a number of clusters that showed food businesses with similar characteristics. SWOT analysis was performed in each cluster to determine the appropriate strategy. SWOT analysis is conducted by considering company's internal factors, consist of strengths and weaknesses, and company's external factors, consist of opportunities and threats. Business strategies formulation were conducted by determining weight and ranking from expert.

3. Result and Discussion

According to data from related office, there was 61 food businesses have been curated. However, there were only 45 micro businesses were willing to become respondents. A total of 45 questionnaires were distributed to owner of food businesses in Yogyakarta city. Nevertheless, only 38 completed questionnaires were returned.

3.1. Cluster Analysis

Result of the questionnaires were processed by k-means method using SPSS 23.0. Standardization process was conducted first because the data used has different scale. This process was carried out by changing the original data into z-score data. According to Rosenthal [14], z-score is a raw score calculated by its distance from the mean. It was measured in standard deviation units. Cluster analysis was conducted after z–score was obtained. Based on discussion with consultant of business and management for MSMEs in Yogyakarta city, micro businesses will be divided into 2 clusters, namely prepared to level up and developing cluster. Cluster analysis was conducted by two steps of iteration, as shown in Table 1. According to the result, the minimum distance between cluster centres was 8.939.

Table 1. Iteration stage.

Iteration	Change in cluster centres	
	1	2
1	3.749	3.618
2	0.000	0.000

Table 2 displays final results of cluster centres. Positive z-score indicated data were above its total average, and vice versa. Cluster 1 was contained micro businesses whose characteristics had scores above the total average score of micro businesses studied, except for variable of business life. Micro businesses included cluster 2 have characteristics against with cluster 1. Scores for all variables in cluster were below the total average score of micro businesses studied, except for variable of business life. Therefore, it was concluded that cluster 1 was contained micro businesses who were prepared to level up and cluster 2 was contained developing micro business.

Table 2. Final result of cluster centres.

Variables	Cluster	
	1	2
Z score(X1_Capital)	0.46153	-0.06993
Z score(X2_Average_gross_profit_permonth)	2.31038	-0.35006
Z score(X3_Average_net_profit_permonth)	2.06170	-0.31238
Z score(X4_Number_of_workers)	1.26477	-0.19163
Z score(X5_Business_life)	-0.02548	0.00386
Z score(X6_Certification_and_legality)	0.32171	-0.04874
Z score(X7_Production_capacity_permonth_inrupiah)	0.87535	-0.13263
Z score(X8_Increasing_for_production_capacity)	0.11199	-0.01697
Z score(X9_Management_and_organization)	0.06358	-0.00963
Z score(X10_Packaging)	0.47811	-0.07244
Z score(X11_Branding)	0.01825	-0.00276
Z score(X12_Modern_market)	0.67182	-0.10179
Z score(X13_Financial_inclusion_and_literacy)	0.05271	-0.00799

ANOVA was used to test the effect of variables composed clusters [15]. Table 3 displays ANOVA results with significance level $\alpha = 0.05$. These results indicated that 4 variables have p value<0.05. They were X2, X3, X4, and X7. These four variables were proven to have significant differences between clusters. It could be concluded that there were four significant variables as cluster forming. The other variables X1, X5, X6, X8, X9, X10, X11, X12, and X13 were known to have p value>0.05. This showed that there was no significance difference between clusters based on these nine variables.

Table 3. ANOVA table.

Variables	Cluster		Error		F	Sig.
	Mean Square	df	Mean Square	df		
Zscore(X1_Capital)	1.226	1	0.994	36	1.234	0.274
Zscore(X2_Average_gross_profit_permonth)	30.733	1	0.174	36	176.547	0.000
Zscore(X3_ Average_net_profit_permonth)	24.473	1	0.348	36	70.331	0.000
Zscore(X4_Number_of_workers)	9.210	1	0.772	36	11.931	0.001
Zscore(X5_Business_life)	0.004	1	1.028	36	0.004	0.952
Zscore(X6_Certification_and_legality)	0.596	1	1.011	36	0.589	0.448
Zscore(X7_Production_capacity_permonth_inrupiah)	4.412	1	0.905	36	4.873	0.034
Zscore(X8_Increasing_for_production_capacity)	0.072	1	1.026	36	0.070	0.792
Zscore(X9_Management_and_organization)	0.023	1	1.027	36	0.023	0.881
Zscore(X10_Packaging)	1.316	1	0.991	36	1.328	0.257
Zscore(X11_Branding)	0.002	1	1.028	36	0.002	0.966
Zscore(X12_Modern_market)	2.599	1	0.956	36	2.719	0.108
Zscore(X13_Financial_inclusion_and_literacy)	0.016	1	1.027	36	0.016	0.901

According to final result, it was found that cluster 1 consisted of five food businesses and cluster 2 consisted of 33 food businesses. There was only 13.16% of food businesses that could be prepared to level up judged by its characteristics. The remaining 86.84% are considered to be MSMEs in developing cluster. This proves that Cooperative, SMEs, Manpower and Transmigration Office of Yogyakarta city still has heavy duty to guide these food businesses in developing cluster to able to level up. Table 4 and Table 5 display list of food business members included in cluster 1 and 2.

Table 4. Members of cluster 1 'prepared to level up cluster'.

No	Food business	Distance
1	Soklat Es Coklat	3.749
2	Selera Roti dan Kue	2.937
3	Premio Jogja Berdikari	2.604
4	Warung Pecel Ndeso	3.237
5	Melinda Brownies Kukus	4.452

Table 6 displays summary of characteristics for each cluster based on significant variables composing clusters, namely average gross profit per month, average net profit per month, number of workers, and production capacity per month in rupiah.

Table 5. Members of cluster 2 'developing cluster'.

No	Food business	Distance	No	Food business	Distance
1	Saung Pepes	2.462	18	Juice Nyess	2.403
2	Dapur Azza	2.083	19	Telur Asin Mentes	3.056
3	Ji-Pong	3.565	20	Blessing Snack	3.17
4	Pawon Sentono	5.043	21	Chie Cik Snack	2.885
5	Maxichic	1.68	22	Andien Food	2.351
6	Sumringah	1.981	23	DK Wingko	3.059
7	Rosemary Herbs	2.515	24	Bakpia Disty	2.914
8	Jahe Joss	3.087	25	Irin Catering	3.871
9	Berkah Sabar Sejahtera	3.79	26	Amiratih	3.385
10	Srikandi	2.963	27	Kedai Jiro	3.018
11	Siomay dan Batagor Tugu	3.708	28	Wedang Rejdo	3.474
12	Java Rainbow	3.618	29	Mie Ayam Bakso	3.887
13	Dya Snack & Catering	4.197	30	Glak Gluk Thai Tea	2.941
14	Hanabi Snack	2.242	31	Shinta Pakualaman The Modern Kluwo	2.595
15	Surya Mandiri	4.19	32	Telo Keju Mas Tejo	2.737
16	Bertina Catering & Cakes	2.509	33	Bakpia Erma	3.971
17	Sambel Granat	4.55			

Table 6. Characteristics of each cluster.

Significant variables forming clusters	Cluster	
	Prepared to level up	Developing
X2 Average gross profit per month	Above 15 million	10-15 million
X3 Average net profit per month	5-7 million	1-5 million
X4 Number of workers	4-5 workers	2 workers
X7 Production capacity per month in rupiah	10-15 million	1-10 million

Explanation for each cluster were listed as follows

1. Cluster 1: Prepared to level up
MSMEs prepared to level up were expected to improve their business competence so that they would become highly competitive MSMEs. The government initiated campaign of one million MSMEs to be prepared to level up. This was an acceleration program conducted by government to improve MSMEs competitiveness. This program was formulated into National Medium-Term Development Plan for 2015-2019. This was aimed to make MSMEs ready to deal with ASEAN Economic Community [16]. Generally, prepared to level up cluster have superior characteristics than developing cluster. Average gross income per month reached more than 15 million rupiah or 300 million rupiahs in a year. Average net income per month reached 5-7.5 million rupiah. According to law no 20 of 2008, an enterprise could be categorized as a small enterprise when the annual sales was between 300,000,000 rupiahs and 500,000,000 rupiahs. Food businesses in this cluster, mostly spent 10-15million per rupiah per month for production cost. This corresponded to large income achieved per month. These food businesses could be prepared to level up due to their good financial management. They also had products according to customers want. Therefore, members inside prepared to level up cluster were able to maximize their business financial turnover.
2. Cluster 2: Developing
Developing cluster was considered to have opportunities to develop. Nevertheless, they had not able to follow provisions from government to level up. Food business in this cluster have to improve their internal management. Average gross income per month was still less than ten million while average net income per month less than five million. The products were very simple and there was no

innovation yet. Therefore, product sales were not optimal. These food businesses had not been able to carry out their financial management properly due to their inability to separate between business fund and private fund. As a result, sometimes business fund was used for personal purposes. This caused business circulation could be faltered. Average number of workers in developing cluster was 2-3 persons. This was accordance with provisions stated micro business has less than five workers. Most of food businesses in this developing cluster still employs their family members as workers. As a result, their management and organization structure was very simple. Sometimes there was a lack of professionalism because of working with fellow of family members. This was in accordance with study by Gedajlovic et. al. [17].

3.2. SWOT Analysis

SWOT analysis was conducted to determine strategies for each cluster. SWOT analysis facilitates to build useful framework to identify internal and external factor of organization. Therefore, this tool includes both environmental analysis and the organizational analysis [18]. Identification for each cluster was carried out on internal factor, called strength and weaknesses, and external factor, called opportunities and threats. Those factors were based on the result of characteristics from each cluster and also conducted discussions with the chosen experts. After identified those factors, expert assessment was conducted to obtain appropriate strategies for each cluster. The chosen experts were a business consultant who handled MSMEs issues in Yogyakarta city and her technical staff. The result of assessment was carried out through IFE (Internal Factor Evaluation) and EFE (External Factor Evaluation) matrices. Table 7 and Table 8 show the IFE result from each cluster.

Table 7. IFE matrix for 'prepared to level up cluster'.

No	Strengths	Weight	Rating	Weight x Rating
1	High sales revenue	0.108	3.67	0.39
2	High production capacity	0.097	3.00	0.29
3	Fairly good financial management	0.118	3.00	0.35
4	Effective utilization of manpower	0.108	2.67	0.29
5	Having desirable products for consumers	0.129	3.00	0.39
6	Already promoting and selling products online	0.097	3.33	0.32
No	**Weaknesses**	**Weight**	**Rating**	**Weight x Rating**
1	Small number of workers (4-5 persons)	0.065	3.33	0.22
2	Having no partnership	0.086	2.67	0.23
3	Products didn't have business license, authorization from National Agency of Drug and Food Control, and halal certification	0.108	3.67	0.39
4	Structure of business management was still low	0.086	3.33	0.29
	Total	**1.00**	**32**	**3.16**

Table 8. IFE matrix for 'developing cluster'.

No	Strengths	Weight	Rating	Weight x Rating
1	Not much spending in labor cost	0.073	2.67	0.19
2	Products were dominated by traditional characteristics	0.064	3.33	0.21
3	Frequently attended to exhibitions	0.091	3.33	0.30
4	Participating in various communities	0.073	2.00	0.15
5	Already promoting and selling products online	0.073	2.33	0.17
No	**Weaknesses**	**Weight**	**Rating**	**Weight x Rating**
1	Low sales revenue	0.093	3.00	0.28
2	Low production capacity	0.084	3.33	0.28
3	Labor used were from family (1-2 persons)	0.093	3.00	0.28
4	Poor financial management	0.103	2.00	0.21
5	Unable to focus on sold product	0.075	2.67	0.20
6	Products didn't have business license, authorization from National Agency of Drug and Food Control, and halal certification	0.084	3.33	0.28
7	Having no partnership	0.084	2.67	0.22
	Total	**1.00**	**34**	**2.80**

High sales revenue was the biggest strength for prepared to level up cluster while frequently participated in exhibition was the highest strength for developing cluster. The products didn't have business license, authorization from National Agency of Drug and Food Control, and halal certification, was the biggest weakness for prepared to level up cluster. This was a critical factor because consumers have begun to consider these things in making a product purchased. The result for assessment of EFE for each cluster were shown in Table 9 and Table 10.

Table 9. EFE matrix for 'prepared to level up cluster'.

No	Opportunities	Weight	Rating	Weight x Rating
1	Getting facilities from local government, such as training, assistance, consultation, and monitoring	0.100	3.67	0.37
2	Getting assistance capital	0.091	2.67	0.24
3	Many exhibitions were available to be joined	0.100	3.67	0.37
4	Promoting and selling online were getting easier	0.073	2.67	0.19
5	Many sales partnership schemes were available	0.082	2.33	0.19
6	Society were in good social condition	0.082	3.00	0.25
No	**Threats**	**Weight**	**Rating**	**Weight x Rating**
1	Many food business competitors	0.073	3.33	0.24
2	The products were easy to imitate	0.091	3.00	0.27
3	Other products have made innovation s	0.091	3.00	0.27
4	Price of raw materials was fluctuating, raw materials were difficult to obtain	0.073	2.67	0.19
5	Need of adaptation to face digitalization era	0.073	3.67	0.27
6	Lack of trust to cooperate with business actors	0.073	3.33	0.24
	Total	**1**	**37.00**	**3.10**

Table 10. EFE matrix for 'developing cluster'.

No	Opportunities	Weight	Rating	Weight x Rating
1	Getting facilities from local government, such as training, assistance, consultation, and monitoring	0.096	3.33	0.32
2	Getting assistance capital	0.088	2.67	0.23
3	Many exhibitions were available to be joined	0.096	3.67	0.35
4	Promoting and selling online were getting easier	0.061	2.67	0.16
5	Many sales partnership schemes were available	0.070	2.00	0.14
6	Society were in good social condition	0.070	3.33	0.23
No	**Threats**	**Weight**	**Rating**	**Weight x Rating**
1	Many food business competitors	0.088	3.33	0.29
2	The products were easy to imitate	0.096	3.67	0.35
3	Other products have made innovations	0.070	3.00	0.21
4	Price of raw materials was fluctuating, raw materials were difficult to obtain	0.096	3.67	0.35
5	Need of adaptation to face digitalization era	0.096	3.33	0.32
6	Lack of trust to cooperate with business actors	0.070	2.67	0.19
	Total	**1**	**37.33**	**3.17**

According to table 9 and 10, the highest score for both clusters was on many opportunities to participate in various exhibition for MSMEs, including food businesses. This could be used for each of food business to introduce their products to consumers. Meanwhile, the biggest threat for both clusters was product ease to be imitated. The score was 0.35 for developing cluster and 0.27 for prepared to level up cluster. Food businesses need to innovate their product to overcome this problem, especially for food businesses in developing cluster due to their higher score.

After evaluating internal and external factors, strategy was formulated for each cluster by plotted their total score as shown in Figure 1.

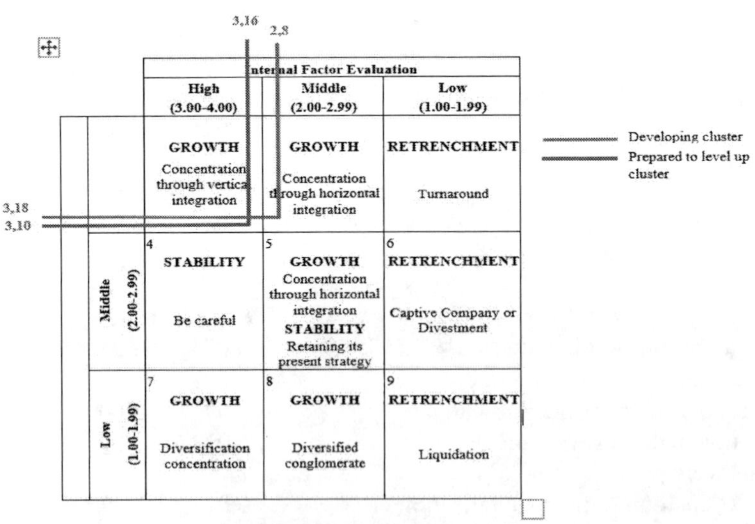

Figure 1. IFAS-EFAS matrix.

Strategy for each cluster were determined by its quadrant position as shown in Figure 1. Strategies for each cluster were explained as follows:

1. Strategy for 'prepared to level up cluster'
   The characteristics from prepared to level up cluster were superior to developing cluster. According to figure 1, the appropriate strategy to be applied was growth strategy with concentration through vertical integration. Adeleke [19] stated that this strategy could be conducted by taking over the function of suppliers (backward integration) and distributors (forward integration). Food businesses in this cluster already have high market share and strong competitiveness. This corresponded to the results of SWOT analysis showed that these food businesses already have high sales revenue because their products were preferred by consumers. In addition, their internal management were also quite good. Therefore, this cluster could begin to focus on other strategies, one of which was taking over the role of suppliers. This aimed to minimize total operational costs, such as raw material costs. In addition, constraints related to raw materials such as availability and quality could be minimized. This was due to direct controlling from food businesses to suppliers. Another strategy was to take over function of distributors. It was expected to minimize total distribution cost through this direct controlling.
2. Strategy for 'developing cluster'
   Growth strategy with concentration through horizontal integration was appropriate to be applied in developing cluster. According to Athoillah [20], integration horizontal strategy could be conducted by expanding its business or coverage area. It was intended to increase its sales or profit. To realize this strategy, food businesses in this cluster needs to improve their production capacity because they still have a small production capacity. Strengths of this cluster such as participating frequently in exhibitions and being registered in various MSMEs communities, could help this cluster to identify potential location for their business expansion.

4. Conclusion

Two clusters for food businesses in Yogyakarta were constructed by applying cluster analysis. They were 'prepared to level up cluster' and 'developing cluster'. 'Prepared to level up cluster' was considered to be more potential than 'developing cluster'. It was expected could develop their business competence to improve competitiveness. Meanwhile, developing cluster have to improve their internal management before being prepared to become a highly competitive MSMEs. Each of cluster was analysed to determine their appropriate strategy. Their strategy was formulated by conducting SWOT analysis. Internal and external factors were identified for each cluster before formulated the strategy. The appropriate strategy for 'prepared to level up cluster' was growth strategy through vertical integration. This could be conducted by taking over the function of suppliers or distributors. Meanwhile, growth strategy through horizontal integration was appropriate to be applied for 'developing cluster'. This cluster needs to expand its business or coverage area to increase market share. Further research can be conducted in the handicraft industry due to the large number MSMEs engaged.

References

[1] Central Bureau of Statistics. Tabel Perkembangan UMKM pada Periode 1997 -2013. 2016. Retrieved from https://www.bps.go.id/statictable/2014/01/30/1322/tabel-perke.

[2] Jaswal SS. Problems and Prospects of Micro, Small & Medium Enterprises (MSME's) in India. *Int. J. Innov. Res. Stud.* 2014;**3**(5):140–161.

[3] Sugiarto I. Obstacles and Challenges in the Development of MSMEs: Case Study. *BIRCI-Journal.* Budapest International Research and Critics Institute; 2018 Dec 22;**1**(4):93–8. doi.org/10.33258/birci.v1i4.97.

[4] Tambunan TTH. Development of Micro, Small and Medium Enterprises and Their Constraints: A Story from Indonesia. *GamaIJB.* Universitas Gadjah Mada; 2011 Feb 12;**13**(1):21. doi.org/10.22146/gamaijb.5492.

[5] Rindrasih E. Tourist's Perceived Risk and Image of the Destinations Prone to Natural Disasters: The Case of Bali and Yogyakarta, Indonesia. *Jurnal Humaniora*. Universitas Gadjah Mada; 2018 Jun 8;**30**(2):192. doi.org/10.22146/jh.v30i2.32239.

[6] Roostika R. The Role of City and Host University Images on Students' Satisfaction with the Assigned Destination. *Rev. Integr. Bus. Econ. Res.* 2017;**6**(1):250–261.

[7] Kurniawati D, Yuliando H. Productivity Improvement of Small Scale Medium Enterprises (SMEs) on Food Products: Case at Yogyakarta Province, Indonesia. *Agriculture and Agricultural Science Procedia*. Elsevier BV; 2015;**3**:189–94. doi.org/10.1016/j.aaspro.2015.01.037.

[8] Roostika R. Citra Merek Tujuan Wisata Dan Perilaku Wisatawan: Yogyakarta Sebagai Daerah Tujuan Wisata. *J. Manaj. dan Akunt.* 2012;**1**(1):41–54.

[9] Fathania V. Strategi Pengembangan Klaster POKLAHSAR (Kelompok Pengolah dan Pemasar) Kerupuk Bandeng dengan Metode K-Means Clustering dan Fuzzy AHP di Kecamatan Rejoso, Kabupaten Pasuruan. *Universitas Brawijaya*. 2017.

[10] Setyaningsih S. Using Cluster Analysis Study to Examine the Successful Performance Entrepreneur in Indonesia. *Procedia Economics and Finance*. Elsevier BV; 2012;**4**:286–98. doi.org/10.1016/s2212-5671(12)00343-7.

[11] Asmaul Mustaniroh S. Development Strategy for Cluster Apple Chips with K-means Clustering and Analytical Hierarchy Process. Industria: *Jurnal Teknologi dan Manajemen Agroindustri*. Brawijaya University; 2016 Aug 30;**5**(2):67–74. doi.org/10.21776/ub.industria.2016.005.02.3.

[12] Hatta IH, Riskarini D, Ichwani T. SME BUSINESS DEVELOPMENT STRATEGY: SWOT AND EFE-EFI ANALYSIS. *JAM*. Brawijaya University; 2018 Sep 1;**16**(3):537–43. doi.org/10.21776/ub.jam.2018.016.03.19.

[13] Raval UR and Jani C. Implementing & Improvisation of K-means Clustering Algorithm. *Int. J. Comput. Sci. Mob. Comput.* 2016;**55**(5):191–203.

[14] Rosenthal JA. *Statistics and Data Interpretation for Social Work*. New York: *Springer Publishing Company*. LLC. 2012.

[15] Concepts TF. Sawyer et al. Analysis of Variance. *JMMT*. 2009;**17**(2):27–38.

[16] Pangestu ME. Competitiveness Towards Asean Economic Community. *J. Ekon. Bisnis Indones, Fakultas Ekon. dan Bisnis Univ. Gadjah Mada*. 2015;**24**(1):22–32.

[17] Gedajlovic E, Carney M, Chrisman JJ, Kellermanns FW. The Adolescence of Family Firm Research. *J Manage*. SAGE Publications; 2012 Jan 5;**38**(4):1010–37. doi.org/10.1177/0149206311429990.

[18] Ghazinoory S, Abdi M, Azadegan-Mehr M. SWOT METHODOLOGY: A STATE-OF-THE-ART REVIEW FOR THE PAST, A FRAMEWORK FOR THE FUTURE / SSGG METODOLOGIJA: PRAEITIES IR ATEITIES ANALIZĖ. *J. Bus. Econ. Manag*. Vilnius Gediminas Technical University; 2011 Apr 12;**12**(1):24–48. doi.org/10.3846/16111699.2011.555358.

[19] Onodugo VA, Adeleke BS, Ike RN. Ethnocentric Behaviour and Business Performance of Multinational Enterprises (MNES): Evidence From South-West Nigeria. *IJAEMS*. AI Publications; 2017;**3**(2):98–106. doi.org/10.24001/ijaems.3.2.17.

[20] Athoillah AS, Firdaus M, Sanim B. Competititve strategy of taxi company in facing environmental changes. *BISMA*. Universitas Negeri Surabaya; 2019 Oct 18;**12**(1):66. doi.org/10.26740/bisma.v12n1.p66-87.

IOP Conf. Series: Materials Science and Engineering 1072 (2021) 012012 doi:10.1088/1757-899X/1072/1/012012

In the search of oil and gas incident causal factors: The Gaussian Fuzzy Analytic Hierarchy Process approach

F D Wicaksono[1] and Y B Arshad[2]

[1]Fakulti Pengurusan Teknologi Dan Teknousahawan, University Technical Malaysia Melaka, Durian Tunggal, Melaka, 76100, Malaysia
[2]Center of Technopreneurship Development (C-TED), University Technical Malaysia Melaka, Durian Tunggal, Melaka, 76100, Malaysia

E-mail: [1]fermidadang@gmail.com, [2]ayusri@utem.edu.my

Abstract. Learning from events is a crucial measure to reduce the number of catastrophic incidents on board the oil and gas industry. It is critical to have a better understanding of the root causes behind the incidents. A better understanding of the incident causal factors can be accomplished by investigating thoroughly previous incident reports and determine the most notable factors in contributing to the incidents. This paper mainly elaborates and analyzes incident reports produced by the International Oil and Gas Producer Association's (IOGP's) reports from 2010 until 2018 located in the worldwide operation. The most contributing factors lead to the oil and gas incidents are examined in this paper utilizing Gaussian fuzzy analytic hierarchy process, an improved methodology of fuzzy analytic hierarchy process (FAHP) by embedding Gaussian fuzzy number within the evaluation process. Gaussian fuzzy number is obtained by random simulation corresponding to the Gaussian probability density function. The result of this paper reveals the more accurate and realistic representation of incident causal factors determination. This point of view helps in better explaining oil and gas incidents causal factors, where precaution measures should be directed and efficiently managed.

1. Introduction

The oil and gas industry is one of the highest-risk operation process industries. The risks involved in the operation shall originate from the complex operation, large manpower employment, and severe incident consequences [1]. The oil and gas industry is one of the sectors with a high incident rate, involving the probability from one incident to escalate to a catastrophic or disastrous event [2]. During 2010 – 2018, there are 987 High potential event incidents (HIPO-incidents), which could escalate to the higher level of consequences, or possess higher probability to result one or more fatalities [3]. These incidents are produced from a number of causal factors, underlying from the Organizational system factor, Process factor leading to unsafe working condition, and Human-behavioral factor.

A number of researches have been performed to identify the underlying causal factors which significantly contribute to the incidents. An example of the study was conducted by Abdelhamid TS & Everett [4] to reveal the root causes of the construction industry accidents. The study concludes that the incidents are likely to occur as a cause of both unsafe conditions and unsafe acts. Zhou & Lei [5] investigate China railway incidents by using the Human Factors Analysis and Classification System and found that higher organizational levels in the railway system result in errors in accidents/ incidents. Managerial position of the organization bears the ultimate responsibility to generate organizational safety culture that significantly prevents incidents occurrence [6]. Organizational system factors involving corporate safety culture, applicable procedure, and adequate competency are considered as the underlying cause of incidents, through potentially influencing the human-behavioral factor and generating unsafe working condition [7] .

In regard to the identified factors, the research question is raised: how significant causal factor could contribute to the incidents, and how is the quantified degree of importance to an incident cause. The conclusion to the questioned would generate not only reveal the specific weighting values to the causal factors but also to elaborates the most effective incidents prevention schemes. The author's contribution to the body of research in the oil and gas safety should a holistic, quantitative model, based on robust methodology, and capable of approximating HIPO-incidents frequency [8].

Analytic hierarchy process (AHP) is firstly introduced by Saaty [9] as the breakthrough of Multi-criteria decision-making methodology. AHP has been implemented in various applications, including business selection problems, energy sector evaluation, and optimization of location selection. The remarkable implementation of AHP is addressed as a structured scientific solution to the Israeli-Palestinian conflict [10]. To overcome this limitation, fuzzy AHP is introduced by integrating fuzzy numbers in the evaluation process [11]. Fuzzy AHP has been rapidly used in the field of Multi-criteria decision analysis [12]. In many study cases, Fuzzy AHPs are utilized based on triangular fuzzy number to represent linguistic variable and approximate reasoning [13]. Yang et al. [14] implement triangular fuzzy number in AHP to evaluate the prioritization of environmental issues in offshore oil and gas operations. Montazar et al. [15] utilized triangular fuzzy number in AHP evaluation to assess the performance of irrigation projects.

In recent studies, Fuzzy AHP is improved by executing Gaussian fuzzy numbers to eliminate the case of zero weights and lack of consistency for the decision matrices [16]. The adoption of Gaussian fuzzy number is also demonstrated by Wicaksono et al. [17] to represent more realistic fuzzy environment in the evaluation of benefit, cost, and risk analysis in the oil and gas processing area. The main capability of Gaussian fuzzy number compared to triangular fuzzy number is the capability to evaluate first fuzzy number which is wrongly weighted as 0 in the TFN [16] [18].

Current paper examines 987 High potential event incidents (HIPO-incidents) occurring from 2010 until 2018 investigated by the International Oil and Gas Producers Association (IOGP). This study aims to develop a quantified degree of importance model for oil and gas incident causal factors by utilizing Gaussian fuzzy analytic hierarchy process. Sensitivity analysis is performed to measure the proposed methodology's accuracy and consistency. The novelty of this study covers the adoption Gaussian fuzzy number in the Fuzzy AHP evaluation which is generated based on random numbers following the Normal distribution density function. This methodology is capable of approximating the randomness and uncertainty that may arise from the decision-making process.

2. Literature Review

The incident causes analysis is merely based on the studies of which performed to identify the root causes of incidents. Abdelhamid & Everett [4] define that the incidents are merely caused by two factors, i.e., Human factor models, and Non-human factor or unsafe conditions. Halim et al. [1] perform the study literature based on the Bureau of Safety and Environmental Enforcement and confirm that incidents are caused by 1. Equipment failure, 2. Human error, and 3. Other significant findings such as, lacked proper isolation, and procedure deviation. Theophilus et al. [19] investigate the Human factors analysis and classification system for the oil and gas industry. They identify that the incidents are dominant to be caused by Human and Organizational factors in the oil and gas industry. The study performed implies that oil and gas incidents is caused by 1. Unsafe acts, 2. Preconditions for unsafe acts, 3. Unsafe supervision, and 4. Organizational influences. Tang et al. [20] perform a study to review offshore oil and gas safety indices. They conclude that Organizational factor is the key success to build the safety culture within the organization. Skogdalen & Vinnem [21] conclude the precursor incidents in oil and gas industry are caused by 1. Human factor, 2. Organizational factor, and 3. Technical factor. The precursor incidents are, therefore generate consequences and harmful effects. Ferjencik [22] defines that the incidents would have occurred due to these factors: 1. Implementation (1st level) causes, 2. Organizational (2nd level) causes, 3. Management attitude (3rd level) causes, and 4. Societal (4th level) causes. Motarjemi & Wallace [6] investigates several incidents in the field of food safety management and conclude that the identified root cause are: 1. Human error or equipment failure, 2. Workplace condition, and 3. Organizational culture and management decision. Bea [23] investigates the human and organizational factor in reliability assessment and management of offshore structures. He implies that the incidents are mainly linked to the factor of Human and Organizational.

Figure 1 illustrates the main variables used to identify the possibility of High potential incident causes. Based on those studies, the author formulates several factors as the main variables of this research, i.e. (1) Human-behavioral factor; (2) Job factor leading to unsafe working condition; and (3) Organizational system factor.

2.1. High Potential Event Incidents (HIPO-Incidents)

Investigating incidents is essential, as effective root-cause and the recommendation of improvements, to make every effort to avoid future incidents. The international Oil and Gas Producer (IOGP) highly recommends carrying out an incident investigation for which is expected to cause fatality or major injuries. High potential event incidents (HIPO-incidents) are defined as any incident or near-miss event which potentially or would have realistically caused in catastrophic events, such as fatalities, major pollution, and/or loss of production facility. HIPO-incidents are also defined as any incident which is expected, in other circumstances, result in catastrophic and disastrous consequences regardless of the actual severity. In other words, HIPO-incidents are probable to escalate from lower consequences incident tor the higher level. Skogdalen & Vinnem [21] identify that HIPO-incidents are precursor incidents which have the potential to lead to or develop into a catastrophic accident and they should be thoroughly investigated in order to prevent them from reoccurring.

The type of HIPO-incidents can be summarized as [2]: (1) Fire or explosion (2) Flash fire in a hazardous are; (3) Significant and Major Hydrocarbon leakage; (4) Emergency shutdown caused by the detection of gas leak or fire; (5) Incidents involving personnel caught in, under or between; (6) Falls from height or person slips and trips (7) Falling objects capable of causing a fatality; and (8) Incident involving confined space working.

2.2. IOGP's Incident Investigation Report

IOGP has displayed information regarding investigated HIPO-incidents in the reports every year on their website. The report contains detailed information on the incidents, including the immediate cause and the underlying cause of HIPO-incidents. As per 31 December 2019, the report shows a total of 987 HIPO-incidents, which are assigned to the causal factors. IOGP classified the HIPO-incidents report based on the working functions involved in the incidents. The working functions are the entities or concerned disciplines holding the responsibility of HIPO-incidents. As in most of upstream oil and gas company, the working functions are classified as in Figure 1.

Figure 1 illustrates the number of HIPO-incidents in accordance with the working functions. It is evident that Production concern discipline contributes the highest number of HIPO-incidents among others. Production entity possesses 452 cases from total of 957 cases (47.2%) in the worldwide operation. Based on this graphical statistic, Production, as the main operation entity, involves the highest risks in the oil and gas industry. In the second place, there are 274 HIPO-incident cases in the drilling and well services working function. This number shows that this concerned disciplines involves high risk and is also considered as the main operating business in oil and gas companies. Construction working functions places at the third rank, following 134 HIPO-incident cases. As the nature of these disciplines is to build offshore platforms and facilities, construction shall possess inherent risks due to the massive manpower involved. Logistics and Explorations working functions lie at the fourth and fifth places with 89 and 38 HIPO-incident cases respectively. Logistics may contribute to the HIPO-incident case through their supporting activities involving land, air, and sea/marine transportation. The number of HIPO-incidents in this concerned discipline are mainly contributed by the lifting activity and material/personal transportation. Exploration may involve lesser risk among other working functions due to its fewer working man-hours.

Table 1. The possibility factors of high potential incident based on literature's variables.

Author	Journal	Incidents categories	Root causes variables
Abdelhamid & Everett [4]	*Journal of construction engineering and management*	Construction	. Human factor model . Unsafe condition
Halim et al. [1]	*Journal of loss prevention in the process industries*	Oil and gas	. Equipment failure . Human error . Other significant findings
Theophilus et al. [19]	*Reliability rngineering and system safety*	Oil and gas	. Unsafe acts . Preconditions for unsafe acts . Unsafe supervision . Organizational influence
Tang et al. [20]	*Safety science*	Oil and gas	. Organizational factor
Skogdalen & Vinnem [21]	*Reliability engineering and system safety*	Oil and gas	. Human factor; . Organizational factor; . Technical factor
Ferjencik [22]	*Safety science*	General	. Implementation (1st level) causes . Organizational (2nd level) causes . Management attitude (3rd level) causes . Societal (4th level) causes
Motarjemi & Wallace [6]	*Encycolopedia of food safety*	Food industry	. Human error or equipment failure . Workplace condition . Organizational culture and management decision
Bea [23]	*Risk analysis*	Oil and gas	. Human factor . Organizational factor
Madigan et al. [24]	*Accident analysis and prevention*	Transportation	. Organizational incluences . Precondition for unsafe acts . Unsafe acts

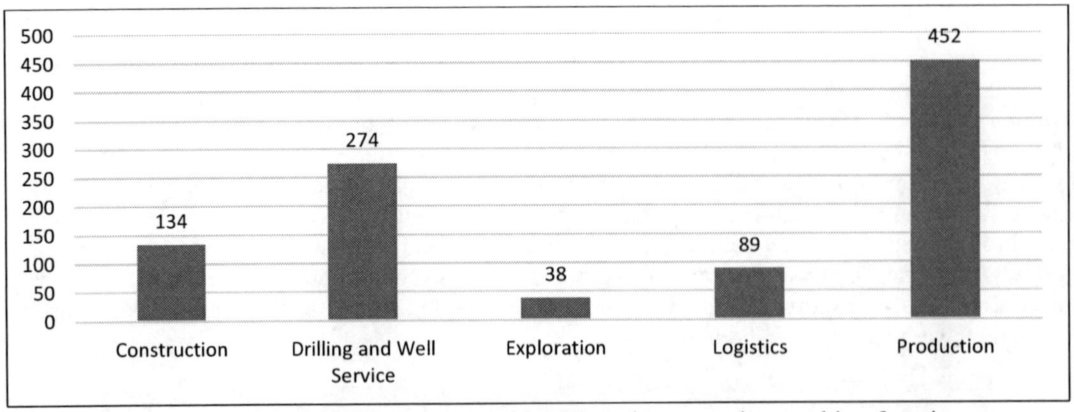

Figure 1. The HIPO-incidents occurred based on the respective working functions.

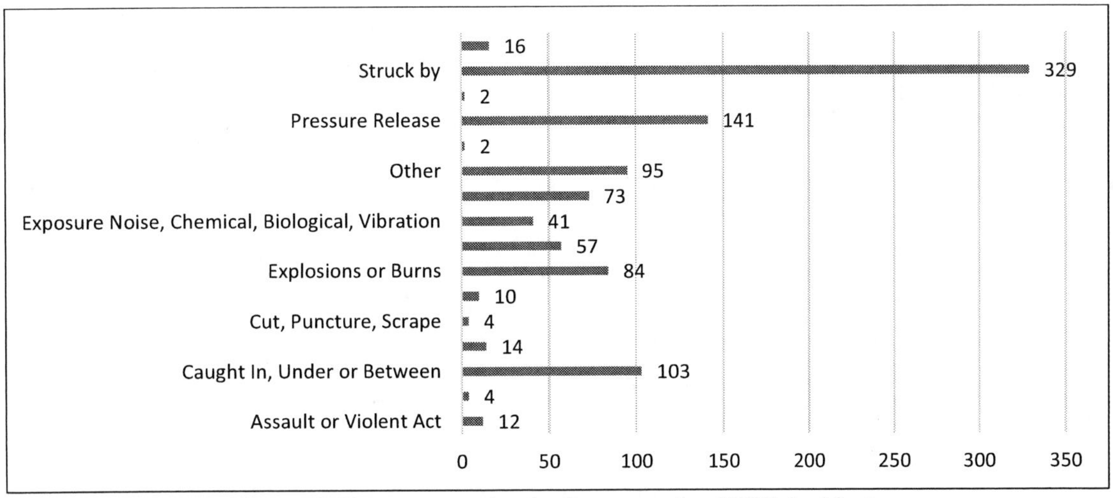

Figure 2. Event categories for the respective HIPO-incidents.

All HIPO-incidents listed in the IOGP's report are categorized based on their event category. The event category describes the activities associated with HIPO-incidents and the nature of incidents. In the IOGP's report, event categories are defined as: water-related incident and drowning; struck by, slips and trips at the same height; pressure release or hydrocarbon leaks; overexertion or strain; falls from height, exposure noise, chemical, biological, vibration; exposure electrical; explosion or burns; dropped objects; cut, puncture, scrape; confined space; caught in, under or between; aviation incident; assault or violent act; and other type of incidents. Figure 2 illustrates the event categories of HIPO-incidents provided by the IOGP's database. The investigation report lists each HIPO-incident case as a single event category despite multiple event categories may also occur within the same HIPO incident cases.

According to graphical statistics in Figure 2, event category "struck by" contributes the most numerous HIPO-incident occurrence (34.4%). This event category explains the conditions where an injury is resulted from being hit by moving equipment and machinery, or by flying or falling objects. This event category also includes vehicular incidents where the vehicle is struck or collided against another object. In the second place, event category "pressure release" lists 141 HIPO-incident cases (14.7%). This type of incident is described as the release of gas, liquid, or object under pressure from pressurized system. Hydrocarbon gas leakage and oil spill are two of the most often case in the operations. Event category "caught in, under or between" is recorded in the third place following 103 (10.8%) HIPO-incident cases occurrence. This event category elaborates an incident where an injured person is crushed between moving parts equipment or machinery. This also includes an incident where injured person caught between rolling tubular or objects being moved, crushed between a boat and a bollard, and also includes vehicle incidents involving rollover. There are 95 HIPO-incidents (9.9%) are categorized as "other" type of incident. This event category explains an incident where it cannot be logically classified under any other category. This event category includes natural incidents and the incidents which cannot be identified nor investigated due to lack of factual record.

2.3. HIPO-Incident Causal Factors

IOGP reported 987 HIPO-incidents based on root cause analysis methods and revealed 36 root causes which are investigated as the causal factors of the incidents. IOGP provides the definition for each causal factor which are identified along with their explanations. All of causal factors are mentioned as three main categories, i.e., Organizational system factor; Process factor leading to unsafe working condition; and Human-behavioral factor. Figure 3 explains the causal factors of HIPO-incidents from 2010 until 2018.

IOP Conf. Series: Materials Science and Engineering 1072 (2021) 012012 doi:10.1088/1757-899X/1072/1/012012

Organizational system factor is identified as the group of root causes associated with the failure of management decision or management system dysfunction. This causal factor is linked to the failure of an organization to provide safe working conditions and adequate health and safety culture. Oftentimes, organizational system failures within incidents are required more in-depth investigation as they may not be revealed directly by the management. Several examples of Organizational system factors are written as follows: inadequate work standards/procedures; inadequate to provide training/competence; inadequate supervision; inadequate communication; poor leadership/organizational culture; and failure to report/learn from events.

Process factor leading to unsafe working condition is a causal factor of incidents which is marked with the existence of physical hazards or high-risk situation. Process condition factor may lead to unsafe working condition due to insufficient control measures. In other circumstances, unsafe working condition may also be resulted from inadequate safety system or poor engineering design. There are four classification categories of process condition factor leading to unsafe working condition. Those are: inadequate protective equipment; failure of preventive maintenance or inspection; insufficient engineering design; and workplace hazards.

The last of incidents causal factor classified is Human-behaviour factor. This causal factor involves either the violating human behaviour or a person actions which incorrectly performed as required. Several examples of this causal factor are: violation intentional or unintentional by individual or group; work or motion at improper parameter; improper use of equipment; failure to warn of hazard; personal Protective Equipment not used or used improperly; acts of violence; use of drugs or alcohol; and fatigue.

Of the 987 HIPO-incidents that IOGP investigated, there are a total of 5124 causal factors appear as the root cause of the incidents. From these numbers, it is confirming that a HIPO case should have resulted from multiple causal factors. Figure 3 demonstrates the names of specific causal factors according to the classifications. Inadequate hazard identification or risk assessment, inadequate work standards/procedures, and improper decision making has been identified as the top of causes and appeared to be influenced by other causal factors. This section thus identifies the essential contributing factors which emerge in conjunction with organizational system factor, process factor, and human-behavioral factor.

In Organizational system factor classification, Figure 3 illustrates 523 (10.2%) causal factors is identified as inadequate hazard identification or risk assessment. This finding becomes critical because failure to manage and control risks shall create a higher-severity incident. As often in many cases, HIPO-incidents involving fire and gas leak could rapidly escalate and produce catastrophic consequences. At the second point, inadequate work standards and procedures contribute 418 (8.2%) of the total causal factors. This finding includes maintaining obsolete procedures/documents, misplaced work instructions, and incompatible organization standards. It is essential finding that inadequate procedure shall impact to the failure of operation management and equipment defect. It is concluded that an improper or insufficient way of conducting a task had initiated an equipment failure that eventually led to the fire [1]. Therefore, providing correct and updated working procedures and organizational documents plays critical roles in creating safer working conditions.

In terms of Human-behaviour factor, it is investigated that 326 causal factors are improper decision making or lack of safety judgment. This terminology is observed where the risky situations were wrongly judged, and the error decision were made, or the personnel involved in the inappropriate activities. This finding includes taking shortcuts, horse-playing, and bypassing safety standards. Another essential factor within Human-behaviour factor is the violation of safety rules unintentionally by the personnel (228 cases). This happens because of the lack of safety knowledge and poor competence of assigned tasks. Personnel may violate safety standards and endanger themselves unintentionally due to insufficient safety awareness, which is resulted from organizational failure to promote safety culture. This finding is also correlated with providing inadequate training/competence in the Organizational system factor classification.

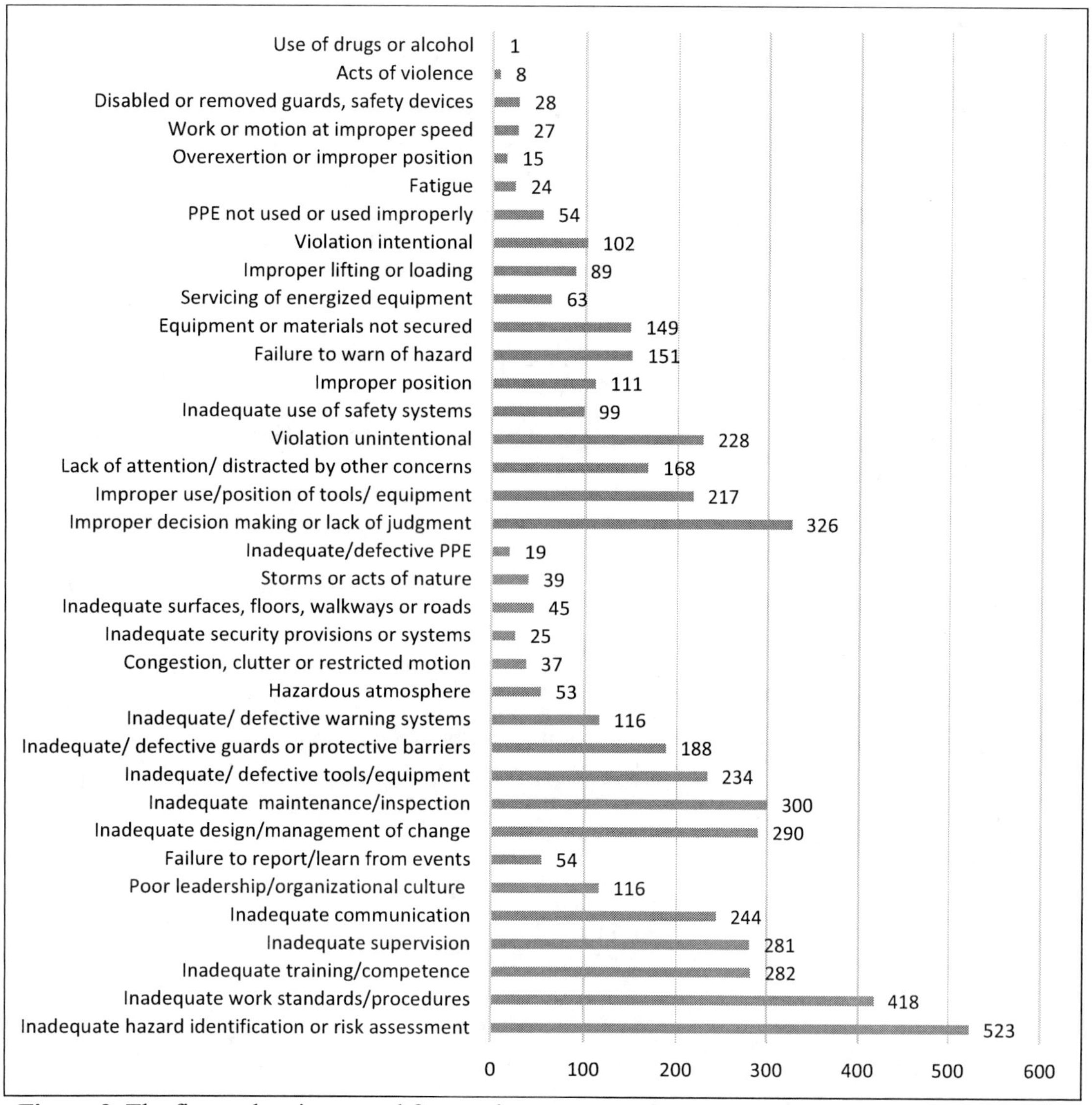

Figure 3. The figure showing causal factors that appear as root cause to the HIPO-incidents. Orange color indicates the human-behavioral factor, green color indicates the process factor leading to unsafe working condition, and blue colour defines the organizational system factor.

The high-risk situation in the process factor, which leads to unsafe working conditions, is mainly caused by inadequate preventive maintenance and inspection (5.9%). Most of the equipment defect and malfunction are attributed to the failure of maintenance program, or it was not performed correctly. Due to this fact, it is obvious that inadequate maintenance and inspection contributes significant factor to the HIPO-incidents occurrence. Another important factor identified is the inadequate design/specification or management of change. There are 290 (5.7%) findings attributed to this causal factor. Such a design flaw may result in equipment degradation due to operational vibrations, heats, and pressures. This causal factor includes improper installations, incompatible materials and products, and operating beyond equipment specification. From these investigations, it is expected that the organization could understand the causes of incidents and provide adequate measures to control the operational risks.

3. Research Methodology

3.1. Development of Gaussian Fuzzy Analytic Hierarchy Process

This paper proposes the Gaussian fuzzy analytic hierarchy process (Gaussian FAHP), a multi-criteria decision analysis methodology that can realistically compensate uncertainty and ambiguity during preference selection. This methodology is advantageous due to its capability to approximate uncertainty and risk of incomplete information due to scattered data. The Gaussian membership function in this paper is generated based on Normal distribution function by simulating Monte-Carlo random variables. The final evaluation value is formed as a single crisp number by calculating centre of gravity defuzzification method. The detailed computation process of Gaussian FAHP in this paper is explained as follows.

Table 2. Conversion table of Gaussian fuzzy number from Triangular fuzzy number in correspondent to Saaty's scale.

Linguistic variables	Saaty's Scale	Triangular fuzzy number	Gaussian fuzzy number
Equally important	1	Triangular (x,1,1,1)	Gaussian (x: 1, rand (σ_i))
Moderately important	3	Triangular (x,1,3,5)	Gaussian (x: 3, rand (σ_i))
Significantly important	5	Triangular (x,3,5,7)	Gaussian (x: 5, rand (σ_i))
Strongly important	7	Triangular (x,5,7,9)	Gaussian (x: 7, rand (σ_i))
Extremely important	9	Triangular (x,1,1,1)	Gaussian (x: 9, rand (σ_i))

Step-1, transform a triangular fuzzy number into Gaussian fuzzy number, based on the linguistic preference scale of the Saaty scale based on Table 2. The conversion of fuzzy membership function in terms of lower, middle, and upper values as Equation 1-5. The value of rand (σ_i) is obtained by simulating Monte-Carlo random variables of Normal distribution function.

$$\tilde{1} = T\left[x_{a_1}, x_{b_1}\right] = G[x\colon \mu_1, \sigma_1];$$
$$l_1 = m_1 = \mu_1$$
$$m_1 = \mu_1$$
$$u_1 = \mu_1 + (Z_{\frac{\alpha}{2}}) \frac{\sigma_1}{\sqrt{n}} \tag{1}$$

$$\tilde{3} = T\left[x_{a_3}, x_{b_3}\right] = G[x\colon \mu_3, \sigma_3];$$
$$l_3 = \mu_3 - (Z_{\frac{\alpha}{2}}) \frac{\sigma_3}{\sqrt{n}}$$
$$m_3 = \mu_3$$
$$u_3 = \mu_3 + (Z_{\frac{\alpha}{2}}) \frac{\sigma_3}{\sqrt{n}} \tag{2}$$

$$\tilde{5} = T\left[x_{a_5}, x_{b_5}\right] = G[x\colon \mu_5, \sigma_5];$$
$$l_5 = \mu_5 - (Z_{\frac{\alpha}{2}}) \frac{\sigma_5}{\sqrt{n}}$$
$$m_5 = \mu_5$$
$$u_5 = \mu_5 + (Z_{\frac{\alpha}{2}}) \frac{\sigma_5}{\sqrt{n}} \tag{3}$$

$$\tilde{7} = T\left[x_{a_7}, x_{b_7}\right] = G[x\colon \mu_7, \sigma_7];$$
$$l_7 = \mu_7 - (Z_{\frac{\alpha}{2}}) \frac{\sigma_7}{\sqrt{n}}$$
$$m_7 = \mu_7$$
$$u_7 = \mu_7 + (Z_{\frac{\alpha}{2}}) \frac{\sigma_7}{\sqrt{n}} \tag{4}$$

$$\tilde{9} = T\left[x_{a_9}, x_{b_9}\right] = G[x\colon \mu_9, \sigma_9];$$
$$l_9 = \mu_9 - (Z_{\frac{\alpha}{2}}) \frac{\sigma_9}{\sqrt{n}}$$

$$m_9 = \mu_9$$
$$u_9 = m_9 = \mu_9 \tag{5}$$

where

$$x_a = \tilde{\mu}_i - \tilde{\sigma}_i\sqrt{-Ln(\alpha)}$$
$$x_b = \tilde{\mu}_i + \tilde{\sigma}_i\sqrt{-Ln(\alpha)} \tag{6}$$

and the value of α is chosen as 0.001 to approximate triangular fuzzy number. $\tilde{\mu}_i$are the mean value of Gaussian fuzzy number, it is typically = m_i, $\tilde{\sigma}_i$ is the standard deviation of the Gaussian fuzzy number, it is obtained by solving Eq. 6.

Step-2, calculate Gaussian fuzzy pairwise comparison matrices for all criteria and alternatives judgments. Construct Gaussian fuzzy judgment matrix $\tilde{A}$ based on the value of (l_i, m_i, u_i) which one of criteria is more important to another.

$$\widetilde{A_\alpha} = \begin{bmatrix} 1 & \tilde{a}_{\alpha 12} & \cdots & \tilde{a}_{\alpha 1n} \\ \tilde{a}_{\alpha 21} & 1 & \cdots & \tilde{a}_{\alpha 2n} \\ \cdots & \cdots & \ddots & \cdots \\ \tilde{a}_{\alpha n1} & \tilde{a}_{\alpha n2} & \cdots & 1 \end{bmatrix}$$

$$= \begin{matrix} \\ C_1 \\ C_2 \\ C_n \end{matrix} \begin{matrix} C_1 & C_2 & & C_n \\ \left[\begin{matrix} 1 \\ \tilde{a}_{\alpha 12}^{\;-1} \\ \cdots \\ \tilde{a}_{\alpha 1n}^{\;-1} \end{matrix}\right. & \begin{matrix} \tilde{a}_{\alpha 12} \\ 1 \\ \cdots \\ \tilde{a}_{\alpha 2n}^{\;-1} \end{matrix} & \begin{matrix} \cdots \\ \cdots \\ \ddots \\ \cdots \end{matrix} & \left.\begin{matrix} \tilde{a}_{\alpha 1n} \\ \tilde{a}_{\alpha 2n} \\ \cdots \\ 1 \end{matrix}\right] \end{matrix} \tag{7}$$

Where $\widetilde{A_{a_{ij}}}$ is the fuzzy pairwise comparison value *i* compare to *j*. The value of (l_i, m_i, u_i) are the value of lower, middle, and upper properties of the fuzzy pair wise comparison matrices $\widetilde{A_{a_{ij}}}$,

Step-3, Determine the fuzzy weight value by implementing geometric mean for each criterion and alternative judgment. The fuzzy weight of each criterion is calculated as follows [25].

$$l_n = \left[\prod_{j=1}^{n} l_{ij}\right]^{\frac{1}{n}}; m_n = \left[\prod_{j=1}^{n} m_{ij}\right]^{\frac{1}{n}};$$
$$u_n = \left[\prod_{j=1}^{n} u_{ij}\right]^{\frac{1}{n}} \tag{8}$$

Step-4, identify the rank for all alternatives by applying defuzzification process of Gaussian fuzzy numbers (l_n, m_n, u_n). Defuzzification process is important because it defines the crisp value from the fuzzy number to determine the most optimum alternative. The defuzzification method of (l_n, m_n, u_n) is applied by center of gravity of the Normal (Gaussian) distribution probability density function.

$$U_n^* = \frac{\int_{l_n}^{u_n} x.G(x:\tilde{\mu}_n, \tilde{\sigma}_n)dx}{\int_{l_n}^{u_n} G(x:\tilde{\mu}_n, \tilde{\sigma}_n)dx}$$

$$U_n^* = \frac{\int_{l_n}^{u_n} \frac{x}{\sqrt{2\pi\tilde{\sigma}_n^{\,2}}} e^{-\frac{(x-\tilde{\mu}_n)^2}{2\tilde{\sigma}_n^{\,2}}} dx}{\int_{l_n}^{u_n} \frac{1}{\sqrt{2\pi\tilde{\sigma}_n^{\,2}}} e^{-\frac{(x-\tilde{\mu}_n)^2}{2\tilde{\sigma}_n^{\,2}}} dx} \tag{9}$$

Where $\tilde{\mu}_n$ and $\tilde{\sigma}_n$ are the mean value and standard deviation of Normal (Gaussian) probability distribution function after performed by Geometric mean operation. It is explained that:

$$\tilde{\mu}_n = m_n \tag{10}$$

$$\tilde{\sigma}_n = \frac{\tilde{\mu}_n - l_n}{3} \quad for\ m_n > l_n\ and\ m_n = u_n$$

$$\tilde{\sigma}_n = \frac{u_n - \tilde{\mu}_n}{3} \quad for\ u_n > m_n\ and\ m_n = l_n \tag{11}$$

Step-5, Perform consistency analysis. After having defuzzified value, it is important to check the consistency of the pairwise comparison matrices. The crisp value is then formed into a comparison matrix similar with conventional AHP. The values of defuzzification U are developed into crisp value comparison matrix.

$$\tilde{U}_n = \begin{bmatrix} U_{11} & U_{12} & \cdots & U_{1n} \\ U_{21} & U_{22} & \cdots & U_{2n} \\ \cdots & \cdots & \ddots & \cdots \\ U_{n1} & U_{n2} & \cdots & U_{nn} \end{bmatrix} \tag{12}$$

As stated by Zheng et al. [26] "When the conventional comparison matrix U is consistent, it means that fuzzy comparison matrix $\tilde{U}$ is also consistent". Consistency analysis is the performed by implementing consistency ratio method similar with those in AHP method.

Step-6, determine the weight of fuzzy value by normalizing each criterion.

$$U_{sum} = \sum_{i=1}^{n} U_i \tag{13}$$

$$\tilde{U}_{ij} = \frac{U_i}{U_{sum}} \quad ; for\ i = 1,2,3,\dots,n \tag{14}$$

where $\tilde{U}_{ij}$ is the fuzzy weight value of the *i*-th criterion, and therefore, obtain fuzzy final value by calculating hierarchical layer sequencing.

$$\widetilde{W_i} = \sum_{j=1}^{n} \tilde{U}_i * \widetilde{U_{ij}} \tag{15}$$

where $\widetilde{U_{ij}}$ is the fuzzy weight value of the *j*-th criteria to the *i*-the alternatives.

4. Results and Discussion

This section analyses the methods implemented in the research. As defining the most contributing causal factor of HIPO-incidents is the ultimate objective, the methods should comply with systematic research framework. For an improved fuzzy AHP research to be precise, its criteria, sub-criteria, and alternatives should be developed based on applicable research methodology. The criteria involved in the proposed NMCFAHP method is determined based on work functions. Whereas the alternatives lie upon the identified root causes, i.e., Organizational system factor; Process factor leading to unsafe working condition; and Human-behaviour factor.

4.1. Evaluation of Gaussian FAHP Criteria Variables

The development of criteria variables evaluation begins with the process of listing all HIPO-incidents from IOGP's report into the table of working function series. This table contains all HIPO in accordance with their specific working function and is used for developing the evaluative criteria. Table 3 defines the series of working functions from 2010 until 2018.

To evaluate the criteria variables, we develop pairwise comparison tables according to the working function in Table 3. All working function is compared from one to another and displayed in a year-series. For instance, in 2010, there are 14 HIPO-incident cases occurred in construction, whereas in drilling there are 32 cases. Therefore, in relative comparison drilling is more important than construction with scale 32:14 ≈ 2.3:1, or equivalently in Saaty's scale 3:1. The result of the pairwise comparison of all criteria is written in Table 4. The reciprocal approach is implemented for reflecting the dominance of second criteria compared with the first. Afterward, from Table 4, we develop the pairwise comparison matrices. The analysis for criteria variables is calculated based fuzzy pairwise comparison matrix. The fuzzy judgment matrix $\tilde{A}$ is developed based on Gaussian fuzzy number (Gaussian (x: μ_i, σ_i)). The value of mean (μ) and standard deviation (σ) is obtained from the average and standard deviation value listed in the years' arrear, as specified in Table 4. We then generate a random Gaussian fuzzy number by applying the Monte-Carlo simulation following the values of μ and σ for each criteria evaluation. This paper employs 1,000 random normal variables. The Monte-Carlo simulation is obtained by entering a spreadsheet formula, i.e. "$= NORMINV(RAND(), \bar{X}, s)$". From the Monte-Carlo simulation, we obtain the result of Monte-Carlo Normal distribution mean (μ_i) and standard deviation (σ_i). Based on the random number generation for the criteria pairwise comparison, we determine the value of l, m, and u as the most probable, lower value, and upper value of the fuzzy pairwise comparison matrices $\tilde{A}$. The value of l, m, and u for all criteria evaluation is determined by solving Equation 1 until 4. After having the value of l, m, and u, the pairwise comparison is modified and written in Table 5.

Table 3. HIPO-incidents in correspond to working function.

Work Function	2010	2011	2012	2013	2014	2015	2016	2017	2018
Construction	14	7	30	8	15	15	12	19	14
Drilling	32	23	42	36	31	28	21	38	23
Exploration	3	6	13	5	2	2	2	1	4
Logistics	17	6	15	15	6	7	12	4	7
Production	31	27	73	60	32	46	56	69	58

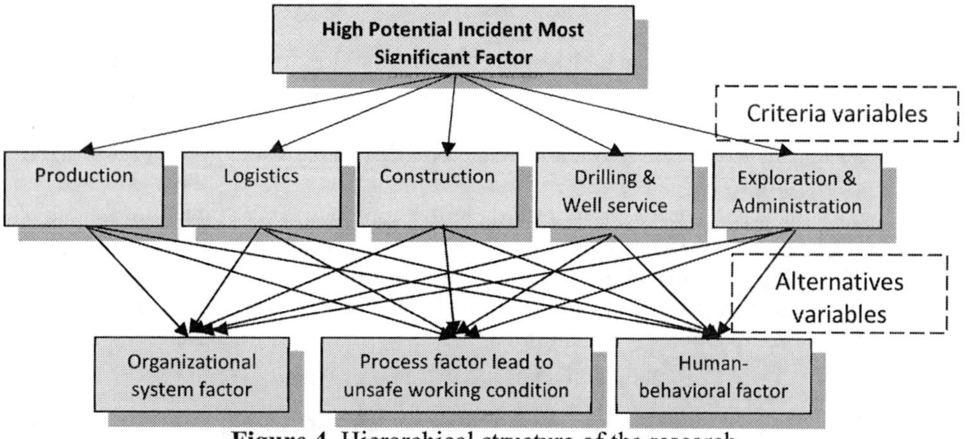

Figure 4. Hierarchical structure of the research.

Based on Figure 4 it can be seen that the researcher uses five criteria and three alternatives. The result of criteria evaluation is plotted as Gaussian probability distribution function. In corresponding to the value of $\tilde{\mu}_n$ and $\tilde{\sigma}_n$ in Equation 10 and 11. For obtaining the final criteria weight values, we perform defuzzification method based on Equation 4. The result of Gaussian distribution plot of criteria is demonstrated in Figure 5, and the result of defuzzification process is written in Table 7.

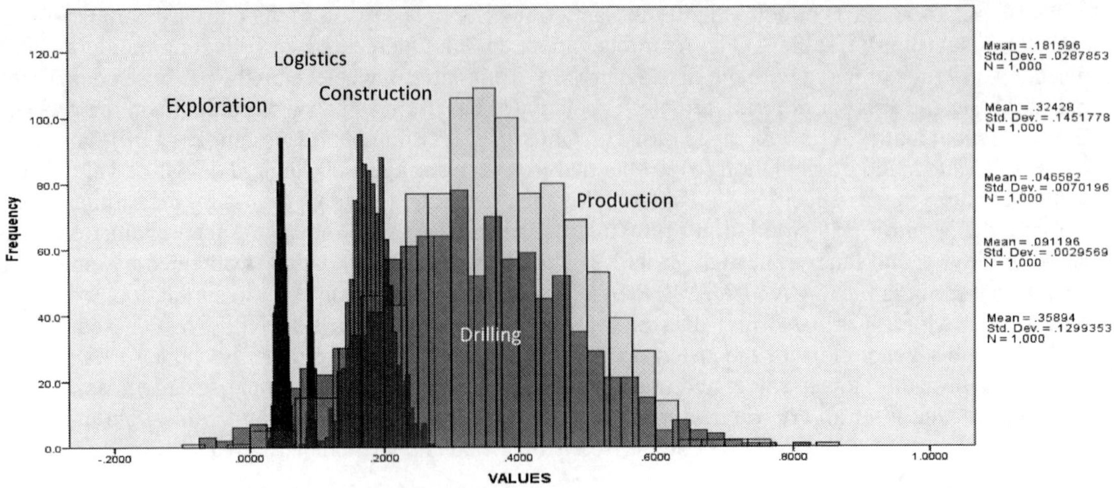

Figure 5. Gaussian probability density function plot in accordance to all working functions (criteria).

4.2. Evaluation of Gaussian FAHP Alternatives Variables

In this research, a similar approach with the criteria evaluation is implemented for evaluating alternatives. The alternatives are assessed by the respective criterion. Therefore, there will be five pairwise comparisons to score the alternatives based on criteria evaluated. Firstly, based on the IOGP's report, the list of causal factors according to the working function is displayed in series arrear. To shorten the writing style, the alternative variables are determined as follows,

$A1 = Organizational\ system\ factor$

$A2 = Process\ factor\ leading\ to\ unsafe\ working\ condition$

$A3 = Human - behavioral\ factor$

Afterward, the pairwise comparison tables are developed for alternatives based on each criterion. For instance, in 2010 there are 27 cases of Organizational system factor, 5 cases of process factor, and 22 cases of Human-behavioral factor within Construction working function. Therefore, pairwise comparison tables and matrices can be obtained based on this ratio. Table 8. until table 13. define the pairwise comparison for the alternatives. Due to space limitation, we only display the causal factor cases in correspond with Construction working function. Similar approaches are applied for the other working function or criteria.

In order to have the alternatives final fuzzy values, multiplication values between criteria and alternatives should be performed. To accommodate this problems, we used the fuzzy hierarchical layer sequencing method. By implementing Equation 16, the fuzzy final value is obtained.

$$\widetilde{U_\iota} = \left((\widetilde{w_1} \otimes \widetilde{w_{12}}) \oplus (\widetilde{w_2} \otimes \widetilde{w_{22}}) \oplus \dots \oplus (\widetilde{w_J} \otimes \widetilde{r_{JJ}})\right) \emptyset \left(\widetilde{w_1} \oplus \widetilde{w_2} \oplus \dots \oplus \widetilde{w_J}\right) \quad (16)$$

The results then is plotted to the Normal distribution probability density function and this computation are shown as a triplet number which follows as a mathematical equation in Equation 17 until 19.

$$u_i{}^L = \tilde{\mu} - (Z_{\frac{\alpha}{2}}) \frac{\tilde{\sigma}}{\sqrt{n}} \quad (17)$$

$$u_i{}^M = \tilde{\mu} \quad (18)$$

$$u_i{}^U = \tilde{\mu} + (Z_{\frac{\alpha}{2}}) \frac{\tilde{\sigma}}{\sqrt{n}} \quad (19)$$

where $w_i{}^L$, $w_i{}^M$, and $w_i{}^U$ are the lower value, most probable value, and upper value of the calculated fuzzy pairwise comparison matrices; $\tilde{\sigma}$ is the fuzzy standard deviation; and $\tilde{\mu}$ is the mean of fuzzy calculated values. Random variables for all alternatives are generated in accordance with the values of $\tilde{\mu}$ and $\tilde{\sigma}$. The result of final fuzzy values is illustrated in Figure 6.

As seen in Figure 6 There are overlapping figure from one alternatives to others. To clarify the alternatives' rank, final crisp value should be obtained. The final crisp values are then calculated by applying a defuzzification using the centre of gravity method (COG).

$$u_i^M(A1) = \tilde{\mu} = 0.473112$$

$$\tilde{\sigma}(A1) = \frac{max|u_i^U - \tilde{\mu}|}{(Z\frac{\alpha}{2})} \sqrt{n} = 0.0619811$$

$$w_i^*(A1) = COG = \frac{\int_{u_i^L}^{u_i^U} \frac{x}{\sqrt{2\pi(0.0619811)^2}} e^{-\frac{(x-0.473112)^2}{2(0.0619811)^2}} dx}{\int_{u_i^L}^{u_i^U} \frac{1}{\sqrt{2\pi(0.0619811)^2}} e^{-\frac{(x-0.1613)^2}{2(0.0619811)^2}} dx} \tag{20}$$

By solving complex integration in Eq. 20, the fuzzy final values for alternative-1 are:

$$w_i^*(A1) \approx \frac{0.01969}{0.041497} = 0.474540$$

The rest of calculations are implemented for A2, A3, and A4 respectively. The final alternatives' rank and score is then displayed in Table 15.

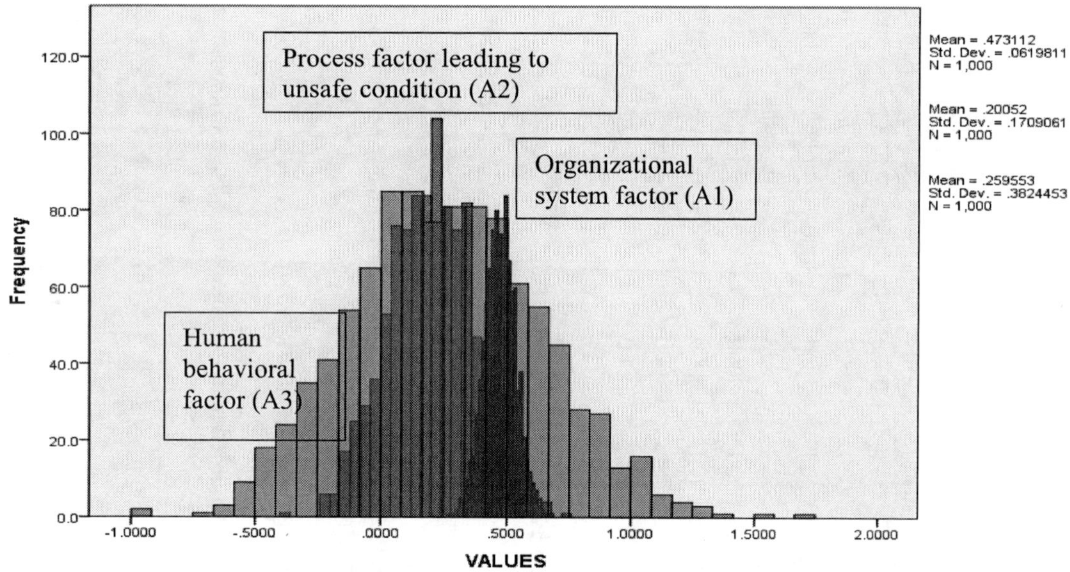

Figure 6. The result of probability density function for alternative final values.

Table 4. Pairwise comparison table of all criteria variables.

Pairwise comparison	**2010**	**2011**	**2012**	**2013**	**2014**	**2015**	**2016**	**2017**	**2018**	**Mean**	**Std. dev.**
Construction vs Drilling	0.3333	0.3333	1.0000	0.2000	0.3333	0.5000	0.5000	0.5000	0.5000	0.4667	0.2273
Construction vs Exploration	5.0000	1.0000	3.0000	2.0000	7.0000	7.0000	7.0000	9.0000	5.0000	5.1111	2.6667
Construction vs Logistics	1.0000	1.0000	3.0000	0.5000	3.0000	3.0000	1.0000	5.0000	3.0000	2.2778	1.4814
Construction vs Production	0.3333	0.3333	0.3333	0.1111	0.5000	0.3333	0.2000	0.2000	0.2000	0.2827	0.1159
Drilling vs Exploration	9.0000	5.0000	5.0000	7.0000	9.0000	9.0000	9.0000	9.0000	7.0000	7.6667	1.7321
Drilling vs Logistics	2.0000	5.0000	3.0000	3.0000	5.0000	5.0000	2.0000	9.0000	5.0000	4.3333	2.1794
Drilling vs Production	1.0000	1.0000	0.5000	0.5000	1.0000	0.5000	0.3333	0.5000	0.3333	0.6296	0.2860
Exploration vs Logistics	0.1429	1.0000	1.0000	0.3333	0.3333	0.3333	0.1429	0.2000	0.5000	0.4429	0.3352
Exploration vs Production	0.1111	0.2000	0.1111	0.1111	0.1111	0.1111	0.1111	0.1111	0.1111	0.1210	0.0296
Logistics vs Production	0.5000	0.2000	0.2000	0.3333	0.1429	0.1429	0.2000	0.1111	0.1429	0.2192	0.1236

Table 5. Pairwise comparison matrix for all criteria variables.

	Construction	Drilling	Exploration	Logistics	Production
Construction	1	G(*x: 0.4667, 0.2273*)	G(*x: 5.1111, 0.2273*)	G(*x: 2.2778, 1.4814*)	G(*x: 0.2827, 0.1159*)
Drilling	G(*x: (1/0.4667), 0.2273*)	1	G(*x: 7.6667, 1.7321*)	G(*x: 4.333, 2.1794*)	G(*x: 0.6296, 0.2860*)
Exploration	G(*x: (1/5.1111), 0.2273*)	G(*x: (1/7.6667), 1.7321*)	1	G(*x: 0.4429, 0.3352*)	G(*x: 0.1210, 0.0296*)
Logistics	G(*x: (1/2.2778), 1.4814*)	G(*x: (1/ 4.333), 2.1794*)	G(*x: (1/0.4429), 0.3352*)	1	G(*x: 0.2192, 0.1236*)
Production	G(*x: (1/0.2827), 0.1159*)	G(*x: (1/0.6296), 0.2860*)	G(*x: (1/0.1210), 0.0296*)	G(*x: (1/ 0.2192), 0.1236*)	1

Table 6. Modified pairwise comparison matrix in accordance to the value of l, m, and u.

	Construction	Drilling	Exploration	Logistics	Production
Construction	(1,1,1)	(0.4481, 0.4620, 0.4758)	(5.0236, 5.1887, 5.3537)	(2.2488, 2.3418, 2.4349)	(0.3330, 0.3912, 0.4493)
Drilling	(2.1015, 2.1647, 2.2318)	(1,1,1)	(7.5803, 7.6911, 7.8020)	(4.1609, 4.2965, 4.4320)	(0.6527, 0.6987, 0.7448)
Exploration	(0.1868, 0.1927, 0.1991)	(0.1281, 0.1300, 0.1319)	(1,1,1)	(0.4541, 0.4787, 0.5033)	(0.1390, 0.1604, 0.1818)
Logistics	(0.4107, 0.4270, 0.4447)	(0.2256, 0.2327, 0.2403)	(1.9869, 2.0890, 2.2021)	(1,1,1)	(0.2477, 0.2777, 0.3078)
Production	(2.2255, 2.5563, 3.0026)	(1.3427, 1.4312, 1.5321)	(5.5009, 6.2345, 7.1939)	(3.2489, 3.6005, 4.0374)	(1,1,1)

Table 7. The value of normalized pairwise comparison matrix.

	Normalized value	Fuzzy weight *(Gaussian (x: $\tilde{\mu}$, $\tilde{\sigma}$))*	Defuzzified value
Construction	(0.1792, 0.1810, 0.1814)	*Gaussian (x: 0.163706, 0.028223)*	0.1788
Drilling	(0.3329, 0.3251, 0.3161)	*Gaussian (x: 0.308160, 0.021542)*	0.3244
Exploration	(0.0469, 0.0473, 0.0474)	*Gaussian (x: 0.042082, 0.002416)*	0.0470
Logistics	(0.0888, 0.0890, 0.0888)	*Gaussian (x: 0.083045, 0.010977)*	0.0904
Production	(0.3521, 0.3576, 0.3662)	*Gaussian (x: 0.403007, 0.016230)*	0.3595

Table 8. Causal factors in correspond to construction criteria.

Construction	2010	2011	2012	2013	2014	2015	2016	2017	2018
Organizational system factor (A1)	27	12	63	18	30	25	24	38	20
Process factor leading to unsafe working condition (A2)	5	6	19	8	10	7	13	22	11
Human-behavioral factor (A3)	22	12	36	15	29	33	33	24	30

Table 9. Pairwise comparison table of alternatives in correspond to construction criteria.

	2010	2011	2012	2013	2014	2015	2016	2017	2018	Mean	Std. Dev.
A1 VS A2	7	2	5	3	3	5	2	2	3	3.5556	1.7401
A1 VS A3	2	1	3	2	1	0.5	1	2	0.5	1.4444	0.8457
A2 VS A3	0.2	0.5	0.5	0.5	0.333333	0.2	0.333333	1	0.333333	0.4333	0.2432

Table 10. Pairwise comparison matrix of alternatives in correspond to construction criteria.

	A1	A2	A3	Wi
A1	(1.0000, 1.0000, 1.0000)	(3.4147, 3.5202, 3.6256)	(1.3881, 1.4423, 1.4964)	(0.5072, 0.5024, 0.5069)
A2	(0.2758, 0.2841, 0.2929)	(1.0000, 1.0000, 1.0000)	(0.4147, 1.4423, 0.4438)	(0.1466, 0.2171, 0.1461)
A3	(0.6682, 0.6933, 0.7204)	(2.2534, 0.6933, 2.4115)	(1.0000, 1.0000, 1.0000)	(0.3461, 0.2290, 0.3468)

Table 11. Pairwise comparison matrix of alternatives in correspond to drilling criteria.

	A1	A2	A3	Wi
A1	(1.0000, 1.0000, 1.0000)	(2.8567, 2.9378, 3.0189)	(1.3178, 1.3792, 1.4405)	(0.4655, 0.4611, 0.4670)
A2	(0.3312, 0.3404, 0.3501)	(1.0000, 1.0000, 1.0000)	(0.5977, 0.6296, 0.6615)	(0.1744, 0.1731, 0.1757)
A3	(0.6942, 0.7251, 0.7588)	(1.5118, 1.5884, 1.6732)	(1.0000, 1.0000, 1.0000)	(0.3041, 0.3032, 0.3098)

Table 12. Pairwise comparison matrix of alternatives in correspond to exploration criteria.

	A1	A2	A3	Wi
A1	(1.0000, 1.0000, 1.0000)	(2.8848, 2.9906, 3.0963)	(2.3682, 2.5357, 2.7032)	(0.5678, 0.5682, 0.5809)
A2	(0.3230, 0.3344, 0.3466)	(1.0000, 1.0000, 1.0000)	(0.8058, 0.8572, 0.9087)	(0.1911, 0.1907, 0.1947)
A3	(0.3699, 0.3944, 0.4223)	(1.1005, 1.1666, 1.2411)	(1.0000, 1.0000, 1.0000)	(0.2218, 0.2233, 0.2307)

Table 13. Pairwise comparison matrix of alternatives in correspond to logistics criteria.

	A1	A2	A3	Wi
A1	(1.0000, 1.0000, 1.0000)	(3.7390, 3.9137, 4.0885)	(1.2519, 1.3147, 1.3775)	(0.5006, 0.4993, 0.5091)
A2	(0.2446, 0.2555, 0.2675)	(1.0000, 1.0000, 1.0000)	(0.5609, 0.5818, 0.6028)	(0.1543, 0.1532, 0.1557)
A3	(0.7259, 0.7606, 0.7988)	(1.6590, 1.7187, 1.7830)	(1.0000, 1.0000,. 1.0000)	(0.3184, 0.3162, 0.3219)

Table 14. Pairwise comparison matrix of alternatives in correspond to production criteria.

	A1	A2	A3	Wi
A1	(1.0000, 1.0000, 1.0000)	(2.1822, 2.2767, 2.3711)	(1.4634, 1.5245, 1.5855)	(0.4407, 0.4379, 0.4449)
A2	(0.4217, 0.4392, 0.4582)	(1.0000, 1.0000, 1.0000)	(0.9166, 0.9304, 0.9442)	(0.2180, 0.2146, 0.2164)
A3	(0.6307, 0.6560, 0.6833)	(1.0591, 1.0748, 1.0909)	(1.0000, 1.0000, 1.0000)	(0.2616, 0.2574, 0.2594)

Table 15. The result of alternatives' rank and score.

Rank	Alternative	Fuzzy weight *(Gaussian (x: $\widetilde{\mu}$, $\widetilde{\sigma}$))*	Modified fuzzy weight (u_i^L, u_i^M, u_i^U)	Defuzzified weight
1st	A1	*Gaussian (x: 0.46807, 0.04404)*	(0.46960, 0.47246, 0.47633)	0.474540
2nd	A3	*Gaussian (x: 0.20052, 0.17961)*	(0.27577, 0.29857, 0.30028)	0.289450
3rd	A2	*Gaussian (x: 0.27545, 0.36673)*	(0.18460, 0.18514, 0.19585)	0.187740

From the result of this research, it is exposed that Organizational system factor (A1) is the most significant contributing factor to the HIPO-incidents occurrence. Besides that, Organizational system factor possesses the narrowest Gaussian probability density function among other causal factors. The calculation of standard deviations for Organizational system factor is less than 0.5% of their means. Based on this figure, it is implied that the incidents statistics are complete and consistent in deciding the fuzzy weights of judgment.

In the second and third rank lies the causal factor of Human behaviour factor (A3) and Process factor leading unsafe working condition (A2). In contrast with A1, A2 and A3 display a wide range of standard deviation (>0.5% of mean). Especially in A3, the value of standard deviation is even more than its mean. This phenomenon shows that in the Human-behaviour causal factor, there is a wide range of fluctuation in the yearly series arrear. Based on this analysis, Human-behaviour factor comes as the most unstable causal factor, which identifies the fragility of personnel/workers involved. This circumstance also indicates that Human-behaviour factor should be consistently taken into account in incident prevention programs.

5. Performance Analysis

5.1. Sensitivity Analysis

The sensitivity analysis is performed to measure the proposed evaluation model's performance. It is necessary to evaluate how input change could lead to output difference. This means that if the input system is changed by adding or lowering the criteria values, how far the expected alternatives scores will change. The sensitivity analysis is performed by replacing the degree of importance of working function criteria. In such manner, we change the least important criteria, e.g., Exploration and Logistics to become the most important criteria. This analysis previses if there is a change in incident statistics or different data approach as the baseline of the research. The result of the sensitivity analysis shall be displayed in descriptive graphics and charts.

According to the evaluation of criteria in section 4.1., the most significant criteria is Production, followed by Drilling. Whereas the lease significant contributing criteria are Exploration and Logistics. The sensitivity analysis is performed by lowering 50% weight of Production and Drilling criteria and adding 50% weight to the least significant criteria, Exploration and Logistics. The result of this change is demonstrated in Figure 7.

In Figure 7, it is observed that there is no ranking change of all alternatives. The result is still consistent, figuring that Organizational system factor (A1) as the most significant causal factor for the HIPO-incidents occurrence. In addition, it is obviously displayed that Human-behaviour factor (A3) and Process factor leading to unsafe working condition (A2) come as the second and third significant factor. Therefore, the evaluation is consistent and robust.

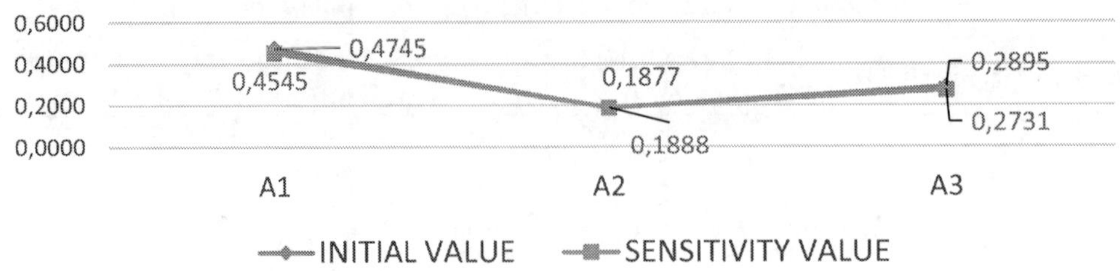

Figure 7. The result of alternatives' defuzzified value after having the sensitivity analysis.

6. Conclusion

This paper has presented the development of Gaussian fuzzy analytic hierarchy process based on the random values generated by Monte-Carlo simulation. The proposed methodology evaluates the most significant causal factor corresponding to the HIPO-incidents occurrence. Based on the analysis, it is concluded that Organizational system factor (A1) is the most contributing causal factor to the HIPO-incidents. In current condition, oil and gas company records the HIPO-incidents' occurrence and defines its frequencies as the notable factor. The detail of root causal factors is almost never adhered to. Oil and gas company has not prioritized its resource to enhance the safety aspect of the organizational system factor. On the other hand, the Human-behaviour causal factor (A3), perches as the second place of significant causal factor and displays a wide range of standard deviation. As observed, the standard deviation is even more than the mean value. This concludes that Human-behaviour factor indicates unstable causal factor and identifies the fragility of personnel/workers involved.

Based on the sensitivity analysis, the observed result demonstrates consistent and accurate analysis. There is no change in terms of alternatives' final ranking and significance. The analysis figures that Organizational system factor (A1) still as the most significant causal factor among the other alternatives. Therefore, it is inferred that the propose methodology is robust and acceptable as the reference for oil and gas companies in determining their incident preventive programs.

References

[1] Halim SZ, Janardanan S, Flechas T, Mannan MS. In search of causes behind offshore incidents: Fire in offshore oil and gas facilities. *J Loss Prev Process Ind*. Elsevier BV; 2018 Jul;**54**:254–65. doi.org/10.1016/j.jlp.2018.04.006

[2] Wicaksono FD. Implementation of Analytic Hierarchy Process for Determination of Incident Causes in Oil and Gas Industry. *Proc Indon Petrol Assoc*. Indonesian Petroleum Association (IPA). doi.org/10.29118/ipa.50.17.155.f.

[3] IOGP Report 2018s. *Safety Performance Indicators – 2018 Data*. London. 2019.

[4] Abdelhamid TS, Everett JG. Identifying Root Causes of Construction Accidents. *JCEM*. American Society of Civil Engineers (ASCE); 2000 Jan;**126**(1):52–60. doi.org/10.1061/(asce)0733-9364(2000)126:1(52).

[5] Zhou J-L, Lei Y. Paths between latent and active errors: Analysis of 407 railway accidents/incidents' causes in China. *Saf Sci*. Elsevier BV; 2018 Dec;**110**:47–58. doi.org/10.1016/j.ssci.2017.12.027.

[6] Motarjemi Y, Wallace CA. Food Safety Assurance Systems: Root Cause Analysis of Incidents. *Food Res. Int*. Elsevier; 2014;331–9. doi.org/10.1016/b978-0-12-378612-8.00442-x.

[7] Attwood D, Khan F, Veitch B. Occupational accident models—Where have we been and where are we going? *J Loss Prev Process Ind*. Elsevier BV; 2006 Nov;**19**(6):664–82. doi.org/10.1016/j.jlp.2006.02.001.

[8] Attwood D, Khan F, Veitch B. Offshore oil and gas occupational accidents—What is important? *J Loss Prev Process Ind*. Elsevier BV; 2006 Sep;**19**(5):386–98. doi.org/10.1016/j.jlp.2005.10.006.

[9] Saaty RW. The analytic hierarchy process—what it is and how it is used. *Appl. Math. Model.* Elsevier BV; 1987;**9**(3-5):161–76. doi.org/10.1016/0270-0255(87)90473-8.

[10] Saaty TL, Vargas LG, Zoffer HJ. A structured scientific solution to the Israeli–Palestinian conflict: the analytic hierarchy process approach. *Decision (Wash D C).* Springer Science and Business Media LLC; 2015 Sep 18;**2**(1). doi.org/10.1186/s40165-015-0017-3.

[11] Buckley JJ. Fuzzy hierarchical analysis. *FUZZY SET SYST.* Elsevier BV; 1985 Dec;**17**(3):233–47. doi.org/10.1016/0165-0114(85)90090-9.

[12] Ho W, Ma X. The state-of-the-art integrations and applications of the analytic hierarchy process. *Oper Res.* Elsevier BV; 2018 Jun;**267**(2):399–414. doi.org/10.1016/j.ejor.2017.09.007.

[13] Chang D-Y. Applications of the extent analysis method on fuzzy AHP. *Oper Res.* Elsevier BV; 1996 Dec;**95**(3):649–55. doi.org/10.1016/0377-2217(95)00300-2.

[14] Yang M, Khan FI, Sadiq R. Prioritization of environmental issues in offshore oil and gas operations: A hybrid approach using fuzzy inference system and fuzzy analytic hierarchy process. *Process Saf Environ Prot.* Elsevier BV; 2011 Jan;**89**(1):22–34. doi.org/10.1016/j.psep.2010.08.006.

[15] Montazar A, Gheidari ON, Snyder RL. A fuzzy analytical hierarchy methodology for the performance assessment of irrigation projects. *Agric Water Manag.* Elsevier BV; 2013 Apr;**121**:113–23. doi.org/10.1016/j.agwat.2013.01.011.

[16] Sahin B, Yip TL. Shipping technology selection for dynamic capability based on improved Gaussian fuzzy AHP model. *Ocean Eng.* Elsevier BV; 2017 May;**136**:233–42. doi.org/10.1016/j.oceaneng.2017.03.032.

[17] Wicaksono FD, Arshad Y, Sihombing H, Baihaqi I. GAUSSIAN FUZZY ANALYTIC HIERARCHY PROCESS FOR THE EVALUATION OF BENEFIT, COST, AND RISK ANALYSIS IN THE INDONESIAN OIL AND GAS PROCESSING AREA. *Sci Eng.* Penerbit UTM Press; 2019 Sep 22;**81**(6). doi.org/10.11113/jt.v81.13623.

[18] Hefny HA, Elsayed HM, Aly HF. Fuzzy multi-criteria decision making model for different scenarios of electrical power generation in Egypt. *Egypt Inf.* Elsevier BV; 2013 Jul;**14**(2):125–33. doi.org/10.1016/j.eij.2013.04.001.

[19] Theophilus SC, Esenowo VN, Arewa AO, Ifelebuegu AO, Nnadi EO, Mbanaso FU. Human factors analysis and classification system for the oil and gas industry (HFACS-OGI). *Reliab. Eng. Syst. Saf.* Elsevier BV; 2017 Nov;**167**:168–76. doi.org/10.1016/j.ress.2017.05.036.

[20] Tang KHD, Md Dawal SZ, Olugu EU. A review of the offshore oil and gas safety indices. *Saf Sci.* Elsevier BV; 2018 Nov;**109**:344–52. doi.org/10.1016/j.ssci.2018.06.018.

[21] Skogdalen JE, Vinnem JE. Combining precursor incidents investigations and QRA in oil and gas industry. *Reliab. Eng. Syst. Saf.* Elsevier BV; 2012 May;**101**:48–58. doi.org/10.1016/j.ress.2011.12.009.

[22] Ferjencik M. An integrated approach to the analysis of incident causes. *Saf Sci.* Elsevier BV; 2011 Jul;**49**(6):886–905. doi.org/10.1016/j.ssci.2011.02.005.

[23] Bea RG. Human and Organizational Factors in Reliability and Design of "Minimum" Offshore Structures. *OMAE*, ASMEDC; 2002 Jan 1;**2**. doi.org/10.1115/omae2002-28198.

[24] Madigan R, Golightly D, Madders R. Application of Human Factors Analysis and Classification System (HFACS) to UK rail safety of the line incidents. *Accid Ana Prev.* Elsevier BV; 2016 Dec;**97**:122–31. doi.org/10.1016/j.aap.2016.08.023.

[25] Krejčí J, Pavlačka O, Talašová J. A fuzzy extension of Analytic Hierarchy Process based on the constrained fuzzy arithmetic. *Fuzzy Optim. Decis. Mak.* Springer Science and Business Media LLC; 2016 Mar 11;**16**(1):89–110. doi.org/10.1007/s10700-016-9241-0.

[26] Zheng G, Zhu N, Tian Z, Chen Y, Sun B. Application of a trapezoidal fuzzy AHP method for work safety evaluation and early warning rating of hot and humid environments. *Saf Sci.* Elsevier BV; 2012 Feb;**50**(2):228–39. doi.org/10.1016/j.ssci.2011.08.042.

IOP Conf. Series: Materials Science and Engineering 1072 (2021) 012013 doi:10.1088/1757-899X/1072/1/012013

PCA effect on the 3D face recognition system speed

A Kusnadi[1,2], Wella[1,3], R Winantyo[1,4] and I Z Pane[1,5]

[1]Department of Informatics, Universities Multimedia Nusantara, Tangerang, Indonesia

E-mail: [2]adhi.kusnadi@umn.ac.id, [3]wella@umn.ac.id, [4]rangga.winantyo@umn.ac.id, [5]izpane@gmail.com

Abstract. In this paper, a system of three-dimensional (3D) face recognition is not done through 3D face reconstruction method but directly uses the data retrieved from Kinect Xbox camera system. From a previous study, there exists a possibility to increase the speed and accuracy of the system. In order to accelerate the recognition speed, a single step in the said study is eliminated, which is the reconstruction of 3D face data. The algorithms used in this research are Backpropagation Neural Network and PCA. Testing is done in two ways. The first test uses a combination of Backpropagation and PCA, while the second test uses Backpropagation only. PCA has the function of simplifying the amount of data, thereby reducing the amount of computing that can increase speed system. The results show that the system that uses a combination of Backpropagation and PCA achieved an increase in speed up to 34.2 times, but a reduced accuracy of 8.5%.

1. Introduction

Recent technologies such as the internet, digital television, mobile phones, household appliances, and smart electronics have entered mainstream use[1]. This caused security concerns, which has become more difficult to maintain than it previously ever been. Authentication is therefore a crucial element of a computer security system. One of the available authentication methods is biometric authentication which has many advantages over traditional credential-based authentication mechanism. Among divers biometric authentication methods, face recognition is a one the most popular technique that is continuously improving over the past decade [2]. Each person face should have some differences, to make face recognition system work [3].

To date, most face recognition systems are based on the evaluation of 2D characteristic points or colour images, and still has weakness due to all kinds of adverse factors such as illumination variation, pose changes, makeup, or facial expressions [4] [5]. In view of the mentioned shortcomings, 3D face recognition is used to solve them. There are many studies of 3D face recognition [6] and at research [7] [8], the data of the face was taken using depth camera such as Tof PMD Cam-Cube 3.0 and Kinect Xbox. It was then processed by reconstructing the data into a 3-dimensional model, before being inputted into the recognition algorithm i.e. Backpropagation, Eigen face, LBP, LDA, and CMA. In research [9] [10], it did not focus on increasing speed. However, it is possible to increase the recognition speed and accuracy of a 3D face recognition. In a study [7] the recognition speed is 2189 Ms and the accuracy is 95.12%, and in a research [8], four algorithms has been tested with recognition speed variation of 260.5 Ms to 623 Ms, and accuracy rate variation of 87.5% to 92.5%.

The basic difference of this research compared to the previous researches is in the number of steps in data process. By eliminating the reconstruction step, the recognition speed of the system thus can be increased. Another notable difference is in the recognition algorithm. This research utilizes a combination of PCA and Backpropagation algorithms. The benefit of Principal Component Analysis (PCA) is to cut down the dimension of the data. No data redundancy is found as the components are orthogonal. With PCA, the complexity of grouping the images can be reduced. PCA is a dimensionality reduction technique which is used in order to simplify the face identification and recognition process, thus increasing the speed of the system. PCA is also known as Eigen space projection or Karhunen-

Loeve transformation [11]. The implementation of both algorithms resulted in a high level of accuracy of the system [12].

In the experiment, we conducted two scenarios. The first test used a combination of PCA and Backpropagation, while the second test excluded the PCA and used only the Backpropagation algorithm. It gets the best results for without using a PCA with a speed of 58.49 Ms and an accuracy of 90%. When compared with previous studies, the increase in recognition speed was up to 34.2 times but failed to improve accuracy as it decreased by 8.5%. This research increased the speed of 3D face recognition. This is very useful in increasing the reliability of applications that use systems like this.

2. Method

The flowchart in the Figure 1 describes the process of the system in this research:

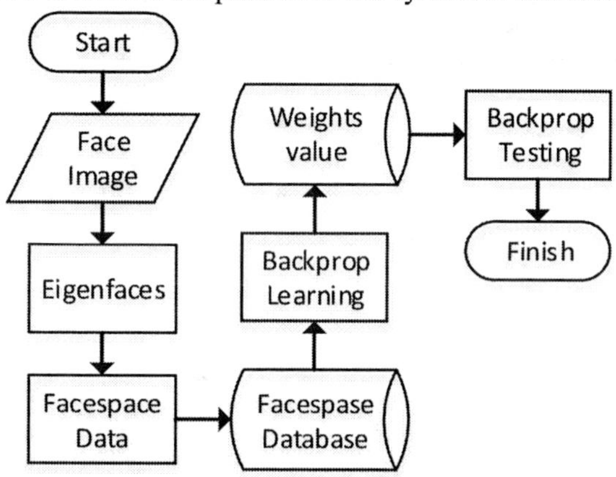

Figure 1. System flowchart.

Our procedure of implementation for the face recognition process is divided into three main steps of process: (1) Image acquisition, (2) Image processing, and (3) Recognition.

2.1. Image Acquisition

The background is plainly colorad and does not contain any other objects to reduce the noise level or disturbance to the depth data obtained using ToF cameras [13]. The face is in a neutral facial expression, not wearing glasses, and facing straight to the camera to improve face recognized accuracy.

(a) (b) (c)

Figure 2. Image samples acquired by application: (a) RGB image, (b) infrared image, and (c) depth map image.

Under these conditions, RGB and depth images can be captured using the application for Kinect. Figure 2. shows a sample of RGB and depth images captured using Xbox One Kinect system. For experimentation, 10 faces are used, and each face is captured 10 times. 80 faces are used as learning data, and 20 faces are used as training data.

2.2. Image Processing

We only used 3D images within our face recognition system. An image data is solely an image with depth information as shown in Figure 5. In other words, an image data is an array of numbers where the numbers quantify the distances between the focal plane of the sensor to the surface of objects within the field of view along the rays emanating from a regularly spaced grid. The image data/depth maps have a resolution of 512 x 424 pixels with each pixel containing the decimal number from eleven binary number. These numbers can vary from 0 to 2^{11} or 2048. The depth maps are normalized to 80 x 120 pixels, represented as 9600 vectors. Next, the mean image is calculated using equation (2) by summing the entire database ID vectors together and then dividing it with the number of faces in the database. Each face is then centred by subtracting the mean image from each image applying equation (1) [11].

$$x^{-i} = x^i - m \tag{1}$$

Where

$$m = \frac{1}{p}\sum_{i=1}^{p} x^i \tag{2}$$

Next, the data matrix is created by merging the centred database image side-by-side. The covariance matrix is then calculated by multiplying the data matrix with its transpose, using equation (3) .

$$\Omega = \bar{X}\bar{X}^T \tag{3}$$

This is continued by the calculation of the Eigenvalues and Eigenvectors for covariance matrix using equation (4).

$$\Omega \mathrm{V} = \lambda V \tag{4}$$

Where V is the Eigenvectors set and λ is corresponding Eigenvalues.

By calculating the eigenvalues and eigenvectors of the covariance matrix, we able to get that the eigenvectors with the largest eigenvalues related the dimensions that have the strongest correlation in the dataset, this is namely the principal component. PCA is a beneficial statistical technique that has been implemented in fields application such as face recognition, image compression and finding patterns in data of high dimension [14] [15].

2.3. Recognition

In the recognition step, we used Backpropagation Neural Network, is the most foundational classic establishing block in a neural network [16]. The algorithm can efficiently train a neural network, after each forward undergo through a network, backpropagation carry out a backward pass while adjusting the model's parameters (weights and biases). This algorithm has been effectively applied in a lot of fields with excellent results such as in these studies [17] [18] [19] [20].

The scheme has two major steps, which are training and testing. During the training step, Backpropagation algorithm will try to discover the right weight by modifying the weight between neurons with speed according to the learning rate and momentum rate predetermined to gain an appropriate outcome, i.e. a network that could recognize the expected data input. The next step is testing the algorithm to recognize the input data, where the weight is coming from the training phase. Both aforementioned stages needed the data from the camera. Below is the condition in designing an artificial neural network used in the system:

1. The neural network is design in three layers: the input layer, hidden layer, and output layer. The input layer is passive, which means that it did not modify the data (Figure 3).
2. Number of output node is 1 for each recognized person.
3. Tolerance number is 0.01.
4. Threshold number is 0.04.
5. Number of hidden layer is 1.
6. Nguyen and Widrow optimization [21] will use a layer between input layer and hidden layer.
7. Momentum factor.
8. The initial weights range between -0.5 to 0.5

There isn't any similar architecture for each different research object. Neural network can have any number of layers and any number of nodes per layer. Most application uses three-layer structure with a maximum of a few hundred input nodes. This explains why an experiment is required to discover an optimal design.

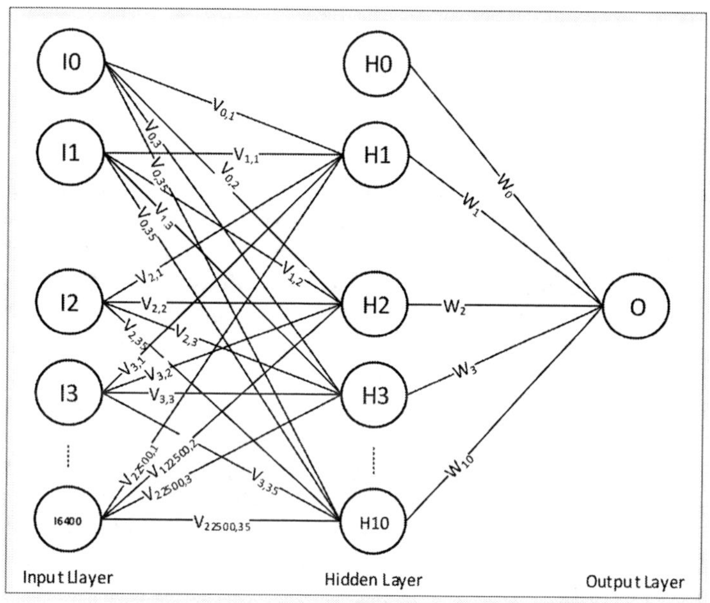

Figure 3. Neural network architecture design.

3. Implementation and Results

Table 1 shows the results of architecture experiments on various combinations of hidden node number and learning rate value. The training is continuously repeated until the smallest MSE values are obtained and stopped when the MSE values started to increase. The termination of training equals with the maximum number of epoch. Two tests have been done. In the first test, the system used a combination of PCA and Backpropagation. In the second test, the system used Backpropagation without using PCA. In Table 1, if a number passes the maximum value that the system can calculate (in this case, the number is too small), it will appear as NaN [22]. It appears in systems that has hidden nodes below 50 and is using PCA. This meant that the measurement of speed and accuracy is impossible in such architecture. Measurement of the accuracy level is based on the tolerance number. If the difference between the output and the target is outside of the tolerated number, no accuracy number is displayed, which means that the architecture cannot recognize the image data. These combinations of architecture are therefore ignored.

Table 1. Architectures of the backpropagation tests.

Hidden Node	Learning Rate	MSE		Epoch		Training Time (second)		Accuracy		Recognition Speed (millisecond)	
		Without PCA	Using PCA	Without PCA	Using PCA	Without PCA	Using PCA	Without PCA (%)	Using PCA (%)	Without PCA	Using PCA
1	0.05	1.895	NAN	10144	NAN	2812	NAN	-	-	-	-
1	0.01	1.901	NAN	10716	NAN	2937	NAN	-	-	-	-
1	0.005	0.009	NAN	28929	NAN	7168	NAN	85	-	42.1446	-
5	0.05	1.894	NAN	30430	NAN	13635	NAN		-		-
5	0.01	0.009	NAN	4981	NAN	2529	NAN	90	-	58.49674	-
5	0.005	0.009	NAN	19861	NAN	25117	NAN	90	-	73.89407	-
10	0.05	1.894	NAN	19765	NAN	13695	NAN	-	-		-
10	0.01	0.009	NAN	3568	NAN	1902	NAN	90	-	51.77404	-
10	0.005	0.009	NAN	14227	NAN	22688	NAN	90	-	52.35587	-
20	0.05	1.894	NAN	5387	NAN	6640	NAN	-	-	-	-
20	0.01	1.896	NAN	4198	NAN	2222	NAN	-	-	-	-
20	0.005	1.896	NAN	6408	NAN	3457	NAN	-	-	-	-
50	0.05	1.894	NAN	10730	NAN	6529	NAN	-	-	-	-
50	0.01	0.0126	NAN	3372	NAN	14520	NAN	-	-	-	-
50	0.005	0.009	2.999	5060	2771	4350	131.007	90	75	81.47682	56.7258
100	0.05	4.999	NAN	235	NAN	3778	NAN	-	-	-	-
100	0.01	0.009	2.999	1836	917	11268	68.6040	85	80	475.89723	65.3427
100	0.005	0.009	2.999	7321	837	17153	62.1808	85	80	451.87912	61.8076

In Table 1, if a number passes the maximum value that the system can calculate (in this case, the number is too small), it will appear as NaN [22]. It appears in systems that has hidden nodes below 50 and is using PCA. This meant that the measurement of speed and accuracy is impossible in such architecture. Measurement of the accuracy level is based on the tolerance number. If the difference between the output and the target is outside of the tolerated number, no accuracy number is displayed, which means that the architecture cannot recognize the image data. These combinations of architecture are therefore ignored.

In this study, if we compare the average time of training speeds between the systems that used PCA and the systems that excluded PCA, an increase in training speed by 125.18 times is achieved by the systems that used PCA, while also increased the recognition time by 5.49 times. This is caused by the smaller amount of data [23]. For the recognition speed compared to the previous study [4], the system in this study is faster by 34.21 times and 7 times respectively for the systems that used PCA. This is due to a smaller number of processes the system had to go through. Maximum speed increases if the system used a combination of PCA and Backpropagation. This is very useful if the system is applied to a device with low-end specifications.

However, while the recognition speed increases, the accuracy rate attained is the opposite for the systems that used a combination of PCA and Backpropagation. Figure 4 illustrates the different rate of accuracy between the previous research and the system in this study.

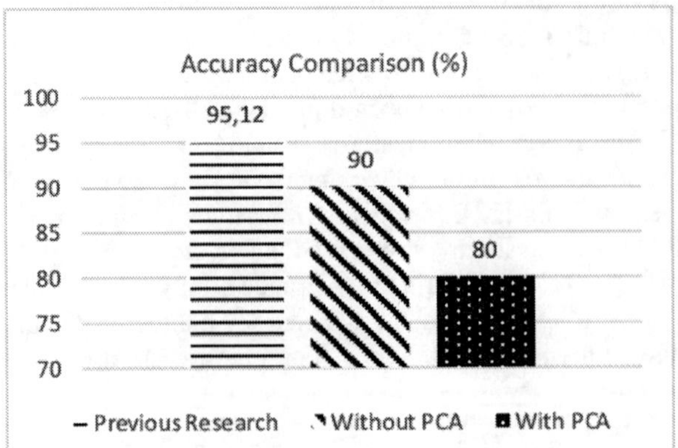

Figure 4. Comparison of accuracy results.

The accuracy rate of is lower compared to the previous research [5]. This is caused by the elimination of 3D model reconstruction process, which can be described as the data normalization process. Accuracy decreases further by 8.5% after the inclusion of PCA, which can be attributed to the amount of data depth reduced by PCA process. This is probably because a way to distinguish a human face from another is through texture/number of the depth of data. However, this theory requires further research.

4. Conclusion

This paper puts forward a new 3D face recognition technique to improve its recognition speed. Backpropagation and PCA can be implemented together for a 3-dimensional face recognition without going through model reconstruction process. 3D face image is taken by Kinect Xbox One time-of-flight depth camera. In order to accelerate the speed of recognition, one data-processing step in the previous research is eliminated. The eliminated process is the reconstruction of 3D face data from Kinect. Instead, the data is directly used as the data input into PCA before being processed using Backpropagation algorithm. It resulted in an increase of recognition speed. Two tests are done. The first test used a combination of PCA and Backpropagation, while the second test excludes the PCA and used only the Backpropagation algorithm. The result from the tests shows that the system with PCA yielded an increase in recognition speed up to 34.2 times but failed to improve accuracy as it decreased by 8.5%.

5. Acknowledgement

The authors gratefully acknowledge the financial support from by "DIKTI" in the Scheme of "PDUPT 2020".

References

[1] Apăvăloaie E-I. The Impact of the Internet on the Business Environment. *Procedia Economics and Finance*. Elsevier BV; 2014;**15**:951–8. doi.org/10.1016/s2212-5671(14)00654-6.

[2] Chen S, Pande A, Mohapatra P. Sensor-assisted facial recognition. *MobiSys*. ACM Press; 2014. doi.org/10.1145/2594368.2594373.

[3] Erwin, Saparudin, Fachrurrozi M, Tamaarsa A, Zhang A. Facial Recognition Technique with Combination Method of Local Binary Pattern Histogram and Image Euclidean Distance. *JPCS*. IOP Publishing; 2019 Mar;**1196**:012009. doi.org/10.1088/1742-6596/1196/1/012009.

[4] Zhang L, Ding Z, Li H, Shen Y, Lu J. 3D Face Recognition Based on Multiple Keypoint Descriptors and Sparse Representation. Zhang Y, editor. PLoS ONE. *PLoS*. 2014 Jun 18;**9**(6):e100120. doi.org/10.1371/journal.pone.0100120.

[5] Zhou S, Xiao S. 3D face recognition: a survey. *hum-cent comput info*. Springer Science and Business Media LLC; 2018 Nov 24;**8**(1). doi.org/10.1186/s13673-018-0157-2.

[6] Alexander L, Kusnadi A, Wella W, Winantyo R, Pane IZ. Authentication System Using 3D Face With Algorithm DLT and Neural Network. *SCIS and ISIS*. IEEE; 2018 Dec. doi.org/10.1109/scis-isis.2018.00039.

[7] Zhang D, Lu G, *3D biometrics: Systems and applications*. New York: Springer New York Heidelberg Dordrecht London, 2013. doi: 10.1007/978-1-4614-7400-5_2.

[8] Cheng Z, Shi T, Cui W, Dong Y, Fang X. 3D face recognition based on kinect depth data. *ICSAI*. IEEE; 2017 Nov; doi.org/10.1109/icsai.2017.8248353.

[9] Nagat A-A, Caleanu C-D. Face identification using kinect technology. 2013 *IEEE* 8th *SACI*. IEEE; 2013 May. doi.org/10.1109/saci.2013.6608961.

[10] Jonathan, Kusnadi A, Julio D. Security system with 3 dimensional face recognition using PCA method and neural networks algorithm. *CONMEDIA*. IEEE; 2017 Nov. doi.org/10.1109/conmedia.2017.8266048.

[11] Turk MA, Pentland AP. Face recognition using eigenfaces. Proceedings 1991 *IEEE Comput. Sco.CVPR*. IEEE Comput. Sco. Press. doi.org/10.1109/cvpr.1991.139758.

[12] Zhao X, Wang K, Yang K, Hu W. Unconstrained face detection and recognition based on RGB-D camera for the visually impaired. Wang Y, Pham TD, Vozenilek V, Zhang D, Xie Y, editors. *ICGIP* 2016. SPIE; 2017 Feb 8. doi.org/10.1117/12.2266122.

[13] Fuchs S, Hirzinger G. Extrinsic and depth calibration of ToF-cameras. 2008 *IEEE CVPR*. IEEE; 2008 Jun. doi.org/10.1109/cvpr.2008.4587828.

[14] Alahmadi A, Hussain M, Aboalsamh HA, Zuair M. PCAPooL: unsupervised feature learning for face recognition using PCA, LBP, and pyramid pooling. *Pattern Anal Appl*. Springer Science and Business Media LLC; 2019 May 10;**23**(2):673–82. doi.org/10.1007/s10044-019-00818-y.

[15] Zhang Y, Xiao X, Yang L-X, Xiang Y, Zhong S. Secure and Efficient Outsourcing of PCA-Based Face Recognition. *IEEE*. Institute of Electrical and Electronics Engineers (IEEE); 2020;**15**:1683–95. doi.org/10.1109/tifs.2019.2947872.

[16] Brunel A, Mazza D, Pagani M. Backpropagation in the simply typed lambda-calculus with linear negation. *PACMPL*. Association for Computing Machinery (ACM); 2020 Jan;**4**(POPL):1–27. doi.org/10.1145/3371132.

[17] Whittington JCR, Bogacz R. Theories of Error Back-Propagation in the Brain. *Trends Cogn. Sci.* Elsevier BV; 2019 Mar;**23**(3):235–50. doi.org/10.1016/j.tics.2018.12.005.

[18] Ramesh VP, Baskaran P, Krishnamoorthy A, Damodaran D, Sadasivam P. Back propagation neural network based big data analytics for a stock market challenge. *Commun. Stat. Theory Methods*. Informa UK Limited; 2018 Nov 17;**48**(14):3622–42. doi.org/10.1080/03610926.2018.1478103.

[19] Lillicrap TP, Santoro A, Marris L, Akerman CJ, Hinton G. Backpropagation and the brain. *Nat. Rev. Neurosci*. Springer Science and Business Media LLC; 2020 Apr 17;**21**(6):335–46. doi.org/10.1038/s41583-020-0277-3.

[20] PAOLA JD, SCHOWENGERDT RA. A review and analysis of backpropagation neural networks for classification of remotely-sensed multi-spectral imagery. *Int J Remote Sens*. Informa UK Limited; 1995 Nov 10;**16**(16):3033–58. doi.org/10.1080/01431169508954607.

[21] Nguyen D, Widrow B. Improving the learning speed of 2-layer neural networks by choosing initial values of the adaptive weights. *IJCNN*. IEEE; 1990. doi.org/10.1109/ijcnn.1990.137819.

[22] Works M. *NaN (MATLAB Functions) 205AD*. 2020 Retrieved from http://matlab.izmiran.ru/help/techdoc/ref/nan.html.

[23] Zhang G, Tang B, Chen Z. Operational modal parameter identification based on PCA-CWT. *Meas*. Elsevier BV; 2019 Jun;**139**:334–45. doi.org/10.1016/j.measurement.2019.02.078.

IOP Conf. Series: Materials Science and Engineering 1072 (2021) 012014 doi:10.1088/1757-899X/1072/1/012014

Analysis of the eco-efficiency level in the dining table production process using life cycle assessment method to increase industry sustainability

R Purwaningsih[1,2], Susanto N[1,3], D A Adiaksa[1,4] and A A A Putri[1,5]

[1]Industrial Engineering Department, Faculty of Engineering, Diponegoro University

E-mail: [2]ratna.tiundip@gmail.com, [3]nophie.susanto@gmail.com, [4]damarazis30@gmail.com, [5]aisyaharifna@student.undip.ac.id

Abstract. This study aims to measure the environmental costs (eco-cost) and the level of eco-efficiency of the production of the dining table in the Jepara furniture industry and formulates recommendations to improve eco-efficiency. The measurement of eco-efficiency was performed using life cycle assessment (LCA) with the eco-cost method. The furniture manufacturer produces a waste of 30 % and the product reject 15 %. The result of the data processing shows that the eco-costs of IDR 134,000 per unit product and net value IDR 208,000. The eco-efficiency index or EEI is 1.55 which means that the products categorized as affordable and sustainable. The eco-efficiency level is 35% means that the environmental performance is low and the furniture production needs improvements in the production process. Strategy to improve the eco-efficiency of production is increasing the utilization of wood waste and increasing the capability of technicians and operators to reduce the product defect.

1. Introduction
The industry's development has brought many changes to environmental damage. The environmental impacts occur throughout the life cycle of a product, starting from the extraction of natural resources as raw materials, the production process, the use and reuse of products and finally to the disposal stage [1]. To create a sustainable industry and product, the use of renewable natural resources must be combined with a large reduction in environmental impacts such as greenhouse gases. Sustainable Development tries to give a new discourse about the importance of preserving the natural environment for the future. Jepara is a furniture industrial cluster located at Jepara, Central Java province. The furniture industries in this area already used the wood legality verification system certification. The regulation requires companies to use only the legal wood in their production process.

Dining table production per month reaches 246 units with a finished product volume of 15.8 m^3. LCA considers materials taken from nature as production inputs and materials released into nature as output from the production process. The use of wood as an input material according to LCA contributes to climate change because trees are producing oxygen and absorbing carbon dioxide. The industries also produce waste in the form of small pieces of wood, sawdust and wood chips. The production process used electrical machinery, LCA takes into account coal burning that occurs to produce electricity as CO2 production and has an impact on climate change. The use of chemical materials for the finishing process contribute to human toxicity in certain amount define by LCA characterization. The interview with the industry management said that they have not planned how to handle the waste to reduce the impact of it to the environment. Therefore, measurement of eco-efficiency of the production process is needed to determine the level of the environmental impact of the production proses.

Eco-efficiency is a strategy that combines the concepts of economic and ecological efficiency in the use of natural resources. Eco-efficiency is defined as a production process that minimizes the use of production resources such as raw materials and energy to reduce the environmental impact per unit of product. The eco-efficiency indicators for the industry are based on the World Business Council for

Sustainable Development (WBCSD) in the form of economic indicators and environmental indicators. [2]. WBCSD defines eco-efficiency as "increasing the production of goods and services with lower resource use, waste, and pollution" [3]. WBCSD describes the eco-efficiency as a concept of reducing resource consumption, reduce environmental impact, and increase the value of products and services. Adoption of the principles of eco-efficiency in production patterns in public policymaking will increase the technological capabilities [4]. Environmental Accounting since 1990 has been focusing on publishing The Greening of Accountancy by the Chartered Association of Certified Accountants (ACCA). Based on US Environmental Protection Agency (US EPA), Green Environmental Accounting has defined a function that environmental costs should be considered by the responsible party to reduce or avoid costs and improve environmental quality [5]. According to ESCAPE, eco-efficiency is a comparison between the economic value of the product indicated by the net value and the environmental cost to produce the product [6].

Nowadays, the eco-efficiency measurement is using the life cycle assessment (LCA) method which has indicators to measure the environmental impact category of a product. LCA is a compilation and evaluation of inputs, outputs and the potential environmental impact of a product or service based on the life cycle [7]. This research aims to measure the environmental impact of the production process as the eco cost of the product resulted, identify the most significant environmental impact category, measure the environmental efficiency performance of the production process and formulate recommendations to improve the efficiency performance of the production process. Method of Life Cycle Assessment is conducted to evaluate the environmental impact and provide an alternative improvement to reduce environmental impact [8].

2. Research Methodology

2.1. Life Cycle Assessment Method

Today, LCA began to be applied in government policy, marketing, strategic planning, process improvement, and product design and also used as the basis for eco-labeling and consumer education program [9]. The eco cost consists of four groups, eco cost of resource scarcity, eco cost of carbon footprint, eco cost of ecosystem and eco cost of human health. Each group consists of some impact categories [10]. The group of eco cost and their impact category describe in Figure 1.

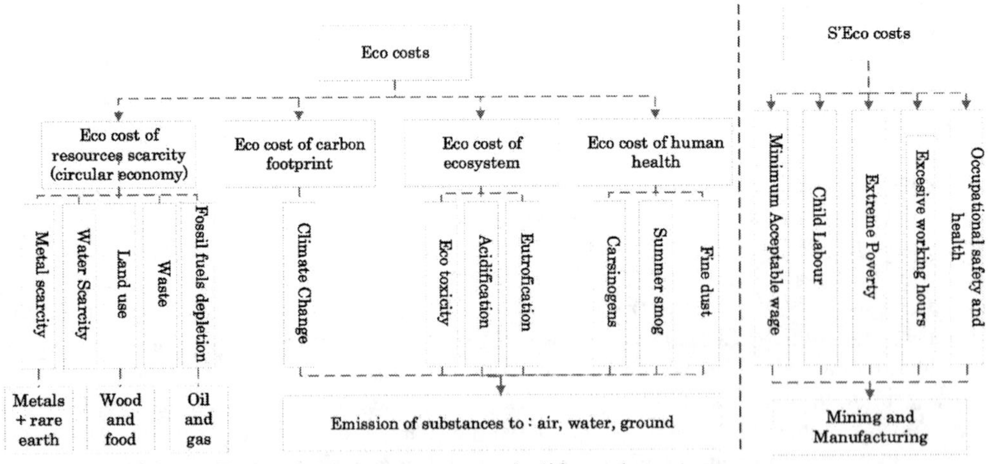

Figure 1. Eco cost category in life cycle assessment.

The LCA method has three phases. The first phase is to define the purpose and scope to determine the direction of the research. LCA goals and scope aim to formulate and describe the objectives, the boundaries, and the assumptions related to impacts throughout the life cycle of the system being evaluated. The scope of the assessment of the dining table starts from the resources entry the manufacture, the production process and ended in the finishing product package. We can say the scope is a gate to gate type. The second phase is the analysis of inventory (life cycle inventory assessment /LCIA) which identifies and collects data of the input and output of the production system and did some data processing through Classification, Characterization, Normalization, Weighting and define a Single Score. LCIA analyzes the type and magnitude of the value of each impact category. The third phase is the environmental impact statement which is intended to get the output of data processing in the form of a potential category to the damage of the environment. The output shows as Output of Impact Category Diagram. The diagram shows the environmental category impact of the production process [11].

2.2. *Life Cycle Assessment Indicators*

LCA has some indicators of the economic aspect and environmental aspect. First, we did the cost-benefit analysis to obtain the net value. The net value is the selling price reduced by the cost of production. Cost-benefit analysis can also determine whether the product is eligible to be sold or not.

$$\text{Net Value} = \text{Selling price} - \text{Production Cost} \quad (1)$$

Then, the net value used to obtain the *Eco-efficiency index* (EEI) which determines the feasibility of the product in terms of ecological efficiency (sustainability) as well as economic efficiency (profitable). Eco-efficiency index value obtained by dividing the net value to the eco-cost [4].

Eco-Efficiency Index (EEI) is used to categorize the product whether the product is affordable and sustainable or not. Products categorized as affordable and sustainable if the value of EEI is more than 1, while affordable but not sustainable if the value of EEI is between 0 – 1. The product categorizes as not affordable and not sustainable if the value of EEI is less than 1 [12],[13].

The EEI value used to calculate the EVR (*Eco-Cost Value Ratio).* EVR is an indicator that compares the eco cost to produce a product that represents the ecologic value to the net value of a product that represents the economic value.

$$\text{EVR} = \text{Eco cost/ Net value} \quad (2)$$

The final calculation is the *Eco-Efficiency Ratio Rate* as the measurement of the eco-efficiency of production [14]. The formulation to calculate this indicator are state below. The EER defines how efficient the environmental performance of the production process in the form of a percentage.

$$\text{EER} = (1 - \text{EVR}) \times 100\% \quad (3)$$

2.3. *The Input of the Life Cycle Assessment*

Inventory flow includes the input of raw materials, auxiliary materials, natural resources, energy use and everything released to land, water, and air. To develop LCI, a system flow model is built using input and output data. We must input to SimaPro all of the materials used in the production process and the materials outputs resulted in process. The raw material for the production of Square Dining Table is Teak wood. The wood must go through a kiln-drying process to reduce the moisture content before processing. The process also used bolts to connect components one another, and other materials for assembly and packaging process. The product and the specification (size) is described in Figure 2. The input and output of the production process are shown in Table 1.

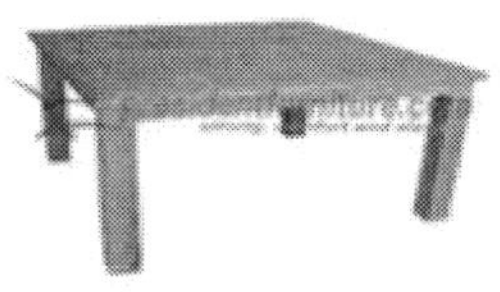

Code	PF14010
Description	Square Dining Table
Length	150 cm
Width	150 cm
Height	75 cm
Volume	0.29 m^3
Weight	30 kg

Figure 2. Square dining table.

Input and Output of the production process are needed to build life cycle inventory. The input and output of the production process are shown in Table 1.

Table 1. Input-output of the production process.

Input	Quantity	Unit
Teak wood	24.2	m^3
Energy	4910	kwh
Output	**Quantity**	**unit**
Dining Table	15.8	m^3
Wood scrap &	8.4	m^3

3. Results and Discussion

The second phase of LCA is LCIA calculations which consist of (1) characterization, (2) normalization, (3) weighting and (4) define a single score of the environmental impact. LCA calculation use the 2012 eco-costs method. The result of the characterization phase and normalization phase are shown in Table 2 as an output of data processing using SimaPro software after input the value of material volume in table 1. To change the environmental impact in characterization column to environmental cost in euro, the characterization value is multiplied by the normalization factors value. Normalization value is the price of environmental impact in euro for one unit of material that taken from and release to the environment. The calculation was done by characterization process on SimaPro. Continue by the weighting process, here the process is ignored by using the same weight for all impact factors because all factors assumed to have the same influence to environment and to human life. The result of characterization, normalization and weighting is shown in Table 2.

Table 2 shows that the highest value of the environmental impact category is climate change which reaches 45.9 kg CO2 resulted from energy consumption and teak wood consumption. Life Cycle Assessment can be used to evaluate the environmental impact of the use of raw materials such as coal and biomass fuels which the removal of these can lead to a problem of how big the emissions and other impacts resulted from the use of biomass fuels [15].

We also see that the highest normalization factors are Fine Dust which reaches € 34 / kg. Fortunately, the fine dust output only has impact category value 0.0012 and the highest eco cost here is still climate change. There are 21 machines used in the production process and the total time of machinery processing is 2073 seconds or 34.5 minutes per unit and consumes 4.910 kwh electrical energy. The electrical energy consumed in the production process is calculated by SimaPro as a component that releases the CO2 emission from the coal burn in generating power. While using 24.2 m3 of teak wood which exploited from the forest also gives an impact on climate change. The total of eco cost for production of one unit of dining table is 9,38 euro or IDR used the exchange rate IDR 14,285 per euro rounded off to 134,000.

The net value defines by selling price reduced by the production cost. The selling price of a product is IDR 700,000 and the production cost is IDR 492,000. The production cost consists of material cost, energy cost, labour cost, and overhead cost. The production cost and other value to obtain net value are given in Table 3.

Table 2. Recapitulation of SimaPro output.

Impact Category	Unit	Characterization	Normalization Factors	Normalization Result (Euro)
Climate Change	kg CO_2 eq	45.9	€ 0.135 / kg CO_2 eq	6.2
Acidification	kg SO_2 eq	0.408	€ 8.25 / kg SO_2 eq	3.36
Eutrophication	kg PO_4-- eq	0.0295	€ 3.9 / kg PO_4 eq	0.115
Photochemical Oxidant Formation	kg C_2H_4 eq	0.00581	€ 9.7 / kg C_2H_4 eq	0.0564
Fine Dust	kg Pm 2.5 eq	0.0012	€ 34 / kg Pm 2.5 eq	0.0408
Human Toxicity	CTUh	1.22 x 10^{-8}	€ 1.030.000 / CTUh	0.0125
Ecotoxicity (Freshwater)	CTUe	12.6	€ 0,0014 / CTUe	0.018
Metals Depletion	euro	0	€ 1	0
Oil&Gas Depletion Excl Energy	kg oil equ	0	N/A	0
Waste	MJ	-37.8	€ 0.012 / MJ	-0.45
Land-Use	euro	0	€ 1	0
Water Stress Indicator	WSI factor	0.0125	€ 0.76 / WSI factor	0.0949
Eco cost				9.38

Table 3. The LCIA indicators for 1 batch or 250 products.

No	Cost	Component	Total (IDR)
1	Material	Teak wood, bolts, and assembly components, packaging	48,400,000
2	Direct labor	Carpenter and machine operators	57,600,000
3	Overhead	supporting materials, electricity and administration	17,000,000
4	Total Production cost	1 + 2+ 3	123,000,000
5	Production cost/unit	123.000.000/ 250	492,000
6	Selling Price	50 USD, exchange rate IDR 14.000.	700,000
7	Net value	5 - 4	208,000
8	Eco cost from table 1	9,38 Euro x IDR 18,500 per unit product	134,000
	EEI	Net value / Eco Cost	1.55
	EVR	Eco cost/ Net value	0.644
	EER	(1 – EVR) 100 %	35 %

We can define the net value as IDR 217,958. The net value used to obtain the eco-efficiency index (EEI) of a dining table from the result of the net value divide by the eco cost. The EEI is 1.55. The value of EEI is greater than 1 and said that the product is affordable and sustainable. This condition shows that dining table products are environmentally friendly and financially quite affordable. Dining table products are said to be sustainable means that products do not harm the environment, and said affordable because the value of eco-costs is smaller than the net value. The result of EEI, EVR and EER value are shown in Table 3.

The EVR value is obtained by dividing eco-costs with net value as the economic value of each product. From these calculations, it can be seen that the EVR of the dining table products is 0.64. The EVR value defines from the net value, the greater net value will result in the smaller EVR value. Small EVR value means better or more feasible for the product to be produced. EVR value is used as input to determine the level of eco-efficiency of the product. The value of EER is 35 % are categorized as less efficient.

3.1 Recommendations for Improvement

There are several problems faced by the wood furniture industry in Jepara with the environment, which are illegal logging and solid waste deposits [16]. Based on the data processing and analysis that has been done, the recommendations were formulated to improve the efficiency of the production process. The environmental impacts category which gives the highest contribution to eco cost is Climate Change and Acidification. Reducing the impact on climate change can be done by the reuse of wood waste. The industry can use wood blocks in the production process of products that require smaller sized parts. Also, wood waste can be used as valuable goods such as shoe racks. The wood scraps produce from the production process usually burned and impacts on air pollution. Workers health complaint are on eyes pain and respiratory organ which caused by air pollution from the smoke of burning wood scraps [17].

The second recommendation is replacing the material used in the finishing process to reduce health impacts on labor. Replace the use of water-based paint by wood paint will be much safer for the worker and reduce the impact on the environment. The substitution of dyes is often an alternative in efforts to reduce the impact on the environment and the health of workers. If possible, it would be better to use bio-degradable raw materials in which the disposal of raw material waste can be directly broken down by nature without causing negative effects on the environment [18].

4. Conclusion

Measurement of environmental impact categories with the eco-costs method finds out 9 categories of environmental impacts arising from the production of dining tables. The highest impact category is climate change with an eco-cost value of IDR 88,114 and the second is acidification with an eco-cost value of IDR 4,752. The eco cost to produce 1 unit product is IDR 134,000, while the product net value is IDR 208,000. Then, the EEI calculation result is 1.55 means that the dining table is categorized as an affordable and sustainable product. Affordable means that the products have an economic value where the selling price is greater than the cost of producing. Sustainable means that products do not harm the environment. The eco-efficiency ratio rate EER of the production process is 35 % which means that the furniture production process is still categorized as less efficient and needs improvement. Recommendations given to the furniture industry to improve the environmental performance of production are to re-use wood waste to reduce the climate change cost and replacing paint of finishing materials with safer materials to reduce the acidification cost.

References

[1] Zhu Q, Sarkis J, Lai K. Green supply chain management implications for "closing the loop."*Transp Res Part E: Log. Transp. Rev*. Elsevier BV; 2008 Jan;44(1):1–18. doi.org/10.1016/j.tre.2006.06.003.

[2] Park H-S, Behera SK. Methodological aspects of applying eco-efficiency indicators to industrial symbiosis networks. *J. Clean. Prod.* Elsevier BV; 2014 Feb; 64:478–85. doi.org/10.1016/j.jclepro.2013.08.032.

[3] Lehni, M. Eco-Efficiency creating more value with less impact. Switzerland: *World Bussiness Council for Sustainable Development* (WBCSD). 2000.

[4] Ichimura, M., Nam, S., Bonjour, S., Rankine, H., Carisma, B., Qiu, Y., & Khrueachotikul, R. Measuring Resource-use Efficiency and the Impact of Economic Activities on the. Bangkok: United Nations Economic and Social Commission for Asia and the Pacific. 2009.

[5] Basuki B. Eco-Efficiency and Sustainable Development as Efforts to Produce Environmentally Friendly Product: An Exploratory Case Study. *ISEA*. Indonesian Center for Social and Environmental Accounting Research and Development (ICSEARD); 2015 Sep 30;9(3):199. doi.org/10.22164/isea.v9i3.105.

[6] UNO, ESCAP. Eco-efficiency Indicators: Measuring Resource-use E_ciency and the Impact of Economic Activities on the Environment. *United Nations Economic and Social Commission for Asia and the Pacific*. 2009.
[7] Lehtinen, H., Saarentaus, A., Rouhiainen, J., Pitts, M., & Azapagic, A. A Review of LCA Methods and Tools and their Suitability for SMEs. Manchester: *Europe Innova Biochem*. 2011.
[8] Pringgajaya, K. A., & Ciptomulyono, U. Implementasi Life Cycle Assessment (LCA) dan Pendekatan Analytical Network Process (ANP) untuk pengembangan Produk Hetric Lamp yang Ramah Lingkungan. *Jurnal Teknik ITS*. 2012.
[9] LEWIS H, DEMMERS M. Life Cycle Assessment and Environmental Management. Australasian *J. Environ. Manage*. Informa UK Limited; 1996 Jan;3(2):110–23. doi.org/10.1080/14486563.1996.10648348.
[10] TuDelf. The Model of the Eco-costs / Value Ratio (EVR). 2020. https://www.ecocostsvalue.com.
[11] Purwaningsih. R., Simanjuntak C, F., Rosyada Z. F. Analysis of Eco-Efficiency ratio of Pencil Production Using Life Cycle Assessment (LCA) to Increase the Manufacture Sustainability. *JATI*. 2020. Vol 22 No 1.
[12] Sari DP, Hartini S, Rinawati DI, Wicaksono TS. Pengukuran Tingkat Eko-efisiensi Menggunakan Life Cycle Assessment untuk Menciptakan Sustainable Production di Usaha Kecil Menengah Batik. *J@ti Undip*. Petra Christian University; 2011 Jun 11;14(2). doi.org/10.9744/jti.14.2.137-144.
[13] Hendriks CF, Vogtländer JG, Janssen GMT. The eco-costs/value ratio. *Int J Ecodyn*. WITPRESS LTD.; 2006 Aug 29;1(2):136–48. doi.org/10.2495/eco-v1-n2-136-148.
[14] Vogtlander J. LCA-Based Assessment of Sustainability: The Eco-Costs/Value Ratio (EVR). *Netherland: Delft University of Technology)*. 2010.
[15] Harjanti, T. R., Fahrurrozi, M., & Bendiyasa, I. M. Life Cycle Assessment Pabrik Semen PT Holcim Indonesia Tbk. Pabrik Cilacap: Komparasi antara Bahan Bakar Batubara dengan Biomassa. *Jurnal Rekayasa Proses*. 2012.
[16] Likumahwa FM, Purwaningsih R, Handayani NU. The Influence of Green Supply Chain management on Company's Performance and Competitiveness in Wood Furniture Industry: An Overview of Conceptual Model. IOP Conference Series: *Mater. Sci. Eng*. IOP Publishing; 2019 Sep 6; 598:012076. doi.org/10.1088/1757-899x/598/1/012076.
[17] Purwaningsih R, Cahyantari AE, Ariyani Z, Susanty A, Arvianto A, Santoso H. Product-service system method to measure sustainability level of traditional smoked fish processing industries. Ma'mun S, Tamura H, Purnomo MRA, editors. *MATEC*. EDP Sciences; 2018; 154:01062. doi.org/10.1051/matecconf/201815401062.
[18] Purwaningsih R, Susanto N, Yudha MC. Penilaian Keberlanjutan UKM Batik Kota Semarang dengan Metode Product Service System (PSS). *J@ti Undip*. Petra Christian University; 2016 Sep 29;18(1). doi.org/10.9744/jti.18.1.31-42.

Process improvement using value stream mapping and lean methodology: a case study application in batch chemical process industry

S R Lie[1,2] and R D Kusumastuti[1,3]

[1]Department of Management, Faculty of Economics and Business, Universitas Indonesia, Depok, 16424, Indonesia

E-mail: [2]sudono.riady@ui.ac.id, [3]ratih.dyah@ui.ac.id

Abstract. Intensifying competition in the global chemicals market is increasing the pressure on companies to adopt lean manufacturing and improve their operational efficiency to remain competitive. Companies operating in emerging economies are no exception. This study looks at the application of value stream mapping (VSM) in the production process of a batch chemical manufacturing company based in Indonesia. The application of value stream method is a multistep approach. A value stream scope is determined in the beginning, followed by a data collection step to construct a current state map. The current state is analysed to gain a deep understanding of how things currently operate and provides the foundation for the ideal future state of the production process. The production process is then redesigned following lean manufacturing principles, and the future state map is drawn. Analysis is then conducted to estimate the potential cost reduction from implementing the proposed future state. The results indicate a more efficient production system as shown by decreased production lead time and reduced production costs.

1. Introduction

The challenge to feed the world is an increasingly difficult task as the world population continues to grow, especially in Africa and Asia. In 2050, it is estimated that the world population will have grown by roughly 30%. This means from now to then, nearly 83 million people will be added every year [1]. On the other hand, it is estimated that 12 million hectares of agricultural land are lost annually due to urbanization among other causes, and by 2050, there would remain only 0.16 hectares of farmland per capita compared to 0.52 hectares per capita in the 1950s.

This situation highlights the need for a more productive farming technique that utilizes technology to increase yield per hectare of available farmland. One of the many tools available to farmers is the use of herbicide to combat the growth of nutrient sapping weeds. In Indonesia, the agricultural sector is an important source of employment as it represents approximately 7.78% of the gross domestic product in 2019. An efficient and reliable herbicide product supply is therefore impactful towards the nation's economic stability and vital in ensuring food security in the country.

Increasing globalization has led to intensifying price competition in the herbicide industry. This situation necessitates industry players to increase their operational efficiency and push towards the productivity frontier in order to thrive in the industry. This case study looks at an herbicide manufacturing company that is facing challenges from fluctuating sales volume. It sees demand surges at certain times of the year, overwhelming production lines and leading to overtime costs increase.

Ace Chemicals (the real name of the company is not disclosed for confidentiality reasons) is a wholly-owned subsidiary of a global chemical conglomerate that is amongst the world's largest agrochemical producers. Ace Chemicals produces herbicide products for both local and export markets. The company has a large footprint in Indonesia with a significant market share in both local and export markets. Local market sales account for roughly 75% of the company's total sales, and sales to various overseas markets including Australia, New Zealand, Korea, and Japan make up the remaining sales volume. Among the global chemical organization's core strengths are customer centricity and process excellence, and one of its most far-reaching initiatives to reflect those core strengths is the launch of a production system with the aim of standardizing its supply chain operations worldwide and increasing its operational efficiency. The initiative was launched in mid-2019, and Ace Chemicals became the organization's first site in Asia to implement the new production system. The organization's production system follows the pattern set by other manufacturing companies and can be traced back to the Toyota Production System (TPS) [2].

Due to intensifying competition in the industry, and in order to align with the parent organization's long-term business sustainability goals, the company needed to improve its production process and reduce its production costs. The research question in this study is on how to utilize the value stream mapping (VSM) and lean principles to analyze and improve the production process of a batch chemical plant producing an herbicide product family for the local Indonesian market. VSM is used as it allows the identification of wastes' sources and offers a systematic approach to realize improvements that takes into consideration the constraints and opportunities from the perspective of the whole process [3]. VSM applied in the context of lean manufacturing has been successfully deployed in many different manufacturing settings from its original application in discrete manufacturing to continuous process manufacturing [3][4]. More recently there have been cases of researchers applying VSM in software development labs to the hospital environment [5][6], demonstrating the versatility of this approach. Additionally, the discussion on VSM and lean in a complex manufacturing environment such as batch chemical manufacturing is still limited [7][8], and this paper aims to explore this topic further to test the applicability of VSM and lean production in the context of batch chemical manufacturing.

2. Literature Review

Lean manufacturing methodology has its origin in the Toyota Production System (TPS) developed by the Toyota Motor Corporation starting in the 1940's in post-World War II Japan. The person largely credited with the invention of TPS is an industrial engineer working for Toyota named Taiichi Ohno [4][9][10]. TPS has two guiding principles. The first is the principle of Just-in-Time production (JIT) that mandates the production of only what was needed, and only when it was needed. Any divergence from the true needs of manufacturing a product is condemned as waste or *muda* in Japanese. The second principle is that of *jidoka* where any production problems are instantly self-evident, and when problems are detected, production is immediately stopped to allow the issue to be resolved. In other words, *jidoka* insisted on built-in quality in the production process and condemned any deviation from value-addition as waste. In the Toyota Production System book, Taiichi Ohno classified waste into seven categories, which are briefly described below [11]:

- Transportation: unnecessary movement of materials or people within a process causing additional needs for maintenance, increasing lead time and introducing risk of damaged product
- Inventory: including raw material, work-in-progress (WIP), and finished goods. Inventory increases operational costs (including conveyance costs, storage costs, and risk of obsolescence)
- Motion: unnecessary movement or motion of people or machine that does not add value to the product
- Waiting: time wasted by people or machines waiting for the completion of a work cycle, which includes waiting for other people, machines or material
- Over processing: this includes all activities or operations that are not required to meet customer demand and does not add value for the customer
- Overproduction: Producing sooner, faster, or in greater quantities than is required by the customer. Overproduced items end up as excess inventory or may even be scrapped
- Defects: Producing parts or products that do not meet customer specification

There are various tools that are commonly deployed during implementation of lean including *kanban*, visual control, *poka yoke*, 5S (i.e. a systematic and sustainable method to organize the workplace), TPM (i.e. Total Productive Maintenance), DDS (i.e. Daily Direction Setting), and value stream mapping [7]. VSM is used for detailed analysis of the production process to identify waste and allows for process improvements to lower production costs, improve response time to satisfy customer demand and produce higher quality products [5][7]. VSM is indispensable for an organization seeking to implement lean. It forms an essential part of lean manufacturing implementation. Mapping the organization's value stream is one of the five principles of lean [12]:

- Define value: understand what value is by discovering what customers find valuable, in other words, value is what the customer is willing to pay for in a product
- Map the value stream: value stream consists of all actions required to develop and manufacture a product or deliver a service
- Create flow: once the value stream has been mapped, each step in the process is then analyzed to maximize its efficiency, reduce, and ideally, eliminate waste to create a state where products flow between production steps
- Establish pull: this means letting customers pull (i.e. order) the product according to their needs instead of constantly producing and pushing unwanted products regardless of customer's demand
- Pursue perfection: this part emphasizes that lean is a cyclical approach to the process of waste reduction to bring the greatest value for the customer

Implementation of company specific production systems built on lean principles has been found to provide competitive parity in the manufacturing industry using the resource-based VRIO (Valuable, Rare, Inimitable, and Organized) business analysis framework [2]. Lean manufacturing has been demonstrated to produce higher levels of quality and productivity and better customer responsiveness. Lacerda, Xambre, and Alvelos [4] demonstrated that they were able to identify lean wastes in an automobile component manufacturing plant using VSM methodology and proposed fourteen solutions to eliminate those wastes, of which eleven were implemented. As a result, the authors were able to reduce the cycle time and improve the efficiency and productivity of the plant [4]. Abdulmalek and Rajgopal [3] presented the application of VSM in process industry by combining VSM with simulation of lean tools at a steel mill and found that VSM is universally applicable in the process industry [3]. Schmidtke, Heiser, and Hinrichsen [7] applied VSM to batch chemical process manufacturing. In their study, the authors observed that there were some challenges related to the application of VSM in batch chemical processes due to its origins in the automotive industry and mechanical production processes requiring some adaptation for use in chemical manufacturing environment. The authors found several limitations in VSM methodology due to process variability, complex process flow, and conflicting cost factors often found in the chemical industry. Nevertheless, the authors were able to apply VSM methodology coupled with Discrete Event Simulation (DES) tool to gain satisfactory cost reduction in their case study application in the production process of exhaust gas purification catalysts [7].

3. Methodology

The objective of this research was to study the application of value stream mapping in a batch chemical production process that include analysing the current state of the company's main production process to identify wastes, designing an improved process following lean manufacturing principles to achieve a more efficient production system shown by decreased production cycle time and reduced production costs. The study began by clarifying the research question. Here the main problem to be addressed was the need by an herbicide manufacturer for an improved production process of its main product to remain competitive in the market through the application of value stream mapping (VSM) methodology complemented by lean manufacturing approach. Next, a literature review focusing on the application of VSM and lean manufacturing methodology was conducted to gain a thorough understanding of the topic.

Data were collected through observations and time study of related processes to gain a true picture of how things are in the current state. Field observations were conducted directly at the company's manufacturing facility during real production activities whenever possible. Next, VSM of the current production process was drawn based on collected data, followed by the creation of a future state map with improved manufacturing process. Then, analysis was carried out to quantify the potential savings from implementing the future state map in the manufacturing process. In the final stage of this study, conclusions were drawn from the results of the analysis made in the previous stages to highlight the implications of this study. Research flowchart in this study is shown in Figure 1.

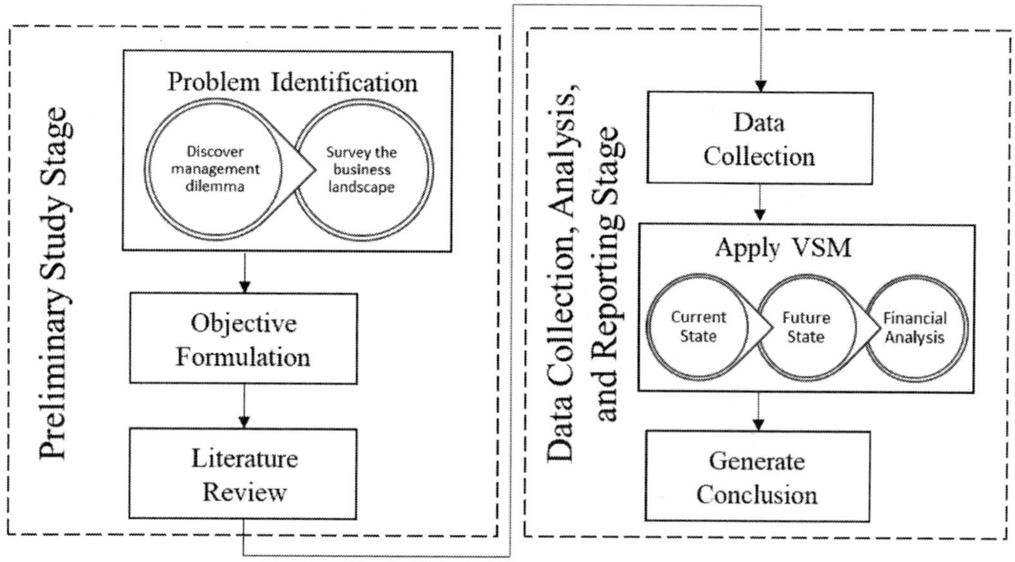

Figure 1. Research flowchart.

Value stream maps were developed for the current and targeted future states. The identification of the dominant "product family" in a production process served as the starting point for the creation of a value stream map. A product family is defined by Rother and Shook as a group of products that pass through similar processing steps and over common equipment in the downstream processes. Next, customer requirements were determined including required product quality, quantity, and tack time. The value stream maps were drawn using paper and pencil on an A3 sized paper to record information and material flows in Ace Chemicals. For each process step, the Process Time (P/T), Delay Time (D/T), and Percent Complete and Accurate (CA%) data were collected and shown on each data box. Inventory and Work-in-Progress (WIP) were counted and recorded on the current state map. Variables used in value stream mapping in this study are shown in Table 1.

Data collection in this study began with an initial sampling of 15 observations [4] for each process step to calculate their mean and standard deviation. The number of required sample size was calculated based on accuracy level of 5% and confidence level of 95% using the following formula:

$$\textbf{Required Sample Size } (n) = \left(\frac{zs}{h\bar{x}}\right)^2 \quad (1)$$

Where z is the standard deviation from the desired confidence level, s is standard deviation from initial sample, h is desired accuracy level, and x-bar is the mean from initial sample. Additional observations were then gathered to reach the required number of samples, and standard time for each process step was then calculated. Performance rating factor was set at 100% since production was running at normal load and all observed operators were trained to do the work and have more than one year of working experience at the plant. An allowance factor of 10% was used in calculating the normal time to consider worker fatigue, bathroom breaks, and delays. The future state map was drawn according lean principles to implement pull and reduce inventory level to achieve lower lead time and production costs [4, 7]. In addition, the future state map aimed to achieve increased production capacity to meet seasonal demand for the product. Finally, financial analysis of the proposed future state was carried out to estimate the expected costs savings.

Table 1. VSM variables used in capturing process data.

Term	Unit	Description	Formula	Data collection method
Processing Time (P/T)	s	The time spent actually performing work and does not include delay time within a process. It can be further broken down into Value Added, incidental and Non-Value-Added time	N/A	Site observation
Delay Time (D/T)	s	The time waiting within a process step. There is no value is being added in delay time	N/A	Site observation
Cycle Time (C/T)	h	The time span between repetitions of the same task. Thus, it is the time taken by all the slowest station or operator in the process	$Cycle\ Time\ (C/T) = Processing\ Time\ (P/T) + Delay\ Time\ (D/T)$	Calculation
Percent Complete and Accurate (CA%)	%	The percentage of time customer (internal or external) requirements are fully met first time in a process step	N/A	Site observation
Takt Time (T/T)	s	The pace at which the manufacturer has to produce its products to meet the demand of the client. This metric synchronizes the rate of production to the rhythm of sales	$Takt\ Time\ (T/T) = \frac{Available\ work\ time\ per\ shift}{Customer\ demand\ rate\ per\ shift}$	Calculation
Wait Time (W/T)	h	The time when no work is being done as a result of the delay between process steps (i.e. sitting in the inventory)	$Wait\ Time\ (W/T) = Cycle\ Time\ (C/T) \times Inventory$	Calculation
Lead Time (L/T)	h	The time it takes for a product to pass through the value stream or the process from the very beginning to the end	$Lead\ Time\ (L/T) = \sum_{i=1}^{n} C/T_i + W/T_i$	Calculation
Rolled Throughput Yield (RTY)	%	The product of CA% at each process step. RTY is the % of time that the process delivers a product or service completely and accurately from intake to the delivery to the end customer	$Rolled\ Throughput\ Yield\ (RTY) = \prod_{i=1}^{n} CA\%_i$	Calculation

4. Results and Analysis

This paper focused on the main product family that makes up around 75% of Ace Chemical's annual production volume. The main difference within this product family is the pack sizes of the final product. The product is delivered in three pack sizes: 1-liter bottles, 4-liter jugs, and 20-liter jars. The normal average demand per week is calculated to be 270,000 litters with 173,000 litters in 1L bottles, and 97,000 litters in 4L pack size. However, during peak production months, demand reaches an average of 472,500 litters comprised of 202,500 litters of 20L jars, 173,000 litters of 1L bottles, and 97,000 litters of 4L pack size. The production process is shown in Figure 2.

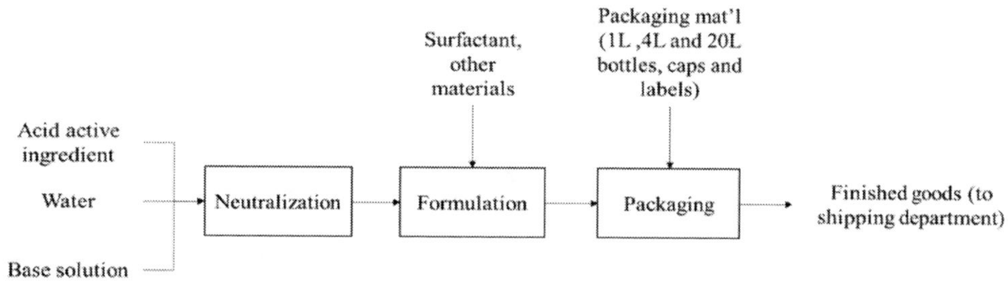

Figure 2. Block flow diagram of case study process.

The production process begins with the neutralization step where acid active ingredient in the form of wet cake in FC bags (i.e. 1-ton bags) are transported from the raw material storage area by forklift and hoisted to the neutralization reactor where it is reacted with base solution in water. The product from this neutralization process is called the active ingredient salt and is stored in surge tanks while waiting for further processing. Next, the active ingredient salt is transformed to finished bulk product in the formulation tank by mixing the salt with surfactant and other materials. The finished bulk product is stored in storage tanks before being transferred to the packaging lines. The packaging machines are run continuously during normal work shift (i.e. 8-hour shift in a 5-day workweek). A process pump transfers the finished bulk to the packaging lines consisting of Line 1 (1L) with 80 L/min capacity, Line 2 (4L) with 48 L/min capacity, and Line 3 (20L) with 94 L/min capacity. Line 1 (1L) and Line 2 (4L) each requires five operators to run, while Line 3 (20L) is run by four operators. Finished goods are pelleted and transported by two forklift operators to the finished goods warehouse for storage before shipping to the customer.

Data gathered from each process step is shown in Table 2. Forklift 1 showed the largest variability compared to other processes due to raw material storage areas that were situated at varying distances from the next process (i.e. neutralization). The decentralized nature of the plant's raw material storage locations was partly due to the high raw material inventory requiring storage space that was more than the capacity of any one storage location. The neutralization step required a sample size of 22 due to variation introduced by rework causing longer batch time. High sample variability was also observed in the shipping process step. This variability was the result of varying truck sizes observed during the sampling process (i.e. large delivery trucks require longer load time per pallet, and vice versa).

Table 2. Current state map process data.

Process step	Mean (s)	Std Dev (s)	Required sample size	Actual sample size	Normal time (s)	Standard time (s)
Forklift 1	1.311	157	30	31	1.311	1.442
Neutralization	4.425	522	22	23	4.425	4.868
Formulation	17.126	1.963	18	18	17.126	18.839
Line 1 (1L)	705	66	13	15	705	775
Line 2 (4L)	1.301	125	14	15	1.301	1.431
Forklift 2	113	9	9	15	113	124
Shipping	299	30	28	28	299	329

The factory operates on an 8-hour shift schedule for 5 days a week, and customer demand is 270,000 litters per week; therefore, the tact time is 533 seconds per kL product. The current state map is shown in Figure 3. The map shows high raw material inventory, push production arrangement, and process bottleneck at the neutralization step as indicated by the lowest P/T to D/T ratio of 0.12. Current lead time was calculated to be 754.5 hours or roughly 31 days with only 0.93% value added time in the whole process.

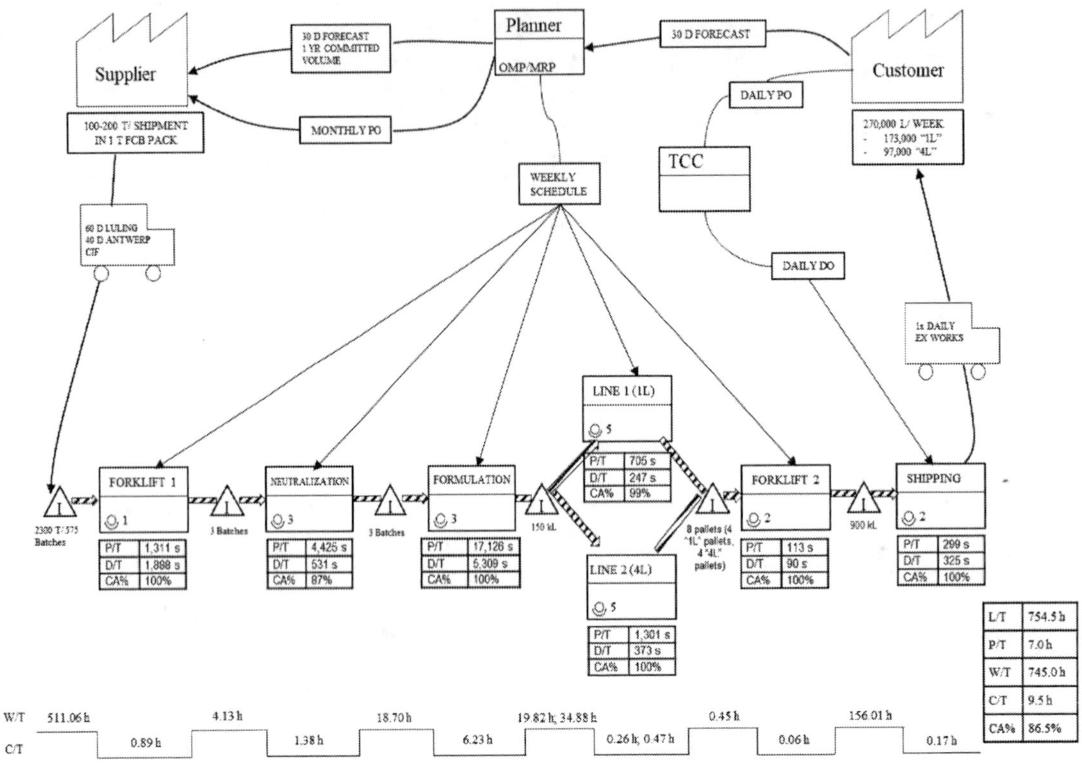

Figure 3. Current state map.

An opportunity assessment analysis that included several rounds of interviews and workshops with each area personnel was conducted to gain insights on possible improvements to the current state with the aim of eliminating waste and applying lean principles to the herbicide production process. The list of potential improvements was narrowed down by considering potential benefits and ease of implementation to the following items:

- Implement pull and inventory supermarket
- Debottleneck neutralization reactor by reactivating an idle reactor with existing neutralization personnel sharing the manning responsibility (i.e. no additional headcount requirement)
- Reduce acid active ingredient inventory to 75D supply
- Reduce finished bulk inventory by one-third
- Reduce finish goods inventory to a week's worth of stock
- Increase formulation capacity to balance with new neutralization capacity
- Bring Line 3 (20L) machine online
- Continuous improvement/ operational excellence initiatives: 5S, TPM, DDS

During peak season, demand reaches 472,500 litters per week, and tact time becomes 305 seconds per kL product. The future state map was drawn up to reflect the proposed changes and shown in Figure 4.

The main features of the future state are the establishment of supermarkets to enable the downstream "pull" of materials at their desired rate to prevent inventory build-up, issuance of weekly production schedule to a single point at the packaging line, and increased production capacity to enable the plant to change output rate to match demand.

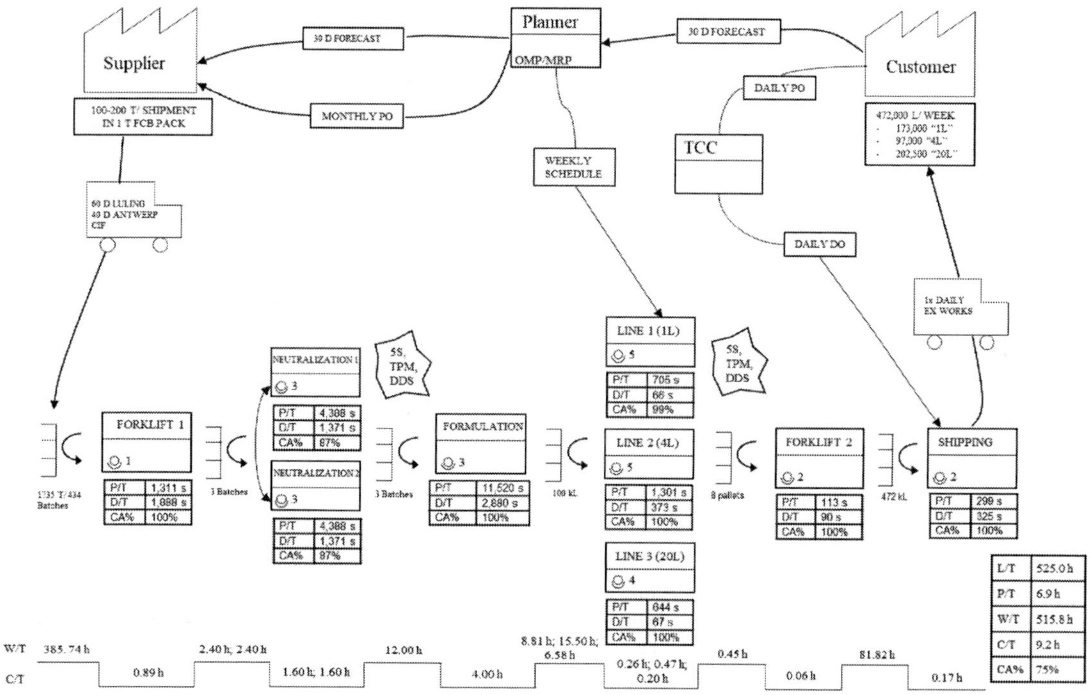

Figure 4. Future state map.

Table 3 shows a comparison between the present state and future state. The analysis shows that production lead time could potentially reduce by 30% and value-added time percentage could increase from 0.93% to 1.31% by applying the future state map to current production process.

Table 3. Current state and future state lean metrics comparison.

		Current state	Future state
Lead time		754.5 h	525.0 h
Value added time		0.93%	1.31%
Inventory	RM	575 batches	434 batches
	WIP	150 kL	100 kL
	FG	900 kL	472 kL
Capacity		54 kL/ shift	94.5 kL/ shift

The potential savings from implementing the proposed changes are shown in Table 4. The savings are achieved from reduction of inventory costs and overtime costs during peak demand season. Due to confidentiality issues, the numbers shown have been masked, but still nevertheless indicate the potential cost impact from the improvement initiatives.

Table 4. Estimated costs and savings from proposed solution.

	Formula		Result
Capital investments (C_0)	N/A	$	(150,000)
Workforce savings (C_1)	$\left(600\frac{USD}{month} \times \#\ of\ personnel/shift \times 360\frac{hour}{year} \times 1.92\right) \times \frac{1}{173}$	$	16,800/ year
Inventory savings (C_2)	$\left(\sum_{i=1}^{n} \Delta\ inventory_i \times cost\ of\ inventory_i\right) \times 10\%\ p.a.$	$	20,600/ year
Estimated total savings	$NPV = C_0 + \sum_{i=1}^{20} \frac{C_i}{(1+0.15)^i}$	$	247,518

5. Conclusions

The objective of this research was to study the applicability of VSM in a batch chemical production process of an herbicide manufacturing company in Indonesia. In this paper, the production process of an herbicide product family was studied by gathering the information regarding current production process. Process improvements were identified following lean principles and a new redesigned process was presented with a reduction of 30% in manufacturing lead time. An analysis of potential savings from implementing the changes was also presented showing a positive return on investment. To implement the proposed changes, Ace Chemicals will have to renegotiate its raw material purchase agreement with its suppliers to synchronize the raw material shipment to customer tact time, engage with engineering contractors to execute its reactor turnaround plan, and formalize its lean manufacturing initiatives to sustain employee participation in the new production system.

This study demonstrated the ability of VSM as a tool to show and analyse the end-to-end view of a manufacturing process in a batch chemical plant. The method was shown to be effective for achieving balanced production rate and meeting customer demand. Managerial implications of this study include highlighting the value of continuous improvement using lean methodology even in a mature plant that has been in operation for years and the applicability of VSM outside of its discrete manufacturing sweet spot (i.e. in batch chemical process industry). Limitation of this research is the single case study at a single production site in Indonesia that makes up part of the parent organization's global presence, which may limit some generalizations to other companies with different levels of resources, process complexity, and readiness for change. Future research may address this limitation by conducting larger study involving more companies in the process industry. Other opportunities that may be explored further include studying the impact of changeover to productivity and lead time and studying the application of simulation to model customer demand variability to further reduce production costs and lead time by optimizing inventory level and formulation batch size.

References

[1] World Population Prospects. *UN*. 2016 Apr 14. doi.org/10.18356/a3bacd57-en.

[2] Netland HT, Aspelund A. Company-specific production systems and competitive advantage. Clegg, Jillian MacBryde and Prasant B, editor. *J. Oper. Prod. Manag*. Emerald; 2013 Nov 11;**33**(11/12):1511–31. doi.org/10.1108/ijopm-07-2010-0171.

[3] Abdulmalek FA, Rajgopal J. Analyzing the benefits of lean manufacturing and value stream mapping via simulation: A process sector case study. *INT J PROD ECON*. Elsevier BV; 2007 May;**107**(1):223–36. doi.org/10.1016/j.ijpe.2006.09.009. .

[4] Lacerda AP, Xambre AR, Alvelos HM. Applying Value Stream Mapping to eliminate waste: a case study of an original equipment manufacturer for the automotive industry. *Int. J. Prod. Res.* Informa UK Limited; 2015 Jun 18;**54**(6):1708–20. doi.org/10.1080/00207543.2015.1055349.

[5] Henrique DB, Rentes AF, Godinho Filho M, Esposto KF. A new value stream mapping approach for healthcare environments. *Prod. Plan. Control*. Informa UK Limited; 2015 Jul 17;**27**(1):24–48. doi.org/10.1080/09537287.2015.1051159.
[6] Ali NB, Petersen K, de França BBN. Evaluation of simulation-assisted value stream mapping for software product development: Two industrial cases. *Inf Softw Technol*. Elsevier BV; 2015 Dec;**68**:45–61. doi.org/10.1016/j.infsof.2015.08.005.
[7] Schmidtke D, Heiser U, Hinrichsen O. A simulation-enhanced value stream mapping approach for optimisation of complex production environments. *Int. J. Prod. Res.* Informa UK Limited; 2014 May 16;**52**(20):6146–60. doi.org/10.1080/00207543.2014.917770.
[8] Braglia M, Carmignani G, Zammori F. A new value stream mapping approach for complex production systems. *Int. J. Prod. Res.* Informa UK Limited; 2006 Sep 15;**44**(18-19):3929–52. doi.org/10.1080/00207540600690545.
[9] Rand G, Womack J, Jones DT. Lean Thinking-Banish Waste and Create Wealth in Your Corporation. *J. Oper. Res. Soc.* JSTOR; 1997 Nov;**48**(11):1148. doi.org/10.2307/3010314.
[10] Melton T. The Benefits of Lean Manufacturing. *Chem Eng Res Des*. Elsevier BV; 2005 Jun;**83**(6):662–73. doi.org/10.1205/cherd.04351.
[11] Ohno T, Bodek N. *TPS*. Productivity Press; 2019 Dec 17. doi.org/10.4324/9780429273018.
[12] Havn E. J.P. Womack, D.T. Jones, and D. Ross,The Machine that Changed the World, Rawson Associates, New York, 1990, 323 PP., $24.95. Int J Hum Factors Manuf. Wiley; 1994;**4**(3):341–3. doi.org/10.1002/hfm.4530040310.

The expired period of Liquefied Petroleum Gas (LPG) rubber hose supports consumer safety aspects

A B Mulyono[1,2] and E H Purwanto[1,3]

[1]Center for Research and Human Resource Development, National Standardization Agency, South Tangerang, Banten, Indonesia

Email: [2]arybudimulyono1988@gmail.com, [3]endi@bsn.go.id

Abstract. Liquefied Petroleum Gas (LPG) rubber hose has the corrosive characteristic with hydrocarbon compounds (butane and propane). This makes LPG rubber hose has an expired period. Standards or technical specifications are needed to guarantee the quality of the product and increasing product safety. Until now, the expired period of LPG hose has not been regulated in the Indonesian National Standard (SNI). This study aims to analyse expired period of LPG rubber hose for gas stoves. In this study, the descriptive qualitative method is used based on primary data (questionnaire and interview) and secondary data (literature study). The result of this study concluded that the determination of LPG rubber hose expired period based on shelf life and service life. The service life of rubber hose will vary due to different raw materials, additives, and usage conditions (time, temperature, pressure, and environment). Seventy-five percent (75%) of LPG rubber hose industry respondents conducted research on LPG rubber hoses service life and recommended service life period is one year.

1. Introduction

The government has launched Kerosene to Liquefied Petroleum Gas (LPG) conversion program that aims to diversify energy supplies in order to reduce dependence on fuel oil and efficiency of the government budget. Before the conversion program began, the government budgeted a trillion rupiah to subsidize kerosene [1]. The kerosene to LPG conversion program is based on Law No.22/2001 concerning Oil and Gas, Presidential Regulation No.5/2006 concerning the National Energy Policy, and Presidential Regulation No.104/2007 concerning the Supply, Distribution, and Determination of Liquefied Petroleum Gas (LPG) 3 kg cylinders. Until 2017, Bank of Indonesia (BI) stated that kerosene to LPG conversion program carried out by the government since 2007 has provided savings of 197 trillion rupiah [2].

Until now, the government continues Kerosene to Liquefied Petroleum Gas (LPG) conversion program. Based on the Minister of Energy and Mineral Resources Regulation, Pertamina has a mandate to carry out the supply and distribution of 531,131 three kilos LPG in 2018. The kerosene to Liquefied Petroleum Gas (LPG) conversion program has successfully increased percentage of households using LPG from 6.4% to 91.1% (2007-2013) [3].

This program has several problems in its implementation. Problems that arise include leakage and explosion of LPG cylinders [[4]. Most of the fires caused by LPG gas cylinders explode are due to leaks in gas distribution channels (including valves, hoses, regulators, and clamps) [5] [6]. The leakage of LPG rubber hoses influenced by several factors including rat bites, sharp object incisions, flames exposure, and direct contact with the environment that causes physical properties changes of LPG rubber hoses.

Rubber hose serves to deliver LPG with a certain pressure from the gas cylinder to gas stove. A good quality rubber hose must have good resistance to LPG [7]. Rubber hose has three main parts, there are rubber hose tube, reinforcement layer, and cover layer. Rubber hose tubes are used to withstand the pressure caused by gas. Reinforcement layers are used to resist mechanical stresses received by the hose, such as internal pressure, vacuum, or tensile pressure. The cover layer is used to protect the amplifier from external pressure, weather and ozone, chemical pressure, and thermal stress.

Standards are technical requirement, including procedures and methods that are prepared based on consensus in accordance with requirements of safety, security, health, environment, science and technology development, experience, as well as current and future developments [8]. Standards are based on the needs and results of the consensus of stakeholders to achieve various economic, social, and environmental aspects supporting sustainable development [9]. Indonesia National Standard (SNI) is standard established by National Standardization Agency of Indonesia (BSN) and applies in Indonesia[9]. National Standardization Agency of Indonesia BSN) has established 2 (two) SNI for LPG hoses, SNI 7213: 2014 Rubber Hoses for LPG Gas Stoves, and SNI 8022: 2014 Elastomer thermoplastic hoses for LPG gas stoves. Both SNIs are mandatory regulated by the Ministry of Industry to protect consumers.

Technical standards or specifications are needed to guarantee product quality and improve LPG hose product safety. In 2009, research conducted by The Centre of Research and Development – National Standardization Agency of Indonesia showed the level of conformity of LPG gas stove rubber hose products to Indonesian National Standard (SNI) requirements is still low. There are indications that the age factor also affects the quality of rubber hose characteristics. It is known that rubber has corrosive susceptibility to hydrocarbon compounds including butane and propane, so that each component such as rubber hose, regulator, and rubber seal have a service life (expired period). As the warranty period expires, the component quality will automatically continue to decline.

Until now, only LPG gas cylinders have a procedure to determine the expired period by retesting LPG gas cylinders every five years after manufactured, while the determination of the expired period of the LPG hose has not been regulated. The purpose of this study was to analyse the expired period for LPG rubber hoses.

2. Method

This study uses qualitative descriptive method. Descriptive analysis is used to collect information by giving attention as much as possible aspects to obtain the overall actual situation. The qualitative method addresses the social aspect of research. The researcher uses open-ended questions and interviews subjects in a semi-structured fashion [10]. The qualitative method also emphasizes data sampling conducted purposively and the research results emphasize in meaning than generalization [11]. Primary data collection is done by interview using a questionnaire, while secondary data obtained through a literature study. Primary data collected are from LPG rubber hose manufacturers and LPG rubber hose consumers. The survey respondents were carried out by purposive sampling (nonprobability sampling) in Medan, Bogor, Depok, Bekasi, Tangerang, and Surabaya.

3. Result and Discussion

3.1. Standardization of LPG Rubber Hose Expired Period

Liquefied Petroleum Gas (LPG) rubber hose is one of the products used to complement the LPG gas stoves. This product must meet not only quality requirements but also technical age. If the product has exceeded its technical age (expired period) but is still being used, then it could cause a fire hazard. The expired period can be divided into [12] [13]:

- Shelf life: the period from receipt or delivery by distributor until the product begins to be used.
- Service life: the period from installation to product replacement.

In this study, identification of standards related to the procedure for rubber and elastomeric products storage can be seen in Table 1.

In these four standards, there are correct procedure recommendations for rubber and elastomeric products storage, so that shelf life of rubber and elastomeric products can be optimized. Recommended procedures for rubber products and elastomers storage can be summarized and seen in Table 2.

Table 1. Standards relating to procedures for rubber and elastomeric products storage.

No	Standard Number	Standard Title
1	DIN 7716: 1982	Rubber products; requirements for storage, cleaning and maintenance
2	BS 4F 68: 2002	Controlled storage of vulcanized rubbers for use in aerospace applications
3	ISO 2230: 2002	Rubber products - Guidelines for storage
4	DIN 9088: 2002	Aerospace Series - Storage life of rubber products

Table 2. Recommended procedures for rubber products and elastomers storage (based on Standards).

Storage Parameters	Storage recommendation
Temperature	Storage temperature between 5 °C to 25 °C. Avoid contact with heat sources such as boilers, radiators, and direct sunlight
Humidity	Storage humidity must below 70%. Avoid conditions too humid or too dry. Condensation should not occur.
Light exposure	Must be protected from light sources, especially direct sunlight or artificial light that contains ultraviolet. Storage containers offer the best protection as long as their packaging is ultraviolet resistant. It is recommended to close all windows in the storage room with red/orange coating screen.
Radiation	Must be protected from all types of ionizing radiation sources which can cause damage to the product.
Oxygen Dan Ozone	Elastomeric material must be protected from air circulation by wrapping packages or airtight packages. Because Ozone is destructive to elastomers, storage space must not contain equipment that produces ozone.
Deformation	Elastomeric material should be stored as far as possible under conditions of tension, compression or other deformation.
Liquid or semi-solid contact	Avoid contact with solvents, oils, greases, or other semi-solid materials during storage
Contact with metals and non-metals	Avoid direct contact with certain metals, such as manganese, iron, brass, copper, and its alloys. These material components are known to have adverse effects on some rubber and elastomeric products.

The shelf life of rubber products varies depending on the type of rubber used as raw material. International standards have been identified and reviewed related to the shelf life of LPG hoses. These standards can be seen in Table 3.

The shelf life of rubber hoses is influenced by many factors. The recommended shelf life of rubber and elastomers products can be seen in Table 4.

Table 3. International standard for rubber hose shelf life.

No	Standard Number	Standard Title
1	ISO 2230 : 2002	Rubber products — Guidelines for storage
2	MIL-HDBK-695 (2017)	Rubber Products: Recommended Shelf Life
3	SAE ARP 5316 (2014)	Storage of Elastomer Seals and Seal Assemblies Which Include An Elastomer Element Prior To Hardware Assembly

Table 4. The recommended shelf life of elastomeric rubber hose products.

Chemical Names	Chemical Abbreviation	Recommended Shelf Life (Year)		
		ISO 2230: 2002	**MIL-HDBK-695 (2005)**	**SAE ARP 5316 (2002)**
Polyacrylate	ACM, ANM	--	20	--
Polyester Urethane	AU	--	3-5	5
Chlorobutyl	CIIR	--	10	--
Chlorinated Polyethylene	CM	--	10	--
Polychloroprene	CR	7-10	5-10	15
Chlorosulfonated Polyethylene	CSM	--	5-10	--
Epichlorohydrin H, EO	CO, ECO	7-10	5-10	--
Polysulfide	EOT, T	--	20	20
Ethylene/Propylene/Diene	EPDM	10-15	5-10	unlimited
Polyether Urethane	EU	--	5-10	5
Tetrafluoroethylene/Propylene	FEPM	--	--	unlimited
Perfluorocarbon	FFKM	--	20	unlimited
Fluorocarbons	FKM	--	20	unlimited
Fluoro Methyl Vinyl Silicone	FVMQ	--	20	unlimited
Polypropylene oxide	GPO	--	5-10	--
Hydrogenated Nitrile	HNBR	7-10	15	--
Isobutylene Isoprene	IIR	7-10	5-10	15
Isoprene	IR	--	3-5	--
Butadiene Acrylonitrile	NBR	7-10	3-5	15
Natural Rubber	NR	--	3-5	--
Butadiene Styrene	SBR	--	3-5	--
Methyl Vinyl Silicone	VMQ, Q	10-15	20	--

Based on the literature study, some experts in Australia (ELGAS Australia) and United Kingdom (UK LPGenius) recommend LPG rubber hose to be replaced every five years. The service life of a polymer is difficult to predict because [14] [15]:

- The characteristics of polymers depend on time, temperature, environment, and pressure.
- The property value at which the polymer fails is often unknown.
- The usage conditions of polymers may vary, and usage errors also occur frequently.
- For complex uses, it is not possible to be the same as the accelerated usage conditions test
- The failure mechanism is not always sustainable - gradual changes in characteristic can occur.
- Polymers are inherent variables.
- Difficulties in conducting long-term testing

The service life of rubber hose will vary due to different raw materials, additives, and usage conditions (time, temperature, pressure, and environment) [13] [15] [16]. While SNI does not state the raw materials used in the manufacture of LPG rubber hoses. Some additives which are added include heat stabilizer, ultraviolet stabilizer, antioxidant, and colorant.

3.2. Survey Results Supports Determination of LPG Rubber Hose Expired Period

To find out data/information needed to develop expired period of LPG rubber hoses, a survey was conducted using interviews with LPG rubber hoses manufacturer (4 manufacturers). The results are:

- All survey respondents (100%) stated that LPG rubber hose complies with SNI 7213: 2014 (Mandatory SNI). respondents stated that they did not experience any obstacles in fulfilling SNI specifications. Respondents also stated that they did not use other standards as a reference in LPG rubber hoses production.

- 50% of respondents stated that fulfilment testing of SNI requirements on internal companies.
- 75% of respondents stated that they had LPG rubber hose expired period for 1 year as a safe limit approach. Determination of service life is based on results of internal research using the aging test method. However, manufacturers stated the service life was very dependent on storage and usage.
- 25% of respondents stated that they monitored products that were still circulating in the market related to the expired period (shelf life). While 75% of respondents said they did not monitor the products circulating in the market. This shows that the level of supervision of the company's products is still low. It is necessary to increase supervision so that the quality of products received by the public can be further improved.
- All survey respondents (100%) stated that the determination of the LPG rubber hoses expired period needs to be included in SNI 7213: 2014. Respondents stated that the expired period is very important so that the quality of LPG rubber hose produced remains guaranteed when the product is used by consumers. However, respondents stated that the determination of the LPG rubber hose expired period must also consider the economic calculation of the company's production.

In addition to the primary data collection in LPG rubber hose manufacturer, the survey was also conducted on LPG gas stoves consumers (22 respondents). This survey was conducted to determine problems encountered in rubber hoses usage. The results show frequent damages are bloated hose, melted hose, cracks, stretchy hose, blocked regulator, and leakage (rat).

4. Conclusion

Based on the results, it can be concluded that the determination of the LPG hose expired period includes shelf life (storage life) (see Table 4) and service life. Rubber hose storage conditions are the most important factor in determining the shelf life. The service life of rubber hose will vary due to different raw materials, additives, and usage conditions (time, temperature, pressure, and environment). The results show 75% LPG rubber hose manufacturer conducted research on LPG rubber hoses service life and obtained recommended service life period is 1 year. The shelf life of rubber hose needs to be stated in SNI 7213:2014 based on raw material specifications to ensure the product quality and increasing product safety.

Acknowledgment

This research was supported by National Standardization Agency of Indonesia. We would like to thank Mr. Mangasa Ritonga for his guidance in this research.

References

[1] Mentang SX. PERSEPSI, SIKAP DAN PARTISIPASI MASYARAKAT TERHADAP KONVERSI MINYAK TANAH KE GAS LPG (Studi di Kelurahan Taratara, Kecamatan Tomohon Barat, Tomohon). Center for Open Science; 2019 Apr 1. doi.org/10.31227/osf.io/4vnfs.

[2] Asyari H, Umar U. Desain Prototipe Kompor Listrik Tenaga Surya. *Emitor*. Universitas Muhammadiyah Surakarta; 2019 Mar 31;**19**(1):6–9. doi.org/10.23917/emitor.v19i1.6997.

[3] Andadari RK, Mulder P, Rietveld P. Energy poverty reduction by fuel switching. Impact evaluation of the LPG conversion program in Indonesia. *Energy Policy*. Elsevier BV; 2014 Mar;**66**:436–49.doi.org/10.1016/j.enpol.2013.11.021.

[4] Kusriyanto M, Firdaus, Yulianto A, Kurniawan S. Early detection of LPG gas leakage based Wireless Sensor Networking. Ma'mun S, Tamura H, Purnomo MRA, editors. *MATEC Web of Conferences*. EDP Sciences; 2018;**154**:01045. doi.org/10.1051/matecconf/201815401045.

[5] Dewi L, Somantri Y. Wireless Sensor Network on LPG Gas Leak Detection and Automatic Gas Regulator System Using Arduino. *Mater. Sci. Eng. C*. IOP Publishing; 2018 Jul;**384**:012064. doi.org/10.1088/1757-899x/384/1/012064.

[6] Rahman A, Istighfarin R. Augmented reality based learning media for virtual instruction of LPG cylinder assembly. EXPLORING RESOURCES, PROCESS AND DESIGN FOR SUSTAINABLE URBAN DEVELOPMENT. *ICETIA*. AIP Publishing; 2019. doi.org/10.1063/1.5112480.
[7] Handayani H, Faturrohman MI, Kinasih NA, Falaah AF. KARET ALAM EPOKSI SEBAGAI BAHAN BAKU PEMBUATAN KOMPONEN KARET PADA KATUP TABUNG DAN REGULATOR LPG. *JPK*. Riset Perkebunan Nusantara; 2018 Jan 11;**35**(2):199. doi.org/10.22302/ppk.jpk.v35i2.416.
[8] Sutanto J, Sulisworo P. Improvement of Nuclear Science Standards (SNI) to Meet Market Needs and Harmonization. *J. Phys*. IOP Publishing; 2019 Apr;**1198**(2):022010. doi.org/10.1088/1742-6596/1198/2/022010.
[9] Saleh AR, Sumarni E, Nihayati N, Bahrudin M. DISTRIBUSI STANDAR NASIONAL INDONESIA KEPADA MASYARAKAT MELALUI LAYANAN PERPUSTAKAAN. J*urnal Standardisasi*. Center for Research and Development of Standardization; 2018 May 11;**18**(2):139. doi.org/10.31153/js.v18i2.706.
[10] MOHAJAN HK. QUALITATIVE RESEARCH METHODOLOGY IN SOCIAL SCIENCES AND RELATED SUBJECTS. *JEDEP*. Fundatia Romania de Maine; 2018 Mar 30;**7**(1):23. doi.org/10.26458/jedep.v7i1.571.
[11] Etikan I. Comparison of Convenience Sampling and Purposive Sampling. *AJTAS*. Science Publishing Group; 2016;**5**(1):1. doi.org/10.11648/j.ajtas.20160501.11.
[12] Das SS, Schroeder LW. Estimation of shelf life of natural rubber latex exam-gloves based on creep behavior. *J Biomed Mater Res B*. Wiley; 2008;**85B**(2):398–408. doi.org/10.1002/jbm.b.30958.
[13] Boyun BM, Rhoads JE. Elastomer Shelf Life: Aged Junk or Jewels? *IEEE*. Institute of Electrical and Electronics Engineers (IEEE); 1989 Jun;**9**(6):45–45. doi.org/10.1109/mper.1989.4310754.
[14] Hulme A, Cooper J. Life prediction of polymers for industry. *Seal Tech*. Elsevier BV; 2012 Sep;**2012**(9):8–12. doi.org/10.1016/s1350-4789(12)70398-7.
[15] Woo CS, Choi SS, Lee SB, Kim HS. Useful Lifetime Prediction of Rubber Components Using Accelerated Testing. *IEEE*. Institute of Electrical and Electronics Engineers (IEEE); 2010 Mar;**59**(1):11–7. doi.org/10.1109/tr.2010.2042103.
[16] Woo CS, Park HS. Useful lifetime prediction of rubber component. *ENG FAIL ANAL*. Elsevier BV; 2011 Oct;**18**(7):1645–51. doi.org/10.1016/j.engfailanal.2011.01.003

IOP Conf. Series: Materials Science and Engineering 1072 (2021) 012017 doi:10.1088/1757-899X/1072/1/012017

Understanding user feedback on learning management system of SIPEJAR by using text mining techniques

P Rahmawati[1,3], **A Larasati**[2,4], **M Farhan**[1,5], **A M Hajji**[2,6], **and N A Fanani**[1,7]

[1]Department of Industrial Engineering, University Negeri Malang, Jl. Semarang No.5, Malang, Indonesia
[2]PUI-PT Disruptive Learning Innovation (DLI) University Negeri Malang, Jl. Semarang No.5, Malang, Indonesia

E-mail: [3]Pujirahma147@gmail.com, [4]aisyah.larasati.ft@um.ac.id, [5]muhfarhanaan11@gmail.com, [6]apif.miptahul.ft@um.ac.id, [7]nazzahraf@gmail.com

Abstract. SIPEJAR (*Sistem Pengelolaan Pembelajaran*) as a learning management system can organize various activities in e-learning such as assignments, exams, learning materials, etc. This study aimed to use text mining techniques to extract information from the answers given by respondents toward questions about SIPEJAR. Text mining is one of the techniques used to extract and analyze useful information from the unstructured data. The data were collected using an online survey, and the total data collected is 10,456 but only 9,315 were used. The results show that course material, assignment, and search for information are the three most important factors that motivate students to access SIPEJAR. In the contrary, the factors that lead students to reluctant using SIPEJAR is bad quality network, difficulty to use, and frequently server down. In addition, the factors that need to be improved from SIPEJAR is a user interface, features, service, notification, and simplicity.

1. Introduction

The popularity of information technology has been increased recently. There is a various application of information technology that has been used, for example in business, education, healthcare, etc. In education, one example of the application of information technology is a *learning management system* (LMS). LMS is environments that meet a set of features for creating and structuring of courses in the distance and the objective is to centralize and simplify administration and management of teaching and learning through e-learning [1].

University Negeri Malang has applied a learning management system, which is named as SIPEJAR (*Sistem Pengelolaan Pembelajaran*). SIPEJAR is a website created by University Negeri Malang that used as a forum for students to collect lecture assignments, to do quizzes, as well as to discuss course material. SIPEJAR also functions as a forum for lecturers to upload assignments and lecture material (e-books, PowerPoint, etc) as well as to assess learning achievement on the website. The goal of SIPEJAR is to reduce the homework assignment in physical form and change them into digital, so the teaching and learning process becomes more effective and easier for both students and lecturers.

SIPEJAR website is designed and created in well-planned processes. However, there still remains flaws that have to be improved. So far, many students still have not used SIPEJAR services during learning processes, even though University Negeri Malang has provided these services for free. This is due to lack of training or introduction for students to use SIPEJAR, so when they are introduced to the new system, students are not aware of the process of using SIPEJAR, which eventually students in this case as users SIPEJAR not responding to the existing of the new system. Besides, there is no guidebook on the website regarding the use of SIPEJAR, especially regarding the functionality of each of the features provided therein. For students who have used SIPEJAR, there are some features are quite confusing to use, and not all of the features that have been provided work in accordance with their functions. Only a few features are frequently used for instance giving or collecting tasks. In addition, the user interface is not interactive, and errors often occur when accessing the website are some other things that must be improved. Therefore, to avoid the risks of decreasing visitors and to increase the quality of the website because of these problems, the research was carried out using questionnaires about SIPEJAR also as a suggestion for college to improve the performance of its services. The questionnaire's purpose is to find out how satisfied website visitors are to the existing features, also to see if there are certain suggestions from visitors. The questionnaire is not only contained by multiple-choice (ratios) only, but also contained by essay questions. Multiple choice questions are easy to assess because the responses are quantifiable [2]. Meanwhile, the responses to essay questions are text data, and objectively understanding each student's preference is quite hard. Besides, it is a challenge to avoid subjective data interpretation. Therefore, to avoid these risks as much as possible, the text-mining method might be useful.

Text mining is a technique to extract new or previously unknown information and knowledge from unstructured text or textual data [3]. In this case, there has been researched on the response or user's behaviour toward the learning management system, but so far, no one has used text mining methods to obtain information on about it. So, it is hoped that by using the text mining method, the information will be obtained more easily. Generally, it's easy for people to understand an unstructured text, but it's a challenge for a computer program. The purpose of text mining is to investigate data with simpler methods for experimentation, which have been linked but previously separate in order to form new theories or actuality [3]. Text mining has been used in various fields such as market trends analysis, social media analysis, healthcare, business analysis, fraud detection, risk management, etc. [4] [5]. In education, text mining applied to discover students' performance, feedback to users, evaluate essays and online assignments, etc. then with a variety of model approaches that have been used such as topic modelling approach, sentiment analysis, and others [5]. However, there is still not much research reviewing the use of text mining in terms of improving the quality of services from the Learning management system (SIPEJAR). One of the text mining software that is available for free is *rapid miner*. *Rapid miner* is the most popular free data science platform that supports data mining, text mining, and other data mining techniques that are applied in combination with text mining [5]. Rapid miner has an easy-to-use graphical user interface consisting of processes and many operators that can be used in data mining. Rapid miner text analytics extension using an easily understandable example set and provided operators that can assist in doing text mining [5]. The method used in this text mining process is Word Frequency. Words from the survey's result are counted to identify the words that appear most often. Words often appear in the respondent's answer indicating that an analysis of the word needs to be done, so the survey can be turned into actionable insight. This paper presents the use of text mining to analyse the questionnaire's essay to get the best conclusion to improve the features of SIPEJAR in order to increase the visitor.

2. Method

This study aims to analyse student opinions about SIPEJAR using text mining. The text mining method consists of three steps. The flowchart process is shown in Figure 1.

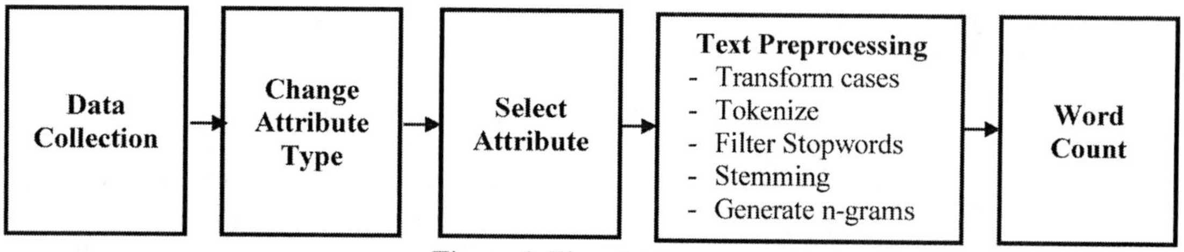

Figure 1. Flowchart process.

The first step is to collect student data about SIPEJAR in University Negeri Malang. Data is collected using an online survey. There are four attributes used in the questionnaire, for more details, see Table 1.

Table 1. Question and respondent profile.

Attribute	Item
Respondent Profile	Name, faculty, department, semester
1	What motivated you to use SIPEJAR?
2	What makes you reluctant to use SIPEJAR?
3	What do you think needs to be improved from SIPEJAR services?

The total data obtained is 10,456 data, but only 9,315 can be processed because many students send data repeatedly so that some data is duplicated, and the language used respondent in answering questions is Bahasa Indonesia.

Data obtained usually cases are unstructured, informal, usually long documents written quickly and containing anomalous or dirty text: typos, misspellings, Adcock abbreviations, code, core dumps, cryptic tables, ambiguous and missing punctuation, and bad use of grammar [6]. Text preprocessing aims to clean up the data so that the data used can be processed in the next step. Text preprocessing makes the input documents more consistent to facilitate text representation, which is necessary for most text analytics tasks [8].

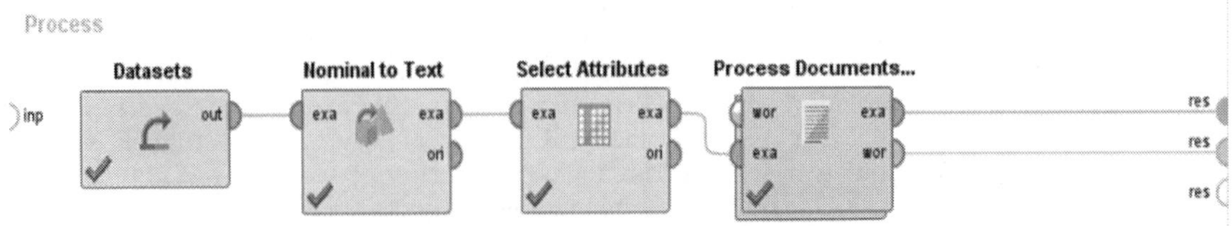

Figure 2. Text mining process.

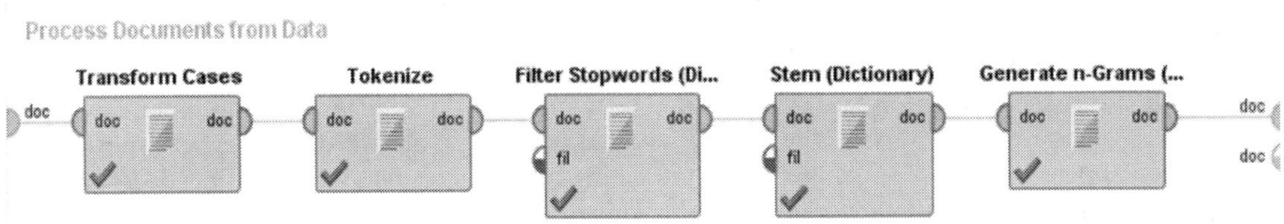

Figure 3. Text preprocessing process.

Text mining process and Text Preprocessing is shown in Figure 2 and Figure 3. These steps using Rapid miner to process the data. In-text mining process, Nominal to Text operator functions to process the survey data so that it can be processed, and the word frequency calculation is performed. While the select attribute aims to select the questions to be processed, the questions are processed one by one so that the meanings and intentions of the respondents are not ambiguous.

Text preprocessing is a sub-operator of the Process Document from Data in Text Mining Process (Figure 1). In-text preprocessing, five algorithms used. First, transform cases algorithm aims to convert text into lowercase letters to make it easier in processing data, because the answers from respondents are different. Second, tokenize aims to break sentences into words called tokens. Then filter stop words aim to eliminate words that are not meant for analysis because it can make look heavier for the text such as in, a , an , and etc. [9]. And stemming aims to change a word into its basic words, to reduce the number of words in the text to save space and memory, and suffixes can be removed [9]. Specifically, for the parameters used in the stop words and stemming filter operators, it is necessary to write a list of words from stop words and stemming itself, because there are no special Indonesian operators on software. After doing the stemming process, we need to use the *generate n-gram* operator. If we break sentences into words, sometimes a word has a different meaning when it becomes a phrase. *Generate n-*

gram functions to record words that appear in a row so that there is no difference in meaning when analyzed. The parameter used to generate n-grams is *max. length: 2*.

Word Count is done using an operator *process document from data*. The parameters used are *prune method: perceptual, prune below percent: 1*. After having done the text mining, then an analysis is performed on the data based on the frequency of the data appearing.

3. Result and Discussion

The aim of this research is to find the response of students using SIPEJAR to improve the services using Text Mining. Text Mining used to help discover information from three questions about SIPEJAR. Attributes are grouped based on the similarity of meaning and relevant to support the answers to each question, so that the following results are obtained the result shown in the table and graph below.

Table 2. Total number of appearance (Question 1).

Attribute	Appearance of Word	Appearance of Document
Assignments from lecture/collect assignment/deadline assignment	2111	2096
Lecturer	2055	2022
SIPEJAR	1000	897
Easy	852	831
Material assignment	803	782
Course	759	736
Searching_for_information	596	589
Demands/instructed_by_lecturer	518	518
Study	359	351
Access	215	215
Needs	171	171

Table 2 and Figure 4 show the total number of document appearance in question 1 ("what motivated you to use SIPEJAR?"). The top five words with the most mentioned in the answer given by the respondent which is related to the question are: **First**, an assignment from lecture/collect assignment/deadline assignment with total 2111 appearance of word and 2096 appearances of document. **Second**, lecturer with total 2055 appearances of word and 2022 appearances of document. **Third**, easy with total appearances of word: 852 and appearances of document: 831. **Fourth**, material assignment with total 803 appearances of word and 782 appearances of document. And **fifth**, searching for information with total appearances of word: 596 and appearances of document: 589. And the last two words that appear the least are access with total 215 appearances of word and 215 appearance of document, then needs with total 171 appearances of word and 171 appearances of document.

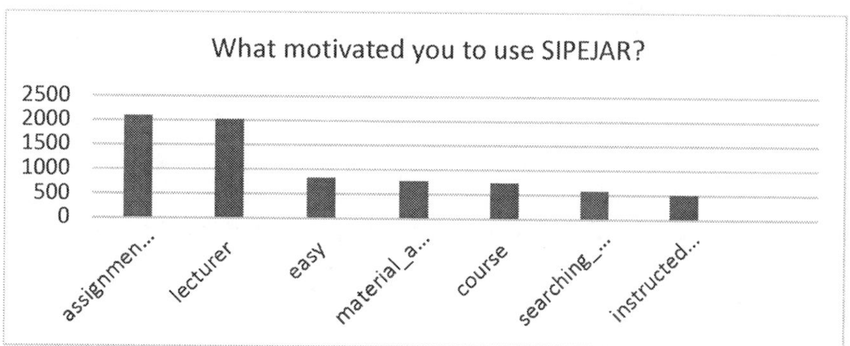

Figure 4. Graph of the total number appearance (Question 1).

Assignment from lecture/collect assignment/deadline assignment, lecturer, easy, material assignment, and searching for information are the words that appear most often. The use of technologies has modified the ways lecturers distribute course materials to students, rarely do university students transcribe notes from blackboards/whiteboards [11]. *Learning Management System* (LMS) changes the behaviour of lecturers and students, starting from the distribution of assignments, materials, quizzes, communication, and even information for can be done in SIPEJAR. It is also the same as the statement from the previous study [12] which state that the instructor, in this case the lecturer is very influential on the behaviour of students in using learning technology, if the lecturer has an interest and willingness to embrace with information technology, then it also affects the success of the application of information technology. Similarly with study in [13] to find academics' behavioural intention to use LSM said that ease of use can be one of the factors that influence the attitude toward using LMS and intention or motivation to use LMS. So, the things that must be improved so that motivation to use SIPEJAR increases are: (1) How the assignment or material assignment can be distributed and collect by adding reminder alarms to students regarding assignment deadlines; (2) Interest of lecturer to introduce and explain about SIPEJAR by giving instructions to the lecturer to be obliged to explain how the operation of SIPEJAR itself; and (3) Ease of use it by increase the type of services and user interface.

Table 3. Total number of appearance (Question 2).

Attribute	Appearance of Word	Appearance of Document
Internet/Connection/network/connection_internet/ signal/network_internet/server	946	939
Server/down/loading/error/slow	648	645
Difficult/complicated/complex/	591	583
Features/display/interesting/interesting display	460	445
Confused/Ease	177	174
Material	145	132
Access	127	126
Task lecturer/task_SIPEJAR	114	113
Utilization	56	53
Socialization	55	54
Services	47	46

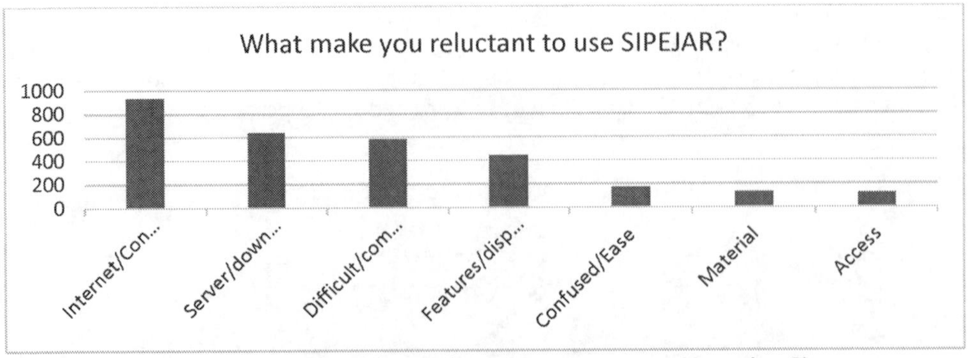

Figure 5. Graph of the total number appearance (Question 2).

Table 3 and Figure 5 show for the total number of appearances in question 2 ("what make you reluctant to use SIPEJAR?"). The table provides the top five words with the most mentioned in the answer given by respondent that are relevant with question 2: **First,** internet/connection/network/connection network/network internet/ server with total 946 appearances of word and 939 appearances of document. **Second**, server/down/loading/error/slow with total 648 appearances of word and 645 appearances on documents. **Fourth,** difficult/ complicated/complex with the total appearances of word: 591 and appearances of document: 583. **Next**, feature/display/interesting/display interesting with total number appearances of words: 460 and total number appearances of document: 445. **Fifth**, material with total number appearances of words: 145 and total number appearances of document: 132. And the last wo words that appear the least are service with total 47 appearances of word and 46 appearances of document, then utilization and socialization with 56 and 55 total appearances of word and 53 and 54 total appearances of document.

The result indicates that users are reluctant to use SIPEJAR because of the influence of the internet connection, especially if the connection is interrupted, because to use SIPEJAR must be connected to the internet. Inversely proportional to the study [14] which shows that the respondents are quite comfortable when using the internet to access the site. This is possible because the level of connection in that country is better than in here. In addition, the operation of the functions offered by SIPEJAR is difficult and complicated, so users will find it difficult to use it then make them reluctant to use it. This is similar to the study described [15] [16] about the LMS survey. The result said that users does not fully understand and be fully aware of the functions offered from the LMS used. Because the user must know how to access it, utilize, navigate, and convenience when using the system like the results obtained in this research [17] so that it is successfully implemented. In that study [15] also shows the same in terms of website performance, if the website server is often error or loading, it also makes users reluctant to use it, especially when uploading assignments, seeing feedback and quizzes, or download the materials, of course it is frustrating for the users. For display design or feature in SIPEJAR, it also gives an impact on users' behavior towards SIPEJAR. If SIPEJAR features is not interesting for users such as inconsistent display, or colors and icons used, this will make the users not interested in using it. Thus, in order to improve services of SIPEJAR, some actions that can be considered are: (1) Internet connection must be improved by providing Wi-Fi with wider access area coverage and increased speed, also to avoid errors and loading when accessing SIPEJAR; and (2) The process of using SIPEJAR is quite difficult or complicated, for that service providers must make the process of operation easier and more attractive user interface design. The last is about material provided, SIPEJAR must improve the types of material and also the information provided on the website.

Table 4. Total number of appearance (Question 3).

Attribute	Appearance of Word	Appearance of Document
Display / Interesting display / Sipejar display	1154	1149
Feature / feature / It's feature	951	861
Notification / Task notification / Notif	532	509
service	468	454
Simplicity	349	349
socialization	205	204
forum discussion / forum	158	146
deadline task	49	48

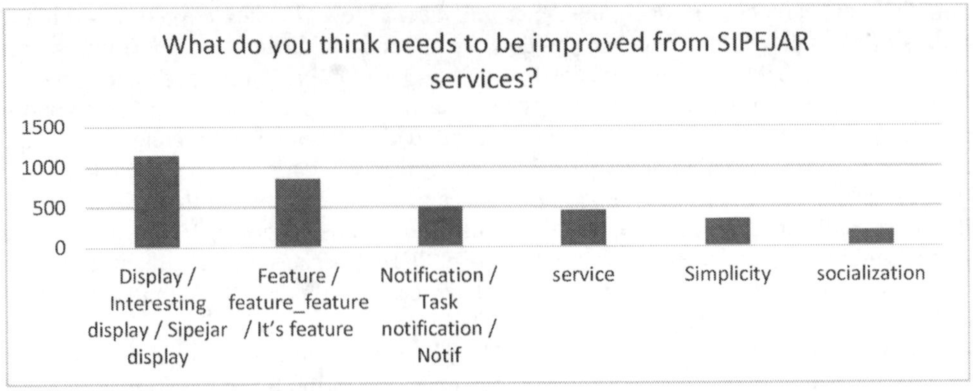

Figure 6. Graph of the total number appearance (Question 3).

Table 4 and Figure 6 show the total number of appearances in question 3 (what do you think needs to be improved from SIPEJAR services?). The top five words with the most mentioned in answer given by respondent which is related to the question are: **First**, display/interesting display/ SIPEJAR display with total 1154 appearances of word and 1149 appearances of document. **Second**, feature/feature feature/ it's feature with total 951 appearances of word and 861 appearances of document. **Third**, service with total appearances of word: 468 and appearances of document: 454. **Fourth**, notification /task notification/notif with total 532 appearances of word and 509 appearances of document. **Fifth**, simplicity with total appearances of word: 349 and appearances of document: 349. And the last two words that appear the least are forum discussion / forum with total 158 appearances of word and 146 appearances of document, then deadline task with total 49 appearances of word and 48 appearances of document.

From the result shows above, there are several things need to be improved, namely the display is made to be more attractive, features such as those offered can be suitable with the needs of the user such as automatic task notifications/reminders, and also the ease of service offered. This is also supported by research conducted by Carleton University (2018) which provides recommendations on some actions that can be done to improve the usefulness of NGOs that are used based on answers from respondents, for example by paying attention to instructions given on the website, improving the search capabilities of the website, navigation, features tips and tricks as well as instructional videos to be able to support the website, so users can also learn about the use of the website through these media. Besides, the user interface needs to be upgraded, it must be consistent in terms of the use of design, color, etc. SIPEJAR features will facilitate students and make students feel that accessing SIPEJAR became their needs. By knowing the material that will be discussed in class students can access further information to read outside of the course material and conduct research through the use of online journals and databases [11].

The growth in the world of information technology give a large volume of data to be processed to see actually what the respondent needs. There have been several techniques used to improving the learning management system besides text mining. For example in [18] and [13], they are use the technology acceptance model (TAM) to assessing the learning management system with attribute perceived ease of use, perceived usefulness, and attitude toward using it. Also, there is some paper related to this research are [14] that examines learner satisfaction when using learning management system and from the user's response is known what should be improved in the system, Carleton University report [15] that doing a research about how the feedback from LMS users about their experience when using it, Ohliati and Abbas [19] that measuring student satisfaction from private university when using the learning management system, and Hussein [20] that examine the attitude of a user in Saudi university toward using their LMS. In this paper, data was collected using a questionnaire, then processed using SPSS. The results generated using SPSS describe of how the user responds, but not as a whole. While, text mining provides results based on the words most often expressed in the response given by the user, and the analyzer must determine which words have an influence on the objectives. SIPEJAR must improve the design to make it more interesting, add new features, make notifications about deadline assignments to be collected or notification that there are new assignments given by lecturers, and simplify the process of its operation.

4. Conclusion

This paper presented SIPEJAR as a learning management system and text mining used to see the respond of the users. Text mining is one of the techniques used to extract meaningful information effectively from a large amount of unstructured data. This paper provided an overview of processing the data using text mining that tries to find information by seeing which words appeared the most from the answers given by respondents. From the result, the most mentioned word for question 1: what motivated you to use SIPEJAR? is because assignment from lecture/collects assignment/deadline assignment, lecturer, ease, material assignment, and searching for information. For question 2: what makes you reluctant to use SIPEJAR? is because of the internet connection, difficulties, and complexity when using it, server error, and interesting display and features. Then, for question 3: What do you think needs to be improved from SIPEJAR services? it is an interesting display, the features provided, services, notification, and simplicity when using it. So, from the results that have been obtained, it is expected that training or workshops will be conducted to improve the understanding and ability of users in operating SIPEJAR, as well as utilizing existing resources to improve services, features, also the user interface of SIPEJAR. Later with the recommendation given in this study, it is hoped that this will help provide useful information to improve the services of SIPEJAR in University Negeri Malang.

References

[1] Agaçi R. Learning Management Systems in Higher Education. UBT International Conference. University for Business and Technology; 2017 Oct 28. doi.org/10.33107/ubt-ic.2017.190.

[2] Takamatsu K, Kozaki Y, Kishida A, Bannaka K, Mitsunari K, et al. ANALYZING STUDENTS' COURSE EVALUATIONS USING TEXT MINING: VISUALIZATION OF OPEN-ENDED RESPONSES IN A CO-OCCURRENCE NETWORK. PEOPLE: *Int. J. Soc. Sci.* Global Research & Development Services; 2018 Nov 15;**4**(3):142–53. doi.org/10.20319/pijss.2018.43.142153.

[3] Maheswari MU. Text Mining : Survey On Techniques And Applications. International Journal of Science and Research. 2017.

[4] Sukanya M, Biruntha S. Techniques On Text Mining Proc. Ieee Int. Conf. Adv. Commun. Control Comput. Technol. 2012;(978):269–271.

[5] Ferreira-Mello R, André M, Pinheiro A, Costa E, Romero C. Text mining in education. *Wiley Interdiscip. Rev. Data Min. Knowl. Discov.* Wiley; 2019 Aug 4;**9**(6). doi.org/10.1002/widm.1332.

[6] Hofmann M, Klinkenberg R. *Rapid Miner Data Mining Use Cases And Business Analytics Applications*. Chapman & Hall/CRC Data Mining and Knowledge Discovery Series. 2014.

[7] Miner GD, Elder J, Nisbet RA. *Practical Text Mining And Statistical Analysis For Non-Structured Text Data Applications*. Computing and Information Sciences from the Association of American Publishers. 2012.

[8] Berry MW, Castellanos M, editors. *Survey of Text Mining II.* Springer London; 2008. doi.org/10.1007/978-1-84800-046-9.

[9] Aggarwal CC, Zhai C, editors. *Mining Text Data.* Springer US; 2012. doi.org/10.1007/978-1-4614-3223-4.

[10] Vijayarani S, Ilamathi MJ, Nithya M, Professor A, Research Scholar MP. *Preprocessing Techniques For Text Mining -An Overview*. 2015;**5**(1): 7–16.

[11] Moyle K, Wijngaards G, editors. *Student Reactions to Learning with Technologies*. IGI Global; 2012. doi.org/10.4018/978-1-61350-177-1.

[12] Teo T. Modelling technology acceptance in education: A study of pre-service teachers. *Comput Educ*. Elsevier BV; 2009 Feb;**52**(2):302–12. doi.org/10.1016/j.compedu.2008.08.006.

[13] Alharbi S, Drew S. Using the Technology Acceptance Model in Understanding Academics' Behavioural Intention to Use Learning Management Systems. *Int. J. Adv. Comput. Sci. Appl.* The Science and Information Organization; 2014;**5**(1). doi.org/10.14569/ijacsa.2014.050120.

[14] Sunkara VM, Kurra RR. An Analysis Of Learner Satisfaction And Needs On E-Learning Systems. *Int. J. Comput. Intell. Res*. 2017;**13**(3):433–444.

[15] University C. Lms Survey. *Educ. Dev. Cent*. 2018:1–16.

[16] Reid DL. Learning Management Systems: The Game Changer for Traditional Teaching and Learning at Adult and Higher Education Institutions. *GJHSS*. Global Journals; 2019 Jul **23**;1–14. doi.org/10.34257/gjhssgvol19is6pg1.

[17] Naveh G, Tubin D, Pliskin N. Student satisfaction with learning management systems: a lens of critical success factors. Technology, Pedagogy and Education. Informa UK Limited; 2012 Sep 28;**21**(3):337–50. doi.org/10.1080/1475939x.2012.720413.

[18] Wichadee S. Factors Related To Faculty Members' Attitude And Adoption Of A Learning Management System. *Turkish Online J. Educ. Technol.* 2015;**14**(4):53–61.

[19] Ohliati J, Abbas BS. Measuring students satisfaction in using learning management system. *Int. J. Emerg. Technol. Learn.* 2019;**14**(4):180–189.

[20] Hussein HB. Attitudes of Saudi universities faculty members towards using learning management system (JUSUR). *Turkish Online J. Educ. Technol.* 2011;**10**(2):43–53.

Determine effective pedestrian paths as evacuation routes for landslide disaster (case study: manyaran and kembangarum villages)

D W Latuihamallo[1,2], N Susanto[1,3] and A Susanty[1,4]

[1]Departement of Management and Industrial Engineering, Faculty of Industrial Engineering, Diponegoro University, Indonesia 50275

E-mail: [2]delianalatuihamallo066@gmail.com, [3]nophie.susanto@gmail.com, [4]ariessusanty@gmail.com

Abstract. Central Java is one of the provinces in Indonesia that is vulnerable to natural disasters, such as earthquakes, volcanic eruptions, and landslides. Landslide incidents in Central Java from 2011-2015 have the highest frequency compared to other disasters, i.e. 568 incidents. Primarily, for landslide, the data shows that the incidents in the city of Semarang have a high frequency. The previous study about the readiness index of several villages in the city of Semarang in facing the landslide disaster has been conducted and the result indicated that 2 out 19 surveyed villages belong to the category not ready for facing the landslides, namely Manyaram and Kembangarum. One of the causes of the lack of readiness is they don't have an evacuation route. So, based on this condition, the purpose of this study is to determine an effective pedestrian path as evacuation routes for landslide disaster for Manyaram and Kembangarum Villages. In this case, the effective pedestrian path will be determined based on the shortest, fastest, and inexpensive route parameters. The method used in this study is Network Analysis with ArcGIS10.3 applications. The result of this study will display the route map and trip descriptions of the shortest, fastest, and inexpensive route.

1. Introduction

Central Java is one of the provinces in Indonesia that is vulnerable to natural disasters, such as earthquakes, volcanic eruptions, and landslides. Of the five disasters that most often occur during the last few years is a landslide disaster. Landslide incidents in Central Java from 2011-2015 have the highest frequency compared to other disasters, i.e. 568 incidents. Primarily, for landslide, the data shows that the incidents in the city of Semarang have a high frequency. According to [1] Semarang City has a high enough potential for landslides in almost all regions. Regions in the city of Semarang that have the potential for landslides are divided into three classes of vulnerability, namely: quite vulnerable, vulnerable, and very vulnerable found in the sub-districts [2]. The condition of the bare mountain slopes and the fragility of the rocks and the unstable soil conditions make these lands unable to hold water in times of heavy rainfall with a high level of the village. Besides being caused by high rainfall, the geological conditions which include the morphology of the area such as hills and slopes, the ~existence of geological structures (faults) that control, as well as the lithology/rock conditions that make up the area. Based on the results of research conducted by [3] the district of West Semarang is dominated by residential areas. Most of the settlements in disaster-prone areas are 63.8% of the total residential area or 45.82% of the total area of the District of West Semarang. This indicates that the risk of disasters in the West Semarang Sub-district is high because the extent of settlements in areas prone to disasters is greater than those that are not prone to disasters.

The previous study about the readiness index of several villages in the city of Semarang in facing the landslide disaster has been conducted and the result indicated that 2 out 19 surveyed villages belong to the category not ready for facing the landslides, namely Manyaram and Kembangarum [4]. This study also described several causes of the low index of preparedness of citizens against landslides in the

Kembang Arum and Manyaram villages which are basically the same. One of the causes of the lack of readiness is they don't have an evacuation route.

The query that requires to be resolved in this study is how to determine some alternative evacuation routes for the community through the pedestrian route in mitigating landslides and how to choose an effective evacuation route for the community through the pedestrian route in mitigating landslides. In this case, the effective pedestrian path will be determined based on the shortest, fastest, and inexpensive route parameters.

Efforts to mitigate the risk of natural disasters with various preventive measures to minimize the negative impacts of the disaster, one of them is by the existence of an effective and efficient evacuation route map based on several available variables so that the community can be helped in finding routes/paths to get to safe and fast evacuation sites via pedestrian evacuation routes. The approach used in determining the evacuation route in this study is the Network Analysis approach which has strong potential for solving pathway problems [5] in the application of Geographic Information Systems (GIS).

2. Methodology

The approach used in this study is a quantitative research approach with a descriptive approach that provides an overview of the object under study through data or samples that have been collected and processed. This research was conducted in Manyaran and Kembang Arum villages, which were conducted from February-March 2020. In this research, a method of data collection in the form of techniques and methods commonly used by researchers to collect data is needed, for example with field activities for primary data collection through interviews, and observations to determine the condition of the problem that is relevant to the purpose of the initial research.

Secondary data, it can be obtained by studying road data / official archives obtained from several agencies related to disaster management, such as the Semarang local government, in this case, BNPB for data related to disaster and regional monographs as well as several articles, journals from research studies with the same topic. The following are some secondary data needed in conducting this research, namely:

- Road data is an important factor in determining the route, in this case, a road network map in the Kembang Arum and Manyaram districts.
- Landslide Disaster Location Data of Kembang Arum and Manyaram villages from Semarang City BPBD which will be used as a reference in determining some shelters or temporary evacuation sites in the event of a natural disaster.
- Digital Map Data on Land Use 1: 25.000 from Bappeda Kota Semarang.
- Data on Land Transportation Map of Manyaram and KembangArum village
- Shelter Location Data is a temporary evacuation site for people affected by the landslide.

Data analysis techniques in this study using the *Network Analysis* method that will produce an effective evacuation route from several existing route alternatives, based on parameters by looking at the shortest route, the fastest route, the low-cost route, and also the pedestrian path with a path safest means that the evacuation paths formed later can also be used by pedestrian communities, as well as the qualitative descriptive analysis used to explain evacuation routes generated in the Geographic Information System application to determine these pathways. Map data that has been collected is then analyzed by scoring, overlapping, and description methods.

The scoring method is a way of analyzing data by giving a rating or score to each indicator in accordance with the parameters used. Furthermore, manual calculation with operation research tools precisely analyses the network with the Shortest Route Algorithm to see how far the comparison of numbers is generated. The results are said to be accurate if both results are close to or the same [6] Research methodology is shown in Figure 1.

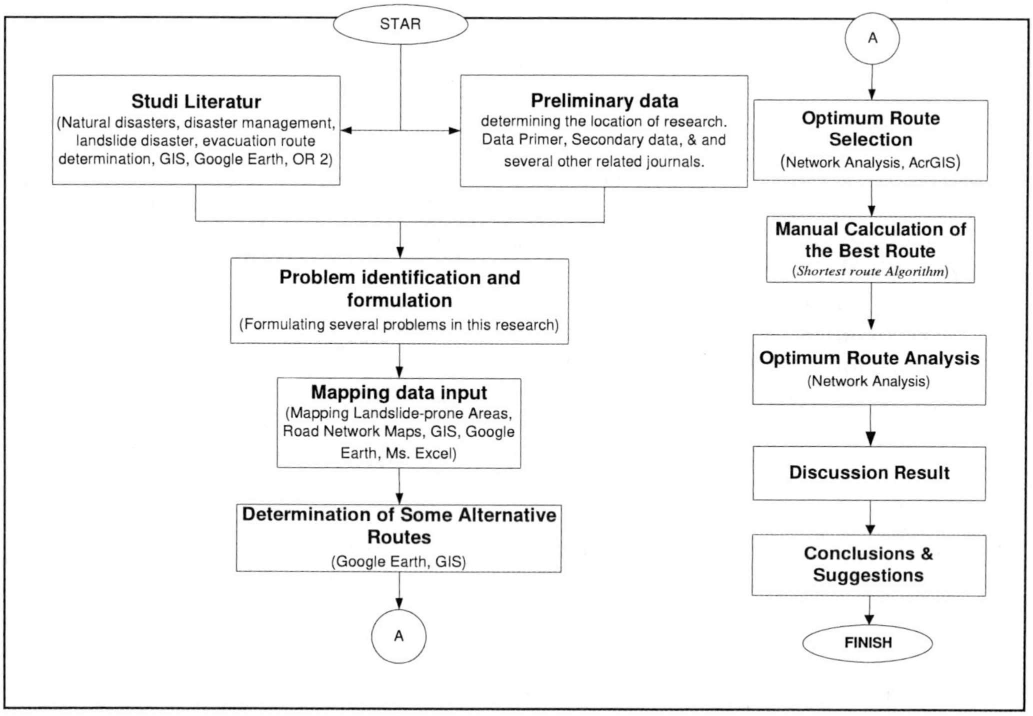

Figure 1. Research methodology.

3. Result and Discussion

3.1. Determination of Evacuation Paths

Evacuation means displacement or resettlement from dangerous areas, such as war hazards, flood hazards, landslides, volcanic eruptions, safe areas; it can also be interpreted as evacuating, moving away, evacuating people from the danger area. Evacuation Route is a special route that connects all areas to a safe area (Gathering Point). In a city, evacuation routes are very important to evacuate people to a safe place if something unexpected happens (in disaster). Therefore, evacuation route signs must be installed in all disaster-prone areas. Disaster evacuation routes are fundamental information needed in disaster-prone areas. Evacuation scenarios need to be supported by the readiness of the government and communities in flood-prone areas as noted in Law No. 24 About disaster management. Evacuation route planning is included in the preparedness stage before natural disasters occur. For the standard width of the evacuation route, there is actually no general provision, which must be considered whether this route can be traversed properly and quickly, and for the evacuation route, it should be able to load two vehicles so that if each other does not obstruct the evacuation process, there is a fixed gathering point with easy and wide access, away from sources of threats and the effects of threats. The determination of the evacuation route must at least consider a number of important things such as the Disaster Preparedness Training Handbook in [7] which has outlined some of the important things, namely:

- Evacuation route with the fastest and safest route for refugees to the refugee camp.
- There are alternative routes besides the main route.
- Suitability of the time needed to reach the shelter.
- Completeness of resources including the availability of vehicles that can be used in the evacuation process.
- It is also important to consider the position of the vehicle and the minimum amount of load if needed.

- Evacuation maps are based on survey and design results that inform evacuation routes, places of refuge, and time to reach them, alternative routes, disaster safe locations, and the position of evacuation team standby posts.

As for some principles in regulating earthquake evacuation locations from the perspective of urban planning according [8] namely :

- Safety, means that potential evacuation sites must avoid potential risk areas such as fault lines, melting risks, and chemical warehouses.
- Land use control, means that prospective evacuation sites can only be selected from plots of land where the use rights are controlled by the government.
- Nearest evacuation, means that when an earthquake occurs, the shelter needs to be close to the residential area, and it must be easy for residents to find the earthquake evacuation shelter.
- Economy, for example, the cost of building a shelter from the government.
- Suitability of distance and time, means that evacuating must consider the distance and time to get to the shelter, and this can increase the likelihood of successful evacuation to the shelter.
- Maximum coverage, for example, a single shelter that has been determined must be with a standard of no more than 2 km.
- Population capacity, meaning that each shelter/evacuation place must have a population limit.

In the evacuation route and to get to an evacuation point very necessary for the role of GIS. The ability of GIS can be used in disaster management as a function of data displays, Land Information Storage and Retrieval, Zone and District Management, Site Selection, Hazard Impact Assessment, and Development / Land Suitability Modelling [9].

3.2. Route Determination with ArcGIS (Network Analysis)

In general, the definition of GIS is a component consisting of hardware, software, human resources, and data that work together effectively to enter, store, improve, update, manage, manipulate, integrate, analyze and display data in a geographically based information. GIS Component is shown in Figure 2.

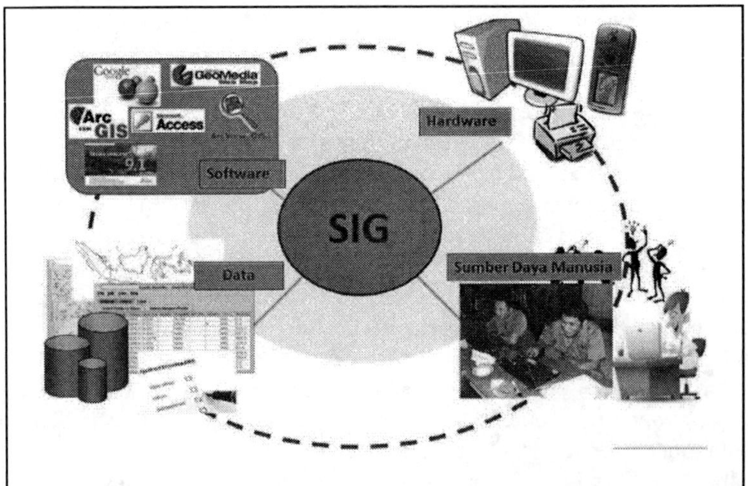

Figure 2. GIS component.

Broadly speaking, a Geographic Information System consists of four main stages or four subsystems that need to be understood earlier by each user. GIS subsystem is shown in Figure 3.

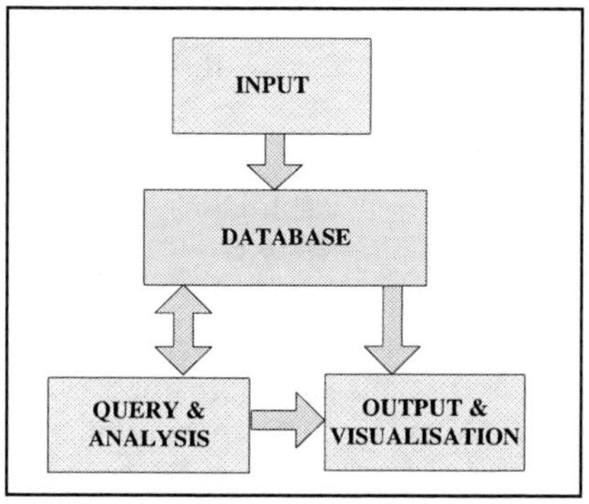

Figure 3. GIS sub-system.

A network is a collection of points called nodes, which are connected by arcs or branches. Figure 4. Show the link between start node and last node. In network analysis, there are several terms that are often used to describe a particular object such as events (events) and activities (activities). Activities are a work assignment, where the completion requires a certain period of time, costs, and facilities. Usually given the arrow symbol. Events are the beginning or end of an activity. Usually given a circle symbol. Many things can be presented with networks, communication systems, distribution systems, project planning, and others. Figure 5. and figure 6. explain the relationship between events and activities in a network system and a simple example of the relationship between the event and the activity above, it can be explained briefly.

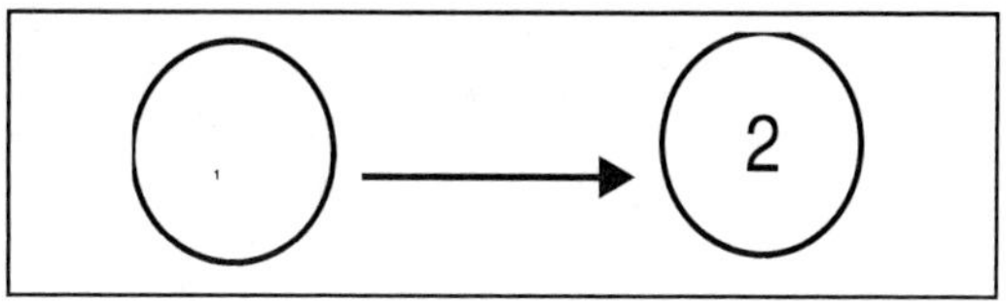

Figure 4. The link between the start node and the last node.

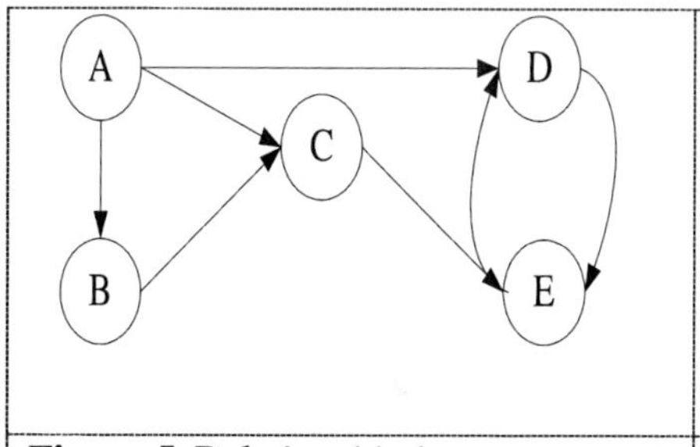

Figure 5. Relationship between events and activities in a network system.

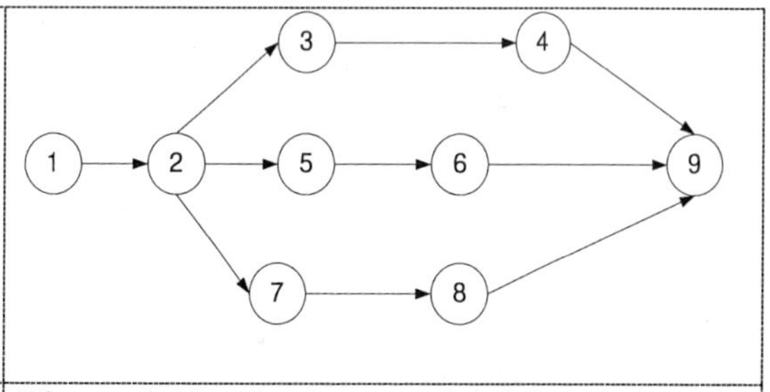

Figure 6. The simple example of the relationship between the event and the activity above, it can be explained briefly.

The simple system that can be generated from this research is as follows, Land Transportation as the Network System, landslide disaster location as the initial node, the evacuation site/shelter as the final node and road/lane as arrows, types of currents in the form of vehicles, pedestrians. Or more clearly seen in table 1. below this:

Table 1. A simple overview of the system in this research.

System Name	Network System	First Node	Lastly Node	Arrows	Current Type
Evacuation Route	Land Transportation	Landslide Disaster Location	Evacuation Location	Rood lane/ Route	Vehicle/On foot

One of the tools that will be used in this research is network analysis or known as Network analyst which has been provided in the ArcMAP GIS application that aims to analyze the transportation network, where in analyzing the network Analyst network will find the path of the least impedance. The analysis of this toolbox can help users to find the shortest route between two or more locations, place a source right in the middle, or find the most efficient route connecting several locations. The location can be determined interactively by placing points on the layer, by entering addresses, or by using points in features that are in the class features. To analyze the Geographical Information System begins with image data as basic data. From the image data in the form of physical data, scanning is performed to obtain digital data. The layering process is the process of separating the layer categories that exist in digital data. The result of this layering is vector data that has been divided into several data layers according to their respective covers. For example, the Semarang city map data has been divided into sub-district boundary layers, road layers, river layers, and others. Each layer will get information from tabular data on the related process. The output of this process is a shape file (file with the extension shp), where this file is ready to be used to display geographically based information. Figure 7. Show the block analysis diagram model [10].

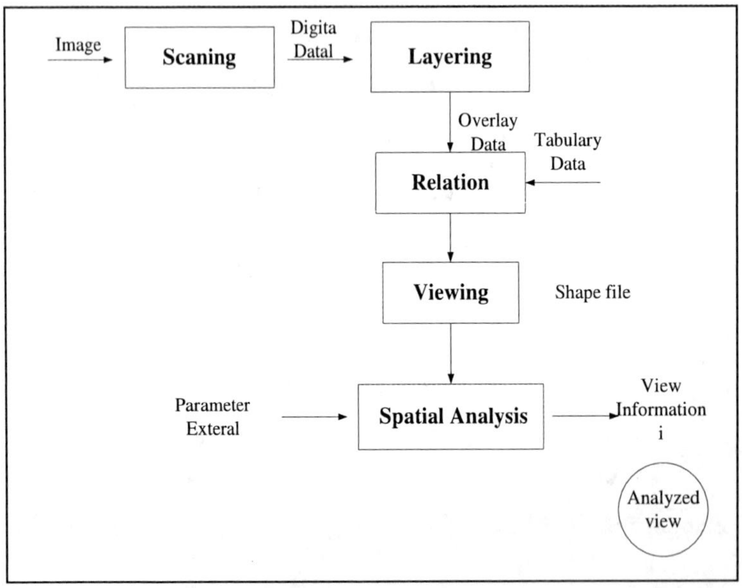

Figure 7. Block analysis diagram model.

In this process, information can only be displayed to users. To get the analyzed display data, then from the view out of the view, a spatial analysis process is carried out by entering input categories called external parameters. The results of the spatial analysis process are data views that have been analyzed that can be used as a user reference in making choices, in this case, the route for selecting disaster evacuation routes [10]. The method of solving problems in finding effective pathways is divided into 4 parts, namely the shortest path, fastest path, cheapest path, safest path. The manual of the analysis model is as follows :

1) Determination of the Shortest Route.
To analyze the shortest path, the shortest path view is prepared. The external parameter used to find the shortest path is the distance/length parameter. Distance is denoted by L, which consists of several alternative roads (from start to finish) so that the notation is (L1, L2, L3 ... Lk). Segments are determined from each intersection. To determine the total cost field can be modeled as follows:
Cost Field (CF) = length
Total Cost Field =
$(D1+D2+D3+\ldots D_N) = \sum_{k=1}^{n} D$ (1)
where : D = Selected Path
N = Number of Alternatives
2) Determination of the Fastest Route.
The second problem is determining the fastest path, from a point/start point to the destination point (finish point). In determining the fastest path the external parameters that also determine the weighting (Cost Field) are the distance and condition of the vehicle's rate. Distance is denoted by L, which consists of several alternative roads (from start to finish) so that the notation is (L1, L2, L3 ... Lk). Segments are determined from each intersection. Vehicle speed conditions are represented by speed V, each segment. So there are (V1, V2, V3, ... VK). To determine the total cost field can be modeled as follows:
Fastest Cost Field (CF) = C = Speed / Distance, L/V
Total Cost Field =
$(E1+E2+E3+\ldots E_N) = \sum_{k=1}^{n} E$ (2)
where: E = Chosen paths in the fastest analysis
N = Number of alternatives.
In the existing model Length is the distance based on the unit (not yet representing the original distance). To adjust the scaling, the length is multiplied by 60 units, this is intended to make the actual distance value. Length x 60 = meters. Because of the scale of 1: 60.
Speed data is in km, so it needs to be adjusted. To find the cost field in meters, the formula is applied:
Fastest Cost Field (CF) = Minutes = (distance / speed * 1000) / 60 (3)
The results of this weight are used as the basic for the analysis.
3) Determination with Low Cost
For the analysis of determining the cheapest route, in this study using the transport route approach. This approach is used because this public transport model is able to trace every analysis road without recognizing road/route boundaries. For the analysis of the cost path, the cheapest route is determined by calculating the fare of the transport. With the distance, he traveled. The results of the multiplication are added with the first start rate. The modeling in this analysis is as follows, Distance is denoted by L, which consists of several alternative paths (from start to finish) so that the notation is (L1, L2, L3 ... Lk). Segments are determined from each intersection. To determine the total cost field can be modeled as follows :
Fastest Cost Field (CF) for D = Distance * Cost, /100 (4)
The unit used is the meter.
Total *Cost Field* =
$(D1+D2+D3+\ldots D_N)$ + Cost/ Starting Point $= \sum_{k=1}^{n} D + Tsp$ (5)
Where : D = Chosen Paths in the Fastest Analysis
N = Number of alternatives.
Tsp = first star rates
Of the three categories above, to determine the best path / the most appropriate path is to model the results of the analysis above to get an effective output in terms of time and cost. From the results of the three categories above displayed in a View, so the effective path can be determined by the user.
4) Determination of the safest path.
For the analysis of the determination of the path by considering the safest path will be done by looking at the environmental conditions around the landslide disaster site. The criteria used are

divided into 2 parts, namely the criteria for determining safe evacuations and the criteria for evacuation routes, shown in the following table 2. :

Table 2. Criteria of the safest path.

Evacuation Route	Evacuation Path
Retaining capacity	Included in the disaster affected zone
Availability of basic needs	Slope level
Toilet availability	Road width / road section
	Pedestrian Path
	Road Conditions
	The direction of the road

3.3. Landslide Hazard Index

Based on reports on the results of natural disaster studies in Indonesia in [11] the landslide hazard index is divided into three levels, namely low, medium and high, and the City of Semarang is high. which is equal to 184. Based on the results of research conducted by [3] the district of West Semarang is dominated by residential areas. Most of the settlements in disaster-prone areas are 63.8% of the total residential area or 45.82% of the total area of the District of West Semarang. This indicates that the risk of disasters in the District of West Semarang is high because the extent of settlements in areas prone to disasters is greater than those that are not prone to disasters. Landslide-prone location data shown in table 3.

Table 3. Landslide prone location data.

Village	Landslide Disaster Point
Kembang Arum	RW 1, RW 7, RW 8, RW 9, RW 10, RW 11
Manyaran	RW 3, RW 4, RW 5, RW 6, RW 9, RW 10

Based on the landslide-prone location data [12],[13] the data collection and research carried out in the RT / RW that experienced the landslide, and the following are the location points that frequently occur in landslides in both villages. Map of landslide disaster point in kembang Arum village and Manyaran village can be shown in Figure 8. and Figure 9.

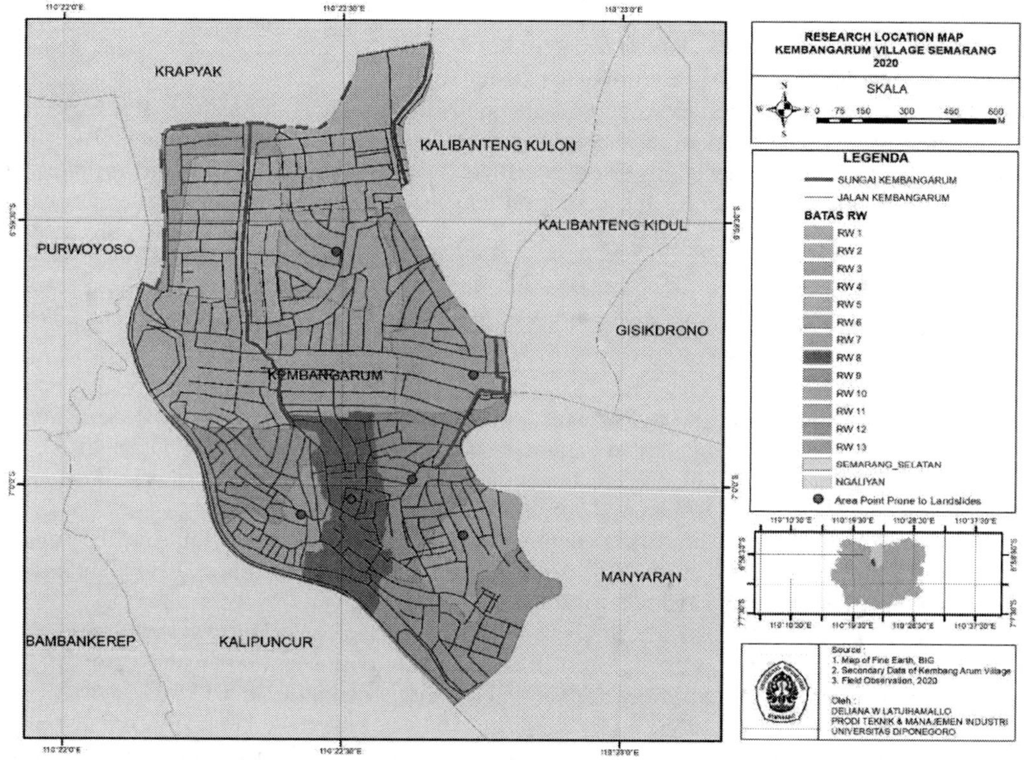

Figure 8. Map of landslide disaster point in Kembang Arum village.

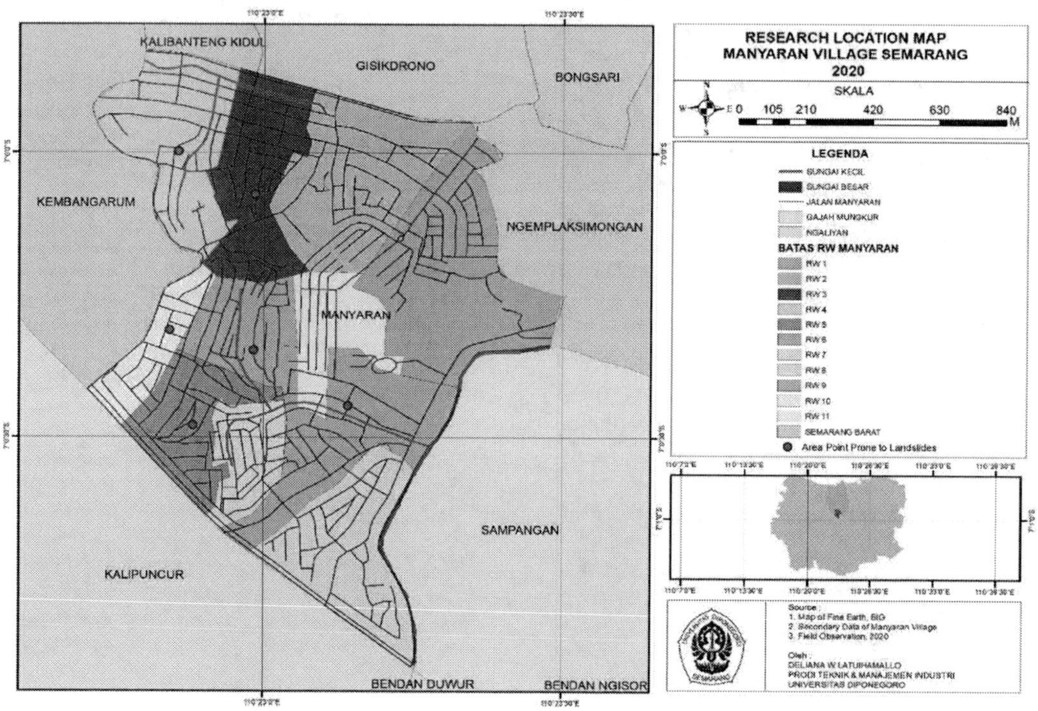

Figure 9. Map of landslide disaster point in Manyaran village.

4. Conclusion

The use of algorithms in network analysis to determine the best evacuation routes produced will then be compared with manual calculations according to existing formulas so that the results of both calculations will be used to determine the best evacuation route points from several alternatives. It is expected that the evacuation route determination can help the local community in evacuating themselves when a landslide disaster comes. Evacuation Pathway Model is made by connecting the location of the evacuation space and the location of residential settlements that have a distance close enough to the location that is considered to have a high level of risk.

References

[1] Susanto N, Putranto TT. Analisis Level Kesiapan Warga Menghadapi Potensi Bencana Longsor Kota Semarang. *UEJS*. Institute of Research and Community Services Diponegoro University (LPPM UNDIP); 2016 Dec 31;**37**(2):54. doi.org/10.14710/teknik.v37i2.9815.

[2] Purba J, Subiyanto S, Sasmito B. Pembuatan Peta Zona Rawan Tanah Longsor Di Kota Semarang Dengan Melakukan Pembobotan Parameter. *Jurnal Geodesi Undip*. 2014;**3**(2): 40-52.

[3] Arief M, Pigawati B. Kajian Kerentangan di Kawasan Pemukiman Rawan Bencana Kec. Semarang Barat, Kota Semarang. *Jurnal PWK Undip*. 2015;**4**(2).

[4] Susanto N, Putranto TT, Ulfa EA. Designing Cognitive Intervention to Improve the Awareness Index of the Residents in the Landslide Area. *EES*. IOP Publishing; 2017 Feb;**55**:012043. doi.org/10.1088/1755-1315/55/1/012043.

[5] Akay AE, Wing MG, Sivrikaya F, Sakar D. A GIS-based decision support system for determining the shortest and safest route to forest fires: a case study in Mediterranean Region of Turkey. *Enviro. Monit. Assess.* Springer Science and Business Media LLC; 2011 Apr 21;**184**(3):1391–407. doi.org/10.1007/s10661-011-2049-z.

[6] Pramudya RA, Subiyanto S. Penggunaan algoritma dijkstra dalam perencanaan rute evakuasi bencana longsor di kota semarang. *UEJS*. Institute of Research and Community Services Diponegoro University (LPPM UNDIP); 2015. doi.org/10.14710/geoplanning.2.2.93-102.

[7] BNPB. *Jurnal Penanggulangan Bencana*. BNPB:Jakarta. 2012;**3**(1).

[8] Liu Y, Yin K, Chen L, Wang W, Liu Y. A community-based disaster risk reduction system in Wanzhou, China. *IJDRR*. Elsevier BV; 2016 Oct;**19**:379–89. doi.org/10.1016/j.ijdrr.2016.09.009.

[9] Levine J, Landis JD, Klosterman R. Geographic Information Systems for Local Planning. *J Am Plann Assoc*. Informa UK Limited; 1989 Jun 30;**55**(2):209–20. doi.org/10.1080/01944368908976022.

[10] Aziz MM. Aplikasi Penentuan Rute Terbaik Scanning data Layering data overlay Relation Theme / Shape File Viewing Analisis Spasial Parameter External. *J Ilm Teknol Inf Din*. 2005;**10**(2):76-83.

[11] BNBP, BIG. *Buku Atlas Kebencanaan Indonesia*.2011 Retrieved from https://bnpb.go.id/buku/atlas-kebencanaan-indonesia.

[12] Secondary data from Kembang Arum Village. 2020.

[13] Secondary data from Manyaran Village. 2020.

Optimization in job shop scheduling problem using Genetic Algorithm (study case in furniture industry)

S L Aquinaldo[1,2], N R Cucuk[1,3] and Yuniaristanto[1,4]

[1]Industrial Engineering Department of University Sebelas Maret, Jl. Ir. Sutami, 36 A, Surakarta, Indonesia

E-mail: [2]saralaurensia@student.uns.ac.id, [3]cucuknur@gmail.com, [4]yuniaristanto@gmail.com

Abstract. Job shop scheduling problem belongs to a class of NP-Hard problems. We solve a scheduling problem in a job shop based furniture company. The company produces several products such as chair, table, home decorations, and home accessories. Currently, the company schedules the order using Earliest Due Date (EDD) and First Come First Serve (FCFS) methods. The best schedule resulted from those methods is then chosen and used as the initial solution for Genetic Algorithm (GA) method. The proposed algorithm is implemented in MATLAB 2019a to minimize the makespan. Parameters used in the GA formation of new generations are done by crossover using the Precedence Preservative Crossover (PPX) method and mutations using job-pair exchange mutations. The selection of chromosomes for regeneration in the crossover process is chosen by two chromosomes that have the best fitness and for the mutation process, one chromosome that has the worst fitness is chosen. Solution from genetic algorithm is better than EDD for the case study. From the results, GA produces shorter makespan compared to EDD and FCFS methods. The EDD method gives a makespan of 104,280 minutes and the FCFS method gives a makespan of 118,440 minutes, while GA provides a makespan of 81,780 minutes.

1. Introduction

Production scheduling in a manufacturing industry has an important role in minimizing production lead time. Decisions in the scheduling that are defined in the assignment are in the form of job sequencing intern of time to start and finish the operations. Scheduling also deals with the allocation of resources that exist at a certain period, with the purpose of optimally used such resources [1]. Finally, production scheduling is useful for increasing productivity and minimizing operating costs [2]. Good scheduling will also minimize delays that occur in a company.

The job shop scheduling problem is one of the scheduling problems that have task processing sequence constraints, and each task must go through each machine exactly once. Each job scheduled has a specific process through a specific set of machinery [3]. Job shop scheduling is included in complex combinatorial optimization scheduling problems and categorized as one of the Np-hard problems, a problem whose search for solutions will rise exponentially as the size of the problem increases linearly [4].

Genetic algorithm method has been able to solve various problems related to optimization, including job shops schedule. In this paper, GA is used to reduce the idle time of the machine. The initial solution is result from EDD method to reduce late delivery problems in a furniture company. Furniture company located in Trangsan RT 01 RW 05 Gatak Sukoharjo, established since April 2002. This company produces products in the form of furniture and home decor with a workforce of 80 people. Focuses on the international market by exporting to several countries including United States, Belgium, France, Netherlands, and Spain. Company makes products based on orders so processing between one order to another is different or can be called as job shops. The objective of the scheduling in this paper is to minimize the makespan.

2. Methods

The production scheduling system used by the company results in the late order completion. Delayed completion of orders and the buildup of orders is a major factor in the emergence of problems with customers. This study uses a heuristic EDD method to solve the delayed completion of orders in a job order furniture company. The resulted EDD schedule still has several idle times. Therefore, GA is used to reduce the idle time based on the EDD schedule. Methodology starts with calculation production scheduling using EDD method to minimize tardiness. After that do the optimization on EDD scheduling by implementing GA using matlab software. Finally, do a comparison between the company's initial method and proposed method. The research methodology is shown in Figure 1.

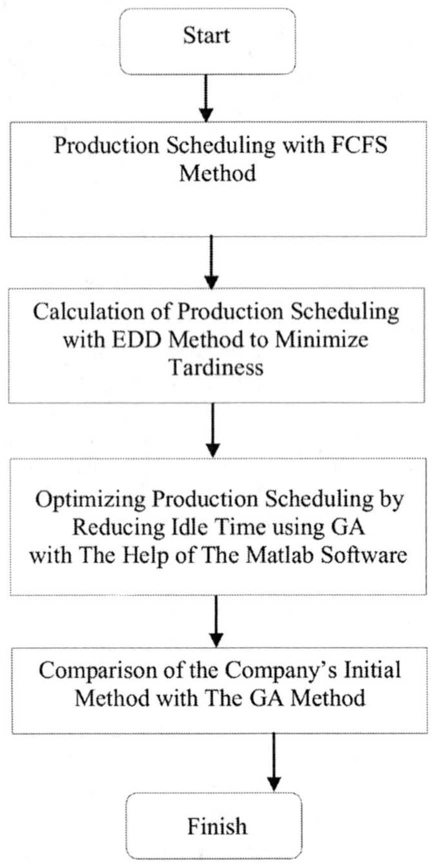

Figure 1. Research methodology.

2.1. Genetic Algorithm

GA is an effective meta-heuristic to solve combinatorial optimization problems and has been successfully adopted to solve the job shops scheduling problem. Recently, many papers discuss about this topic. [5] developed the scheduling algorithm for job shop scheduling problems with parallel machines with re-entrant processes. [6] applied a product grouping system, so that there are groups of jobs that must be done in the same vulnerable time based on the same product group. [7] formulated Global Selection (GS) and Local Selection (LS) which designed to generate high-quality initial population in the initialization stage.

Genetic algorithm is one of effective metaheuristic methods to solve optimization problems. This method was developed based on idea that exist in Darwin genetics and theory. The optimum value is found by a searching technique which carried out simultaneously on several possible solutions known as population. Individuals in a population are called chromosomes. This chromosome is a solution consisting of some genes. The initial population is built randomly, while the next population is the result of the evolution of chromosomes through an iteration called generation. In each generation, chromosomes will undergo an evaluation process using a measuring instrument called a fitness function. Fitness value of a chromosome will indicate the quality of chromosomes in the population.

The next generation is known as offspring, formed from a combination of two chromosomes of the current generation that act as parents using crossover operator. Apart from crossing operations, chromosomes can also be modified using another genetic operator called mutation. New generation population is formed by selecting the fitness value of the parent chromosome and the fitness value of the offspring chromosome and rejecting the other chromosomes (chromosomes with a small fitness value) so that the size of the population is constant. After going through several generations, this algorithm will converge to the best chromosome that will be the solution of the problems.

3. Result and Discussion

3.1. Process Operations Data

The EDD method ranks jobs according to their due date and this sequence is then used as a basis to determine the production schedules. Data recap of the processing time of each operation for six products coded as J_1, J_2, J_3, J_4, J_5, and J_6 are shown in Table 1.

Table 1. Data processing time.

Operation	Suar join pedestal (minute)	Suar join stool (minute)	Wood stool only (minute)	Donat suar stool (minute)	Stool hocker black (minute)	Stool hocker white (minute)
1	$T_{1,1,1(1)}=780$	$T_{2,1,1(1)}=1200$	$T_{3,1,1(1)}=420$	$T_{4,1,2(2)}=960$	$T_{5,1,1(1)}=1620$	$T_{6,1,1(1)}=1560$
2	$T_{1,2,2(1)}=720$	$T_{2,2,2(1)}=900$	$T_{3,2,2(1)}=300$	$T_{4,2,3(3)}=840$	$T_{5,2,3(3)}=300$	$T_{6,2,3(3)}=300$
3	$T_{1,3,2(1)}=840$	$T_{2,3,2(1)}=1320$	$T_{3,3,2(1)}=360$	$T_{4,3,5(5)}=180$	$T_{5,3,5(5)}=180$	$T_{6,3,5(5)}=180$
4	$T_{1,4,2(2)}=540$	$T_{2,4,2(2)}=780$	$T_{3,4,2(2)}=240$	$T_{4,4,1(1)}=240$	$T_{5,4,1(1)}=480$	$T_{6,4,1(1)}=420$
5	$T_{1,5,1(1)}=180$	$T_{2,5,1(1)}=180$	$T_{3,5,1(1)}=180$	$T_{4,5,1(1)}=360$	$T_{5,5,1(1)}=360$	$T_{6,5,1(1)}=360$
6	$T_{1,6,2(1)}=540$	$T_{2,6,2(1)}=720$	$T_{3,6,2(1)}=60$	$T_{4,6,1(1)}=360$	$T_{5,6,1(1)}=300$	$T_{6,6,1(1)}=300$
7	$T_{1,7,1(1)}=240$	$T_{2,7,1(1)}=360$	$T_{3,7,1(1)}=180$	$T_{4,7,1(1)}=240$	$T_{5,7,1(1)}=360$	$T_{6,7,1(1)}=360$
8	$T_{1,8,2(1)}=540$	$T_{2,8,2(1)}=720$	$T_{3,8,2(1)}=600$	$T_{4,8,1(1)}=360$	$T_{5,8,1(1)}=360$	$T_{6,8,1(1)}=360$
9	$T_{1,9,1(1)}=1320$	$T_{2,9,1(1)}=2040$	$T_{3,9,5(3)}=180$	$T_{4,9,5(5)}=720$	$T_{5,9,1(1)}=180$	$T_{6,9,1(1)}=180$
10	$T_{1,10,2(1)}=360$	$T_{2,10,2(1)}=360$	$T_{3,10,1(1)}=180$	$T_{4,10,1(1)}=2700$	$T_{5,10,1(1)}=360$	$T_{6,10,1(1)}=360$
11	$T_{1,11,3(1)}=60$	$T_{2,11,3(3)}=120$	$T_{3,11,1(1)}=360$	$T_{4,11,1(1)}=420$	$T_{5,11,1(1)}=300$	$T_{6,11,1(1)}=300$
12	$T_{1,12,5(5)}=60$	$T_{2,12,5(5)}=120$	$T_{3,12,3(3)}=360$	$T_{4,12,1(1)}=120$	$T_{5,12,1(1)}=420$	$T_{6,12,1(1)}=420$
13	$T_{1,13,1(1)}=240$	$T_{2,13,1(1)}=300$	$T_{3,13,5(5)}=240$	$T_{4,13,1(1)}=1440$	$T_{5,13,1(1)}=1440$	$T_{6,13,1(1)}=1440$
14	$T_{1,14,1(1)}=360$	$T_{2,14,1(1)}=360$	$T_{3,14,1(1)}=360$	$T_{4,,14,1(1)}=60$	$T_{5,14,1(1)}=60$	$T_{6,14,1(1)}=60$
15	$T_{1,15,1(1)}=240$	$T_{2,15,1(1)}=360$	$T_{3,15,5(2)}=180$			
16	$T_{1,16,1(1)}=240$	$T_{2,16,1(1)}=300$	$T_{3,16,1(1)}=60$			
17	$T_{1,17,1(1)}=360$	$T_{2,17,1(1)}=360$	$T_{3,17,1(1)}=1440$			
18	$T_{1,18,1(1)}=240$	$T_{2,18,1(1)}=360$	$T_{3,18,1(1)}=60$			
19	$T_{1,19,1(1)}=240$	$T_{2,19,1(1)}=300$				
20	$T_{1,20,1(1)}=1440$	$T_{2,20,1(1)}=1440$				
21	$T_{1,21,1(1)}=60$	$T_{2,21,1(1)}=60$				

3.2. EDD Results

To solve the delayed completion of orders in a job order using the EDD method. Table 2 below shows the schedule results of EDD method.

Table 2. The schedule EDD results.

Product	Processing Time		Start	Due date	Finish
	(minute)	(days)			
Suar Join Pedestal	8,880	21	08/04/2019	11/06/2019	03/05/2019
Suar Join Stool	17,220	41	08/04/2019	11/06/2019	27/05/2019
Wood Stool Only	18,720	38	15/05/2019	17/07/2019	08/07/2019
Donat Suar Stool	17,880	48	06/05/2019	13/07/2019	10/07/2019
Stool Hocker Black	19,980	46	06/05/2019	13/07/2019	08/06/2019
Stool Hocker White	21,600	51	06/05/2019	13/07/2019	13/07/2019
Total	104,280	245			

From the table above, it can be seen that by using the EDD method, the problem of delayed can be overcome for all products where each product can be completed on time but still have several idle times. Hence, the results of makespan with EDD method is 104,280 minutes.

GA is used to reduce the idle time based on the EDD schedule. In this paper we use 400 generations with a population of 80 and the problem is represented using encoding permutation. For problem representations, chromosomes are represented as integers, where the time calculation is based on two matrices, namely the order process matrix and the time matrix. Maximum operating limit is 21 with 19 workstations. The generation of chromosomes is done by generating random numbers. The size of chromosomes is based on the number of jobs multiplied by the number of machines (n x m). The initial population chromosome is 80 x 102, where 80 expresses the population size and 102 expresses the length of the chromosome. Chromosomes that have been raised will be determined based on the fitness value. In this paper, we use crossover probability of 0.45 and probability of mutation is 0.01 [8]. The results of GA are shown in Table 3.

Table 3. The schedule GA results.

Product	Processing Time		Start	Due date	Finish
	(minutes)	(days)			
Suar Join Pedestal	15,600	44	08/04/2019	11/06/2019	31/05/2019
Suar Join Stool	19,080	41	08/04/2019	11/06/2019	27/05/2019
Wood Stool Only	8,400	38	15/05/2019	17/07/2019	04/07/2019
Donat Suar Stool	14,220	30	06/05/2019	13/07/2019	19/06/2019
Stool Hocker Black	14,280	24	06/05/2019	13/07/2019	31/05/2019
Stool Hocker White	10,200	17	06/05/2019	13/07/2019	22/05/2019
Total	81,780	194			

Total of makespan for the six products is 81,780 minutes. The comparison of the makespan between the GA method and EDD is shown in Figure 2.

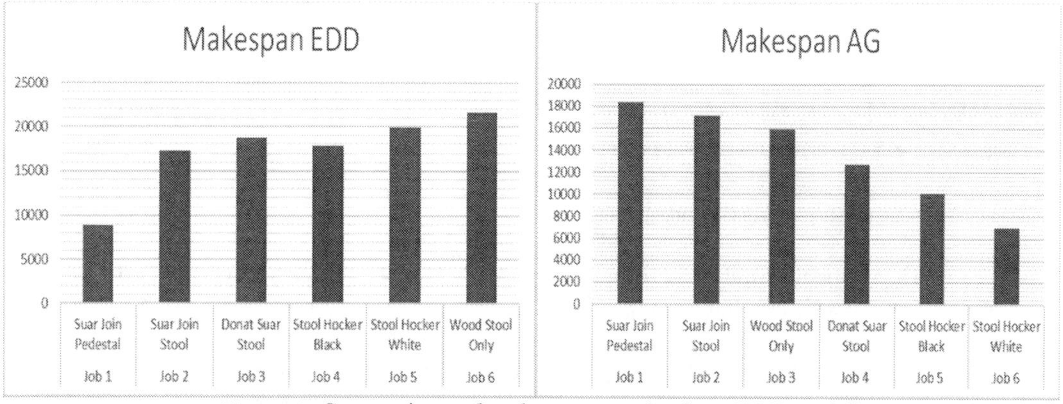

Figure 2. Comparison of makespan results from EDD and GA.

From the figure, we can see that the EDD method results a makespan of 104,280 minutes and the GA method results a makespan of 81,180 minutes. Hence, the GA gives 20.5% better results than EDD rule.

4. Conclusions

In this paper, GA is used to reduce the idle time of machines based on schedule resulted from EDD rule. The results of GA give total makespan of 81,780 minutes while EDD rule gives total makespan of 104,280 minutes. The efficiency obtained by genetic algorithms is 20.5%. Suggestion for further research that work insertion can be done.

References

[1] MCKAY K, PINEDO M, WEBSTER S. PRACTICE-FOCUSED RESEARCH ISSUES FOR SCHEDULING SYSTEMS*. *Prod Oper Manag*. Wiley; 2009 Jan 5;**11**(2):249–58. doi.org/10.1111/j.1937-5956.2002.tb00494.x.

[2] Mohamed Y, Borrego D, Francisco L, Al-Hussein M, AbouRizk S, Hermann U. Simulation-based scheduling of module assembly yards: case study. *Eng. Constr. Archit. Manag*. Emerald; 2007 May 8;**14**(3):293–311. doi.org/10.1108/09699980710744926.

[3] Applegate D, Cook W. A Computational Study of the Job-Shop Scheduling Problem. *ORSA Journal on Computing*. Institute for Operations Research and the Management Sciences (INFORMS); 1991 May;**3**(2):149–56.doi.org/10.1287/ijoc.3.2.149.

[4] Aryawan B. PT. PENERAPAN ALGORITMA TABU SEARCH DALAM PENJADWALAN JOB SHOP. *MAKARA of Technology Series*. Universitas Indonesia, Directorate of Research and Public Service; 2010 Oct 14;**7**(3). doi.org/10.7454/mst.v7i3.197.

[5] Chen JC, Wu C-C, Chen C-W, Chen K-H. Flexible job shop scheduling with parallel machines using Genetic Algorithm and Grouping Genetic Algorithm. *Expert Syst. Appl*. Elsevier BV; 2012 Sep;**39**(11):10016–21. doi.org/10.1016/j.eswa.2012.01.211.

[6] Krisnanti, Ria, Sudiarso A. Penjadwalan Mesin Bertipe Job Shop untuk Meminimalkan Makespan dengan Metode Algoritma Genetika (Studi Kasus Pada PT X). Simposium Nasional RAPI XI FT UMS. 2012:60-65.

[7] Zhang G, Gao L, Shi Y. An effective genetic algorithm for the flexible job-shop scheduling problem. *Expert Syst. Appl*. Elsevier BV; 2011 Apr;**38**(4):3563–73. doi.org/10.1016/j.eswa.2010.08.145.

[8] Grefenstette J. Optimization of Control Parameters for Genetic Algorithms. *IEEE Trans. Syst. Man Cybern*. Institute of Electrical and Electronics Engineers (IEEE); 1986 Jan;**16**(1):122–8. doi.org/10.1109/tsmc.1986.289288.

IOP Conf. Series: Materials Science and Engineering 1072 (2021) 012020 doi:10.1088/1757-899X/1072/1/012020

Designing supply chain models with blockchain technology in the fishing industry in Indonesia

S Larissa[1] and J Parung[1,2]

[1]Department of Industrial Engineering, University of Surabaya, Raya Kalirungkut, Surabaya 60293, Indonesia

E-mail: [2]Jparung@staff.ubaya.ac.id

Abstract. The fishing industry contributes to the national economy and has a high risk of being traded because it has a long shipping time, limited time to consume, and the amount traded varies depending on the season. This risk can reduce the potential profit of players in the fishing industry. To reduce this risk and maximize added value, blockchain technology is used to change the supply chain of the fishing industry. Blockchain technology is a ledger database that is distributed to record transactions among parties that can be verified and permanently. This paper proposes a blockchain technology framework in the fishing industry. This paper will explain the definition and benefits of blockchain, current applications, and design a new model of the supply chain. The potential benefits of the blockchain are transparency and traceability, reducing costs and time, and preventing illegal or unethical fishery products from entering the supply chain. Moreover, this paper will also explore how to build this blockchain technology in the supply chain.

1. Introduction

Indonesia is known as an archipelagic country with two-thirds of the sea area larger than land which provides benefits for fish production in Indonesian, given the ease in obtaining raw materials. A fish caught from various locations can change hands dozens of times and pass many processing and packaging steps before reaching its destination. However, there are several challenges in the fishery supply chain, including non-transparent file recording, limited production infrastructure and supporting facilities, high logistics costs, and long shipping time that leads to low product quality [1].

The lack of an open supply chain allows many fishery business players to access and manipulate data; this is undoubtedly a violation of recognized legal and ethical standards. Illegal, unreported, and unregulated (IUU) activities, such as overfishing, human rights violations, and fraud, often occur in the fishing industry [2]. Fully transparent and traceable fishery supply chains are essential to solve this problem and can be facilitated by blockchain technology [2]. Blockchain has the potential to achieve end-to-end transparency.

Blockchain has beneficial potential in the supply chain field but has limited attention in the literature. There is an essential gap in the literature for examining blockchain opportunities in the fishery supply chain. The purpose of this paper is to provide a blockchain supply chain model for the fishing industry in Indonesia. At the same time, it provides the potential and positive contribution of the blockchain in the fishery supply chain.

2. Literature review

Several papers were reviewed to know the current application of blockchain. Table 1 shows a comparison between the application of blockchain in many industries from previous studies and this study.

Table 1. Comparison of blockchain applications in various industries.

	(Perboli et al.) [3]	(Wang et al.) [4]	(Tonnissen et al.) [5]	(This research, 2020)
Topic	Logistics costs	Blockchain applications and technical challenges	Various effects of blockchain	Designing blockchain in the supply chain
Application	Fresh food delivery	Financial service, e-business, education, web community, privacy protection	Logistics, food trade, supply chain management and transportation	Fishery industry
Platform	IBM	Ethereum, Hawk	-	Treum
Conclusion	Blockchain provides benefits to all players in the process	Blockchain can transform the traditional industry	Blockchain is very dynamic and leads to permanent changes	Blockchain increases traceability, trust, and reduces cost

2.1 Global fishery supply chain

Fishery trade is a complex supply chain with each player possesses its own problems; thus, the role of distribution is critical because of the very perishable nature of fisheries (Figure 1). Fishery products pass through various stages in the supply chain, especially at the primary level between producers and first buyers or main processors, then between distributors and end consumers [6-7].

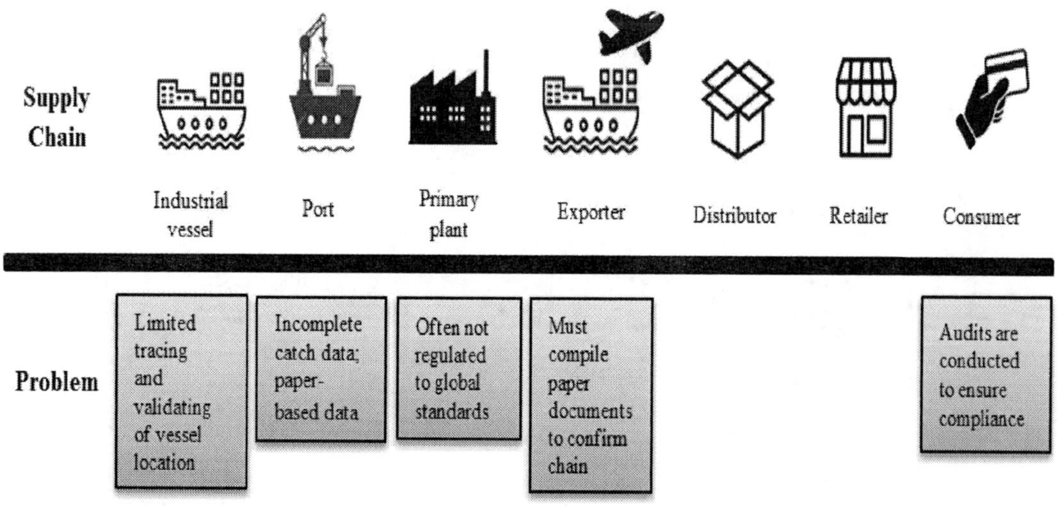

Figure 1. Supply chain in the fishing industry and the problems of each player.

2.2 Blockchain Technology

Blockchain is a trusted and decentralized network where nodes are anonymous and ledgers are owned by each node to maintain transaction history [8-10]. Figure 2 shows that blockchain records all transactions with encrypted, decentralized, verifiable, secure, and lower costs [11-13], thereby minimizing verification and network costs.

Figure 2. A simplified blockchain in the supply chain.

2.3 The WWF Pilot Project – Fiji Tuna Supply Chain

The World Wildlife Foundation (WWF) declared the Blockchain Tracking Project in the Supply Chain in January 2018 [2], to eradicate illegal tuna fishing. WWF partnered with Viant and Sea Quest Fiji Ltd. The objective of this project was to create a truly transparent and traceable supply chain by applying blockchain technology. Every part of the fish will be marked and verified at the arrival stage and brought to processing. In processing, the tag remains with the fish through the entire process to ensure the continuity of the fish's history. This tag will be removed in the packaging process and replaced with a unique QR code that matches the fish when it is sent to the market.

3. Designing a supply chain model with blockchain for the Indonesian fishing industry

WWF's success in implementing blockchain technology with Sea Quest Fiji Ltd could be a supporting factor for implementing blockchain in the Indonesian fishing industry. This paper will provide a fishery supply chain design model with blockchain. In-depth interviews and survey methods with many suppliers and employees from the fishing industry in Tuban, East Java were carried out to collect data about their problems in tracking fish products.

3.1 Fishery supply chain

An example of a fishery supply chain in Tuban, East Java, is shown in Figure 3. Data is recorded manually from the beginning to delivery through paper and book expeditions. It is certainly less efficient and difficult to track and monitor fish history at all times. Therefore, the authors propose the use of blockchain technology in the fishing industry.

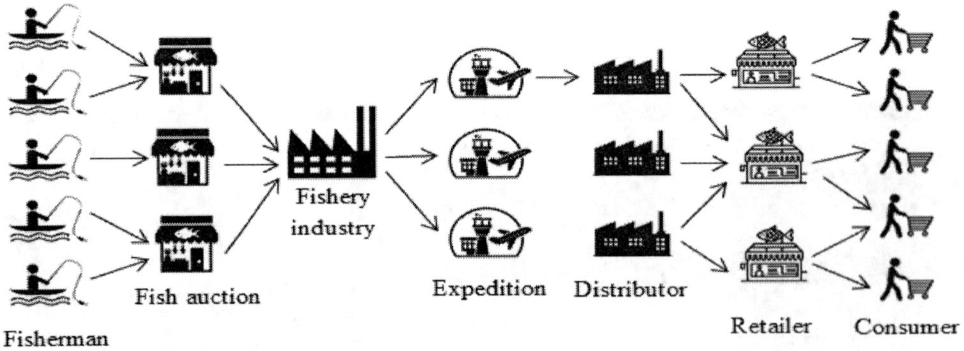

Figure 3. Fishery supply chain in Tuban, East Java.

3.2 Supply chain with blockchain

The design of the supply chain using blockchain for the fishing industry in Indonesia is shown in Table 2. Data recording is carried out by creating a QR code on each player, and tracking history will be checked by scanning the QR code.

Table 2. Design of supply chain with blockchain.

Player	Role	Tools	QR code	Tracking code
	Record the catch and create a product ID	GPS logbook IOT		**AA01HKSA** AA : Fisherman ID 01 : Fishing location HK : Fishing gear SA : Fish types
	Record, verify, track, and give label	IOT package		**XX01** XX : Supplier ID 01 : Fish auction ID
	Record, verify, track, and give label	IOT package		**SSAPTK100** SSA : Fishing Industry ID PTK : Product ID 100 : Weight of fish
	Record, verify, track, and give label	IOT package		**RHY** RHY : Shipper ID
	Record, verify, track, and give label	IOT package		**ABC** ABC : Distributor ID
	Record, verify, track, and give label	IOT package		**XYZ** XYZ : Retailer ID
	Check product history	IOT package		**AA01HKSAXX01SSAPTK100 RHYABCXYZ** History of fish product

Customers are able to scan the QR code and see all fish product history. Thus, they will know about the product journey and how the product is processed, which eventually will increase traceability and trust in the customer while buying the product.

3.3 What are the benefits of implementing blockchain?

Blockchain provides many benefits because it can provide openness and trust among players [14]. This technology facilitates fish tracking as fish enters customers through real-time monitoring. This process will increase traceability in the supply chain; decentralized ledgers help connect inputs from suppliers, manufacturers, distributors, to remote retailers [15]. Blockchain technology can ensure the quality products along the supply chain and increase the company's reputation among the consumers [15].

In this paper, an example was taken in the fishing industry in Tuban, East Java. The industry is experiencing difficulties in tracking down the causes of rotten fish products. At present, most of the rejected fish will be returned directly to the supplier and are assumed to come from fishing sources. In fact, fish damage can occur when washing and cutting time is too long, cooling equipment is damaged, and loading time is too long, causing a decrease in the quality of fish products. With blockchain technology, contaminated products can be easily tracked to find out the cause of the damage.

3.4 What cost should be considered to apply blockchain in the fishery supply chain?

As this technology is only in the early stages of development, the full cost structure cannot be determined at this time. However, it is possible to assess some basic costs that are an early indication when the system is implemented while thinking about more complex future costs. Table 3 shows the estimated costs for the fishing industry [2].

Table 3. Cost components for blockchain applications.

Type	Cost components
Hardware	QR code tags (per trip)
	QR code scanner – Handheld (on off)
	Tablet - Vessel (on off)
	Tablet - Landing & Processing (on off)
Software (per month)	Treum
Resources (manhours)	Training (on off)
	Support (per trip)
Miscellaneous	Fish tag printing (per trip)
	Label printing (per trip)
	Mobile data (per tablet)
	Internet bandwidth (per month)

4. Conclusion

Blockchain is a promising technology in providing a transparent fish product supply chain, but many obstacles and challenges still hinder its popularity in the fishery supply chain. However, blockchain with its main characteristics: decentralization, anonymity, transparency, and traceability, can change the traditional fishing industry for the better. Blockchain can be combined with other emerging technologies (big data, robotics, IoT, RFID, NFC, etc.), for higher automation of the fishery supply chain, resulting in full transparency and tracking.

This paper provides knowledge about blockchain technology, including its definition and potential, current applications, and designing new supply chain models. In addition, this paper lists some basic costs that can be an indication of the implementation of the blockchain in the fishery supply chain. Blockchain also has economic benefits for all users because it can make production more efficiently at lower costs.

Acknowledgments

We thank the University of Surabaya through the Master of Industrial Engineering Study Program of University of Surabaya that gives funding for this research.

References

[1] I. Seafood and S. Chain, "Chapter 6," no. 5, 2019.
[2] B. Cook, "Blockchain : Transforming the Seafood Supply Chain," pp. 1–41, 2018.
[3] G. Perboli, S. Musso, and M. Rosano, "Blockchain in Logistics and Supply Chain: A Lean Approach for Designing Real-World Use Cases," *IEEE Access*, vol. 6, no. c, pp. 62018–62028, 2018, doi: 10.1109/ACCESS.2018.2875782.
[4] H. Wang, Z. Zheng, S. Xie, H. N. Dai, and X. Chen, "Blockchain challenges and opportunities: a survey," *Int. J. Web Grid Serv.*, vol. 14, no. 4, p. 352, 2018, doi: 10.1504/ijwgs.2018.10016848.
[5] S. Tönnissen and F. Teuteberg, "Analysing the impact of blockchain-technology for operations and supply chain management: An explanatory model drawn from multiple case studies," *Int. J. Inf. Manage.*, no. January, pp. 0–1, 2019, doi: 10.1016/j.ijinfomgt.2019.05.009.
[6] C. Leong, T. Viskin, and R. Stewart, "Tracing the Supply Chain - How blockchain can enable traceability in the food industry," *Accenture*, pp. 1–63, 2019, [Online]. Available: https://www.accenture.com/us-en/insights/blockchain/food-traceability.
[7] M. C. Leal, T. Pimentel, F. Ricardo, R. Rosa, and R. Calado, "Seafood traceability: Current needs, available tools, and biotechnological challenges for origin certification," *Trends Biotechnol.*, vol. 33, no. 6, pp. 331–336, 2015, doi: 10.1016/j.tibtech.2015.03.003.
[8] S. Aggarwal, R. Chaudhary, G. S. Aujla, N. Kumar, K. K. R. Choo, and A. Y. Zomaya, "Blockchain for smart communities: Applications, challenges and opportunities," *J. Netw. Comput. Appl.*, vol. 144, no. February, pp. 13–48, 2019, doi: 10.1016/j.jnca.2019.06.018.
[9] Kairos Future, "Blockchain use cases for food traceability and control Blockchain use cases for food traceability and control A study to identify the potential benefits from using blockchain technology for food traceability and control," 2017.
[10] Z. Zheng, S. Xie, H. N. Dai, X. Chen, and H. Wang, "Blockchain challenges and opportunities: A survey," *Int. J. Web Grid Serv.*, vol. 14, no. 4, pp. 352–375, 2018, doi: 10.1504/IJWGS.2018.095647.
[11] M. Andoni *et al.*, "Blockchain technology in the energy sector: A systematic review of challenges and opportunities," *Renew. Sustain. Energy Rev.*, vol. 100, no. October 2018, pp. 143–174, 2019, doi: 10.1016/j.rser.2018.10.014.
[12] A. Kamilaris, A. Fonts, and F. X. Prenafeta-Boldú, "The rise of blockchain technology in agriculture and food supply chains," *Trends Food Sci. Technol.*, vol. 91, pp. 640–652, 2019, doi: 10.1016/j.tifs.2019.07.034.
[13] B. Bhandari, "Supply Chain Management, Blockchains and Smart Contracts," *SSRN Electron. J.*, pp. 1–22, 2018, doi: 10.2139/ssrn.3204297.
[14] K. Christidis and M. Devetsikiotis, "Blockchains and Smart Contracts for the Internet of Things," in IEEE Access, vol. 4, no. , pp. 2292-2303, 2016.," *IEEE Access*, vol. 4, pp. 2292–2303, 2016, doi: doi: 10.1109/ACCESS.2016.2566339.
[15] J. Parung, "The use of blockchain to support sustainable supply chain strategy," *IOP Conf. Ser. Mater. Sci. Eng.*, vol. 703, no. 1, 2019, doi: 10.1088/1757-899X/703/1/012001.

IOP Conf. Series: Materials Science and Engineering 1072 (2021) 012021 doi:10.1088/1757-899X/1072/1/012021

Efficiency measurements of Diponegoro University faculty using the Data Envelopment Analysis hierarchical network model

B M Hendrawan[1,2], B Purwanggono[1,3], and H Suliantoro[1,4]

[1]Department of Industrial Engineering, Faculty of Engineering, Diponegoro University, Semarang, Indonesia

E-mail: [2]bagusmaulana30@gmail.com, [3]b.purwanggono@gmail.com, [4]suliantoro_hery@yahoo.com

Abstract. Diponegoro University (UNDIP) is a state university located in Semarang city. UNDIP is aiming to become a world-class university. To become a world-class university, UNDIP has to maintain its quality in academic, research, and community service. But UNDIP was facing some challenges, like has to compete with other universities that growing rapidly. To achieve the target, the efficiency of every faculty in UNDIP should be continuously monitored and improved. Measuring the overall efficiency of every faculty is the aim of this research. The samples used in this research were all faculty of undergraduate programs in UNDIP. The hierarchical network data envelopment analysis model measuring the efficiency level of every faculty at UNDIP for systems with a hierarchical structure. In this research, the input variables used were the number of lecturers, the number of students, and amount of operational expenses. While the output variables were the number of graduates, the number of researches, the number of publications, the number of researches, the number of guidance & consultation, and the number of activities. The results of this study would help the management of university especially the dean of faculty to identify the inefficiency at every faculty in UNDIP and then plan for improvement.

1. Introduction

Measuring efficiency has an important role in the management of an organization. Measuring efficiency in an organization could indicate some directions for a few necessary developments in the future, and to list some achieved accomplishments in the past [1]. Not only applied in the sectors of business and organization, measuring efficiency, in this case, should be applied to the field of education in the higher learning institutions (university) [2]. An effective technique to measure a set of decision-making units (DMUs) relative efficiency that can be applied with using multiple inputs to produce multiple outputs is data envelopment analysis (DEA) [3]. In measuring the relative efficiency of an organization, data envelopment analysis is a suitable method to be used to measure the relative efficiency of an organization that uses multiple inputs to produce multiple outputs [2]. As a linear programming technique, to define the results of measuring efficiency using the data envelopment analysis results are easy to understand compared with the other linear programming methods because the efficiency scores result of data envelopment analysis are the result based on the inputs and outputs data that more specified [4].

The conventional data envelopment analysis model is not considering the internal structure of DMUs. Conventional data envelopment analysis model evaluates the system and treats this as a black box, only the inputs were supplied to and the outputs produced from it are considered. The internal processes in the black-box data envelopment analysis model ignoring all of the operations and its interdependence processes, thus it produces a result that the system with its component processes are assessed as efficient, even though it is not [1]. Meanwhile, organizations have various structures, as an example like parallel structure, two-stage structure, and hierarchical structure.

To achieve a relevant efficiency measurement with the framework of network data envelopment analysis, taking consideration of the component processes and the availability of data must be considered [1]. Most research focused on forming a single stage of data envelopment analysis model by combining all the functions of faculty in the university when it's applied to a faculty, each function of its internal activities were ignored. Because of this, the faculty having some difficulties in determining the overall inefficiencies of research, teaching, and services done by some individuals in the faculty [2]. This type of structure in the black-box data envelopment analysis model could produce a misleading result when it's applied to measuring both of the efficiency of DMU's and the internal efficiency units couldn't be measured, hence the hierarchical network data envelopment analysis model should be used [1].

The organizational structure of faculty at university has three main functions, those are research, teaching, and service, therefore a suitable model to measure the efficiency of the faculty in university is the hierarchical network data envelopment analysis [2]. A first previous study conducted by [1] measuring the efficiency of the Department of Physics at a university using a hierarchical structure, Kao's hierarchical system dividing teaching function into two activities: activities of teaching undergraduate, and masters. The second previous study conducted by [2] extended Kao's study about the faculty hierarchical system at university. In [2], Kao's hierarchical system was applied as the efficiency measurement and then extended in the university. Three activities of teaching function were further divided into teaching undergraduate program students, teaching master program students, and teaching Ph.D. program students. Two activities of service were divided into consultation activities and cooperation activities. In this paper, [1] and [2] hierarchical system is applied in the measurement of efficiency and then extended, based on Indonesia's policy to higher education institutions to apply *Tri Dharma Perguruan Tinggi* (Tri Dharma of higher education institutions). The purpose of achievement to be done is the tri dharma of higher education institutions, it's a responsibility for every college in Indonesia whether its state university or private university. Tri dharma of higher education institutions consists of education and teaching, research and development, and devotion/service to the community. Based on the tri dharma of higher education institutions, the function of research was divided into research function and service function. The function of research was further divided into two activities, those are work on research projects, and publications, meanwhile for the services function divided into two activities consists of consultation activities and academic activities.

In this paper, the number of first-level units on each DMU's hierarchical system have particular purposes. In the first level of units, if a unit has subordinate units at a lower level, all DMU must have the same number of subordinate units. Every DMU in the first level performing the same number of subordinate units and performing the same functions. Hence, several of one DMU of the unit have the same resemblance with the other DMUs of the unit [1]. Assumed that the intermediate unit's outputs were the outcome from the subordinate units, where the intermediate unit's output doesn't generate the outputs by itself. Every intermediate unit's inputs were distributed to the subordinate units. Pre-specified proportions were used to distribute the shared inputs to the subordinate units, or it could be said that the data reflecting the most advantageous proportions [1].

Aim of this paper is to compare the measurement of the overall efficiency at every faculty in Diponegoro University using the constant return to scale data envelopment analysis and the efficiency measurement of every faculty in Diponegoro University without disregarding its internal structure using the hierarchical network data envelopment analysis model for systems with a hierarchical structure [4]. This paper has five sections, the first section is the introduction, the second section contains efficiency measurement, the third section explains the hierarchical structure of a university faculty with its components, the fourth section is measurement model with constant return to scale data envelopment analysis and hierarchical network data envelopment analysis to calculate the efficiency of a university faculty, the fifth section is conclusions.

2. Literature

Different techniques and approaches can be used to measure efficiency. Two main methods in research are parametric method and non-parametric method, the appropriateness of the method depends heavily on the level of data used in the analysis [2]. The regression analysis approach is not suitable to be used in the education sector, because institutions of education producing multiple results of output using numerous inputs, while performance indicator constructed using regression model is derived from a production function, which is the average line through data, not frontiers around the data [2]. The reason for using the data envelopment analysis was because of due to its high flexibility in the educational sector. Instead of a single number of final inputs, a sequence of intermediate outputs and hardly not just a range of inputs was the function of production capability to adapt the processes [2].

2.1. Data Envelopment Analysis (DEA)

Data envelopment analysis (DEA) is an approach of the non-parametric method to measure the efficiency of a set of units to be measured called DMU for example, DEA mainly used to measure the efficiency of private firm, state agency, higher education institute, etc. [5]. Charnes, Cooper, and Rhodes (CCR) were the first to introduce the DEA in 1978. A technique of linear programming is the base of DEA. DEA has been widely used as an effective technique to measure the efficiency of a certain number of DMUs by using some specific number of inputs to produce a certain amount of output [4]. DEA is a non-parametric method for measuring efficiency when some many inputs and outputs must be used, while there are no generally acceptable weights to be able to combine several of these inputs and outputs in the aggregate [5]. DEA is very appropriate when used in evaluating efficiency by converting several inputs to several outputs. The advantage of DEA is that it can identify alternative configurations of some inputs that can produce a higher number of outputs without the need to increase the use of existing resources [5]. A linear programming technique like DEA could make the management to be able to assess each DMU and did a benchmarking process to a DMU with the best score of performance and provide estimates of potential improvements for inefficient DMUs. DEA also can make an approximate improvement for DMUs with inefficient assessment [5].

DEA recognized as a good method to analyze an achieved accomplishment, create an ideal model for the organization, and the process of an organization operational [6]. To evaluate education sector system, it is not the same as to use an evaluation mechanism on market (example: benefit assessment, inputs, and outputs the economic value, etc.) to determine DMU performance, DEA creates a highlight for the university to target for a certain number of inputs and a certain amount of output choice, and then create some option of quality to the system about indicators of input and output [6]. A set of boundaries including performers with the most efficient assessment produced by DEA, making it possible to compare among the performers directly. The parameters of assumed functional form evaluated by a method of statistic regression only with a process of enhancement to all of the DMU's where the DEA uses optimization for dissimilar DMUs without an assumption of priority as the basis for the forms of functional. *W*ith features that are owned by *the* DEA that have been proven unique, various organization with a different background that needs to evaluate its efficiency has been applied using DEA [6].

2.1.1. Data Envelopment Analysis Constant Return to Scale Model

Data envelopment analysis constant return to scale model also can be called DEA Charnes-Cooper-Rhodes (CCR) model is the most basic, it's treated as a black-box (single-stage DEA) model. In the Single-stage model, only inputs are supplied, thus outputs were produced and then to be calculated. Otherwise, the operation and interdependence of internal processes are ignored, resulting in the system can be assessed as efficient when each of the components that were measured was not.

2.1.2. Data Envelopment Analysis Hierarchical Network Model

Data envelopment analysis hierarchical network model was used to get a more accurate measurement of efficiency, it must consider whether the data is available or not in the system's internal structure [2]. Network DEA has been introduced by Färe & Grosskopf in 2000. The DEA hierarchical network model approach under the framework of DEA considers all the process components. The internal process of the DEA model has been explored and the systems can have several structures such as two-stage, parallel, hierarchical, series, and multistage [1]. Some organizations have several units at the first level, causing several subunits are set up at the second level, and it is possible large subunits are split up to some number of sub-subunits with its functions that different with others at the third level, the next level can be continued if necessary. Compared to the DEA constant return to scale, the model of DEA hierarchical network was less flexible because of the total internal weight in each category of each level of the hierarchy have to be the same as one. It's causing a result in the score of efficiency for every DMU that has been calculated from this network model will not surpass the one that calculated from the one level model, the DEA discriminating power will increase to a certain amount [2]. The structure of hierarchy was described that has different functions/activities in the model and its weight on one level from the other levels could be treated as different.

2.1.3. Previous Research

Related research about the measurement of efficiency defined in Table 1. From previous research and literature, variables of input and output to measure in this paper were referred from it. The input of a faculty in system were the number of students, the number of lecturers, and the operational expenses [1] [6]. The output of a faculty in system were the number of graduates, the number of research, the number of publication, the number of consultations, and the number of activities [1] [2] [5] [6]. Input and output used in this paper for measurement in efficiency can be seen in Table 2.

The description of the input variables and output variables in this paper was explained below:

- *The number of lecturers*

Referring to the literature from [1], [2], [5], and [6], researchers determined the number of lecturers as input variables in the study. The number of lecturers is the number of permanent lecturers who have the status as the permanent lecturers that have teaching experience for more than ten years in each faculty at Diponegoro University.

- *The number of students*

Referring to the literature from [6], researchers determined the number of active students as input variables in the study. The number of students is the number of active students is the number of students who have active lecture status, are not on leave or drop out status in each faculty at Diponegoro University.

- *Amount of operational expenses*

Referring to the literature from [1], [2], [6], and [7], researchers determined the amount of operational expenses as input variables in the study. The amount of operational expenses is the large amount of funds used by each faculty at Diponegoro University which is used for the operational activity of the faculty itself.

- *The number of graduates*

Referring to the literature from [1], [2], [5], [6], and [7], researchers determine the number of graduates is the number of students who graduate on time as output variables in research. The number of graduates is the number of students who graduate on time at each faculty in Diponegoro University who can complete the study period within a predetermined period of time or faster.

- *The number of researches*

Referring to the literature from [2] and [6], researchers determine the number of researches as output variables in research. The number of researches is accumulation of all research done by each faculty at Diponegoro University.

- *The number of publications*

Referring to the literature from [1], [2], and [5], researchers set the number of publications as output variables in the study. The number of publications is the accumulation of the total number of publications (books, articles, journals, papers, newspapers, magazines, etc.) that have been carried out by each faculty at Diponegoro University.

Table 1. Previous research.

Author	Sample	Input Variable	Output Variable	Empirical Methodology
[2]	14 faculties at a university	- Number of professors - Number of associate professors - Number of senior lecturers - Number of lecturers - Number of foreign academic staff - Number of non-academic staff - Operating costs	- Number of graduates from undergraduate level study programs - Number of graduates from S-2 level study programs - Number of graduates from PhD level study programs - Number of publications - Amount of the grant - Number of key researchers based on various types of grants - Number of experienced lecturers - Number of collaborative activities carried out under the MoU	DEA Hierarchical Network Model
[7]	28 departments of a university	- Operating expenses - Training resources - Credit hours	- Number of graduates - Promotions - Public services activities	DEA CCR model and DEA BCC Model
[1]	20 hypothetic universities physics department	- Number of personnel including academic faculty and administrative staff - Expenses including the procurement, maintenance of the equipment and facilities, salaries for the part-time workers, and the other types of expenditures	- Number of graduating undergraduate students in a year - Total credit-hours course for undergraduate - Number of graduating undergraduate students in a year - Total credit-hours taught for graduate courses - Total number of publications by faculty of the department in o year - Grants acquired from government funding agencies and research achievements - The income received from such contracts and extension programs	DEA CCR model and DEA Hierarchical Network Model
[5]	26 faculties in a university	- Outdoor and indoor area of university (m^2) - Number of academic staff - Number of administrative staff	- Number of publications - Number of graduate students	DEA CCR model and DEA BCC Model
[6]	21 departments of 2 university	- The number of registered students - The number of teaching staff	- The number of graduates - The number of passed students to higher levels - The performed research work	DEA CCR model

- The number of consultations

Referring to the literature from [7], researchers set the number of consultations as an output variable in the study. The number of consultations is the amount of guidance and consultation that have been done by the faculty in forms of mentoring activities including guidance and consultation about academic, community service program (KKN), industrial work practices, student creativity programs (PKM), and others.

- The number of academic activities

Referring to the literature from [2], researchers set the number of academic activities as output variables in the study. The number of academic activities is the accumulation of all academic activities held in or by the faculty which can be in the form of seminars, training, visiting professors, workshops, service, student exchanges, and other activities in accordance with the tri dharma of higher education institutions.

Table 2. Variables used as input and output for measuring efficiency in faculty.

INPUT	OUTPUT
1. The number of lecturers	1. The number of graduates
2. The number of students	2. The number of researches
3. Amount of operational expenses	3. The number of publications
	4. The number of consultations
	5. The number of activities

3. Hierarchical Structure of University Faculty

Faculty in University could be created as a structure in the form of a hierarchical structure since the faculty in university has three major functions, those functions are teaching functions, research functions, and service functions, which utilizes the resources to a certain amount of outputs [2]. The first previous study directed by [1] used the hierarchical structure to measure the efficiency of the Department of Physics at a university, Kao's hierarchical system dividing teaching function into two activities: activities of teaching undergraduate, and masters. The second previous study conducted by [2] extended Kao's study to the system of faculty level in university. In [2], a hierarchical system conducted by Kao was implemented in university. The function of teaching was further divided into several activities: teaching undergraduate, masters, and Ph.D. students, while the function of service was further divided into consultation activities and cooperation activities.

In this paper, hierarchy system in [1] and [2] was adopted, modified, and extended based on Indonesia's policy to institutions of higher education to apply *Tri Dharma Perguruan Tinggi* (Tri Dharma of higher education institutions). Tri Dharma of higher education institutions is the purpose of achievement to be done, it is a responsibility for every college in Indonesia. Tri Dharma of higher education institutions consists of education and teaching, research and development, devotion/service to the community. Based on Tri Dharma of higher education institutions the research function divided research function and services function into two activities to each function. Research function divided into two activities: work on research projects, and publications. Services function divided into two activities: consultation activities and academic activities. Teaching functions and research functions have been recognized as two of the primary tasks in the university in most studies on the performance of universities [8]. Difficult was faced to measure the performance of these two primary tasks, it is important to achieve some specific indicators that can be used to symbolize of achievements from those tasks, also it was taken into consideration of the resources that have been used by the faculty in bearing out these duties [2]. Based on the description before, the hierarchical structure can be illustrated in Figure 1.

From hierarchical structure as shown on Figure 1, creating a hierarchical structure with mathematical symbol will make the creating of formula about efficiency measurement in faculty at Diponegoro University a lot easier to understand. Figure 2 shows the hierarchical structure with mathematical symbol.

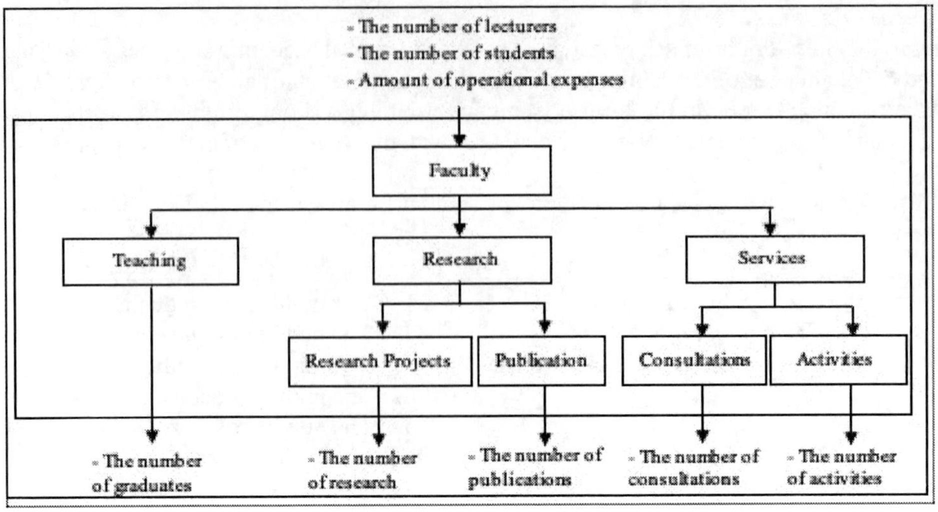

Figure 1. The hierarchy structure of a system in faculty.
(Source: A combination of a hierarchical structure from [1] and [2] that has been adapted to the actual conditions and research needs in the Diponegoro University)

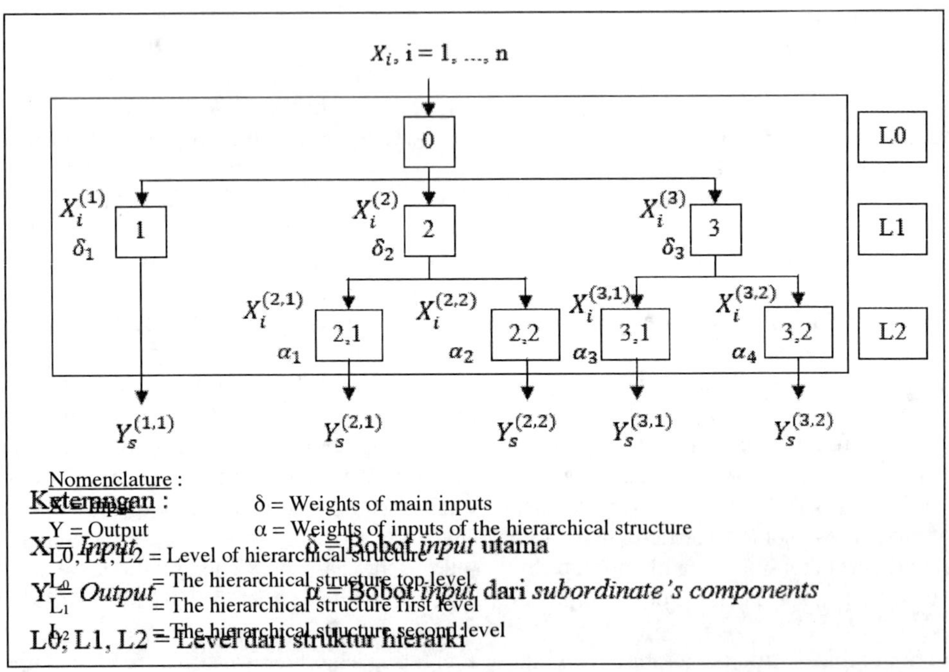

Figure 2. A hierarchical system of two level in faculty.
(Source: A combination of a hierarchical structure from [1] and [2] that has been adapted to the actual conditions and research needs in the Diponegoro University)

4. Efficiency Measurement

As can be seen at Figure 2, it is the hierarchy structure of DEA hierarchical network model as the basis of formulation to measure the efficiency of eleven faculties of undergraduate program at Diponegoro University. The 11 faculties whose efficiency was measured were Faculty of Law, Faculty of Economy and Business, Faculty of Social Science and Political Science, Faculty of Cultural Studies, Faculty of Psychology, Faculty of Public Health, Faculty of Medicine, Faculty of Animal Husbandry and Agriculture, Faculty of Science and Mathematics, Faculty of Fisheries and Marine Science, and Faculty of Engineering.

This study uses three inputs and five outputs. Between the different components, the input in every faculty were shared in the hierarchical process [2].

4.1. Measuring Efficiency Using DEA Constant Return to Scale (CRS) Model

The collected data then calculated using the DEA constant return to scale model as a black box to determine the overall efficiency value of each university faculty that is represented as the DMU. A basic hierarchic structure of system was assumed to be every T-th component of the DMU j (j = 1, 2, ..., m) $X_{ij}^{(T)}$ having *input* i (i = 1, 2, ..., n) and $Y_{sj}^{(T)}$, *output* s (s = 1, 2, ..., r). V_i is input weights and U_s are output weights. In general, the DEA method black-box model was used for efficiency measurements of DMU j [1]. The formula of the CRS DEA model is:

$$E_{d}^{CCR} = \max \frac{\sum_{s=1}^{r} Us\left(Y_{sd}^{(1,1)} + Y_{sd}^{(2,1)} + Y_{sd}^{(2,2)} + Y_{sd}^{(3,1)} + Y_{sd}^{(3,2)}\right)}{\sum_{i=1}^{n} V_i X_{id}} \quad (1)$$

Subject to:

$$\sum_{s=1}^{r} Us\left(Y_{sd}^{(1,1)} + Y_{sd}^{(2,1)} + Y_{sd}^{(2,2)} + Y_{sd}^{(3,1)} + Y_{sd}^{(3,2)}\right) - \sum_{i-1}^{n} V_i X_{ij} < 0, j = 1, \ldots, m \quad (2)$$

$$U_s, V_i > 0, s = 1, \ldots, r, i = 1, \ldots, n \quad (3)$$

Nomenclature:

$E^{CCR}{}_{d}$ = Efficiency measurement using DEA CCR model of DMU d
U_s = *Output weights*
$Y_{sd}^{(T)}$ = The s-th input of the T-th unit of DMU d
V_i = Input weights
X_{id} = The i-th input of the T-th unit of DMU d

DMU (faculty) which gets a value of 1 can be interpreted that the DMU is efficient, if the DMU has a value of less than 1 then the DMU has a low efficiency. DMU that can be used as a benchmark is DMU with a value of 1. The results of calculations using DEA CRS models that have been carried out at each DMU are then made in a tabular form to facilitate the evaluation of the efficiency values of each DMU.

4.2. Measuring Efficiency Using Network DEA Model for System with Hierarchical Structures

As can be seen in Figure 2 is the DEA hierarchical network model hierarchy structure in this paper which is the basis for calculating the efficiency of eleven faculties of the undergraduate program at Diponegoro University with three inputs and five outputs. In the hierarchy process, the input in each faculty is divided between different components [2]. For example, lecturers still have responsibilities in lecturing, conducting research, conducting publications, conducting services, and providing academic activities, so it is difficult to describe the amount of effort allocated by permanent lecturers to achieve each goal, so it must determine about the weight of the function [1]. In this study, it is assumed that the weight of the faculty for every input is allocated at 44% for lectures, 39% for research, and 17% for services [2]. Weights proportion used for every input (δ1, δ2, δ3) allocated to each function, then the relationship between these inputs is assumed to be δ1=1.128 δ2, δ2=2.294 δ3, δ1 + δ2 + δ3 = 1 [2]. Because the relationship (δ) is an estimate, the proportions are expressed in ranges : 0.5 δ2 ≤ δ1 ≤ 2 δ2, δ3 ≤ δ1 ≤ 4 δ3, δ3 ≤ δ2 ≤ 4 δ3 [1].

There are two categories in research: research conducted, and publications. It is assumed that each faculty allocates the same amount of input to both categories (research conducted, and publications) as follows: α1 = α2, α1 + α1 = 1 and expressed in the form: 0.5 α2 ≤ α1 ≤ 2 α2.

There are two categories of services: guidance & consulting, and academic activities. It is assumed that each faculty allocates the same amount of input in both categories (guidance & consultation, and academic activities) as following: α3 = α4, α3 + α4 = 1 and expressed in the form: 0.5 α4 ≤ α3 ≤ 2 α4 [2].

The efficiency measurement at every faculty using the data envelopment analysis hierarchical network system model can be solved by the following equation:

$$E_d^{HNW} = max \sum_{s=1}^{r} Us \left(Y_{sd}^{(1,1)} + Y_{sd}^{(2,1)} + Y_{sd}^{(2,2)} + Y_{sd}^{(3,1)} + Y_{sd}^{(3,2)}\right) \quad (4)$$

Subject to:

$$\sum_{i-1}^{n} V_i X_{ij} = 1 \quad (5)$$

$$\sum_{s=1}^{r} U_s Y_{sj}^{(1)} - \sum_{i=1}^{n} V_i \delta_1 X_{ij}^{(1)} < 0, f = 1, \dots m \quad (6)$$

$$\sum_{s=1}^{r} U_s Y_{sj}^{(2,1)} - \sum_{i=1}^{n} V_i \delta_2 \alpha_1 X_{ij}^{(2,1)} < 0, f = 1, \dots m \quad (7)$$

$$\sum_{s=1}^{r} U_s Y_{sj}^{(2,2)} - \sum_{i=1}^{n} V_i \delta_2 \alpha_2 X_{ij}^{3} < 0, f = 1, \dots m \quad (8)$$

$$\sum_{s=1}^{r} U_s Y_{sj}^{(3,1)} - \sum_{i=1}^{n} V_i \delta_3 \alpha_3 X_{ij}^{(3,1)} < 0, f = 1, \dots m \quad (9)$$

$$\sum_{s=1}^{r} U_s Y_{sj}^{(3,2)} - \sum_{i=1}^{n} V_i \delta_3 \alpha_4 X_{ij}^{(3,2)} < 0, f = 1, \dots m \quad (10)$$

$$0.5\ \delta_2 \leq \delta_1 \leq 2\ \delta_2,\ \delta_3 \leq \delta_1 \leq 4\ \delta_3,\ \delta_3 \leq \delta_2 \leq 4\ \delta_3,\ \delta_1+\delta_2+\delta_3 = 1 \quad (11)$$

$$0{,}5\alpha_2 \leq \alpha_1 \leq 2\alpha_2,\ \alpha_1 + \alpha_2 = 1 \quad (12)$$

$$0{,}5\ \alpha_4 \leq \alpha_3 \leq 2\ \alpha_4,\ \alpha_3 + \alpha_4 = 1,\ U_s, V_i \geq 0,\ s = 1, \dots, r,\ i = 1, \dots, n \quad (13)$$

Nomenclature:

$E^{HNW}{}_d$ = Efficiency measurement using hierarchical network model of DMU d
U_s = *Output weights*
$Y_{sd}^{(T)}$ = The s-th input of the T-th unit of DMU d
V_i = Input weights
$X_{ij}^{(T)}$ = The i-th input of the T-th unit of DMU j
$Y_{sj}^{(T)}$ = The s-th output of the T-th unit of DMU j

δ_k = The proportions of each inputs allocated amounts of resources to teaching, research, and services.
α_k = The proportions of each inputs allocated amounts of resources to research projects, publications, consultations, and services.

The efficiency of each component in the faculty was calculated using the formula below when the optimal solution (U *, V *) has been achieved.

$$E_d^{(Teaching)} = \frac{\sum_{s=1}^{r} U_s{}^* \, Y_{sd}^1}{\sum_{i=1}^{n} V_i{}^* \delta_1 X_{id}^1} \tag{14}$$

$$E_d^{(Research)} = \frac{\sum_{s=1}^{r} U_s{}^* \, Y_{sd}^{(2,1)}}{\sum_{i=1}^{n} V_i{}^* \delta_2 \alpha_1 X_{id}^{(2,1)}} \tag{15}$$

$$E_d^{(Publication)} = \frac{\sum_{s=1}^{r} U_s{}^* \, Y_{sd}^{(2,2)}}{\sum_{i=1}^{n} V_i{}^* \delta_2 \alpha_2 X_{id}^{(2,2)}} \tag{16}$$

$$E_d^{(Consultation)} = \frac{\sum_{s=1}^{r} U_s{}^* \, Y_{sd}^{(3,1)}}{\sum_{i=1}^{n} V_i{}^* \delta_3 \alpha_3 X_{id}^{(3,1)}} \tag{17}$$

$$E_d^{(Activities)} = \frac{\sum_{s=1}^{r} U_s{}^* \, Y_{sd}^{(3,2)}}{\sum_{i=1}^{n} V_i{}^* \delta_3 \alpha_4 X_{id}^{(3,2)}} \tag{18}$$

<u>Nomenclature:</u>
$E_d^{(T)}$ = Efficiency of the T-th unit of DMU d
U_s = *Output weights*
V_i = Input weights
$X_{id}^{(T)}$ = The i-th input of the T-th unit of DMU d
$Y_{sd}^{(T)}$ = The s-th output of the T-th unit of DMU d
δ_k = The proportions of each inputs allocated amounts of resources to teaching, research, and services.
α_k = The proportions of each inputs allocated amounts of resources to research projects, publications, consultations, and services.

To identify the relative efficiency scores of each DMU and its components in the hierarchical system, the model was run a couple of times. If each DMU and its components in the system obtain an efficiency score of 100%, thus the DMU can be considered as an efficient DMU [2]. Theorem 1 = DMU can be said to be efficient if all the components are efficient, then it can happen if all DMUs are inefficient which is contrary to the results of conventional DEA models [2]. From the results of calculations using the DEA Hierarchical Network Systems model that has been done at each DMU, tables are made to facilitate the evaluation of efficiency. A table containing the efficiency scores calculated by DEA hierarchical network model from each faculty is made with the aim that the evaluation process is easy.

5. Conclusions

This paper has a purpose to measure the efficiency of faculty at undergraduate program in Diponegoro University using hierarchical network data envelopment analysis. Diponegoro University previously never taken efficiency measurements in each of its faculty, so this research can contribute to Diponegoro University to be able to carry out an efficiency evaluation process to develop better through the internal efficiency processes of the organization. This paper can help the management of University especially the dean of faculty to measure the overall and each components efficiency of each faculty, and to create improvement strategy and policy in the future, therefore the target of Diponegoro University as a World Class University can be achieved. The DEA conventional black-box (constant return to scale) model is selected as a comparative method to DEA Hierarchical network model with a hierarchical structure. Both of method is comparable, with the same underpinning theory but different structures. The DEA black-box model and DEA hierarchical network model as proposed in the paper will be extended further in the future by exploring more new cases and its new components in the hierarchy system.

References

[1] Kao C. Efficiency measurement for hierarchical network systems. *Omega*. Elsevier BV; 2015 Mar;51:121–7. doi.org/10.1016/j.omega.2014.09.008

[2] Kashim R, Mat Kasim M, Abd Rahman R. MEASURING EFFICIENCY OF A UNIVERSITY FACULTY USING A HIERARCHICAL NETWORK DATA ENVELOPMENT ANALYSIS MODEL. Journal of Information and Communication Technology. *Int. J. Inf. Commun. Technol.* UUM Press, Universiti Utara Malaysia; 2018 Oct;17(4). doi.org/10.32890/jict2018.17.4.3

[3] Liu, J.S, Lu LYY, Lu WM, Lin BJY. A Survey of DEA Applications. Omega; 2013; 41(5): 893-902. https://doi.org/10.1016/j.omega.2012.11.004

[4] Ku-Mahamud KR, Ghani NFA, Kasim MM, Abdullah F. Efficiency Measurement Model of Projects Using DEA. 2010 *Int. Conf. Comput. Intell. Model. Simul*. IEEE; 2010 Sep; doi.org/10.1109/cimsim.2010.32

[5] Goksen, Yilmaz, Onur Dogan, Bilge Ozkarabacak. A Data Envelopment Analysis Application for Measuring Efficiency of University Departments. *Procedia Economics and Finance*. Elsevier BV; 2015; 19: 226-237

[6] Rayeni, M. M, Saljooghi, F. H. Benchmarking in the Academic Departments using Data Envelopment Analysis. American Journal of Applied Sciences; 2010 Nov 8;7(11): 1464-1469 https://doi.org/10.3844/ajassp.2010.1464.1469

[7] Agha, R., Salah, Ibrahim, Kuhail, Nader Abdelnabi, Mahmoud Salem, Ahmed Ghanim. Assessment of Academic Depatments Efficiency Using Data Envelopment Analysis. Journal of Industrial Engineering and Management; 2011; 4(2): 301-325 https://doi.org/10.3926/jiem.2011.v4n2.p301-325

[8] Saniee Monfared MA, Safi M. Network DEA: an application to analysis of academic performance. *J. Ind. Eng. Int.* Springer Science and Business Media LLC; 2013 May 26;9(1). doi.org/10.1186/2251-712x-9-15

[9] Kao C. Efficiency decomposition for parallel production systems. *J. Oper. Res. Soc*. Informa UK

Limited; 2012 Jan;63(1):64–71. doi.org/10.1057/jors.2011.1
[10] Kao C. Efficiency measurement and frontier projection identification for general two-stage systems in data envelopment analysis. *Eur. J. Oper. Res*. Elsevier BV; 2017 Sep;261(2):679–89. doi.org/10.1016/j.ejor.2017.03.016
[11] Purwaningsih R, Pratiwi CG, Susanto N, Santosa H. Measurement of Intermediary Trader Efficiency in Poultry Distribution Using Data Envelopment Analysis Method. IOP Conference Series: *Mater. Sci. Eng*. IOP Publishing; 2019 Sep 6;598:012009. doi.org/10.1088/1757-899x/598/1/012009

Bi-objective optimization model for integrated planning in container terminal operations

D N Prayogo[1], A Hidayatno[2,3], and Komarudin[2,4]

[1]Department of Industrial Engineering, University of Surabaya, Raya Kalirungkut, Surabaya, 60293, Indonesia
[2]Department of Industrial Engineering, University of Indonesia, Margonda Raya, Depok, 16424, Indonesia

Email: [1]dnprayogo@staff.ubaya.ac.id, [3]akhmad@eng.ui.ac.id, [4]komarudin74@ui.ac.id

Abstract. In this paper, we studied an integrated planning optimization model in container operations consisting of tactical and operational level decisions. At the tactical level decision, integration planning includes berth allocation problem, quay crane assignment problem and yard allocation planning. These tactical level decisions are integrated with quay crane scheduling which is an operational level decision at the container terminal. Integration of planning in container terminals is needed because the decision on berth allocation, namely berthing position and berthing time in seaport, is highly dependent on the decision of reserved sub-blocks positions in container yard allocation, the number of quay cranes assigned to each vessel and its scheduling on each vessel bays. This integrated planning decision optimization has two objective functions; namely: (i) minimizing total operational cost, and (ii) maximizing service level. Numerical experiments have been carried out to obtain total cost efficiency and maximum service level resulting from the proposed optimization model.

1. Introduction

In line with the rapid development of international trade in the past decade, the utilization of container terminals to support the efficiency of global transportation costs is very important. There are around 793.26 million TEUs managed by container ports around the world and have a growth rate of 4.7% in 2019 with 64% of the total container port traffic occurring in Asia. This is a huge challenge for container terminal managers to have a facility planning, i.e.: quay wharf and quay cranes in seaside ports, and at yard storage area in landside as container terminal resources in efficiently and effectively to increase the competitiveness of seaports.

Tactical level planning decisions in container terminal consists of Tactical Berth Allocation Problem (TBAP), Quay Crane Assignment Problem (QCAP), Specific Quay Crane Assignment Problem (SQCAP) and Tactical Yard Allocation Planning (TYAP). While operational level planning decisions includes Quay Crane Scheduling Problem (QCSP), Stowage Planning, Internal Transport Scheduling Problem (ITSP) and Yard Crane Scheduling Problem (YCSP).

TBAP decisions are berthing start and end time and berthing position, which are interrelated to the number of assigned QCs and the handling time of each vessel which are the decisions of QCAP and the location of reserved sub blocks for each vessel in the container yard storage area. However, intertwined tactical planning decisions of QCAP and TYAP also depend on the TBAP decisions, those are berthing time and position. Moreover, QCSP operational level decisions related to the sequence of tasks to be processed by each quay crane (QC) also depend on tactical decisions of TBAP and SQCAP. Therefore, solving the problem of tactical and operational planning at the container terminal should be carried out in integrated manner and considering the interests of all parties involved, namely container terminal managers and shipping liner owners.

In this study, we proposed a bi-objective optimization model for integration of tactical and operational planning decisions at container terminal operations. Furthermore, the organization of this paper has the following structure. Section 2 briefly discusses the literature review regarding various types of integration of container terminal planning. Explanation of problems related to the operational processes contained in container terminals at seaside and landside is described in Section 3. Section 4 describes the formulation of the bi-objective optimization model for the proposed integrated planning in container terminal operations. The solution method discussed in Section 5 and continues with the numerical experiments and analysis of the results of applying the integration model and the proposed solution method in Section 6. Finally, Section 7 presents the conclusion remarks and further research directions.

2. Literature Review

Several integration planning models of tactical and operational level at container terminal have been developed by researchers. The methods for solving the integration model can be classified into three categories, namely: (1) Sequential integration; an integration model is divided into several sub-problems and each sub-problem is solved in stages, where the optimization results from one stage will be input data for the next stage, such that the obtained results do not guarantee optimal overall; (2) Deep integration: the integration model is solved as a single model. If the integration model is a NP-Hard problem, then the optimal solution will hardly be obtained from the deep integration method, and (3) Functional integration: the integration model is divided into several sub-problems that are interrelated to each other, and solved by a mechanism feedback loop output from one sub-problem as an input parameter for another sub-problem. This mechanism is done iteratively until a convergent solution condition obtained in accordance with the specified requirements.

The most developed integration planning model at the container terminal were the integration model between TBAP and QCAP, which is solved by sequential integration as by Chan and Chung [1] using deep integration by Meisel and Bierwirth [2], Imai et al. [3], Zhang et al. [4], Giallombardo et al., [5], Meisel and Bierwirth [6], Liang et al. [7], Vacca et al. [8], Shang et al. [9], Iris et al. [10], [11], Wang et al. [12], Krimi et al. [13], Zheng et al. [14], Prayogo et al., [15], Wang et al. [16], Xie et al. [[17]. The integration models of TBAP, QCAP and SQCAP as had been done by Karam and Eltawil [18] using functional integration The integration models of TBAP, QCAP and QCSP had been proposed by Park and Kim [19], Agra and Oliveira, [20], Kasm et al. [21], and solved using sequential integration, functional integration was used by Li et al.[22], Meisel and Bierwirth [23], and Türkogulları et al. [24]. The integration models between TBAP and TYAP are Tao and Lee [25], Ma et al. [26]. The integration models between QCAP and QCSP as in the model from Fu et al. [27], Fu and Diabat [28], Theodorou and Diabat [29], and Alsoufi et al. [30]. The integration models of TBAP, QCAP and TYAP as had been done by Zhen et al. [31], Prayogo et al. [32], and Liu [33] and were solved by using functional integration.

This study proposed an integration model between tactical planning decisions, namely TBAP, TYAP, QCAP, and SQCAP, and QCSP as operational level decisions in container terminal operations. The optimization model of container terminal planning integration had been solved by using bi-objective optimization with point of views from all involved parties, those are container terminal manager and shipping liner owners, which objectives to minimize the total operating costs and maximize service level to increase customer satisfaction. This bi-objective optimization for integrated planning model is solved by applying the Fuzzy Multi-Objective Programming approach.

3. Problem Statement

This study solves problem related to berth allocation for every arrival of ships on quay wharf with continuous berth layout dynamically. Determination of berthing time and position of each vessel depends on the number of assigned QCs and specific QC assignment decisions aims to arrange the sequence of tasks to be processed by each quay crane in QC scheduling problems.

In this problem, we assumed a dynamic berth allocation problem, each vessel can arrive at any time within the planning horizon, but their feasible and expected arrival times are deterministic. Each vessel can moor at any position along the quay wharf, which is called as continuous berth allocation problem. The berth is divided into several berth segments with equal section length. Likewise, the time horizon, which is divided into several time steps in equal time interval. The integrated planning is cyclical berth and yard planning without service priorities.

In this problem also considered a time-variant assignment of quay cranes, which is, an assigned quay crane in a vessel could be moved to another vessel while the previous vessel was still being serving by other quay cranes if this resulted in a more efficient solution overall. The moving time of QCs from one vessel to another is not considered. QCs can move in bi-directional following a single track with safety margin between QCs and have an identical working rate. In QC assignment and scheduling, it's also considered the safety space among QCs and non-crossing constraints, such that QCs should be ordered from left to right along quay wharf.

For TYAP, consignment strategy is applied in yard allocation to cope of traffic congestion at container yard storage area. Yard storage areas are divided into transit, import and export container yard areas. Loading and unloading activities in the yard area consist of direct and indirect transhipment, import and export operations.

4. Mathematical Formulation

4.1. Notation

In the following sections, detailed indices, model parameters and decision variables were used to develop the integration of tactical planning and operational planning model in container terminal operations. The design of the integrated container terminal planning model had adopted from the models in [21], [27] [33], [34], and [35].

4.1.1. Indices

i, j: indices of vessels, $i, j \in V, i \neq j$, V: set of vessels
b: index of berth segments, $b \in B$, B: set of berth segments
q: index of quay cranes (QCs), $q \in Q$, Q: set of quay cranes
m, n: indices of bays, $m, n \in M, m \neq n$, M_i: set of bays in each vessel i
k: index of sub-blocks, $k \in K$, K: set of sub-blocks, $K = K^T \cup K^I \cup K^E$
K^T: set of transshipment sub-blocks in container yard area
K^I: set of import sub-blocks in container yard area
K^E: set of export sub-blocks in container yard area
n: index of neighborhood sub-block pairs in container yard area, $n \in N$, N: set of all the neighbor sub-block pairs in container yard area
g: index of all blocks, which each block consists of five sub-blocks in the export and transhipment container areas, $g \in G$, G: set of all the blocks in container yard area
t: index of time step, $t \in T$, which is $T = H + E$
H: length of planning horizon expressed in number of time steps
E: the longest handling time of all vessels

4.1.2. Model parameters

L: the quay wharf length in meters
l: a berth segment length in meters
l_i: vessel length i including safety distance between two adjacent mooring vessels
$[a_i^f, b_i^f]$: range of feasible service time for vessel i; $i \in V$ and $a_i^f, b_i^f \in T | t \leq H$
$[a_i^e, b_i^e]$: range of expected service time for vessel i; $i \in V$ and $a_i^e, b_i^e \in T | t \leq H$
f: Maximum start time step difference between two vessels conducting direct transhipment activity
v_{ij}^T: number of transshipment containers from vessel i to vessel j; $i, j \in V, i \neq j$
v_i^I: number of import containers should be unloaded from vessel i; $i \in V$
v_i^E: number of export containers should be loaded from vessel i; $i \in V$
m_i: total workload for vessel i; $i \in V$ in terms of QC – time steps, required for discharging and loading activities
Q_t: number of available quay cranes at time step t
λ: interference exponent for productivity loss the QCs
q_i^{min}: minimum number of QCs that should be assigned to vessel i; $i \in V$
q_i^{max}: maximum number of QCs that can be assigned to vessel i; $i \in V$
pq: productivity of the quay crane in TEUs per hour

l_{im}: location of bay m in vessel i; $i \in V, m \in M_i$
w_{im}: workload (in QC-time steps) to be done for bay m in vessel i; $i \in V, m \in M_i$
u_{imt}: set to 1 if a QC is operating on bay m of vessel i at time step t; $\forall i \in V,\ m \in M_i, t \in T$
R: minimum of safety distance among quay cranes assigned in a vessel
s_i^T: number of transshipment sub-blocks in container yard area must be reserved for vessel i; $i \in V$
s_i^I: number of import sub-blocks in container yard area must be reserved for vessel i; $i \in V$
s_i^E: number of export sub-blocks in container yard area must be reserved for vessel i; $i \in V$
D_{bk}^U: distance between berth segment b and sub-block k for unloading activities; $b \in B, k \in K^T \cup K^I$
D_{kb}^L: distance between sub-block k and berth segment b for loading activities; $b \in B, k \in K^T \cup K^E$
$D_{b_i b_j}^D$: distance between berth segment of vessel i (b_i) and berth segment of vessel j (b_j) for direct transshipment activities; $i, j \in V,\ i \neq j,\ b_i, b_j \in B,\ b_i \neq b_j$
C_i^1: penalty cost for waiting of berthing start time of vessel i from its expected start service time; $i \in V$
C_i^2: penalty cost for tardiness of completion time of vessel i from its expected departure time; $i \in V$
C^T: transportation cost per container-meter in container yard area
M: a sufficiently large positive number

4.1.3. Decision variables
α_i: the berthing start time of vessel i; $i \in V$
β_i: the berthing end time of vessel i; $i \in V$
τ_i^{a+}: waiting time before berthing start time of vessel i; $i \in V$
τ_i^{a-}: earliness of berthing start time of vessel i; $i \in V$
τ_i^{b+}: tardiness of berthing end time of vessel i; $i \in V$
τ_i^{b-}: earliness of berthing end time of vessel i; $i \in V$
b_i: berthing position for vessel i; $i \in V$ and $bp_i \in [1, \dots, L]$
ω_{ib}: set to 1 if the middle point of vessel length i is located at berth segment b and 0 otherwise; $i \in V, b \in B$
ξ_{ib}: set to 1, if vessel i occupies berth segment b; $i \in V, b \in B$
π_{it}: set to 1, if the start berthing time of vessel i is $t = \alpha_i$ and 0 otherwise; $i \in V, t \in T \mid t \leq H$
θ_{it}: set to 1, if vessel i is berthing in time step t; $i \in V, t \in T$
ht_i: number of time steps for handling time of vessel i; $\forall i \in V$
δ_{ij}^x: set to 1, if vessel i is completed before vessel j starts service and 0 otherwise; $i, j \in V, i \neq j$
δ_{ij}^y: set to 1, if vessel i is berthed on the left-side of vessel j along the quay wharf and 0 otherwise; $i, j \in V, i \neq j$
s_b: the start time of berth segment b is occupied; $b \in B$
e_b: the finish time of berth segment b is occupied; $b \in B$
x_{ipt}: set to 1, if there is p QCs assigned to vessel i at time step t and 0 otherwise; $i \in V, p \in [0, q_i^{min}, \dots, q_i^{max}], t \in T$
nq_{it}: number of QCs should be assigned to vessel i at time step t; $i \in V, t \in T$
ρ_t: number of assigned QCs in time step t; $t \in T$
Z_{iqt}: set to 1, if QC q is assigned to vessel i at time step t and 0 otherwise; $i \in V, q \in Q, t \in T$
L_{iqt}: set to 1, if QC q is the leftmost QC assigned to vessel i at time step t and 0 otherwise; $i \in V, q \in Q, t \in T$
R_{iqt}: set to 1, if QC q is the rightmost QC assigned to vessel i at time step t and 0 otherwise; $i \in V, q \in Q, t \in T$
y_{imqt}: set to 1, if QC q is assigned to bay m of vessel i at time step t and 0 otherwise; $i \in V, m \in M_i, q \in Q_t, t \in T$
u_{imt}: set to 1 if bay bay m of vessel $i$is fully handled at time step t and 0 otherwise; $i \in V, m \in M_i, t \in T$
φ_{ik}: set to 1, if sub block k is assigned to vessel i and 0 otherwise; $i \in V, k \in K$

γ_{ikt}: set to 1, if sub block k is assigned to vessel i at time step t and 0 otherwise; $i \in V, k \in K, t \in T$
v_{ij}^{D}: set to 1, if there is direct-transhipment activity from vessel i to vessel j and 0 otherwise; $i, j \in V, i \neq j$
v_{ij}^{I}: set to 1, if there is indirect-transhipment activity from vessel i to vessel j and 0 otherwise; $i, j \in V, i \neq j$
ψ_{jkib}: set to 1, if there is indirect transshipment activities from vessel i at berth segment b to vessel j which has reserved sub block k and 0 otherwise, $i, j \in V,\ i \neq j, k \in K^{T}, b \in B$
κ_{ibk}: set to 1, if vessel i at berth segment b and reserved sub block k and 0 otherwise, $i \in V, b \in B, k \in K$
$\varpi_{ib_ijb_j}$: set to 1, if bething position of vessel i at berth segment b_i and berthing position of vessel j at berth segment b_j and 0 otherwise; $i, j \in V,\ i \neq j,\ b_i, b_j \in B,\ b_i \neq b_j$
θ_{ib}^{L}: set to 1, if berth segment b is on the left side of vessel i and 0 otherwise; $i \in V, b \in B$
θ_{ib}^{R}: set to 1, if berth segment b is on the right side of vessel i and 0 otherwise; $i \in V,\ b \in B$
ι_{ij}^{L}: set to 1, if vessel i has berthing start time before berthing start time of vessel j and 0 otherwise; $i, j \in V, i \neq j$
ι_{ij}^{R}: set to 1, if vessel j has berthing start time before berthing start time of vessel i and maximum allowance time of difference start times f and 0 otherwise; $i, j \in V,\ i \neq j$

4.2. Objective functions
This integrated planning model has two objective function, those are OBJ_1, that minimize the total operational cost, consist of cost of delay in berthing start time, penalty cost of tardiness of berthing completion time, QCs operation costs and total transportation cost of import, export and transit containers in container yard areas.

$$Obj_1 = Min\ TC = \sum_{i\in V} c_i^1\,\tau_i^{a+} + c_i^2\tau_i^{b+} + C^q \sum_{i\in V}\sum_{t\in T} nq_{it} + c^T \sum_{i\in V}\sum_{\substack{j\in V\\ j\neq i}}\sum_{b\in B}\sum_{k\in K} v_{ij}^{T}\,\frac{D_{bk}^{U}\psi_{jkib}+D_{kb}^{L}\kappa_{jbk}}{s_j^{T}} + c^T \sum_{i\in V}\sum_{b\in B} v_i^{I}\, D_{bk}^{U}\omega_{ib} + c^T \sum_{i\in V}\sum_{k\in K}\sum_{b\in B} v_i^{E}\,\frac{D_{kb}^{L}\kappa_{ibk}}{s_i^{E}} \qquad (1)$$

The second objective function (Obj_2) is to find the maximize of minimum service level among all vessels.

$$Obj_2 = Max\ SL \qquad (2)$$

Service level of each vessel is correlated to delay completion time of its expected service end time as presented in constraint (3)

$$SL \leq 1 - \frac{\tau_i^{b+}}{b_i^{e}-a_i^{e}} \qquad \forall i \in V \qquad (3)$$

4.3. Constraints
Constraints of the tactical and operational planning model at the container terminal were divided into three sub problems, those are the constraints related to the integration models: (1) TBAP, QCAP and SQCAP; (2) TBAP and TVAP, and (3) SQCAP and QCSP which would be explained in detail as follows.

TBAP, QCAP and SQCAP related constraints:
Constraints (4) guarantee that the middle point of each vessel length is positioned at a berth segment

$$\sum_{b\in B}\omega_{ib} = 1 \qquad \forall i \in V \qquad (4)$$

Constraints (5) and (6) ensure that berthing position of vessel i should be located within the range $[l \times (b-1), l \times b - 1]$ on the quay wharf.

$$l.\sum_{b\in B}(b-1).\,\omega_{ib} \leq b_i \qquad \forall i \in V \qquad (5)$$

$$b_i \leq l.\sum_{b\in B} b.\,\omega_{ib} - 1 \qquad \forall i \in V \qquad (6)$$

Constraint (7) ensures that for any two vessels i and j, if $\delta_{ij}^{y} = 1$, then the leftmost berthing position of vessel j is not less than the rightmost berthing position of vessel i.

$$b_i + l_i/2 \leq b - l_j/2 + M\left(1 - \delta_{ij}^{y}\right) \qquad \forall i,j \in V, i \neq j \tag{7}$$

Constraint (8) defines that for any two vessels $i$and j, if $\delta_{ij}^{x} = 1$, then the end time of vessel i must less than the start time of vessel j.

$$\beta_i \leq \alpha_j + M\left(1 - \delta_{ij}^{x}\right) \qquad \forall i,j \in V, i \neq j \tag{8}$$

Constraint (9) defines that there is no overlap for any two vessels in berth plan graph.

$$\delta_{ij}^{x} + \delta_{ji}^{x} + \delta_{ij}^{y} + \delta_{ji}^{y} \geq 1 \qquad \forall i,j \in V, i \neq j \tag{9}$$

Constraint (10) ensures that the rightmost berthing position of vessel i should be within the quay wharf.

$$b_i + l_i/2 \leq L \qquad \forall i \in V \tag{10}$$

Constraint (11) defines that each vessel is scheduled to start handling at a time step in planning horizon.

$$\sum_{t=1}^{H} \pi_{it} = 1 \qquad \forall i \in V \tag{11}$$

Constraint (12) relates the start time-step α_i and binary variable π_{it} for each vessel i .

$$\alpha_i = \sum_{t=1}^{H} t.\pi_{it} \qquad \forall i \in V \tag{12}$$

Constraint (13) connects the start and end time-steps of each vessel.

$$\alpha_i + \sum_{t \in T} \theta_{it} - 1 = \beta_i \qquad \forall i \in V \tag{13}$$

Constraint (14) shows the total handling time of each vessel.

$$ht_i = \sum_{t \in T} \theta_{it} \qquad \forall i \in V \tag{14}$$

Constraints (15) and (16) relate variables π_{it} and θ_{it}, θ_{it} and Z_{iqt}.

$$\pi_{it} \geq \theta_{it} - \theta_{it-1} \qquad \forall i \in V, t \in T | t \leq H \tag{15}$$

$$\theta_{it} \geq Z_{iqt} \qquad \forall i \in V, q \in Q_t, t \in T | t \leq H \tag{16}$$

The following four constraints (17) - (20) show the relation between $\zeta_{ib}$and b_i. and

$$b_i - l_i/2 \leq s.b - 1 + M.\theta_{ib}^{L} \qquad \forall i \in V, b \in B \tag{17}$$

$$b_i - l_i/2 \geq s.b - M\left(1 - \theta_{ib}^{L}\right) \qquad \forall i \in V, b \in B \tag{18}$$

$$b_i + l_i/2 \geq s.(b-1) - M.\theta_{ib}^{R} \qquad \forall i \in V, b \in B \tag{19}$$

$$\beta_i + l_i/2 \leq s.(b-1) - 1 + M\left(1 - \theta_{ib}^{R}\right) \qquad \forall i \in V, b \in B \tag{20}$$

Constraint (21) guarantees that if both $\theta_{ib}^{L} = 0$ and $\theta_{ib}^{R} = 0$, then $\zeta_{ib} = 1$.

$$\zeta_{ib} \geq 1 - \theta_{ib}^{L} - \theta_{ib}^{R} \qquad \forall i \in V, b \in B \tag{21}$$

Constraints (22) and (23) define start and end time berth b occupied by vessels.

$$s_b \leq \alpha_i + M(1 - \zeta_{ib}) \qquad \forall i \in V, b \in B \tag{22}$$

$$e_b \geq \beta_i - M(1 - \zeta_{ib}) \qquad \forall i \in V, b \in B \tag{23}$$

Constraints (24) guarantees that the time window of berth b are occupied by vessels within the planning horizon.

$$e_b - s_b + 1 \leq H \qquad \forall b \in B \tag{24}$$

Constraint (25) relates nq_{it} and θ_{it}

$$nq_{it} \leq M.\theta_{it} \qquad \forall i \in V, t \in T | t \leq H \tag{25}$$

Constraints (26) and (27) enforce that the number of used QCs in each time step should not exceed the available QCs in the cyclical berth planning.

Constraints (26) and (27) ensure that in cyclical berth planning, the number of QCs used in each time step does not exceed the number of available QCs in container terminal.

$$\sum_{i \in V} nq_{it} \leq Q_t \qquad \forall t \in T | E + 1 \leq t \leq H \tag{26}$$

$$\sum_{i \in V} nq_{it} - \sum_{i \in V} nq_{it+H} \leq Q_t \qquad \forall t \in T | t \leq E \tag{27}$$

Constraint (28) – (30) ensure vessel operation carry out to completion.

$$X_{i0t} + \sum_{c=q_i^{min}}^{q_i^{max}} X_{ict} = 1 \qquad \forall i \in V, t \in T | t \leq H \tag{28}$$

$$\sum_{c=q_i^{min}}^{q_i^{max}} c.X_{ict} = nq_{it} \qquad \forall i \in V, t \in T | t \leq H \tag{29}$$

$$\sum_{t=1}^{H} \sum_{c=q_i^{min}}^{q_i^{max}} c^{\lambda} X_{ict} \geq m_i \qquad \forall i \in V \tag{30}$$

Constraint (31) ensures that each QC can only serve maximum one vessel at every time step.

$$\sum_{i\in V} Z_{iqt} \leq 1 \qquad \forall q \in Q_t, t \in T|t \leq H \qquad (31)$$

Constraint (32) and (33) and (34) and (35) define L_{iqt} and R_{iqt} , respectively.

$$1-(Z_{iqt}-Z_{iq-1,t}) \leq M(1-L_{iqt}) \qquad \forall i \in V, q \in Q_t, t \in T|t \leq H \qquad (32)$$

$$Z_{iqt}-Z_{iq-1,t} \leq ML_{iqt} \qquad \forall i \in V, q \in Q_t, t \in T|t \leq H \qquad (33)$$

$$1-(Z_{iqt}-Z_{iq+1,t}) \leq M(1-R_{iqt}) \qquad \forall i \in V, q \in Q_t, t \in T|t \leq H \qquad (34)$$

$$Z_{iqt}-Z_{iq+1,t} \leq MR_{iqt} \qquad \forall i \in V, q \in Q_t, t \in T|t \leq H \qquad (35)$$

Constraints (36) – (38) define the leftmost and rightmost QCs, that are assigned to a vessel.

$$\sum_{q=1}^{Q_t} L_{iqt} = \theta_{it} \qquad \forall i \in V, t \in T|t \leq H \qquad (36)$$

$$\sum_{q=1}^{Q_t} R_{iqt} = \theta_{it} \qquad \forall i \in V, t \in T|t \leq H \qquad (37)$$

$$\sum_{q=1}^{Q_t} Z_{iqt} = \theta_{it} \qquad \forall i \in V, t \in T|t \leq H \qquad (38)$$

Constraints (39) and (42) enforce each vessel must be served within its feasible service time window.

$$\alpha_i \geq a_i^f \qquad \forall i \in V \qquad (39)$$

$$\beta_i \leq b_i^f \qquad \forall i \in V \qquad (40)$$

$$a_i^e - \alpha_i = \tau_i^{a+} - \tau_i^{a-} \qquad \forall i \in V \qquad (41)$$

$$\beta_i - b_i^e = \tau_i^{b+} - \tau_i^{b-} \qquad \forall i \in V \qquad (42)$$

TBAP and TYAP related constraints:

Constraints (43) guarantees that each sub-block can be reserved for maximum one vessel.

$$\sum_{i\in V} \varphi_{ik} \leq 1 \qquad \forall k \in K \qquad (43)$$

Constraints (44) defines number of sub-blocks are reserved for vessel i.

$$\sum_{k\in K} \varphi_{ik} = s_i \qquad \forall i \in V \qquad (44)$$

Constraints (45) shows the relation among $\psi_{jkib}, \varphi_{jk}$ and ω_{ib} as follow

$$\psi_{jkib} = \varphi_{jk} + \omega_{ib} - 1 \qquad \forall\, i,j \in V, i \neq j, k \in K^T, b \in B \qquad (45)$$

Constraints (46) defines the relation among $\kappa_{jkb}, \varphi_{jk}$ and ω_{jb} as follow.

$$\kappa_{jkb} = \varphi_{jk} + \omega_{jb} - 1 \qquad \forall j \in V, k \in K^I \cup K^E \cup K^T, b \in B \qquad (46)$$

If there is a transhipment activity, then this transhipment task must be carried out in either direct-transhipment or indirect-transhipment as presented in constraint (47).

$$v_{ij}^D + v_{ij}^I = A_{ij}^T \qquad \forall\, i,j \in V, i \neq j \qquad (47)$$

Constraints (48) and (49) ensure that the number of sub-blocks in yard transhipment area should be reserved for the vessel whose carry out indirect-transhipment activity.

$$\sum_{k\in K^T} \varphi_{jk} \geq s_j^T - M\left(1 - \sum_{\substack{i\in V \\ i\neq j}} v_{ij}^I\right) \qquad \forall\, j \in V \qquad (48)$$

$$\sum_{k\in K^T} \varphi_{jk} \leq s_j^T + M\left(1 - \sum_{\substack{i\in V \\ i\neq j}} v_{ij}^I\right) \qquad \forall\, j \in V \qquad (49)$$

The following three constraints (50) – (52) guarantee that for direct or indirect transhipment activity from vessel i to vessel j are presented as follow.

$$\alpha_j \geq \alpha_i - M(1-\iota_{ij}^L) \qquad \forall i,j \in V, i \neq j \qquad (50)$$

$$\alpha_j \leq \alpha_i + f - 1 + M(1-\iota_{ij}^R) \qquad \forall\, i,j \in V, i \neq j \qquad (51)$$

$$v_{ij}^I \geq A_{ij}^T(2-\iota_{ij}^L-\iota_{ij}^R) \qquad \forall i,j \in V, i \neq j \qquad (52)$$

Constraint (53) defines the links among variables ξ_{ikt} to φ_{ik} and θ_{it} for each sub-block in yard export area.

$$\xi_{ikt} \geq \varphi_{ik} + \theta_{it} - 1 \qquad \forall i \in V, k \in K^E, t \in T \qquad (53)$$

Constraint (54) shows the links between variables ξ_{jkt} to φ_{jk}, θ_{jt} and v_{ij}^I for each sub-block in yard transhipment area.

$$\xi_{jkt} \geq \varphi_{jk} + \theta_{jt} + \sum_{\substack{i\in V \\ i\neq j}} v_{ij}^I - 2 \qquad \forall\, j \in V, k \in K^T, t \in T \qquad (54)$$

Constraints (55) and (56) define that at most only one sub-block in each adjacent sub-block pairs can carry out loading containers to vessels at each time step in yard export area and transhipment area.

$$\sum_{k\in n}\sum_{i\in V}\xi_{ikt}+\sum_{k\in n}\sum_{i\in V}\xi_{ikt+H}\leq 1 \qquad \forall t\in T|t\leq E, n\in N \tag{55}$$
$$\sum_{k\in n}\sum_{i\in V}\xi_{ikt}\leq 1 \qquad \forall t\in T|E+1\leq t\leq H, n\in N \tag{56}$$

Constraints (57) and (58) ensure that at most one sub-block in each block can carry out loading containers to vessels at each time step in yard export area and transhipment area.

$$\sum_{k\in g}\sum_{i\in V}\xi_{ikt}+\sum_{k\in g}\sum_{i\in V}\xi_{ikt+H}\leq 1 \qquad \forall t\in T|t\leq E, g\in G \tag{57}$$
$$\sum_{k\in g}\sum_{i\in V}\xi_{ikt}\leq 1 \qquad \forall t\in T|E+1\leq t\leq H, g\in G \tag{58}$$

SQCAP and QCSP related constraints:

Constraints (59) and (60) ensure that each QC can only be placed in a single bay at any time and only one QC can be placed in a bay at any time.

$$\sum_{i\in V}\sum_{m\in M_i}y_{imqt}\leq 1 \qquad \forall q\in Q_t, t\in T \tag{59}$$
$$\sum_{q\in Q_t}y_{imqt}\leq 1 \qquad \forall i\in V, m\in M_i, t\in T \tag{60}$$

Constraint (61) defines that number of operating QCs must less than the number of available QCs at any time step.

$$\sum_{i\in V}\sum_{m\in M_i}\sum_{q\in Q_t}y_{imqt}\leq Q_t \qquad \forall t\in T \tag{61}$$

Constraint (62) is a QC non-crossing constraint with the location in increasing order from left to right and minimum safety space R bays

$$\sum_{i\in V}\sum_{m\in M_i}l_{im}.y_{imq+1,t}-\sum_{i\in V}\sum_{m\in M_i}l_{im}.y_{imqt}\geq R \qquad \forall q\in Q_t, t\in T \tag{62}$$

Constraint (63) ensures that all containers are handled in the planning horizon.

$$\sum_{q\in Q_t}\sum_{t\in T}y_{imqt}= w_{im} \qquad \forall i\in V, m\in M_i \tag{63}$$

Constraint (64) defines the value of binary variable u_{imt} is set to 1 if all containers in bay m of vessel i are completed.

$$\frac{\sum_{q\in Q_t}\sum_{h=1}^{t}y_{imqh}}{w_{im}}\geq u_{imt} \qquad \forall i\in V, m\in M_i, t\in T \tag{64}$$

Constraint (65) ensures that a vessel is finished served after the task for every bay of vessel i is completed.

$$u_{imt}\geq \theta_{it} \qquad \forall i\in V, m\in M_i, t\in T \tag{65}$$

Constraint (66) links the berthing status to the service status. A vessel remains berthed until the required service is finished handled.

$$M.\theta_{it}\geq \theta_{it}-1-\sum_{m=1}^{M_i}w_{im}\left(1-\sum_{p=1}^{t}\pi_{ip}\right)+\sum_{m=1}^{M_i}w_{im}-\sum_{m=1}^{M_i}\sum_{h=1}^{t-1}u_{imp} \quad \forall i\in V, t\in T \tag{66}$$

Constraint (67) enforces that before the vessel starts berthing there are no quay cranes operating.

$$u_{imt}\leq \sum_{p=1}^{t}\pi_{ip} \qquad \forall i\in V, m\leq M_i, t\in T \tag{67}$$

Constraint (68) guarantees that no quay crane will be assigned to vessel that is not moored

$$\sum_{m=1}^{M_i}u_{imt}\leq M.\theta_{it} \qquad \forall i\in V, t\in T \tag{68}$$

Constraint (69) ensures that all containers are finished handled

$$\sum_{t\in T}u_{imt}=w_{im} \qquad \forall i\in V, m\leq M_i \tag{69}$$

Constraint (70) ensures that the position between the quay cranes meets the safety clearance distance requirements

$$\sum_{a=m}^{m+R}u_{iat}\leq 1 \qquad \forall i\in V, m\leq M_i-R, t\in T \tag{70}$$

Constraint (71) ensures that the number of QCs operating on vessel i does not exceed the number of QCs assigned to vessel i as resulting in QCAP.

$$\sum_{m=1}^{M_i}u_{imt}\leq nq_{it} \qquad \forall i\in V, t\in T \tag{71}$$

Each decision variable domain is stated in (72) – (81).

$$b_i,\alpha_i,\beta_i,\tau_i^{a+},\tau_i^{a-},\tau_i^{b+},\tau_i^{b-},s_b,e_b,\rho_t\geq 0\ \&\ integer \qquad \forall i\in V, b\in B, t\in T \tag{72}$$
$$\omega_{ib},\varphi_{ik},\xi_{ikt},\zeta_{ib}\in\{0,1\} \qquad \forall i\in V, b\in B, k\in K^I\cup K^E\cup K^T, t\in T \tag{73}$$
$$\pi_{it},\theta_{it}\in\{0,1\} \qquad \forall i\in V, t\in T \tag{74}$$
$$\delta_{ij}^x,\delta_{ij}^y,v_{ij}^D,v_{ij}^I\in\{0,1\} \qquad \forall i,j\in V, i\neq j \tag{75}$$
$$\psi_{jkib}\in\{0,1\} \qquad \forall i,j\in V, i\neq j, k\in K^T, b\in B \tag{76}$$
$$\kappa_{jkb}\in\{0,1\} \qquad \forall j\in V, k\in K^I\cup K^E\cup K^T, b\in B \tag{77}$$

$$\varpi_{ib_ijb_j} \in \{0,1\} \quad \forall i,j \in V, i \neq j, b_i, b_j \in B, b_i \neq b_j \tag{78}$$
$$\theta^L_{ib}, \theta^R_{ib}, \iota^L_{ij}, \iota^R_{ij} \in \{0,1\} \quad \forall i,j \in V, i \neq j, b \in B \tag{79}$$
$$u_{imt} \in \{0,1\} \quad \forall i \in V, m \in M_i, t \in T \tag{80}$$
$$y_{imqt}\{0,1\} \quad \forall i \in V, m \in M_i, q \in Q_t, t \in T \tag{81}$$

5. Solution method

In this study, the tactical and operational decisions of integration model at the container terminal consists of integration models for tactical level planning, namely: TBAP, TYAP, QCAP, and SQCAP, and QCSP at the operational level planning. The integrated planning models for container terminal were solved using a combination of functional integration with feedback loop structure and deep integration methods. Functional integration method had been used to solve the integration model of TBAP, QCAP and SQCAP, as well as the integration model between TBAP and TYAP. While deep integration was used to solve the integration model between SQCAP and QCSP. The proposed procedure for solving the integration model of tactical and operational planning at container terminal is shown in Figure 1. The solution method for bi-objective optimization model uses a fuzzy multi-objective programming approach to obtain a tradeoff between the interests of the container terminal manager to achieve total operational cost efficiency and the interests of the shipping liner owners to maximize service level in order to improve customer satisfaction.

The following model is a fuzzy multi objective programming that was used to solve the integrated planning model at the container terminal [36].

$$Max.\lambda \tag{82}$$
S.t:
$$\mu_1 \leq \lambda \tag{83}$$
$$\mu_2 \leq \lambda \tag{84}$$
$$\mu_1 = \frac{f_1^- - f_1}{f_1^- - f_1^+} \tag{85}$$
$$\mu_2 = \frac{f_2 - f_2^-}{f_2^+ - f_2^-} \tag{86}$$
$$0 \leq \lambda \leq 1 \tag{87}$$

Where: λ: fuzzy membership function
μ_1: Membership function of objective function minimization-mode Obj_1
μ_2: Membership function of objective function maximization-mode Obj_2
f_1: Objective function value of Obj_1
f_2: Objective function value of Obj_2
f_1^+: the best value of objective function Obj_1
f_1^-: the worst value of objective function Obj_1
f_2^+: the best value of objective function Obj_2
f_2^-: the worst value of objective function Obj_2

6. Numerical Experiments

The proposed bi-objective optimization model of integrated planning in container terminal was applied to 12 cases which the model parameters are combination of number of calling vessels, length of quay wharf, number of available QCs, number of sub-blocks in yard area as shown in Table 1. Quay wharf is divided into intervals of berth segment with a length of 50 meters. Planning time horizon is divided into time steps for 4 hours and carried out of periodic planning in 1 week. Each sub-block has length of 50 meters and contains of 240 containers. For each case, we generated randomly 10 datasets. For each instance, three vessel types consist of Feeder vessels, Medium vessels and Jumbo vessels are generated with composition of 90%, 30% and 10%, respectively. The vessel length and the number of containers for each vessel type, as well as the composition of the number of transit, import and export containers; and number of bays in each vessel follow the distribution of model parameters in [33] and [27]. Python programming was applied to Gurobi optimization to solve proposed bi-objective optimization model of integrated planning in container terminal and using the Fuzzy Multi Objective Programming approach.

Table 1. shows the average of minimum total operational costs and maximum service levels for each case that consist of 10 instances.

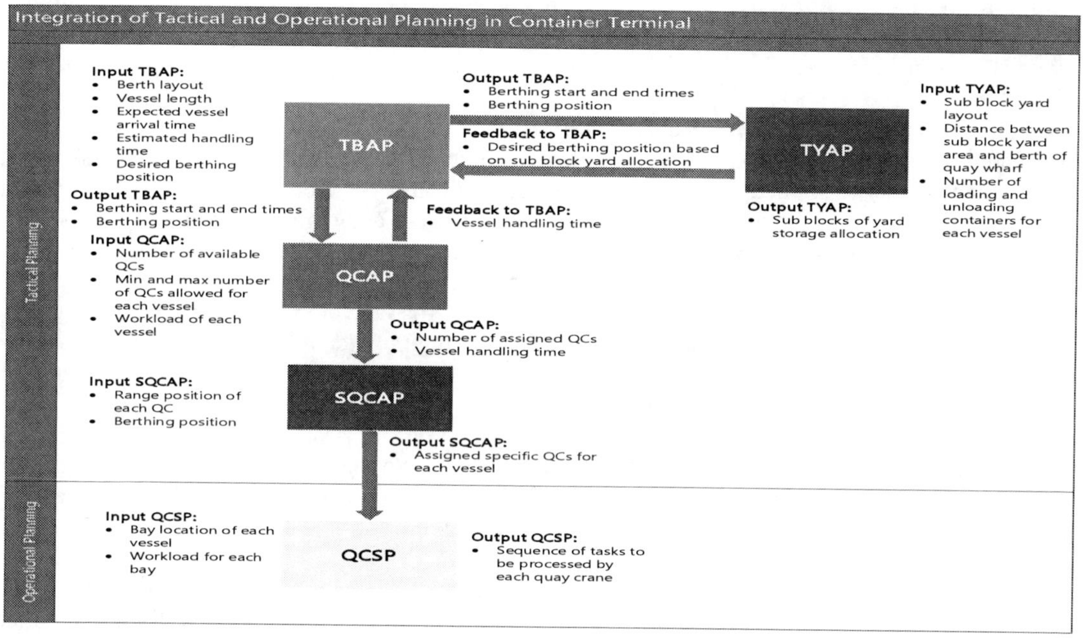

Figure 1. Solution method for integrated planning model in container terminal operations.

Table 1. Parameters and optimal solutions for each case.

Case	V [a]	L [b]	QC [c]	K [d]	H [e]	SL [f]	TC [g]	Run-time (s)
C6-600-80	6	600	8	80	42	100.00%	52.35	0.66
C9-600-80	9	600	8	80	42	97.93%	245.29	0.74
C12-600-80	12	600	8	80	42	93.03%	523.64	2.57
C15-1000-120	15	1000	16	120	42	97.83%	358.12	3.15
C18-1000-120	18	1000	16	120	42	92.69%	477.06	6.74
C21-1000-120	21	1000	16	120	42	95.57%	561.49	8.03
C24-1400-160	24	1400	24	160	42	96.13%	285.30	12.72
C27-1400-160	27	1400	24	160	42	94.07%	829.74	33.12
C30-1400-160	30	1400	24	160	42	90.78%	1175.89	68.68
C33-1800-200	33	1800	30	200	42	95.69%	742.54	84.48
C36-1800-200	36	1800	30	200	42	94.92%	909.93	97.57
C40-1800-200	40	1800	30	200	42	91.72%	1510.49	176.10

a. V: number of vessels
b. L: length of quay wharf in meters
c. QC: number of available quay cranes
d. K: number of sub-blocks in yard area
e. H: number of planning horizon in time step.
f. SL: average of maximize service level
g. TC: average of minimize total service cost in container terminal

Based on the optimal solutions in Table 1, shown that the increasing in number of vessel arrivals at the case of the same quay wharf length, the number of QCs availability and the number of sub-blocks in yard area have an effect on increasing total costs and decreasing service level. Runtime is proportionally increasing with respect to the number of calling vessels, berths segments, and sub-blocks in yard storage area.

7. Conclusions and Further Research Direction

This study has developed a bi-objective optimization integration model of tactical and operational planning in container terminal operations that consist of tactical berth allocation problem, specific quay crane assignment problem, tactical yard allocation planning and quay crane scheduling problem. The optimization model has two objective functions, namely minimization of the total operational cost and maximization of the overall service level. The integration model had solved using the Fuzzy Multi Objective Programming approach. As a solution method, we had developed solution method by using functional integration with feedback loop structure for solving the integration model of TBAP, QCAP and SQCAP and integration of TBAP and TYAP. The deep integration was used to solve the integration of QCAP and QCSP using Python Programming.

For further research, it might be considered uncertainty environment in container terminal operational planning such as vessel arrival times, handling times of each vessel and productivity rate of quay cranes. In addition, for solving the integration model effectively, the development of metaheuristic algorithms could be considered to solve large scale problems.

Acknowledgement

The authors would like to thank the anonymous reviewers for all constructive suggestions and editors for their useful comments in improving this paper.

References

[1] Ma H, Chan FT, Chung S. A fast approach for the integrated berth allocation and quay crane assignment problem. *Proc Inst Mech Eng B J Eng Manuf.* SAGE Publications; 2014 Aug 14;229(11):2076–87. doi.org/10.1177/0954405414544555

[2] Meisel F, Bierwirth C. Integration of Berth Allocation and Crane Assignment to Improve the Resource Utilization at a Seaport Container Terminal. *Oper. Res. Proc. 2005.* Springer Berlin Heidelberg; 105–10. doi.org/10.1007/3-540-32539-5_17.

[3] Imai A, Chen HC, Nishimura E, Papadimitriou S. The simultaneous berth and quay crane allocation problem. *Transp. Res. Part E. Logist. Transp. Rev.* Elsevier BV; 2008 Sep;44(5):900–20. doi.org/10.1016/j.tre.2007.03.003

[4] Zhang C, Zheng L, Zhang Z, Shi L, Armstrong AJ. The allocation of berths and quay cranes by using a sub-gradient optimization technique. *Comput. Ind. Eng.* Elsevier BV; 2010 Feb;58(1):40–50. doi.org/10.1016/j.cie.2009.08.002.

[5] Giallombardo G, Moccia L, Salani M, Vacca I. Modeling and solving the Tactical Berth Allocation Problem. *Transp. Res. Part B Methodol.* Elsevier BV; 2010 Feb;44(2):232–45. doi.org/10.1016/j.trb.2009.07.003.

[6] Meisel F, Bierwirth C. Heuristics for the integration of crane productivity in the berth allocation problem. *Transp. Res. Part E. Logist. Transp. Rev.* Elsevier BV; 2009 Jan;45(1):196–209. doi.org/10.1016/j.tre.2008.03.001.

[7] Liang C, Huang Y, Yang Y. A quay crane dynamic scheduling problem by hybrid evolutionary algorithm for berth allocation planning. *Comput. Ind. Eng.* Elsevier BV; 2009 Apr;56(3):1021–8. doi.org/10.1016/j.cie.2008.09.024

[8] Vacca I, Salani M, Bierlaire M. An Exact Algorithm for the Integrated Planning of Berth Allocation and Quay Crane Assignment. *Transp. Sci.* Institute for Operations Research and the Management Sciences (INFORMS); 2013 May;47(2):148–61. doi.org/10.1287/trsc.1120.0428

[9] Shang XT, Cao JX, Ren J. A robust optimization approach to the integrated berth allocation and quay crane assignment problem. *Transp. Res. Part E. Logist. Transp. Rev.* Elsevier BV; 2016

Oct;94:44–65. doi.org/10.1016/j.tre.2016.06.011

[10] Iris Ç, Pacino D, Ropke S, Larsen A. Integrated Berth Allocation and Quay Crane Assignment Problem: Set partitioning models and computational results. *Transp. Res. Part E. Logist. Transp. Rev*. Elsevier BV; 2015 Sep;81:75–97. doi.org/10.1016/j.tre.2015.06.008

[11] Iris Ç, Pacino D, Ropke S. Improved formulations and an Adaptive Large Neighborhood Search heuristic for the integrated berth allocation and quay crane assignment problem. *Transp. Res. Part E. Logist. Transp. Rev.* Elsevier BV; 2017 Sep;105:123–47. doi.org/10.1016/j.tre.2017.06.013

[12] Wang T, Wang X, Meng Q. Joint berth allocation and quay crane assignment under different carbon taxation policies. *Transp. Res. Part B Methodol.* Elsevier BV; 2018 Nov;117:18–36. doi.org/10.1016/j.trb.2018.08.012

[13] Krimi I, Benmansour R, el Cadi AA, Duvivier D, Elhachemi N, Deshayes L, et al. The integrated multi-quay Berth Allocation and Crane Assignment Problem: Formulation and case study. 2018 *7th Int. Conf. Ind. Technol. Manag.* (ICITM). IEEE; 2018 Mar; doi.org/10.1109/icitm.2018.8333938

[14] Zheng F, Li Y, Chu F, Liu M, Xu Y. Integrated berth allocation and quay crane assignment with maintenance activities. *Int. J. Prod. Res.* Informa UK Limited; 2018 Nov;57(11):3478–503. doi.org/10.1080/00207543.2018.1539265

[15] Prayogo DN, Hidayatno A, Komarudin. Developing a robust optimization model for seaside operations in container terminal under uncertainty environment. *IOP Conf. Ser. Earth Environ Sci.* IOP Publishing; 2019 Feb 20;235:012067. doi.org/10.1088/1755-1315/235/1/012067.

[16] Hu H, Li M, Wang T. Berth allocation and quay crane-yard truck assignment considering carbon emissions in port area. *Int. J. Shipp. Transp. Logist.* Inderscience Publishers; 2019;11(2/3):216. doi.org/10.1504/ijstl.2019.10020672..

[17] Xie F, Wu T, Zhang C. A Branch-and-Price Algorithm for the Integrated Berth Allocation and Quay Crane Assignment Problem. *Transp. Sci.* Institute for Operations Research and the Management Sciences (INFORMS); 2019 Sep;53(5):1427–54. doi.org/10.1287/trsc.2019.0894.

[18] Karam A, Eltawil AB. Functional integration approach for the berth allocation, quay crane assignment and specific quay crane assignment problems. *Comput. Ind. Eng.* Elsevier BV; 2016 Dec;102:458–66. doi.org/10.1016/j.cie.2016.04.006.

[19] Park Y-M, Kim KH. A scheduling method for Berth and Quay cranes. *Contain. Termin. Autom. Transp. Syst. Logist. Control Issues Quant. Decis. Support*. Springer-Verlag; 159–81. doi.org/10.1007/3-540-26686-0_7

[20] Agra A, Oliveira M. MIP approaches for the integrated berth allocation and quay crane assignment and scheduling problem. *Eur. J. Oper. Res.* Elsevier BV; 2018 Jan;264(1):138–48. Available from: http://dx.doi.org/10.1016/j.ejor.2017.05.040.

[21] Abou Kasm O, Diabat A, Cheng TCE. The integrated berth allocation, quay crane assignment and scheduling problem: mathematical formulations and a case study. *Ann. Oper. Res.* Springer Science and Business Media LLC; 2019 Jan 2;291(1-2):435–61. doi.org/10.1007/s10479-018-3125-3

[22] Li F, Sheu J-B, Gao Z-Y. Solving the Continuous Berth Allocation and Specific Quay Crane Assignment Problems with Quay Crane Coverage Range. *Transp. Sci.* Institute for Operations Research and the Management Sciences (INFORMS); 2015 Nov;49(4):968–89. doi.org/10.1287/trsc.2015.0619

[23] Meisel F, Bierwirth C. A Framework for Integrated Berth Allocation and Crane Operations Planning in Seaport Container Terminals. *Transp. Sci.* Institute for Operations Research and the Management Sciences (INFORMS); 2013 May;47(2):131–47. doi.org/10.1287/trsc.1120.0419

[24] Türkoğulları YB, Taşkın ZC, Aras N, Altınel İK. Optimal berth allocation, time-variant quay crane assignment and scheduling with crane setups in container terminals. *Eur. J. Oper. Res.* Elsevier BV; 2016 Nov;254(3):985–1001. doi.org/10.1016/j.ejor.2016.04.022

[25] Tao Y, Lee C-Y. Joint planning of berth and yard allocation in transshipment terminals using

multi-cluster stacking strategy. *Transp. Res. Part E. Logist. Transp. Rev.* Elsevier BV; 2015 Nov;83:34–50. doi.org/10.1016/j.tre.2015.08.005

[26] Ma HL, Chung SH, Chan HK, Cui L. An integrated model for berth and yard planning in container terminals with multi-continuous berth layout. *Ann. Oper. Res.* Springer Science and Business Media LLC; 2017 Jul 13;273(1-2):409–31. doi.org/10.1007/s10479-017-2577-1.

[27] Fu Y-M, Diabat A, Tsai I-T. A multi-vessel quay crane assignment and scheduling problem: Formulation and heuristic solution approach. *Expert Syst. Appl.* Elsevier BV; 2014 Nov;41(15):6959–65. doi.org/10.1016/j.eswa.2014.05.002

[28] Fu Y-M, Diabat A. A Lagrangian relaxation approach for solving the integrated quay crane assignment and scheduling problem. *Appl. Math. Model.* Elsevier BV; 2015 Feb;39(3-4):1194–201. doi.org/10.1016/j.apm.2014.07.006.

[29] Theodorou E, Diabat A. A joint quay crane assignment and scheduling problem: formulation, solution algorithm and computational results. *Optim. Lett.* Springer Science and Business Media LLC; 2014 Sep 20;9(4):799–817. doi.org/10.1007/s11590-014-0787-x.

[30] Alsoufi G, Yang X, Salhi A. Combined quay crane assignment and quay crane scheduling with crane inter-vessel movement and non-interference constraints. *J. Oper. Res. Soc.* Informa UK Limited; 2018 Jan 17;69(3):372–83. doi.org/10.1057/s41274-017-0226-3.

[31] Zhen L, Chew EP, Lee LH. An Integrated Model for Berth Template and Yard Template Planning in Transshipment Hubs. *Transp. Sci.* Institute for Operations Research and the Management Sciences (INFORMS); 2011 Nov;45(4):483–504. doi.org/10.1287/trsc.1100.0364.

[32] Prayogo DN, Hidayatno A, Komarudin. Development of integrated tactical level planning in container terminal. *Int. Conf. Ind. Eng. Eng. Manag.* IEEE; 2017 Dec; doi.org/10.1109/ieem.2017.8290045

[33] Liu M, Lee C-Y, Zhang Z, Chu C. Bi-objective optimization for the container terminal integrated planning. *Transp. Res. Part B Methodol.* Elsevier BV; 2016 Nov;93:720–49. doi.org/10.1016/j.trb.2016.05.012

[34] Wang K, Zhen L, Wang S, Laporte G. Column Generation for the Integrated Berth Allocation, Quay Crane Assignment, and Yard Assignment Problem. *Transp. Sci.* Institute for Operations Research and the Management Sciences (INFORMS); 2018 Aug;52(4):812–34. doi.org/10.1287/trsc.2018.0822

[35] Liu C. Iterative heuristic for simultaneous allocations of berths, quay cranes, and yards under practical situations. *Transp. Res. Part E Logist. Transp. Rev.* Elsevier BV; 2020 Jan;133:101814. doi.org/10.1016/j.tre.2019.11.008.

[36] Mohamed RH. The relationship between goal programming and fuzzy programming. *Fuzzy Sets Syst.* Elsevier BV; 1997 Jul;89(2):215–22. doi.org/10.1016/s0165-0114(96)00100-5

Analysis of sugar import policy effects on sugar cane farmer's income in East Java: A system dynamic approach

A F Fudhlaa[1,2], W Rachmawati[1,3], and D Retnowati[1,4]

[1]Industrial Engineering, Universitas Maarif Hasyim Latif, Raya Ngelom Megare Taman Sidoarjo 61257

E-mail: [1]fatih_fudhla@dosen.umaha.ac.id, [2]windirachmawati.97@gmail.com, [3]dini_retnowati@dosen.umaha.ac.id

Abstract. Besides rice, sugar becomes a national strategic agricultural commodity in Indonesia. Until 2019, since local sugar production has not been able to meet national needs, import is required. However, the uncontrolled entry of imported sugars could hit the presence of the local sugar. This can smite particularly the traditional sugar cane farmers who are the backbone of national sugar production. Based on the observation in East Java as a province with the largest sugar production in Indonesia, in recent years, when they are depressed with the increase of living and farming costs, the entry of imported sugar in large volumes will make their conditions worse. This study develops a system dynamics model that can evaluate the impact of sugar import policies on the fluctuations in the welfare of traditional sugar cane farmers in East Java. After the validation process, the result of simulation shows that if there are imported sugars in volume 10% of local productions, sugar cane farmers' incomes decrease 38% on average per year. When imported sugars reach 60% of local production, sugar cane farmers' annual income dropped drastically tends to go bankrupt. When there are no imported sugars, the income crept up an average of 4.35% per year. The Model could be used by policy makers as an early warning mechanism on this issue.

1. Introduction

Sugar is one of the agriculture strategic commodities in Indonesia [1]. National sugar consumption in 2018 reaches 5.5 million tons which domestic production could fulfill only 38% [2]. As the largest sugar-producing province in Indonesia [3], East Java has an important role in national sugar needs fulfillment [4]. Its sugarcane farmers have a key role in the supply of sugar raw material [5]. Nevertheless, since 2012 sugar production has decreased [2] and affected the depression in sugar businesses, especially sugarcane farmers. During the sugarcane planting, sugar price has increased, but when harvesting it has decreased. it gets worse when imported sugar came into East Java. Imported sugar is a national policy intended to meet the lack of supply to national sugar demand [2]. However, the entry of refined sugar/imported sugar into East Java which has a surplus in sugar production could cause huge problems [3]. Big traders or broker were the sides who benefit the most, sugar cane farmers welfare were the worst, so an analytical study is needed to relieve this problem.

Furthermore, the sugarcane was not a simple system, many complex interactions between system components [6]. One analytical method that can accommodate them is system dynamic approach [7]. This study developed an instrument of system dynamics modeling that could evaluate how traditional sugar cane farmers' welfare fluctuation in East java affected by sugar import policies.

2. System Dynamics Modelling

System dynamics is a simulation methodology to describe and analyze the complex feedback system that the behaviors are changing over time [8]. According to Feofilovs et.al [9], System dynamics is a simulation model that used to describe the complexity of system behavior which includes social aspects by using causal loops and stock and flow in its interactions.

Simulation dynamic system modelling was initially developed at the United State by JW Forrester in 1956, which is used to analyse complex behaviour in social science, especially in management science. While according to Forrester [10], system dynamics is a simulation methodology used to analyse complex systems that change from time to time, especially in management science and modern control theory using computer simulations.

In the modelling, the initial step is "fully understanding" the real system behaviour, then identifying variables and relations among them using the Causal Loop Diagram (CLD). The feedback relationships generate the loops that will make the system behave simultaneously and dynamically like a real system [11] [12]. The main concept of system dynamic theory is stock and flow. Stock is an element that can change and can describe the process of storing materials. Stock can change due to the influence of flow. While flow shows the pattern of changes in stock and adds or reduces some entities or information. In the model, flow is addressed as a variable that can affect a stock level [12] [13].

Model validation is performed to determine the accuracy of a model based on mathematical equations. To show the real problem, the model has to be simulated to obtain outputs based on system behaviour [14]. If model behaviour similar to the real system, then the model is valid [15].

2.1. Previous Research (System Dynamic in Agriculture)
System dynamic approaches in various fields of agriculture have been carried out in previous studies, including such as; research on production and soybean demand [16], supply chain perspective of soybean sufficiency [17], research that discusses the hydrological process in plantations [18], studies that discuss the welfare of farmers and developing scenarios how to reduce poverty of farmers in Central America [19], and in New Zealand [20]. While previous system dynamic research related to sugarcane farming or sugar systems included; research that discusses the exploration of sugarcane yields that affect the production process and ethanol consumption in Brazil [21], research for enhance availability of sugar in East Java [1], and research examining sugarcane logging in the Bondowoso Indonesia [22].No research discusses the welfare of sugarcane farmers in System Dynamics before.

3. Result and Discussion
3.1. Research Object Description
East Java is the largest sugar-producing province in Indonesia. In 2018, it produced 1.04 million tons of sugar, which up to 450 thousand tons were absorbed by local needs, and the remaining were absorbed by national needs. Based on data from Indonesia's Central Bureau of Statistics/ "Badan Pusat Statistik" (BPS), East Java has the largest sugarcane area up to two hundred thousand hectares. In this province stand 33 plants of sugar cane refineries, the largest number in Indonesia, owned by state and private companies. Based on data from BPS 98% of sugarcane land in East Java is owned and managed by farmers, others by state and private companies. These show that most sugar raw material supply is obtained from the sugar farmers' harvest. There are two large groups of farmers, namely landowners/farmers and farmworkers. In this study, these two groups' income will be highlighted as a very vulnerable welfare class when there are imported sugars got into East Java Province.

3.2. Sugarcane Farmers Welfare System Dynamic Model
This research begins with the identification of the variables involved in the income of sugar cane farmers and farmworkers, and then it will be outlined in the Causal Loop Diagram (CLD). By using *Vensim* (Ventana Simulation), the relationship between variables in the income system of Sugar Cane Farmers in East Java is shown in Figure 1.

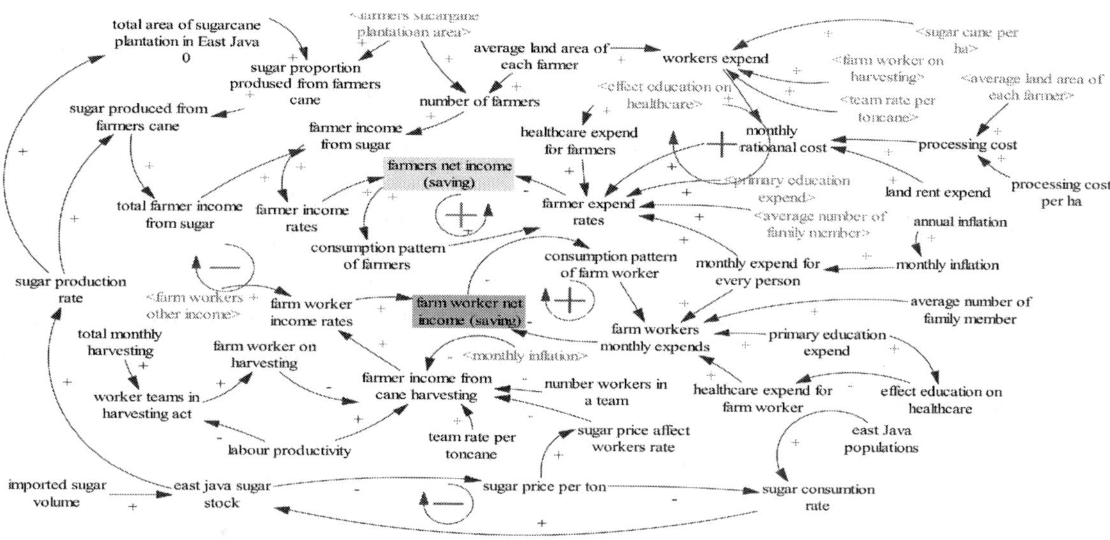

Figure 1. Causal loop diagram East Java sugarcane farmer's income.

Figure 1 shows the interaction among variables. Positive arrows indicate a positive relationship and negative arrows indicate a negative relationship. The different colours highlight two central variables, "Farmer Net income (saving)" and "farmworkers net income (saving)". The Farmer is influenced negatively by farmers expend rate and positively by income rates. It's means that "farmers net income" will decrease when the expend rise and increase when the income ascend. Imported sugars affect price through the stock/storage. "Sugar consumption rate" will decrease when the price increases.

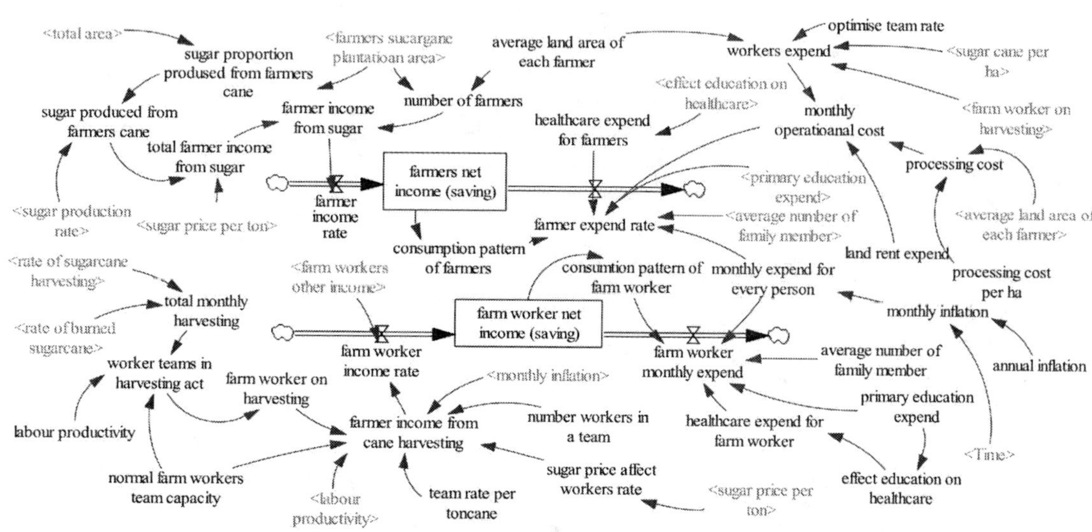

Figure 2. Main model (stock and flow diagram) of sugarcane farmers' income in East Java.

Figure 2 is a Stock and flow diagram of the main model. This model is based on Causal Loop Diagram before with net income of two farmers as the main variable. Net income is the total net income of the farmer, i.e. the total amount of income that has been reduced by the total amount of expenditure per period. Expenditure is derived from the total cost of daily needs, health costs, education costs, and operational costs for farmers. Operational costs are formed from processing costs, land rental costs, and labour costs. Labour costs are the costs for farmers who own the land, but the income for farm laborers. When it is not in a harvesting season, farmworkers also get income from other freelance jobs, such as *ojek*, construction workers, and land or field workers. "Farmers Net income (saving)" is established from the reduction of "farmer income rate" and "farmer expends rate" added by the initial value 55 million rupiah (Rp). Several variable formulas have shown in Table 1.

Table 1. Several examples of variable formulas configuring the model.

No	Variable	Type	Formula	Units
1	Farmers net income (saving)	level	INTEG (+farmer income rate - farmer expend rate) with initial value: 55000000	Rp
2	Farmer income rate	rate	farmer income from sugar	Rp/month
3	Farmer expend rate	rate	monthly operational cost + primary education expends × average number of family member/2 + monthly expend for every person × average number of family member × consumption pattern of farmers + healthcare expends for farmers	Rp/month
4	Farm worker net income	level	INTEG (farm worker income rate- farm worker monthly expend) with initial value: 1500000	Rp
5	Farm worker income rate	rate	farmer income from cane harvesting + farm workers other income	Rp/month
6	Farm workers other income	attribute	(income from others freelance + other income from construction freelance + others income from "ojek") × IF THEN ELSE (farm worker on harvesting=0, 1 , 0)	Rp/month

Figure 3 below represents the sub model of sugar and cane production. This sub model describes that sugar production rate depends on both sugarcane volume planted and sugar yield (*rendemen*) of cane. There are probabilities of cane burned every year. And volumes of cane depend on total land/farm area.

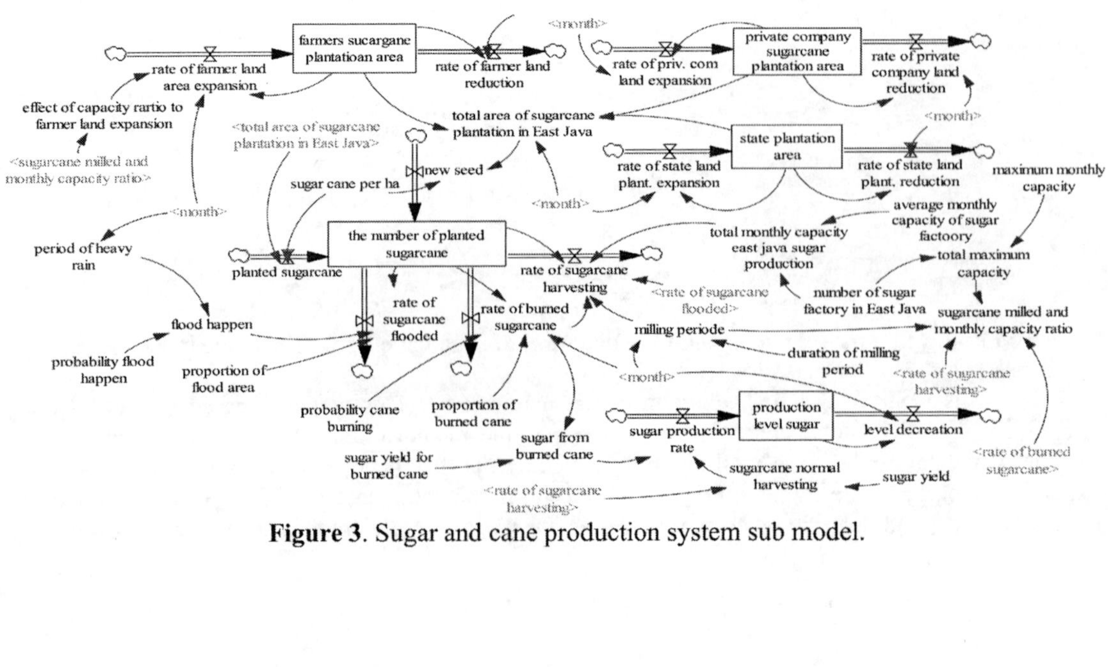

Figure 3. Sugar and cane production system sub model.

Sugar stock volume in East Java besides local consumption it is also used to fulfil national needs. The stock gets supply from "sugar production rate" and imported sugar that gets into East Java. "Local consumption rate" is affected by the population of East Java. All of them are represented in Figure 4. All sub-models are integrated into large models to support the main model. They are not running separately.

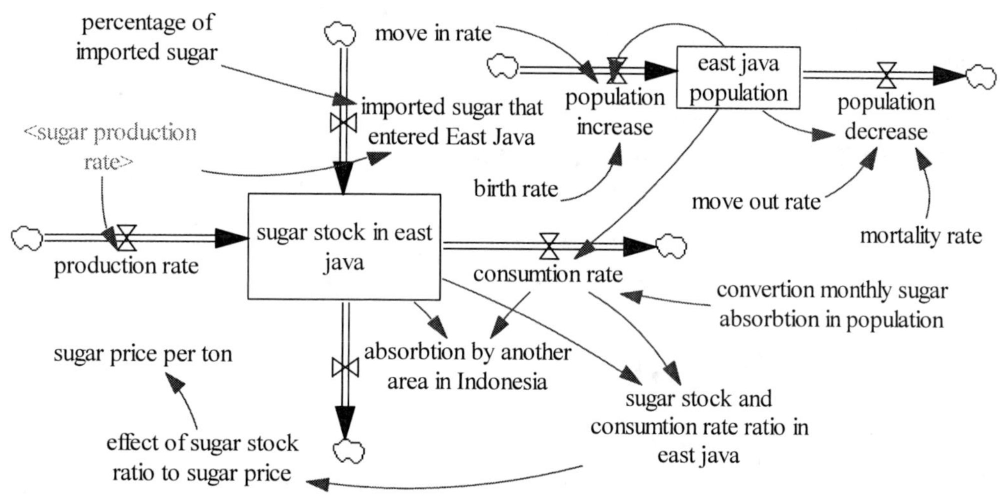

Figure 4. Sugar market absorption sub model.

3.3. *Validation*

After the modelling process, the next steps are simulating and validating the model. The first validation is performed by parameter testing. The parameter tested was variable of "total sugarcane area". The model is simulated for 100 months. And recapitulation obtained a total of 8 annual sugarcane area data. The results of this simulation will be compared statistically with data sourced from BPS. 12 historical data of sugarcane land area in East Java are obtained from 2006 to 2017 as shown in Figure 5 below.

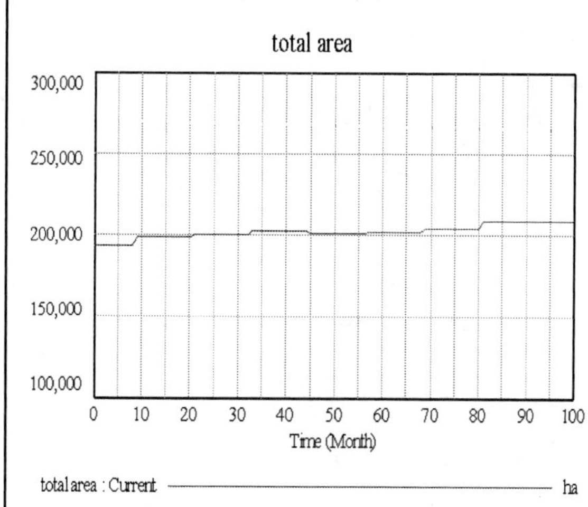

Simulation result		historical data (BPS)	
year period	sugar farms area (ha)	year	sugar farms in East Java (ha)
1	193,817	2006	173,829
2	197,951	2007	198,343
3	199,820	2008	200,822
4	202,821	2009	186,026
5	201,215	2010	193,396
6	201,740	2011	197,762
7	204,087	2012	203,483
8	208,242	2013	205,805
		2014	218,706
		2015	175,259
		2016	200,702
		2017	193,940

Figure 5. Recapitulation of simulation data and historical data for model validation.

Two-sample t-test was applied to perform a hypothesis testing that computes whether two populations have statistically significant differences or not in a certain confidence interval [23]. The hypothesis is formulated as follow:

$$H_0: \mu_1 - \mu_2 = \delta_0 \quad \text{versus} \quad H_1: \mu_1 - \mu_2 \neq \delta_0$$

where μ_1 and μ_2 are the population means of “simulation result” and “historical data of total area planted by sugarcane”. And δ_0 is the hypothesized difference between the two equals to zero. The null hypothesis states that the two populations are similar or not different. Meanwhile, the alternative hypothesis states that differences exist.

Two-Sample T-Test and CI: simulate; historical data

```
Two-sample T for simulate vs historical data

                   N    Mean  StDev  SE Mean
simulate           8  201,21   4,27      1,5
historical data   12   195,7   12,6      3,6


Difference = mu (simulate) - mu (historical data)
Estimate for difference:  5,53887
95% CI for difference:  (-2,92141; 13,99916)
T-Test of difference = 0 (vs not =): T-Value = 1,40  P-Value = 0,182  DF = 14
```

Figure 6. Two-sample t-test result for validation.

Based on the result of the two-sample t-test, P-value 0.182 was obtained. It means in $\alpha = 0.05$, H_0 accepted, there were no differences between the average simulation result and historical data. a 95% confidence interval is (-2.92, 13.99) which includes zero, thus suggesting that the difference doesn’t exist (statistically similar). Validation processes were obtained by the boundary adequacy test and model behaviour test as well. The model was valid.

3.4. Simulation and Scenario Analysis

The following are simulation results from models that have passed the validation. The simulation was run in a monthly time step. Simulations run for 100 months. Variables of “Farmers net income” and “Farmworker net income” obtained the results of the simulation as shown consecutively in Figure 7 (a) and 7 (b). Each graph fluctuates depending on the period in the harvesting season or not. In variable of “Farmworker net income” it is shown that farm workers biggest income was in the sugarcane harvesting season, they could reach more than 10 million rupiahs in saving. If it was not in harvesting season, they worked any freelance jobs, so that their income dropped dramatically.

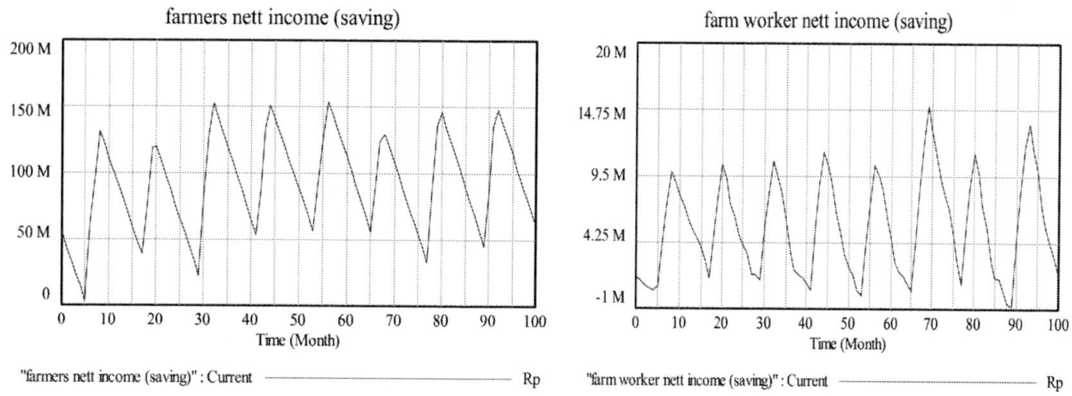

(a) (b)

Figure 7. (a) Simulation result graph for "Farmers net income" variable.
(b) Simulation result graph for "Farm workers net income" variable.

There are gaps between the two main variables. But both of them tend to increase. As shown in Figure 8(a) and 8(b), the results of the two variables were plotted on trend analysis. This was performed to find out how big the change was over time. The linear approach was obtained. For the "Farmers net income" variable, at the early period fits 77.407.900 and at the 100th period fits 106.119.300. It means the income increased 28.711.400 in 100 months, 0.37% a month, or 4.5% a year. For "Farmworkers net income" variable, at first period fit to 4.655.200 and 100th fits 6.268.600. it means the income increased 1.613.400 in 100 months, 0.35% a month, or 4.2% for a year. Both increased by 4.3% a year on average.

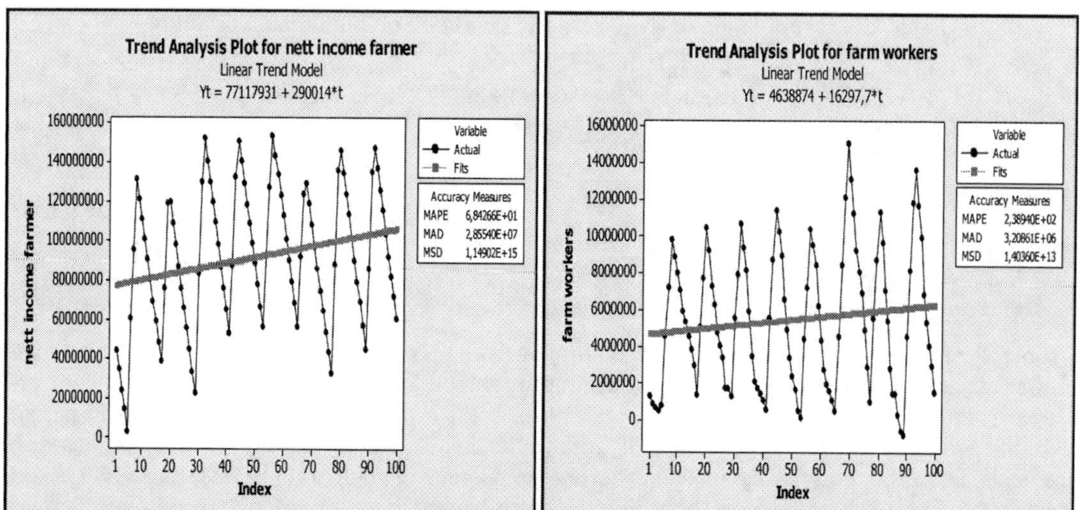

(a) (b)

Figure 8. (a) Trend analysis for "Farmer net income" variable.
(b) Trend analysis for "Farm workers net income" variable.

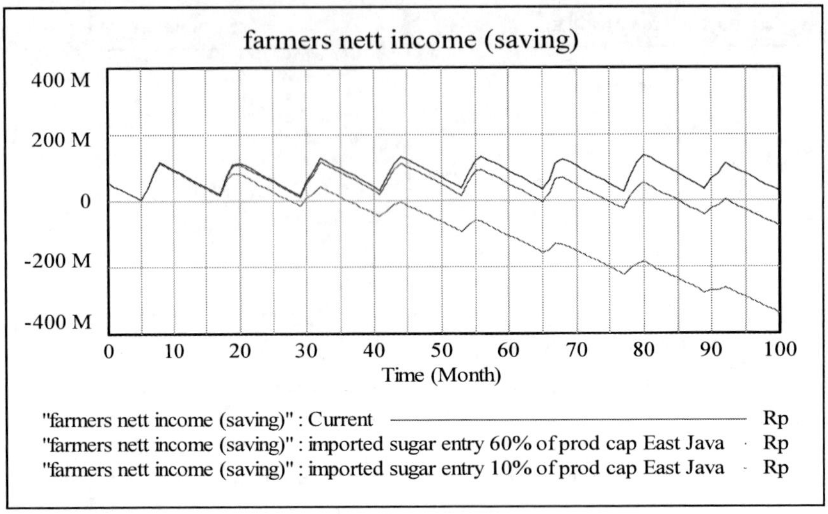

Figure 9. Result of "Farmer net income" variable from three scenario of the entry of imported sugar in East Java.

After no imported sugar entry, the next step is simulating the model with the entry exist. The first scenario is carried out by a volume of 10% of the level of local sugar production and the second by an extreme volume in 60% of the level of local sugar production. 100 months running period are performed as well. Result and Comparison of them are represented in Figure 9 above. Farmer's net income from the results of scenario 1 looks lower than scenario 0 (no imported sugar entry), but the second scenario shows drastically drop in net income. Three of them show a similar level at the beginning period, but after two harvesting seasons, they look different.

Table 2. Recapitulation of three scenario condition on both main variables.

Simulation scenario	Imported Sugar (% Volume of East Java Sugar Production Rate)	Average Of "Farmers Net Income (Saving)" (Rp)	Gap from scenario 0 (%)	Average Of "Farm worker Net Income (Saving)" (Rp)	Gap from scenario 0 (%)	Average Gab (%)
0	0%	76,050,500	0	5,249,400	0	0%
1	10%	39,641,800	-47.9%	3,764,050	-28.3%	-38%
2	60%	(83,738,000)	-210.1%	1,776,300	-66.2%	-138%

Table 2 represents the average value of both main variables in three scenarios. All of them show that the entry of imported sugar into a province that surplus in local production as East Java harms all sugarcane farmers' welfare. Farmers gain greater losses. As a landowner, pre-planting, planting, and harvesting processes require large costs. Misfortune for them the entry of imported sugar causes the price deals below the total production costs during the post milling auction process. Sugar sales could not cover the costs. The scenarios show that uncontrolled entry of imported sugar affects farmers' income badly. If needed by industry, the government must make strict regulations and ensure that the imported sugar does not leak to be sold freely to end in the public market.

4. Conclusion

From the analysis of simulation model results, sugarcane farmers' income threatened with the entry of imported sugar in east java. The simulation results that if there are imported sugar get into East Java with 10% of local sugar volume production, sugarcane farmers' income will decrease 38% per year on average. When the entry of imported sugar into East Java goes out of control, for example, 60% of local sugar production, the average annual income of sugar cane farmers will drastically decrease. But when imported sugar does not enter, the income of smallholder sugarcane farmers creeps up 4.3% on average per year. Therefore, it is hereby recommended for Government especially the Local Government of East Java Province to tighten the entry of imported sugar into East Java to protect the welfare of sugar cane farmers.

Acknowledgement
The authors would like to acknowledge colleagues in the Industrial Engineering Laboratory of Universitas Maarif Hasyim Latif Sidoarjo for the facilitations in developing the models.

References
[1] Putra A B and Suryani E,. (2014). Skenario Kebijakan Industri Gula Untuk Meningkatkan Ketersediaan Gula Dipasaran Dengan Menggunakan Pendekatan Sistem Dinamik. *Sisfo.* 05, 01 p. 51–61. doi.org/10.24089/j.sisfo.2014.03.013
[2] Bagus S, (2019). Petani Tebu Dukung Kebijakan Gula Nasional, *Media Indoensia E-Paper (https://mediaindonesia.com/read/detail/221902-petani-tebu-dukung-kebijakan-gula-nasional).* [Online]. Available: https://mediaindonesia.com/read/detail/221902-petani-tebu-dukung-kebijakan-gula-nasional. [Accessed: 20-Feb-2020].
[3] Kumalasari A D Budiraharjo K and Setiadi A,. (2019) Komparasi Produksi dan Pendapatan Petani Tebu MItra dan Non Mitra Pabrik Gula Rendeng di Kabupaten Kudus. *Agrisocionomics: Jurnal Sosial Ekonomi Pertanian.* 3, 1 p. 28. doi.org/10.14710/agrisocionomics.v3i1.4021
[4] Sadikun S,. (2019). Sejauh Mana Jawa Timur Wujudkan Swasembada Gula Nasional 2019?, *Mediajatim.com.* [Online]. Available: https://mediajatim.com/2019/10/15/sejauh-mana-jawa-timur-wujudkan-swasembada-gula-nasional-2019/. [Accessed: 20-Feb-2020].
[5] Asrol M, Marimin M, Machfud M, Yani M, Taira E. Supply Chain Fair Profit Allocation Based on Risk and Value Added for Sugarcane Agro-industry. *Oper. Supply Chain Manag Int. J.* OSCM Forum; 2020 Mar 12;150–65. doi.org/10.31387/oscm0410259
[6] Shongwe M I and Bezuidenhout C N,. (2018). A heuristic for the selection of appropriate diagnostic tools in large-scale sugarcane supply systems. *AIMS Agric. Food* 4, 1 p. 1–26. doi.org/10.3934/AGRFOOD.2019.1.1
[7] Ima Dudin M D Wiranatha A S and Sadyasmara C A B,. (2020). Simulasi Model Sistem Dinamik Ketersediaan Bawang Putih (Allium sativum, L.) di Provinsi Bali. *Jurnal Rekayasa Dan Manajemen Agroindustri.* 8, 1 p. 114. doi.org/10.24843/JRMA.2020.v08.i01.p12
[8] Seeler K A, (2014). Introduction to System Dynamics. *Syst. Dyn.*, (New York, NY: Springer New York), p. 1–44. doi.org/10.1007/978-1-4614-9152-1_1
[9] Feofilovs M Gravelsins A Pagano A J and Romagnoli F,. (2019). Increasing resilience of the natural gas system with implementation of renewable methane in the context of Latvia: A system dynamics model *Energy Procedia* 158 p. 3944–3950. doi.org/10.1016/j.egypro.2019.01.848
[10] Forrester J W and Collins F, Dec. (1972). World Dynamics *J Dyn Syst Meas Control.* 94, 4 p. 339–339. doi.org/10.1115/1.3426619
[11] Dewi N, Miharja M and Yudoko G,. (2015). Analisis Kebijakan Distribusi Bahan Baku Rotan Dengan Pendekatan Dinamik Sistem Studi Kasus Rotan Indonesia *Jurnal Perencanaan Wilayah dan Kota* 26, 3 p. 177–191. doi.org/10.5614/jpwk.2015.26.3.3
[12] Forrester J W,. (1994) System dynamics, systems thinking, and soft OR *System Dynamics Review* 10, 2–3 p. 245–256. doi.org/10.1002/sdr.4260100211
[13] Budi Supangat A, Sudira P, Supriyo H and Poedjirahajoe E,. (2018). Simulasi Model Dinamik

Pengaruh Legume Cover Crops (LCC) Terhadap Limpasan Dan Sedimen Di Lahan Hutan Tanaman *Jurnal Penelitian Pengelolaan Daerah Aliran Sungai* 2, 1 p. 17–34. doi.org/10.20886/jppdas.2018.2.1.17-34

[14] Forrester J W,. (2016). Learning through System Dynamics as Preparation for the 21st Century *Syst. Dyn. Rev.* 32, 3–4 p. 187–203. doi.org/10.1002/sdr.1571

[15] Shreckengost RC. Dynamic simulation models: how valid are they? *PsycEXTRA Dataset*. American Psychological Association (APA); 1985; Available from: http://dx.doi.org/10.1037/e496952006-007

[16] Hasan N, Suryani E, Hendrawan R. Analysis of Soybean Production and Demand to Develop Strategic Policy of Food Self Sufficiency: A System Dynamics Framework. *Procedia Comput. Sci.* Elsevier BV; 2015;72:605–12. doi.org/10.1016/j.procs.2015.12.169

[17] Oktyajati N, Hisjam M and Sutopo W,. (2018). The dynamic simulation model of soybean in Central Java to support food self sufficiency: A supply chain perspective in *AIP Conference Proceedings* p. 030015. doi.org/10.1063/1.5024074

[18] Ouyang Y, Xu D, Leininger TD, Zhang N. A system dynamic model to estimate hydrological processes and water use in a eucalypt plantation. *Ecol. Eng.* Elsevier BV; 2016 Jan;86:290–9. doi.org/10.1016/j.ecoleng.2015.11.008

[19] Marín-González O, Parsons D, Arnes-Prieto E, Díaz-Ambrona CGH. Building and evaluation of a dynamic model for assessing impact of smallholder endowments on food security in agricultural systems in highland areas of central America (SASHACA). *Agric. Syst.* Elsevier BV; 2018 Jul;164:152–64. doi.org/10.1016/j.agsy.2018.02.005

[20] Rich K M, Rich M and Dizyee K, (2018) Participatory systems approaches for urban and peri-urban agriculture planning: The role of system dynamics and spatial group model building *Agric. Syst.* Elsevier BV; 2018 Feb; 160 p. 110–123. doi.org/10.1016/j.agsy.2016.09.022

[21] Demczuk A and Padula A D,. (2017). Using system dynamics modeling to evaluate the feasibility of ethanol supply chain in Brazil: The role of sugarcane yield, gasoline prices and sales tax rates *Biomass Bioenergy* 97 p. 186–211. doi.org/10.1016/j.biombioe.2016.12.021

[22] Yunitasari D, Lestari E K and Istiyani N,. (2018). Analisis Potensi Tebu dalam Mendukung Pencapaian Swasembada Gula di Kabupaten Bondowoso *Buletin Tanaman Tembakau, Serat & Minyak Industri* 10, 1 p. 13. doi.org/10.21082/btsm.v10n1.2018.13-20

[23] Gerald B,. (2018). A Brief Review of Independent, Dependent and One Sample t-test *Int. J. Appl. Math. Theor. Phys.* 4, 2 p. 50. doi.org/10.11648/j.ijamtp.20180402.13

Structural modeling for usability attributes on technology acceptance model for smart parking mobile application

A Amiruddin[1,2], R S Dewi[1,3] and E Widodo[1,4]

[1]Institut Teknologi Sepuluh Nopember,Surabaya, Indonesia

E-mail: [2]amaliahamiruddin@gmail.com, [3]ratna.sari.dewi80@gmail.com, and [4]erwin@ie.its.ac.id

Abstract. Smart parking is a mobile application that is created to overcome parking problems, although there are many smart parking applications developed in the world, only a few have a good interface design. One of the methods to assess the quality of the interface design is by measuring the usability of the application. Also, several studies have explored the relationship between usability and TAM (Technology Acceptance Model). However, is still lacking, especially about smart parking as an object. Due to that, this study focuses on determining usability attributes in smart parking applications and also modelling that usability attributes to TAM. This study using SEM (Structural Equation Model) method for modelling the relationship between Usability and TAM. The goal is the model could be used as a guide by application developers to see what factors must be considered when designing a smart parking mobile application, and for future research, it is expected that the developed model can be used as a guide when measuring the usability of a smart parking application. This research proposed ten hypotheses and findings two hypotheses have no significant effect that would be explained furthermore in the result and discussion section.

1. Introduction

In this world, there are many countries that have developed smart parking application to overcome parking problems that arise. In Canada, smart parking is developed to make the citizen be able to order a parking lot that is still empty, such as Best Parking, Parking Panda, and Spot Hero. In Taiwan, there are iTaipei Parking, Taoyuan Easy GO, Tainan Easy Parking, and many more. And in Indonesia itself there is Go Parking owned by the government of Surabaya, Cari Parkir owned by PT. Astra and in Sidoarjo Regency there is SPON (Sistem Parkir Online) that is still developed. Although there are many varieties of parking applications that have developed, only a few have a good interface design. Most of these applications are difficult to use because of poor interaction between applications and users.

One of the ways to evaluate the quality of the interface is by conducting usability evaluation. Usability is a well-known and well-defined concept in human-computer interaction (HCI) research, referring to the extent the user and the system can communicate clearly without misunderstanding through the interface [1]. According to the International Organization for Standardization (ISO) defines usability is the extent to which an application can be used by specified users to achieve specified goals with effectiveness, efficiency, and satisfaction in a specified context of use [2]. Neglecting usability may result in losing the users' interest, and the users will choose another application that serves its purpose better. Therefore, usability needs to be considered when developing smartphone applications.

In recent years, various studies have used different attributes or variables when evaluated the usability of products. Depends on the device used (smartphone or computer screen) will affect the model and the usability attributes to be selected. In 2013, Harrison, Flood, and Duce [3] studies developed a new usability model named PACMAD (People at the Centre of Mobile Application Development), where the attributes were effectiveness, efficiency, satisfaction, learnability, memorability, errors, and cognitive load. In 2015, Shawgi and Noureldien [4] developed a new usability model to measure website usability that named UMM (Usability Measurement Model). The usability variables consist of accessibility, understandability, learnability, operability, attractiveness, and navigability. In the following year, Hoehle, Aljafari, and Venkatesh [5] developed and conceptualized mobile application usability by analysing Microsoft mobile usability guidelines to be a model that could be followed by other researchers in the future. The usability model developed is how the relationship between mobile application usability (the attributes were aesthetic graphics, colour, obviousness control, entry points, fingertip-sized controls, fonts, gestalt, hierarchy, subtle animation, and transition) with the outcome variable is continued intention to use a mobile application.

In addition, usability is one of the determining factors of the TAM (Technology Acceptance Model). Where when creating an application, the developer also needs to see how the user's acceptance of the application. Based on previous research, it is known that new technology will not be easily accepted and used by humans. The process of accepting new technology is influenced by several factors. The TAM model is a valid model that is widely accepted and trusted to predict the acceptance of new technology by end-users. The main determinants of the TAM model are Perceived Usefulness (PU), which refers to an individual's belief that a particular system will improve work performance, and Perceived Ease of Use (PE), which refers to an individual's belief that using a particular system will free from effort [6]. Perceived usefulness and perceived ease of use are influenced by relevant external determinants that have an influence on Behavioural Intention to Use (BI) [6]. However, very limited research has been conducted to develop the model between usability attributes and TAM. Some studies that try to see the relationship between the usability attribute and TAM then find a relationship between them are [7] and [8].

Hence, this study intends to contribute a structural design model for the usability attribute to the Technology Acceptance Model (TAM) for smart parking mobile applications. Based on several existing studies, researchers see the absence of structural modelling for usability attributes on TAM for smart parking mobile applications in Indonesia. While in the future the smart parking application can be used by the Indonesian government in solving parking problems that arise and can be an opportunity for application developers in Indonesia to create and design a good smart parking application and have an impact on the desire of the Indonesian people to be interested in using it. Also, the model can be a valuable reference for other researchers when conducting research in the same field.

2. Methodology

This study has a purpose to determine the usability attributes in the smart parking mobile application then explore the relationship between usability attributes and Technology Acceptance Model (TAM). Based on these objectives, the main object in this study is the smart parking mobile application "Cari Parkir" owned by PT. Astra Indonesia. This study conducted several stages, including the research model and hypotheses development, data collection, measurement model, and interpretation of research results.

2.1 Research Model and Hypotheses Development

At this phase, researchers establish the model by study literature about smart parking, usability, Technology Acceptance Model (TAM), and Partial Least Square-Structural Equation Modelling (PLS-SEM). These study literature, [3-9] help researchers to determine the variables and indicators of usability and technology acceptance model (TAM), then connecting each latent variables to construct the model. See Figure 1 about the reflective-formative model proposed.

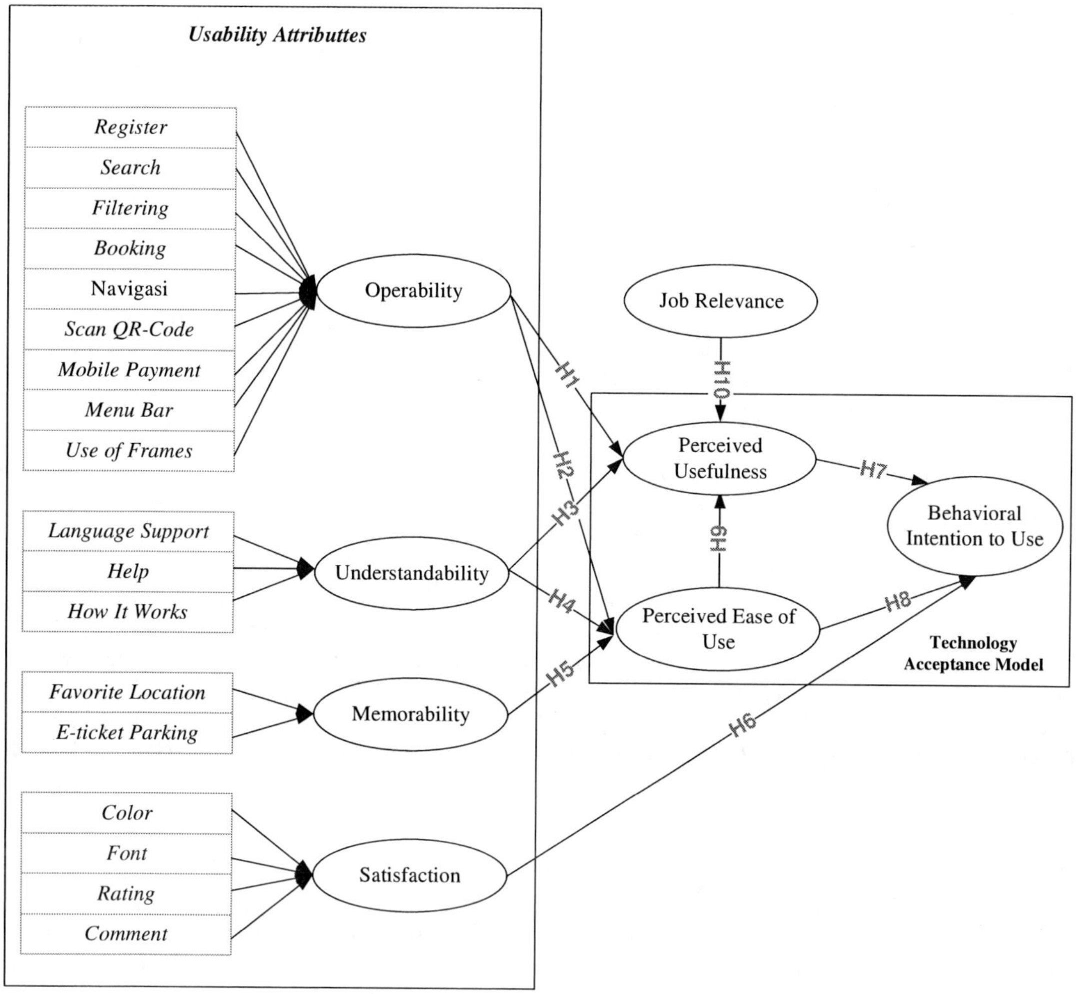

Figure 1. Proposed research model.

Based on Figure 1 can be seen that researcher proposed 10 hypotheses, they are:

H1: Operability has a positive and significant impact on Perceived Usefulness
H2: Operability has a positive and significant on Perceived Ease of Use

Operability is "the capability of the software product to enable the user to operate and control it" [10]. Operability in this model as a latent variable that contains nine indicators: register [11], search, filtering, booking, navigation, scan QR code, mobile payment, menu bar and, use of frames [4]. Perceived usefulness is defined as "the degree to which a person believes that using a particular system would enhance his or her job performance" [6]. And perceived ease of use refers to "the degree to which a person believes that using a particular system would be free of effort" [6]. Based on the definition of each variable, we assume that operability affects on perceived usefulness and perceived ease of use.

H3: Understandability has a positive and significant impact on Perceived Usefulness
H4: Understandability has a positive and significant impact on Perceived Ease of Use

Understandability is "the capability of the software to allow the user to understand whether the software is suitable, and how it can be used for particular tasks and conditions of use" [10].

H5: Memorability has a positive and significant impact on Perceived Ease of Use

Memorability is the ability of a user to retain how to use an application effectively [3]. In [7-8] studies also try to prove that memorability affects on perceived ease of use variable.

H6: Satisfaction has a positive and significant impact on Behavioural Intention to Use

Satisfaction is freedom from discomfort and positive attitudes towards the use of the application [3]. Behavioural intention to use is the degree to which a person has formulated a conscious plan to perform or not perform some specified behaviour [12]. Based on the definition of each variable, we assume that satisfaction affects behavioural intention to use.

H7: Perceived Usefulness has a positive and significant impact on Behavioural Intention to Use
H8: Perceived Ease of Use has a positive and significant impact on Behavioural Intention to Use
H9: Perceived Ease of Use has a positive and significant impact on Perceived Usefulness
H10: Job Relevance has a positive and significant impact on Perceived Usefulness

The hypotheses for Technology Acceptance Model (TAM) variables in this study are the following TAM 2 model. Where Perceived Usefulness (PU) and Perceived Ease of Use (PE) are the two main variables used to see how people's acceptance of new technology. Whereas the Behavioural Intention to Use (BI) variable is influenced by both main variables. Job Relevance (JR) as an external variable that affects the variable Perceived Usefulness (PU). The next hypothesis is Perceived Ease of Use (PE) influences Perceived Usefulness (PU) [13-14].

2.2 Data Collection Phase
The researcher conducted an online survey by using google docs from 1 May 2020 till 18 May 2020. The number of questionnaires filled in was 100 questionnaires but after we checked the data only 90 questionnaires that we could analyse. We removed 10 respondent because had straight-line pattern responses (dominant answer point 1 or point 6). The data is enough to process and analyse using SmartPLS, based on this rule saying that the minimum sample size should be 10 times the largest number of formative indicators used to measure a single construct [9]. The respondent in this study who is lived in Jabodetabek (Jakarta, Bogor, Depok, Tangerang, and Bekasi), Surabaya, Bandung, Makassar, and has used or familiar using "Cari Parkir" application. The characteristic of the respondent is displayed in Table 1.

2.3 Measurement Model
In this study, researchers applied Partial Least Square – Structural Equation Modelling (PLS-SEM) to assess the measurement and structural model. The measurement model is a model that describes the relationship between latent variables and indicators. Measurement model evaluation conducted to assess the validity and reliability of the model. The measurement model types in this study are divided into two called formative and reflective. For operability, understandability, memorability, and satisfaction variables are formative types. Where the indicators cause the variables. As for the variable perceived usefulness, perceived ease of use, job relevance, and behavioural intention to use are reflective types in which the indicators are caused by the variable. After testing reliability and validity (measurement model), the next step is to evaluate the results of the structural model. The aim is to examine the relationship between structural models represented by hypothetical relationships between constructs that have been built [9].

Table 1. Characteristic of Respondent (n=90).

Item	Demographics	Sample	Percentage
Gender	Female	48	53%
	Male	42	47%
Age	17-20	4	5%
	21-25	53	59%
	26-30	28	31%
	31-35	3	3%
	36-41	2	2%
Residence	Jabodetabek	54	60%
	Surabaya	14	15%
	Makassar	9	10%
	Bandung	7	8%
	others	6	7%
Job	Civil servant	10	11%
	Public company	8	9%
	Private company	41	45%
	Entrepreneur	6	7%
	College student	23	26%
	Housewife	2	2%
Parking frequency	Everyday	25	28%
	2 - 3 times in a week	35	39%
	Once in a week	30	33%

3. Result and Discussion

3.1 Measurement Model Evaluation

Firstly, we evaluate the formative type measurement model (usability variables) from the value of Variance Inflation Factor (VIF) and the significance of the outer weight indicator. Variance inflation factor (VIF) is used to assess the collinearity of the formative type measurement model (the value of VIF should be less than 5.00). While outer weight is used to assess how important formative indicators contribute to forming a construct or variable (level significance 0.05).

In Table 2 we could see that the register, search, navigation, scan QR code, mobile payment, use of frames, language support, help, color, and comment indicators are not significant. However based on reference [9] said that: 1) When the outer weight (w) of an indicator is not significant but the value of outer loading (l) is relatively high ($l \geq 0.50$), it is statistically significant (the indicator must be retained). 2) If the value of outer weight (w) is not significant and outer loading (l) is relatively low ($l < 0.5$), the researcher must strongly consider removing formative indicators from the model. Based on that statement, then register, search, navigation, scan QR code, mobile payment, use of frames, help, and color are still significant forming the variables. While the language support and comment indicators are not significant and will be removed from the model.

Based on the observations of researchers, the comment indicator has no effect on the satisfaction variable because the indicator is indeed not available in the "Cari Parkir" application and has been represented by a rating indicator that is both used to provide an assessment of the quality of the application and parking attendant services. Whereas for language support indicators that do not have an influence on the understandability variable, researchers see this because even though the language support feature is not available in the "Cari Parkir" application, the user does not mind it because the language used in the application uses Bahasa, where the respondents in this study are all Indonesian. Maybe the results will be different if there are respondents who are not Indonesian then the application does not provide universal language support, in this case, the English language.

Table 2. Formative measurement.

Formative variable	Formative indicator	Outer Weight (Outer Loading)	*t* Value	*p* Value	Significance ($p < 0.05$?)	VIF[a]
Operability	Register	0.007 (0.535)	0.081	0.935	No	1.781
	Search	0.093 (0.685)	0.900	0.368	No	1.936
	Filtering	0.245 (0.809)	1.992	0.046	Yes	2.896
	Booking	0.340 (0.783)	2.995	0.003	Yes	1.994
	Navigation	0.033 (0.652)	0.283	0.777	No	1.963
	Scan QR Code	0.060 (0.607)	0.523	0.601	No	2.130
	Mobile Payment	-0.090 (0.550)	0.892	0.372	No	2.037
	Menu Bar	0.393 (0.876)	2.971	0.003	Yes	2.738
	Use of Frames	0.154 (0.751)	1.274	0.203	No	1.995
Understandability	Language Support	-0.077 (-0.374)	0.798	0.425	No	1.108
	Help	0.080 (0.674)	0.460	0.645	No	1.648
	How it Works	0.922 (0.995)	6.913	0.000	Yes	1.695
Memorability	Favorite Location	0.638 (0.886)	5.663	0.000	Yes	1.284
	E-ticket Parking	0.526 (0.827)	5.533	0.000	Yes	1.284
Satisfaction	Color	0.197 (0.528)	0.959	0.337	No	1.329
	Font	0.449 (0.796)	2.337	0.019	Yes	1.514
	Rating	0.623 (0.841)	4.040	0.000	Yes	1.222
	Comment	-0.046 (-0.319)	0.249	0.803	No	1.087

[a] VIF: Variance Inflation Factor

Table 3. Reflective measurement.

Reflective variable	Kode	Cronbach's alpha	CR[a]	AVE[b]
Perceived Usefulness	PU	0.865	0.908	0.712
Perceived Ease of Use	PE	0.859	0.899	0.643
Job Relevance	JR	0.922	0.950	0.864
Behavioural Intention to Use	BI	0.789	0.876	0.703

[a] CR: Composite reliability
[b] AVE: Average Variance Extracted

Secondly, for the reflective type measurement model (TAM variables) we calculated internal consistency value: Cronbach's alpha and composite reliability, and convergent validity: Average Variance Extracted (AVE). The purpose of evaluating the reflective measurement model is to ensure the reliability and validity of the relationship of the variable with its indicators [9]. From the results of running internal consistency: Cronbach's alpha value and composite reliability of each TAM variable (perceived usefulness, perceived ease of use, job relevance and behavioral intention to use) have values between the range 0.70–0.90 [9]. This means that each TAM variables in the model are reliable. For the results of running convergent validity, it can be seen that the AVE value is greater than 0.50, so it can be interpreted that the four TAM variables are well represented the original data [9].

Last, to examine discriminant validity, we conducted cross-loading criteria and the results are summarized in Table 4. The cross-loading value of each indicator to the variable is must be greater than the value of the cross-loading of other variables. In Table 4 we could see the cross-loading value also obtained good results for each indicator of the TAM variable. It means that the model has good discriminant validity.

Table 4. Discriminant validity (cross-loading).

	BI[a]	JR[b]	PE[c]	PU[d]
BI1	**0.799**	0.518	0.516	0.672
BI2	**0.875**	0.455	0.734	0.759
BI3	**0.839**	0.387	0.507	0.620
JR1	0.493	**0.932**	0.469	0.539
JR2	0.417	**0.928**	0.407	0.436
JR3	0.577	**0.928**	0.566	0.585
PE1	0.663	0.390	**0.828**	0.654
PE2	0.554	0.475	**0.873**	0.645
PE3	0.491	0.590	**0.725**	0.633
PE4	0.481	0.231	**0.729**	0.508
PE5	0.634	0.405	**0.842**	0.667
PU1	0.574	0.503	0.669	**0.827**
PU2	0.633	0.425	0.593	**0.813**
PU3	0.755	0.544	0.660	**0.853**
PU4	0.787	0.442	0.703	**0.881**

[a] BI: Behavioral intention to use
[b] JR: Job Relevance
[c] PE: Perceived ease of use
[d] PU: Perceived usefulness

3.2 Structural Model Evaluation

In structural model evaluation, the researcher inspected the path coefficient to see the accepted and rejected hypothesis. The criteria used to determine hypotheses accepted or rejected are t-value ≥ 1.65 with p-value 0.1.

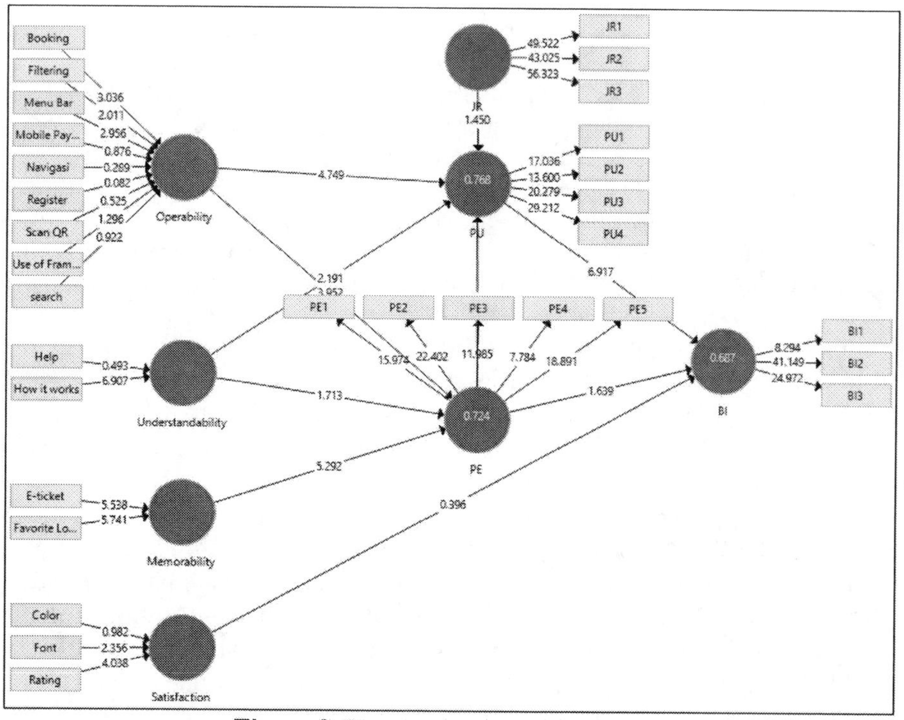

Figure 2. The structural model (t-value).

In Figure 2 we could see the result of our test of the structural model. The accepted and rejected hypotheses are:

- Operability has a positive and significant effect on Perceived Usefulness with t-value 4.749 (H1 accepted). It means that users can use and operate smart parking applications properly, thus speed up the driver's parking process.
- Operability has a positive and significant effect on Perceived Ease of Use with t-value 3.952 (H2 accepted). It means that users can use and operate smart parking properly, thus drivers feel the convenience to do the parking process.
- Understandability has a positive and significant effect on Perceived Usefulness with t-value 2.191 (H3 accepted). It means that users feel smart parking application easy to understand so that it speeds up the driver's parking process.
- Understandability has a positive and significant effect on Perceived Ease of Use with t-value 1.713 (H4 accepted). It means that the smart parking application has clear guidelines so that it makes the driver easy to do the parking process.
- Memorability has a positive and significant effect on Perceived Usefulness with t-value 5.292). By this result, we can conclude that smart parking applications make the users easy to remember their frequently visited parking, the last location of parking, the arrival time, and the parking fees so that users feel the convenience during the parking process.
- Satisfaction has a positive effect but not a significant relationship on Behavioral intention to use with t-value 0.396 (H6 rejected). We can conclude that that application users do not care much about the satisfaction factor (color, font, and rating) as a factor for using the smart parking application, this can be seen from the results, although the positive effect is 0.037 the effect is not significant. These results are the same as the study [9] where the color and font elements of the usability variable for mobile applications do not significantly affect the user's intention to continue using a mobile application.
- Perceived Usefulness has a positive and significant effect on Behavioral intention to use with t-value 6.917 (H7 accepted). It means that people who often parked their vehicles felt the benefits of using the smart parking application so they were interested in using smart parking applications.
- Perceived Ease of Use has a positive and significant effect on Behavioral Intention to Use with t-value 1.757 (H8 accepted). It means that people who often parked their vehicles felt the ease of using the smart parking application so they were interested in using smart parking applications.
- Perceived Ease of Use has a positive and significant effect on Perceived Usefulness with t-value 2.255 (H9 accepted). It means that people who always parked their vehicles felt the ease of using the application so that users felt the benefits when using it.
- Job relevance has a positive effect but not a significant relationship on Perceived Usefulness with t-value 1.450 (H10 rejected). We can conclude that the users do not really feel the usefulness and relevance of smart parking application to their work or activities.

Meanwhile, H6 and H10 have no significant relationship (t-value less than 1.65) we considered to remove satisfaction and job relevance from the model. After we removed the variables and rerunning the model, we get the final model (See Figure 3). This final model has Standardized Root Mean Square Residual (SRMR) value 0.093. It means the model has fit based on the references [15] [16] that said value of SRMR must less than 0.10. Next, in figure 3 we could see all the hypotheses finally accepted. H1 (Operability has a positive and significant effect on Perceived Usefulness with t-value 5.299), H2 (Operability has a positive and significant effect on PE with t-value 3.884), H3 (Understandability a positive and significant effect on PU with t-value 2.057), H4 (Understandability a positive and significant effect on PE with t-value 1.683), H5 (Memorability has a positive and significant effect on PU with t-value 5.184), H7 (PU has a positive and significant effect on BI with t-value 7.546), H8 (PE has a positive and significant effect on BI with t-value 1.757) and H9 (PE has a positive and significant effect on PU with t-value 2.415).

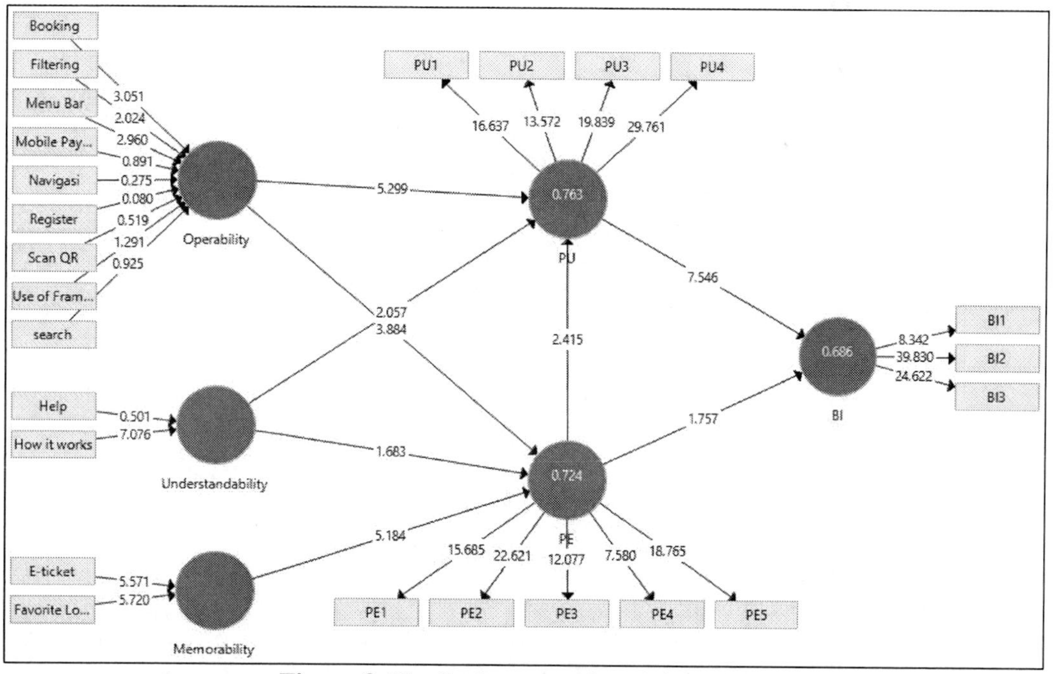

Figure 3. The final structural model (t-value).

4. Conclusion

This study combines usability attributes and TAM to evaluate smart parking mobile application in Indonesia. The objective of this study is to determine variables and indicators in usability attributes with smart parking as an object. After that, we develop and test a model that explores the relationship between usability attributes and TAM. The data we were collected to test our proposed hypotheses. From running the data, the test result indicates that operability, understandability, and memorability have a positive relationship with TAM variables (see figure 3). And the highest total effect value is the relationship between operability to behavioural intention to use (0.496). It means that drivers need a smart parking application that has more information and easy to operate. The limitation of this study is the participant fairly small. This is due to the lack of users who are familiar with the smart parking application. Therefore for future research, the researcher can use a larger sample or the researcher can focus on quantitative usability testing (i.e. eye-tracking) by using the usability attribute in this model. For the application developer can pay attention to features or indicators that exist in the operability variable in this model. We suggest the operation of the registration feature must be improved because it has the lowest coefficient value in this model. This could be an indication of why this application is not yet familiar because users have difficulty log in to the application.

References

[1] Chou J-R, Hsiao S-W. A usability study on human–computer interface for middle-aged learners. *Comput. Hum. Behav.* Elsevier BV; 2007 Jul;23(4):2040–63. doi.org/10.1016/j.chb.2006.02.011
[2] Venkatesh, Ramesh. Web and Wireless Site Usability: Understanding Differences and Modeling Use. MIS Q. JSTOR; 2006;30(1):181; pp. 181–206. doi.org/10.2307/25148723
[3] Harrison R, Flood D, Duce D. Usability of mobile applications: literature review and rationale for

a new usability model. Journal of Interaction Science. Springer Science and Business Media LLC; 2013;1(1):1. doi.org/10.1186/2194-0827-1-1

[4] N. A. Shawgi, E., & Noureldien, "Usability measurement model (UMM): a new model for measuring websites usability," *Int. J. Inf. Sci.*, vol. 5, no. 1, pp. 5–13, 2015.

[5] Hoehle H, Aljafari R, Venkatesh V. Leveraging Microsoft's mobile usability guidelines: Conceptualizing and developing scales for mobile application usability. *Int. J. Hum. Comput. Stud.* Elsevier BV; 2016 May;89:35–53. doi.org/10.1016/j.ijhcs.2016.02.001

[6] Davis FD. Perceived Usefulness, Perceived Ease of Use, and User Acceptance of Information Technology. *MIS Quarterly*. JSTOR; 1989 Sep;13(3):319. doi.org/10.2307/249008

[7] Lin C-C. Exploring the relationship between technology acceptance model and usability test. Inf. Technol. Manag. Springer Science and Business Media LLC; 2013 Apr 17;14(3):243–55. doi.org/10.1007/s10799-013-0162-0

[8] S. Aqil Burney, S. Asim Ali, P. Afshan Ejaz, and F. Ahmed Siddiqui, "Discovering the Correlation between Technology Acceptance Model and Usability," *IJCSNS Int. J. Comput. Sci. Netw. Secur.*, vol. 17, no. 11, p. 53, 2017.

[9] J. F. Hair, G. T. M. Hult, and C. M. Ringle, *A Primer on Partial Least Squares Structural Equation Modeling (PLS-SEM)*, 2nd ed. (Thousand Oaks: Sage), 2017.

[10] Moumane K, Idri A, Abran A. Usability evaluation of mobile applications using ISO 9241 and ISO 25062 standards. *Springerplus*. Springer Science and Business Media LLC; 2016 Apr 29;5(1). doi.org/10.1186/s40064-016-2171-z

[11] Nazrul Islam M, Tétard F. Exploring the impact of interface signs' interpretation accuracy, design, and evaluation on web usability. *J. Syst. Inf. Technol.* Emerald; 2014 Nov 4;16(4):250–76. doi.org/10.1108/jsit-03-2014-0019

[12] Warshaw PR, Davis FD. Disentangling behavioral intention and behavioral expectation. *J. Exp. Soc. Psychol.* Elsevier BV; 1985 May;21(3):213–28. doi.org/10.1016/0022-1031(85)90017-4

[13] Chu PY, Wu TZ, Lee CH. Technology acceptance in public sector: an empirical study of a Knowledge Management System in Kaohsiung City Government. *Int. J. Manag. Decis. Mak.* Inderscience Publishers; 2009;**10**(5/6):341-58. doi.org/10.1504/ijmdm.2009.026682

[14] Kianpisheh A, Mustaffa N, Mei Yean See J, Keikhosrokiani P. User Behavioral Intention toward Using Smart Parking System. *Commun. Comput. Inf. Sci.* Springer Berlin Heidelberg; 2011;732–43. doi.org/10.1007/978-3-642-25453-6_61

[15] Hu L, Bentler PM. Fit indices in covariance structure modeling: Sensitivity to underparameterized model misspecification. *Psychol. Methods.* American Psychological Association (APA); 1998;3(4):424–53. doi.org/10.1037/1082-989x.3.4.424

[16] Henseler J, Dijkstra TK, Sarstedt M, Ringle CM, Diamantopoulos A, Straub DW, et al. Common Beliefs and Reality About PLS. *Organ. Res. Methods.* SAGE Publications; 2014 Apr;17(2):182–209. doi.org/10.1177/1094428114526928

Waste transportation route garbage using network analysis method, a research method design

B J Maspaitella[1]*, A Susanty[1] and R Purwaningsih[1]

[1]Depertemenof Management and Industrial Engineering, Faculty of Engineering, Diponegoro University, Jl. Prof. Soedarto, SH, Kampus UndipTembalang, Semarang, Indonesia 50275

E-mail: *maspaitella_billy@yahoo.com

Abstract. Semarang city is one of the cities in Indonesia and is the centre of the economy, the centre of education and it is also not free from the garbage problem. Garbage production in Semarang City is 1,200 tons per day and not fully transported. The increasing accumulation of garbage can cause pollution of soil, ground water and air. The volume of garbage that has been produced for the Banyumanik district area is 27% of all the total Semarang waste production. It is important to optimizing garbage transportation routes that can minimize transport distances and maximize transportation routes with constrain all transportable points being passed. The method used in this research is network analysis with ArcGIS10.3 application. The results of this study will display the route map of the garbage transportation which minimizes the distance of transportation from the origin to the final disposal site and a description of the trip of the shortest, fastest and cheapest route.

1. Introduction

Semarang is the capital of the Central Java Province and, the fifth largest metropolitan city in Indonesia. Central Java Province is also the centre of the economy and education centre [1]. The problems faced by big cities is garbage handling and processing. Sometimes, the community does not realize the danger effects of garbage which is not handles properly. Indonesia is the second largest producer of plastic waste after China [2]. Every day, the production of plastic waste in Indonesia reaches 175,000 tons. Every year plastic waste in Indonesia reaches 63.9 million tons. Indonesia have about 500 Final Disposal Sites [3].Waste production in the city of Semarang reaches 1,200 tons per day and the volume of waste produced, in the process of transportation is not fully transported. The last five years it was recorded that Semarang City experienced an increase in waste production of 1.67% every year and caused various kinds of problems, one of which was a problem in the social environmental aspect [4].

The Semarang City Environment Agency has been releasing a program on improving city cleanliness responsibility of the government and, the community. Just imagine if the city is inhabited by 5,000 people, and every day one person throws away at least one garbage. So every day there is 5,000 rubbishes produced and more and more garbage will accumulate [1].Banyumanik is one of the large sub-districts in Semarang City which located south at south and one of the densely populated area. Resident in the area only moves waste to landfill without any processing at the source. Based on the 2010 sampling result on waste production, the per capita waste heap in the Banyumanik 325.55 m3/day. The waste managed by the Semarang City Sanitation and Landscaping Service by preparing Temporary Disposal Sites and Final Disposal Sites [5].

Jatibarang is one of the Final Disposal Sitesin Semarang, the existence of the Jatibarang has begun to be full capacity. It means that the Jatibarang disposal capacity has been declining because the area not increased while the quantity of waste production has increased every year [6]. Kadafa, A.A, explained that open disposal and management of solid waste had a detrimental impact on the financing aspect because of the relatively high use of public funds [7].

Research conducted by Setiawan et all, 2018 uses LINGO 11 software where the purpose of the study is to look at alternative points of household waste transfer, and the results obtained in research in Surakarta are alternative points of disposal of household waste including Sodakan Grave, Norowangsa, SPSA, Rail Length, Bonoloyo, and Kedung Tungkul [8].Another study was also conducted by Setiawan et al, 2019 in Boyolali using LINGO 11 software to see the transfer point activity to serve all waste generators, and the results obtained were from 37 selected 13 alternative transfer points, this is because the volume of waste capacity increased to 50% [9].

The urgency of this research is that in Banyumanik District has problems in the collection and transportation of waste. The residents of Banyumanik increases every year and, automatically the number of piles or volume of waste production also increase linear to the increasing of population. In 2012 waste which transported from Banyumanik reached 76.40% of the total waste produced, which amounted to 394.34 cubic meters per day [10]. The purpose of this study is to design the shortest route in the process of transporting waste from Banyumanik to Jatibarangus analysis network application ArcGIS 10.3.

Geographic Information Systems (GIS) is a tool that consist of hardware, software, human resources and data that working together effectively to enter, store, improve, update, manage, manipulate, integrate, analyse and display data in a geographically based information [11]. ArcGIS is one of the tools used in this study as a network analysis that has been provided in a GIS application to analyse transportation networks [12]. Based on the background of the problem, the research is "Routes of transporting waste using the network analysis method in the Banyumanik District of Semarang".

2. Methodology

This research uses descriptive analysis to describe the circumstances that occur based on facts. The stage of the research is the preparation phase, the data collection phase, system analysis and model development phase, data processing phase, and continued by conclusions and formulate recommendation. Data collection uses primary data and secondary data. The variable in this study is determining the shortest route of transportation of waste from the origin to the point of final disposal. In this study data processing using a computerized system using the application ArcGIS 10.3.

2.1. Define the Shortest Route

The search of shortest route used Dijkstra algorithm, one of the greedy algorithms. This algorithm is used to solve the shortest path problem with one source on a graph that has no negative side cost and produces the shortest path[13].Broadly speaking, Dijkstra's algorithm divides the node into two, then it is entered into different table, namely permanent and temporal tables. The permanent table contains the initial node and nodes that have gone through the inspection process and the labels have been changed from temporal to permanent. The data analysis stage begins with a thorough review of the data collection stage, then analyses the data using the Dijkstra algorithm. Can be seen in Figure 1.

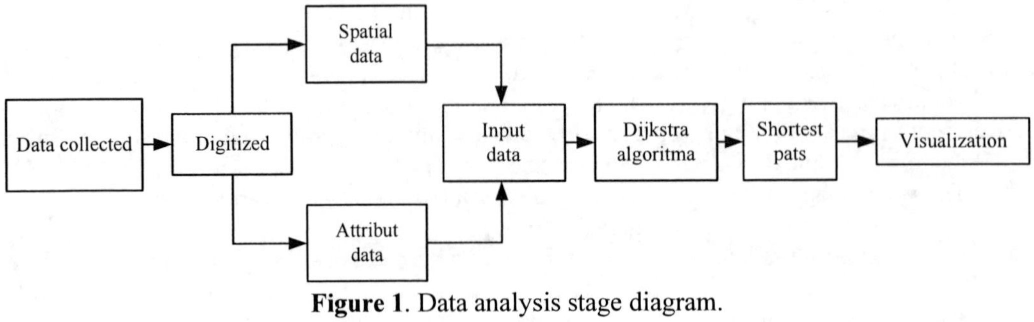

Figure 1. Data analysis stage diagram.

3. Result and Discussion

Data which becomes an input to determine the shortest path is the location data of Temporary Disposal Sites and Final Disposal Sites, and the status road of Banyumanik in Semarang City. There are two kinds of input data for the Dijkstra algorithm, which is of the starting point location and location Final Disposal Sites that have been determined based on the Operational System of the Sanitation Office. The output of the Dijkstra algorithm process is the shortest path from the starting point location head Final Disposal Sites location. Details analysis process for determining the shortest path can be seen in Figure 2.

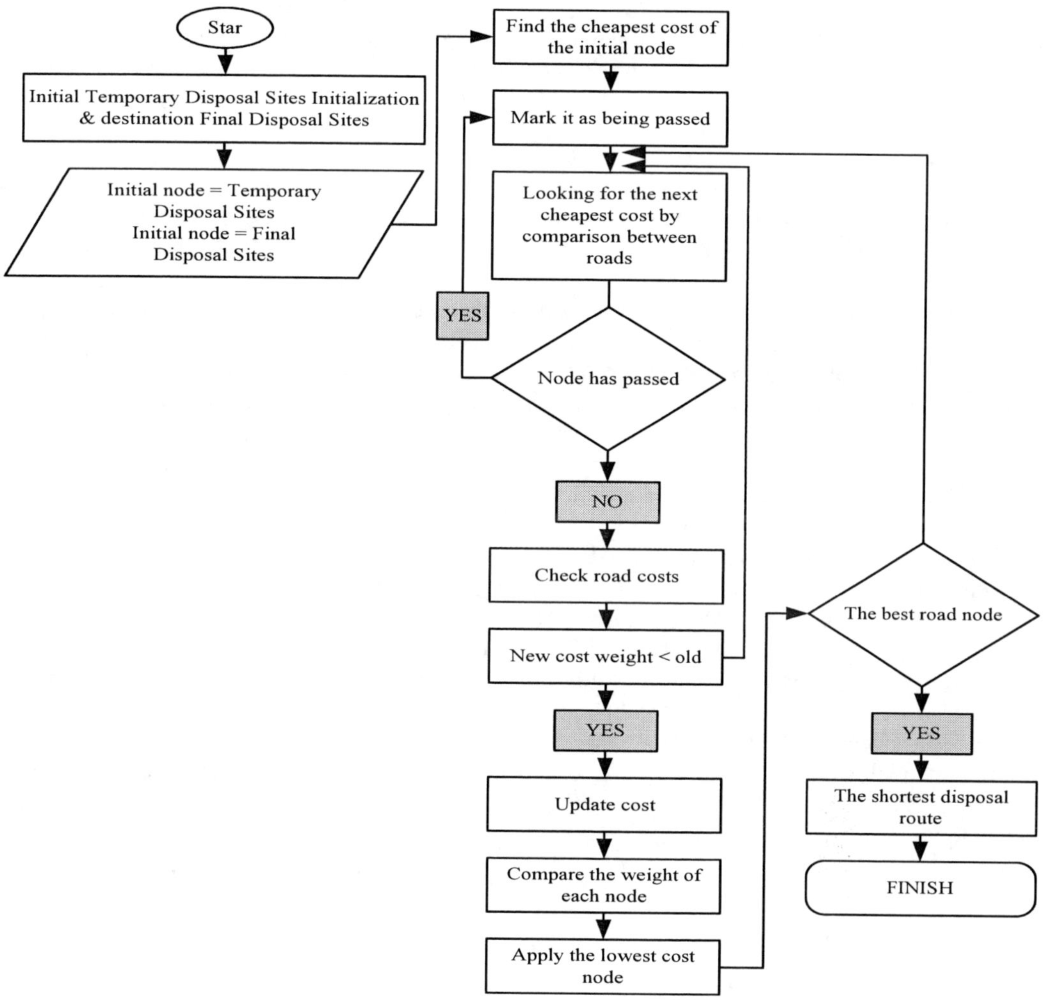

Figure 2. Shortest path determination flowchart [14].

The results of the study are looking at designed the shortest route that is appropriate and can be applied in the process of transporting waste garbage from the Temporary Disposal Sites to the Final Disposal Sites Jatibarang in Semarang. it's hoping that this research can be a suggestion in the process of changing the construction bee better transportation route waste garbage and expected that with the results that will be obtained later, the waste problem in Semarang city will more be resolved, especially in Banyumanik. The following is an overview of the results of the research to be carried out or the target of this study.

Simple system that can be produced from this research as follow: land transportation as the network system, the location of the Temporary Disposal Sites as the initial node, Final Disposal Sites location and the road/trajectory of the transport as arrows. Types of grooves in the form of vehicles can bee see at Table 1.

Table 1. A simple overview of the system in this.

System Name	Network System	Node		Arrows	Current Type
		Initial	Finish		
Transport line	Land transportation	Temporary Disposal Sites location	Final Disposal Siteslocation	Road/trajectory Transportation	Transport

3.1. Transport Route Garbage Search Algorithm
The waste transportation garbage route in this system is divided into four types, that is the closest distance from the garbage transport pool to the Temporary Disposal Sites, the closest distance from the Temporary Disposal Sites to the Temporary Disposal Sites, the closest distance from the Final Disposal Sites to the garbage transport pool. Route selection process on the four types of garbage transportation, the distance from all the nodes will be searched and to be connected, and determine the points of Temporary Disposal Sites and Final Disposal Sites. For the process of finding a route from garbage transport pool to Temporary Disposal Sites, only process just pool node as the initial node, to find a route from Final Disposal Sites to Temporary Disposal Sites and from Final Disposal Sites to Pool, the initial node used is Final Disposal Sites node. While, for route search from Final Disposal Sites, Temporary Disposal Sites to Final Disposal Sites, should be included in all Temporary Disposal Sites like initial nodes and only Final Disposal Sites routes are stored [15].

3.2. Route Garbage Trash Transport Search
The garbage transportation route consists of branching points, pools, Temporary Disposal Sites, and Final Disposal Sites. There are 240 Temporary Disposal Sites Based report on the masterplan activity Semarang City, and 28 Temporary Disposal Sites in Banyumanik. The activity of transporting waste garbage from Temporary Disposal Sites to Final Disposal Sites uses Armroll and Dump truck transportation. The waste garbage transportation route from Banyumanik to the Final Disposal Sites Jatibarang is as depicts in Figure 3 [4].

Figure 4 explains that 28 Temporary Disposal Sites are located in the Banyumanik sub-district and the locations can be seen in Table 2. Meanwhile, Final Disposal Sites is found in the Mijen sub-district in Gedungpane Village.

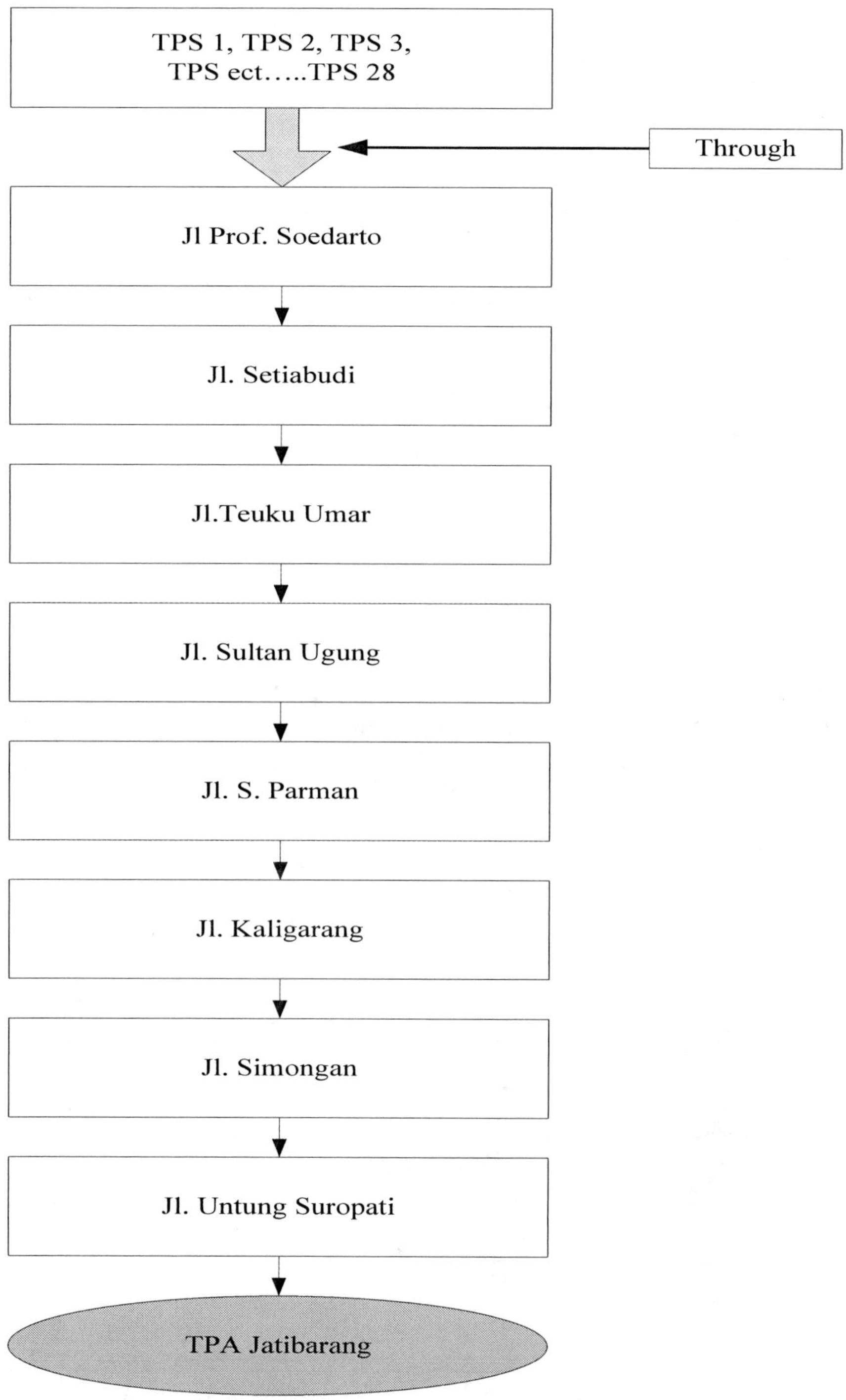

Figure 3. Trash transport trip routes.

Table 2. Temporary disposal sites.

No	Disposal site
1.	TPS Murbei
2.	TPS Hotel Serata
3.	TPS Meranti
4.	TPS PerumahanSetiabudi
5.	TPS PasarRasamala
6.	TPS Brimob
7.	TPS SrondolAsri
8.	TPS RW 1
9.	TPS Hotel Plasa
10.	TPS PerumBrigrif
11.	TPS PasarSrondol
12.	TPS DiklatSrondol
13.	TPS Ada Swalayan
14.	TPS Bukit Regensi
15.	TPS PudakPanyung
16.	TPS Kantor BPK
17.	TPS Trangkil
18.	TPS PasarBanyumanik
19.	TPS Bukit Sari
20.	TPS Resto Panorama
21.	TPS Alam Indah
22.	TPS RW 2
23.	TPS Swiss Roti
24.	TPS PasarDamar
25.	TPS Ulin
26.	TPS PasarJatingaleh
27.	TPS GrahaEstetika
28.	TPS Gedawang

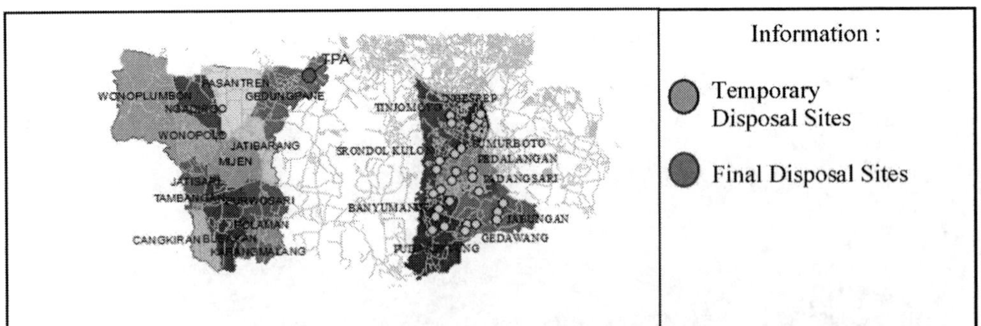

Figure 4. Point map temporary disposal sites and Jatibarang final disposal sites.

3.3. Route Determination with ArcGIS (Network Analysis)

The network is a point called nodes, which are connected by arcs, several terms are often used to describe a particular object in network analysis such as events and activities. Activities are a work assignment, where the completion requires a certain time, costs, and facilities. Usually given the arrow symbol. Events are the beginning or end of an activity. Usually given the circle symbol, seen in Figure 3. ArcGIS is one of the tools used in this study to analyse network that has been provided in a GIS application to analyse transportation network, wherein conducting network analysis will find the path of least impedance. Geographic Information Systems can be divided into two classifications, that is manual and automatic. The difference between the two lies in the processing method. In the manual processing method by combining data in the form of maps transparent sheet media for overlapping. Data can be processed and analysed manually without using a computer. Different case with automated systems that can process data through the digitization process. Data sources can be in the form of satellite imagery or aerial photographs that have been digitized and thematic maps that have been digitized.

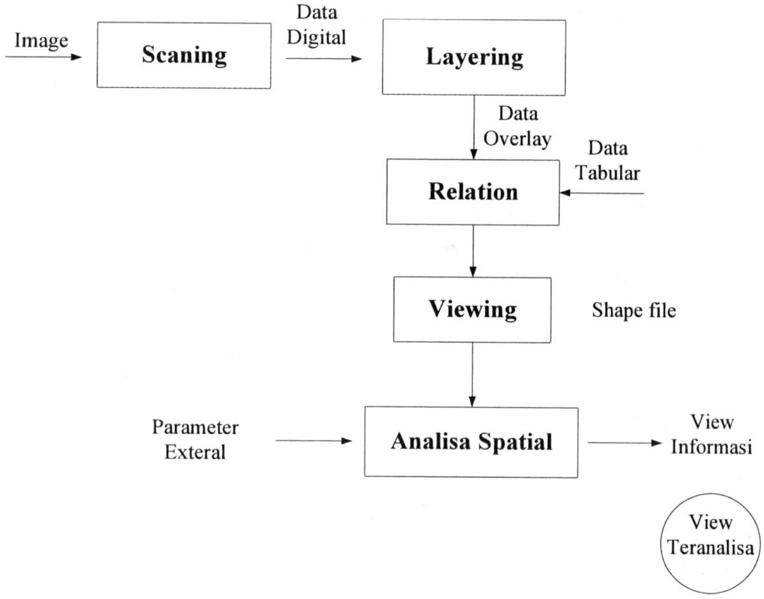

Figure 5. Block analysis model diagram [12].

Figure 5 above explains that to analyse the Geographic Information System begins with image data as basic data. From the image data in the form of physical data, scanning is performed to obtain digital data. The layering process is the process of separating the layer categories that exist in digital data. The result of this layering is vector data that has been divided into several data layers according to their respective covers. For example, the Semarang city map data has been divided into sub-district boundary layers, road layers, river layers, and others. Each layer will get information from tabular data on the related process. The output of this process is a shapefile, where this file is ready to be used to display geographically based information. The parameters in this study were seen based on Aziz 2005 as written in Table 3 [12].

Table 3. Research parameters.

Effective Path	Parameters
Shortest Path	Distance
Cheapest Path	Cost

Based on Table 3 above, the modelling analysis is as follows:

3.3.1. Shortest Route Determination
The external parameter used to find the shortest path is the distance parameter. Segments are determined from each intersection, to determine the total cost field can be modelled as follows:

$$Total\ Cost\ Field\ =\ (D_1 + D_2 + D_3 + \cdots D_N) = \textstyle\sum_{k=1}^{n} D \qquad (1)$$

Where: D = Selected Path
N = Number of Alternatives

3.3.2. Low Cost Determination
This study analyses the determination of the cheapest route using the transport route approach. The analysis of the cost path of the cheapest route is determined by considering the tariff of the transport. Segments are determined from each intersection. To determine the total cost field can be modelled as follows:

$$Fastest\ Cost\ Field\ (CF)\ forD\ =\ \frac{Distance \times Rates}{100} \qquad (2)$$

The unit used is meter.

Total *Cost Field* =

$$(D_1 + D_2 + D_3 + \cdots D_N) + Tarrif/First\ start\ = \textstyle\sum_{k=1}^{n} D + Tsp \qquad (3)$$

Where: D = Chosen Path in Fastest Analysis
N = Number of Alternatives
Tsp = First Star Rates

By using these equations, the route that provides the shortest distance can be determined after the input data obtained the distance between the point of disposal site and available routes.

4. Conclusion

The route development uses the Dijkstra algorithm to determine the route shortest waste garbage transporting from Temporary Disposal Sites to Final Disposal Sites with manual calculations following existing formulas so that the results of both calculations can be used to determine the shortest route in waste garbage transportation. Hope, whit result from research in determining the shortest route of garbage transportation can help the Semarang City government, especially in Banyumanik can be apply the shortest route waste garbage transporting, so that can get more effective and efficient of waste garbage transporting results in the process that is from Temporary Disposal Sites to Final Disposal Sites.

Reference

[1] Pertiwi A, Kiky SMP, Wiwik B, Ratna P, Budi PS, Arya R. Preliminary Study on Plastic Waste Handling in Semarang City - Indonesia: Estimated Generation and Existing Management. Hadiyanto, Maryono, Warsito B, editors. *E3S Web of Conferences*. EDP Sciences; 2018;73:07008. doi.org/10.1051/e3sconf/20187307008.

[2] Synthesis Report. Marine Debris "Rapid Assessment". 2018. 1–40.

[3] Farizal F, Aji R, Rachman A, Nasruddin N, Mahlia TMI. Indonesia's Municipal Solid Waste 3R and Waste to Energy Programs. *Makara Journal of Technology*. Universitas Indonesia, Directorate of Research and Public Service; 2018 Jan 9;21(3):153. doi.org/10.7454/mst.v21i3.3536.

[4] Semarang City Environmental Agency Management Report Final Waste Disposal of Jatibarang. 2017. 1–20.

[5] Hapsari I, Zaman B, Dwi Siwi Handayani Perencanaan Sistem Pengolahan Sampah Terpadu

Kecamatam Banyumanik Semarang. 2012.
[6] Kementrian Pekerjaan Umum DIrektorat Jenderal Cipta Karya Jawa Tengah Laporan Perencanaan Pembangunan Daerah Kota Semarang. 2012.
[7] Kadafa, A. A., Latifah, A. M., Abdullah, H. S., & Sulaiman WN Comparative Assessment of the Municipal Solid Waste Management Services. Life Sci. J. 2013.
[8] Setiawan E, Nugrahadi B, Widiyastuti Y, Djunaidi M, Suranto. Positioning Household Waste Transfer Points: A Municipality Government-Organized Waste Perspective. *Adv. Sci. Lett.* American Scientific Publishers; 2018 Dec 1;24(12):9128–32. doi.org/10.1166/asl.2018.12109.
[9] Setiawan E, Zuhri N, Nugroho YA, Djunaidi M, Suranto, Astuti SP A Bi-Perspective Positioning of Household Waste Transfer Points for Boyolali Regency , Central Java , Indonesia. 2019.
[10] Nudiana DF, Yuningsih T, Maesaroh Studi kualitatif pengelolaan sampah di kecamatan banyumanik kota semarang. 2013.
[11] Wibowo KM, Indra K, Jumadi J Sistem Informasi Geografis (SIG) Menentukan Lokasi Pertambangan Batu Bara di Provinsi Bengkulu Berbasis Website. J Media Infotama. 2015. 11:51–60.
[12] Aziz M.M. Aplikasi Penentuan Rute Terbaik Berbasis Sistem Informasi Geografis. J Ilm Teknol Inf Din Vol X; 2005. 76–83.
[13] Fakhri Penerapan Algorirma Dijkstra Dalam Pencarian Solusi Maximum Flow Problem. 2008.
[14] Hapsari L.I. Penerapan Algoritma Djikstra Dalam Pencarian rute Terpendek Truk Pengangkutan Sampah : Studi Kasus Kabupaten Jember. Universitas Jember. 2017.
[15] Sakaria S. Sistem Informasi Pengangkutan Sampah Dengan Rute Terpendek Menggunakan Algoritma Dijkstra Pada Dinas Kebersihan Kota Malang. 2013. 1–10.

IOP Conf. Series: Materials Science and Engineering 1072 (2021) 012026 doi:10.1088/1757-899X/1072/1/012026

Supply-demand optimization in the product distribution process through the branching method

V Lukitosari[1,3] and A P Subriadi[2]

[1] Mathematics Department, Faculty of Science and Data Analytics, Sepuluh Nopember Institute of Technology (ITS), Surabaya 60111, Indonesia
[2] Department of Information System, Faculty of Intelligent Electrical and Informatics Technology, Sepuluh Nopember Institute of Technology (ITS), Surabaya 60111, Indonesia

Email: [3]valerianalukitosari@gmail.com

Abstract. Interaction and relationship between seller-buyer for the availability of certain products always involve the distribution process from suppliers to a number of customers so that demand at each destination is fulfilled. If the product cannot be fulfilled, the customer will find another supplier to meet the product's shortcomings. Thus the supplier is trying to prevent the supply of other competitors of the same type. Effective logistical arrangements in the distribution of goods need to pay attention to the behavior characteristics of shipping costs. The purpose of this study is to minimize shipping costs with a distribution scheme that can optimize product loads. By using the Balinski linear approach, the problem of sending fixed costs can be solved by the branching method. The branching method gradually produces new problem-solving branches with different objective functions and constraints. Each value branch could be compared so that the optimal solution for the cost is found.

1. Introduction

The product distribution process is very important for a company. In distribution management, distribution managers are responsible for managing company logistics, such as product inventory being delivered on time and economically. Another responsibility is to determine whether distribution services must use private vehicles or must rent. Successful companies realize that there is no cheap distribution to develop their business. Unless distribution is managed effectively and efficiently, so that procurement, manufacturing, and distribution performance can meet expectations.

In the problem of product distribution and delivery, there are fixed costs and variable costs. The problem of fixed costs is modeled as an integer programming problem 0 - 1. Fixed costs arise when investment or rental of transport vehicles. Variable costs in the distribution process include fuel costs and engine maintenance costs. Product distribution and shipping costs will change according to the distance that must be traveled product loads. This distance will be aligned with the cost of fuel, so the further the distance will provide a greater variable cost. The problem of product distribution and delivery will be simplified by linear formulation. Branching method is a very useful method of solving fixed cost distribution problems [1]. Distribution formulation is completed periodically by updating the upper and lower limits so that the optimal solution is obtained. The solution obtained is very useful for solving large-scale fixed cost problems.

Distribution problems in distribution problems can be solved by approximation techniques [2]. The problem of redistribution in the Navy supply system is carried out on small samples. This technique behaves well when fixed costs are higher and few routes are used, at present, no improvement at this lower threshold will come. The approximation technique described is very strong and the general conditions produce an optimal solution to the distribution problem with a fixed cost.

Several studies have discussed the branching method [1, 9], the inventory problems [2, 3, 6], and minimizing cost in logistics [4, 5, 7, 8]. The branching method provides the lower and upper limits for fixed cost distribution problems so that each branch can be considered a solution. In contrast to previous studies, this study can encourage the linear nature of finding optimal solutions through the Balinski approach. There is at least one optimal point to guarantee that the problem has a solution.

2. Preliminaries

The distribution process is a distribution problem that involves the distribution of products originating from m suppliers to n customers (customers). Inventory at each supplier i is unit a_i and demand for each agent j is unit b_j. Suppose x_{ij} is the number of products sent from supplier i to customer j with the shipping cost per unit of the product being c_{ij} and a fixed cost of f_{ij}. Distribution problems in this study use fixed costs. Binary integer programming for fixed costs is y_{ij}. In general, a mathematical model for distribution with fixed costs can be formulated as follows:

$$P: \quad Min\, Z(P) = \sum_{i=1}^{m} \sum_{j=1}^{n} \left(c_{ij}x_{ij} + f_{ij}y_{ij}\right) \tag{1}$$

subject to:

$$\boldsymbol{\sum_{i=1}^{m} x_{ij} = a_i \qquad \text{for } i = 1, 2, \ldots, m} \tag{2}$$

$$\sum_{j=1}^{n} x_{ij} = b_j \qquad \text{for } j = 1,2, \ldots, n \tag{3}$$

$$x_{ij} \geq 0 \qquad \text{for all } (i,j)$$

$$y_{ij} = \begin{cases} 0, & \text{if } x_{ij} = 0 \\ 1, & \text{if } x_{ij} > 0 \end{cases}$$

We assume that the inventory owned by source $\boldsymbol{a_i}$ is balanced by the demand owned by the destination $\boldsymbol{b_j}$.

$$\sum_{i=1}^{m} a_i = \sum_{j=1}^{n} b_j \text{ and } a_i, b_j, c_{ij}, f_{ij} \geq 0$$

Product distributions in connection with demand will be met according to available inventory. At the time of distribution of the product will be considered matters relating to the number of loads and total distribution costs.

3. The product distribution process through the branching method

Development of a linear approach of the fixed cost distribution problem [1] by relaxing the integer constraint on y_{ij} with that property as follows:

$$y_{ij} = x_{ij}/m_{ij} \tag{4}$$

where

$$m_{ij} = min\left(a_i, b_j\right) \tag{5}$$

The problem of distribution is relaxed into a fixed cost distribution problem that is transformed into a classic distribution problem without fixed costs. Distribution unit costs are formulated as follows:

$$C_{ij} = c_{ij} + f_{ij}/m_{ij} \tag{6}$$

This relaxation distribution problem can be called PB and this formulation is used to develop a feasible initial solution for the fixed cost distribution problem. The optimal solution of the PB distribution problem is $\left(x^B{}_{ij}\right)$, then a feasible solution is obtained $\left(x^B{}_{ij}, y^B{}_{ij}\right)$ from P with the following conditions as follows:

$$y^B{}_{ij} = \begin{cases} 0, & \text{if } x^B{}_{ij} = 0 \\ 1, & \text{if } x^B{}_{ij} > 0 \end{cases} \tag{7}$$

The optimal value $Z(PB)$is the lower limit on the optimal value $Z^*(P)$ of the fixed cost distribution problem. So the solution $\left(x^B{}_{ij}, y^B{}_{ij}\right)$ becomes a feasible solution that gives an upper limit on $Z^*(P)$. Thus $Z(PB) \leq Z^*(P) \leq Z(P)$ can be written as follows:

$$\sum\sum C_{ij}\, x^{B}{}_{ij} \le Z^{*}(P) \le \sum\sum\left(c_{ij}x^{B}{}_{ij} + f_{ij}y^{B}{}_{ij}\right). \quad (8)$$

The solution of a relaxation distribution (PB) problem is integer, the lower limit of inequality in Equation (8) of the optimal (PB) solution can be rounded up to the nearest interval.

4. Supply-demand optimization

Supply and demand allocation is a matter of distribution. Problem-solving uses the formulation of distribution problems with the Balinski linear approach. Certain cells are selected according to certain selection criteria. This method branches off progressively by excluding selected cells to find the optimal solution. We assume that the distribution problem formulation uses fixed costs, with the coefficient of costs c_{ij}, fixed costs f_{ij} and m_{ij}.

The branching process of fixed cost distribution problems are as follows:

The first step : Formulate Balinski relaxation distribution matrix problem with equation (6)

Second step : Solve it as a classic distribution problem and identify the charge as $x^{B}{}_{ij}$.

Third step : Check the termination conditions for certain branches with the following conditions:

a. $\sum\sum C_{ij}\, x^{B}{}_{ij} = \sum\sum\left(c_{ij}x^{B}{}_{ij} + f_{ij}y^{B}{}_{ij}\right)$ or

b. $\sum\sum C_{ij}\, x^{B}{}_{ij} \ge Z$ value, so far stop and end this branch.

Fourth step : Select cells that are partially loaded, i.e. cells with $0 < x^{B}{}_{ij} < m_{ij}$.

Fifth step : For each cell identified in Step 4, by counting $\Delta = f_{ij} - (f_{ij}/m_{ij})x^{B}{}_{ij} = f_{ij}(1 - x^{B}{}_{ij}/m_{ij})$.

The value $\Delta = f_{ij}(1 - x^{B}{}_{ij}/m_{ij})$ is the differentiation of fixed costs between the problem of relaxation distribution and the problem of distribution of fixed costs to load $x^{B}{}_{ij}$ in cells (i,j).

Sixth step : Select the cell (s,t) with the highest Δ among those identified in fifth step. Break the bond by choosing Δ with the largest f_{ij}. If there is more than one Δ, then choose one randomly.

In sixth step produces two branches, namely: (a) the Y branch, the selected cell is given a charge. The associated fixed costs are summarized as Z_{Ycost} to be combined with the objective function and $f_{st} = 0$ in this fixed cost distribution problem; (b) in the selected branch N, is excluded by setting costs related to a very large amount.

Branch (s,t): contains cells (s,t).

Seventh step : Note $Z_{Ycost} = f_{st}$, where Z_{Ycost} represents fixed costs summarized along branch $Y(s,t)$. Set $f_{st} = 0$ in the fixed cost distribution problem.

Eighth step : Repeat Steps 1-6 with the distribution problem of fixed costs that have been adjusted.

Branch (s,t): except cell (s,t).

Ninth step : Set $c_{st} = M$ in the fixed cost distribution problem.

Tenth step : Repeat Steps 1-6 with the fixed cost distribution problem that has been adjusted.

The branching method utilizes the relationship presented in inequality in Equation (8), which provides the lower and upper limits for fixed cost distribution problems.

This method sequentially looks for fixed costs f_{ij} given by distribution problems. This method can encourage the linear nature of finding optimal solutions by sequentially increasing the lower limit obtained from the Balinski formulation. This method guarantees that there is at least one optimal point, identified by the lowest upper limit.

5. Numerical example

We apply the problem of product distribution to the allocation of supply and demand in companies that have three branch factories that operate together. The three branches are called Plant 1, Plant 2, and Plant 3. Product distribution is carried out to meet the market supply based on product demand data provided by Agent 1, Agent 2, and Agent 3 in the maximum and minimum demand periods.

The vehicle used for product distribution is a truck with a load capacity of 30 tons with a fixed cost of IDR 45,000,000 per unit per month. The cost of fuel needed is calculated starting from departing to returning to the warehouse of each plant at a price of IDR 5,500 per liter for a distance of 2 km. The cost distribution matrix is prepared by formulation (f_{ij}, c_{ij}) where $f_{ij}$is a fixed cost that is the cost vehicle rental, and $c_{ij}$are variable costs, i.e. fuel costs.

5.1. Maximum demand period

The distribution cost for the maximum demand period can be seen in Table 1.

Table 1. The distribution cost (f_{ij}, c_{ij}) on maximum demand period (million).

	Agent 1	Agent 2	Agent 3	Supply (ton)
Plant 1	(720,343)	(720,375)	(630,402)	14.130
Plant 2	(720,79)	(720,206)	(630,192)	14.130
Plant 3	(720,510)	(720,628)	(630,404)	14.330
Demand (Ton)	13.890	16.130	12.570	42.590

The relaxation distribution cost C_{ij} from the Balinski linear approach in Equation (6) when the maximum demand period can be seen in Table 2.

Table 2. Relaxation distribution cost C_{ij} on maximum demand period (million).

	Agent 1	Agent 2	Agent 3	Supply (ton)
Plant 1	343,052	375,051	402,05	14.130
Plant 2	79,052	206,051	192,05	14.130
Plant 3	510,052	628,05	404,05	14.330
Demand (Ton)	13.890	16.130	12.570	42.590

Based on Balinski relaxation and branching processes, the optimal solution is found in Table 3. Table 3 shows each plant 14.130/30 tons in one month (30 days). Each plant is sent a load of 16 trucks so that 48 trucks per day is needed on a maximum period.

Table 3. Optimal solutions to problems of distribution fixed costs maximum demand period.

	Agent 1	Agent 2	Agent 3	Supply (ton)
Plant 1	0	14.130	0	14.130
Plant 2	13.890	240	0	14.130
Plant 3	0	1,760	12.570	14.330
Demand (Ton)	13.890	16.130	12.570	42.590

5.2. Minimum demand period

The distribution cost for the minimum demand period can be seen in Table 4.

Table 4. The distribution cost (f_{ij}, c_{ij}) on minimum demand period (million).

	Agent 1	Agent 2	Agent 3	Supply (ton)
Plant 1	(405,193)	(585,305)	(585,373)	7.650
Plant 2	(495,55)	(585,167)	(585,178)	13.703
Plant 3	(495,350)	(585,511)	(585,375)	11.603
Demand (Ton)	9.540	11.386	12.030	32.956

The relaxation distribution cost C_{ij} from the Balinski linear approach in Equation (6) when the minimum demand period can be seen in Table 5.

Table 5. Relaxation distribution cost C_{ij} on minimum demand period (million).

	Agent 1	Agent 2	Agent 3	Supply (ton)
Plant 1	193,053	305,076	373,076	7.650
Plant 2	55,052	167,051	178,049	13.703
Plant 3	350,052	511,051	375,05	11.603
Demand (Ton)	9.540	11.386	12.030	32.956

Based on Balinski relaxation and branching processes, the optimal solution is found in Table 6. Table 6 shows Plant 1 for 1.7650 / 30 = 255 in one month (30 days). Plant 1 distribution has sent a load of 9 trucks per day. The same way for Plant 2 (16 trucks) and Plant 3 (13 trucks). Total product distribution is done 38 trucks per day is needed on a minimum period.

Table 6. Optimal solutions to problems of distribution fixed costs minimum demand period.

	Agent 1	Agent 2	Agent 3	Supply (ton)
Plant 1	0	7.650	0	7.650
Plant 2	9.540	3.736	427	13.703
Plant 3	0	0	11.603	11.603
Demand (Ton)	9.540	11.386	12.030	32.956

5.3. Results and Discussion

We use two data scenarios in the numerical example, which are the maximum demand period and the minimum demand period. This is to show the difference between demand in the high season and low season. Two distribution request scenarios are made to look in more detail at resolving the average demand problem.

The fixed costs observed in this study are vehicle rental costs. Fuel costs are variable costs that change according to changes in distribution. Both costs affect the costs the company budgeted for product distribution. The company packages products weighing 40 kg per bag to fulfil supplies according to the agent's request. Product distribution uses trucks with a payload capacity of 30 tons. The shipping amount and weight of the package are integers.

The branching method can solve problems related to integers. This method combined Balinski's formulation and completion by selecting certain cells to be loaded maximally and providing minimal distribution costs.

Numerical simulations during the period of maximum demand obtained minimal results from the distribution of products from 3 plants with 48 trucks. The distribution of each plant for 16 trucks spent Rp.421,244,600,000 per month. While optimization of product distribution in the minimum period consists of 9 trucks for Plant 1, 16 trucks for Plant 2, and 13 trucks for Plant 3. The cost required is Rp.239,924,100,000 per month using 38 units of trucks, the vehicle is carrying 814,900 bags.

6. Conclusion and Future Research

Factors and characteristics of distribution cost are developed as important factors in logistical arrangements. One factor that influences this is the behavior of distribution costs. In this study, the distribution costs are considered fixed costs. The fixed costs include distribution problems that are difficult to solve. The distribution costs are minimized by using the branching method. The numerical example shows the problem of fixed costs on the maximum and minimum period that have different quantities of supply and demand. We use the branching method in the iteration process to get the upper and lower limits that correspond to the constraints. The two limits can be compared to get the optimal solution which is the minimum distribution costs needed to maintain optimal inventory and can meet every customer demand. The next research agenda, fixed costs, and variable costs will be tested for the dominance or importance/weight of both.

Acknowledgments

The authors are grateful to the referees for providing valuable comments and suggestions.

References

[1] Adlakha V, Kowalski K, Lev B. A branching method for the fixed charge transportation problem. *Omega*. Elsevier BV; 2010 Oct;**38**(5):393–7. doi.org/10.1016/j.omega.2009.10.005

[2] Balinski M L. Fixed cost distribution problem. *Naval Research Logistics Quarterly*. 1961;**2928** pp 41–54.

[3] Bowersox D J, Closs S J, and Cooper M B. *Supply Chain Logistic Management.* (New York: The McGraw-Hill Companies Inc). 2002

[4] Kowalski K. On the Structure of the fixed charge distribution problem. *International Journal of Mathematical Education in Science and Technology*. 2005;**36** 8 pp 879–88

[5] Lukitosari V, Subriadi AP, Siyamah I. Minimizing cost and backorder risk through customer priority on spares inventory. *J Phys Conf Ser*. IOP Publishing; 2019 May;1218:012061. doi.org/10.1088/1742-6596/1218/1/012061

[6] Lukitosari V, Subriadi AP. Inventory models for short life cycle clothing products use a logistic growth model. *J Phys Conf Ser*. IOP Publishing; 2020 Mar;1490:012060. doi.org/10.1088/1742-6596/1490/1/012060

[7] Lukitosari V, Simanjuntak TF, Utomo DB. A game-theoretic model of marketing strategy using consumer segmentation. *J Phys Conf Ser*. IOP Publishing; 2020 Mar;1490:012026. doi.org/10.1088/1742-6596/1490/1/012026

[8] Lukitosari V, Suparno, Pujawan IN, Widodo B. Inventory strategy for spare parts redundancy to support server operations during production processes. *Production & Manufacturing Research*. Informa UK Limited; 2019 Jan 1;**7**(1):395–414. doi.org/10.1080/21693277.2019.1630681

[9] Palekar US, Karwan MH, Zionts S. A Branch-and-Bound Method for the Fixed Charge Transportation Problem. *Manage Sci*. Institute for Operations Research and the Management Sciences (INFORMS); 1990 Sep;**36**(9):1092–105. doi.org/10.1287/mnsc.36.9.1092

Human error probability analysis using Success Likelihood Index Method (SLIM) approach in grinding activities

I Santiasih[1,2*], R A Ratriwardhani[3]

[1]Department of Safety Engineering, Politeknik Perkapalan Negeri Surabaya, Keputih, Sukolilo, 60111 Surabaya, Indonesia

E-mail: [2]indri.santiasih@ppns.ac.id; [3]indri.santiasih@gmail.com

Abstract. Grinding activities produced 31% accident cases that occurred repeatedly and had the same cause. Most of them were investigated and were proved to be caused by human error. The study aims to obtain the value of Human Error Probability (HEP), thus, it could be employed as prevention and controlling consideration. HEP value was obtained by finding the value of. Success Likelihood Index (SLI). SLI value could be obtained using questionnaires and Performance Shaping Factor (PSF) assessment which had been filled by expert judgment. The scenarios were developed to predict and reduce the occurrence of human error as applied to Success Likelihood Index Method (SLIM). This research illustrated the factors that affected errors in the grinding process were procedure, fatigue, complexity, training, and experience. The task that contained the highest error value was in the first task (use PPE following predetermined). While the task which included the lowest HEP was task 2 (prepare documents according to pre-defined). As part of efforts to reduce the error probability values grinding process, it recommended reducing the probability of error by increasing the value of PSF, ranging from improved procedures, reduction of fatigue of the workers, the reduction of the complexity of the job, increased training for workers, and increased workers' qualification through work experience history.

1. Introduction

The "human factor" plays an important role to predict the safe operation of a facility. Hence, the information about human capacities and behaviors should be applied precisely to improve the safety of a systematic process. Norman [1] and Reason [2] investigate an "error" arise in situations where an act is devoted both purposely and un-purposely; however, the error itself and the original purpose of the act are often described separately. Sanders and McCormick [3] illustrate human errors as inappropriate decisions that obtain a negative effect on the effectiveness of the safety system and performance. Providing a system classification may facilitate to organize human error data and provide insight into how errors can be prevented. Several studies have determined that such errors are a major cause of accidents in grinding [4–6].

Reduction in human errors will naturally lead to a reduction in costs [7]. Human error has been well known as the most contributing factor to accidents. Several factors contribute to human errors including personal characteristics, managerial or organizational aspects, the complexity of work methods, environmental conditions, machine design, training methods, supervision methods, presence, and/or absence of work instruction [8].

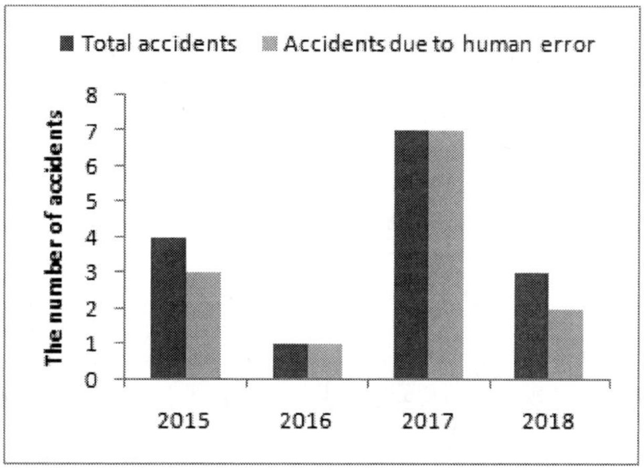

Figure 1. The number of accidents in grinding activities.

The previous studies are conducted to investigate the contribution of human factors into accidents. Heinrich concludes that unsafe acts (88%), unsafe conditions (10%), and unpredictable factors (2%) account for these accidents are analyzed in 7500 accidents [9]. In another study conducted in Australia, 83% of 2000 accidents are due to human error. A similar study is carried out at Berlin Technical University illustrates that the cause of 64% of all accidents is a human failure [10, 11]. It can be argued that human error is very complicated not only due to individual mistakes but also the conditions of human error.

Human Reliability Analysis (HRA) always has an essential concern of safety engineers and risk assessment analysts. The main reason for that is the subjectivity of the methods employed to estimate human reliability and the uncertainty of the data concerning human factors, along with the complexity of human behavior. For an HRA technique to be valid, the model specifies the likely human error modes and the process whereby numerical estimates of HEP are obtained must both be valid. A discussion of validity in the human-machine system context is available in Hollnagel [12], and Cronbach and Meehl [13] consider the validity issue within the context of psychological testing.

Several methods are developed as the teamwork of engineers and psychologists, to assist the analysis of human error and human reliability. Most of them require expert judgment, statistical data, and simulation proofs including Success Likelihood Index Methods (SLIM) approach.

The SLIM approach is a technique applied in the Human Reliability Analysis (HRA) and aims to analyze the possibility of human error that occurs when conducting a job. A likely basis for the evaluation of the results from the SLIM test is convergent validity. SLIM is also an HRA approach with an approach that considers various PSF value recommendations and sociotechnical factors using a mathematical formula that produces a Human Error Probability (HEP) value, as long as the weight and rating of the PSF are known. Therefore the results of the analysis can be employed to provide recommendations to reduce the occurrence of errors, with the expectation that the same accident will not be repeated [14]. Excess SLIM compared to other methods in HRA analysis among others is the errors can be measured at any level of job, sub-work, even on any each task of sub-work, thus it can be known which task has the highest risk. Moreover, SLIM uses a mathematical formula to generate HEP value, as long as the weighting and the rating of PSF are known. Consequently, the parameters that affect an individual's ability to perform a given job can be recognized [15].

Grinding is an activity to produce a smooth surface and can achieve high accuracy. Grinding can also be employed to create workpieces such as tidying the results of cutting, smoothing out welds, curves on angled workpieces, preparing workpiece surfaces to be welded, refine, and make accurate measurements on the surface of the workpiece (finishing), and others. On a grinding machine, the sharpening stone rotation at workpiece slicing requires a very high rotation speed. Grinding activities can be dangerous, among others, if there is a lack of operator expertise, operations are not following SOPs and tools are not checked before the operation. 31% of accidents occur in international companies in Surabaya engaged in manufacturing steam boiler manufacturing are accidents in the grinding process (Figure 1). Grinding is the type of work that produces the most accident cases, and 66.67% - 100% in the last five years due to human error. Therefore, it is necessary to have an in-depth study of the SLIM approach as one of the techniques of human reliability analysis to study human error in grinding activities.

2. Material and Methods

2.1. Selection of Expert Judgment

Expert judgment consisted of a group leader, Environmental Health and Safety (EHS) inspector and supervisor. The criteria for determining expert judgment referred to Skjong and Wentworth's [16] included participating in the design or evaluation of a grinding process security system, having participated in a risk assessment in the grinding field, having experience in the grinding field, not filling in the questionnaire simultaneously with other judges, willing to spend time during working hours to be asked for information related to grinding work, had a lot of knowledge about grinding work, and also contained a good reputation in the company, neutral, honest, and confident.

2.2. The Determining of Task Analysis

The task analysis was identified as the working step of the job in the details of grinding. Task analysis on the work of portable grinding machines prepared based on the work instructions grinding work, then developed in more detail. Task analysis verified by EHS manager, a grinding operator, and expert judgment. The task was determined to produce a questionnaire of Performance Shaping Factors (PSF) weighting and an assessment questionnaire of PSF. Error probabilities were calculated for each task and subtask.

2.3. The Determining of Suitable Performance Shaping Factors (PSF)

The PSF was the factors that affect the probability of an error occurrence. It served to produce the PSF weighting questionnaire, the PSF assessment questionnaire, and also to calculate the Success Likelihood Index (SLI). PSF determined based on accident data and the factors that cause errors to be identified and verified by expert judgment.

2.4. The Development of A Weighting Questionnaire

The questionnaire was filled out by expert judgment, which aimed to find out how much influence in each PSF generated errors in the grinding work by providing a weight for each PSF in each task. Weights are given on a range of 1 to 10. Weight 10 had the greatest effect and 1 contained the least effect. It meant that the greater the number of PSF showed the increasingly influential in the emergence of errors in grinding work. Inversely, the smaller the number of PSF had a small influence on the incidence of errors in grinding work compared to other PSFs.

2.5. The Assessment of PSF

The determining of the value of PSF determined the quality of each PSF in each task. It could be utilized to calculate the Success Likelihood Index (SLI). Ratings were verified through the results of the PSF assessment questionnaire that had been filled out by expert judgment.

2.6. The Rating of Performance Shaping Factors (PSF) on Each Task

This questionnaire aims to determine the quality of each PSF in each task. The quality is described in the form of weight. The rating scale starts on a scale of 0 to 100. The range of PSF rating scale in Table 1 referred to DiMattia [17].

Table 1. Description of the PSF rating scale for each task.

PSF (*Performance Shaping Factors*	Scoring Scale 100	50	0
Procedure	The procedure in this task is almost perfect	Rather perfect procedure	No procedure
Fatigue	Workers are not tired of this task	Rather tired	Very tired
Complexity	Not a complicated task	A bit tricky	Very complicated
Training	A lot of training so workers are highly trained in this task	There is some training	No training
Experience	Experienced in this task	Rather experienced	inexperienced

2.7. The Calculating of Success Likelihood Index (SLI) value

The SLI value is used to calculate Human Error Probability (HEP). SLI could be applied as a performance indicator and could be employed as an aspect of monitoring the Occupational Health and Safety (OHS) management system as well. SLI values were calculated using Equation 1, referred to Embrey and Humpreys [18].

$$SLI_j = \sum R_{ij} W_i \tag{1}$$

SLI_j = SLI value of $task_j$
R_{ij} = PSF_i rating of $task_j$
W_i = PSF_iNormalization Weight ($\sum W_i = 1$)

The SLI value in each task was determined by the rating values and normalization weights obtained.

2.8. The Converting of SLI to Human Error Probability (HEP)

The SLI value was converted to HEP value to obtain the probability of human error in grinding work. It referred to Embrey and Kontogiannis [19] using Equation 2.

$$\log(HEP) = a\ SLI + b \tag{2}$$

a = Constant
SLI = Success Likelihood Index
b = Constant

The Probability of Success (POS) value was acquired from the SLI value referred to Embrey and Kontogiannis [19] using Equation 3.

$$POS = 1 - HEP \tag{3}$$

a and b value was determined by identifying the error probability minimum in 2 tasks. The error probability value could be recognized from the accident data in task 1 and task 4. These tasks were selected because there were the highest accident cases compared to other tasks. 5 accidents occurred for

5 years in task 1 (task A) due to lack of presence of task 1. While the absence of task 4 (task B) produced 2 accidents for 5 years, referred to Equation 4.

$$\textit{Task}\ \text{A or Task B} = \frac{Total\ accidents\ for\ 5\ years}{\frac{hours}{days} x \frac{days}{weeks} x \frac{weeks}{years} x\ years} \quad (4)$$

3. Result and Discussion

3.1. The Determining of Performance Shaping Factors (PSF)

The PSF identified as influencing the occurrence of errors was the procedure, fatigue, illumination, complexity, and work shift. verification results from the judges found that from the 5 PSFs determined. The term of PSF was used to denote both of human traits and conditions of work settings that were perceived by judges to have a predominant influence on success likelihood in the scenario being evaluated [18]. It turned out that lighting had no probability of meaningful consequences regarding the expert suggestion. Subsequently, there were 2 additional PSFs suggested by the judges, namely training and the experience factor. Hence, there were 5 PSFs that affected the grinding work namely procedure, fatigue, complexity, training, and experience.

3.2. The Weighing of Performance Shaping Factors (PSF)

The weight of the PSF was determined through a questionnaire by expert judgment. The weighted results of the judges were averaged. The weight was then normalized through each value divided by the total value overall. The total normalized weight should be 1.00. The weighting results showed a comparison of the level of importance of the PSFs to the grinding work (Figure 2).

Table 2. Weighting the weighting of performance shaping factors (PSF) questionnaire results.

No	Task	PSF Value				
		Procedure	Fatigue	Complexity	Training	Experience
1	Use PPE following requirement	8.75	6,50	7	7.25	8.50
2	Prepare a document following requirement	5.50	3,25	5	5.25	5.75
3	Perform autonomous maintenance	5.75	4,50	4	5.50	5.25
4	Select the grinding stone according to the type of material would be grinded and the work should be conducted	5.75	4	4.75	6	6.75
5	Install grinding stones	5.75	3.25	3.25	5.50	5
6	Turn on the grinding machine	5.50	5.50	5.50	5.50	4.75
7	Grinding the material	6.50	5.75	6.25	6	7.75
8	Turn off the grinding machine	5.50	5.50	5.50	5.50	4.75
9	Check the results of the grinding whether following the document	7	3.75	4.50	5	5.75
10	Perform autonomous maintenance	5.75	5	3.50	4.75	4.75
11	Return the grinder to its original place	3.75	4	2.75	2.75	3.25
12	Carry out 5S	5.25	5.25	5.25	6.25	5
	Total	70.75	56.25	57.25	65.25	67.25
	Average (Weight)	5.90	4.69	4.77	5.44	5.60

The level of importance of PSFs in order from the largest was procedure, experience, training, complexity, and fatigue. The procedure was one of PSF on grinding work if it was not carried out properly would be produced the most initiating errors. Therefore, it was followed by work experience factors that were less possibility of generating an error to the experienced worker. The lack of training could affect as well to the level of accidents. The complexity in the grinding process also influenced the occurrence of errors. The complex workpiece or the difficulty of grinding positions increased the probability to generate an error. The fatigue decreased worker concentration, which did not infrequently produce sleepy, and other signs of exhausted. There have been many cases of accidents that have occurred due to this factor. These five factors are the most influential in grinding work.

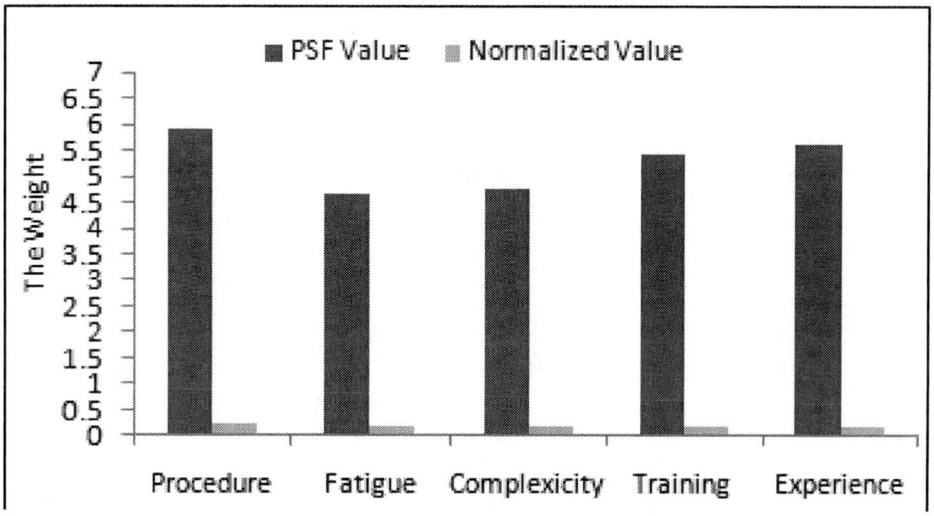

Figure 2. The PSF and normalized value. Data presented in mean of PSF and normalized value.

3.3. The Determining of Performance Shaping Factors (PSFs) Value for Each Task

Ratings were determined through the results of the PSF assessment questionnaire that had been filled out by expert judgment. Afterward, the rating results of the judges were averaged. The quality of procedures on grinding work that had been made by the company was 92.5. It meant that the quality of procedures in this task was almost perfect because it was closed to a rating of 100. While the fatigue factor in the first task was 80, meaning that the task of using PPE as specified, the workers did not feel tired of this task. For complexity, the first task was not a complicated task to execute. The quality of training in the first task was 100, meaning that there was a lot of training in the task. Hence, workers were already highly trained in the task. The experience factor of the first task was 92.5. It meant that the workers were very experienced.

3.4. The Calculation of Success Likelihood Index (SLI)

The rating values and normalization weights were determined, hence SLI value could be identified in each task, shows in table 3.

3.5. Converting the value of Success Likelihood Index (SLI) to Human Error Probability (HEP) Value

After the SLI value was identified, the next step was to find the error probability (Equation 2).To find out the values of a and b (Equation 2) at least the probability of error in 2 tasks should be recognized.

The error probability value could be identified from the accident data in task 1 and task 4. These 2 tasks were selected because there were the highest accident cases compared to other tasks. In task 1 (task A), the number of accidents that occurred due to lack of task 1 fulfillment was 5 accidents for 5 years. While the number of accidents that occurred due to ignore task 4 (task B) was 2 accidents for 5 years. The probability of error is calculated according to Equation 2.

Hence that the equation was obtained as follows:

$\log(\text{HEP}) = a\ \text{SLI} + b$

$\log(\text{HEP}) = 0{,}0066323\ \text{SLI} - 3{,}831614$

While the POS value was obtained from Equation 3. Table 4 illustrates the calculations of the SLI, HEP, and POS.

Table 3. The weight of PSF and the value of SLI.

Task	PSF Value					SLI
	Procedure	Fatigue	Complexity	Training	Experience	Value
Use PPE following requirement	92.50	80	82.50	100	92.50	90.03
Prepare a document following requirement	75	40	50	72.5	70	62.63
Perform autonomous maintenance	95	85	72.50	87.50	80	84.43
Select the grinding stone according to the type of material would be grinded and the work should be conducted	97.50	87.50	82.50	97.50	82.50	89.85
Install grinding stones	100	62.50	72.50	80	75	78.85
Turn on the grinding machine	87.50	75	77.50	87.50	77.50	81.35
Grinding the material	92.50	75	77,50	82.50	92.50	84.55
Turn off the grinding machine	95	77.50	77.50	77.50	75	80.83
Check the results of the grinding whether following the document	90	70	67.50	80	67.50	75.53
Perform autonomous maintenance	92.50	57.50	70	92.50	90	81.63
Return the grinder to its original place	77.50	77.50	77.50	65	87.50	76.98
Carry out 5S	60	85	85	70	90	77.40
The normalized weight	0.22	0.18	0.18	0.21	0.21	∑SLI = 964.06

Table 4 illustrates that the increase of SLI value would be followed by the increase of HEP value. Even though the POS value was inversely proportional to HEP value. The HEP value increased would be followed by a decrease in the POS value and vice versa.

Table 4. The value of SLI, HEP, and POS.

No	Task	SLI	HEP	POS
1	Use PPE following requirement	90.03	0.0005828	0.9994172
2	Prepare a document following requirement	62.63	0.0003835	0.9996165
3	Perform autonomous maintenance	84.43	0.000535	0.999465
4	Select the grinding stone according to the type of material would be grinded and the work should be conducted	89.85	0.0005812	0.9994188
5	Install grinding stones	78.85	0.0004913	0.9995087
6	Turn on the grinding machine	81.35	0.0005104	0.9994896
7	Grinding the material	84.55	0.000536	0.999464
8	Turn off the grinding machine	80.83	0.0005064	0.9994936
9	Check the results of the grinding whether following the document	75.53	0.000467	0.999533
10	Perform autonomous maintenance	81.63	0.0005126	0.9994874
11	Return the grinder to its original place	76.98	0.0004805	0.9995195
12	Carry out 5S	77.40	0.0004775	0.9995225

Table 4 shows that the task with the highest error probability was task 1 (use PPE following requirement). Giving evidence that task 1 had the largest accident occurrence with the highest frequency and severity according to accident records. This data illustrates that these accidents occurred due to there was a lack of protection from the proper use of PPE. Balkhyour *et al.* [20] investigated that PPEs are very effective in minimizing occupational injuries, accidents, and other hazards which otherwise result in substantial manpower and financial losses in 37 small-scale industries situated in urban premises of Jeddah. This research illustrates that education correlated with the use of PPE only for safety mats, face masks, and safety glasses. Besides these three types of PPE, education was not related to the use of PPE. Setyaningrumand Saputra [21] studied the correlation between the use of PPE with the occupational accident. This research concluded that PPE was one of the risk management control accidents. The use of PPE can reduce the risk of accidents, moreover for the work with large potential hazards, and high risk.

Table 4 describes the lowest HEP was task 2 (prepare a document following requirement) in which these tasks had the lowest risk both of the accident frequency and severity, thus, they had a lower probability to generate accidents.

As efforts to reduce the error probability value in grinding work, the following recommendations were provided to decrease the error probability value by increased PSF value. It was ranging from improving procedures, reducing fatigue of workers, reducing the complexity of work, developing training and socialization on workers to improve worker knowledge [22], and improving worker qualifications through a history of work experience. Wachter and Yorio [23] found that the effectiveness of a safety management system and its practices in reducing accident rates depended on the levels of safety-focused cognitive and emotional worker engagement.

1. Procedure
   a. Tasks and task descriptions in the work instruction should be more detailed
   The procedure classified into work instruction and presented as job safety analysis (JSA). Job safety analysis conducted individually or as a group. Unsafe conditions prior to and while jobs were being carried out should be eliminated before work began [24]. Work instruction of grinding work using a portable grinding machine that lacked detailed contents, work instructions only contained points that were already common in grinding work. Therefore, it was necessary to create a task analysis for grinding work using a portable grinding machine.
   b. Providing work instruction socialization
   Work instruction socialization was required to provide detailed explanations related to work instruction and to provide a common perception of each workforce, since all this time, work instruction was only at some point in the workshop. The work instruction should be applied at every bay (workplace) thorough the routine company program for improving worker knowledge such as safety briefing, safety talks, safety training, and seminars. Hamdani *et al.* [22] found that the socialization of PPE could significantly improve the worker's knowledge and attitude to frequently use of PPE at any task of work.
2. Increased supervision
   The essential for supervision here by the supervisor or group leader to ensure the implementation of procedures and work instructions. Supervisors had to ensure the control programs arranged by management and the investigation team were employed [24].
3. Arrange work time and rest time for workers
   One of the PSFs that has a high weight to cause errors was fatigue, it was necessary to study to regulate work time and rest periods of workers. The management of working time and rest periods were necessary to ensure the worker obtained the workload regarding their capabilities. The regulation of worker workload referred to the threshold value of the work climate in Standard Nasional Indonesia (SNI) [25].
4. Complexity
   The level of complexity of the work which was one of the PSFs could be minimized by compiling more detailed work instructions and conducting training and re-training. The training was conducted to increase the knowledge and expertise of workers in grinding activities. It was essential as well to refreshing back the workers who had already received the previous training.
5. Experience
   One important PSF that produced errors was experienced with grinding work. Hence, it was important to recruit workers who had qualified skills as a good base for safety behavior. Safety management practices were designed to influence employee knowledge, skills, motivation, decision-making, attitudes, and perceptions [23].

4. Conclusion

Grinding activities contain high risks for potential hazards. Accident analysis data illustrate that the frequent accident due to human errors. The factors that influenced the occurrence of errors in grinding work are the procedure, fatigue, complexity, raining, and experience factors. The task that has the highest error probability value is the usage of the specified PPE with the HEP value is 0.0005828, which means the probability of success (no error) is the lowest compared to other tasks. This result shows that this task most often results in accidents caused by human error. These accidents occur due to there is a lack of protection from the proper use of PPE. While the lowest error probability value is the task of preparing documents under what has been determined with the HEP value of 0.0003835. Recommendations given to reduce the value of error probability are by compiling detailed work instructions as job safety analysis. Moreover, employee knowledge has to be improved through training or socialization programs for accident prevention. The supervision should be increased to ensure the procedures and work instruction in job safety analysis is implemented. The management of work time and rest periods of workers should be conducted to certify the worker obtains the workload according to their capabilities. The worker recruitment is applied to find an appropriate worker who has qualified skills.

References

[1] Norman D A. Categorization of action slips. *Psychol Rev*. American Psychological Association (APA); 1981 Jan;**88**(1):1–15. doi.org/10.1037/0033-295x.88.1.1

[2] J. Reason, Human Error. Retrieved from https://books.google.co.id/books?id=WJL8NZc8lZ8C&printsec=frontcover&dq=Reason,+J.,+1990.+Human+Error*pdf&hl=en&sa=X&ved=0ahUKEwjyivKesP3nAhWFguYKHfDlAHAQ6AEIKTAA#v=onepage&q&f=false (accessed March 3, 2020).

[3] M.S. Sanders, E.J. Mccormick, Human Engineering Factors In And Design, 1993. Retrieved from Http://Books.Google.Com/Books?Id=3d5qpqaacaaj&Pgis=1 (Accessed March 7, 2020).

[4] Ariefiani M, Handoko L, Amrullah HN, Ashari L, Shah M, Hamzah F. Human Error Probability of Grinding Operation in Fabrication and Construction Company Using Fuzzy HEART Method. *Proc. of the 2019 1st Int. Conf. on Engineering and Management in Industrial System*. Atlantis Press; 2019; doi.org/10.2991/icoemis-19.2019.17

[5] D. Ayundha Novianti, A. Maisarah Disrinama, D. Haidar Natsir Amrullah, P. Studi Teknik Keselamatan dan Kesehatan Kerja, J. Teknik Permesinan Kapal, P. Perkapalan Negeri Surabaya. Analisis Probabilitas Human Error Pada Pekerjaan Grinding dengan Metode HEART dan SLIM-ANP di Perusahaan Jasa Fabrikasi dan Konstruksi, 2018.

[6] Maulida ZA, Santiasih I, Handoko L. Human Reliability Analysis dengan Pendekatan Cognitive Reliability dan Error Analysis Method (Cream). *J@Ti Undip : Jurnal Teknik Industri*. Institute of Research and Community Services Diponegoro University (LPPM UNDIP); 2015 Jan 23;**10**(1). doi.org/10.12777/jati.10.1.1-6

[7] S. Rahimi Kamal, J. Nasl Saraji, I. Mohammad Fam, Assessment of human error probability index for gas compressor station musters (region 3 of gas transmission operation), Journal of School of Public Health and Institute of Public Health Research, 2013. Retrieved from http://journals.tums.ac.ir/ (accessed March 3, 2020).

[8] Mohammadian M, Choobineh A, Mostafavi Nave A, Hashemi Nejad N, et al. Human errors identification in operation of meat grinder using TAFEI technique. *J Occup Health Psychol*. CASRP: Center of Advanced Scientific Research and Publications; 2012 Oct 1;**1**(3):171–81. doi.org/10.18869/acadpub.johe.1.3.171

[9] Brauer RL. *Safety and Health for Engineers*. John Wiley & Sons, Inc.; 2005 Nov 4; doi.org/10.1002/047175093x

[10] Kariuki S G, Löwe K. Integrating human factors into process hazard analysis. *Reliab. Eng. Syst. Saf.* Elsevier BV; 2007 Dec;**92**(12):1764–73. doi.org/10.1016/j.ress.2007.01.002

[11] Konstandinidou M, Nivolianitou Z, Kiranoudis C, Markatos N. A fuzzy modeling application of

CREAM methodology for human reliability analysis. *Reliab. Eng. Syst. Saf.* Elsevier BV; 2006 Jun;**91**(6):706–16. doi.org/10.1016/j.ress.2005.06.002

[12] E. Hollnagel, General rights The methodology of man-machine systems: Problems of verification and validation, 1981.

[13] Cronbach LJ, Meehl PE. Construct validity in psychological tests. *Psychol. Bull.* American Psychological Association (APA); 1955 Jul;**52**(4):281–302. doi.org/10.1037/h0040957

[14] D.E. Embrey, P. Humphreys, E.A. Rosa, B. Kirwan, K. Rea, SLIM-MAUD: an approach to assessing human error probabilities using structured expert judgment. Volume II. Detailed analysis of the technical issues, 1984.

[15] Kirwan B. *A Guide to Practical Human Reliability Assessment.* CRC Press; 2017 Dec 14; doi.org/10.1201/9781315136349

[16] R. Skjong, B.H. Wentworth, Expert Judgment And Risk Perception, 2001.

[17] D.G. DiMattia, Human error probability index for offshore platform musters. 2004.

[18] K. Embrey, D.E.; Humphreys, P.; Rosa, E.A.; Kirwan, B.; Rea, SLIM-MAUD: an approach to assessing human error probabilities using structured expert judgment. Volume I. Overview of SLIM-MAUD (Technical Report) | OSTI.GOV, 1984. https://doi.org/AC02-76CH00016.

[19] *Guidelines for Preventing Human Error in Process Safety.* John Wiley & Sons, Inc.; 2004 Aug 1; doi.org/10.1002/9780470925096

[20] M.A. Balkhyour, I. Ahmad, M. Rehan, Assessment of personal protective equipment use and occupational exposures in small industries in Jeddah: Health implications for workers, *Saudi J. Biol. Sci.* 26 (2019) 653–659. doi.org/10.1016/j.sjbs.2018.06.011.

[21] M. Setyaningrum, Ratna; Saputra, Analysis of affecting factors of work accidents and use of personal protective equipment in welders in A. Yani Street Banjarbaru 2016, (n.d.). Retrieved from https://www.researchgate.net/publication/306138372_Analysis_of_affecting_factors_of_work_accidents_and_use_of_personal_protective_equipment_in_welders_in_A_Yani_Street_Banjarbaru_2016 (accessed June 7, 2020).

[22] Hamdani MZ, Rudyarti E, Phuspa SM. The Correlation Of Personal Protective Equipment Socialization Toward The Changing Of Occupational Safety And Health Behavior Of Musical Instrument Craftsmen. *J. Vocat. Heal. Stud.* Universitas Airlangga; 2018 Nov 7;**2**(1):14. doi.org/10.20473/jvhs.v2.i1.2018.14-19

[23] Wachter JK, Yorio PL. A system of safety management practices and worker engagement for reducing and preventing accidents: An empirical and theoretical investigation. *Accid. Anal. Prev.* Elsevier BV; 2014 Jul;**68**:117–30. doi.org/10.1016/j.aap.2013.07.029

[24] Morrish C. Incident prevention tools–incident investigations and pre-job safety analyses. *Int. J. Min. Sci. Technol.* Elsevier BV; 2017 Jul;**27**(4):635–40. doi.org/10.1016/j.ijmst.2017.05.009

[25] SNI, SNI 16-7061-2004 tentang Pengukuran Iklim Kerja (Panas) dengan Parameter, 2004. Retrieved from https://www.slideshare.net/miemamk/sni-1670612004-tentang-pengukuran-iklim-kerja-panas-dengan-parameter-indeks-suhu-basah-dan-bola (accessed June 12, 2020).

IOP Conf. Series: Materials Science and Engineering 1072 (2021) 012028 doi:10.1088/1757-899X/1072/1/012028

Evaluation of the SAP R/3 system implementation using optimized modules (Case study: Chemicals company)

T G Amran[1,2] and S Yuniati[1,3]

[1]University of Trisakti, Kyai Tapa No.1 - Grogol, 11440, Indonesia

E-mail: [2]amran.tiena@gmail.com; tiena@trisakti.ac.id, [3]sriyuniati@yahoo.com

Abstract. This study evaluates the SAP R/3 System Implementation through optimization of modules in resin chemicals industry. The company has implemented SAP R/3 software since 2013, but 62% of it has not been adjusted and need to be evaluated for implementation performance, therefore it is unknown whether SAP already reached its optimum point. Therefore, SAP R/3 evaluation is needed through SY Check List, improvement of forecasting, and Material Requirement Planning (MRP). The chosen product that was calculated was the A category product by ABC analysis. The 6 periods (months) forecasting will be used as a base of Master Production Schedule (MPS) to calculate the raw material requirement. The MRP calculation method used Lot-For-Lot because it does not need price information. SY Check List was used to identify in-optimum modules which SAP provided beforehand. The suggestion made based on the best forecasting calculation was used as other A category products forecasting calculation method. Therefore, SY Check List optimized SAP modules not only as a book data, but also logistic as integrated system which have to be maximized. As actual demand fluctuation forecasting method should be changed. More accurate forecasting will define MRP parameters accuracy.

1. Introduction

Choosing and implementing the ERP-SAP (Enterprise Resource Planning-System Application and Processing) in chemical company from the head office to subsidiaries in Tokyo is aimed to integrate the whole logistic information system. Therefore, the data collection, controlling and monitoring are directly conducted by using global template. A similar and holistic system must be implemented to all subsidiaries around the world as soon as possible in Information System (IT) technology and this information system will process the business into a new one. In the computer-based information system, Enterprise Resource Planning (ERP) is connected to the software of information system technology [1]. ERP is a concept that integrates all function in the business process of the company in a single database, so that each function of each different department in a company can share information to each other in a better way. The main factors of successful ERP implementation are teamwork and composition, Change Management program and Culture, top management support, business plan and vision, business process reengineering with minimum customization, project management, and software development, testing and troubleshooting [2].

Offering a comprehensive information management system for organizations, ERP is aimed to integrate various business processes by eliminating redundant elements while enabling data sharing between business functions. Coordination of planning activities, global and geographic audit of the organization, and the integration of top-level information can be achieved through ERP. ERP is a system that uses the most advanced information technology that responds to all these needs [3]. The ERP system successfully benefits the rationalization, standardization, fault reduction and cost reduction process [4]. In the opposite case, the inability to create an effective ERP system will not only lead to cost and time losses, in addition, the ERP can cause serious losses by causing damage to the company's culture, minimizing production, introducing excessive training needs, and misleading customer demand [5].

The problem is before the use of ERP-SAP the old system BaaN is one part of the production management division of the company, Invensys PLC [4]. This transition period finds that (1) there are inaccuracies in detecting raw material input that is consumed by the system, (2) there is an inability to trace the quantity of inputs and the output of the system, (3) the old system cannot export the data directly but a modifier system named Crystal Report must be available, (4) Purchase Requisition (PR) or request of raw materials to Purchasing section are still using Ms. Excel, and (5) the MRP module is not permitted to be used due to the purchase/rental package. The purpose of this research is to evaluate the SAP R/3 implementation at operational level with SY Check List approach to improve forecasting, MPS and MRP.

2. Methods

The methodology and research framework of this research is shown in Figure 1.

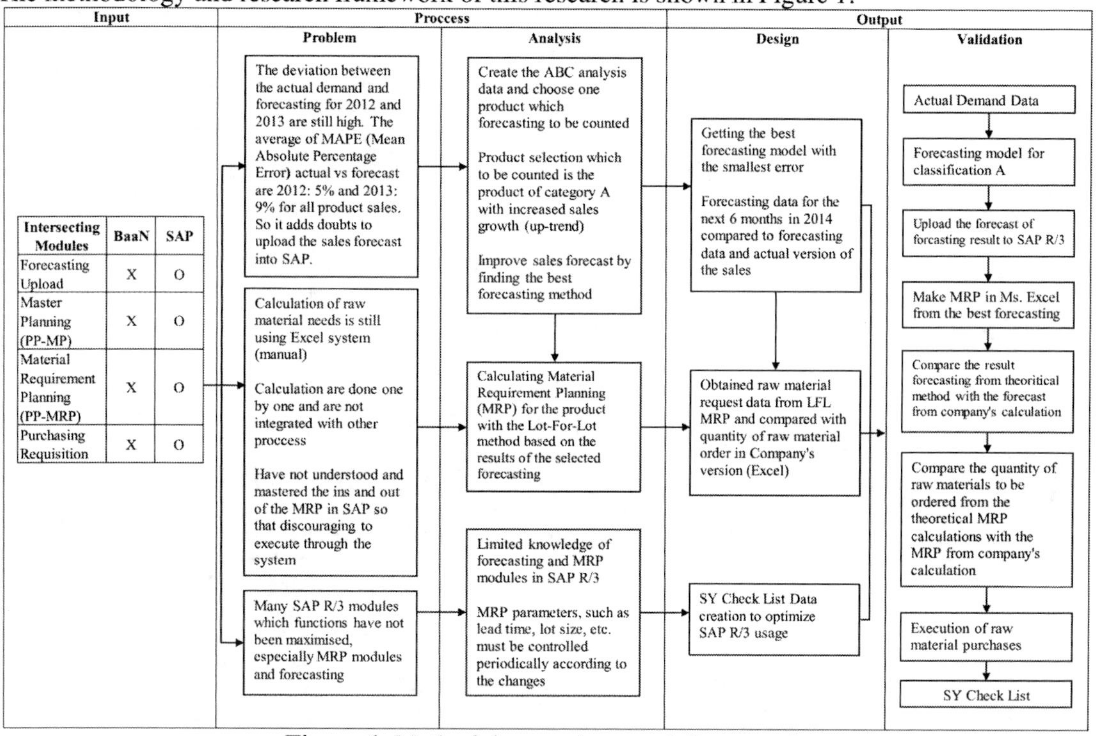

Figure 1. Methodology and research framework.

3. Results and Discussion

The option of switching and using the SAP in Table 1, there are many modules obtained that have not been in BaaN but can be accommodated in SAP R/3. This condition has not been optimally enabled by the company because the sales forecasting accuracy is unsure. Forecasting mistakes can lead to overstock or out of stock. Therefore, it is necessary to calculate the planning of sales more accurately so that the calculation result can be uploaded to SAP R/3. Calculation results of the forecasting are used to calculate the needs of raw materials (MRP).

SAP R/3 system was adopted more frequently compared to BaaN (Figure 2), as for example: (1) Sales which consisted of master data, sales, shipping, billing, (2) PPIC which consisted of basic data, capacity requirement planning, repetitive manufacturing, master planning, reduction orders, material requirement planning, warehouse management, (3) Accounting which consisted of inventory management, general ledger, account receivable, account payable, product cost controlling, (4) Purchasing which consisted of raw material order processing, packaging processing, trading goods, return processing, (5) Technical which consisted of technical trial, material master, and (6) QC which consisted of inspection processing, quality control and quality certificate. Meanwhile, human resource data such as attendance, leave, payroll has yet to use SAP R/3 system.

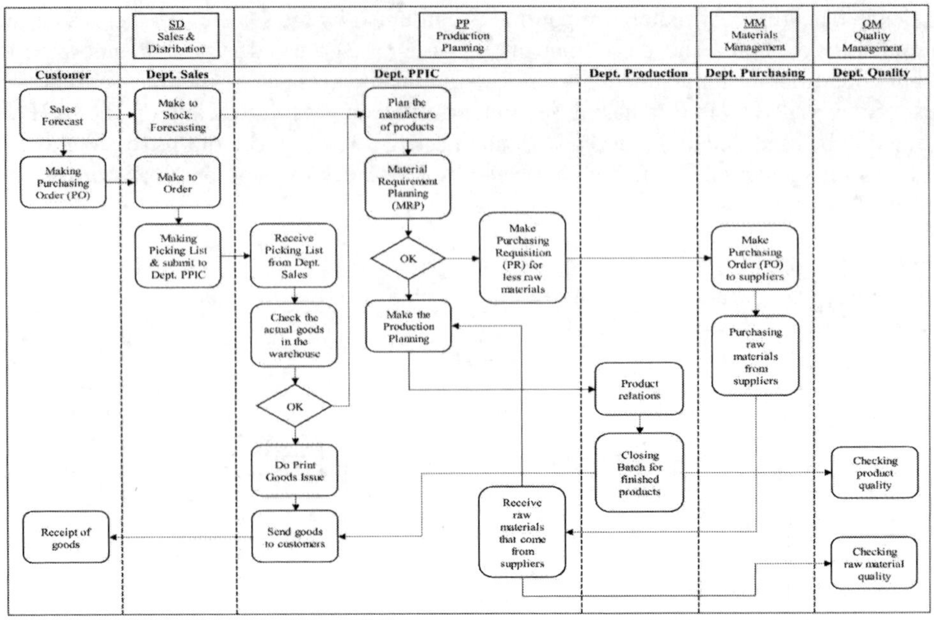

Figure 2. Flow chart process and information system SAP/R3 in chemical company.

Table 1. Comparisons of total modules of transaction coverage between BaaN and SAP.

Process Planning (PP), Material Management (MM), Manufacturing (M)					
Module	BaaN	SAP	Module	BaaN	SAP
Basic Data (PP-BB)	O	O	Material Requirement Planning (MM-MRP)	X	O
Sales and Operation Planning (PP-SOP)	O	O	Purchasing (MM-PUR)	O	O
Master Planning (PP-MP)	X	O	Inventory Management (MM-IM)	O	O
Capacity Requirement Planning (PP-CRP)	X	O	Warehouse Management (MM-WM)	O	O
Material Requirement Planning (PP-MRP)	X	O	Invoice Verification (MM-IV)	O	O
Production Orders (PP-MRP)	O	O	Information System (MM-IS)	O	O
Product Costing (PP-MRP)	O	O	Electronic Data Interchange (EDI) (MM-MRP)	X	X
Kanban/JIT Production (PP-MRP)	X	X	Bill of Material	O	O
Repetitive Manufacturing (PP-REM)	X	O	Cost Price Calculation	O	O
Assembly Orders (PP-ATO)	X	O	Engineering Change Control	X	X
Process Planning for Process Industry (PP-)	O	X	Engineering Data Management	X	X
Plant Data Collection (PP-PDC)	O	O	Hours Accounting	O	O
Information System (PP-IS)	O	X	Product classification	O	O
			Product configuration	O	O
			Production control	X	O
			Production Planning	O	O
			Project Budgeting	X	X
			Project Control	X	X
			Repetitive Manufacturing	X	O
			Routing	O	O
			Shop Floor Control	X	O
			Tool Requirement Planning and Control	X	O
			Capacity Requirement Planning	X	O
			Master Product Scheduling	X	O
			Master Requirement Planning	X	O
	6	3		12	5

SY Check List approach in Production Planning (PP), Material Management (MM), and Manufacturing (M) operations.

Table 1 explained the existing modules within the thirty-eight modules with BaaN and SAP/R3, 38% of the modules was suitable and 62% was unsuitable. In this research, module adjustments that will be evaluated is the Precom Chemicals case in Process Planning (PP). the forecasting improvement from sampling of one chosen product is done with ABC classification approach [6][7].

3.1. Graphs and Calculation of Selected Product Forecasting ABC Classification

Product selection which represent ABC classification approach prioritize materials with high product sales (Table 2).

Table 2. ABC analysis category a for product sales results.

SAP Code	Name	Actual Demand 2017 (Kg)	%	Cumulative	Classification
129999995706	DK0190_BURNOCK ECL-341-PDJ	2,140,603	19%	19%	A
129999995598	CK0018_DICDRY LX-830-PDC	1,018,766	9%	27%	A
129999995752	DK0220_POLYLITE DE-347-IM	954,360	8%	36%	A
129999995628	DK0180_12X0020-P	775,980	7%	42%	A
129999995722	DK0230_POLYLITE FG-208	669,760	6%	48%	A
120000008613	EK0224_PRECOMNBA-13P	660,820	6%	54%	A
129999995630	DK0180_12X0921-AF	483,300	4%	58%	A
129999995746	DK0220_POLYLITE CN-225-SF	474,760	4%	62%	A
129999995754	DK0220_POLYLITE DE-350	438,720	4%	66%	A
129999995783	DK0180_Styrene Monomer	385,200	3%	69%	A
129999995733	DK0230_POLYLITE CS-810-NW	289,800	3%	72%	A
129999995594	CK0018_DICDRY LX-820-PD	279,654	2%	74%	A
129999995558	DK0190_ACRYDIC A-418	250,800	2%	76%	A
129999995678	CK0018_DICDRY LX-420A	240,678	2%	79%	A
129999995682	CK0018_SP-75 D	240,084	2%	81%	A
129999995709	DK0230_POLYLITE BS-109	232,530	2%	83%	A
129999995632	DK0180_12X1153-AF	227,520	2%	85%	A
129999995735	DK0220_POLYLITE KT-114	209,880	2%	86%	A
129999995674	CK0018_SI-75 D	204,736	2%	88%	A
129999995583	DK0190_BECKOSOL PD-7300-S	204,440	2%	90%	A
129999995737	DK0220_POLYLITE KT-114-HV	194,260	2%	92%	A
129999995606	DK0190_BECKOLITE M-6401-50S	188,670	2%	93%	A
129999995516	DK0190_ACRYDIC 52-473	181,654	2%	95%	A
129999995544	DK0190_ACRYDIC FL-829D-50-PD	173,850	2%	96%	A
129999995471	DK0190_ACRYDIC A-801-PHV	149,210	1%	98%	A
129999995666	CK0018_DICDRY LX-405-A D	133,852	1%	99%	A
129999995728	DK0230_POLYLITE TA-130	133,730	1%	100%	A

Example: Precom Product (classification A).

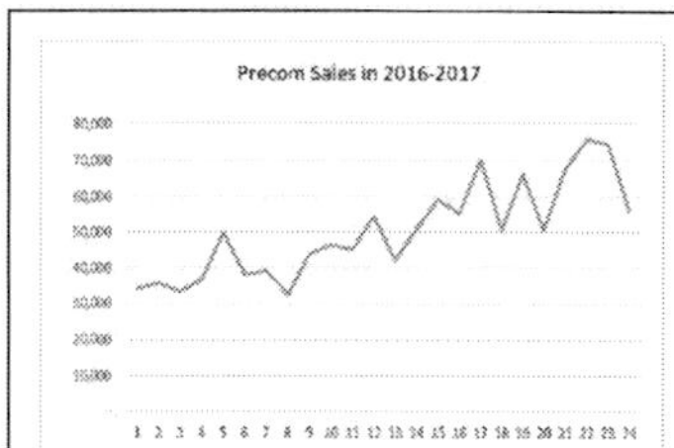

Figure 3. Trend graphs.

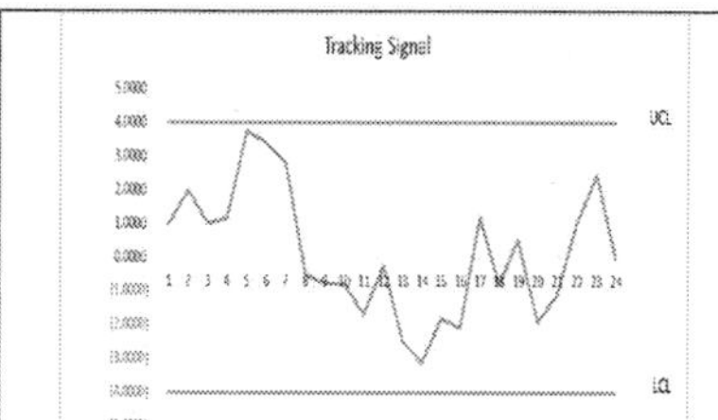

Figure 4. Control of difference between forecasting and actual delivery.

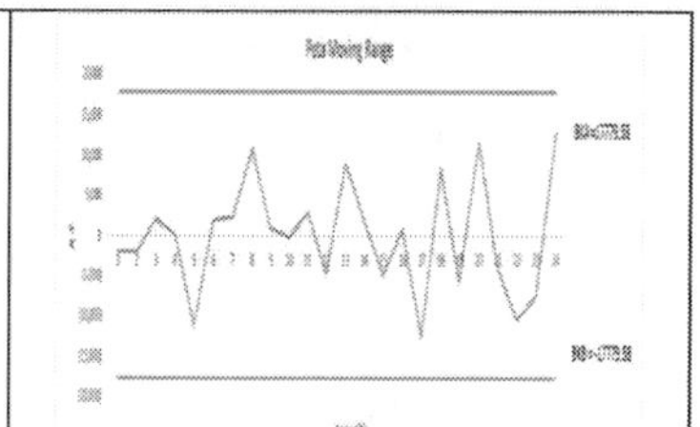

Figure 5. Tracking signal of upper and lower control limits.

Sales Forecasting Method	MAD
Moving Average	8669
Single Exponential Smoothing	9075
Linear	5464

MAD in the Linear method is the smallest error value so that this forecasting method can be used to calculate demand plan in the next 6-month period

MR = 6684,1
UCL = 17779,58
LCL = - 17779,58

Linear method forecasting is chosen and applied to products in classification A

Figure 3 and 4 showed gaps in the actual forecasting of the company's demand and calculation by 110,456 kg. Whereas the linear method of forecasting is no difference, this approach is valid and can be used for Precom products. By assigning this data to the SAP R/3, hence the production needs are calculated automatically, the plan and purchase of raw materials eventually become a completely integrated system without manual calculations and the deviation rate errors resolved.

3.2. Material Requirement Planning (MRP) (Example: Classification A Products)
Precom product in A classification was chosen as sample in MRP forecast calculation (Table 3, Table 4, and Table 5) based on the performance as superior product and the availability of its raw materials must always be maintained. The growth trend of Precom product increases but it has yet to be a commodity, which resulted in lower number of competitors with similar product. Therefore, it is important to keep the product overstocking vice versa.

MRP Calculation.

Table 3. Precom formula.

Product Name	Raw Material	Total (Kg)
Precom	PB-X	18
	PB-Y	25
	A	6
	B	145
	C	30

Table 4. PB-X formula.

Intermediate Name	Raw Material	Total (Kg)
PB-Y	G	3000
	H	1000

Table 5. PB-Y formula.

Intermediate Name	Raw Material	Total (Kg)
PB-Y	G	3000
	H	1000

With the MRP Lot-For-Lot method, in the exploding stage for Level 0 (Precom Product), Level 1 (PB-X and PB-Y), and level 2 is for the raw material of PB-X forming, as in the Table 4. This MRP calculation result should be compared by Purchasing Requisition (PR) in SAP R/3.

3.3. Master Production Schedule (MPS)
Based on Linear method forecast (Figure 5), six future periods demand were calculated. This will be the basis of Master Production Schedule (MPS) arrangement (Table 6).

Table 6. MPS precom 1st semester 2017.

Item	Year 2017					
Month	Jan	Feb	Mar	Apr	May	Jun
Period	1	2	3	4	5	6
EK0224_PRECOM NBA-13P	70.051	71.627	73.203	74.779	76.355	77.931

3.4. Inventory Precom Status
This table shows the information of supply availability at the beginning of period 0.

Table 7. Status inventory precom.

Periode t	0	1	2	3	4	5	6
Gross needs Dt	-	70.051	71.627	73.203	74.779	76.355	77.931
Acceptance of Orders Qt (IOO)	-	71.680	71.680	71.680	71.680	71.680	71.680
Available supplies (IOH)	7.511	9.140	9.193	7.670	4.571	(104)	(6.355)

From the calculation in the Table 7, it is shown that in 5th and 6th period, the supply becomes -104 and -6355. Inventor in 5th month was 4571 – 76355 + 71680 = -104. This resulted in *intermediate PB-Y* shortage in fifth and sixth month and that need to be restocked since fourth or sixth month. For

Precom manufacture, raw material "G" was imported from Japan, which required 2 months of shipment time and "F" from Europe, which required 3 months of shipment. MRP calculation should be compared and proceeded as Purchasing Requisition (PR) in SAP R/3.

3.5. Lot-For-Lot Netting

At this stage, the net needs of Precom is calculated. The Table 8 shows that in the period 5th and 6th, Precom experienced a deficiency, so the net needs in months 5 and 6 is as much as the deficiency itself.

Table 8. LFL netting for precom.

Periode t	1	2	3	4	5	6
Gross needs Dt	70.051	71.627	73.203	74.779	76.355	77.931
Acceptance of Orders Qt (IOO)	71.680	71.680	71.680	71.680	71.680	71.680
Available supplies It (IOH)	9.140	9.193	7.670	4.571	(104)	(6.355)
Net Needs (Rt)	0	0	0	0	104	6.355

3.6. Lot-For-Lot Exploding

Once the net needs are obtained, the size of the reservation lot (lot size), and the lead time, then the next step is exploding for the product structure at level 0 and level 1 and the product level 2, which are *intermediate PB-X* and *PB-Y* (sample calculation). On the table is the exploding on the structure of the product level 2 with raw materials G and H.

Table 9. LFL exploding for level 2 (PB-Y).

PB-Y : Level 1 (25 Kg)	**0**	**1**	**2**	**3**	**4**	**5**	**6**
Gross Needs Dt					12	709	
Order received Qt (IOO)				220			
Available supplies It (IOH)	50	50	50	270	258	-451	-451
Net needs (Rt)						451	451
Lot size (Qt)						451	451
Order Plan - POR					451	451	
G : Level 2 (3000 Kg)	**0**	**1**	**2**	**3**	**4**	**5**	**6**
Gross needs Dt					54106	54106	
Order received Qt (IOO)				9720	9720		
Available supplies It (IOH)	0	0	0	9720	-34666	-88772	-88772
Net needs (Rt)					34666	88772	88772
Lot size (Qt)					34666	88772	88772
Order Plan - POR			34666	88772	88772		
H : Level 2 (1000 Kg)	**0**	**1**	**2**	**3**	**4**	**5**	**6**
Gross Need Dt					18035	18035	
Order received Qt (IOO)				2500	2500	2500	
Available supplies It (IOH)	0	0	0	2500	-13035	-28571	-28571
Net needs (Rt)					13035	28571	28571
Lot size (Qt)					13035	28571	28571
Order Plan - POR		13035	28571	28571			

Table 9 displayed the exploding calculation in Level 2 product structure, which has been chosen as *intermediate* PB-Y.

3.7. SY Check List

On the Table 10 is a check list to record PPQM modules that have and have not been executed along with the information on which parts are responsible and the cycle of the checking.

SY Check List is a table to analyze whether the modules that have been charged to the user are executed properly and correctly according to their functions. From the table, SY Check List to PPQM module is that the process number 1 to number 5, which is the process that the forecasting is obtained from the Sales department, should be uploaded until the raw materials monthly checking has not been done. This is because the PPIC part is still doubtful of the accuracy of the forecasting given by the sales section. Less accurate forecasting will result in inaccuracy of process number 3 and number 5. If the forecast from the Sales section forced to be uploaded to the SAP system, then the number of production plans indicated by the ZPPP_R0007 and the amount of raw materials needed in the aggregate becomes doubted to be used as the basis of execution.

Table 10. SY check list for PPQM module.

No	Transaction Code	Process Definition	Done	Undone	Remark	Dept in Charge	Checking Cycle
1	ZEPP_R0001	Forecast Upload		√		PPIC	Monthly
2	MD40	MPS Run		√		PPIC	Monthly
3	ZPPP_R0007	Monthly production Planning report		√		PPIC	Monthly
4	MD01	MRP Run		√		PPIC	Monthly
5	ZPPP_R0008	Monthly production Planning report		√		PPIC	Weekly
6	ME51N	Create Purchasing Requisition	√		Individual Running	PPIC	Daily
7	ME55	Release Purchasing Requisition	√		Individual Running	PPIC	Daily
8	MD04	Check stock requirement list	√		Individual Running	PPIC	Weekly
9	MDVP	Check Planned order availability		√		PPIC	Daily
10	MD04 / COR7	Convert Planned order to Process order		√		PPIC	Daily
11	COR1	Create Process Order	√		Individual Running	PPIC	Daily
12	ZPPP_E0004	Release Pr. Order + Make ticket & QC ticket printing	√		Individual Running	PPIC	Daily
13	CORK	Process order confirmation	√		Individual Running	Production	Daily
14	COGI	Process order Confirmation errors	√		Individual Running	Production	Daily
15	QA32	Record Inspection Result	√		Individual Running	Quality	Daily
16	QA32	Record UD	√		Individual Running	Quality	Daily
17	COR2	Technically close the Process order (TECO)	√		Individual Running	Production	Weekly

In process number 9th and 10th, it is a direct effect that occurs if process number 1 until number 5 is not executed. If the forecast is not uploaded to SAP, then the planned order will not be automatically generated by SAP. So, to make the used process order as a command for product creation to the production section, it is created from the COR1 module directly. Currently, what takes place in the company is production plans and raw material needs are calculated with Ms. Excel. The planned production will be realized, directly created in COR1. While the raw material plan to be ordered is made directly at ME51N. With the existing MRP facility in SAP R/3 it will work optimally, bearing in mind that there are hundreds of raw materials.

3.8. Design and Modeling of SY Check List

Design and modeling of SY Check List is created to identify which modules have and have not been executed during the application of SAP R/3. This identification is limited only to the Production Planning module. These design parameters of SY Check List include:

1. *Transaction Code* : A transaction code is used to enter the SAP R/3 Module window.
2. *Process Definition* : A detailed description of the transaction name in the opened window.
3. *Done/ Undone* : The marking (check list) of any transaction code that have or have not been executed by company.
4. *Remark* : A description field that contains information if this transaction code is executed individually (individual running).
5. *Dept in Charge* : Certain sections/departments are responsible for transactions charged.
6. *Checking Cycle* : A time cycle that is recommended to always be done before the closing of the end-month book. This is to see which transactions are still hanging and not completed.

In the Remark column, it can be seen that some of the transaction codes that have been implemented are running individually and without any relationship between one module to another. This will be different if the uploaded forecast is already executed, then the transaction order and the relationship between modules will be very visible.

3.9. Modeling of SAP R/3 Implementation Integration system
From the results of the study above, it can be acknowledged that in order to use the SAP R/3 system, especially in the forecast upload module, it is necessary to determine the accuracy of the raw material requirements calculation and overall production plan. After forecasting has been calculated with the right method, the new forecasting data is integrated into the SAP R/3 system. Some other steps that should be done such as calculating the MRP and identifying the implementation of SAP PPQM modules with SY Check List in a given cycle and communicating with related sections for modules that have not been executed properly. Thus, a flowchart was built to design the integration model of SAP R/3 Implementation system (Figure 6):

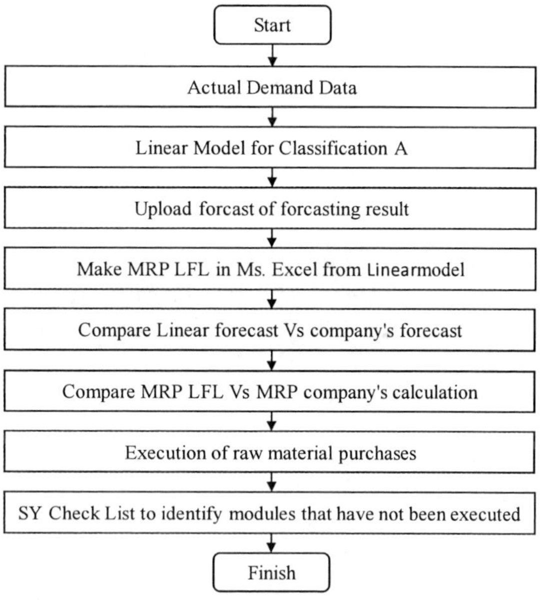

Figure 6. Integration model design of SAP R/3 implementation system.

The Novelty Score is the creation of SY Check List and the flowchart design of SAP R/3 implementation system integration model. On SY Check List, the order of transaction process and its definition is a provision from SAP AG. Meanwhile, the new update is the addition of the Done, Undone, Remark, Department in Charge, and Checking Cycle columns. The flow chart design is created from the research results which effectiveness calculation method has been calculated, proven, verified, and validated. The steps in this flowchart are a sequence of processes that should be used in chemicals companies (who have implemented SAP R/3) before uploading the forecasting to the system to avoid inaccuracies of raw material needs and production plans.

Product forecasting and Material Requirement Planning (MRP) such as safety stock, minimum lot size, lead time are done periodically every 3 months because it will be expired. Because it is imported material then the safety stock, minimum lot size, lead time will affect the automatic calculation of MRP in SAP R/3. The company needs to conduct training of SAP modules to its employees so that the available modules can be optimized for use. SY Check List is used for mapping the most unused modules yet having an outstanding function can greatly help the company's performance.

4. Conclusion

Readiness to use SAP R/3 facility especially the forecast uploading module is simultaneous with the readiness of its logistic information system accuracy. Linear forecasting model is the best method for products in classification A. From the actual demand data, a forecast calculation was made with Linear, Moving Average and Simple Exponential Smoothing model. The smallest Mean Absolute Deviation (MAD) is shown by Linear forecasting and is used to calculate the forecasting in the next 6 periods. The calculation results of the 6 periods is used as a basis of Master Production Schedule (MPS) and with the product structure BOM (Bill of Material), then to calculate the raw material needs as a basis for purchasing.

From the calculation of Material Requirement Planning, the Lot-For-Lot method was chosen. This result is used as a basis to purchase raw materials more accurately because some of the raw materials have to be imported from foreign suppliers like Japan and Europe. The MRP calculation result shows that the company will go through stock deficiency in the fifth and sixth period/month, so that it has to be produced from a month before. The deficient raw materials are PB-X and PB-Y which known as intermediate materials. There are 6 material that is deficient in the fifth and sixth months, which are imported materials. The SY checklist which has been made shows that, all this time the SAP R/3 that never been used optimally, is only because there is a doubt in uploading the forecast received from Sales Department. SY Check List is open for further research development until the SAP R/3 modules increases the company branch performance, in accordance to main company target performance.

References

[1] Yontar E. A Comparative Study to Evaluate of Sap and Logo Erp Software's for Smes and Big Businesses. *Turkish J. Eng. Environ. Sci.* 2019 Jan 1;1–10. doi.org/10.31127/tuje.416678

[2] Fui-Hoon Nah F, Lee-Shang Lau J, Kuang J. Critical factors for successful implementation of enterprise systems. *Bus. Process Manag. J.* Emerald; 2001 Aug;7(3):285–96. doi.org/10.1108/14637150110392782

[3] Amran TG, Azmi N, Surjawati AA. Information system and website design to support the automotive manufacture ERP system. IOP Conference Series: *Mater. Sci. Eng. R Rep.* IOP Publishing; 2017 Dec;277:012007. doi.org/10.1088/1757-899x/277/1/012007

[4] Okrent MD, Vokurka RJ. Process mapping in successful ERP implementations. *Ind. Manag. Data Syst.* Emerald; 2004 Oct;104(8):637–43. doi.org/10.1108/02635570410561618

[5] Fadlilah S, Utama AAGS. The Adaptation of Delone and Mclean's Information System Model to Test The Success of the Erp-Sap Implementation in Middle-Level Management: A Study Case at PT Petrokimia Gresik. *1st International Conference on Islamic Economics, Business, and Philanthropy*. SCITEPRESS - Science and Technology Publications; 2017; doi.org/10.5220/0007081102950299

[6] Su Y, Yang C. A structural equation model for analyzing the impact of ERP on SCM. *Expert.* Syst. Appl. Elsevier BV; 2010 Jan;37(1):456–69. http://dx.doi.org/10.1016/j.eswa.2009.05.061

[7] Millstein MA, Yang L, Li H. Optimizing ABC inventory grouping decisions. *Int. J. Prod. Econ.* Elsevier BV; 2014 Feb;148:71–80. http://dx.doi.org/10.1016/j.ijpe.2013.11.007

Applying the ISO 31000:2009 in livestock industries (Case study: XYZ livestock group)

H Fitria[1] and A Mansur[2*]

[1]Master of Industrial Engineering, Universitas Islam Indonesia
[2]Industrial Engineering Department, Universitas Islam Indonesia

Email: *agusmansur@uii.ac.id

Abstract. XYZ Livestock Group is one of the livestock industries contributing to beef cattle and dairy cattle. However, there are many problems in the livestock industry. Because of the many uncertainties in managing the livestock industry, many cattle ranchers have switched professions to sand miners or residential entrepreneurs. This research is designated as a reference for livestock industry stakeholders to improve their business, especially to handle their risk. ISO 31000:2009 Framework is used as a method in this research. In this study, we found 13 risks with 13 causes of risk that occurred in the XYZ Livestock Group. Further analysis shows that there are three risks in the red zone, namely that livestock cannot be sold (dead/sick), and consumers of XYZ Livestock Group are declining. Furthermore, this research proposes a mitigation strategy to reduce risk. Three mitigations are obtained based on risk priorities, namely taking or increasing risks to get opportunities, eliminate sources of risk, and change the consequences.

1. Introduction

Sleman is a district in the Yogyakarta Special Region that contributes livestock products in Indonesia from existing beef cattle and dairy farmers. There are 2,053 beef cattle and 20 dairy cows managed by the Sleman livestock group. Sleman Regency is ranked fifth after Kalasan Regency [1]. However, dairy farmers in Sleman always complain about the low yield of dairy cows [2]. Another complaint is the emergence of various diseases in dairy cows, such as the Theiler virus, which infects many livestock farmers [3]. As a result, many dairy farmers directly switch their professions to become sand miners or residential entrepreneurs [4]. This study aims to create a comprehensive risk analysis and design risk mitigation strategy at the livestock group using the ISO 31000:2009 approach.

According to Moschini and Hennessy (2001), Agriculture or livestock industry is an industry that contains a lot of risks and uncertainties [5]. Livestock products such as cattle are an essential agricultural commodity because they contribute to global kilocalorie food security by 17% and global protein by 33%. [6] Failure to manage existing risks in a way will cause difficulties in meeting organizational goals [7]. In China, many farmers use two risk control instruments, namely insurance and agricultural cooperatives [5]. Insights into the social, technological, and economic systems, as well as the dynamics of cooperative behavior to regulate the transition in the energy sector and other sectors such as the agriculture industry [8]. This condition makes the livestock industry in Sleman Regency a challenge to find suitable risk management [9].

Risk management will face severe challenges when approaching specific and endogenous needs [10]. According to Knight (2007), in the face of increasing threats posed by climate change and natural disasters, the International Organization of Standardization developed ISO 3100:2009 in November 2009. ISO 31000:2009 provides general guidelines and frameworks in risk management from various countries [11]. The guidelines in ISO 31000:2009 provide an impetus for preventive measures that can be applied in anticipating negative impacts from emerging risks [12].

This study, calculations will be carried out to mitigate risk using the ISO 31000:2009 approach. The purposes of this study are to (1) set goals and context of the company, (2) conduct an assessment of risks, (3) conduct treatment of risks, and (4) analyze mitigation based on identified risks.

2. Methods

Agribusiness is an agricultural business that covers all activities, such as the procurement of agricultural inputs to the marketing of agricultural products or processed agricultural products [13]. The livestock industry also has the same business process in the agribusiness industry. The focus of this research is to find risk assessment in the livestock industry to mitigate identified risks based on ISO 31000:2009. The location of this research is the XYZ Livestock Group, located in Cangkringan, Sleman, DI. Yogyakarta. The framework of this study is to mitigate five risk aspects from the Livestock Industry using ISO 31000:2009. The result of this study will produce mitigating each risk from Livestock Industry, based on a study case in XYZ Livestock Group. The research framework explains as Figure 1.

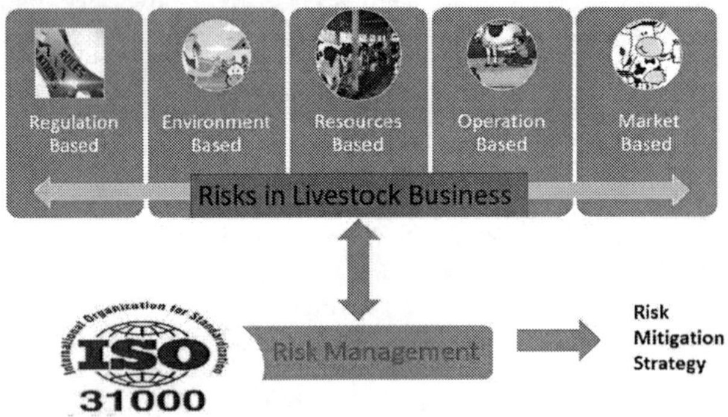

Figure 1. Research framework diagram.

Based on Figure 1, the scope of this research will focus on identifying and mitigating risk potential related to factors regulation, environment, resources, process, and market. Figure 2 explains the flow chart of this research. It consists of problem identification, literature study, determining research objectives, data collection, data processing, processing analysis, discussion, conclusions, and suggestions.

3. ISO 31000:2009

Risk management guidelines based on ISO 31000:2009 require the support of models and theories in each sector, industry, service, organization, and project. Thus, the application of risk management can achieve effectiveness and easy to implement in a real work environment. The essential element of risk management is the uncertainty of human behavior, operated in complex systems, as concluded in different R&D environments [8].

The application of ISO 31000:2009 has been carried out in many industries, as applied to the agricultural and livestock industries, which have a high degree of uncertainty [5]. Unfortunately, risk management in the XYZ Livestock Group in Cangkringan, Sleman, has not well managed.

3.1. Establishing the Context

The initial step in conducting risk management based on ISO 31000:2009 is determining the goals and context to be achieved by the company. The livestock industry is a type of industry that has a fairly complex uncertainty problem. The livestock industry needs to adapt its industry to the weather, climate and range conditions that affect livestock production. The livestock industry must also be prepared to respond to feed prices and livestock market prices [14].

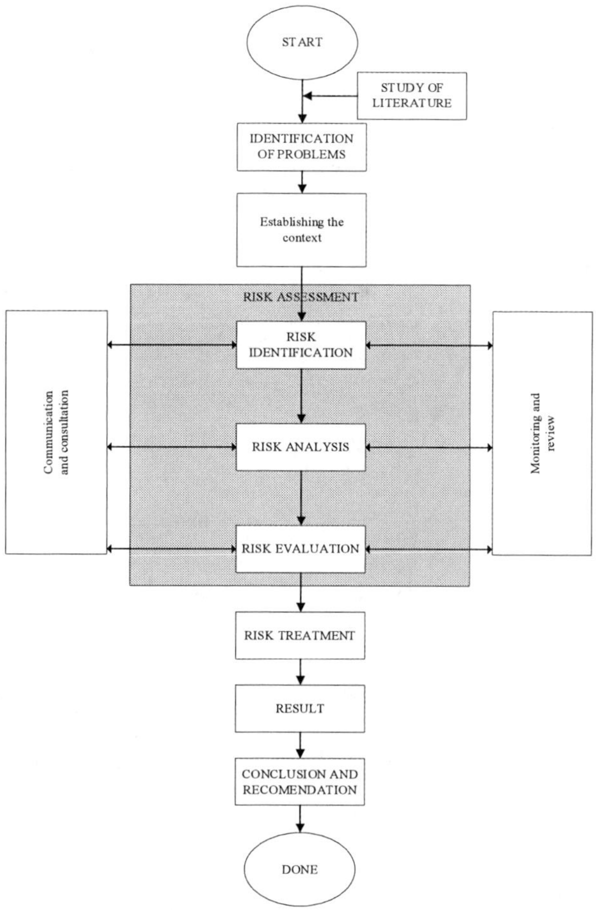

Figure 2. Flowchart diagram.

At the stage of setting objectives and context, it appears that the XYZ Livestock Group wants to improve livestock products such as meat and milk by avoiding the risks faced. The XYZ Livestock Group has yet to have a special SOP on risk management in managing beef cattle and dairy cattle farms.

3.2. Risk Assessment

3.2.1. Risk identification

Referring to the International Organization of Standardization, the organization has developed ISO 3100 guidelines in November 2009. ISO 31000:2009 provides general guidelines and frameworks in risk management of various countries [9]. There are five essential aspects to consider in managing risks in the livestock industry.

For example, the livestock industry will face risks, the first threat of livestock is disease [3], lack of knowledge of institutions, marketing, and capital [15], natural disasters from Mount Merapi, and the unstable price of feed and the selling price of cows [2].

In this case, the livestock industry will face several risks, namely law, environmental, marketing, resource, and operational aspects. Furthermore, the risk register event obtained will be elaborated as follows:

Table 1. Event risk register XYZ livestock group.

No	Risk Area	Sub No.	Risk Event	Risk Agent
1	Law	1a	Loss of consumer confidence	Lost important certificate
		1b	XYZ Livestock Group 's profit has dropped dramatically	Changes in government regulations
		1c	Lost a big profit opportunity	Production limits in the livestock industry
2	Environment	2a	Livestock cannot be sold (dead / sick)	The threat of livestock disease
		2b	Loss of XYZ Livestock Group assets	Natural disasters and human error
		2c	The XYZ Livestock Group office cannot operate	The XYZ Livestock Group organization is opposed by the public
3	Marketing	3a	Lack of public interest in transacting XYZ Livestock Group	The traditional shopping patterns of the community are still attached
		3b	XYZ Livestock Group's profit margin decreased	The decline in livestock market prices
		3c	Consumers of XYZ Livestock Group are declining	The emergence of similar industries
4	Resources	4a	Losses incurred by XYZ Livestock Group	Theft / embezzlement of money / company assets
		4b	Hampered business processes XYZ Livestock Group	Employees who are not compliant with SOP
5	Operational	5a	Lost or changed data on the system	The system code on the XYZ Livestock Group database has changed
		5b	Cessation of the process on the XYZ Livestock Group database	Database XYZ Livestock Group overload

3.2.2 Risk analysis

Risk analysis is the stage in which researchers search for information about the likelihood and impact that can occur in the XYZ Livestock Group. According to Water (2007) the analysis can be done in three ways [16], namely:

- Using knowledge of the situation and calculating the probability
- Employing historical data to see the frequency of events to predict future events
- Asking the risk owner (expert, employee, or leader) to get a view of the possibility of the existing events

The technique that will be used in this research is consequence or probability matrix. According to ISO 31000:2009 consequence or probability matrix is a combination of quantitative and semi-qualitative ranking techniques of impact and likelihood. This assessment is performed by experts. Following are the results of the analysis identified in the XYZ Livestock Group's industry:

Table 2. Risk analysis of XYZ livestock group.

Sub No.	Risk Event	Likelihood	Impact
1a	Loss of consumer confidence	Low	High
1b	XYZ Livestock Group profit has dropped dramatically	Moderate	Moderate
1c	Lost a big profit opportunity	Moderate	High
2a	Livestock cannot be sold (dead / sick)	High	High
2b	Loss of XYZ Livestock Group assets	Low	Moderate
2c	The XYZ Livestock Group office cannot operate	Low	Moderate
3a	Lack of public interest in transacting XYZ Livestock Group	Moderate	Low
3b	XYZ Livestock Group's profit margin decreased	Moderate	Moderate
3c	Consumers of XYZ Livestock Group are declining	High	Moderate
4a	Losses incurred by XYZ Livestock Group	Low	High
4b	Hampered business processes XYZ Livestock Group	Moderate	Low
5a	Lost or changed data on the system	Moderate	Moderate
5b	Cessation of the process on the XYZ Livestock Group's database	Low	High

Based on the results of the risk analysis in table above, it can be recognized that there are some risks that have a very critical level. Further research can be proceeded on evaluating the identified risks.

3.2.3. Risk evaluation

In the risk evaluation stage, the likelihood and impact of risks will be identified and analyzed based on the results of the risk analysis above. This stage will form an evaluation matrix based on likelihood and impact values. The evaluation matrix can also be referred to as a risk map. An evaluation matrix or risk map is formed based on existing risk analysis parameters. Details of the risk evaluation parameters can be seen in Table 4, while the risk evaluation matrix can be seen in Table 3.

Table 3. Risk map.

		Impact		
		Low	Moderate	High
Likelihood	High		3c	2a
	Moderate	3a,4b	1b, 3b, 5a	1c
	Low		2b, 2c	1a, 4a, 5a

Based on the results of the evaluation matrix above, a risk map is obtained, then an evaluation will be carried out as shown in Table 4 below. Possible risks will be sorted from high to low to find out the risk priorities, which are as follows:

Table 4. Risk priority.

Sub No.	***Risk Event***	***Likelihood***	***Impact***	**The value of Risk**
1c	Lost a big profit opportunity	Moderate	High	High
2a	Livestock cannot be sold (dead / sick)	High	High	High
3c	Consumers of XYZ Livestock Group are declining	High	Moderate	High
1a	Loss of consumer confidence	Low	High	Moderate
1b	XYZ Livestock Group 's profit has dropped dramatically	Moderate	Moderate	Moderate
3b	XYZ Livestock Group 's profit margin decreased	Moderate	Moderate	Moderate
4a	Losses incurred by XYZ Livestock Group	Low	High	Moderate
5a	Lost or changed data on the system	Moderate	Moderate	Moderate
5b	Cessation of the process on the XYZ Livestock Group database	Low	High	Moderate
2b	Loss of XYZ Livestock Group assets	Low	Moderate	Low
2c	The XYZ Livestock Group office cannot operate	Low	Moderate	Low
3a	Lack of public interest in transacting XYZ Livestock Group	Moderate	Low	Low
4b	Hampered business processes XYZ Livestock Group	Moderate	Low	Low

4. Result and Discussion

The results of the risk evaluation analysis show that there are three highest risks with high evaluation matrix values. It is the risk of losing substantial profit opportunities, livestock cannot be sold (dead/sick), and consumers of XYZ Livestock Group are declining. Furthermore, there are six risks in the moderate risk evaluation matrix and four risks in the low-risk evaluation.

At the risk treatment stage, a proposal or mitigation will be identified in the hope that the mitigation undertaken can minimize the likelihood of such risks arising. Besides, mitigation is also carried out to minimize the impact that occurs.

The risk of losing a significant opportunity caused by legal factors is the existence of regulations from the local government to limit production in the livestock industry. In the event, the risk factor is unavoidable. XYZ Livestock Group can make other products that can be conducted together with raising beef cattle and dairy cows. Another product that can be produced is farming with an intercropping system.

The other risk is that livestock cannot be sold (dead / sick) caused by livestock disease threat factors. The completion of the risk involves routinely conducting health checks on beef cattle and dairy cattle belonging to the XYZ Livestock Group, periodic vaccination, and routine cage cleaning. Those are mitigation efforts that can be done to eliminate the source of risk.

Due to the emergence of similar industries, XYZ Livestock Group's consumer has decreased due to certain benefits offered to consumers. In this case, XYZ Livestock Group can reduce its impact by increasing the quality of beef cattle and dairy cattle. Quality improvement can be made by ensuring that raw materials (beef cattle) have superior quality.

The risk of losing consumer trust can be caused by the loss of essential certificates, such as halal certificates. So as to avoid these risks, XYZ Livestock Group can improve quality control of its products. The next risk is that the profits of XYZ Livestock Group dropped dramatically due to government regulations. At this risk, the legal nature of XYZ Livestock Group can do the same thing in point one. Managing risk in the profits of XYZ Livestock Group decreased due to marketing factors, namely the decline in livestock market prices. This risk can be avoided by making new preparations in beef and dairy cattle's initial product, thereby increasing selling prices and profits.

The risk of loss borne by XYZ Livestock Group is due to theft/embezzlement of money/company assets. This risk can be avoided by improving the XYZ Livestock Group's security system and imposing strict sanctions on its employees.

The changed system code causes the risk of data missing or changing on the system on the XYZ Livestock Group database. In this case, the risk can be minimized by backing up data on another storage system to avoid losing or changing data.

The risk of process interruption in the XYZ Livestock Group database is due to the XYZ Livestock Group database overload. This risk can be avoided by increasing the database capacity on the XYZ Livestock Group system.

Risk of loss of assets of XYZ Livestock Group caused by natural disasters or human error. In this case, the risk cannot be avoided, so XYZ Livestock Group must face the risk. However, XYZ Livestock Group can minimize its impact and possibilities by applying for insurance and training.

The risk of the XYZ Livestock Group Office not being able to operate is due to the opposition factor that the community does. The root of the problem of this risk is the existence of river pollution due to XYZ Livestock Group's waste, so that, in avoiding the risk of XYZ Livestock Group, it is necessary to treat the waste before it has flowed into the river. Proper treatment of waste can minimize community opposition

The next risk is the lack of public interest in the XYZ Livestock Group transaction due to the community's inherent traditional shopping patterns. In dealing with this risk, XYZ Livestock Group needs to disseminate information to the public to switch it to modern transactions.
The risk of obstruction in the business process of XYZ Livestock Group is caused by employees who do not comply with SOP. This rick could be handled by providing strict sanctions to its employees.

5. Conclusions and recommendations

5.1. Conclusion

This study successfully identifies risks and create mitigation efforts to reduce the XYZ livestock group's level of risk. By using ISO 31000:2009, this research effectively-identified thirteen risks, which were grouped into three categories. Namely: high risk (red zone), moderate risk (orange zone), low risk (green zone). Red zone risk requires more handling intensity than others. There are three risks in the red zone category: six in the orange zone and four in green categories. This study also succeeded in offering a risk mitigation strategy to reduce the impacts of thirteen identified risks.

5.2. Recommendations
This study succeeded in building a risk map and obtained a proposal to reduce risk at each level. This study's weakness is not calculating the magnitude of the risk costs and not considering the time needed to solve the risk factor. Future studies are suggested to consider with resources are needed to handle the risks, such as money, time, and others.

Acknowledgments
Praise and pray of the authors to Allah SWT, for the blessings and grace. So that, the writer can finish this research smoothly. The author also thanks the XYZ Livestock Group in Cangkringan, Sleman, DI. Yogyakarta for its willingness to be the case study material in this research. A gratitude also be expressed to the Master Program in Industrial Engineering, Universitas Islam Indonesia for supporting this research.

References
[1] Badan Pusat Statistika. Direktori Pasar Tradisional dan Pusat Perbelanjaan 2017. Diakses 13 Februari 2020. Retrieved from https://slemankab.bps.go.id/statictable/2018/10/11/354/banyaknya-ternak-menurut-jenisnya-per-kecamatan-di-kabupaten-sleman-2017.html.
[2] Siswanto F, Rubiyatno R, Dwiatmaka Y. IbM Peternak Sapi Perah dan Pengolahan Susu di Pakem Sleman. ABDIMAS ALTRUIS: *Jurnal Pengabdian Kepada Masyarakat*. Sanata Dharma University; 2018 Apr 30;**01**(01):01–7. doi.org/10.24071/altruis.2018.010101
[3] Wijana, E., P., E. Puluhan sapi di sleman diduga terjangkit virus theileria, ada yang mati. Diakses 13 Februari 2020. Retrieved from https://jogja.suara.com/read/2020/01/16/130558/puluhan-sapi-di-sleman-diduga-terjangkit-virus-theileria-ada-yang-mati.
[4] Ermando, A. Banyak beralih profesi, jumlah peternak dan populasi sapi perah di sleman menurun. Diakses 13 Februari 2020. Retrieved from https://jogja.tribunnews.com/2020/01/23/banyak-beralih-profesi-jumlah-peternak-dan-populasi-sapi-perah-di-sleman-menurun.
[5] Zhang Y, Ju G, Zhan J. Farmers using insurance and cooperatives to manage agricultural risks: A case study of the swine industry in China. *J. Integr. Agric.* Elsevier BV; 2019 Dec;**18**(12):2910–8. doi.org/10.1016/s2095-3119(19)62823-6
[6] Rojas-Downing, M., M., Nejadhasemi, A., P., Harrigan, T., and Wozniki, S., A. 2017. Climate change and livestock: impact, adaption, and mitigation. *Climate Risk Management*. Vol. 16, pp. 145-163.
[7] Masár M, Hudáková M, Šimák L, Brezina D. The current state of project risk management in the transport sector. *Transportation Research Procedia*. Elsevier BV; 2019;**40**:1119–26. doi.org/10.1016/j.trpro.2019.07.156
[8] Spijker E, Anger-Kraavi A, Pollitt H, van de Ven D-J. Evaluating integrated impacts of low-emission transitions in the livestock sector. *Environ. Innov. Soc. Transit.* Elsevier BV; 2020 Jun;**35**:482–92. doi.org/10.1016/j.eist.2019.11.003
[9] Sistiaga M, Herrmann B, Forås E, Frank K, Sunde LM. Prediction of size-dependent risk of salmon smolt (Salmo salar) escape through fish farm nets. *Aquac. Eng.* Elsevier BV; 2020 May;**89**:102061. doi.org/10.1016/j.aquaeng.2020.102061
[10] Domingues MSQ, Baptista ALF, Diogo MT. Engineering complex systems applied to risk management in the mining industry. *Int. J. Min. Sci. Technol.* Elsevier BV; 2017 Jul;**27**(4):611–6. doi.org/10.1016/j.ijmst.2017.05.007
[11] Seño A /EnP. SSD. How sustainable are the Philippines-based housing donor programs? A multi-disciplinary perspective. *Procedia Economics and Finance*. Elsevier BV; 2014;**18**:892–9. doi.org/10.1016/s2212-5671(14)01015-6
[12] Muchfirodin M, Guritno AD, Yuliando H. Supply Chain Risk Management on Tobacco Commodity in Temanggung, Central Java (Case Study at Farmers and Middlemen Level). *Agriculture and Agricultural Science Procedia*. Elsevier BV; 2015;**3**:235–40. doi.org/10.1016/j.aaspro.2015.01.046

[13] Antriyandarti E, Ferichani M, Ani SW. Sustainability of Post-Eruption Socio Economic Recovery for the Community on Mount Merapi Slope through Horticulture Agribusiness Region Development (Case Study in Boyolali District). *Procedia Environmental Sciences*. Elsevier BV; 2013;**17**:46–52. doi.org/10.1016/j.proenv.2013.02.010
[14] Shrum TR, Travis WR, Williams TM, Lih E. Managing climate risks on the ranch with limited drought information. *Clim. Risk Manag.* Elsevier BV; 2018;**20**:11–26. doi.org/10.1016/j.crm.2018.01.002
[15] Nugroho H., S. Pengembangan sumber daya manusia kelompok tani ternak sapi manunggal; 2018
[16] Meilani, T. Penerapan iso 31000 dalam pengelolaan risiko pada bank perkreditasn rakyat (studi kasus bank perkreditan rakyat x). *Jurnal Administrasi Bisnis*; 2014;**10**(1):17-32

Lean manufacturing with multilevel Value Stream Mapping

Y Prasetyawan[1,2], **B A Ramadhan**[1,3] **and L Salsabila**[1,4]

[1]Departement of Industrial Engineering, Institut Teknologi Sepuluh Nopember, Kampus ITS Sukolilo – Surabaya 60111, Indonesia

Email: [2]yudhaprase@gmail.com, [3]abdurrahmanbagas@gmail.com, [4]bila.latifah@gmail.com

Abstract. The lean method is a process that helps the companies to produce more products by minimizing time of the process, capital, and other resources. VSM is a tool which is based on the lean manufacturing that used to identify the current scenario of the process in the company to solve the problems and fix the future scenario. As VSM identifies all the process steps, the value added, and non-value-added activities are analysed using this tool to help see the source of waste and the hidden waste. The existing VSM is limited to explain the production process in general and it is less detailed, so the improvements that made are limited to the level of the production line and not every production process. The purpose of this paper is to develop a dynamic mapping method to enable a more detailed analysis of lean problems by making value stream mapping into several levels (multilevel).

1. Introduction

The production process in manufacturing system requires cost reduction to maximize the profit. This time, it is one of the key problems that the companies must face. One of the ways to do the cost reduction is using lean method. The lean method is a process that helps the companies to produce more products by minimizing time of the process, capital, and other resources [1]. Lean is a systematic process that aims to fulfil demands and satisfy performance from the customer without reducing quality by using method that is efficient, optimize costs, and optimize delivery times for making the finish product [2] [3]. To achieve the goals, it is better to act based on the three sources of inefficiency, they are: waste, flexibility, and variability [3]. The measure of the lean level in the company can be seen using a tool called Value Stream Mapping (VSM), it is a method used to identify wastes and improve performance [4]. Not all problems can be easily identified using VSM because the analysis only considers the static aspect of the system and oversimplify problems that exist in manufacturing system, while manufacturing system is influenced by various dynamic problems of the production process and it includes high complexity [5]. Therefore, a method by considering the dynamic system must be applied. In this paper the authors want to develop VSM which considers static aspect to be dynamic based on Quality Function Deployment (QFD) framework with several levels, not based on the QFD matrix but from the dynamic aspects that shown in QFD.

VSM is a tool which is based on the lean manufacturing that used to identify the current scenario of the process in the company to solve the problems and fix the future scenario. VSM focused to reorganizing all of the manufacturing system with lean perspective. It described by information flow that includes all the process from customer order, manufacturing process, and delivers the product to the customer [6]. As VSM identifies all the process steps, the value added, and non-value-added activities are analysed using this tools to help see the source of waste and the hidden waste [7]. VSM can brings out the positive result on process ratio, takt time, process tine and inventory level, total lead time, and reduced the labour power [8]. Not all of the problems can be easily solved by VSM, because it was able to display a static production flow and only suitable for representing production lines with one product or a product family [9]. It is difficult for applying VSM to complex product flow conditions. QFD is a method that translate customer requirements and manage it into the final product [10]. QFD discusses customer demand to improve product quality to match the voice of customer as a whole, which contains several multilevel matrices that are product planning matrix, product design matrix, process design matrix and process control matrix [11].

A publication presents QFD integrated with VSM that used for enabling identification of wastes and tools to eliminate them. The QFD framework has been used for identified scientific prioritization of wastes from the current VSM and techniques to eliminate them, then the result can be implemented for eliminating the wastes that visualized with future VSM [12]. Value stream mapping developed into a dynamic concept that taking both recent digitalization technologies and organization structures into account. The goals are developing static method into DVMM (Dynamic Value Stream Management) as a feedback to the changing status of challenges in the production line in manufacturing system. The concept of the DVMM framework is designed with two dimensions: technical infrastructure facing industry 4.0 and management approach that describe relevant day-a-day activities [13]. The use of value stream mapping with other methods for lean manufacturing has been done in research before, one of them is integration of value stream mapping with dynamic systems in modelling and analysing manufacturing lines [14]. The purpose of the method is to improve waste identification and eliminate it. The method has the main stages for identification waste, the first is creating a value stream mapping as in general, then making a value stream analysis, after which it is simulated using a scenario from a dynamic system analysis. The main purpose of value stream mapping and value stream analysis is to identify the problems and waste, while dynamic system analysis is make it possible to understand how the problem can affect the other system such as inventory levels that change in time and how that it can affect the amount of production and client satisfaction.

The purpose of this research is to develop a dynamic mapping method to enable a more detailed analysis of lean problems by making value stream mapping into several levels (multilevel). There are three levels of multilevel VSM namely previous VSM, current VSM, and future VSM which dynamically shows the improvement of the company's production process.

2. Method

In order to be able to see the detail of wastes that exists in the production process, the authors develop VSM into several levels (multilevel). There are three level of the VSM: (1) level 1 is making detailed VSM for each process that occurs in the production line; (2) level 2 is making previous, current, and future scenarios of VSM for one production line (not each process) to see the improvement scenarios that have been made in a structured way by showing the impact of what would happen if a small change in production process was made; then (3) level 3 is a combination of level 1 and level 2 by combining the detailed process with each scenario (previous, current, future). To apply multilevel VSM, the authors takes an example of a case study in industrial and building paint company [6]. Below is the result of the multilevel VSM for the case study that has been modified by the authors.

2.1. VSM Level 1

The production process consists of first mixing process, big mixer process, testing, and packaging. There are some wastes for each process that is rarely known if we make direct VSM for one production line, therefore VSM can be made more detailed for each process or in this case referred as VSM level 1. The VSM has shown the details of processes that occur within a process, so detailed wastes can be identified. Below is the example of VSM level 1 in first mixing process and big mixer process. VSM we use as example is from [15].

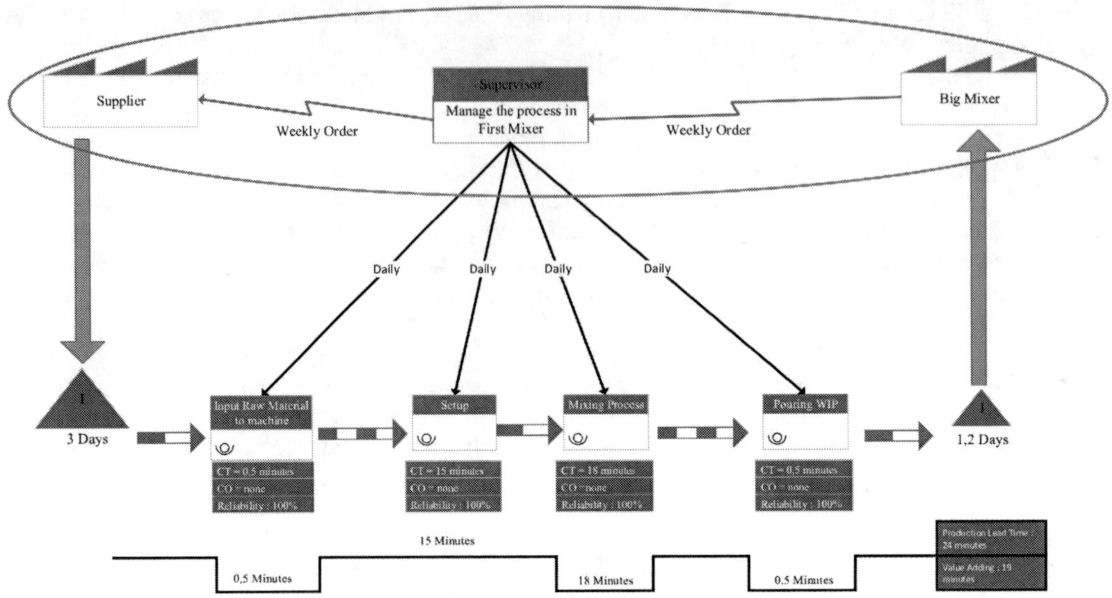

Figure 1. VSM level 1 process 2 (first mixer).

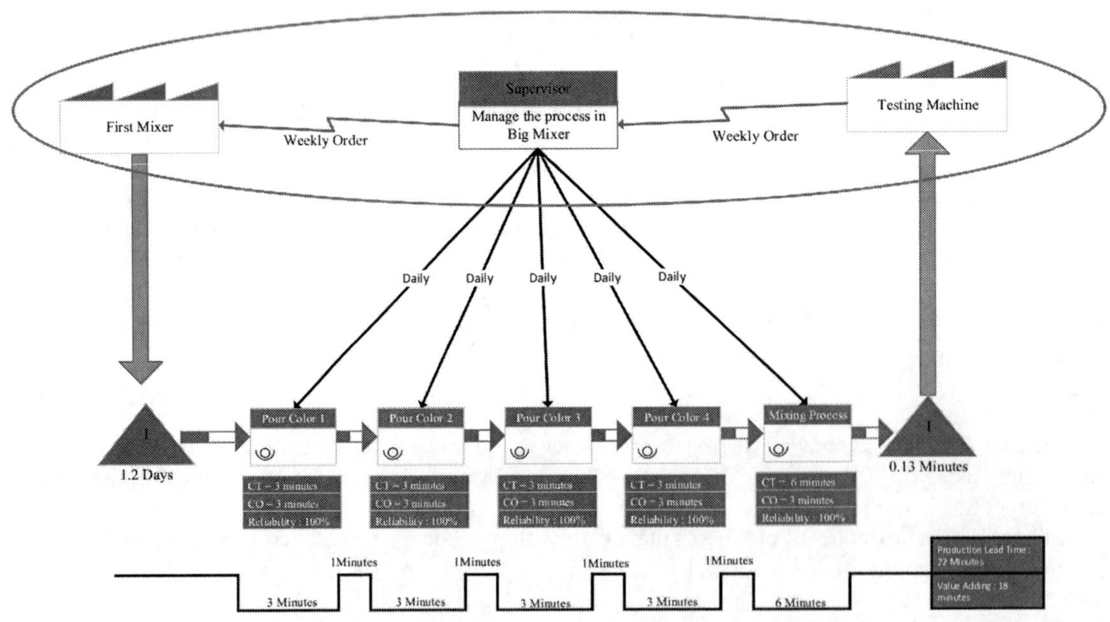

Figure 2. VSM level 1 process 2 (big mixer).

2.2. VSM Level 2

This level illustrates previous, current, and future scenarios of VSM for one production line. By making VSM for each production process at level 1, it can be identified waste that occurs in each production process. The wastes can be reduced using several tools, for example: the use of Kanban, single minute exchange dies (SMED), and all available lean manufacturing tools. At level 2, the previous VSM is shown before the improvement process is performed. After the company makes improvements by eliminating waste, current VSM is made to describe the company's current condition. Then the company makes another improvement to become future VSM.

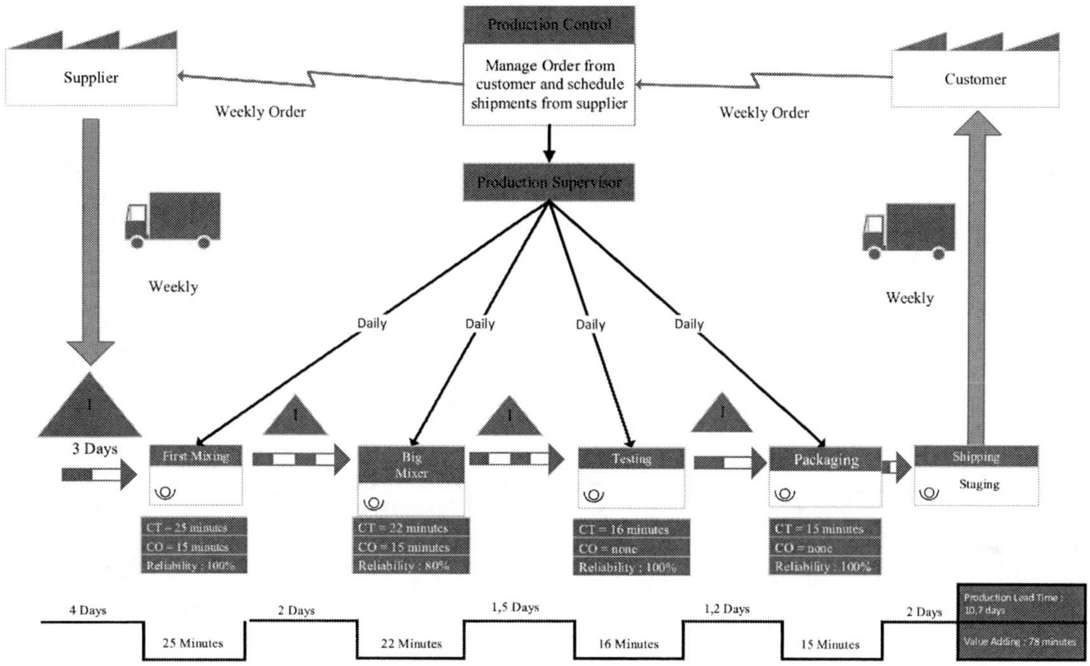

Figure 3. VSM level 2 (previous VSM).

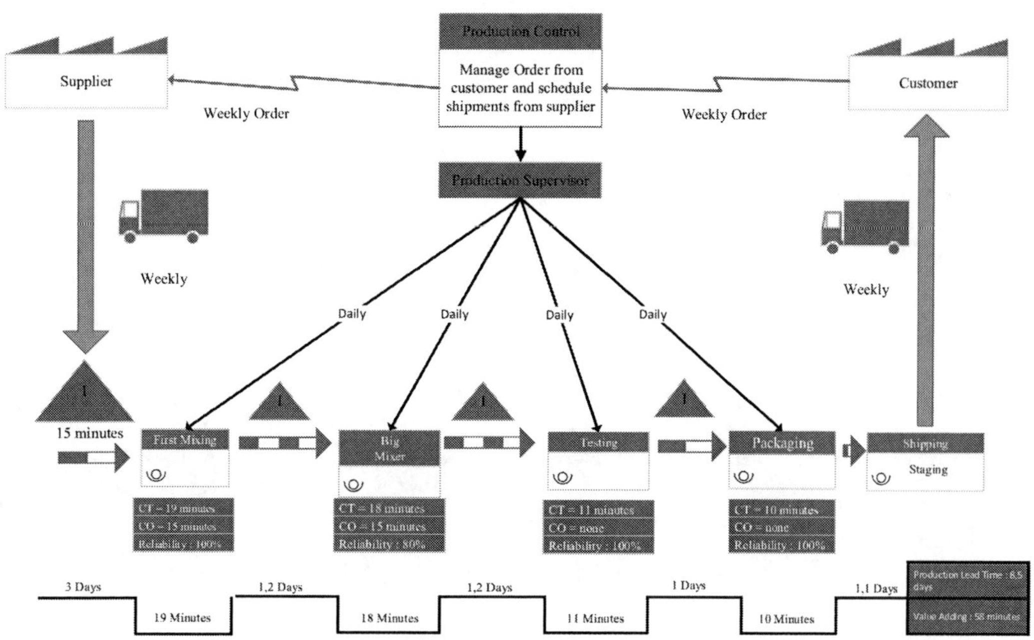

Figure 4. VSM level 2 (current VSM).

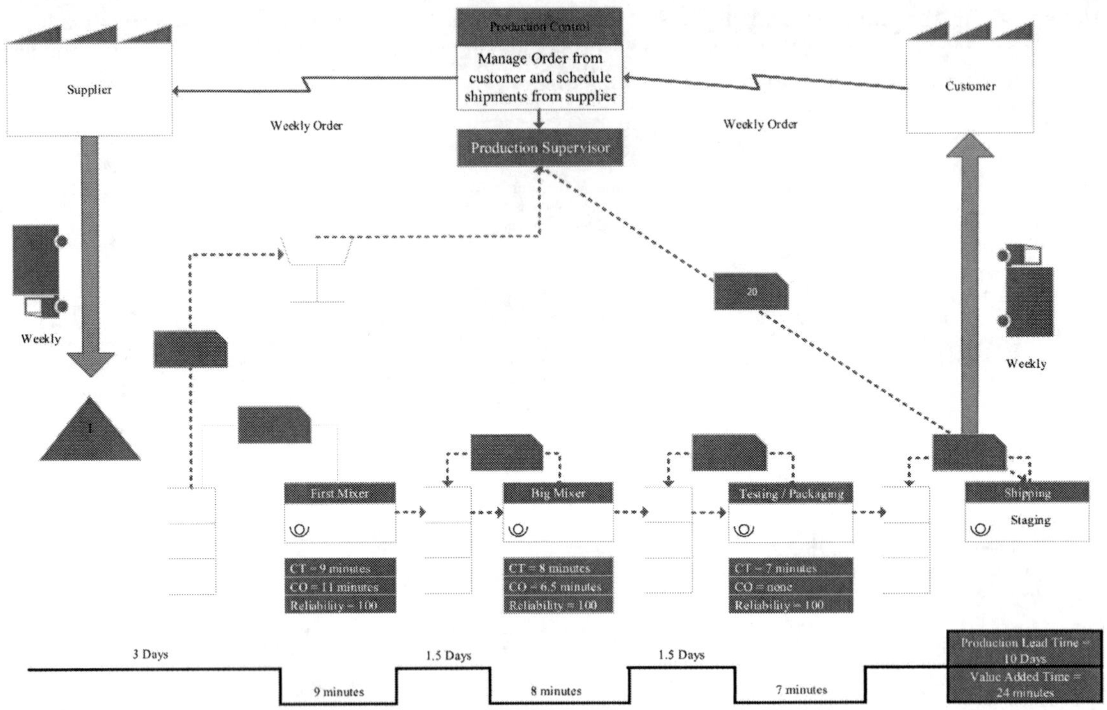

Figure 5. VSM level 2 (future VSM).

2.3. VSM Level 3

VSM level 3 is a combination of level 1 and level 2 where the previous, current, and future VSM are described in detail for each process, so the VSM can be seen in detail what process that occur, and it wastes. Figure 6, Figure 7, and Figure 8 show VSM level 3 for previous, current, and future VSM.

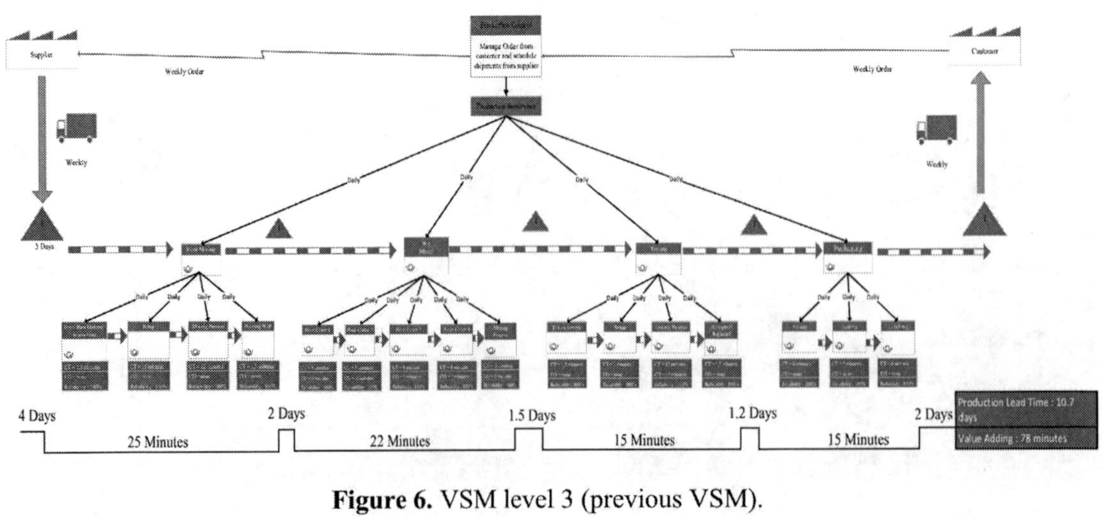

Figure 6. VSM level 3 (previous VSM).

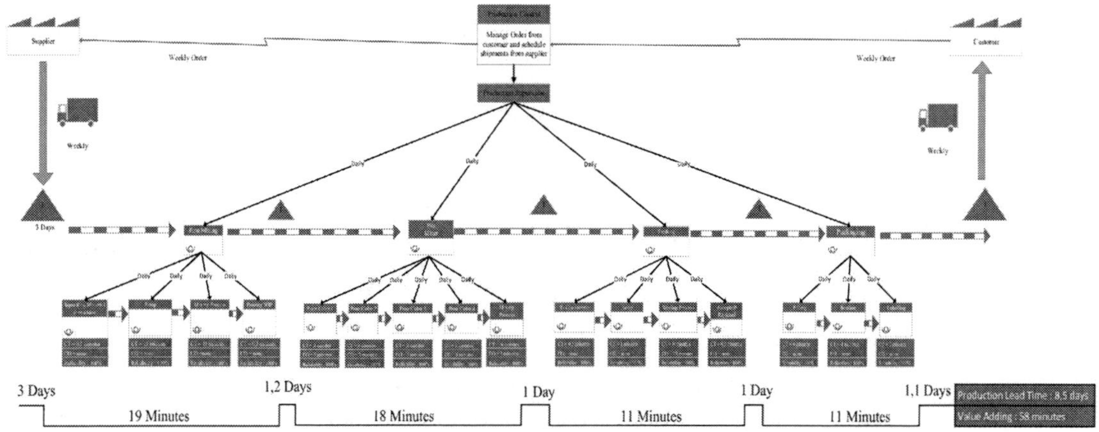

Figure 7. VSM level 3 (current VSM).

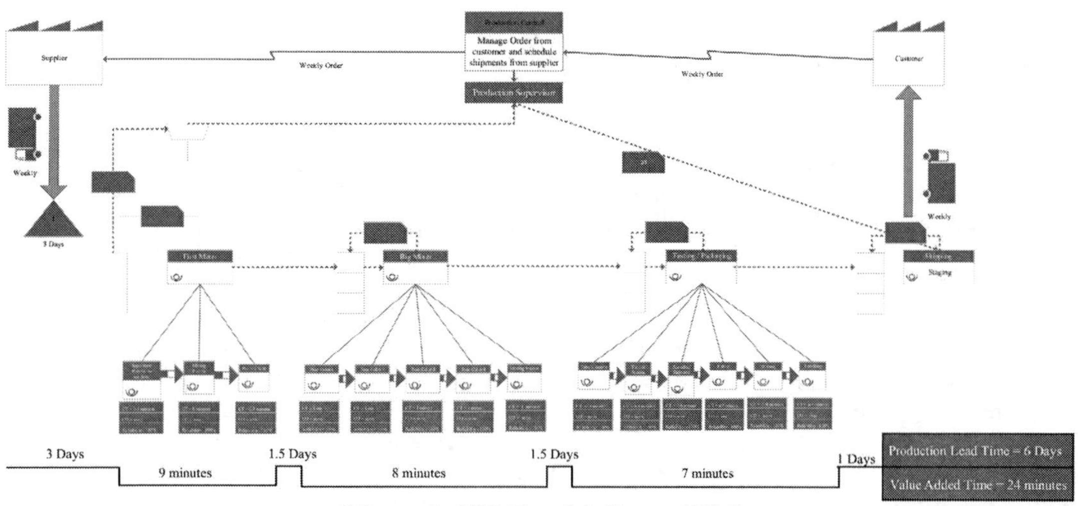

Figure 8. VSM level 3 (future VSM).

By describing VSM up to level 3, the company can find out where the waste occurs in more detail so the elimination of the wastes can be right on target. It is not the same as the existing method of VSM that only explains the production process in general and it is less detailed, so the improvements that are made only at the level of the production line and not every production process.

3. Result and Discussion

After making the three levels of Value Stream Mapping, it can be performed an analysis of each operation that occurs in the production line to determine the amount of Value Added (VA), Non-Value Added (NVA), and Necessary but Non-Value Added (NNVA). Table 1 shows an analysis of total activity in previous VSM, current VSM, and future VSM.

It can be seen in Table 1 that if we use existing VSM in general, we can only know the total cycle time and total lead time, more detailed activities cannot be analysed in general VSM. But by using Multilevel VSM, it can be identified the activities that occur in each process in its production line in order to reduce the number of Non-Value Added activities. From Table 1 is known that from previous VSM to current VSM had reduced lead tine from 10.7 days to 8.5 days and the cycle time from 78 minutes into, also current VSM to future VSM had reduced lead time from 8.5 days to 6 days and the cycle time had reduced from 58 minutes into 24 minutes. From that fact there has been an improvement in current VSM abd there has been improvement again in Future VSM, it can be seen that the most visible reduction is in non-value added, total cycle time, and lead time.

Table 1. Total activity in previous VSM, current VSM and future VSM.

Previous VSM			Current VSM			Future VSM		
Activity	**Qty**	**Total Time**	**Activity**	**Qty**	**Total Time**	**Activity**	**Qty**	**Total Time**
Operation	10	61.5	Operation	10	46.5	Operation	10	18
Transportation	3	1.5	Transportation	3	1.5	Transportation	3	1.5
Inspection	1	15	Inspection	1	10	Inspection	1	4.5
Storage	5	10.7 days	Storage	5	8.5 days	Storage	3	6 days
Delay	2	25	Delay	2	25	Delay	0	-
VA	7	158.72	VA	7	158.72	VA	7	158.72
NVA	6	11.5	NVA	6	11.5	NVA	3	11.5
NNVA	8	36.51	NNVA	8	36.51	NNVA	7	36.51
Total CT		78	Total CT		58	Total CT		24
Total LT		10.7 days	Total LT		8.5 days	Total LT		6 days

4. Conclusion

VSM is a tool which is based on the lean manufacturing that used to identify the current scenario of the process in the company to solve the problems and fix the future scenario. Multilevel VSM is divided into three levels and multilevel VSM is developed to identify wastes for each process that occurs in the production line, so the waste elimination can be used on time especially for reducing non-value added activity. Level 1 is making detailed VSM for each process that occurs in the production line. Level 2 is making previous, current, and future scenarios of VSM. The previous VSM is shown before the improvement process is performed. After the company makes improvements by eliminating waste, current VSM is made to describe the company's current condition. Then the company makes another improvement to become future VSM. Then level 3 is a combination of level not as the same as existing and level 2 by combining the detailed process with each scenario (previous, current, future). It is not one method of VSM that only explains the production process in general and it is less detailed, so the improvements that are made only at the level of the production line and not every production process.

5. References

[1] B. Dighe, A. Kakirde, Lean manufacturing implementation using value stream mapping: a case study of pumps, Int. J. Sci. Res. (IJSR) **3**. 2014; 2319–7064

[2] Deshkar A, Kamle S, Giri J, Korde V. Design and evaluation of a Lean Manufacturing framework using Value Stream Mapping (VSM) for a plastic bag manufacturing unit. *Materials Today: Proceedings*. Elsevier BV; 2018;**5**(2):7668–77. doi.org/10.1016/j.matpr.2017.11.442

[3] El Kihel Y, Amrani A, Ducq Y, Amegouz D. Implementation of Lean through VSM modeling on the distribution chain: Automotive case. *International Colloquium on Logistics and Supply Chain Management (LOGISTIQUA)*. IEEE; 2019 Jun. doi.org/10.1109/logistiqua.2019.8907282

[4] Lacerda AP, Xambre AR, Alvelos HM. Applying Value Stream Mapping to eliminate waste: a case study of an original equipment manufacturer for the automotive industry. *Int. J. Prod. Res.* Informa UK Limited; 2015 Jun 18;**54**(6):1708–20. doi.org/10.1080/00207543.2015.1055349

[5] Lian Y-H, Van Landeghem H. Analysing the effects of Lean manufacturing using a value stream mapping-based simulation generator. *Int. J. Prod. Res.* Informa UK Limited; 2007 Jul;**45**(13):3037–58. doi.org/10.1080/00207540600791590

[6] Masuti PM, Dabade UA. Lean manufacturing implementation using value stream mapping at excavator manufacturing company. *Materials Today: Proceedings*. Elsevier BV; 2019;19:606–10. doi.org/10.1016/j.matpr.2019.07.740
[7] Rahani AR, al-Ashraf M. Production Flow Analysis through Value Stream Mapping: A Lean Manufacturing Process Case Study. *Procedia Engineering*. Elsevier BV; 2012;**41**:1727–34. doi.org/10.1016/j.proeng.2012.07.375
[8] Vamsi Krishna Jasti, N., Sharma, A. Lean manufacturing implementation using value stream mapping as a tool. *Int. J. Lean Six Sigma*; 2014;**5**:89–116. DOI : 10.1108/ijlss-04-2012-0002
[9] Gurumurthy A, Kodali R. Design of lean manufacturing systems using value stream mapping with simulation. *J. Manuf. Technol. Manag.* Emerald; 2011 May 3;**22**(4):444–73. doi.org/10.1108/17410381111126409
[10] Apparao K C, Birru A K. QFD-Taguchi based hybrid approach in die casting process optimization. *Transactions of Nonferrous Metals Society of China*. Elsevier BV; 2017 Nov;**27**(11):2345–56. doi.org/10.1016/s1003-6326(17)60260-7
[11] Karimi Gavareshki MH, Abbasi M, Rostamkhani R. Application of QFD & VE & lean approach for control tests in a product design. *Arch. Mater. Sci. Eng.* Index Copernicus; 2017 Apr 2;**84**(2):65–78. doi.org/10.5604/01.3001.0010.0980
[12] Mohanraj R, Sakthivel M, Vinodh S. QFD integrated value stream mapping: an enabler of lean manufacturing. *Int. J. Product. Qual. Manag.* Inderscience Publishers; 2011;**7**(4):501. doi.org/10.1504/ijpqm.2011.040546
[13] Lugert A, Völker K, Winkler H. Dynamization of Value Stream Management by technical and managerial approach. *Procedia CIRP*. Elsevier BV; 2018;**72**:701–6. doi.org/10.1016/j.procir.2018.03.284
[14] Stadnicka D, Litwin P. Value stream mapping and system dynamics integration for manufacturing line modelling and analysis. *Int. J. Prod. Econ.* Elsevier BV; 2019 Feb;**208**:400–11. doi.org/10.1016/j.ijpe.2018.12.011
[15] Rohani JM, Zahraee SM. Production Line Analysis via Value Stream Mapping: A Lean Manufacturing Process of Color Industry. *Procedia. Manuf.* Elsevier BV; 2015;**2**:6–10. doi.org/10.1016/j.promfg.2015.07.002

Proposing predictive maintenance strategy to increase OEE through system upgrade scenarios and AHP

Y Prasetyawan[1, 2], **I Rachmayanti** [1, 3]

[1]Department of Industrial Engineering, Institut Teknologi Sepuluh Nopember, Kampus ITS Sukolilo-Surabaya 60111, Indonesia

E-mail: [2]yudhaprase@gmail.com, [3]irindria@gmail.com

Abstract. Maximizing the value of OEE is a crucial thing to do for a company in order to increase their productivity. OEE is one of the most common technique to measure equipment's effectiveness. To enhance the number of OEE, it is very important to do a maintenance task to keep the equipment in a good condition. Predictive maintenance task is one of the maintenance strategies to see any potential failure in an equipment and do a preventive way to prevent it from a total stop/break down. As a result, understanding the needs to do a maintenance upgrades for better performance and proposing a good predictive maintenance strategy will enhance the value of OEE and result in cost savings. The objective of this paper is to propose the necessary predictive maintenance task using analytical hierarchy process (AHP) using three respondents to decide the AHP process and consistency ratio to get the most suitable maintenance task for fertilizer company as small medium enterprise.

1. Introduction

In the current era of globalization where there are no boundaries between nations in various matters throughout the world including competition in the economic and business, manufacturing sector is one of the most crucial field in helping to improve the economic growth of a country. Based on this fact, many companies are competing to be able to provide the best manufacturing performance and continuous improvement by maximizing the effectiveness of the manufacturing process [1]. Companies are starting to look for some techniques in order to enhance the excellence of the manufacturing process and improving the overall performance. Thus, it will help reducing the cost to gain more profit to the company. One of the things that needs to be considered in improving the effectiveness of the manufacturing process is maximizing the utilization of the equipment. The most common technique to measure the effectiveness of equipment is Overall Equipment Effectiveness (OEE) which is a part of Total Productive Maintenance (TPM) [2]. TPM strives to improve productivity and reduces losses by doing an appropriate maintenance [3]. OEE is a calculation tool to provide information about the process using multiple data and issues in manufacturing. OEE delivers a framework to improve manufacturing process in a company. Efficiencies can be improved by using the six big losses as in OEE concept. The Component in OEE are Availability, Performance Rate and Quality [4]. By improving quality product, time and efficiencies will help to gain customer's satisfaction [5].

The other things to be considered in maximizing the overall equipment effectiveness is by doing maintenance task. Unplanned downtime in a machine can cause delays and reduce the productivity of an equipment. Thus, it will also reduce the OEE value. One of the maintenance actions that can be applied is predictive maintenance. Predictive Maintenance is a maintenance task to see the potential problem of an equipment by looking at the historical data and analysing the pattern. By detecting the potential failure earlier, it can also reduce the cost for maintenance.

By knowing the performance issues, the predictive maintenance strategy can be formed by integrating all the system and doing scenario upgrades for better improvement in enhancing the value of OEE and predicting how long the strategy will last.

Fertilizer company is one of the most important part of Indonesia as an agricultural country. It takes part to help farmer on providing the best food for the society. Based on the data of the fertilizer producer association, the need of fertilizer increased 5% along 2018 for about 6,27 million from the initial number 5,97 million in 2017. It is necessary to keep the production going with minor interruption. Maintenance task could be a very important and beneficial role to increase machine's ability and enhance the number of OEE. Maintenance strategy for such a big fertilizer company maybe already good, but how to choose a good maintenance task which fit with the machine's characteristic for small medium fertilizer companies? Most of them still not familiar with the expert maintenance strategy by extending the maintenance activities or doing the cost reduction for breakdown maintenance.

The objective of this paper is to propose a predictive maintenance strategy in order to enhance the value of overall equipment effectiveness using system upgrade scenarios and analytical hierarchy process (AHP) method to select some maintenance alternatives to be implemented in a company based on some important criteria. AHP is a method used to make an order of decision and alternatives to get the best decision. The main input of AHP is human perception. AHP would be beneficial to know which task is better than another task by using expert judgment and consistency ratio calculation. It's easier for small medium fertilizer company to be applied rather than focusing on detecting maintenance task using reliability cantered maintenance (RCM), the well-known method for maintenance. Since there is no any quantitative procedure in AHP, it is very necessary to choose an expert for doing the AHP process to make it more relevant. AHP is commonly used in any sector since it is not a new method in industry.

2. Methods

As stated previously, the method used in making this paper is by analysing the upgrades needed by understanding the preparation for upgrades scenarios, then, mapping the predictive maintenance task and linked it with six major losses in Total Productive Maintenance (TPM) concept. By using Analytical Hierarchy Process (AHP) as the multicriteria decision tools, we try to propose the best predictive maintenance task importance priority for company. These two methods is not a new method, but we combine of two existing method.

2.1. Analysing Upgrade Scenarios

An upgrade is necessary for a company especially if it affect the productivity. The first thing to do is understanding the reason to do an upgrade and why is it necessary to be applied to the company. For this step, a focus group discussion involving all the stakeholder can be taken as an alternative. Planning the upgrade strategy and scenarios for the company should be depend on the business case. In this paper, it is about the performance issues and in order to enhance the OEE value of an equipment. The strategy should be made after we analyse the benefit of implementing updates by comparing the cost for upgrade and its profit for the company. The next phase is determining the readiness and the path for upgrade and scope the project. This paper aim to scope the project on predictive maintenance task upgrade by selecting some necessary task to minimize the cost but still fulfilling all the needs for predictive maintenance. Since it's also a qualitative research, the respondent must be someone who really takes part and involve in maintenance activity of the company. It should be a head of maintenance division, senior field staff of maintenance, head of production division and the machine's operator. They are responsible for the production and maintenance. The combination for these four respondents will help us to use the AHP to choose the best maintenance task to enhance the production. These all four experts who can make this research become more relevant since it's a qualitative research. Thus, the combination of technical procedure and expert judgment will give the expected optimum result of cost reduction and maximizing performance.

2.2. Mapping the Predictive Maintenance Task

Mapping the predictive maintenance task by understanding its importance and involving all stakeholder is one of the most crucial things to do. The predictive maintenance task should consist of plan, do, check, and action concept. Action proposed in predictive maintenance task shows in Table 1.

IOP Conf. Series: Materials Science and Engineering 1072 (2021) 012031 doi:10.1088/1757-899X/1072/1/012031

Table 1. Predictive maintenance task & maintenance task cost in a fertilizer company.

No	Predictive Maintenance Task	Cost Spend/year
1	Vibration monitoring	$ 1,500
2	Thermographic inspection	$ 1,200
3	Oil analysis	$ 3,000
4	Visual inspection	-
5	Shock pulse	$ 2,700
6	Ultrasonic leak detector	$ 1,700
7	Electrical insulation	$ 2,900
8	Performance testing	$ 3,200
9	Wear and dimensional measurement	$ 700
10	Signature analysis, time, and frequency domain	$ 600
11	Non-destructive testing by using ultrasonic	$ 800

By knowing the predictive maintenance task, we can link it with six major losses in TPM based on the value of OEE. Table 2 shows the relation between maintenance task and their effect on six big losses. It shows whether the activity give big impact to the losses or not. The six big losses including unplanned stops, planned stops, small stops, slow cycle, production reject and start-up reject [6]. Each loses represent the element for overall equipment effectiveness which are availability loss, performance loss, and quality loss.

Table 2. OEE components.

Overall Equipment Effectiveness	Six Big Losses	Conventional Six Big Losses
Availability Loss	Unplanned Shutdown	Machine Failure
	Planned Shutdown	Machine Adjustments
Performance Loss	Idle Time	Machine Idle
	Slow Speed	Speed Reducing
Quality Loss	Product Rejection	Product Defect
	Start-up Rejection	Yield Reducing
OEE	Total Productive Time	Valuable Operational Time

Source: Nakajima, 1990

2.3. Using Analytical Hierarchy Process

The analytic hierarchy method (AHP), is a powerful multicriteria decision-making tool. It was developed by Saaty and has been applied in many field including economics, engineering and even politics [7]. This method will help the decision maker to assign some value/weight which represent preferences of some given alternatives so that one best possible alternative will be chosen as the most weighted alternatives. This method allows decision maker to make some decisions in an objective way based on the decision maker point of view by the appropriate ranking for each alternative with some criteria. The importance level of each criteria is not the same. Thus, the next step to do is proposing the relative weight for all criteria [8]. Using paired comparison, the decision maker will be able to express their preference for each criterion [9]. The AHP process started by forming the hierarchy in form of tree and connecting some decision element with other component such as main objectives and criteria of alternatives solution [10]. The next phase is doing the comparison between the criteria importance and variants of decision element [10]. Table 3 explain the dependencies between criteria, and explanation for each scale shows in Table 4.

Table 3. Linked between maintenance task and six big losses.

	Maintenance Task	Six Big Losses					
		Equipment failure	**Setup and Adjustment**	**Idling and Minor Stop**	**Reduced speed**	**Process defect**	**Reduces yield**
1	Vibration monitoring	V		V		V	
2	Thermographic inspection	V		V			V
3	Oil analysis	V	V	V			
4	Visual inspection	V	V	V		V	
5	Shock pulse	V		V			
6	Ultrasonic leak detector	V		V			
7	Electrical insulation	V		V			
8	Performance testing	V	V	V	V	V	V
9	Wear and dimensional measurement		V		V		V
10	Signature analysis, time, and frequency domain	V		V	V	V	
11	Non-destructive testing by using ultrasonic	V		V			

Table 4. Explanation for each scale.

Scales	Degree of Preferences	Explanation
1	Equally	Two activities contribute equally to the objectives
3	Moderately	Experience and judgement slightly to moderately favour one activity over another
5	Strongly	Experience and judgement strongly or essentially favour one activity over another
7	Very Strongly	An activity is strongly favoured over another and its dominance is showed in practise
9	Extremely	The evidence of favouring one activity over another is of the highest degree of possible of an affirmation
2,4,6,8	Intermediate Values	Used to represent compromises between the preferences in weights 1, 3, 5, 7, 9
Reciprocals	Opposite	Used for inverse comparison

Since this method contains high subjectivity, it is necessary to count Consistency Ration (CR) using this formula

$$CR = \frac{CI}{RI} \tag{1}$$

Where

$$CI = \frac{\lambda max - n}{n-1} \tag{2}$$

While RI represents parameter based on the number of matrix element. If the number of CR is less than 0,1 or equal to 10% of average inconsistency, the judgement is usually acceptable. If it is more than 0.1 or 10%, the judgement should be revised [11]. Table 5 shows weights of the criteria based on pairwise comparison.

Table 5. AHP results.

No	Predictive Maintenance Task	Priority	Rank
1	Vibration monitoring	0.117	3
2	Thermographic inspection	0.1	5
3	Oil analysis	0.119	2
4	Visual inspection	0.6	10
5	Shock pulse	0.104	4
6	Ultrasonic leak detector	0.098	7
7	Electrical insulation	0.099	6
8	Performance testing	0.126	1
9	Wear and dimensional measurement	0.062	9
10	Signature analysis, time, and frequency domain	0.065	8
11	Non-destructive testing by using ultrasonic	0.005	11

*Consistency ratio: 9.3%

Depend on the comparison above, the consistency ratio result is 9.3%. It means that the judgement made is acceptable because it is still under 10%.

3. Result and Discussion

Doing maintenance task is a very important things to do for a company. But there are several maintenance tasks which is not really relevant to detect a problem in an equipment. Besides, it is sometimes too expensive especially for a small and medium enterprises which still trying to survive in market. After mapping the maintenance task and find the preferences of each alternatives, we found several maintenance tasks which is necessary and give high impact to detect the equipment's problem. Based on the analytical hierarchy process, we found that the maintenance task with the highest rank is performance testing. It is shown on the table. that it can affect all the six big losses. By doing performance testing we can understand clearly whether the equipment is still in a good condition or not. This task represents all testing for the equipment. If the performance decrease, it should be that the equipment is on trouble and decrease on its reliability. The company should choose the best and most relevant maintenance task.

In the table above, we can see the rank of each predictive maintenance task. The three highest position are doing performance testing, oil analysis and vibration analysis. This is the cost need supposed if we want to do an upgrade by implementing just the five highest rank.

Implementing only several maintenance task which represent the whole maintenance for an equipment will able to save the cost for maintenance and also enhance the OEE value. As we know this task already suit the predictive maintenance strategy which is scheduled on condition and failure finding task.

Table 6. Maintenance task rank based on pairwise comparison.

No	Maintenance Task	Rank
1	Performance tasting	1
2	Oil analysis	2
3	Vibration analysis	3
4	Shock pulse	4
5	Thermographic inspection	5
6	Electrical insulation	6
7	Ultrasonic leak detector	7
8	Signature analysis, time, and frequency domain	8
9	Wear and dimensional measurement	9
10	Visual inspection	10
11	Non-destructive testing by using ultrasonic	11

Table 7. Cost saving after AHP.

No	Maintenance Task	Cost Spend/Year
1	Performance testing	$ 3,200
2	Oil analysis	$ 3,000
3	Vibration analysis	$ 1,500
4	Shock pulse	$ 2,700
5	Thermographic Inspection	$ 1,200

Table 8. Cost comparison before and after the upgrades.

Cost Before Upgrade	Cost After Upgrade	Cost Savings
$ 18,300	$ 11,600	$ 6,700

Doing an upgrade can spend around $6,700 each year for the company's maintenance task. Since we got the top five maintenance task which is relatable with the machine's characteristic based on expert judgement using AHP. As we seen before the connection between each maintenance task with the OEE component, we justify that the number of every losses will be minimum. This maintenance task is enough to prevent systems from unplanned breakdown and enhance the number of system's availability. Thus, we hope by increasing the number of availabilities, it will also enhance the number of OEE for the system. Since by the formula of OEE we can see that there's three factors affecting OEE which are availability, performance and quality. We hope by providing this simple model will be helpful and can be applied for small medium enterprises to enhance their capability in minimizing the unplanned breakdown.

4. Conclusion

Based on the result of AHP, it is shown that performance testing is the most important task to do in predictive maintenance. Implementing upgrades and proposing necessary predictive maintenance strategy can save cost of maintenance around $ 6,700 each year for a company and still helping in predicting potential failure in order to enhance the value of OEE. But it's necessary to choose the most suitable respondent to make it relevant. Due to the lack of information, the real implementation and experiment cannot be taken because it will consume much time and cost.

5. References

[1] Meddaoui A. A New Manufacturing Improvement Model Based on Overall Equipment Effectiveness and Lean Maintenance. *International Journal of Engineering Management*. Science Publishing Group; 2018;**2**(2):15. doi.org/10.11648/j.ijem.20180202.11

[2] Nakajima,S. Introduction to TPM: Total Productive Maintenance. *Cambridge, MA, Productivity Press.* 1988

[3] Agustiady TK. Total Productive Maintenance. *CRC Press*; 2016 Feb 3; doi.org/10.1201/b18641

[4] Almeanazel, O.T.R. 2010. Total Productive Maintenance Review and Overall Equipment Effectiveness Measurement. *Jordan Journal of Mechanical and Industrial Engineering*,Vol. 4, No. 4. ISSN 1995-6665, Department of Industrial Engineering, Hashemite University, Zarqa, Jordan.

[5] Díaz-Reza JR, García-Alcaraz JL, Martínez-Loya V. Impact Analysis of Total Productive Maintenance. *Springer International Publishing*; 2019; doi.org/10.1007/978-3-030-01725-5

[6] Reichhart A, Holweg M. Lean distribution: concepts, contributions, conflicts. *Int. J. Prod. Res.* Informa UK Limited; 2007 Aug 15;**45**(16):3699–722. doi.org/10.1080/00207540701223576

[7] Cabala P., Using the Analytic Hierarchy Process in Evaluating Decision Alternatives. Operations Research and Decisions, 2010;**1**:5- 23.
[8] Mu, E. , Milagros M.R. 2017. Pratical Decision Making : An Introduction to the Analytic Hierarchy Process (AHP) Using Super Decision v2. *Springer Nature*
[9] Saaty, T.L. Deriving The AHP 1-9 Scale From First Principles. *J. ISAHP*. 2001
[10] Balubaid M, Alamoudi R. Application of the Analytical Hierarchy Process (AHP) to Multi-Criteria Analysis for Contractor Selection. *American Journal of Industrial and Business Management*. Scientific Research Publishing, Inc.; 2015;**5**(9):581–9. doi.org/10.4236/ajibm.2015.59058
[11] Leal JE. AHP-express: A simplified version of the analytical hierarchy process method. *MethodsX*. Elsevier BV; 2020;7:100748. doi.org/10.1016/j.mex.2019.11.021

Workstation and posture improvement in cutting machine process using virtual modelling

L Studiyanti[1], W Septiani[1,3], and N Aulia[2]

[1]Universitas Trisakti, West Jakarta, Indonesia
[2]Graduated from Univesitas Trisakti, West Jakarta, Indonesia

Email: [3]winnie.septiani@trisakti.ac.id

Abstract. The activity of squatting and bending can cause an awkward working body posture and the risk of Musculoskeletal Disorders (MSD). This is mainly happening in the cutting machine process as a workstation that needs high precision so that it requires close eye contact with the workpiece. Through an initial assessment using an NBM questionnaire, it is obtained that almost all of the Cutting Operator's body parts are uncomfortable after finish working. A further investigation is done towards two healthy males (height 169 ± 1.4 cm and weight 73.75 ± 2.5 kg) by capturing the current working body posture and then to be translated to a virtual model and virtual environment, by using JACK 8.2 software. The MSD risk parameters are using PEI by processing LBA, OWAS, and RULA data. Improvement of the posture is done by enhancing the workstation through virtual simulation so that the PEI on the two operators are decreasing from 2.81 and 2.43 to 1.52 and 1.12. Unexpected things happen in this semi-manual cutting workstation, the awkward posture is inevitable as seen in the RULA output of the improved workstation that did not show if the posture is safe from MSD risks. However, MSD risks can be reduced.

1. Introduction

The activity of 'cutting' requires a high level of precision and most of the time, it comes with squatting or bending posture. Muscles hold a static burden repeatedly for a long period. In many cases, it will cause potential injuries for the operators. The symptoms are stiffness in muscles, limited joint motion, and swelling [1]. If the injuries come to the long term, it potentially causes Musculoskeletal Disorders (MSD). Another research related to cutting machines noted that too much time spent by human labour can cause an increase in workload [2]. Therefore, some companies also tried to change the process of cutting to make production faster and easier by setting up the full automation process [3][4]. Similar research has designed the cutting processes that applied in leather which has a huge dimension by using die cutting and collaborated with electro-mechanical technology [5].

Based on field observation towards the Cutting Division in one of the leather tanning automotive company in Jakarta City, Indonesia, it can be acknowledged that the operator's posture is very awkward. To be able to put a leather workpiece that is very thin and wide on the working table, the operator must squat up at the top of the working table. The symptom of MSD is found as 90% of operators that complained about sustained stiffness so that they are absent at work. Besides the awkward posture, the physical workload of an operator is very high because they have 8 hours/day working hours, for 5 working days/week. Therefore, as the initial confirmation, 18 operators are completing the Nordic Body Map (NBM) and from that, it confirmed that the most uncomfortable body part is the back, left shank, and right shank. This is related to the squatting posture the operators do when working which applied another research that analysed the influence of work design to psychological strain labour [6]-[8].

Modelling using a virtual human and virtual environment simulation has been used many times to investigate MSD problems [9]-[13]. The existing MSD risks from postures can be identified through symptoms that can be measured afterward through simulation so improvement in posture can be applied to minimize the impact of MSD risks that could happen in the future. Simulation can facilitate to know whether the improvement has done was accurate or not, considering that the MSD risks are only can be seen in a long-time frame.

In doing the simulation, this research is using JACK 8.2 software, with some advantages as it makes easy for users to manage human model based on Biomechanics principles as possible and as accurate in term of a virtual environment, virtual human, and analyse their performance [14]. Also, the Anthropometry data can be adjusted with the real system. Parameters used in this software are Posture Evaluation Index (PEI) that is obtained from processing the data from Low Back Analysis (LBA), Ovako Working Posture Analysis (OWAS), and Rapid Upper Limb Assessment (RULA). Another information obtained is the Static Strength Prediction (SSP) that can biomechanically identify the percentage of operators who can do the posture with sufficient energy needed and associated Anthropometrics factors that allow the body to move [10].

There was some earlier study that connects work body posture improvement and virtual simulation, but there has been no research that specifically analyses the impactful relationship between process in cutting activity, squatting posture, and MSD risks. Therefore, this research can complement prior researches about virtual modelling in analysing MSD risks in cutting process activity. The purpose of the present study is to improve the work body posture in Cutting Machine Operator so that the risks of MSD can be minimized, by doing a virtual human and virtual environment modelling.

2. Methods

The condition of the existing workstation can be viewed in Figure 1. In the picture, sheets of the very large leather seat are put on the top of the working table manually by the operator. After that, the cutting machine will cut the leather seat according to the pattern. Two healthy males are chosen as the representation from 18 Operators. Overall, the operator's tasks can be divided into two job descriptions as Operator 1 and 2 did. Operator 1 is responsible for putting the pattern on the table, then the Operator 2 is responsible to give working objects in the form of leather seat to the Operator 1 and then to move the pieces that already been cut to the next workstation. Due to the huge size and the height of the table, to reach the centre of the table, the Operator must climb to the top of the working table and squat to adjust the leather seat to be in an accurate position.

Figure 1. Operator's working posture (existing).
Left for operator 1 and right for operator 2.

In this research, Task Analysis Toolkit (TAT) in the JACK software is chosen to evaluate the human virtual posture that is working in the virtual environment as Caputo framework [14]. The algorithm of this research can be seen in Figure 2.

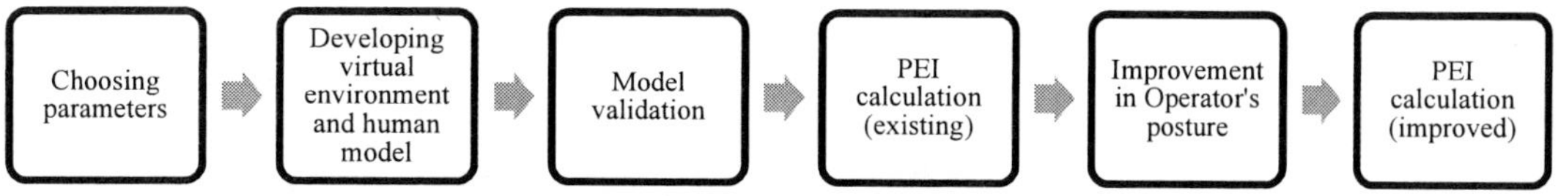

Figure 2. Research algorithm.

2.1. Parameters
At the end of the research, the Posture Evaluation Index (PEI) score in the first posture and final posture will be compared. PEI parameters are chosen because they can comprehensively provide posture analysis information that includes LBA, OWAS, and RULA in various population percentiles [14]. Following are the data needed to count the PEI score:

1. Low Back Analysis (LBA)
   This parameter is used to evaluate the force taken by the Operator's spine L4/L5, especially in the squatting and bending posture. The allowed LBA score according to the NIOSH standard is 3400 N at the maximum [10][14][15][16].
2. Ovako Working Posture Analysis (OWAS)
   This analysis is used to obtain the level of discomfort caused by the working posture of the Operators. Four things to be analyzed by OWAS such as the back, arm, leg, and the load of the body undertaken by the operator. OWAS output that is made by JACK is in the form of code that explains all the four matters.
3. Rapid Upper Limb Assessment (RULA)
   The use of RULA is to measure the upper body posture risks of the operator that can cause MSD. In the JACK software, the RULA calculation is divided into group A (upper arms, lower arms, wrists, and hand rotation) while group B consists of the torso and neck.

Another data that can be considered in the model analysis is the %SSP that is used to figure out the percentage of the operator that can perform certain work postures. The SSP score expected is higher than 90% which means 90% of workers can perform the job in a static strength simulation [10]. The evaluated body parts are wrist, elbow, shoulder, torso, hip, knee, and ankle.

2.2. Developing the Virtual Environment and Virtual Human Model
The virtual environment is developed by measuring the dimension of the real workstation and then translated into objects in the AutoCAD software. The objects then imported into the JACK as a virtual environment. The workstation is referred to as a combination of 6 working tables that are used by the cutting machine operator.

To make a human model (Operator), the Advanced Human Scaling function is chosen in JACK because the function generates a better detail by inputting 14 Anthropometry parameters of the cutting operator. The two healthy operators (average height 169±1.4 cm and average weight 73.75±2.5 kg) that have an average performance, are chosen to be the subjects to represent human in JACK. The Anthropometry data that has been inserted such as stature, head breadth, head height, head length, bi-deltoid, hip breadth, arm length, shoulder-elbow, elbow rest height, elbow-fingertip, buttock-knee, thigh clearance, ankle height, and foot length. Human models that have been formed will be the virtual human in this research. Besides inputting the Anthropometry data, the angles of posture were adjusted in the form of a virtual human and virtual environment. The model is shown in Figure 3.

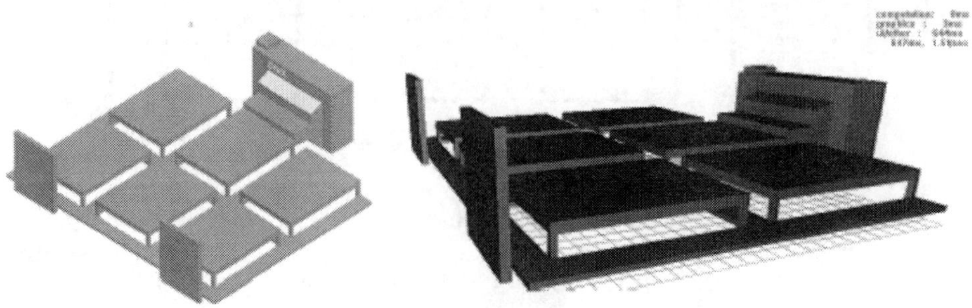

Figure 3. A virtual environment in AutoCAD (left) and JACK (right).

2.3. Model Validation

Before simulating, validation is needed to ensure that the virtual human and virtual environment model has represented the real system. Validation is conducted by comparing between LBA, OWAS, and RULA scores that are obtained from the manual calculation, and a score that is obtained by TAT output from JACK software. The model validation is shown in Figure 4.

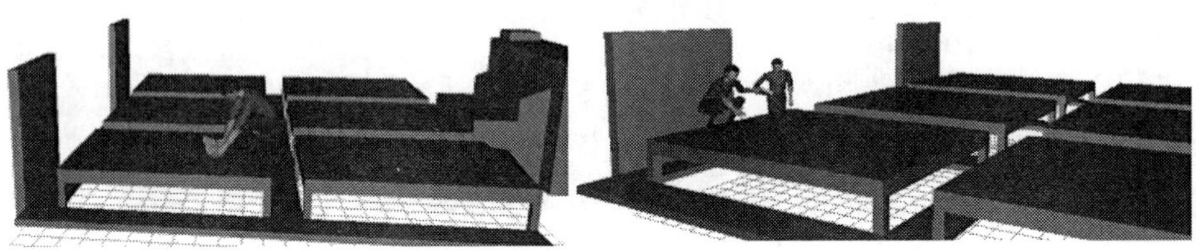

Figure 4. Working posture of cutting operator in JACK.

2.4. Improvement in Operator's posture

A better posture is expected from the improvement of the existing workstation. To design the new workstation, Anthropometry data from the operators will be processed so that the improved workstation (by AutoCAD) can be re-inserted to be the virtual environment in JACK.

2.5. Posture Evaluation Index (PEI) Calculation

Analysis using the PEI is to acknowledge which working posture that gives comfort to as many populations as possible [14]. The lower the PEI score, the more comfortable the posture is so that the MSD risks will also be decreased [10][14]. PEI calculation is conducted by using the formula below [14]:

$$PEI = I_1 + I_2 + (I_3 \times mr) \tag{1}$$

Where:

$$I_1 = \frac{LBA}{3400\ N} \tag{2}$$

$$I_2 = \frac{OWAS}{4} \tag{3}$$

$$mr = amplification\ factor = 1.42 \tag{4}$$

3. Results

3.1. Parameter Evaluation of the Existing Workstation

It proven that there is an MSD risk, the first analysis is done by using the Static Strength Prediction (SSP) parameter in JACK. From the SSP result, Operator 1 and 2 has an SSP score that is lower than 90%, with the highest MSD risk of Operator 1 is on the hip and knee, while in Operator 2 on the hip and sole of the feet. The low score of SSP shows that the current posture is unsafe to be performed for a long period.

The next test is the Low Back Analysis (LBA) in JACK software, with a result that scores of Operator 1 and 2 are 2169 N and 1735 N or still below the NIOSH standard which is 3400N. Therefore, the squatting and bending posture of the Operator are not potentially causing back spine L4/L5 injury, because the pressure point of the Operator is on another body parts.

The OWAS score for Operator 1 and 2 in consecutive were 3 and 2, with code 2141 and 4121. Operator 1 with the code 2141 means that the back posture (2) forward flexion, hand (1) is on the level below shoulder height, feet (4) bent and focused on the feet, and weight (1) that is focused is below 10 kg so that the final score is 3, which shows that the posture of Operator 1 will cause a harmful level of stress on MSD.

The OWAS score for Operator 2 with code 4121 means that the back posture (4) is bent to the side/ forward and accompanied by rotation or movement to the side, hands (1) located below the shoulder height, legs (2) standing with both straight legs, and weight (1) less than 10 kg, so the final score is 2. Therefore Operator 2 has a better posture than Operator 1 but still may harm MSD. This is because Operator 2 is not only doing a static posture in which his job responsibility is to put the pattern on the working table, but also to occasionally stand up to help to move the pattern from the working table to the next workstation.

In contrast to the JACK output for SSP, LBA, and OWAS parameters, the JACK software generates RULA output that is similar to the whole operators, in which the scores are 7. Or in other words, the score reaches the maximum limit. Therefore, according to this result, the MSD risks for all the operator's posture has to be investigated and improved as soon as possible. A recapitulation of MSD risks potential for all parameters can be viewed in Table 1.

Table 1. Evaluation of MSD risks (existing posture).

Operator	Parameters			
	SSP	**LBA**	**OWAS**	**RULA**
Operator 1	<90%	Safe	Will cause harmful level of stress on MSD	Investigation and changes are required immediately
Operator 2	<90%	Safe	May harm MSD	Investigation and changes are required immediately

The validation of the JACK output can be done for OWAS and RULA parameters by using a manual calculation, while for LBA it cannot be calculated manually because to make it sure it requires observation using x-rays. The manual calculation results for OWAS output towards Operator 1 and 2 in consecutive are 3 and 2 or corresponded and matched to the OWAS score from JACK. Besides, the RULA score for the manual calculation to all Operator is 7, which means it has already corresponded to the output on the software. Thus, it can be confirmed that the human model and the human environment is valid for representing the posture in the real workstation.

Table 2. Result of PEI calculation (existing posture).

Operator	I_1	I_2	I_3	Mr	PEI
1	0.64	0.75	1	1.42	2.81
2	0.51	0.5	1	1.42	2.43

As seen in Table 2, the PEI score of Operator 2 has a smaller PEI score which is 2.43. Through the PEI score of Operator 1 and 2, the existing posture must be improved because the PEI score approaches 3.

3.2. Workstation Improvement

Improvement of the workstation is done by considering the Anthropometry size from the working table that is used by the operators. This anthropometric data was chosen to be more fit with current workers. Although currently the cutting process is done in a semi-manual manner, in the future this process will lead to a fully automatic process, so that the current process is a transition process.

The existing working table has a length of 350 cm and 295 cm wide so that the operators are having difficulties to reach the centre of the table and must climb up the top of the table and do a squat posture. Therefore, improvement is done to the workstation that consists of 6 cutting working tables, by re-designing the cutting worktable so that it can improve the operator's work body posture.

Seven types of Anthropometry data are needed to re-design the new cutting worktable. Each dimension was averaged to consider the size of the 50 percentiles from all of the operators to accommodate more. The heights and wideness of the new working table are adjusted by considering that, during manual work where workers often need space for equipment, material, and containers, the height of the working platform just below the height of the standing elbow [17].

According to the anthropometry measurement result on all cutting operators, the height of the standing elbow is on average 101.86 cm so that the height of the working table must be adjusted lower than that height. The wideness of the table is reduced by 50 cm and become 345 cm so that the operator can reach an object in the centre of the table. Therefore, the squatting posture is no longer needed and the stairs to climb the top of the table can be removed as shown in Figure 5 below is the dimension comparison of the existing and improved workstation.

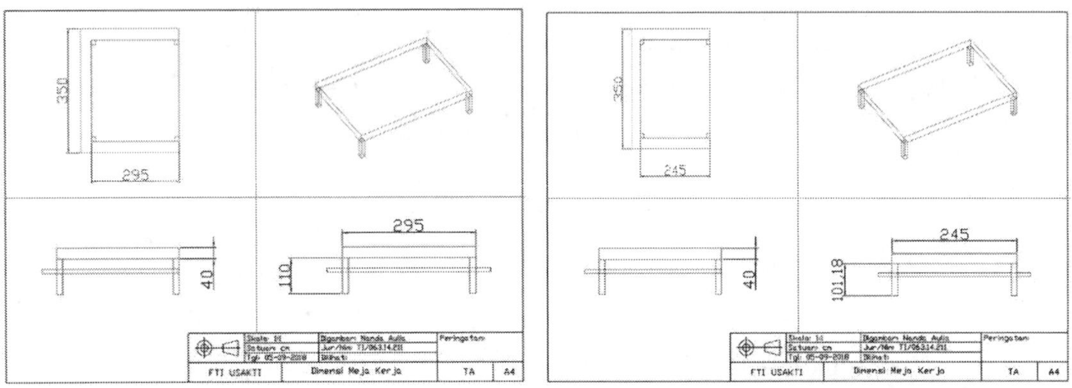

Figure 5. The dimension of existing cutting workstation (left) and improvement (right).

The virtual environment of the model is improved, and work posture is simulated by using the same virtual human with the existing model. The result from the virtual human and virtual environment modelling in the improved workstation can be seen below, the posture of the operator is better than the awkward posture. However, verification is conducted by recalculating the final PEI score for operators that work in the improved workstation. Figure 6 shows an improved operator's workstation in JACK.

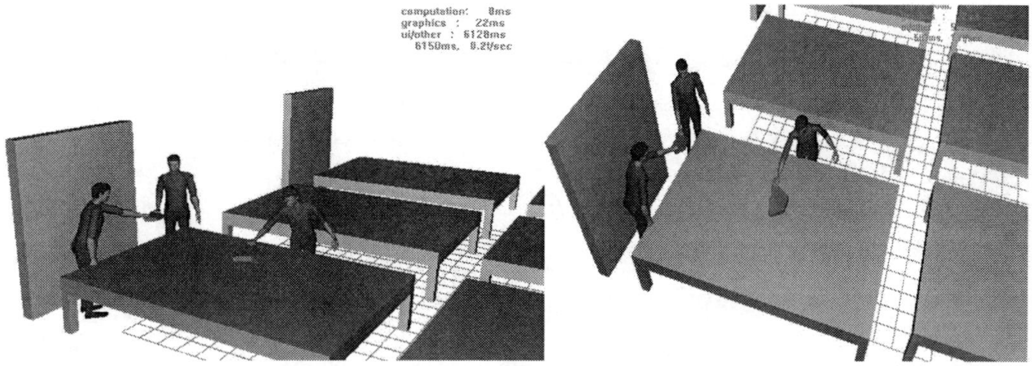

Figure 6. Improved workstation.

3.3. PEI Scores of Improved Posture

TAT calculation is done with the same steps as the calculation made to the existing workstation. The SSP output on JACK coloured grid. It means that the posture has been able to accommodate the posture capability of more than 90% of operators. Furthermore, the LBA score for improvement posture on Operator 1 dropped from 2169 N to 1288 N and for Operator 2 it dropped from 1735 N to 883 N. Thus, the pressure on the operators L4/L5 decreased and better. The OWAS score for Operator 1 has dropped from 3 to 2 which means that the posture of Operator 1 might cause a risk of MSD so further investigation must be carried out again. For Operator 2, the OWAS score dropped from 2 to 1 so that the posture safe. Finally, the RULA score for all operators dropped from 7 to 3. However, although the score has dropped, the implication of score 3 still requires further investigation and improvement. According to the output produced, the comparison between the PEI output of Operator 1 and 2 improved posture can be seen in Table 3 and Table 4. The PEI score for Operator 1 is 1.52 and Operator 2 is 1.12.

Table 3. Evaluation of MSD risks (improved posture).

Operator	Parameters			
	SSP	LBA	OWAS	RULA
Operator 1	>90%	Safe	May harm MSD	Further investigation needed. Changes may be required.
Operator 2	>90%	Safe	The working posture seems normal and natural	Further investigation needed. Changes may be required.

Table 4. Result of PEI calculation (improved posture).

Operator	I_1	I_2	I_3	Mr	PEI
1	0.41	0.50	0.43	1.42	1.52
2	0.26	0.25	0.43	1.42	1.12

4. Discussion

4.1 Output Analysis

According to the reference [14], the score at I_1 indicates the level of posture pressure on the L4 and L5 lumbar with a valid value that is a score below value 1. The scores of I_1 on Operator 1 and Operator 2 respectively on the existing and improved posture are 0.64; 0.41 (Figure 7) and 0.51; 0.26 (Figure 8). As such, this output meets the requirements of the PEI calculation.

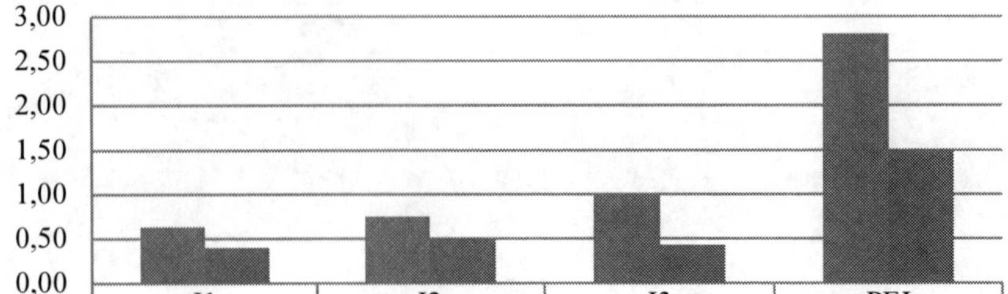

	I1	I2	I3	PEI
Existing	0,64	0,75	1	2,81
Improved	0,41	0,50	0,43	1,52

Figure 7. Changes in PEI scores (operator 1).

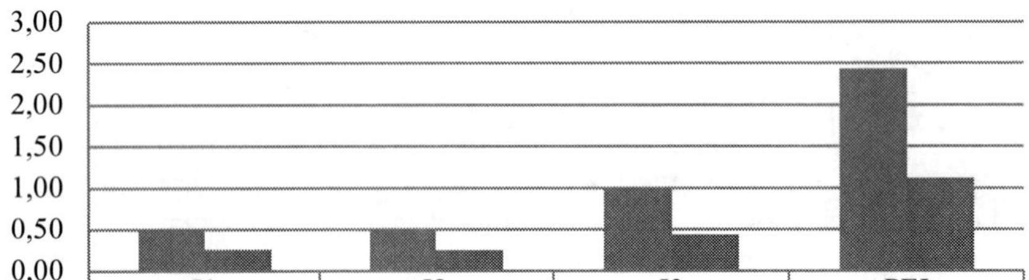

	I1	I2	I3	PEI
Existing	0,51	0,50	1	2,43
Improved	0,26	0,25	0,43	1,12

Figure 8. Changes in PEI scores (operator 2).

An I_2 score indicates the level of discomfort in posture [14]. From the following data, the discomfort score falls on the posture of improvement with the value for Operator 1 is 0.5 (Figure 7) and Operator 2 is 0.25 (Figure 8).

I_3 score indicates the level of fatigue in the upper body part [14]. In Figure 7 and Figure 8 for both operators can be seen, the level of fatigue dropped from 1 to 0.43.

The minimum score for PEI is 0.47 while the maximum score is 3 [14]. The maximum PEI value is very dependent on the I_1 score because when the PEI score reaches 3, it means that the pressure on L4 and L5 reaches a maximum limit of 3400 N. According to the calculation results obtained, the PEI score on the existing posture is close to the maximum value but after improvement, the score drops for Operator 1 of 2.81 to 1.52 for Operator 2 and 2.43 to 1.12.

From Table 3, we can see, even the overall scores have decreased but the OWAS scores for Operator 1 still may harm MSD and also the RULA scores for both operators still noted 'Further investigation needed, and changes may be required. This becomes natural because the activities carried out by the operators are manual, so it's difficult to get an ideal posture that has no MSD risk for the operators especially if the work is carried out repeatedly over a long time.

4.2 Standing Posture for Cutting Workstation

Improvements to the operator's workstation has been reduced the risk of MSD in the future according to the analysis in section 4.1. However, some similar research has observed that almost all the cutting activity is performed with bending posture even if it can be done by sitting, squatting, or standing. To be noted that to cut or to position a workpiece it needs an approximate close distance from the eyes. The smaller the workpiece, the required distance is also smaller.

As Operator 2 tasks, it known that the Operator did squatting and standing by the period of time. Squatting is harmful for the risk of MSD and it suggested to choose the standing position because it is more alert and good for mental state. However, another reference [18] stated that the energy needed by standing position is higher by 10-15% compared to sitting position. So that, how to overcome this problem are periodical stretching and minimizing subjective complaints in reaching, bending, and making awkward head posture [19].

Nevertheless, not all activities that require a level of detail are better done with standing postures. The activity that is best done with standing posture has the following four criteria [18]: (1) There is no space available for legs and knees, (2) Must hold objects which weight more than 4.5 kg, (3) Often carried upwards, bottom, and sides, and (4) This work is carried out frequently and is kept downward.

5. Conclusion

The present study is aimed to improve the working body posture of the Cutting Machine operator process so that the risk of MSD can be minimized by doing a virtual human and virtual environment modelling.

From this study, the enhancement of working body posture is done successfully by improving the workstation. The MSD risks can be seen in the size of the PEI parameter by calculating the value of LBA, OWAS, and RULA. At the end of the study, the PEI value decreases from 2.81 and 2.43 to become 1.52 and 1.12. So, it means that the improvement of the workstation can lower the MSD risks in Cutting Operators. This result supported by SSP value that was bigger than 90%.

Future research on virtual modelling will be carried out not only by using one software but also by several integrated software to get a more realistic model. The next trend research might include aspects of human behaviour analysis, a combination of modelling with AR, workplace design, and also collaboration between humans and robots [20] but in addition, that research will need much more anthropometry data.

References

[1] Tayyari F, Smith JL. *Occupational Ergonomics: Principles and Applications*. London: Chapman and Hall; 1997;**11**;54-9

[2] Baseganni C, Suhel S, Pujar H. Design and analysis of bamboo stick cutting machine. *Adv. J. Grad. Res*. 2018 Jan 1;**3**(1):34–40. doi.org/10.21467/ajgr.3.1.34-40

[3] Jagushte et al. Design and automation of bar cutting machine. *Int. J. Eng. Sci. Res. Techno.* 2017;**6**(4);658-61. doi.org/10.5281/zenodo.557203

[4] Narendra Patil S. Design and development of automatic wire cutting machine: a case study in small scale industry. *Int. J. Res. Appl. Sci. Eng. Technol*. 2017 Mar 28;**5**(3):951–6. doi.org/10.22214/ijraset.2017.3176

[5] Pathan AS and Borlepwar. Development of automated leather cutting machine. *Int. J. Mod. Trends Eng. Res*. 2016 Sep 5;**3**(8):236–42. doi.org/10.21884/ijmter.2016.3029.z6w6x

[6] Panatik SAB. Impact of work design on employee psychological strain among malaysian technical workers. *Procedia. Soc. Behav. Sci.* 2012;**40**:404–9. doi.org/10.1016/j.sbspro.2012.03.207

[7] Van den Broeck A, Parker SK. Job and work design. *Oxford Research Encyclopedia of Psychology.* 2017 May 24. doi.org/10.1093/acrefore/9780190236557.013.15

[8] Parker SK, Andrei DM, Van den Broeck A. Poor work design begets poor work design: Capacity and willingness antecedents of individual work design behavior. *J. Appl. Psychol.* 2019 Jul;**104**(7):907–28. doi.org/10.1037/apl0000383

[9] Purwaningsih R, P. DA, Susanto N. Desain stasiun kerja dan postur kerja dengan menggunakan analisis biomekanik untuk mengurangi beban statis dan keluhan pada otot. *J@ti Undip: Jurnal Teknik Industri.* 2017 Mar 22;**12**(1):15. doi.org/10.14710/jati.12.1.15-22
[10] Muslim E, Nurtjahyo B, Ardi R. Analisis ergonomi industri garmen dengan posture evaluation index pada virtual environment. *MAKARA of Technology Series.* 2011 Sep 22;**15**(1). doi.org/10.7454/mst.v15i1.860
[11] Akhmad H, Aisyah IN, Gagas HP. Analisis ergonomi desain ruang kerja penembak pada kendaraan tempur armoured personnel carrier dalam virtual environment. *Prosiding Seminar Nasional Ergononomi dan Kongres Nasional PEI.* 2012
[12] Szombathyova E. Projecting of working activity by use of technomatic jack. *International Scientific Journal about Simulation.* 2015;**1**(2);9-13
[13] Pekarcikova M. Utilization of the Software Product Tecnomatic JACK in Optimizing of Working Activities. *International Scientific Journal About Simulation.* 2015;**1**;5-11
[14] Caputo F et al. Ergonomic optimization of a manufacturing system work cell in a virtual environment. *Acta Polytechnica: Journal of Advanced Engineering.* 2006;**46**(5);21-7
[15] Badiru AB, Racz L, editors. *Handbook of Emergency Response*. CRC Press; 2013 Aug 22; doi.org/10.1201/b15372
[16] Iqbal BM, Moeis AO, Renalda Krissalam RK. An analysis of the ergonomic design of the tactical commander console (tacco) in the virtual environment of medium-range twin-engine maritime patrol aircraft (MPA). *Int. J. Technol.* 2017 Dec 26;**8**(6):1097. doi.org/10.14716/ijtech.v8i6.704
[17] Clark D. Workstation Evaluation and Design. *Occup. Ergon.* 2012 Mar 8;293–320. doi.org/10.1201/b11717-13
[18] Iftikar ZS. Duduk, berdiri, dan ketenagakerjaan Indonesia. *Prosiding Seminar Nasional Ergonomi dan Kongres Nasional PEI.* 2000
[19] Babur MP. *Fundamentals of Industrial Ergonomics.* New Jersey: Hall International; 1992
[20] Zhu W, Fan X, Zhang Y. Applications and research trends of digital human models in the manufacturing industry. *Virtual Reality & Intelligent Hardware.* 2019 Dec;**1**(6):558–79. doi.org/10.1016/j.vrih.2019.09.005

Aggregate production planning model under demand uncertainty: A case study in an Indonesian cement company

I Gozali[1,2], I N Pujawan[1,3], and N I Arvitrida[1,4]

[1]Department of Industrial Engineering, Institut Teknologi Sepuluh Nopember, Surabaya, Indonesia

E-mail: [2]gozali.imron@gmail.com, [3]pujawan@ie.its.ac.id, [4]niniet@ie.its.ac.id

Abstract. In this paper, a multi-product, multi-stage and multi-period aggregate production-planning model is proposed to deal with demand uncertainty in an Indonesian Cement Company. The demand uncertainty is modelled as a number of discrete scenarios with known probabilities. This problem consists of determining the optimal production, subcontract and inventory level of clinker and cement in order to fulfil demand at minimum cost within the planning horizon. This research aims to develop an aggregate production-planning model under demand uncertainty in an Indonesian Cement Company. The model constructed as a linear programming problem using scenario-based approach to satisfy demand at minimum cost. The outcome of this research is the development of an aggregate production-planning model that accommodates demand uncertainty in Indonesian Cement Company by considering material balance at each stage of the process, equipment capacity and inventory storage capacity during the planning horizon. The proposed model can be used as a decision-making basis for production and inventory control manager to produce or subcontract clinker and semen in each period to fulfil demand at minimum cost.

1. Introduction

Production planning is a key area in operational management [1]. Aggregate production planning is the medium-term capacity planning that determines production plans required to meet customer demands at minimum cost. Based on uncertainty level, aggregate production planning model classified into deterministic model and uncertain model. In deterministic model, demand is assumed to be known in advance. Researchers very often researched this model but practitioners do not like it. The practitioners prefer to develop a model that accommodates uncertainty in parameter data, demand for example, in order to deal with the real situation [2]. Nowadays, satisfying customers are more challenging due to the increasing of uncertainty [3]. However, uncertainty cannot be removed completely from the supply chain and from every part of the supply chain including production [4]. If the uncertainty and its impact are underestimated, it will lead to bad planning decisions [5].

In production context, there are several forms of uncertainty. It can be categorized into two groups, namely system uncertainty and environmental uncertainty. System uncertainty in the production system can be in the form of yield uncertainty, production lead-time uncertainty, quality uncertainty, and uncertainty of failure in the production system and changes in product structure. Environmental uncertainty, on the other hand, is outside of the production system such as supply uncertainty and demand uncertainty [6]. Supply uncertainty is related to the uncertainty of materials' supply. On the other hand, demand uncertainty refers to the probabilistic nature of demand quantity, types, timing and locations [3].

Demand uncertainty can cause frequent changes in production planning in the planning period [7]. To overcome this uncertainty, there are two methodologies used, scenario-based approach and distribution-based approach. In a scenario-based approach, demand conditions divided into low demand, medium demand and high demand [8]. Gupta et al.[5] argued that uncertainties of demand are explained by a set of discrete scenarios that capture how uncertainty might occur in the future. Each scenario is associated with a level of probability that represents the decision makers' expectations about the occurrence of a particular scenario.

Research about aggregate production planning in cement plant begins as the development of a bi-objective model for production planning in a Thailand cement factory. The model determines the amounts of material input to each major production facility in order to maximize utility and minimize cost [9]. However, the limitation of this model was still in wet process, single period, single product cement type and did not consider the demand uncertainty. An alternate aggregate production-planning model for cement plant by using constant workforce and chase strategy has been proposed [10]. Nevertheless, the model did not consider the subcontract of clinker and cement among cement plant. To handle demand uncertainty in supply chain, a model of multi-echelon supply chain network under demand uncertainty was developed. In this model, scenario-based approach used to interpret the demand uncertainty. The scenario divided into high, moderate and low demand while the probability of each scenario is equal [11]. However, this model is only for single period.

In this paper, we developed linear programming model to solve the model while demand uncertainty modelled as a number of discrete scenarios, high – moderate - low, with known probabilities (equal probability for each scenario). Therefore, this study aims to develop an aggregate production-planning model under demand uncertainty in an Indonesian Cement Company. This paper is divided into 5 section. Section 1 discusses about research background. Section 2 provides cement production process at a glance. Section 3 presents a proposed model. Section 4 provides real case study in an Indonesian Cement Company. Finally, the last section discusses about conclusion and future research.

2. Cement Production

Cement is a substance applied to the solid bodies' surface to make them firmly cohere. Portland cement produced in a series of processes. Limestone and other materials containing appropriate proportions of calcium, silicon, aluminum and iron oxides crushed and milled to raw meal. In a kiln, the calcium carbonate of raw meal dissociates to calcium oxide and generates carbon dioxide. Then, the calcium oxide reacts with the other components to form calcium silicates and aluminates, which partially melt at material burning temperature up to 1450 °C. The kiln product is clinker, a black nodular material. At cement mill, the clinker added with a small proportion of gypsum and other additive material yielding a fine-finished product called cement [12]. Figure 1 shows cement production process.

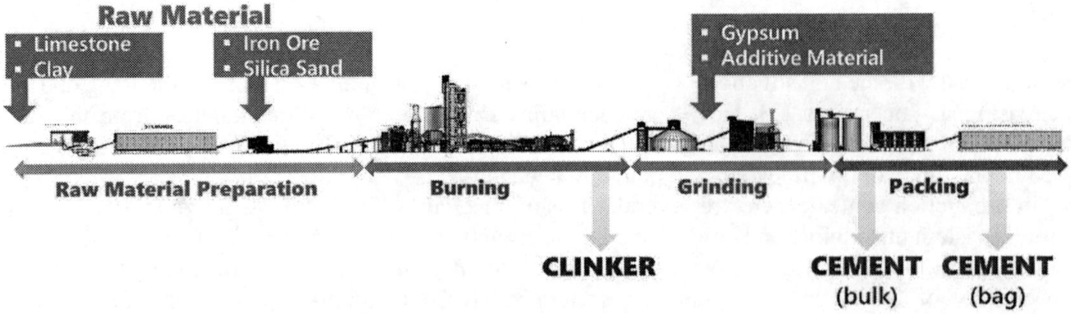

Figure 1. Cement production process.

Based on raw material preparation, cement production process classified into wet process and dry process. Wet process consumed more energy than dry process. Therefore, nowadays cement manufacturer use dry process to save energy [13].

3. The Proposed Model

The problem statement for the proposed model is described as follows: to determine amount of clinker and cement produced, subcontracted, and hold in inventory storage in order to satisfy demand. The research objective is to minimize total cost that consist of production cost (represented by raw material, fuel and additive material cost), subcontracted cost and holding inventory cost. The model is restricted by some assumptions. Uncertainty only happened at demand parameter. Demand uncertainty modelled as a number of discrete scenarios, high moderate low, with known probabilities (equal probability for each scenario). System boundary of this model is from Kiln to Release Facilities. Raw material always available and unlimited. Workforce level assumed constant and ignored. The proposed conceptual model shown in Figure 4 while the mathematical model provided below.

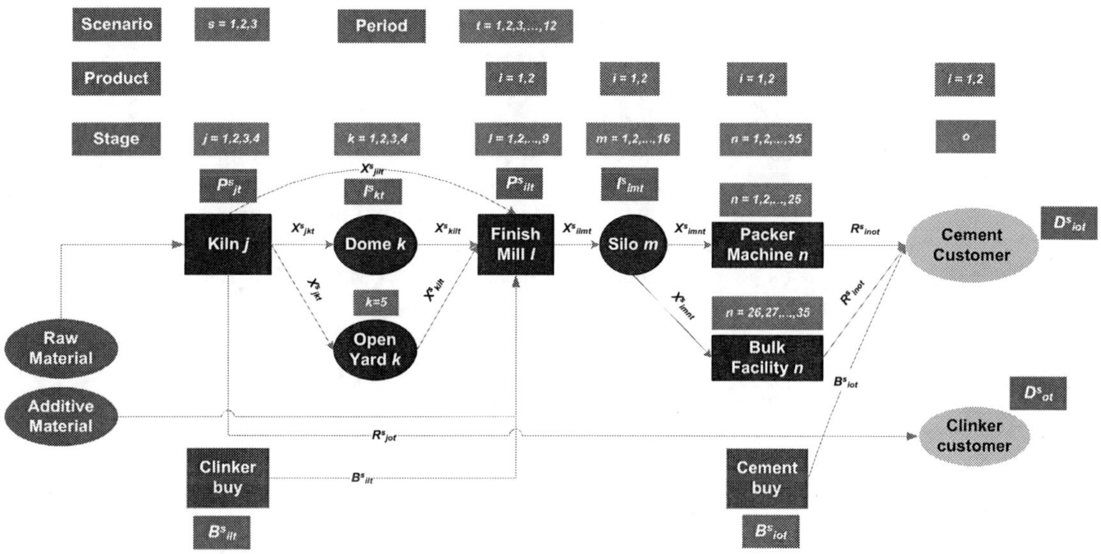

Figure 2. Conceptual model.

3.1. Indices, Parameters and variables

3.1.1. Indices

- e : Raw material e, $\forall_e \in \{limestone,\ clay,\ silica\ sand,\ copper\ slag\}$
- f : Fuel f, $\forall_f \in \{coal\}$
- g : Additive material e, $\forall_g \in \{gypsum,\ trass, limestone, fly\ ash, GBFS\}$
- i : Cement type i, $\forall_i \in \{OPC, PCC\}$
- j : Kiln j, $\forall_j \in \{1,2,3,4\}$
- k : Clinker storage k, consist of clinker dome $k \in \{1,2,3,4\}$ and open yard $k \in \{5\}$
- l : Finish mill l, $\forall_l \in \{1,2,\dots,9\}$
- m : Cement silo m, $\forall_m \in \{1,2,\dots,16\}$
- n : Cement release facilities n, consist of packer machine $n \in \{1,2,\dots,25\}$ and bulk facility $n \in \{26,27,\dots,35\}$
- o : Customer o, $o \in \{1\}$
- t : Monthly period t within planning horizon, $\forall_t \in \{1,2,\dots,12\}$
- s : Demand scenario s, $\forall_s \in \{1,2,3\}$

IOP Conf. Series: Materials Science and Engineering 1072 (2021) 012033 doi:10.1088/1757-899X/1072/1/012033

3.1.2. Parameter

D^s_{iot}	: Demand of cement type *i* for customer *o* at period *t* at scenario *s*
D^s_{ot}	: Demand of clinker for customer *o* at period *t* at scenario *s*
ρ_s	: Probability of scenario *s*
CF_i	: Clinker factor of cement type *i* (show the amount of clinker required to produce 1 ton cement *i*)
RMI_e	: Raw material index *e* (show the amount of raw material required to produce 1 ton clinker)
$FuelI_f$	: Fuel index *f* (show the amount of fuel required to produce 1 ton clinker)
AMI_{gi}	: Additive material index g (show the amount of add. material required to produce 1 ton cement *i*)
crm_e	: Raw material unit price *e*
cf_f	: Fuel unit price *f*
cam_g	: Additive material unit price g
cb	: Cost of buying 1 ton clinker from another cement plant
cb_i	: Cost of buying 1 ton cement type *i* from another cement plant
ch	: Holding cost of 1 ton clinker for each month
ch_i	: Holding cost of 1 ton cement type *i* for each month
F_{jk}	: The presence (1) or absence (0) of material direct transfer facilities from kiln *j* to clinker storage *k*
F_{kil}	: The presence (1) or absence (0) of material direct transfer facilities from clinker storage *k* to finish mill *l* for producing cement type *i*
F_{jil}	: The presence (1) or absence (0) of material direct transfer facilities from kiln *j* to finish mill *l* for producing cement type *i*
F_{il}	: The presence (1) or absence (0) facilities of finish mill *l* to produce cement type *i*
Fb_{il}	: The presence (1) or absence (0) of receiving buying clinker facilities of finish mill *l* to produce cement type *i*
F_{lm}	: The presence (1) or absence (0) of material direct transfer facilities from finish mill *l* to cement silo *m*
F_{ilm}	: The presence (1) or absence (0) of facilities to transfer material cement type *i* from finish mill *l* to cement silo *m*
F_{mn}	: The presence (1) or absence (0) of material direct transfer facilities from cement silo *m* to release facilities *n*
F_{imn}	: The presence (1) or absence (0) of facilities to transfer material cement type *i* from cement silo *m* to release facilities *n*
F_{in}	: The presence (1) or absence (0) facilities of release facilities *n* to release cement type *i*
F_{ino}	: The presence (1) or absence (0) facilities of release facilities *n* to release cement type *i* to customer *o*
F_{jo}	: The presence (1) or absence (0) of material direct transfer facilities from kiln *j* to customer *o*
F_{jot}	: The presence (1) or absence (0) of material direct transfer facilities from kiln *j* to customer *o* at period *t*
$CSCap_k$	: Clinker storage *k* capacity to store clinker
$SCap_m$	: Cement silo *m* capacity to store cement
$KilnCap_{jt}$	: Kiln *j* capacity to produce clinker at period *t*
$FMCap_{ilt}$	: Finish Mill *l* capacity to produce cement type *i* at period *t*
$RCap_{nt}$	: Release Facilities *n* capacity to release cement at period *t*
$RCap_{int}$	: Release Facilities *n* capacity to release cement type *i* at period *t*
CD_t	: Calendar days at period *t*
ADT_{jt}	: Allowable downtime days of kiln *j* at period *t*

ADT_{lt} : Allowable downtime days of finish mill l at period t
ADT_{nt} : Allowable downtime days of release facilities n at period t
α_k : Minimum percentage of clinker inventory level at clinker storage k
β_m : Minimum percentage of cement inventory level at cement silo m
I_{im0} : Initial inventory level of cement type i at cement silo m
I_{k0} : Initial inventory level of clinker at clinker storage k

3.1.3. Decision variables

OD_{jt}^s : Operating days of kiln j at period t at scenario s
OD_{lt}^s : Operating days of finish mill l at period t at scenario s
OD_{ilt}^s : Operating days of finish mill l to produce cement type i at period t at scenario s
OD_{nt}^s : Operating days of release facilities n at period t at scenario s
OD_{int}^s : Operating days of release facilities n to release cement type i at period t at scenario s
P_{ilt}^s : The amount of cement type i produced at finish mill l at period t at scenario s
P_{lt}^s : The amount of cement produced at finish mill l at period t at scenario s
P_{jt}^s : The amount of clinker produced at kiln j at period t at scenario s
R_{nt}^s : The amount of cement released at release facilities n at period t at scenario s
R_{int}^s : The amount of cement type i released at release facilities n at period t at scenario s
R_{inot}^s : The amount of cement type i released at release facilities n to customer o at period t at scenario s
R_{jot}^s : The amount of clinker released at kiln j to customer o at period t at scenario s
B_{iot}^s : The amount of cement type i bought from another cement plant to customer o at period t at scenario s
B_{ilt}^s : The amount of clinker bought from another cement plant to produce cement type i at finish mill l at period t at scenario s
Cl_{lt}^s : The amount of clinker required to produce total cement at finish mill l at period t at scenario s
Cl_{ilt}^s : The amount of clinker required to produce cement type i at finish mill l at period t at scenario s
RM_{ejt}^s : Raw material e usage to produce clinker at kiln j at period t at scenario s
$Fuel_{fjt}^s$: Fuel f usage to produce clinker at kiln j at period t at scenario s
AM_{gilt}^s : Additive material g usage to produce cement i at finish mill l at period t at scenario s
X_{jkt}^s : The amount of clinker out from kiln j to clinker storage k at period t at scenario s
X_{jilt}^s : The amount of clinker out from kiln j to finish mill l to produce cement type i at period t at scenario s
X_{kilt}^s : The amount of clinker out from clinker storage k to finish mill l to produce cement type i at period t at scenario s
X_{ilmt}^s : The amount of cement type i out from finish mill l to cement silo m at period t at scenario s
X_{imnt}^s : The amount of cement type i out from cement silo m to release facilities n at period t at scenario s
I_{imt}^s : Inventory level of cement type i at cement silo m at period t at scenario s
I_{kt}^s : Inventory level of clinker at clinker storage k at period t at scenario s

3.2. Objective function

$$Min\ Z = \sum_{s=1}^{3} \rho_s \left(\sum_{t=1}^{12}\sum_{j=1}^{4}\sum_{e=1}^{4} crm_e . RM_{ejt}^s + \sum_{t=1}^{12}\sum_{j=1}^{4}\sum_{f=1}^{1} cf_f . Fuel_{fjt}^s + \sum_{t=1}^{12}\sum_{l=1}^{9}\sum_{i=1}^{2}\sum_{g=1}^{5} cam_g . AM_{gilt}^s \right.$$
$$+ \sum_{t=1}^{12}\sum_{l=1}^{9}\sum_{i=1}^{2} cb . B_{ilt}^s + \sum_{t=1}^{12}\sum_{o=1}^{1}\sum_{i=1}^{2} cb_i . B_{iot}^s + \sum_{t=1}^{12}\sum_{k=1}^{5} ch . I_{kt}^s \quad (1)$$
$$\left. + \sum_{t=1}^{12}\sum_{m=1}^{16}\sum_{i=1}^{2} ch_i . I_{imt}^s \right)$$

Eq. 1 shows the objective function: to minimize expected cost that consist of raw material, fuel and additive material cost that represent production cost, clinker and cement subcontract cost and inventory holding cost. Expected cost calculated from the sum-product of probability and cost for each scenario.

3.3. Constraints

3.3.1. Cement release

$$OD_{nt}^s \le CD_t - ADT_{nt}, \forall_{n,t,s} \quad (2)$$
$$OD_{nt}^s = \sum_{i=1}^{2} OD_{int}^s \cdot F_{in}, \forall_{n,t,s} \quad (3)$$
$$R_{int}^s = OD_{int}^s \cdot RCap_{int} \cdot F_{in}, \forall_{i,n,t,s} \quad (4)$$
$$R_{nt}^s = \sum_{i=1}^{2} R_{int}^s \cdot F_{in}, \forall_{n,t,s} \quad (5)$$
$$R_{inot}^s = R_{int}^s \cdot F_{ino}, \forall_{i,n,o,t,s} \quad (6)$$

3.3.2. Cement production

$$OD_{lt}^s \le CD_t - ADT_{lt}, \forall_{l,t,s} \quad (7)$$
$$OD_{lt}^s = \sum_{i=1}^{2} OD_{ilt}^s, \forall_{l,t,s} \quad (8)$$
$$P_{ilt}^s = OD_{ilt}^s \cdot FMCap_{ilt} \cdot F_{il}, \forall_{i,l,t,s} \quad (9)$$
$$P_{lt}^s = \sum_{i=1}^{2} P_{ilt}^s \cdot F_{il}, \forall_{l,t,s} \quad (10)$$

3.3.3. Clinker production

$$OD_{jt}^s \le CD_t - ADT_{jt}, \forall_{j,t,s} \quad (11)$$
$$P_{jt}^s = OD_{jt}^s \cdot KilnCap_{jt}, \forall_{j,t,s} \quad (12)$$

3.3.4. Clinker required to produce cement

$$Cl_{ilt}^s = CF_i . P_{ilt}^s, \forall_{i,l,t,s} \quad (13)$$
$$Cl_{lt}^s = \sum_{i=1}^{2} Cl_{ilt}^s \cdot F_{il}, \forall_{l,t,s} \quad (14)$$

3.3.5. Additive material required to produce cement

$$AM_{gilt}^s = AMI_{gi} . P_{ilt}^s, \forall_{g,i,l,t,s} \quad (15)$$

3.3.6. Raw material and fuel required to produce clinker

$$RM_{ejt}^{s} = RMI_{e}.P_{jt}^{s}, \forall_{e,j,t,s} \tag{16}$$

$$Fuel_{ft}^{s} = FI_{e}.P_{jt}^{s}, \forall_{e,j,t,s} \tag{17}$$

3.3.7. Total cement released and cement bought must satisfy cement demand

$$\sum_{n=1}^{35} R_{inot}^{s}.F_{ino} + B_{iot}^{s} \geq D_{iot}^{s}, \forall_{i,o,t,s} \tag{18}$$

3.3.8. Total clinker released must satisfy clinker demand

$$\sum_{j=1}^{4} R_{jot}^{s}.F_{jot} \geq D_{ot}^{s}, \forall_{o,t,s} \tag{19}$$

3.3.9. Total cement released must not exceed release facility capacity to release cement

$$R_{inot}^{s}.F_{ino} \leq CD_{t}\ .\ RCap_{nt}, \forall_{i,n,o,t,s} \tag{20}$$

3.3.10. Total cement produced must not exceed finish mill capacity to produce cement

$$P_{lt}^{s} \leq CD_{t}\ .\ FMCap_{lt}, \forall_{l,t,s} \tag{21}$$

3.3.11. Total clinker produced must not exceed kiln capacity to produce clinker

$$P_{jt}^{s} \leq CD_{t}\ .KilnCap_{jt}, \forall_{j,t,s} \tag{22}$$

3.3.12. Total cement produced is equal to the amount of cement out from finish mill

$$P_{ilt}^{s} = \sum_{m=1}^{16} X_{ilmt}.F_{ilm}\ , \forall_{i,l,t,s} \tag{23}$$

3.3.13. Total clinker produced is equal to the amount of clinker out from kiln

$$P_{jt}^{s} = \sum_{k=1}^{5} X_{jkt}^{s}.F_{jk} + \sum_{i=1}^{2}\sum_{l=1}^{9} X_{jilt}^{s}.F_{jil} + R_{jot}^{s}.F_{jot}, \forall_{j,o,t,s} \tag{24}$$

3.3.14. Release facilities material balance

$$\sum_{m=1}^{16} X_{imnt}^{s}.F_{imn} = R_{inot}^{s}.F_{ino}, \forall_{i,n,t,s} \tag{25}$$

3.3.15. Finish mill material balance

$$\sum_{k=1}^{5} X_{kilt}^{s}\ .F_{kil} + \sum_{j=1}^{4} X_{jilt}^{s}\ .F_{jil} + B_{ilt}^{s}.Fb_{il} = \sum_{m=1}^{16} CF_{i}.X_{ilmt}^{s}.F_{ilm}\ , \forall_{i,l,t,s} \tag{26}$$

3.3.16. Cement silo inventory balance

$$I_{imt}^{s}.F_{im} = I_{im(t-1)}^{s}.F_{im} + \sum_{i=1}^{2}\sum_{l=1}^{9} X_{ilmt}^{s}.F_{ilm} - \sum_{n=1}^{35} X_{imnt}^{s}.F_{imn}\ , \forall_{i,m,t,s} \tag{27}$$

3.3.17. Clinker storage inventory balance

$$I_{kt}^{s} = I_{k(t-1)}^{s} + \sum_{j=1}^{4} X_{jkt}^{s}.F_{jk} - \sum_{l=1}^{9} X_{kilt}^{s}.F_{kil}\,, \forall_{k,t,s} \quad (28)$$

3.3.18. Cement inventory range at cement silo

$$\beta\,.SCap_{m} \leq I_{mt}^{s} \leq SCap_{m}, \forall_{m,t,s} \quad (29)$$

$$\sum_{m=1}^{16} I_{mt}^{s} \geq 150{,}000, \forall_{t,s} \quad (30)$$

3.3.19. Clinker inventory range at clinker storage

$$\alpha\,.CSCap_{k} \leq I_{kt}^{s} \leq CSCap_{k}, \forall_{k,t,s} \quad (31)$$

$$\sum_{k=1}^{4} I_{kt}^{s} \geq 200{,}000, \forall_{t,s} \quad (32)$$

3.3.20. Total cement inventory balance

$$\sum_{m=1}^{16} I_{mt}^{s} = \sum_{m=1}^{16} I_{m(t-1)}^{s} + \sum_{l=1}^{9} P_{lt}^{s} - \sum_{n=1}^{35} R_{inot}^{s}.F_{ino}\,, \forall_{t,s} \quad (33)$$

3.3.21. Total clinker inventory balance

$$\sum_{k=1}^{5} \boldsymbol{I_{kt}^{s}} = \sum_{k=1}^{5} \boldsymbol{I_{k(t-1)}^{s}} + \sum_{j=1}^{4} \boldsymbol{P_{jt}^{s}} - \sum_{j=1}^{4} \boldsymbol{R_{jot}^{s}.F_{jo}} - \sum_{i=1}^{2}\sum_{l=1}^{9} \boldsymbol{Cl_{ilt}^{s}} + \sum_{i=1}^{2}\sum_{l=1}^{9} \boldsymbol{B_{ilt}^{s}.Fb_{il}}\,, \forall_{t,s} \quad (34)$$

3.3.22. Non negativity

$$\boldsymbol{OD_{jt}^{s}, OD_{lt}^{s}, OD_{ilt}^{s}, OD_{nt}^{s}, OD_{int}^{s}, P_{ilt}^{s}, P_{lt}^{s}, P_{jt}^{s}, R_{int}^{s}, R_{inot}^{s}, R_{jot}^{s}, B_{iot}^{s}, B_{ilt}^{s},}$$
$$\boldsymbol{Cl_{ilt}^{s}, Cl_{ilt}^{s}, X_{jkt}^{s}, X_{jilt}^{s}, X_{kilt}^{s}, X_{ilmt}^{s}, X_{imnt}^{s}, I_{imt}^{s}, I_{kt}^{s} \geq 0} \quad (35)$$

Eq. (2) to (6) show cement release constraints that consist of operating day availability and the amount of cement released. Eq. (7) to (10) show cement production consist of operating day availability and the amount of cement produced. Eq. (11) to (12) show clinker production consist of operating day availability and the amount of clinker produced. Eq. (13) to (14) show the amount of clinker required to produce cement. Eq. (15) show the amount of additive material required to produce cement. Eq. (16) to (17) show the amount of raw material and fuel required to produce clinker. Eq. (18) shows cement demand fulfilled both from cement release and from subcontract while eq. (19) clinker demand fulfilled only from clinker release. Eq. (20) to (22) guarantee that the output from equipment are less than or equal to equipment maximum capacity. Eq. (23) to (24) show that the amount of cement (clinker) produced are equal to the amount of cement (clinker) out from finish mill (kiln). Eq. (25) and (26) show material balance of release facilities and finish mill. Eq. (27) and (28) show inventory balance of cement silo and clinker storage. Eq. (29) to (32) show the limitation of inventory range at cement silo and clinker storage. Eq. (33) and (34) show cement and clinker total inventory balance. The last constraint makes sure that all decision variables are non-negative.

4. Case Study

PT X, a cement plant in Indonesia, has 4 Kiln, 5 Clinker Storage, 9 Finish Mill, 16 Cement Silo, and 35 Release Facilities that consist of 25 Packer Machines and 10 Bulk Facilities. As mentioned on section 3, the system boundary observed on this research is from Kiln to Release Facilities and the visualization of conceptual model shown on Figure 4. There are three demand scenarios (high – moderate - low), 12 months of planning horizon, two types of finished product (cement) and an intermediate product (clinker) considered. Cement demand from customer fulfilled from release facilities or subcontract while clinker demand from customer fulfilled directly from kiln. Clinker subcontract from the other cement plant used for finish mill to produce cement.

PT X face demand uncertainty during 2016-2019. Cement demand in Cement Company can be represented with cement release parameter that shown in Figure 2. Based on Figure 2, in 2017 cement actual release is greater than cement planned release while in 2019 cement actual release is less than cement planned release. The average absolute forecast error each year was 7.80% in 2016, 11.33% in 2017, 4.13% in 2018 and 11.64% in 2019. From 2016 to 2019, average absolute forecast error was 8.72% and shown in Figure 3.

The existence of the forecast error certainly affects increasing the idle of the equipment and decreasing the utilization [14]. In 2019, overall idle for Finish Mill and Kiln at PT X was 9,310 hours and 1,018 hours. This idle make overall utilization at PT X was only for 78.58% for Finish Mill and 86.11% for Kiln. In the midst of demand uncertainty at PT X, it is necessary to develop an aggregate production-planning model under demand uncertainty in order to satisfy demand at minimum cost.

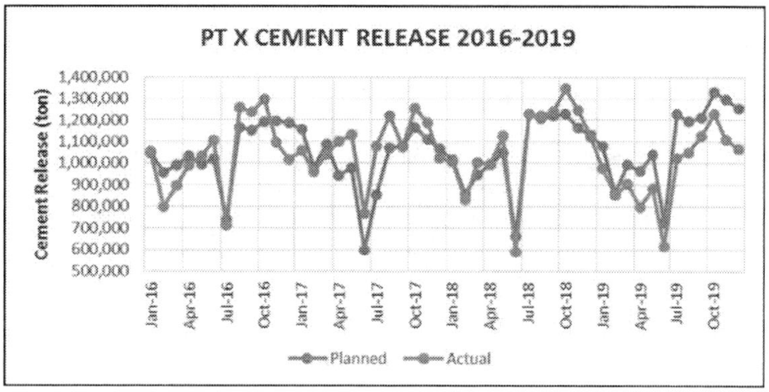

Figure 3. PT X cement release 2016-2019.

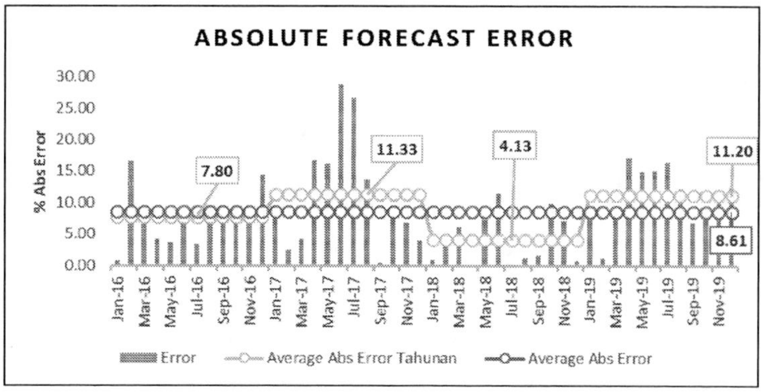

Figure 4. Absolute forecast error PT X 2016-2019.

4.1. Data

The data parameter is taken from PT X and shown in the Table 1 to Table 7 below. The demand and its probability presented in Table 1 and Table 2. Raw material index, fuel index, additive material index and its material unit price shown in Table 4 and Table 5. Equipment capacity and its allowable downtime shown in Table 6. Storage capacity and its inventory shown in Table 7.

Table 1. Demand for each product, each period and each scenario in 2020.

Month	Demand (ton)								
	High			Moderate			Low		
	Clinker	OPC	PCC	Clinker	OPC	PCC	Clinker	OPC	PCC
Jan	125000	327647	709196	110750	330957	732829	101250	303730	683076
Feb	155000	304269	612402	110750	302153	632817	101250	279598	595226
Mar	185000	304482	669928	140750	323827	641976	101250	310300	629958
Apr	165000	287876	664270	110750	296333	625329	131250	291438	611622
May	161000	212429	543123	110750	234055	528474	101250	203976	483075
Jun	155000	259625	720315	110750	260521	713375	101250	246209	664317
Jul	125000	372195	800242	110750	345778	776073	101250	339101	731358
Aug	125000	406301	793595	110750	390934	744635	101250	382382	721360
Sep	125000	413552	819260	140750	405995	775472	101250	397550	738858
Oct	125000	467837	820328	140750	448885	783293	71250	439957	758997
Nov	95000	451164	765418	110750	427302	727606	71250	420364	699370
Dec	125000	398256	722018	140750	377190	681056	101250	369614	644221

Table 2. Probability for each demand scenario.

Scenario	High	Moderate	Low
Probability (ρ)	$^1/_3$	$^1/_3$	$^1/_3$

Table 3. Clinker factor and cost parameter.

Material	Clinker Factor	Subcontract Cost	Inventory Cost
	ton clinker/ton cement	Rp/ton	Rp/ton/month
Clinker	-	430,000	7,644
OPC Cement	0.88	845,000	9,156
PCC Cement	0.65	665,000	6,174

Table 4. Raw material-fuel index and its unit price for clinker production.

Raw Material* / Fuel**	Index	Material Unit Price
	ton material/ton clinker	Rp/ton material
Limestone*	1.3222	22,618
Clay*	0.2927	25,880
Silica Sand*	0.0370	53,000
Copper Slag*	0.0303	195,875
Coal**	0.2185	693,000

Table 5. Additive material and its unit price for cement production.

Additive Material	Index for OPC	Index for PCC	Material Unit Price
	ton material/ton cement	ton material/ton cement	Rp/ton material
Gypsum	0.0360	0.0330	449,573
Trass	0.0008	0.0530	76,750
Limestone	0.0480	0.2400	22,618
Fly Ash	0.0005	0.0240	140,000
GBFS	0.0347	0.0000	317,000

Table 6. Equipment capacity and allowable downtime.

Equipment	Capacity tpd	Allowable Downtime days/month
Kiln	8100	1
Finish Mill	4,320 (FM1-6); 5,040 (FM7-9)	2
Release Facilities	1,200 (RF1-25); 400 (RF27,31); 2,500(RF26,28,33); 5,000 (RF29,30,32,34,35)	2

Table 7. Storage capacity and inventory.

Storage	Capacity	Initial Inventory	Minimum Inventory
	ton	ton	%
Clinker storage	65000 (CS 1-4); 25000 (CS 5)	50000 (CS 1-4); 10000 (CS 5)	40
Cement silo	17500	10000	30

4.2. Results and Discussion

Using LINGO 11 software in Intel (R) Core(TM) i7-7500U CPU @ 2.70 GHz 2.90 GHz RAM 8.00 GB computer, the experimental result show global optimum condition and all constraints that developed fully satisfied. The objective function result presented in Table 8.

Table 8. Objective function result.

Cost (Rp)	Scenario			Expected
	High	Moderate	Low	
Raw material cost	493,955,598,456	477,373,260,180	447,175,594,785	472,834,817,807
Fuel cost	1,648,386,121,210	1,593,048,968,729	1,492,275,918,102	1,577,903,669,347
Additive material cost	354,848,260,063	348,989,999,144	333,335,744,844	345,724,668,017
Clinker buy cost	953,950,081	0	0	317,983,360
Cement buy cost	85,715,445,463	0	0	28,571,815,154
Clinker inventory cost	21,730,138,171	21,519,352,540	19,262,880,000	20,837,456,904
Cement inventory cost	14,922,048,042	13,462,249,632	11,918,959,735	13,434,419,136
Total cost	**2,620,511,561,485**	**2,454,393,830,225**	**2,303,969,097,465**	**2,459,624,829,725**

Based on Table 8, the expected total cost is two trillion and forty-six billion rupiah. The total cost tend to be lower as the demand scenario lower. However, there are clinker and cement subcontracted only occur in high demand scenario. This indicates the higher demand the higher clinker and cement will be probably subcontracted. Table 10 provide clinker and cement production-release-inventory balance on high demand scenario. All demand in all period is satisfied by own production and subcontract form other cement plant. Clinker and cement inventory are kept minimum to minimize inventory cost.

Utilization of equipment is also measured to check whether the production facilities utilized properly or not. As shown on Table 9, the higher demand lead to the higher utilization of equipment. The expected utilization of kiln (87.80%) and finish mill (82.42%) in the planning horizon show better result than its utilization (86.11% and 78.58%) in 2019.

Table 9. Utilization result.

Utilization (%)	Scenario			Expected
	High	Moderate	Low	
Kiln	91.80%	88.72%	83.11%	87.88%
Finish Mill	84.60%	83.20%	79.46%	82.42%
Release Facilities	54.89%	53.98%	51.56%	53.48%

Table 10. Clinker and cement production-release-inventory balance for high demand scenario.

Parameter		Jan	Feb	Mar	Apr	May	Jun	Jul	Aug	Sep	Oct	Nov	Dec
		Period											
Cement													
Demand	OPC	327647	304269	304482	287876	212429	259625	372195	406301	413552	467837	451164	398256
	PCC	709196	612402	669928	664270	543123	720315	800242	793595	819260	820328	765418	722018
Release	OPC	327647	304269	304482	287876	212429	259625	372195	406301	413552	467837	451164	398256
	PCC	709196	612402	669928	664270	543123	720315	800242	793595	819260	758895	697956	722018
Production	OPC	308647	304269	304482	287876	212429	308625	372195	396565	374288	467837	451164	398256
	PCC	718196	612402	669928	664270	555192	789246	800242	793595	774832	722323	697956	722018
Buy	OPC	0	0	0	0	0	0	0	0	0	0	0	0
	PCC	0	0	0	0	0	0	0	0	0	61433	67462	0
Inventory	OPC	21000	21000	21000	21000	21000	70000	70000	60264	21000	21000	21000	21000
	PCC	129000	129000	129000	129000	141069	210000	210000	210000	165572	129000	129000	129000
Clinker													
Demand		125000	155000	185000	165000	161000	155000	125000	125000	125000	125000	95000	125000
Release		125000	155000	185000	165000	161000	155000	125000	125000	125000	125000	95000	125000
Usage to Produce Cement		738,437	665819	703397	685106	547813	784600	847688	864814	833014	881206	850696	819777
Production		863437	820819	888397	850106	783813	939600	972000	972000	939600	972000	939600	944777
Buy		0	0	0	0	0	0	0	0	0	0	2218	0
Inventory		210000	210000	210000	210000	285000	285000	284312	266498	248084	213877	210000	210000

5. Conclusion

Aggregate production planning plays important role in operation management. Due to demand uncertainty, production planning and inventory manager must be able to determine the amount of product produced by his own plant or subcontracted from another plant and to determine how much product kept in inventory to satisfy demand at minimum cost. By using linear programming and scenario-based approach to handle demand uncertainty, aggregate production planning model can be constructed with respect to material balance at each stage of the process, equipment capacity and inventory storage capacity during the planning horizon. The optimization result reached global optimum condition and showed that all demand satisfied from own production and subcontracted from another plant at minimum cost. As the demand higher, the higher also the utilization of equipment and subcontract of product will occur. However, the model still can be developed in future research by considering planned maintenance, lead-time or quality of material.

References

[1] Zanjani MK, Nourelfath M, Ait-Kadi D. A multi-stage stochastic programming approach for production planning with uncertainty in the quality of raw materials and demand. *Int. J. Prod. Res*. 2009 Aug 3;**48**(16):4701–23. doi.org/10.1080/00207540903055727

[2] Cheraghalikhani A, Khoshalhan F, Mokhtari H. Aggregate production planning: A literature review and future research directions. *Int. J. Ind. Eng. Comput*. 2019;309–30. doi.org/10.5267/j.ijiec.2018.6.002

[3] Angkiriwang R, Pujawan IN, Santosa B. Managing uncertainty through supply chain flexibility: reactive vs. proactive approaches. *Prod. Manuf. Res*. 2014 Jan;**2**(1):50–70. doi.org/10.1080/21693277.2014.882804

[4] Mula J, Poler R, García-Sabater JP, Lario FC. Models for production planning under uncertainty: A review. *Int. J. Prod. Econ*. 2006 Sep;103(1):271–85. doi.org/10.1016/j.ijpe.2005.09.001

[5] Gupta A, Maranas CD. Managing demand uncertainty in supply chain planning. *Comput. Chem. Eng*. 2003 Sep;**27**(8-9):1219–27. doi.org/10.1016/s0098-1354(03)00048-6

[6] Ho CJ. Evaluating the impact of operating environments on MRP system nervousness. *Int. J. Prod. Res*. 1989 Jul;**27**(7):1115–35. doi.org/10.1080/00207548908942611

[7] Demirel E, Özelkan EC, Lim C. Aggregate planning with Flexibility Requirements Profile. *Int J Prod Econ*. 2018 Aug;**202**:45–58. doi.org/10.1016/j.ijpe.2018.05.001

[8] Escudero LF, Kamesam PV. On solving stochastic production planning problems via scenario modelling. *Springer Sci. Bus. Media*. 1995 Jun;**3**(1):69–95. doi.org/10.1007/bf02574804

[9] Adulbhan P, Tabucanon MT. A biobjective model for production planning in a cement factory. *Comput. Ind. Eng*. 1979 Jan;**3**(1):41–51. doi.org/10.1016/0360-8352(79)90003-2

[10] Asghar G, Safeen W, Jahanzaib M. *Tech. J. University of Engineering and Technology (UET) Taxila, Pakistan*. 2015;**20**:12-18.

[11] Tsiakis P, Shah N, Pantelides CC. Design of Multi-echelon Supply Chain Networks under Demand Uncertainty. *Ind. Eng. Chem. Res*. 2001 Aug;**40**(16):3585–604. doi.org/10.1021/ie0100030

[12] Alsop PA, Chen H, Chin-Fatt AL, Jackura AJ, McCabe MI and Tseng HH. *Cement Plant Operations Handbook for Dry Process Plants*. 4th ed. Bern: Tradeship Publications Ltd; 2005. 19 p.

[13] Ohunakin OS, Leramo OR, Abidakun OA, Odunfa MK, Bafuwa OB. Energy and Cost Analysis of Cement Production Using the Wet and Dry Processes in Nigeria. *Scientific Research Publishing, Inc.*; 2013;**05**(09):537–50. doi.org/10.4236/epe.2013.59059

[14] Biggs JR, Campion WM. The effect and cost of forecast error bias for multiple-stage production-inventory systems. *Decis. Sci*. 1982 Oct;**13**(4):570–84. doi.org/10.1111/j.1540-5915.1982.tb01183.x

Supply chain network design to adjust with acquisition of competitor: Case study of a cement company in Indonesia

R B Setiyawan[1,2], I N Pujawan[1,3] and N I Arvitrida[1,4]

[1]Department of Industrial and Systems Engineering, Institut Teknologi Sepuluh Nopember, Surabaya, Indonesia.

E-mail: [2]roi.setiyawan@semenindonesia.com, [3]pujawan@ie.its.ac.id, [4]niniet@ie.its.ac.id

Abstract. In this paper, the problems encountered by the largest market share cement manufacturer in Indonesia after acquiring a competitor is considered. These problems consist of determining the right allocation of production facilities to serve a market, the decision to continue to operate or stop the operation of a grinding and or packing plants, determining the number of clinkers sent from the integrated cement plants to the grinding plants and determining of the amount of cement production in each production facility. This research was conducted with the aim of designing the supply chain network in order to minimize total annualized cost in meeting demand in all marketing areas in Indonesia. The supply chain network is mathematically modelled and the optimal solution is solved using mixed integer linear programming (MILP) method by considering the fixed costs of facility infrastructure, clinker and cement bag and bulk unit production and distribution costs. The results provide recommendation that in the current oversupplied market situation to close 21 out of the 31 units of the packing plant and 2 of 3 units of the grinding plant along with the allocation of each facility with a utilization rate of 76.6% of the capacity owned after the acquisition.

1. Introduction

The Indonesian cement market has been experiencing oversupply since 2014. This condition is caused by a higher increase in production capacity compared to a relatively lower increase in demand. Several new players in the cement industry, both local and foreign, also helped enliven the national cement market. Under these conditions, the national cement company which has the largest market share has completed a strategic action in the form of an acquisition of 80.6% of its competitor in the first quarter of last year. This strategic step has significant potential benefit including an increase in market share to more than 53.4% so that it becomes a dominant cement industry player in the region, increased ready mix and aggregate business along with the potential for innovation in the cement derivative business and the potential for creating significant operational synergies with centralize the procurement, logistics and maintenance processes. This of course can improve the company's bargaining position with suppliers such as raw materials, coal, refractories and maintenance spare parts. However, this strategic step also raises new challenges regarding supply chain networks. When an area of demand that was previously a competition area, after this acquisition, the company must be able to operate in synergy in order to obtain the minimum total cost to meet market demand. This is in accordance with the strategic fit of cement products that are classified as functional products so that all activities along the value chain must be efficient to improve the competitiveness of the company. Firm competition in the cement industry is not enough to compete only on the low cost of goods produced but must implement efficiency strategies along the company's value chain. In addition, companies must also continue to improve services to customers in order to remain competitive [1].

The supply chain network design is a strategic decision because it has a long-term impact on supply chain performance. The possibility of network configuration changes only occurs in relatively long-time intervals but operational processes in the network take place continuously. The supply chain network design can provide an optimal platform for efficient and effective supply chain management [2]. In designing supply chain networks, companies have quite a variety of problems even though the aim is generally the same to obtain the most efficient total cost. When a developing company decides to open a new branch store, the decision is about the location, number, capacity and allocation of production and distribution facilities. In general, supply chain network configurations need to be re-evaluated when there are changes in demand patterns, product variations, production processes, procurement strategies or related to operational costs of the facility. Another factor that really needs to be re-evaluated and redesigned the supply chain network configuration is when companies make mergers and acquisitions that require to integrate different logistical networks. The addition of demand and production facility infrastructure from this strategic acquisition step mandates the company to re-evaluate its supply chain network.

Much research has been done to assist practitioners in making the best decisions at both the strategic and operational levels in the supply chain network. The supply chain network which generally consists of a number of suppliers, production facilities, warehouses and distribution centres in an effort to meet demand in several market locations is analysed with the aim of minimizing total annual costs. Several studies have modelled supply chain networks mathematically and used linear programming methods to obtain optimal solutions [3-13]. Other methods used to find optimal solutions are genetic algorithms [2,14,15] and hybrid approaches that combine analytic and simulation methods [16].

In this paper, we design a supply chain network in one of the cement companies in Indonesia that has acquired a competing company. After the acquisition, the company's infrastructure as a group experienced a significant change, namely the addition of similar production facilities in the form of an integrated cement plant, grinding plant and packing plant (Table 1). The locations of some of these production facilities are close each other because previously the two companies geographically competed (Figure 1). Some strategic decisions are analysed which include determining the right allocation of production facilities in serving an area of demand, the decision to continue to operate or stop the operation of a grinding plants and or packing plants, determining the number of clinkers sent from the integrated cement plants to the grinding plants and the determination of the amount of cement production in each production facility in order to obtain the most efficient total annualized cost in meeting demand in all marketing areas in Indonesia. The supply chain network is mathematically modelled and the optimal solution using the mixed integer linear programming (MILP) method by considering the fixed costs of production facility infrastructure, clinker and cement in bag and bulk unit production costs and distribution costs.

This paper is organized as follows. Section 2 presents the theoretical basis and related previous research. Section 3 explains the problem with the proposed mathematical definition and model. Section 4 contains data analysis of the case being discussed. Finally, Section 5 consists of conclusions and suggestions for future research.

2. Literature Review

Several studies on supply chain network design have been developed to help practitioners solve their supply chain planning. Daskin [3] modelled a multi-product multi-echelon production distribution network system consisting of 3 (three) production facilities that produce 3 (three) types of products which are then sent to 2 (two) warehouse locations and then sent to several customer locations. The system is mathematically modelled which has an objective function to minimize the total fixed costs of facilities and variable costs with the assumption that the warehouse is incapacitated and the production capacity of each plant according to the type of product is independent. Melkote and Daskin [4] introduced a combined of facility location and network design problem in which facilities have constraining capacities on the amount of demand they can serve. They called it the capacitated facility location/network design problem (CFLNDP) where the problem was modelled in the formulation of mixed integer programming with the objective function to minimize total costs with some valid inequality constraint functions. The costs considered in this problem ware travel cost per unit flow on link, cost of constructing a facility at some node and cost of constructing link. This introduced model can be applied in many fields such as regional planning, distribution, telecommunication, energy management and others.

Tsiakis and Papageorgiou [5] conducted research on optimizing production and distribution allocations in the global supply chain network by considering operational and financial constraints. The operational limit covers production capacity, warehouse capacity, production allocation, while financial limits cover production costs, transportation costs, and other related costs in the supply chain network related to differences in currency exchange rates. This research also proposed related to the potential of outsourcing production in business decisions when demand exceeds the capacity of the company. Case studies on global companies engaged in chemical manufacturing were carried out using mixed integer linear programming (MILP) methods in explaining this problem. Tsiakis et al. [6] also conducted research on multiproduct, multi-echelon design in supply chain networks consisting of a number of production facilities located in fixed locations, a number of warehouses facilities and distribution centres at locations that are determined from several potential locations and a number of demands in several customer zones in fixed locations. The system was mathematically modelled using a mixed integer linear programming (MILP) method with the aim of determining the number, location and capacity of the warehouse and distribution centre to be setup, the transportation path of the amount of production and the flow of material in the network. This optimization is done to minimize the total annualized cost by considering infrastructure and operational costs.

A similar study was also carried out by Yu et al. [8] In their paper, they developed integrated location, production and distribution planning for supply chain network design that focused on strategic decisions regarding determining the right number and location to build new factories and distribution centres and allocation problems in each echelon. They formulated problems in mathematical models that have an objective function to minimize total costs in the supply chain and implement these problems in LINGO software to find optimal solutions. The mathematical model they proposed was pure integer linear programming (PILP) which has 4 echelons, multiproduct and multi-period characteristics. Numerical test results showed that the model can be applied to small or medium scale problems in manufacturing industries such as automobiles, electronics, furniture and others. The issue of integrated production planning was also discussed by Chen and Wang [12]. In their paper, they conducted a study at a steel manufacturing company in Canada which had one main production facility centre and several finishing plants at another location. They formulate problems in the form of a linear programming model by considering the cost of production, the level of throughput, the cost of purchasing raw and semi-finished materials and the costs of distribution and transportation. Testing of the model was carried out using production data and produced an analysis that high-level financial benefits can be achieved through the use of an integrated planning methodology.

The topic of integrated modelling can also be applied in multinational companies. Bhutta et al. [13] presented a mixed integer linear formulation that extends the model in determining the location of facilities at a multinational company that considers exchange rates and tariff rates and incorporating production, distribution and investment decision. This model provided insight into production levels, facility configurations, distribution strategies and the level of investment required in this facility and explores the factors that influence facility configuration decisions. Several scenarios have been considered based on various facility configurations and exchange rates and tariff rates. The results obtained that the developed model is in accordance with accepted economic theory.

In the next section we present a description of the problems in this case study along with the proposed formulation of a mathematical model that integrates four different echelons of the supply chain. This mathematical model is then written in a programming script in LINGO to get the optimal solution as the answer to some of the problems mentioned in section 1 above.

3. Problem Description and Model Formulation

In this research we develop optimal supply chain network design in a cement company which has the largest market share in Indonesia after acquiring a competitor. When an area of demand that was previously a competition area then after the acquisition, all production facilities and distribution networks of the acquired companies must be integrated into the existing supply chain network that aims to minimize the total annualized cost in meeting demand in all marketing areas in Indonesia. Some strategic decisions after the acquisition include determining the right allocation of production facilities to serve an area of demand, the decision to continue to operate or stop the operations of the existing grinding plants and or packing plants, determining the number of clinkers sent from the integrated cement plants to the grinding plants and the determination of the amount of cement production in each production facility.

The cement company considered in this paper has production facilities (Table 1) in the form of integrated cement plants (CP), grinding plants (GP) and packing plants (PP) located in several locations to reach marketing areas throughout Indonesia (Figure 1). Integrated cement plant is an integrated cement production facility starting from the processing of cement raw materials (limestone, clay, iron sand and silica sand) into a semi-finished product in the form of a clinker which is then followed by the grinding process into the final product in the form of cement. The integrated cement plant location generally approaches the location of the main raw material, the limestone quarry. To save on transportation costs in an effort to reach marketing areas far from the location of the integrated cement plant, a grinding plant and / or packing plant were built. The grinding plant facility is used to process the semi-finished material (clinker) that is sent from the integrated cement plant into cement. The packing plant facility is used to package bulk cement sent from the integrated cement plant to the final product in the form of packaged cement. Thus, there are three types of material flow that may come out of the integrated cement plant, namely clinker, bulk cement and bag cement.

Table 1. Number of facility locations in Indonesia.

No.	Description	Integrated Cement Plant	Grinding Plant	Packing Plant
1	Number of locations of acquiring company facilities.	4	3	25
2	Number of locations of company facilities acquired.	4	2	6
Total location of post-acquisition production facilities.		8	5	31

We develop a mathematical model in the post-acquisition supply chain network which is an amalgamation of two existing supply chain networks to answer some of the strategic decisions mentioned above. Figure 2 shows the multi-echelon network considered in this study. The formulation of the mathematical model is carried out using the concept of mixed integer linear programming which has the objective function to minimize the total cost in the system in meeting customer demand from all provinces in Indonesia with some formulation of limiting functions explained in the following section below. As explained in the paragraph above that there are differences in the form of material flowing in the network, namely clinker and cement, so these differences in calculating units must be converted into equal cement. The conversion factor used is the clinker factor (CF), which is the percentage of the amount of clinker used to form cement. The percentage value of this clinker factor depends on the type of cement to be produced.

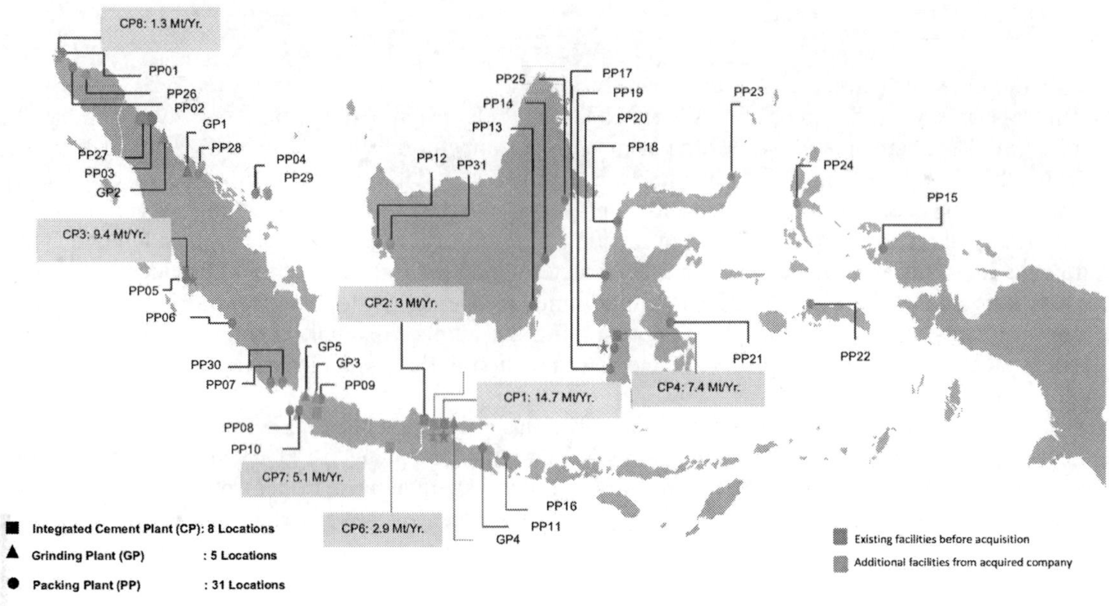

Figure 1. Location of cement production facilities owned after the acquisition.

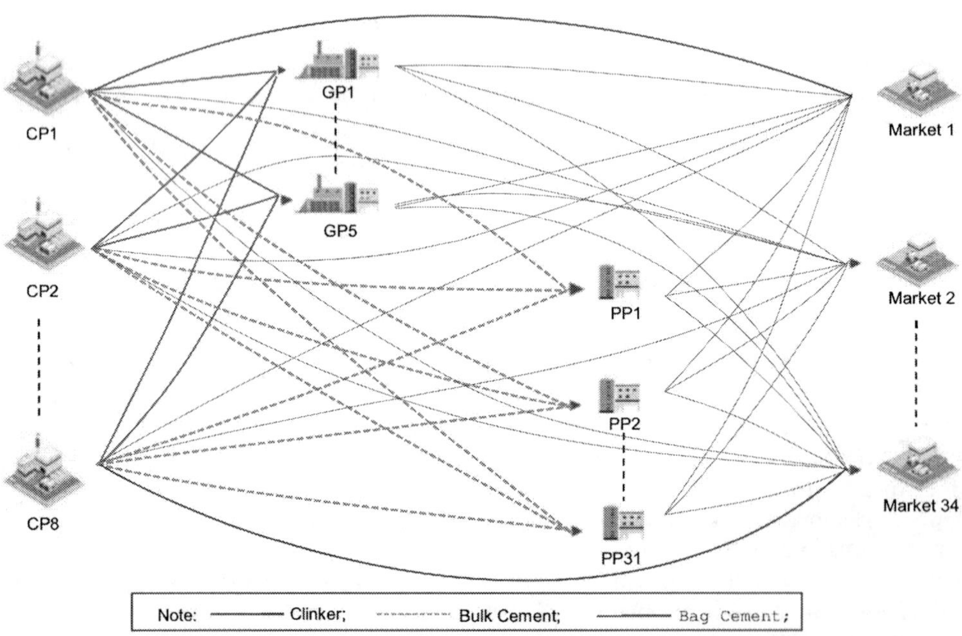

Figure 2. The supply chain network model under consideration.

This proposed model uses the notations shown as follows:

1. Indices:
   1. j Integrated cement plant with an index of 1,2 ... J.
   2. k Grinding plant with an index of 1,2 ... K.

3. l Packing plant with an index of 1,2 … L.
4. m Market at the provincial level with an index of 1,2 ... M.

2. Parameters:
1. f_j^{CP} Fixed operational costs of integrated cement plant j in a year.
2. $f_k^{\text{GP,e}}$ Fixed operational costs of grinding plant k in a year.
3. $f_k^{\text{GP,s}}$ Fixed costs of closing a grinding plant k in a year.
4. $f_l^{\text{PP,e}}$ Fixed operational costs of packing plant l in a year.
5. $f_l^{\text{PP,s}}$ Fixed costs of closing the packing plant l in a year.
6. $C_j^{\text{P,cmt\_zak}}$ Cost of bagged cement production per ton at the integrated cement plant j.
7. $C_j^{\text{P,cmt\_bulk}}$ Cost of bulk cement production per ton at the integrated cement plant j.
8. $C_j^{\text{P},cli}$ Clinker production cost per ton at the integrated cement plant j.
9. C_k^{P} Cost of cement production per ton at the grinding plant k.
10. C_l^{P} Packaging cost per ton at the packing plant l.
11. C_{jm}^{T} Unit transportation cost from the integrated cement plant j to market m.
12. C_{jk}^{T} Unit transportation cost from the integrated cement plant j to the grinding plant k.
13. C_{jl}^{T} Unit transportation cost from the integrated cement plant j to the packing plant l.
14. C_{km}^{T} Unit transportation cost from the grinding plant k to the market m.
15. C_{lm}^{T} Unit transportation costs from packing plant l to market m.
16. Cap_j^{CP} Annual cement production capacity at the integrated cement plant j.
17. $MinOps_j^{\text{CP}}$ Minimum annual cement production in integrated cement plant j.
18. Cap_k^{GP} Annual cement production capacity at the grinding plant k.
19. Cap_l^{PP} Annual cement production capacity at the packing plant l.
20. CF Clinker factor as a conversion factor to calculate the volume of equal cement.
21. D_m Demand in market zone m.

3. Decision variables:
1. Y_k 1 if the grinding plant k is operated; 0 otherwise.
2. Y_l 1 if the packing plant l is operated; 0 otherwise.
3. Q_{jm}^{CP} The volume of cement products shipped from the integrated cement plant j to market m.
4. Q_{jk}^{CP} The volume of clinker products sent from the integrated cement plant j to the grinding plant k. Note: to convert from clinker material to cement, clinker factor (CF) multiplied by clinker volume is required.
5. Q_{jl}^{CP} The volume of cement products shipped from the integrated cement plant j to the packing plant l.
6. Q_{km}^{GP} The volume of cement products sent from the grinding plant k to the market m.
7. Q_{lm}^{PP} The volume of cement products sent from the packing plant l to the market m.

The model to be developed considers the deterministic product demand for aggregate cement product. The mixed-linear integer programming (MILP) problem formulation is explained as follows:

1. Objective function.
   The objective function in this study is to minimize the total annual cost of the cement supply chain network consisting of:
   a. Fixed costs for facility infrastructure.
      This objective function relates to fixed operational costs of the integrated cement plant j and the decision to continue to operate or close the grinding plant k and or packing plant l which is formulated with the following function:

$$\sum_j f_j^{\mathrm{CP}} + \sum_k f_k^{\mathrm{GP,e}} Y_k + \sum_k f_k^{\mathrm{GP,s}} (1 - Y_k) + \sum_l f_l^{\mathrm{PP,e}} Y_l + \sum_l f_l^{\mathrm{PP,s}} (1 - Y_l) \quad (1)$$

The components of these costs include annual operating cost and annual cost of the investment value of production assets which are calculated by parameters of the value of the asset investment, the technical life of the asset and interest using the concept of equal-payment-series capital recovery factor (A = P (A / P), i, n).

b. Facility production costs.
These production costs consist of total cement and clinker production at the integrated cement plant *j*, the cost of cement grinding at the grinding plant *k* and packing costs at the packing plant *l* with the formula of production multiplied by the unit cost of producing per ton as follows:

$$\sum_j C_j^{\mathrm{P,cmt\_zak}} Q_{jm}^{\mathrm{CP}} + \sum_j C_j^{\mathrm{P,cli}} Q_{jk}^{\mathrm{CP}} + \sum_j C_j^{\mathrm{P,cmt\_bulk}} Q_{jl}^{\mathrm{CP}} + \sum_k C_k^{\mathrm{P}} Q_{km}^{\mathrm{GP}} + \sum_l C_l^{\mathrm{P}} Q_{lm}^{\mathrm{PP}} \quad (2)$$

c. Transportation costs.
These transportation costs consist of unit costs of transportation of material in the form of cement from the location of the production facility to the market m and / or clinker from the integrated cement plant j to the grinding plant k multiplied by the volume of material sent. These transportation costs are stated in the following functions:

$$\sum_j C_{jm}^{\mathrm{T}} Q_{jm}^{\mathrm{CP}} + \sum_j C_{jk}^{\mathrm{T}} Q_{jk}^{\mathrm{CP}} + \sum_j C_{jl}^{\mathrm{T}} Q_{jl}^{\mathrm{CP}} + \sum_k C_{km}^{\mathrm{T}} Q_{km}^{\mathrm{GP}} + \sum_l C_{lm}^{\mathrm{T}} Q_{lm}^{\mathrm{PP}} \quad (3)$$

d. The overall objective function.
By combining the cost functions in (1) until (3), the total cost of the supply chain network is stated in the overall objective function as follows:

Min Fixed costs of facility infrastructure + Production costs + Transportation costs

$$\begin{aligned} Min \quad & \sum_j f_j^{\mathrm{CP}} + \sum_k f_k^{\mathrm{GP,e}} Y_k + \sum_k f_k^{\mathrm{GP,s}} (1 - Y_k) + \sum_l f_l^{\mathrm{PP,e}} Y_l + \sum_l f_l^{\mathrm{PP,s}} (1 - Y_l) \\ & + \sum_j C_j^{\mathrm{P,cmt\_zak}} Q_{jm}^{\mathrm{CP}} + \sum_j C_j^{\mathrm{P,cli}} Q_{jk}^{\mathrm{CP}} + \sum_j C_j^{\mathrm{P,cmt\_bulk}} Q_{jl}^{\mathrm{CP}} + \sum_k C_k^{\mathrm{P}} Q_{km}^{\mathrm{GP}} + \sum_l C_l^{\mathrm{P}} Q_{lm}^{\mathrm{PP}} \\ & + \sum_j C_{jm}^{\mathrm{T}} Q_{jm}^{\mathrm{CP}} + \sum_j C_{jk}^{\mathrm{T}} Q_{jk}^{\mathrm{CP}} + \sum_j C_{jl}^{\mathrm{T}} Q_{jl}^{\mathrm{CP}} + \sum_k C_{km}^{\mathrm{T}} Q_{km}^{\mathrm{GP}} + \sum_l C_{lm}^{\mathrm{T}} Q_{lm}^{\mathrm{PP}} \end{aligned} \quad (4)$$

The objective function above is limited by all the constraints described as follows:

2. Capacity constraints of integrated cement plant facilities.
The amount of cement sent from the integrated cement plant *j* goes directly to the market *m* and or packing plant *l*, and the amount of clinker sent to the grinding plant *k* may not exceed the annual production capacity but is greater than or equal to the minimum annual production of the integrated cement plant *j*. The basis of calculation in this model is in units of tons of cement per year so that the clinker material sent from integrated cement plant *j* to grinding plant *k* can be known by means of equal cement divided by clinker factor (CF). The capacity constraint functions of the integrated cement plant facilities are stated as follows:

$$MinOps_j^{\mathrm{CP}} \le \sum_m Q_{jm}^{\mathrm{CP}} + \sum_l Q_{jl}^{\mathrm{CP}} + \sum_k \left(\frac{Q_{jk}^{\mathrm{CP}}}{CF}\right) \le Cap_j^{\mathrm{CP}}, \quad \forall j \quad (5)$$

Clinker factor is deterministic according to the type of cement product. In this paper taking into account aggregation level of single product, the value of clinker factor is obtained from the weighted average clinker factor calculation of several types of general cement products produced by the company.

3. Network structure constraints of the grinding plant facilities.
If the grinding plant *k* is operated, the number of clinker sent from the integrated cement plant *j* to the grinding plant *k* may not exceed the production capacity of the grinding plant *k*. As in constraint

(5) above, the basis for this calculation is in units of tons of cement per year so the number of clinkers sent must be converted first using a clinker conversion factor (CF). The network structure constraint functions of the grinding plant facilities are stated as follows:

$$\sum_j \left(\frac{Q_{jk}^{CP}}{CF}\right) \leq Cap_k^{GP}\, Y_k, \quad \forall k \tag{6}$$

4. Network structure constraints of the packing plant facilities.
   If the packing plant l is operated, the amount of cement product sent from the integrated cement plant j to the packing plant l may not exceed the production capacity of the packing plant l. The basis of this calculation is in units of tons of cement per year. The constraints are stated as follows:

$$\sum_j Q_{jl}^{CP} \leq Cap_l^{PP}\, Y_l, \quad \forall l \tag{7}$$

5. Constraints of demand fulfilment.
   The amount of demand fulfilment in the market m is guaranteed to always be fulfilled. The number of demand in a particular market area is multiplied by the company's target market share in that market area. The constraints of demand fulfilment are stated as follows:

$$\sum_j Q_{jm}^{CP} + \sum_l Q_{lm}^{PP} + \sum_k Q_{km}^{GP} \geq D_m, \quad \forall m \tag{8}$$

6. Constraint of grinding plant synchronization.
   Assuming a zero stock inventory target at the end of the year at the grinding plant facilities, the amount of material entering must be the same as coming out of the grinding plant k. Thus the constraint functions are as follows:

$$\sum_j \left(\frac{Q_{jk}^{CP}}{CF}\right) = \sum_m Q_{km}^{GP}, \quad \forall k \tag{9}$$

7. Constraint of packing plant synchronization.
   This constraint has a similar meaning to the constraint function (9) so the amount of material entering must be the same as coming out of the packing plant l. Thus the constraint functions are as follows:

$$\sum_j Q_{jl}^{CP} = \sum_m Q_{lm}^{PP}, \quad \forall l \tag{10}$$

8. Constraints of non-negativity and integrality.
   Non-negativity constraints mean that the flow of material in the supply chain must not be negative. These constraints are stated in the function as follows:

$$Q_{jm}^{CP} \geq 0,\ \forall j,m;\quad Q_{jk}^{CP} \geq 0,\ \forall j,k;\quad Q_{jl}^{CP} \geq 0,\ \forall j,l;\quad Q_{km}^{GP} \geq 0,\ \forall k,m;\quad Q_{lm}^{PP} \geq 0,\ \forall l,m \tag{11-15}$$

   The integrality constraints explains the decision to keep operating or closing the grinding plant k and or packing plant l which are stated in the following functions:

$$Y_k \in \{0,1\},\ \forall k \in K\ ;\quad Y_l \in \{0,1\},\ \forall l \in L \tag{16-17}$$

4. Computational Results

The case study in this paper is a representation of the real problems faced by a cement company related to the integration of the company's supply chain network that was acquired in the existing supply chain network in order to minimize the total cost of serving customer demand in all provinces in Indonesia. The limitation in this paper is that the product analyzed is aggregated at the level of cement (single product), all integrated cement plants is operated at pre-determined minimum production capacity and facilities that produce special cement such as white cement and granulated blast-furnace slag (GBFS) cement are ignored because those production facilities are dedicated. Thus, in this paper, the number of grinding plant facilities with initial data of 5 locations is adjusted to 3 locations and the number of other facilities remains (J = 8, K = 3, L = 31) and the total market area of 34 provinces (M = 34). To maintain the confidentiality of the original data from the company in this case study, the values in the parameters used in this calculation are not exactly the same but still represent the original conditions.

The parameters used in this case study is explained as follows. Table 2 shows the parameters related to the integrated cement plants (CPs) which includes data on annual production capacity, annual operating fixed costs, unit production costs for clinker, bulk cement and bag cement and unit transportation costs for clinker to the grinding plant k. Table 3 shows the parameters related to grinding plant facilities (GPs) which include annual production capacity, unit production costs, annual operational fixed and closing costs. Table 4 shows the parameters related to packing plant facilities (PPs) which include annual production capacity, unit of packing production cost, annual operational fixed and closing costs. In this paper, except for clinker transportation data from integrated cement plants to grinding plants that have been shown in Table 2, data on bulk cement and cement bag transportation between production facilities and from production facilities to the markets are not displayed in this paper because of the size of the data need more pages if presented in this paper. Cement bag transportation data from the integrated cement plant to the market is represented in an 8x34 matrix. Cement bag transportation data from the grinding plant to the market is represented in a 3x34 matrix. Cement bag transportation data from packing plant to market is represented in a 31x34 matrix. Data on bulk cement transportation from integrated cement plants to packing plants is represented in an 8x31 matrix. These transportation data are represented in units of Rp/ton. Then the target demand fulfilment data in each market zone in all provinces in Indonesia is shown in Table 5.

Table 2. Parameters of capacity, fixed operational costs, transportation costs and unit production costs at cement plants (CPs).

CPs	Capacity *1,000 (Ton/Year)	Annual Operational Fixed Cost (Rp)	Unit Production Cost at Cement Plant j (Rp/Ton)			Transportation Cost Cement Plant j to Grinding Plant k		
			Clinker	Bulk	Bag	GP1	GP2	GP3
CP01	14700	1,874,548,742	362	309	379	295	449	225
CP02	3000	589,791,446	476	413	498	344	497	199
CP03	9400	1,420,793,275	446	398	467	178	185	128
CP04	7400	1,238,723,563	395	337	413	403	556	174
CP05	3600	449,253,368	447	388	468	317	471	248
CP06	3400	469,537,663	450	393	471	372	536	135
CP07	6000	771,028,850	451	396	472	241	405	108
CP08	1800	282,747,598	448	387	469	262	82	248

The clinker factor (CF) parameter used in the calculations in this paper is the weighted average clinker factor of all types of cement produced. Based on production data of several types of cement produced by this company, a CF value of around 0.74 was obtained.

Table 3. Parameters of capacity, unit production cost, fixed operational cost & fixed closing cost at grinding plant (GPs).

GPs	Capacity*1,000 (Ton/year)	Unit Prod. Cost (Rp/Ton)	Fixed Annual Operational Cost (Rp)	Fixed Annual Closing Cost (Rp)
GP01	600	95	96,682,533	76,686,033
GP02	120	115	27,336,507	15,338,757
GP03	540	102	86,014,279	69,017,529

Table 4. Parameters of capacity, unit production costs, fixed operational and closing cost at packing plant (PPs).

PPs	Capacity (Ton/Year)	Unit cost in Rp.		
		Unit Cost (Bag)	Fixed Annual Operational Cost	Fixed Annual Closing Cost
PP 01	300000	43	13,981,746	9,831,746
PP 02	300000	43	13,689,656	9,539,656
PP 03	600000	41	24,684,632	16,884,632
PP 04	300000	43	14,243,612	9,393,612
PP 05	1800000	37	48,034,487	35,709,487
PP 06	300000	45	16,580,326	11,730,326
PP 07	300000	43	14,473,835	10,123,835
PP 08	300000	43	14,462,679	9,612,679
PP 09	600000	41	23,991,245	15,848,745
PP 10	375000	43	13,882,455	9,094,955
PP 11	300000	42	18,012,846	14,112,846
PP 12	300000	43	12,452,232	8,152,232
PP 13	300000	43	15,134,022	10,734,022
PP 14	600000	41	16,154,234	12,104,234
PP 15	300000	44	22,658,379	17,658,379
PP 16	600000	41	22,759,130	15,209,130
PP 17	600000	41	24,419,576	16,669,576
PP 18	300000	44	12,535,463	8,285,463
PP 19	1500000	38	40,633,371	31,208,371
PP 20	600000	41	24,465,621	16,620,121
PP 21	300000	44	13,896,137	9,996,137
PP 22	387500	43	16,023,835	10,370,335
PP 23	600000	41	25,087,933	17,837,933
PP 24	300000	44	13,601,522	9,101,522
PP 25	600000	42	26,926,067	18,276,067
PP 26	180000	46	7,258,317	4,818,317
PP 27	1125000	39	34,145,540	24,370,540
PP 28	462000	42	18,560,549	12,843,049
PP 29	204600	45	9,652,942	6,947,942
PP 30	540000	41	17,174,951	14,412,451
PP 31	180000	46	6,720,183	4,507,683

Table 5. Parameter of demand per market *m*.

D_m	Target demand fulfillment (Ton)
M 01	1184648
M 02	2017189
M 03	1318637
M 04	1567146
M 05	499906
M 06	370127
M 07	734897
M 08	344293
M 09	110644
M 10	1013299
M 11	1350085
M 12	1780831
M 13	3875671
M 14	5328631
M 15	949791
M 16	7474216
M 17	708686
M 18	526280
M 19	363705
M 20	320889
M 21	630391
M 22	107379
M 23	1823403
M 24	284604
M 25	560342
M 26	99038
M 27	325744
M 28	579099
M 29	228411
M 30	122318
M 31	144949
M 32	247752
M 33	273612
M 34	518171

Table 6. Amount of clinkers sent from cement plant *j* to grinding plant *k*.

	Amount of Clinker from Cement Plant j to Grinding Plant k (x1000)			Sub total
	GP1	GP2	GP3	
	-	-	-	-
	-	-	-	-
	444.0	-	-	444.0
	-	-	-	-
	-	-	-	-
	-	-	-	-
	-	-	-	-
	-	-	-	-
Σ	444.0	-	-	444.0

The model developed in section 3 above is solved by the LINGO software. The model formulation is written in three parts of the framework, namely the set of parameters and variables, the mathematical model and the data set that matches the dimensions of the model. Completion of the optimal solution is solved simultaneously using the concept of branch and bound, which is an enumeration process that looks for a series of better and better solutions until the best solution is found.

The optimal solution is obtained as follows. Utilization of production facilities is 76.6% to meet the target demand. Of all the packing plant (PP) facilities that are owned after the acquisition, it is recommended to keep operating 10 units and close the remaining 21 units, including 2 units of the newly acquired facility. Only one grinding plant (GP) should be operated and the remaining two units are recommended to be closed which includes the newly acquired grinding plant. The assignment of GP units that continue to operate as well as the amount of clinker allocation sent from CP is shown in Table 6 (to the right of Table 2) and the total allocation of products serving to which market is shown in Table 9. Table 7 shows the assignment of CPs that directly serve which markets along with the allocation amount cement sent. Table 8 shows the assignment of CPs along with the allocation of the amount of bulk cement sent to PPs that continue to operate as well as the allocation of products to serve the which market is shown in Table 10. A brief summary of the optimal solutions of the post-acquisition network model is shown in Table 11.

Table 7. Amount of cement sent from cement plants (CPs) to markets (tons * 10,000).

Q_{jm}^{CP} (Ton)	Market																																	Sub total
	1	2	3	4	5	6	7	8	9	10	11	12	13	14	15	16	17	18	19	20	21	22	23	24	25	26	27	28	29	30	31	32	33	34	
CPC1	-	-	-	-	-	-	-	-	-	-	-	-	-	338	95	513	71	53	36	32	63	11	-	-	-	-	-	-	-	-	14	25	27	52	**1.330**
CPC2	-	-	-	-	-	-	-	-	-	-	-	-	-	195	-	-	-	-	-	-	-	-	-	-	-	-	-	-	-	-	-	-	-	-	**195**
CPC3	-	-	132	18	2	-	-	-	-	33	-	-	-	-	-	-	-	-	-	-	-	-	-	-	-	-	-	-	-	-	-	-	-	-	**185**
CPC4	-	-	-	-	-	-	-	-	-	-	-	-	-	-	-	-	-	-	-	-	-	-	182	28	56	10	33	58	23	12	-	-	-	-	**402**
CPC5	-	-	-	-	-	-	-	-	-	-	-	-	-	-	-	234	-	-	-	-	-	-	-	-	-	-	-	-	-	-	-	-	-	-	**234**
CPC6	-	-	-	-	-	-	-	-	-	-	-	-	309	-	-	-	-	-	-	-	-	-	-	-	-	-	-	-	-	-	-	-	-	-	**309**
CPC7	-	-	-	-	-	-	-	-	-	-	68	178	79	-	-	-	-	-	-	-	-	-	-	-	-	-	-	-	-	-	-	-	-	-	**324**
CPC8	118	62	-	-	-	-	-	-	-	-	-	-	-	-	-	-	-	-	-	-	-	-	-	-	-	-	-	-	-	-	-	-	-	-	**180**
Σ	**118**	**62**	**132**	**18**	**2**	**-**	**-**	**-**	**-**	**33**	**68**	**178**	**388**	**533**	**95**	**747**	**71**	**53**	**36**	**32**	**63**	**11**	**182**	**28**	**56**	**10**	**33**	**58**	**23**	**12**	**14**	**25**	**27**	**52**	**3.160**

Table 8. Amount of bulk cement sent from cement plant (CPs) to packing plants (tons * 10,000).

Q_{jl}^{CP} (Ton)	Packing Plant																														Sub total
	1	2	3	4	5	6	7	8	9	10	11	12	13	14	15	16	17	18	19	20	21	22	23	24	25	26	27	28	29	30	31	
CP01	-	-	-	-	-	-	-	-	-	-	-	30.0	-	-	-	-	-	-	-	-	-	-	-	-	-	-	-	-	-	-	18.0	**48.0**
CP02	-	-	-	-	-	-	-	-	-	-	-	-	-	-	-	-	-	-	-	-	-	-	-	-	-	-	-	-	-	-	-	**-**
CP03	-	-	60.0	-	110.5	-	-	-	-	-	-	-	-	-	-	-	-	-	-	-	-	-	-	-	-	-	112.5	46.2	-	36.9	-	**366.1**
CP04	-	-	-	-	-	-	-	30.0	11.2	37.5	-	-	-	-	-	-	-	-	-	-	-	-	-	-	-	-	-	-	-	-	-	**78.7**
CP05	-	-	-	-	-	-	-	-	-	-	-	-	-	-	-	-	-	-	-	-	-	-	-	-	-	-	-	-	-	-	-	**-**
CP06	-	-	-	-	-	-	-	-	-	-	-	-	-	-	-	-	-	-	-	-	-	-	-	-	-	-	-	-	-	-	-	**-**
CP07	-	-	-	-	-	-	-	-	48.8	-	-	-	-	-	-	-	-	-	-	-	-	-	-	-	-	-	-	-	-	17.1	-	**65.9**
CP08	-	-	-	-	-	-	-	-	-	-	-	-	-	-	-	-	-	-	-	-	-	-	-	-	-	-	-	-	-	-	-	**-**
Σ	**-**	**-**	**60.0**	**-**	**110.5**	**-**	**-**	**30.0**	**60.0**	**37.5**	**-**	**30.0**	**-**	**-**	**-**	**-**	**-**	**-**	**-**	**-**	**-**	**-**	**-**	**-**	**-**	**-**	**112.5**	**46.2**	**-**	**54.0**	**18.0**	**558.7**

Table 9. Amount of cement sent from grinding plants (GPs) to markets (tons * 10,000).

Q_{jm}^{CP}	Market																																		Sub total
(Ton)	1	2	3	4	5	6	7	8	9	10	11	12	13	14	15	16	17	18	19	20	21	22	23	24	25	26	27	28	29	30	31	32	33	34	
GP01	-	-	-	60	-	-	-	-	-	-	-	-	-	-	-	-	-	-	-	-	-	-	-	-	-	-	-	-	-	-	-	-	-	-	60
GP02	-	-	-	-	-	-	-	-	-	-	-	-	-	-	-	-	-	-	-	-	-	-	-	-	-	-	-	-	-	-	-	-	-	-	-
GP03	-	-	-	-	-	-	-	-	-	-	-	-	-	-	-	-	-	-	-	-	-	-	-	-	-	-	-	-	-	-	-	-	-	-	-
Σ	-	-	-	60	-	-	-	-	-	-	-	-	-	-	-	-	-	-	-	-	-	-	-	-	-	-	-	-	-	-	-	-	-	-	60

Table 10. Amount of cement sent from packing plants (PPs) to markets (tons * 10,000).

Q_{jm}^{CP}	Market																									Sub total
(Ton)	1	2	3	4	5	6	7	8	9	10	11	12	…	22	23	24	25	26	27	28	29	30	31	32	33	34	
PP 01	-	-	-	-	-	-	-	-	-	-	-	-	-	-	-	-	-	-	-	-	-	-	-	-	-	-	-
PP 02	-	-	-	-	-	-	-	-	-	-	-	-	-	-	-	-	-	-	-	-	-	-	-	-	-	-	-
PP 03	-	27.7	-	32.3	-	-	-	-	-	-	-	-	-	-	-	-	-	-	-	-	-	-	-	-	-	-	60.0
PP 04	-	-	-	-	-	-	-	-	-	-	-	-	-	-	-	-	-	-	-	-	-	-	-	-	-	-	-
PP 05	-	-	-	-	-	37.0	73.5	-	-	-	-	-	-	-	-	-	-	-	-	-	-	-	-	-	-	-	110.5
PP 06	-	-	-	-	-	-	-	-	-	-	-	-	-	-	-	-	-	-	-	-	-	-	-	-	-	-	-
PP 07	-	-	-	-	-	-	-	-	-	-	-	-	-	-	-	-	-	-	-	-	-	-	-	-	-	-	-
PP 08	-	-	-	-	-	-	-	-	-	-	30.0	-	-	-	-	-	-	-	-	-	-	-	-	-	-	-	30.0
PP 09	-	-	-	-	-	-	-	34.4	11.1	14.5	-	-	-	-	-	-	-	-	-	-	-	-	-	-	-	-	60.0
PP 10	-	-	-	-	-	-	-	-	-	-	37.5	-	-	-	-	-	-	-	-	-	-	-	-	-	-	-	37.5
PP 11	-	-	-	-	-	-	-	-	-	-	-	-	-	-	-	-	-	-	-	-	-	-	-	-	-	-	-
PP 12	-	-	-	-	30.0	-	-	-	-	-	-	-	-	-	-	-	-	-	-	-	-	-	-	-	-	-	30.0
PP 13	-	-	-	-	-	-	-	-	-	-	-	-	-	-	-	-	-	-	-	-	-	-	-	-	-	-	-
PP 14	-	-	-	-	-	-	-	-	-	-	-	-	-	-	-	-	-	-	-	-	-	-	-	-	-	-	-
PP 15	-	-	-	-	-	-	-	-	-	-	-	-	-	-	-	-	-	-	-	-	-	-	-	-	-	-	-
PP 16	-	-	-	-	-	-	-	-	-	-	-	-	-	-	-	-	-	-	-	-	-	-	-	-	-	-	-
PP 17	-	-	-	-	-	-	-	-	-	-	-	-	-	-	-	-	-	-	-	-	-	-	-	-	-	-	-
PP 18	-	-	-	-	-	-	-	-	-	-	-	-	-	-	-	-	-	-	-	-	-	-	-	-	-	-	-
PP 19	-	-	-	-	-	-	-	-	-	-	-	-	-	-	-	-	-	-	-	-	-	-	-	-	-	-	-
PP 20	-	-	-	-	-	-	-	-	-	-	-	-	-	-	-	-	-	-	-	-	-	-	-	-	-	-	-
PP 21	-	-	-	-	-	-	-	-	-	-	-	-	-	-	-	-	-	-	-	-	-	-	-	-	-	-	-
PP 22	-	-	-	-	-	-	-	-	-	-	-	-	-	-	-	-	-	-	-	-	-	-	-	-	-	-	-
PP 23	-	-	-	-	-	-	-	-	-	-	-	-	-	-	-	-	-	-	-	-	-	-	-	-	-	-	-
PP 24	-	-	-	-	-	-	-	-	-	-	-	-	-	-	-	-	-	-	-	-	-	-	-	-	-	-	-
PP 25	-	-	-	-	-	-	-	-	-	-	-	-	-	-	-	-	-	-	-	-	-	-	-	-	-	-	-
PP 26	-	-	-	-	-	-	-	-	-	-	-	-	-	-	-	-	-	-	-	-	-	-	-	-	-	-	-
PP 27	-	112.5	-	-	-	-	-	-	-	-	-	-	-	-	-	-	-	-	-	-	-	-	-	-	-	-	112.5
PP 28	-	-	-	46.2	-	-	-	-	-	-	-	-	-	-	-	-	-	-	-	-	-	-	-	-	-	-	46.2
PP 29	-	-	-	-	-	-	-	-	-	-	-	-	-	-	-	-	-	-	-	-	-	-	-	-	-	-	-
PP 30	-	-	-	-	-	-	-	-	-	54.0	-	-	-	-	-	-	-	-	-	-	-	-	-	-	-	-	54.0
PP 31	-	-	-	-	18.0	-	-	-	-	-	-	-	-	-	-	-	-	-	-	-	-	-	-	-	-	-	18.0
Σ	-	140.2	-	78.5	48.0	37.0	73.5	34.4	11.1	68.5	67.5	-	-	-	-	-	-	-	-	-	-	-	-	-	-	-	558.7

Table 11. Summary of the calculation results of the optimal solution.

	Parameter	Amount	Parameter	Amount
A	Clinker Factor (CF)	0.74	Total cement plant capacity	49,300,000
B	Obj. Function Min. Total Cost (Rp)	28,627,677,645.6	Total targeted demand	37,784,802
1	Infrastructure cost (Rp)		Total demand fulfilment	37,784,802
	Establish GP	96,682,532.5	% Cement plant utilization	76.6%
	Shutdown GP	84,356,285.8	Number of GPs Opened	1
	Establish PP	214,108,954.8	Number of GPs Closed	2
	Shutdown PP	270,569,570.4	Number of PPs Opened	10
2	Production cost (Rp)		Number of PPs Closed	21
	Cement bag	13,461,989,992.5	Vol. of cement bag delivered from CP to market	31,597,777.0
	Clinker	198,125,973.3	Vol. of clinker delivered from CP to GP	444,000.0
	Cement bulk	2,130,092,155.3	Vol. of cement bulk delivered from CP to PP	5,587,025.0
	Cement from GPs	56,880,000.0	Vol. of cement bag delivered from GP to market	600,000.0
	Cement from PPs	225,486,195.0	Vol. of cement bag delivered from PP to market	5,587,025.0

All the data above that shows costs are represented in units of one thousand Rupiah (Rp. x 1,000). The results of the above calculation show its relevance to the condition of the national cement market which is still oversupply where the target demand is only around 76.6% of the integrated cement plant production capacity owned after the acquisition.

5. Conclusions and recommendations

It is highly recommended to redesign the supply chain network when a company acquires a competitor's company. This is because the production facilities of the acquired companies must be integrated into the existing supply chain network. Meanwhile total demand also increased from the result of merging demand. Designing supply chain networks can be a fairly complex problem depending on the level of depth of analysis related to the strategic decisions being considered. The more detailed strategic decisions that are considered certainly require large enough input data and must be accurate enough so that the resulting decision can be right on target. However, this effort is proportional to the results of efficient supply chain operations because operations in the network take place continuously.

The results of the optimal solution provide recommendations to the company that in this oversupply market situation to close several units of production facilities in order to obtain the most efficient total cost in meeting the target market share in accordance with the current demand of only 76.6% of the total capacity owned after the acquisition. Packing plants which are still in operation are PP03, PP05, PP08, PP9, PP10, PP12, PP27, PP28, PP30 and PP31 and the recommended amount to be closed is 21 units which include 2 units of packing plant owned by the acquired company. Grinding plant that is still in operation is GP1 and the recommended amount to be closed is 2-unit grinding plant owned by the acquired company. PP12 (existing facility) and PP31 (PP of the acquired company) in the M18 market area, both of which are still operated according to the demand. The cost of closing the production facility, which includes the investment cost of the asset, is included in the annual total supply chain cost. Further policies regarding the closure of these facilities are not described in this paper. The options are to remain on standby until there is an increase in demand, especially government infrastructure projects and an increase in residential building needs.

The integrated supply chain network model proposed in this paper with the results of indicative figures in accordance with the strategic decisions to be taken shows that the model can be applied. Henceforth, this model can be further developed in relation to strategic decisions on how to maximize profits in the holding company that has several operating companies from the acquisition where each operating company has each brand for the same type of product.

References

[1] Thomas DJ, Griffin PM. Coordinated supply chain management. *Eur. J. Oper. Res.* Elsevier BV; 1996 Oct;**94**(1):1–15. doi.org/10.1016/0377-2217(96)00098-7

[2] Altiparmak F, Gen M, Lin L, Karaoglan I. A steady-state genetic algorithm for multi-product supply chain network design. *Comput. Ind. Eng.* Elsevier BV; 2009 Mar;**56**(2):521–37. doi.org/10.1016/j.cie.2007.05.012
[3] Daskin MS, editor. *Network and Discrete Location: Models, Algorithms, and Applications*, 2nd Edition. (John Wiley & Sons, Inc.). 2013 Jun 25. doi.org/10.1002/9781118537015
[4] Melkote S, Daskin MS. Capacitated facility location/network design problems. *Eur. J. Oper. Res.* Elsevier BV; 2001 Mar;**129**(3):481–95. doi.org/10.1016/s0377-2217(99)00464-6
[5] Tsiakis P, Papageorgiou LG. Optimal production allocation and distribution supply chain networks. *Int. J. Prod. Econ.* Elsevier BV; 2008 Feb;**111**(2):468–83. doi.org/10.1016/j.ijpe.2007.02.035
[6] Tsiakis P, Shah N, Pantelides CC. Design of multi-echelon supply chain networks under demand uncertainty. *Ind. Eng. Chem. Res.* American Chemical Society (ACS); 2001 Aug;**40**(16):3585–604. doi.org/10.1021/ie0100030
[7] Taxakis K, Papadopoulos C. A design model and a production–distribution and inventory planning model in multi-product supply chain networks. *Int. J. Prod. Res.* Informa UK Limited; 2016 Mar 21;**54**(21):6436–57. doi.org/10.1080/00207543.2016.1158882
[8] Yu VF, Normasari NME, Luong HT. Integrated location-production-distribution planning in a multiproducts supply chain network design model. *Math. Probl. Eng.* Hindawi Limited; 2015;**2015**:1–13. doi.org/10.1155/2015/473172
[9] Cóccola ME, Zamarripa M, Méndez CA, Espuña A. Toward integrated production and distribution management in multi-echelon supply chains. *Comput. Chem. Eng.* Elsevier BV; 2013 Oct;**57**:78–94. doi.org/10.1016/j.compchemeng.2013.01.004
[10] Melo MT, Nickel S, Saldanha da Gama F. Dynamic multi-commodity capacitated facility location: a mathematical modeling framework for strategic supply chain planning. *Comput. Oper. Res.* Elsevier BV; 2006 Jan;**33**(1):181–208. doi.org/10.1016/j.cor.2004.07.005
[11] Papageorgiou LG, Rotstein GE, Shah N. Strategic supply chain optimization for the pharmaceutical industries. *Ind. Eng. Chem. Res.* American Chemical Society (ACS); 2001 Jan;**40**(1):275–86. doi.org/10.1021/ie990870t
[12] Chen M, Wang W. A linear programming model for integrated steel production and distribution planning. *Int. J. Oper. Prod. Manag.* Emerald; 1997 Jun;**17**(6):592–610. doi.org/10.1108/01443579710167276
[13] Bhutta KS, Huq F, Frazier G, Mohamed Z. An integrated location, production, distribution and investment model for a multinational corporation. *Int. J. Prod. Econ.* Elsevier BV; 2003 Dec;**86**(3):201–16. doi.org/10.1016/s0925-5273(03)00046-x
[14] Ghosh D, Mondal S. An integrated production-distribution planning with transshipment between warehouses. *Int. J. Adv. Oper. Manag.* Inderscience Publishers; 2017;**9**(1):23. doi.org/10.1504/ijaom.2017.10006422
[15] Fahimnia B, Luong L, Marian R. Genetic algorithm optimisation of an integrated aggregate production–distribution plan in supply chains. *Int. J. Prod. Res.* Informa UK Limited; 2011 Jun 14;**50**(1):81–96. doi.org/10.1080/00207543.2011.571447
[16] Lee YH, Kim SH. Production–distribution planning in supply chain considering capacity constraints. *Comput. Ind. Eng.* Elsevier BV; 2002 Jul;**43**(1-2):169–90. doi.org/10.1016/s0360-8352(02)00063-3

Provision model for line balancing in multi-echelon multi-product continuous cement production

N Permana[1,2], S G Partiwi[1,3], and E Widodo[1]

[1]Department of Industrial Engineering, Institut Teknologi Sepuluh Nopember, Surabaya, Indonesia

E-mail: [2]Nallend.permana@gmail.com, [3]sgpartiwi@gmail.com

Abstract. This study investigated model for line balancing in multi eschelon and multi product. A cement company which have 4 integrated plant consist of multi-eschelon 4 rawmill, 4 kiln, 9 finish mill with multi-product OPC and PCC. This paper proposed production cement plant mathematical model using mixed integer linear programming (MILP). The main objective of the model is minimize cost by changing decision variable production rate (ton/hour) and product allocation route. Constraint is demand fulfillment of multi product, inventory and production balancing, inventory and production min/max production. Furthermore, the improvement recommendation also will be declared in order to reduce production cost.The implication practice of proposed model of line balancing in multi-eschelon multi product continuous production is production manager could develop production planning and manage production rate in order to reduce production cost during multi period.

1. Introduction

Integrated cement plant are consist of raw mill, kiln, finish mill and packer. Some integrated cement plant have multiple production line. The production rate of each production line has to regulate to minimize production cost while comply the demand [1]. When Cement demand is low. Finish mill suffer Start-Stop Operational condition due to high level Cement Silo. Dome clinker's inventory stock will rise. There is an option to prevent dome clinker from over load by transport the clinker out of dome using truck for some transport cost. Kiln's operation always needs free space of dome clinker. Kiln needs Industrial Diesel Oil (IDO) to heating up. IDO is expensive fuel will cause high heating up cost of kiln. When kiln feed is low, raw mill also start-stop operational [2]. In this study, a mathematical model for an optimum production, is proposed and analysed for multi-production line, multi-echelons, multi-product integrated cement plant.

This paper is consist of several chapter. In the next chapter, several literature and previous research is reviewed to determine the research gap in production planning and production cost minimization. Chapter 3 is problem description, conceptual model, and formulation of provision model. Chapter 4 discuss about real cased study to apply in provisioned model. Numerical result from case study shown in chapter 5. Finally, chapter 6 conclude this paper and discussed about possible future research.

2. Literature review

The cement production cost optimization model has been discussed in several paper in the last decade. Most of the research discussed and focus in cement supply chain as cement is bulk material that transportation model and cost is big issued.

Model conceptual of integrated cement plant is shown in Figure 1.

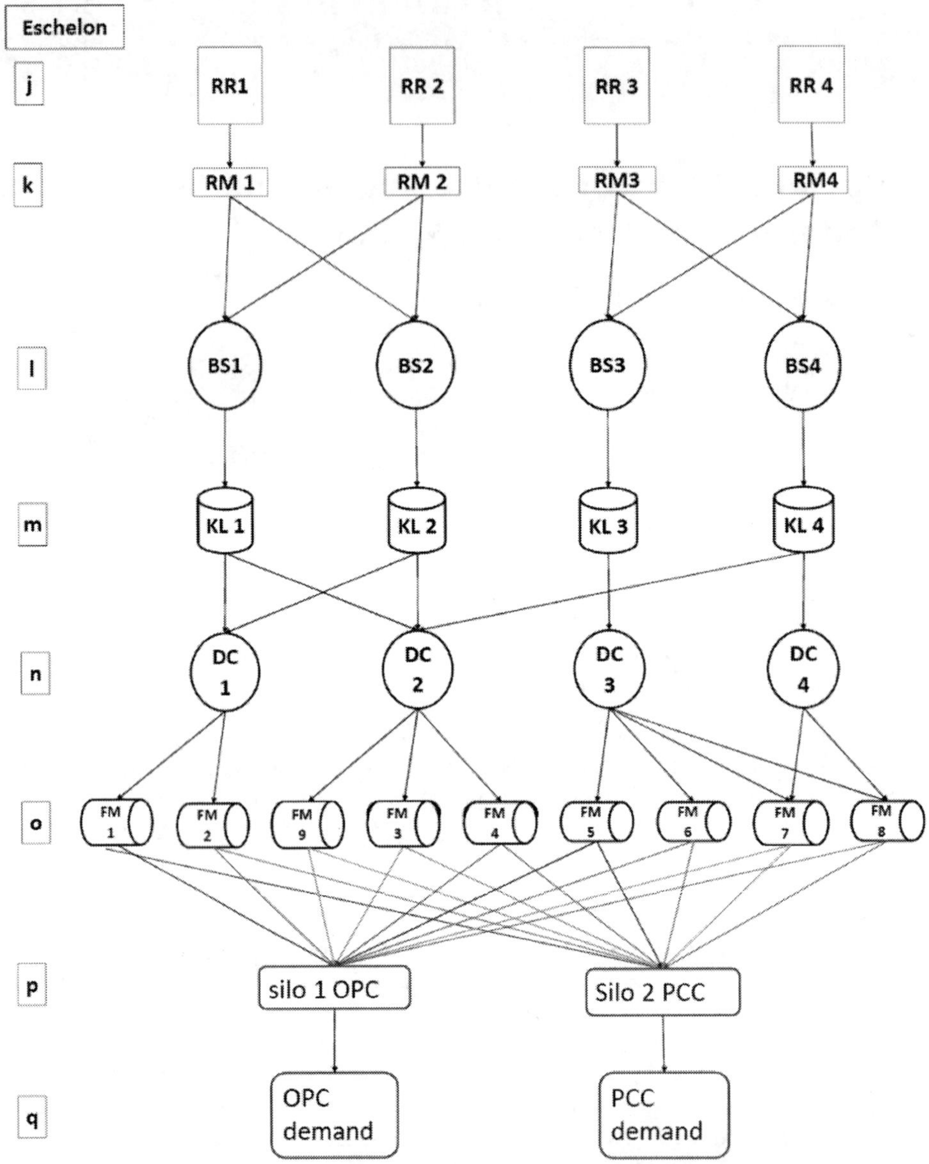

Figure 1. Conceptual model of integrated cement plant.

Current cement manufacture mathematical model investigated production planning and transportation cost minimization. It considers the cement supply chain from mining to distribution of end product [3]. The model is deterministic parameter of demand, production cost, inventory cost and optimal production considering cement and clinker process using linear modelling.

The other mathematical model investigated unit commitment in electric power thermal unit. The cost of starting up an electric power thermal unit after being offline for a couple of days is very high. High starting cost force the operational to carefully planning of the startups and shutdowns of any thermal unit. The electric power thermal unit commitment mathematical model considered the shutdown and startup schedule, fixed cost and variable cost, and also technical and demand security constraint for every electric power thermal unit [4].

Cement production process model has several limitation; irreversible cement production processes, undetailed manufacturing cost, and kiln is a thermal unit etc. literature lacks to provide a model which focus on cost of good manufacture per each cement plant. The deterministic models of the cost of production were taken from annual production report [5]. In this study, a provisioned linear programming model is developed to fill the gap between reviewed paper. Further analyses are presented in chapter 5.

3. Problem description

3.1. Cement production process

Cement is made from limestone, clay, gypsum, and other pozzoland material. Lime stone and clay are mine from open mining. Limestone crush in crusher until 10 cm diameter. Crushed limestone mix together with 80% limestone and 20% clay and stocked in Raw Mix Pile in Reclaimer (RR). Raw Mix Pile in RR will deliver to Raw Mill (RM). Raw Mill will grind Raw Mix pile until fine powder. Fine Powder will be stored in Blending silo (BS). Blending silo supply the kiln (KL) feed. Blending Silo's stock must be confirmed to prevent kiln production slowdown or stop. The raw material are heated in rotary kiln up to 1300-1500°C. Kiln product is Clinker then stored in clinker dome. Clinker stored in clinker dome (CD) supply the finish mill (FM). Clinker will mix with gypsum and any other pozzolan in finish mill then grind until become fine product and stored at cement silo (CS). This complete cement production process called integrated cement plant. Integrated cement plant is multi-echelon and multi product model. Integrated cement plant flow process shown in Figure 2.

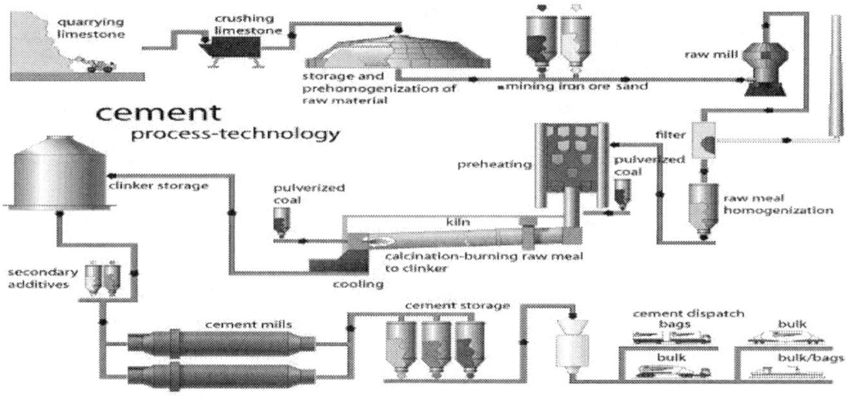

Figure 2. Integrated cement plant flow-process.
Source: cementequipment.org

In this study, a provisioned linear model to minimize production cost and also solve cement production planning problem. For this purpose, the Fixed cost, variable cost, heating up cost, shutdown cost of Raw mill, kiln, and finish mill are taken into consideration [6]. Following, the indices, parameters and the variables of the provisioned model are introduced.

Indices and sets

- k index of rawmill which grind rawmix to fine product, k = 1,2…K
- l index of Blending silo which stored rawmix fine product, l = 1,2…L
- m index of Kiln which process rawmix fine product into clinker, m = 1,2…M
- n index of Dome clinker which stored clinker, n = 1,2…N
- o index of Finish Mill whi grind clinker and mix it with addictive to produce cement, o = 1,2…O
- p index of Cement silo which stored cement, p = 1,2…P
- q index of Type of cement production, 1= OPC, 2 = PPC
- t index of Time period in days a month, t = 1...T, where T typically number days in specific month

Parameter

Ak	where Ak is fixed cost of unit rawmill k
Am	where Am is fixed cost of unit Kiln m
Ao	where Ao is fixed cost of unit Finish mill o
BRMkl	where BRMkl is variable cost of unit rawmill k fill blending silo l
BKLmn	where BKLmn is variable cost of unit kiln m fill clinker dome n
BFMop	where BFMop is variable cost of unit finish mill o fill silo p
Cm	where Cm is start-up cost of unit m (kiln)
Ek	where Ek is shut down cost of unit rawmill k. Minimum power for shutdown (ex. Lighting, oil pump circulation etc)
Em	where Em is shut down cost of unit kiln m. Minimum power for shutdown (ex. Lighting, oil pump circulation etc)
Eo	where Eo is shut down cost of unit finishmill o. Minimum power for shutdown (ex. Lighting, oil pump circulation etc)
Bm5 Dm5t	where Bm[n=5] is variable cost of transport equipment kiln m to n=5 (Clinker open yard) in Rp/ton and Dm[n=5]t is tonage of clinker is being transported from kiln m to clinker open yard n=5 during of period t
B5o D5ot	where B[n=5]o is variable cost of transport equipment n=5 (Clinker open yard) to finish mill o in Rp/ton and D[n=5]ot is tonage of clinker is being transported from open yard to finish mill o during of period t
D1t	Demand for type cement OPC in ton
D2t	Demand for type cement PCC in ton
X1t...X16t	Total tonage product fill in cement silo 1... Cement Silo 16
IBSLt	Inventory Blending Silo L at period t
ICDnt	Inventory clinker dome n at period t
ICSpt	Inventory cement silo p at period t
YCSpqt	Connectivity matrix binary between cement silo p to demand q during period t
YFMopt	Connectivity matrix binary between cement finish mill o to cement mill during period t
YYFMopt	Connectivity matrix maximum production rate finish mill o to cement mill p during period t

3.2. Proposed model

3.2.1 Objective function

Objective function is minimized total cost which are divided into several sub- categories. Raw Mill production Cost, Kiln Production cost and finish mill production cost. Following the expressed production cost formula.

Raw mill production cost

$$\sum_{t=1}^{T}\sum_{k=1}^{4}\sum_{l=1}^{7}[A_k FRM_{kt} + BRM_{kl}RRM_{klt} + E_k(1 - FRM_{kt})] \quad (1)$$

Kiln Production Cost

$$\sum_{t=1}^{T}\sum_{m=1}^{4}\sum_{n=1}^{4}[A_m FKL_{mt} + BKL_{mn}RKL_{mnt} + C_m G_{mt} + E_m(1 - FL_{mt})] \quad (2)$$

Finish mill production cost

$$\sum_{t=1}^{T}\sum_{o=1}^{4}\sum_{q=0}^{1}[A_o FFM_{ot} + BFM_{op}RFM_{opt} + E_o(1 - FFM_{ot})] \quad (3)$$

Transport cost from kiln m to open storage

$$\sum_{t=1}^{T}\sum_{m=1}^{4} B_{m5}D_{m5t} \quad (4)$$

Transport Cost from clinker open storage to finish mill o

$$\sum_{t=1}^{T}\sum_{o=1}^{9} B_{5o} D_{5ot} \tag{5}$$

Following represents the total costs of the system which is the summation of total production cost of every production equipment,

$$Min\ Z = \sum_{t=1}^{T}\sum_{k=1}^{4}\sum_{l=1}^{7}[A_k FRM_{kt} + BRM_{kl} RRM_{klt} + E_k(1 - FRM_{kt})] + \sum_{t=1}^{T}\sum_{m=1}^{4}\sum_{n=1}^{4}[A_m FKL_{mt} + BKL_{mn} RKL_{mnt} + C_m G_{mt} + E_m(1 - FL_{mt})] + \sum_{t=1}^{T}\sum_{o=1}^{4}\sum_{q=0}^{1}[A_o FFM_{ot} + BFM_{op} RFM_{opt} + E_o(1 - FFM_{ot})] + \sum_{t=1}^{T}\sum_{m=1}^{4} B_{m5} D_{m5t} + \sum_{t=1}^{T}\sum_{o=1}^{9} B_{5o} D_{5ot} \tag{6}$$

3.2.2. Constrain

Demand fulfilment constrain

$$\sum_{p=1}^{16} YCS_{pq} * RCS_{pqt} * 24 \geq D_{qt},\ \forall\, q = 1,2\ ,\ t = 1..7 \tag{7}$$

Constraint that maximum release of a cement silo must be lower than cement silo inventory

$$\sum_{q=1}^{2} YCS_{pq} * RCS_{pqt} * 24 \leq ICS_{pt},\ \forall\, p \tag{8}$$

Inventory for next period t+1

Cement Silo

$$\sum_{o=1}^{9} YFM_{op} * RFM_{opt} * 24 - \sum_{q=1}^{2} YCS_{pqt} * RCS_{pqt} * 24 + ICS_{pt} = ICS_{p(t+1)},\ \forall\, p \tag{9}$$

$$\sum_{m=1}^{4} YKL_{mn} * RKL_{mnt} * 24 - \sum_{o=1}^{9} YCD_{not} * RCD_{not} * 24 + ICD_{nt} = ICD_{n(t+1)},\ \forall\, n \tag{10}$$

$$\sum_{k=1}^{4} YRM_{kl} * RRM_{klt} * 24 - \sum_{m=1}^{4} YBS_{lm} * RBS_{lmt} * 24 + IBS_{lt} = IBS_{l(t+1)},\ \forall\, p \tag{11}$$

Dome Clinker

Blending silo

Constraint max inventory

Every Cement silo, Clinker dome, and Blending silo has its unique max capacity, also Constraint max capacity every Finish mill, kiln and raw mill production rate has its own unique maximum capacity rate.

Constraint Production Balance

Finish mill 1 -9

$$\sum_{n=1}^{5} YCD_{no} * RCD_{not} \leq \sum_{p=1}^{16} RFM_{\ opt} .\ YYYFM_{OP}, \tag{12}$$

Kiln 1-4

$$\sum_{l=1}^{7} YBS_{lm} * RBS_{lmt} \geq \sum_{n=1}^{5} RKL_{\ mnt} .* YYKL_{mn}, \tag{13}$$

Constraint of heating up kiln will take 2 days before production

$$\sum_{t}^{t+1} FKL_{mt} \leq 2(1 - G_{mt}), \tag{14}$$

$$FKL_{m(t+1)} - FKL_{mt} \leq M(G_{m(t-1)}) \tag{15}$$

4. Case Study

An Indonesia cement company have a production site consist of four integrated plants. The company produce two types of cement, OPC (ordinary Portland cement) and PCC (Portland composite cement). In this study, the data on cost and demand are extracted from company's production report from

December 2018 to December 2019. Cost parameter are extracted from real data after plant overhaul 2019. This study discusses about monthly production planning. Currently, the operational manager does not have any tool to decide which equipment should be operated, how much cement to be produced, and when to stop for equipment to annually overhaul maintenance. This study is aimed to helping operational manager to make decision operate equipment with production cost minimization.

Demand for January 2019 until December 2019 as indicated in Table 1 initial inventory per 31 December 2018 as indicated in Table 2, Ratio of raw material to clinker and clinker to cement type as indicated in Table 3.

Table 1. Demand data.

Month	January	February	March	April	May	June
OPC	346443	296883	357526	237505	297721	156350
PPC	632226	555620	546883	481343	587416	459157

Month	July	August	September	October	November	December
OPC	319155	286555	329980	336597	319423	243044
PPC	707209	648517	799286	894819	828728	675517

Table 2. Initial inventory cement silo.

Initial Inventory 31 December 2018					
Clinker Dome	**Ton**	**Blending Silo**	**Ton**	**Cement Silo**	**Ton**
1	73443	1	33620	OPC	45244
2	63267	2	31440		
3	44461	3	36859	OPC	147802
4	70193	4	32062		

Table 3. Ratio of raw material to clinker and clinker to cement type.

Raw Material to Clinker Ratio		Clinker to Cement Product Ratio			
Input Kiln	1	Input Clinker	0.85	Output OPC	1
Output Kiln	0.625		0.65	Output OPC	1

5. Numerical Result and Discussion

The mathematical model presented in this study is a deterministic model which optimizes production planning decisions in a cement production industry. It is applied in a real case in Indonesia cement plant to justify the proposed model. In this model, demand is considered to be a deterministic parameter and objective function is representing system's costs, including fixed cost, variable cost, kiln's heating up cost, and idle cost. In this case minimum stock of inventory cement is 50%, 50% for blending silo and clinker dome according to management preference.

Using LINGO 11 software, the result show global optimum with all constrain is satisfied. Table 4, Table 5 and Table 6 are presenting optimal quantities of type cement, clinker production, and raw mill production. Table 7 are presenting cost expressions of the objective function are presented separately for more clarity.

Table 4. Production for finish mill.

Month	Cement Type	Finish Mill								
		FM 1	**FM 2**	**FM 3**	**FM 4**	**FM 5**	**FM 6**	**FM 7**	**FM 8**	**FM 9**
January	OPC	0	133200	0	0	133200	0	0	0	63599
	PCC	139200	6000	139200	139200	6000	0	167040	0	117361
February	OPC	0	0	133200	0	0	0	73797	144000	165600
	PCC	0	139200	6000	139200	0	0	93243	23040	15360
March	OPC	0	0	0	0	0	0	0	0	137812
	PCC	139200	139200	0	0	0	139200	167040	167040	43148
April	OPC	0	0	0	0	0	0	144000	93505	0
	PCC	0	0	0	139300	0	0	23040	73535	180690
May	OPC	0	0	90321	0	0	133200	0	144000	0
	PCC	0	139200	48879	0	6000	600	0	23040	180690
June	OPC	0	0	0	0	0	0	0	0	0
	PCC	139200	139200	139200	0	139200	0	167040	167040	180690
July	OPC	0	0	0	0	0	0	83860	144000	165600
	PCC	139200	0	0	0	139200	139200	83180	23040	15360
August	OPC	0	0	0	0	0	0	0	0	165600
	PCC	0	139200	139200	139200	139200	0	167040	167040	15360
September	OPC	0	115731	0	133200	0	133200	144000	144000	165600
	PCC	0	23469	0	6000	0	6000	23040	23040	15360
October	OPC	0	6269	0	0	0	0	0	144000	0
	PCC	0	132931	139200	139200	133657	139200	167040	23040	180960
November	OPC	0	0	0	0	0	0	0	0	0
	PCC	0	139200	139200	0	0	0	167040	167040	180690
December	OPC	0	0	0	0	0	0	144000	99044	0
	PCC	0	139200	0	139200	0	0	23040	67996	180690

Table 5. Production for kiln.

Date	To dome clinker	From Kiln			
		Kiln 1	Kiln 2	Kiln 3	Kiln 4
Jan	DC1	129503	33902	0	0
	DC2	129696	0	0	147590
	DC3	0	0	210485	0
	DC4	0	0	0	0
Feb	DC1	124505	0	0	0
	DC2	134694	78300	0	145349
	DC3	0	0	259200	0
	DC4	0	0	0	71394
Mar	DC1	295200	114013	0	0
	DC2	0	145186	0	0
	DC3	0	0	199056	0
	DC4	0	0	0	0
Apr	DC1	1129840	0	0	0
	DC2	0	259200	0	0
	DC3	0	0	78300	0
	DC4	0	0	0	259200
May	DC1	0	0	0	0
	DC2	0	259200	0	38731
	DC3	0	0	161393	0
	DC4	0	0	0	137376
Jun	DC1	0	173954	0	0
	DC2	0	85245	0	0
	DC3	0	0	259200	0
	DC4	0	0	0	259200
Jul	DC1	0	35952	0	0
	DC2	259200	223248	0	0
	DC3	0	0	246164	0
	DC4	0	0	0	45054
Aug	DC1	0	0	0	0
	DC2	0	0	0	0
	DC3	0	0	249885	0
	DC4	0	0	0	259200
Sep	DC1	0	0	0	0
	DC2	800759	0	0	187104
	DC3	0	0	259200	0
	DC4	0	0	0	72095
Oct	DC1	0	91733	0	0
	DC2	78300	167466	0	52817
	DC3	0	0	259200	0
	DC4	0	0	0	103782
Nov	DC1	90480	0	0	0
	DC2	168720	39384	0	0
	DC3	0	0	78300	0
	DC4	0	0	0	0
Dec	DC1	0	90480	0	0
	DC2	39384	168720	0	0
	DC3	0	0	6560	0
	DC4	0	0	0	259200

Table 6. Production for rawmill.

Date	**To blending silo**	**From Rawmill**			
		RM 1	**RM 2**	**RM 3**	**RM 4**
Jan	BS1	41470	0	0	0
	BS2	0	54245	0	0
	BS3	0	0	336776	0
	BS4	0	0	0	0
Feb	BS1	414720	0	0	0
	BS2	0	125280	0	0
	BS3	0	0	414720	540000
	BS4	0	0	0	0
Mar	BS1	414720	0	0	0
	BS2	0	414720	0	0
	BS3	0	0	318489	0
	BS4	0	0	0	0
Apr	BS1	207744	0	0	0
	BS2	0	414720	0	0
	BS3	0	0	125280	0
	BS4	0	0	0	0
May	BS1	0	0	0	0
	BS2	0	414720	0	0
	BS3	0	0	258228	258228
	BS4	0	0	0	281771
Jun	BS1	0	0	0	0
	BS2	0	414720	0	0
	BS3	0	0	414720	540000
	BS4	0	0	0	0
Jul	BS1	414720	0	0	0
	BS2	0	414720	0	0
	BS3	0	0	393862	0
	BS4	0	0	0	0
Aug	BS1	0	0	0	0
	BS2	0	0	0	0
	BS3	0	0	399816	0
	BS4	0	0	0	0
Sep	BS1	129215	0	0	0
	BS2	0	0	0	0
	BS3	0	0	414720	414720
	BS4	0	0	0	125280
Oct	BS1	125280	0	0	0
	BS2	0	414720	0	0
	BS3	0	0	414720	540000
	BS4	0	0	0	0
Nov	BS1	414720	0	0	0
	BS2	0	63014	0	0
	BS3	0	0	125280	0
	BS4	0	0	0	0
Dec	BS1	63014	0	0	0
	BS2	0	414720	0	0
	BS3	0	0	10497	10497
	BS4	0	0	0	414720

Table 7. Total cost for the scenario.

Total Cost RM	Rp191,450,000,000
Total Cost KL	Rp789,407,100,000
Total Cost FM	Rp386,816,300,000
Total Cost	Rp1,367,673,400,000

After run the provision model using software, the results are shown in Table 4 answer the operational manager question "which finish mill" have to produce "what product" in "how many tonnage". Table 4 also inform maintenance manager which mill have to stop in specific month and do annual overhaul maintenance. Table 5 kiln production are shown most kiln are running all year with several month, especially in low season, shutdown to maintenance. Table 6 are shown almost only 3 raw mill running. Table 6 indicates that total installed capacity of raw mill is oversupply and consider idling up 1 raw mill and reengineering to make 3 raw mill could supply 4 kiln.

Compared to actual data from company Flash Report 2019, for total energy cost is Rp1,472,824,000,000. The provision linear programming model reduce production cost by 7%.

6. Conclusion

This paper investigates multiple line integrated cement plant production from beginning process raw material to finished product. Considering cement, clinker and raw mill production process, an optimal production plan is driven out from a linear mixed integer linear programming. In the provision model, demand is assumed to be deterministic parameter, Fixed cost, variable cost, kiln thermal heating up cost and idle up cost are taken into consideration. Optimal results indicate that a better total production cost by 7%. The provision model also provides production managers with maintenance schedule.

The main future research avenues are unfolded into the following sub-categorizes as follows: (1) Extending production planning from monthly to daily production planning; (2) Considering Heuristic approach to support more production complexity and faster calculation; and (3) Considering the stochastic and uncertain demand to increase accuracy of the future demand.

References

[1] Adulbhan P, Tabucanon MT. A biobjective model for production planning in a cement factory. *Comput. Ind. Eng.* 1979 Jan;**3**(1):41–51. doi.org/10.1016/0360-8352(79)90003-2

[2] Ozkan NF, Ulutas BH. Efficiency analysis of cement manufacturing facilities in Turkey considering undesirable outputs. *J. Clean. Prod.* 2017 Jul;**156**:932–8. doi.org/10.1016/j.jclepro.2017.04.102

[3] Ivanov D, Tsipoulanidis A, Schönberger J. Global supply chain and operations management. *STBE.* 2017. doi.org/10.1007/978-3-319-24217-0

[4] Sarwani MZ, Mahmudy WF, Naba A. Cost Optimization of Multi-Level Multi-Product Distribution Using An Adaptive Genetic Algorithm. *JITeCS.* 2017 Jan 12;**1**(2):53. doi.org/10.25126/jitecs.20161218

[5] Noche B, Elhasia T. Approach to innovative supply chain strategies in cement industry; analysis and model simulation. *Procedia Soc. Behav. Sci.* 2013 Apr;**75**:359–69. doi.org/10.1016/j.sbspro.2013.04.041

[6] Benati S, García S, Puerto J. Mixed integer linear programming and heuristic methods for feature selection in clustering. *J. Oper. Res. Soc.* 2018 Jan 5;**69**(9):1379–95. doi.org/10.1080/01605682.2017.1398206

IOP Conf. Series: Materials Science and Engineering 1072 (2021) 012036 doi:10.1088/1757-899X/1072/1/012036

Workload analysis using Modified Full Time Equivalent (M-FTE) and NASA-TLX methods to optimize engineer headcount in the engineering services department

Y Rachmuddin[1,2], D S Dewi[1,3] and R S Dewi[1,4]

[1]Department of Industrial and System Engineering Institut Teknologi Sepuluh November, Surabaya, Indonesia

Email: [2]yunita.rahmuddin@gmail.com, [3]dyahsd@gmail.com, [4]ratna.sari.dewi80@gmail.com

Abstract. With the increasing number of project initiatives, requests for engineering technical assistance are also increased. This raises a problem in the Engineering Services department which is uneven project workloads assigned among engineers, while there are no standard tools yet used to calculate accurate workload. To achieve maximum engineering productivity and performance, this study conducted to calculate engineer's physical and mental workload. To calculate physical workload, the modified Full Time Equivalent (M-FTE) method is used, where engineer's performance rating and allowance factors are also used to combine with the basic FTE calculation method. Estimate hours divided according to project's level of complexity are converted into an index of FTE categories, i.e. underload, normal, and overloaded. The NASA-TLX method is used to get mental workload subjectively based on the average weight rating of six indicators. FTE index result shows adequate number of electrical engineers where half of them can still be assign to new project, while most of the instrument engineers already overloaded. Engineer's mental workload from NASA-TLX result shows an average WWL (Weighted Workload) up to 81.72, which categorized as 'Very High' and corresponds to the nature of projects that always expected to be on schedule with high quality result.

1. Introduction

Every profit-oriented organization or company will always try to improve efficiency and reduce operational costs, while still being required to maintain high standards by providing quality results. As is known that human resources are the main assets of the company in improving the quality of products or services. Effective workforce optimization can make a significant difference to the success of a company. For an organization to operate efficiently and effectively, it must know what its workload is. While there is no one exact definition of workload, it is commonly thought to be the amount of work assigned to, or expected to be completed by a worker in a specified time period. Workload that is set too high or too low can negatively affect overall performance. The main objectives of assessing and predicting workload are to achieve an evenly distributed, manageable workload and to accurately determine the resource levels needed to carry out the work.

In supporting the operations of a company, the main task of the Engineering Services department is to provide technical support and manage resources to be able to help the entire area of the company's operations. Technical support provided includes engineering assessment, technical advice and design, cost calculation, and also root cause failure analysis.

In general, the Engineering Services department works on two types of projects:

1. Operating project; costs less than or equal to USD 40,000 taken from the operating expenses of the owner of the area, and not categorized as company asset.
2. Capital project; costs above USD 40,000 funded by capital expenditure and categorized as company asset to be capitalized.

The main key performance indicators (KPI's) in the Engineering Services department are:

1. Develop Engineering Work Package (EWP) which can consist of detailed designs, cost estimates, material specifications, reports and technical recommendations, as well as construction drawings. This EWP is created and targeted to be completed according to the schedule set by each project manager. This KPI is called EWP Compliance.
2. Ensuring that the result is in accordance with the goals, scope and technical design of the project so that it can be handed over to the project sponsor / owner as well as the maintenance team. This KPI is called Design Conformity.

These two KPI's must always exist in each project assign to every engineers. Some projects may have one or more of each KPI's.

With the increasing number of project initiatives in the company that are related to business continuity, improvement ideas, occupational safety and health, or research and development, the number of requests for engineering support are also increased. So that in the Engineering Services department itself several issues and challenges arise, including:

1. Increasing workload, both in operating and capital projects, which are required to deliver results on time with the best quality.
2. Related to human resource management, the unequal workload between one engineer and another.
3. There are no quantitative tools that can provide strong justification for making decisions on the addition of manpower, also to determine whether an engineer can still take an assignment of a new project or not.

2. Research Methods

Effective workload planning by optimizing the capacity of workers must be done to increase productivity [1]. Methods that are widely used to calculate or analyse workload objectively include the Full Time Equivalent (FTE) method. This method is used to determine the optimal amount of labour for a particular task from the workload it receives [2]. The FTE method can calculate the number of hours worked by one full-time employee over a fixed period of time which is considered as one month or one year [3].

The NASA-TLX method is a fairly popular technique used to measure mental workload subjectively. This method was chosen to complement the results of this study because of previous studies found that mental burden also greatly affects employee performance, both excessive and less mental burden. The use of six subscales in the NASA-TLX method to calculate overall workload scores has been found to reduce variability among subjects, relative to unidimensional workload ratings, and also provide diagnostic information about the source of the workload [4].

In general, engineer's tasks, responsibilities and accountability of engineer in the Engineering department are as follows:

- Develop plans, designs, and technical calculations for various elements in electrical and instrumentation projects according to all procedures, standards and policies of Engineering services and HSE departments.
- Prepare engineering construction drawings and ensure the issuance of engineering packages that are comprehensive, understandable, and constructible.
- Ensure project design is completed according to plan and is responsible for safe and reliable results.
- Manage project workflows by providing direct technical support throughout the project and providing reports under the direct supervision of the Electrical / Instrument Supervisor.
- Resolving technical issues during project implementation and completing commissioning after the installation is completed, to ensure the project meets the planned objectives.

IOP Conf. Series: Materials Science and Engineering 1072 (2021) 012036 doi:10.1088/1757-899X/1072/1/012036

- Responsible for quality, technical competence, and professionalism in the engineering efforts undertaken.
- Providing technical assistance to Maintenance and Operations Department when needed.

Table 1 below shows the profile of the engineers who become respondents in this study. The permanent employees consist of 12 (twelve) engineers with three different level:

- Junior Engineer: 5 (five) people
- Engineer: 6 (six) people
- Senior Engineer: 1 (one) person

Table 1. Engineers profile in the electrical/instrument section.

No.	Name	Position	Age (years)	Years of experience
1	Maharani Hasan	Jr. Instrument Engineer	24	2
2	Sofyadi W. Sofyan	Jr. Electrical Engineer	42	2.5
3	Rachmat S. Putra	Jr. Electrical Engineer	34	2.5
4	Christian Yohanis	Jr. Electrical Engineer	38	5
5	A. Betteng Gau	Jr. Electrical Engineer	40	5
6	Busran Mahmud	Instrument Engineer	36	8
7	Leo Agung A. Purnomo	Instrument Engineer	35	8
8	Zulkifli Hamzah	Instrument Engineer	42	16
9	Muhammad Ihsan	Electrical Engineer	36	8
10	Zulfikar Ibnu Mas'ud	Electrical Engineer	37	8
11	Iwan Ignatius Adam	Electrical Engineer	53	22
12	Sofyan	Sr. Electrical Engineer	40	15

2.1. Modified Full Time Equivalent (M-FTE) calculation

Because the position of the engineer is divided into three levels, namely Junior Engineer, Engineer, and Senior Engineer, each of which also has a different service period, it is necessary to distinguish the weight from the estimated time given to complete the same task in advance by determining a certain multiplier factor (performance rating) for each level and tenure. This weight will be added to the factor in the FTE calculation, so it is called Modified Full Time Equivalent (M-FTE).

The types of projects are also differed from one another. Thus, the estimated work time of each task in a project also needs to be distinguished based on the complexity and cost of the project (capital value) using the matrix used in the conceptual project development stage of FEL (Front-End Loading) method, where the project level is divided into 3 (three) categories, namely: Light, Moderate, and Complex project.

Table 2 shows the total number of tasks for each engineer/respondent from January 2020 until May 2020 taken from the IProM (Integrated Project Management) application, along with the division of project levels for each task that has been determined based on a matrix of complexity and project costs. This data will be processed later in the FTE calculations.

Table 2. Total number of tasks for each E/I engineer as of May 2020

No	Name	Position	EWP Compliance			Design Conformity			Total Tasks
			Light	Moderate	Complex	Light	Moderate	Complex	
1	Maharani Hasan	Jr. Instrument Engineer	6	5	1	3	9	1	25
2	Sofyadi W. S.	Jr. Electrical Engineer	1	0	2	2	2	1	8
3	Rachmat S. Putra	Jr. Electrical Engineer	0	9	4	2	2	1	18
4	Yohanis Christian	Jr. Electrical Engineer	3	1	1	2	3	1	11
5	A Betteng Gau	Jr. Electrical Engineer	5	1	0	7	1	0	14
6	Busran Mahmud	Instrument Engineer	4	5	1	5	2	0	17
7	Leo Agung A. P.	Instrument Engineer	0	7	0	2	5	0	14
8	Zulkifli	Instrument Engineer	11	2	4	1	1	1	20
9	Muhammad Ihsan	Electrical Engineer	7	3	0	14	2	1	27
10	Zulfikar I. Masud	Electrical Engineer	10	4	1	3	3	0	21
11	Iwan Adam	Electrical Engineer	3	3	0	2	0	0	8
12	Sofyan	Sr. Electrical Engineer	3	2	0	2	2	0	9

After collecting those datas, then it can proceed with the standard stages of the FTE method, which sets the effective work time used to complete a task. FTE calculation derives into two parts, which are FTE of the company and FTE of the employee. FTE of the company is determining how many hours of work that the employees have (the available time), and FTE of the employee is determining how many hours one full time employee works. The total workload hours are divided by the working hours of one employee.

$$FTE\ of\ Employee = \frac{\text{Workload hours}}{\text{FTE of company}} \tag{1}$$

Workload hour of an activity is determined by how frequent the activity is done in a year (period) and in each period (frequency), how many times the activity is done per unit frequency (quantity or volume), and how long it takes for completing the activity for each unit quantity (process time).

$$Workload\ hours\ = period \times frequency \times quantity \times process\ time \tag{2}$$

The available working hour or FTE of the company is determined by considering working days, off days, standby, and allowances that an employee has in one year working time. The formula is written as Equation 3 as follow.

$$FTE\ of\ company\ = (365\ - off\ days\ - standby\ days\ - allowance)\ x\ number\ of\ working\ hours\ per\ day \tag{3}$$

The FTE of employee becomes:

$$FTE\ of\ employee = \frac{period \times frequency \times quantity \times process\ time}{(365\ - off\ days\ - standby\ days\ - allowance)\ x\ number\ of\ working\ hours\ per\ day} \tag{4}$$

The result of employee FTE calculation can be categorized into three types which are called as FTE index. These types are Overload, Normal, and Underload. The details of FTE index types are shown in Table 3.

Table 3. FTE index.

Index	Definition
0 – 0.99 FTE	Underload
1 – 1.28 FTE	Normal
FTE > 1.28	Overload

The first step to calculate FTE is to define list of task that need to be done by a certain position or subject to be analyse, and then determine how much time needed to complete each task. The duration to complete each task in the KPI is based on estimation data from the Supervisor engineer (expert judgment), by reviewing and analysing historical data for the past three years and also the actual work duration data from the timesheet inputted by each engineer in the existing timesheet online system. From the experience so far, the estimated completion time of a task given by the engineer himself is very vulnerable to bias and cannot be used as an accurate reference.

Table 4 and Table 5 below show list of tasks in each of engineering KPI's (EWP compliance and Design Conformity). As mention before, all projects assign to engineer must have one or more of this KPI. For example, project A required two EWP tasks and one Design Conformity task. Thus, engineer workload should be calculated based on the number tasks and how many hours required to finish each task.

Table 4. EWP compliance tasks list.

List of Task EWP Compliance
Phase: CAR Development / Project Scoping
PSD meeting
Site visit
Develop PSD form and submit for approval
Develop conceptual design (design brief)
Develop Preliminary Design (basic design engineering)
Preliminary Hazop meeting
Develop WBS and cost estimate
Develop project schedule
Phase: Detail Design / Engineering Work Package (EWP)
Site visit
Provide detail engineering calculation (Manual and/or Software)
Develop engineering sketch (design draft)
Provide/review material specification and MR/BoM
Review/revise drawings from designer
Prepare EWP document and submission
Follow up equipment/material procurement - Before PO Issue
PLC / HMI Development (For Instrument discipline only)
Develop/review functional description
Develop/review IO list
Logic programming
HMI programming
Logic & HMI test and simulation

Table 5. Design conformity tasks list.

List of Task Design Conformity
Phase: Material Procurement
Create MIR (Material Input Request) for new or additional material
Follow up equipment/material procurement after PO Issue
Factory Acceptance Test (FAT) & Site Acceptance Test (SAT)
Phase: PLC / HMI Development (For Instrument discipline only)
Logic & HMI test and simulation
Phase: Construction & Commissioning
Provide tender document (pricing schedule, etc.)
Pre-tender meeting and site visit with contractor
Award tender process (clarification and pre-award meeting, kick-off meeting)
Construction progress meeting
Construction assistance, QA/QC inspection, and problem solving
Test and commissioning
Safety Observation & Inspection (SOI)
Phase: Close Out
Prepare punch list and follow up to construction team/others
Prepare mark-up and as-built drawing
Prepare and submit PCC, OCC, CCC and hand over document for approval
Coordination meeting, etc. (if any)

2.2. NASA-TLX Method

This stage is done by collecting and summarizing the results of a questionnaire that has been circled in one of two paired indicators given to each engineer as a respondent, which is felt to be more dominant towards the work performed. The questionnaire is a pairwise comparison to calculate the number of tally of each indicator that has the most influence and will be the weight of the mental load indicator [5]. Furthermore, respondents will rank for six indicators of perceived mental workload. Then the rating multiplication will be performed on each indicator with the weight that has been obtained before, and then an addition is made to get the total value of the product and the distribution is done by fifteen (the number of paired comparisons). Then the results obtained are adjusted to a standard range of values. The paired comparison questionnaire to be given to respondents is given in Table 6, while the questionnaire for the weighting of indicators is shown in Table 7.

Table 6. Paired comparison questionnaire.

No	Indicators	Code	✓	Indicators	Code	✓
1	Physical Demand	PD		Mental Demand	MD	
2	Temporal Demand	TD		Mental Demand	MD	
3	Own Performance	OP		Mental Demand	MD	
4	Effort	EF		Mental Demand	MD	
5	Frustration Level	FR		Mental Demand	MD	
6	Temporal Demand	TD		Physical Demand	PD	
7	Own Performance	OP		Physical Demand	PD	
8	Effort	EF		Physical Demand	PD	
9	Frustration Level	FR		Physical Demand	PD	
10	Own Performance	OP		Temporal Demand	TD	
11	Effort	EF		Temporal Demand	TD	
12	Frustration Level	FR		Temporal Demand	TD	
13	Effort	EF		Own Performance	OP	
14	Frustration Level	FR		Own Performance	OP	
15	Effort	EF		Frustration Level	FR	

Table 7. Indicators weighting.

Score range	Category
0% - 9%	Low
10% - 29%	Medium
30% - 49%	Medium High
50% - 79%	High
80% - 100%	Very High

3. Result and Discussion

3.1. M-FTE result

Table 8 below shows summary of each employee workload based on M-FTE calculation as mention in section 2.1 above. Number of tasks on each KPI (EWP compliance and Design Conformity) are data from January 2020 until May 2020 as shown on Table 2 previously downloaded from the Integrated Project Management (IProM) software, an online system for controlling and monitoring project used by the project team (Project Manager, Project Controller, Engineer, Material Coordinator) in the company.

Table 8. Engineer's Workload Summary with M-FTE Calculation.

No	Name	Position	FTE Index	Category
1	Sofyan	Sr. Electrical Engineer	0.51	Underload
2	Iwan Adam	Electrical Engineer	0.42	Underload
3	Zulfikar Ibnu Mas'ud	Electrical Engineer	1.26	Normal
4	Muhammad Ihsan	Electrical Engineer	1.27	Normal
5	Sofyadi W. Sofyan	Jr. Electrical Engineer	0.74	Underload
6	Rachmat S. Putra	Jr. Electrical Engineer	1.78	Overload
7	A. Betteng Gau	Jr. Electrical Engineer	0.51	Underload
8	Christian Yohanis	Jr. Electrical Engineer	0.79	Normal
9	Zulkifli Hamzah	Instrument Engineer	1.61	Overload
10	Leo Agung A. Purnomo	Instrument Engineer	1.16	Normal
11	Busran Mahmud	Instrument Engineer	1.10	Normal
12	Maharani Hasan	Jr. Instrument Engineer	1.68	Overload

Based on the workload categories found in the FTE calculation, from total of 8 (eight) electrical engineers, there are 4 (four) or half of them are underload compare to other engineers in the same discipline. While one of them in Junior engineer position is considered overload. Whilst for the total of 4 (four) instrument engineers, half of them already overload and the others are normal.

3.2. NASA-TLX survey result

Using NASA-TLX method, data processing until obtaining the level of mental workload are done as follows:

- Calculate the rank comparison between paired indicators, and later add up the results of the comparison for each selected descriptor.
- Calculate the weighted workload (WWL) for each indicator with the equation: $WWL = \Sigma(rating_i \times weight_i)$.
- Calculate the average WWL by dividing it with the total weight, that is 15.

Table 9 below shows the recapitulation of total and average weighted workload taken from the NASA-TLX survey form for each respondent with different level and experiences.

Table 9. Recapitulation of total and average WWL.

Name	Position	Age	Years of Experience	WWL	Average WWL	Category (Simanjuntak, 2010)
Busran Mahmud	Instrument Engineer	36	8	1215	81.00	Very High
Leo Agung A. Purnomo	Instrument Engineer	35	8	1095	73.00	High
Zulkifli Hamzah	Instrument Engineer	42	16	1215	81.00	Very High
Maharani Hasan	Jr. Instrument Engineer	24	2	1295	86.33	Very High
Muhammad Ihsan	Electrical Engineer	36	8	1405	93.67	Very High
Iwan Ignatius Adam	Electrical Engineer	53	22	920	61.33	High
Zulfikar Ibnu Mas'ud	Electrical Engineer	37	8	1200	80.00	Very High
Sofyadi W. Sofyan	Jr. Electrical Engineer	42	2.5	1400	93.33	Very High
Christian Yohanis	Jr. Electrical Engineer	38	5	1365	91.00	Very High
Rachmat S. Putra	Jr. Electrical Engineer	34	2.5	1240	82.67	Very High
A. Betteng Gau	Jr. Electrical Engineer	40	5	1075	71.67	High
Sofyan	Sr. Electrical Engineer	40	15	1285	85.67	Very High

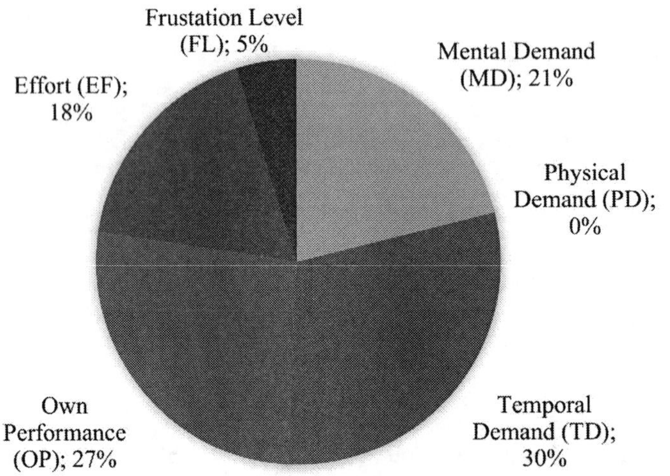

Figure 1. Average workload for each indicator.

Based on the survey result, the most influential indicator that effect engineers' workload as shown on Figure 1 is temporal demand by 30%, followed by own performance and mental demand by 27% and 21% respectively. While physical demand is the lowest to 0%.

Comparing the result of M-FTE calculation and NASA-TLX above, especially for electrical engineer position, the physical workload and mental workload are contradictory, where they still have time for new project assignment, but have high, even very high mental demand. One of similar study done by Fernanda, A [6] in department Engineering and Administration of PT PLN Sidoarjo shows similar result, where in certain position they're doing worker reduction, but the study result shows that position has the highest mental workload.

4. Conclusion

The modified Full Time Equivalent (M-FTE) of each engineer is calculated by determine number of tasks and estimate hours to complete each task. This calculation also must take in to account the performance rating number because of the different in level and experience of each engineer. The method of estimating hours for each task can still be improved perhaps by setting a company standard to record each hour with combining different method such as using stopwatch study.

The M-FTE method can be used to determine each engineer workload from time to time, so it can be the basis for the Supervisor/Manager whether or not he/she can give a new project assignment to an engineer. Because of the nature of the project, it is necessary to differentiate estimate work hours according to the complexity of the project.

Based on the result from the FTE calculation, current electrical engineer headcount is still sufficient to handle current workload, where half of the them considered underload and shall have the assignment for new projects that require electrical discipline to balance the workload among electrical engineers. But for instrumentation discipline, there are only 4 (four) engineers available where half of them already have both high physical and mental workload, which indicate the need of additional manpower for new project requires instrumentation discipline.

Engineer's mental workload based on the result from NASA-TLX survey shows an average WWL (Weighted Workload) up to 81.72, which can be categorized as 'Very High'. The indicator with the highest value of 30% is the Temporal Demand (TD), which is in accordance with the nature of each task carried out by the engineer that always has a completion target determined based on the project schedule. The indicator with the second highest value of 27% is Own Performance (OP), which is also in line with the demands of the engineer who always expected to deliver quality results in accordance with the objectives of the project.

Improvements to physical and mental workload can be done in 3 (three) ways, including: elimination, isolation, and minimization. Elimination is a solution by reducing the activities or jobs that are manual and then replace it with an automatic machine. Isolation is a solution by preventing a job from being handled by untrained or inexperience workers. Related to this research, the project level assigned to each engineer may also consider their level and experience, since currently, even junior engineers are also assigned to a complex project as well. The mental burden experienced by trained or experienced workers doing certain task shall be smaller than inexperienced workers doing the same task. Minimization can be done by increasing the number of workers, adding time rest, changing work processes to avoid excessive physical or mental exertion [7].

From this result, apart from project assignment balancing, it is unavoidable that the challenge of the engineering supervisor/manager nowadays is that they need to start thinking of new ways of working and communicating to support their human resources and have more of their engagement, also maybe start to invest in stress reliefs program. Further study needs to be done to recommend the best improvement to reduce physical and mental workload for the project-based worker.

References

[1] Berseneva KV, Natsubidze AS, Chaschin VV. Human Resource Planning in Implementation of Industrial Enterprises` Strategy. *World Applied Sciences Journal.* 2014;**29**(3):433-437.

[2] Adawiyah W, Sukmawati A. Analisis Beban Kerja Sumber Daya Manusia dalam Aktivitas Produksi Komoditi Sayuran Selada (Studi Kasus: CV Spirit Wira Utama). *Jurnal Manajemen dan Organisasi.* 2016 Jul 18;**4**(2):128. doi.org/10.29244/jmo.v4i2.12619

[3] Tracy B. *Full Time Equivalent (FTE) Calculation Tool.* 2015. Retrieved from http://www.coachmarksteinke.com/sites/default/files/users/FocalPoint_Full_Time_Equivalent.pdf.

[4] Hart SG, Staveland LE. Development of NASA-TLX (Task Load Index): Results of Empirical and Theoretical Research. *Human Mental Workload.* 1988;139–83. doi.org/10.1016/s0166-4115(08)62386-9

IOP Conf. Series: Materials Science and Engineering 1072 (2021) 012036 doi:10.1088/1757-899X/1072/1/012036

[5] Simanjuntak, RA. Analisis Beban Kerja Mental dengan Metoda Nasa-Task Load Index. *Jurnal Teknologi Technoscientia*. 2010;**3**:78-86.
[6] Fernanda A. Analisa Beban Kerja Untuk Menentukan Jumlah Optimal Karyawan (Studi Kasus: Departemen Teknik dan Administrasi PT PLN (Persero) Rayon Sidoarjo Kota). *Thesis Fakultas Teknologi Industri, Institut Teknologi Sepuluh November*. 2014.
[7] Darby F, Walls C. *Stress and Fatigue.* Wellington: Occupational Safety and Health Service of the Department of Labour. 1998.

A model for joint decision between production rate and clinker export proportion using cooperative game theory approach

H T Amrillah[1,2] and E Widodo[1,3]

[1]Department of Industrial and System Engineering, Institut Teknologi Sepuluh Nopember, Surabaya, Indonesia

E-mail: [2]hta1703@gmail.com, [3]erwin@ie.its.ac.id

Abstract. This study aims to provide a decision making model to determine clinker export proportion. Cooperative game theory approach used to build the model. Players in this game are manager of production & material planning, and manager of cement production. Payoff matrix of the game composed using Mixed Integer Linear Programming (MILP) based equation. Expected contribution of this study is to provide a decision making model in determining acceptable proportion between clinker amount to export, and clinker amount to be grinded. Furthermore, this study expected to support revenue maximizing in the company.

1. Introduction

The amount of cement supply and demand in Indonesia has changed since 2015, from an undersupply condition to an oversupply condition. Based on data from the Indonesian Cement Association (ASI), the national cement production capacity currently reaches 113.1 million tons per year. From this capacity, national cement demand is only around 70 million tons per year, so there is an excess production of around 40 million tons per year. Oversupply of cement production in Indonesia is partly due to the entry of foreign cement producers into Indonesia. This condition has forced cement industry to looked at foreign markets to sell their products not only in the form of cement but also clinker.

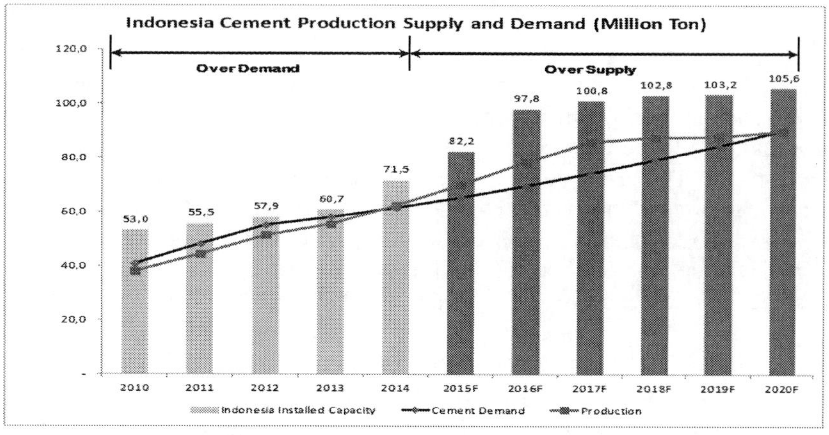

Figure 1. Cement supply and demand in Indonesia.

As one alternative to obtain revenue, one particular cement company in Indonesia has started clinker exports since 2017, but in that year clinker exports were not yet the focus of the company. Only in 2019 will clinker exports become one of the focuses of the company. Clinker production during low season exceeds the amount needed, while the kiln must operate continuously. Therefore there are times when the clinker is forced to be placed in open storage. In 2019 kiln 1 and kiln 2 at one plant of one particular cement company in Indonesia was forced to stop for quite a long time (approximately 20 days) due to improper clinker production scheduling during low season. Therefore, aside from being an alternative to obtain revenue, clinker export becomes a crucial thing for cement industry during cement low demand season.

The proportion of clinker exported with clinker to be grinded into cement mill must be calculated appropriately, so that the demand for cement can still be met but not to the point that excess clinker production is stored in open storage. The decision of clinker amount to be exported, the clinker amount to be grinded into cement mill, the kiln and cement mill production rates involved two parties, namely the material & production planning manager and cement production manager. Both parties have common interest and conflicting interest. The common interest is that they both want to contribute to increase company's revenue. The conflicting interest is that the material & production planning manager wants to export all of the excess from kiln production, whereas the cement production manager wants to use all the clinker produced to be grinded into cement mill. One method that can be used to solve this problem is cooperative game theory. Cooperative game theory can be used because the illustration of the problem that have been explained meets the cooperative game premises, that is involves at least 2 parties, and the benefits are higher when compared to the result from working individually [2].

This paper will propose a model to determine the composition of clinker amount to be exported and clinker amount to be forwarded to cement mill, and also production rates of kiln and cement mill using cooperative game theory, where the payoff matrix is arranged using mixed integer linear programming (MILP), so that optimal revenue associated with the allocation of clinker production can be obtained. The proposed model then demonstrated on one particular cement industry through data experiment.

2. Literature Review

Cooperative game theory is one branch of game theory that models how agents or players compete and work together as a coalition in unstructured interactions to create and capture value [1]. Cooperative game theory is one of mathematical tools developed for the first time by Von Neumann and Morgenstern in 1944. Unlike decision theory where the interests of players are unique or intersect, or zero-sum game where interests between players are opposite, cooperative game theory focuses on situations where mutual benefits occurs between players. One of non-cooperative game example is two person zero sum game, whose equilibrium value is 0, while the example of cooperative game is a two person non-zero sum game.

Research using cooperative game theory approach has been carried out in various fields. Research on human-robot collaborative manufacturing (HRC-Mfg) demonstrated that cooperative game theory can be applied to such system [2]. In another study, cooperative aggregate production planning (Co-APP) introduced as a way to reduce manufacturing plant's operating costs consisting of production costs, labor and inventory [3]. The principle of Co-APP is that when the production planning of 2 or more facilities/plants are combined, they can exchange labor and product inventory so that product demand can be met at a lower cost. Opportunity for cost savings arise from cooperation between plants, which is from reducing inventory and labor levels. Still in the same study, it was also found that job security and worker satisfaction could increase dramatically with collaboration between plants.

Mixed Integer Linear Programming (MILP) is an Integer Programming (IP) which requires several integer variables, while IP is Linear Programming which requires partial or entire of its variables are non-negative integers [4]. MILP has been widely implemented in various studies. Emir, Guler developed the MILP model to plan production at the cement plant by considering changes in electricity consumption prices [5]. This study shows that the amount of obtained savings when the price of electricity consumption is included in the production plan MILP at the cement plant is 11% in average, even reaching 35% for certain months. The amount of obtained savings will be even greater if factory capacity is less utilized.

The implementation of MILP in scheduling steelmaking-continuous casting production has been investigated also. Several constraints considered in the process scheduling of this study are job grouping, technological interdependence, no dead time inside the same group of jobs and dynamic processing time of jobs. The study shows that by using MILP at a continuous casting plant, a significant reduction in production time will be obtained [6].

3. Proposed Model

The problem that will be modeled in this paper is how to find a proportion value that can be accepted by both parties, material & production planning manager and cement production manager in this case, regarding the amount of clinker that will be exported and the amount of clinker that will be grinded into cement mill in 1 month. The approach used to model the problems in this paper is cooperative game theory. Payoff matrix of game is arranged using mathematical equations which will be described below. The proposed model is expected to maximize revenue from clinker production at plant owned by one particular cement company in Indonesia. Several assumptions made for proposed model are: exported clinker always absorbed by market, there is always storage space for clinker and cement that made, and no downtime occurred in kiln and finish mill/cement mill. Proposed model in this paper is limited to one particular cement industry in Indonesia.

A player in a game is an entity that makes decisions [7]. The players of the game in this paper are material & production planning manager (referred to as player 1), and cement production manager (referred to as player 2). Both players have common interest and conflicting interests. The common interest is that both players want to contribute to increase company's revenue, whereas the conflicting interests of the two players are as follows:

- Player 1 wants to export all of the excess from kiln production.
- Player 2 wants to use all the clinkers produced.

3.1. Model Indices, Parameters, Decision Variables

Model indices, parameters and decision variables of the model to be formulated are as follows:

Indices

	i	Kiln; 1, 2, …, 4
	j	Finish mill/cement mill; 1, 2, 3, …, 9
Parameters		
	O_{KL_i}	Kiln *i* operation time (hour) in 1 month
	$H_{E_{CL}}$	Clinker export price ($/ton)
	A	Proportion of clinker exported (%) to the amount of clinker production in 1 month. Maximum 30%
	O_{FM_j}	Finish mill/cement mill *j* operation time (hour) in 1 month
	H_S	Cement price ($/ton)
	Pr_j	Clinker proportion (%) in finish mill/cement mill *j*
	DS	Cement demand (ton) in 1 month
	DC	An amount of clinker needed (ton) in 1 month to produce cement according to demand
	SC	Clinker inventory at the beginning of the month
	(1-A)	Proportion of clinker grinded (%) into finish mill/cement mill to the amount of clinker production in 1 month. Minimum 30%
Decision Variables		
	P_{KL_i}	Production rate (ton/hour) of kiln *i*
	P_{FM_j}	Production rate (ton/hour) of finish mill /cement mill *j*

3.2. Model Formulation

Formulated model from model indices, parameters, and decision variables described above is shown as follows:

$$\text{Max } Z = A(((\textstyle\sum_{i=1}^{4} P_{KL_i} O_{KL_i}) + SC) H_{E_{CL}}) + ((\textstyle\sum_{j=1}^{9} Pr_j\, P_{FM_j}\, O_{FM_j}) H_s)) \quad (1)$$

Subject to:

$$P_{KL_1} \geq 320 \quad (2)$$
$$P_{KL_1} \leq 339 \quad (3)$$
$$P_{KL_2} \geq 320 \quad (4)$$
$$P_{KL_2} \leq 342 \quad (5)$$
$$P_{KL_3} \geq 320 \quad (6)$$
$$P_{KL_3} \leq 350 \quad (7)$$
$$P_{KL_4} \geq 320 \quad (8)$$
$$P_{KL_4} \leq 339 \quad (9)$$
$$(1 - A)((\textstyle\sum_{i=1}^{4} P_{KL_i} O_{KL_i}) + SC) \geq DC \quad \forall i \quad (10)$$
$$P_{FM_1} \geq 126 \quad (11)$$
$$P_{FM_1} \leq 177 \quad (12)$$
$$P_{FM_2} \geq 180 \quad (13)$$
$$P_{FM_2} \leq 238 \quad (14)$$
$$P_{FM_3} \geq 163 \quad (15)$$
$$P_{FM_3} \leq 188 \quad (16)$$
$$P_{FM_4} \geq 179 \quad (17)$$
$$P_{FM_4} \leq 223 \quad (18)$$
$$P_{FM_5} \geq 167 \quad (19)$$
$$P_{FM_5} \leq 202 \quad (20)$$
$$P_{FM_6} \geq 157 \quad (21)$$
$$P_{FM_6} \leq 189 \quad (22)$$
$$P_{FM_7} \geq 201 \quad (23)$$
$$P_{FM_7} \leq 232 \quad (24)$$
$$P_{FM_8} \geq 215 \quad (25)$$
$$P_{FM_8} \leq 234 \quad (26)$$
$$P_{FM_9} \geq 251 \quad (27)$$
$$P_{FM_9} \leq 267 \quad (28)$$
$$\left(\textstyle\sum_{j=1}^{9} P_{FM_j} O_{FM_j}\right) \geq DS \quad \forall j \quad (29)$$
$$\left(\textstyle\sum_{j=1}^{9} Pr_j\, P_{FM_j}\, O_{FM_j}\right) \leq (1\text{-A})\,((\textstyle\sum_{i=1}^{4} P_{KL_i} O_{KL_i}) + SC) \quad \forall i, \forall j \quad (30)$$
$$P_{KL_i}, P_{FM_j} \geq 0 \quad (31)$$
$$P_{KL_i}, P_{FM_j} \text{ integer} \quad (32)$$

Equation (1) is an objective function to maximize revenue, which comes from clinker export and clinker production for 1 month. Revenue from clinker exports is obtained from multiplication of kiln production rates (P_{KL_i}) with kiln operating hours (O_{KL_i}), selling price of exported clinker per ton ($H_{E_{CL}}$), and proportion of clinker exported (A). Revenue from clinker production is obtained by multiplying proportion of clinker in each finish mill/cement mill (Pr_j) with finish mill production rate (P_{FM_j}), finish mill operating hours (O_{FM_j}), and cement selling price per ton (H_S). Proposed model in this paper has several constraints: Kiln production rate constraints (2), (3), …, (9) finish mill/cement mill production rate constraints (11), (12), … (28), clinker amount that is not exported must be equal or greater than clinker demand per month (10), constraints for cement demand fulfilment (29), the proportion of clinker in finish mill/cement mill must be smaller or equal to clinker amount that is not exported (30), and non-negativity and integer constraint (31) and (32).

3.3. Game Strategy

Generally represents an action or action plan to play a game. Each player takes some action or action plan and then its outcome is determined [7]. S_i is the strategy of player i, then $S = (S_1, \ldots, Si)$ is called strategy profile. The strategies chosen by player 1 and player 2 to maximize revenue can be seen in table 1.

Table 1. Player strategy.

	Player 1 Strategy
S1.1	Clinker export proportion high, kiln production rate high
S1.2	Clinker export proportion high, kiln production rate medium
S1.3	Clinker export proportion high, kiln production rate low
S1.4	Clinker export proportion medium, kiln production rate high
S1.5	Clinker export proportion medium, kiln production rate medium
S1.6	Clinker export proportion medium, kiln production rate low
S1.7	Clinker export proportion low, kiln production rate high
S1.8	Clinker export proportion low, kiln production rate medium
S1.9	Clinker export proportion low, kiln production rate low
	Player 2 Strategy
S2.1	Proportion of clinker grinded into finish mill high, finish mill production rate high
S2.2	Proportion of clinker grinded into finish mill high, finish mill production rate medium
S2.3	Proportion of clinker grinded into finish mill high, finish mill production rate low
S2.4	Proportion of clinker grinded into finish mill medium, finish mill production rate high
S2.5	Proportion of clinker grinded into finish mill medium, finish mill production rate medium
S2.6	Proportion of clinker grinded into finish mill medium, finish mill production rate low
S2.7	Proportion of clinker grinded into finish mill low, finish mill production rate high
S2.8	Proportion of clinker grinded into finish mill low, finish mill production rate medium
S2.9	Proportion of clinker grinded into finish mill low, finish mill production rate low

Player 1 strategy is a combination of clinker export percentage with kiln production rate. In this paper the percentage of clinker exported and kiln production rate is categorized into 3: High, medium and low. Player 2 strategy is a combination of percentage of clinker grinded into finish mill with finish mill production rate. The percentage of clinker grinded into finish mill and finish mill production rate also categorized into 3: High, medium and low.

3.4. Payoff Matrix

Payoff is a numerical expression of preference order over outcomes. Each player respectively has a different payoff function [7]. Payoff function for player 1 shown as follows:

$$\text{Max Z} = A(((\sum_{i=1}^{4} P_{KL_i} O_{KL_i}) + SC)H_{E_{CL}}) \tag{33}$$

From the equation above it can be seen that obtained revenue comes from amount of clinker exported. Payoff function for player 2 shown as follows:

$$\text{Max Z} = \left(\sum_{j=1}^{9} Pr_j\, P_{FM_j}\, O_{FM_j})H_s\right) \tag{34}$$

This payoff function calculates clinker proportion in each finish mill, so that the clinker's contribution in cement sales can be calculated. Game strategy and payoff function are used to arrange the payoff matrix shown in table 2 below:

Table 2. Payoff matrix.

		Player 2								
		S2.1	**S2.2**	**S2.3**	**S2.4**	**S2.5**	**S2.6**	**S2.7**	**S2.8**	**S2.9**
	S1.1	Scenario 1	Scenario 2	Scenario 3	Scenario 4	Scenario 5	Scenario 6	Scenario 7	Scenario 8	Scenario 9
	S1.2	Scenario 10	Scenario 11	Scenario 12	Scenario 13	Scenario 14	Scenario 15	Scenario 16	Scenario 17	Scenario 18
	S1.3	Scenario 19	Scenario 20	Scenario 21	Scenario 22	Scenario 23	Scenario 24	Scenario 25	Scenario 26	Scenario 27
	S1.4	Scenario 28	Scenario 29	Scenario 30	Scenario 31	Scenario 32	Scenario 33	Scenario 34	Scenario 35	Scenario 36
Player 2	**S1.5**	Scenario 37	Scenario 38	Scenario 39	Scenario 40	Scenario 41	Scenario 42	Scenario 43	Scenario 44	Scenario 45
	S1.6	Scenario 46	Scenario 47	Scenario 48	Scenario 49	Scenario 50	Scenario 51	Scenario 52	Scenario 53	Scenario 54
	S1.7	Scenario 55	Scenario 56	Scenario 57	Scenario 58	Scenario 59	Scenario 60	Scenario 61	Scenario 62	Scenario 63
	S1.8	Scenario 64	Scenario 65	Scenario 66	Scenario 67	Scenario 68	Scenario 69	Scenario 70	Scenario 71	Scenario 72
	S1.9	Scenario 73	Scenario 74	Scenario 75	Scenario 76	Scenario 77	Scenario 78	Scenario 79	Scenario 80	Scenario 81

A total of 81 scenarios are found in the payoff matrix, obtained from game strategy in table 1. Scenario 1 for example, is obtained from player 1 strategy S1.1 and player 2 strategy S2.1.

4. Data Experiment

The proposed model is demonstrated through data experiment. This demonstration using real data, except for clinker export price and cement price due to confidentiality, from one particular cement industry's production plan and monthly production report. The data shown in table 3 below:

Table 3. Data from production plan and monthly production report.

	Parameter								
i or j	*DS*	*DC*	*OKLi*	*OFMj*	*PRj*	HE_{CL}	H_S	*A*	*SC*
	950,640	703,474				100	200	0.06	235,186
1			480	528	0.92				
2			696	528	0.65				
3			696	528	0.88				
4			696	528	0.65				
5				552	0.88				
6				552	0.65				
7				528	0.65				
8				528	0.65				
9				528	0.65				

Simulation in LINGO 11.0.0.20 was held based on model formulation in section 3.2 and table 3. The result of simulation can be seen below:

Table 4. Simulation result.

	LINGO Result		Actual Data		% Error	
i or j	P_{KLi}	P_{FMi}	P_{KLi}	P_{FMi}	P_{KLi}	P_{FMi}
1	339	177	333.5	169.7	1.65	4.30
2	342	238	325.2	183.6	5.17	29.63
3	350	188	325.7	179.1	7.46	4.97
4	339	223	315.8	191.9	7.35	16.21
5		202		172.6		17.03
6		189		157.9		19.70
7		232		219.6		5.65
8		234		224.2		4.37
9		267		263.5		1.33

Simulation result shows that with clinker export proportion (*A*) 6%, kiln and finish mill production rates should be rolled out with maximum rates to get maximum revenue. These results are different from the actual monthly report data, especially in finish mill 2, 4, 5 and 6 production rate (more than 15%

error). These differences may be caused by several factors such as type of cement grinded in finish mill, consideration of remaining storage space so finish mill is run with a low production rate, and downtime at kiln and finish mill.

The payoff matrix is obtained based on game strategy in table 1, equation (33) & (34), and experiment data. Data used in this experiment shown in table 5 below:

Table 5. Experiment data.

	Experiment Data											
	A			*(1-A)*			P_{KLi}			P_{FMi}		
i or j	High	Medium	Low	High	Medium	Low	High	Medium	Low	High	Medium	Low
	30%	20%	10%	90%	80%	10%						
1							339	330	320	177	152	126
2							342	331	320	238	209	180
3							350	335	320	188	176	163
4							339	330	320	223	201	179
5										202	185	167
6										189	173	157
7										232	217	201
8										234	225	215
9										267	259	251

The payoff matrix obtained is shown as follows:

Table 6. Payoff matrix (using experiment data).

Payoff Matrix		*Player 2*																	
		S2.1		**S2.2**		**S2.3**		**S2.4**		**S2.5**		**S2.6**		**S2.7**		**S2.8**		**S2.9**	
Player 1	**S1.1**	0	0	0	0	0	0	0	0	0	0	0	0	33464460	156167480	33464460	152152280	33464460	147972440
	S1.2	0	0	0	0	0	0	0	0	0	0	0	0	32604060	156167480	32604060	152152280	32604060	147972440
	S1.3	0	0	0	0	0	0	0	0	0	0	0	0	31708380	156167480	31708380	152152280	31708380	147972440
	S1.4	0	0	0	0	0	0	22309640	178477120	22309640	173888320	22309640	169111360	22309640	156167480	22309640	152152280	22309640	147972440
	S1.5	0	0	0	0	0	0	21736040	178477120	21736040	173888320	21736040	169111360	21736040	156167480	21736040	152152280	21736040	147972440
	S1.6	0	0	0	0	0	0	21138920	178477120	21138920	173888320	21138920	169111360	21138920	156167480	21138920	152152280	21138920	147972440
	S1.7	11154820	156167480	11154820	152152280	11154820	147972440	11154820	178477120	11154820	173888320	11154820	169111360	11154820	156167480	11154820	152152280	11154820	147972440
	S1.8	10868020	156167480	10868020	152152280	10868020	147972440	10868020	178477120	10868020	173888320	10868020	169111360	10868020	156167480	10868020	152152280	10868020	147972440
	S1.9	10569460	156167480	10569460	152152280	10569460	147972440	10569460	178477120	10569460	173888320	10569460	169111360	10569460	156167480	10569460	152152280	10569460	147972440

The payoff matrix in table 6 then simulated into Gambit 15.1.1 software to obtain the equilibrium. The result is shown in figure 2:

#	1: 1	1: 2	1: 3	1: 4	1: 5	1: 6	1: 7	1: 8	1: 9	1: 10	1: 11	2: 1	2: 2	2: 3	2: 4	2: 5	2: 6	2: 7	2: 8	2: 9
1	1	0	0	0	0	0	0	0	0	0	0	0	0	0	0	0	0	1	0	0
2	$\frac{1}{8}$	0	0	$\frac{7}{8}$	0	0	0	0	0	0	0	0	0	0	$\frac{1}{3}$	0	0	$\frac{2}{3}$	0	0
3	0	0	0	1	0	0	0	0	0	0	0	0	0	0	1	0	0	0	0	0

Figure 2. Equilibria of game.

Payoff value of equilibria 1 is 189631940, equilibria 2 is 187308019, and equilibria 3 is 200786760. Because objective function in this study is to maximize revenue from clinker production, equilibria 3 is chosen. Game strategy used to obtain equilibria 3 is S1.4 and S2.4. Referring to table 1 and table 5, S1.4 can be interpreted as: Clinker export percentage is 20% with high kiln production rate. Also, S2.4 can be interpreted as: Proportion of clinker grinded into finish mill is 80% with high finish mill production rate.

5. Conclusion

Clinker export is an alternative to obtain revenue in the condition of cement oversupply. In this paper a model has been proposed to determine the proportion of clinker exported and the proportion of clinker grinded into cement using the cooperative game theory approach. Simulation result in table 4 shows that kiln production rate and finish mill production rate should be set to maximum so that maximum revenue can be obtained. Production rate resulted from simulation is different from actual data, especially in finish mill production rate.

The strategy in table 1, equation (33) & (34) and experiment data in table 5 used to compile the payoff matrix in table 6. The payoff matrix is then simulated in Gambit 15.1.1 software to determine proportion of clinker exported and proportion of clinker grinded into finish mill, which can be accepted by player 1 and player 2. Equilibria in figure 2 shows that 20% clinker export proportion and high kiln production rate combined with 80% clinker grinded into finish mill and high finish mill production rate should be accepted by both player 1 and player 2 to maximize revenue. However, this result may differ if: Different clinker export proportion value (A) is used and various categorization in formulating game strategy is applied (i.e., other than 3 categorizations of A, P_{KL_i}, and P_{FM_j}). Future research will consider several things as follows in order to represent actual situation at plant:

- Downtime that occurs in kiln and finish mill.
- The type of cement that is grinded in finish mill.
- Storage capacity and storage space.
- Clinker export demand.

References

[1] Chatain O. *Cooperative and Non-Cooperative Game Theory*. The Palgrave Encyclopedia of Strategic Management. doi.org/10.1057/9781137294678.0132

[2] Liu Z, Liu Q, Xu W, Zhou Z, Pham DT. Human-robot collaborative manufacturing using cooperative game: framework and implementation. *Procedia CIRP*. 2018;**72**:87–92. doi.org/10.1016/j.procir.2018.03.172

[3] Hafezalkotob A, Chaharbaghi S, Lakeh TM. Cooperative aggregate production planning: a game theory approach. *J. Ind. Eng. Int.* 2019 Feb 12;**15**(S1):19–37. doi.org/10.1007/s40092-019-0303-0.

[4] Winston WL. *Operations Research: Applications And Algorithm, Belmont*. Brooks-Cole-Thomson Learning. 2004.

[5] Emir T, Güler MG. *Production Planning Using Day-Ahead Prices in a Cement Plant*. Exergetic, Energetic and Environmental Dimensions. 2018;149–66. doi.org/10.1016/b978-0-12-813734-5.00009-3.

[6] Bellabdaoui A, Teghem J. A mixed-integer linear programming model for the continuous casting planning. *Int. J. Prod. Econ.* 2006 Dec;**104**(2):260–70. doi.org/10.1016/j.ijpe.2004.10.016

[7] Nishino N, Tjahjono B. Game theory approach to product service systems. *Procedia CIRP*. 2018;**73**:304–9. doi.org/10.1016/j.procir.2018.03.326

Selecting plant for joint venture scheme using linear programming and enterprise value conversion

A Y Yulianto[1,2], I N Pujawan[1], and I Vanany[1]

[1]Departement of Industrial and System Engineering, Institut Teknologi Sepuluh Nopember, Surabaya, Indonesia

E-mail: [2]akhmadyaniyulianto@yahoo.com

Abstract. Every company in doing business expansion, especially business outside the core business of the previous business field, generally requires sufficient funding sources to ensure that the expansion goes as expected. Funding according to the needs is often a pretty difficult problem especially when faced with choices that have a significant cost impact. In this study we address the problem faced by a cement company in Indonesia that has extra capacity and potentially can be used to produce other products under a joint venture scheme. The company is faced with 3 choices of funding sources to support expansion into the building materials business, one of which was to sell a certain proportion of shares of the capacity of 8 factories that were not absorbed by the domestic sales target in Indonesia to strategic partners. This study attempt to find the most optimal capacity allocation of all factories in meeting the domestic sales volume target by considering the minimal transportation costs using the linear programming method and compare the remaining plant capacity with the magnitude of the converted plant capacity. The results are expected to support funding proposals to realize the vision of entering the building materials business in Indonesia.

1. Introduction

Supply Chain Management (SCM) research in cement industry is interesting, but relatively are quite a lot of previous research with SCM topics such as supply chain strategic, supply chain innovation, supply chain scheduling, and demand forecasting. Noche et al. [1] developed an innovative approach to supply chain strategies using simulation in the cement industry. Dubey et al. [2] investigated the supply chain innovation in Indian cement manufacturing. Naso et al. [3] applied a genetic algorithm for supply chain scheduling with case studies in the distribution of ready-mixed concrete. A few topics in SCM research in cement industry studied to develop an optimization model for production and distribution from the plant until distribute to marketing area to determine which plant that had excess production capacity. This study develops an optimization model for production and distribution to determine which plant are overproduced and to analyze enterprise value conversion in PT A.

The study of the allocation of production volumes mostly aims to determine the most minimal costs, where the goal stops at minimal costs, even though it is often accompanied by the results of subsequent calculations in the form of unused production capacity. This study provides a solution by offering unused production capacity as a funding tool to get capital by offer to new investors with a joint venture scheme.

2. Problem Description

After acquiring the other cement company in case's company, PT A as case study in this research focused its internal funds only on operational expenditure and capital expenditure on the cement business and in supporting the change in the company's vision to enter the building materials business. The cement company needs funds for capital expenditure of 3.8 trillion rupiahs. The source of the funds is decided by management using an external source with debts from banks, issued new bonds, or selling the proportion of certain factory shares to strategic partners.

The three funding options above are the main considerations in raising funds because the availability of internal funds is focused on the operation and development of the cement business to be more efficient. From the three alternative options above, options number 1 and 2 face many challenges from banks and financial markets because the burden of financial companies is currently very high due to the previous acquisition of PT Z. This is quite logical because the amount of debt has increased sharply in 2019 when compared to the period of the previous year. The nominal debt of this company of 3.9 trillion rupiahs and the leverage ratio of 0.16 in 2014 to 21.4 trillion rupiahs and lever ratio of 0.65 in 2019, an increase of 546%. This condition makes the potential choices number 1 and 2 experience considerable difficulties because they have to convince banks or financial markets to mobilize funds, and it is estimated that the negotiation process is quite long with an uncertain success rate in accordance with expectations, either nominal or interest rate. While the third option is to sell the proportion of shares in accordance with the target funds, it is possible to do so considering that many cement companies abroad want to have cement production units in Indonesia where production costs are considered to be quite low and the quality of raw materials such as limestone and clay are among the quality ranks highest in the world but has challenges in making factory investments from the start both in terms of legal, social and project time that is less predictable on time. This third alternative option, does not have a description of which factories can be cooperated or sold a certain proportion of shares to strategic partners except factory decisions that are not optimal in serving the Indonesian domestic market. The funds that will be obtained from the funding plan above will be used to support the new vision of company entered the building materials business with a focus on investment in light brick/panel factories (Autoclaved Aerated Concrete), cement boards, mortar cement, building material applicator companies (residential specialty contractors) and mega distributor companies.

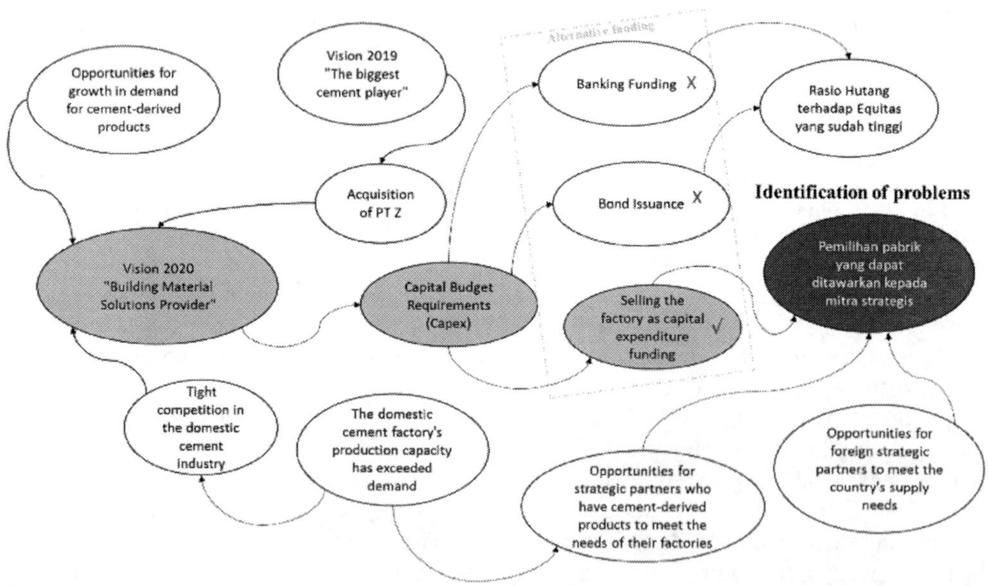

Figure 1. Rich Picture Problem diagram which explains the root causes and impacts that are the focus of research.

Based on the background description explained earlier, the problem discussed in this study is how to determine which factories can be cooperated with strategic partners in supporting funding for the company's capital expenditure in 2020 to 2021. The purpose of this study are optimizing the allocation of supply capacity to each market area, selecting a factory and determining the amount of capacity that is not absorbed by the target demand in the country, calculate the enterprise value and its conversion value with the funds needed, determine the proportion shares of the factory that can be cooperated with a strategic partner as funding in supporting capital expenditure for building materials outside of cement. The benefit of this study is to provide recommendations on the determination of factories that can be collaborated with strategic partners as an alternative funding for the company in supporting the building materials business.

3. Mathematical Model

3.1 Formulation of conversion of enterprise value to factory capacity (ton)

In general, it is easier to judge a public company than a closed company where stock prices provide an actual barometer of the market value of equity. If debt is also traded in the public, then that value can also be calculated from the market. The total value of equity and debt gives the total market value of the enterprise, which is defined to measure the price that the acquirer will pay for the overall value of the enterprise.

There is a strong correlation between calculating the enterprise value open company approach with the calculation of the enterprise value with the discounted cash flow approach.

- Enterprise Value Formulation
  The calculation of Open Enterprise Value can be calculated by the following formula [4]:

$$EV = (Stock\ Quantity \times Avg\ Price) + Debt) - Cash \quad (1)$$
$$\text{EV} = (5{,}931{,}520{,}000 \times 12{,}508) + 21{,}378{,}095{,}604{,}000) - 3{,}436{,}124{,}754{,}000$$
$$\text{EV} = 92{,}133{,}752{,}538{,}888.9$$

Where:

EV : Enterprise Value (Rupiah)
Stock Quantity : Number of shares outstanding on the exchange (Share)
Avg Price : Average price per share on the stock exchange (Rupiah)
Debt : Total amount of interest-bearing debt (Rupiah)
Cash : Amount of cash and cash equivalents (Rupiah)

- Formulation of Enterprise Value compared to Production Capacity
  After knowing the Enterprise Value, the next stage is determined the Enterprise Value formulation compared to the production capacity (Ton).

$$EV\ per\ ton = \left(\frac{EV}{Overall\ Plant\ Capacity}\right) \quad (2)$$
$$EV\ per\ ton = \left(\frac{92{,}133{,}752{,}538{,}888.9}{46{,}461{,}000}\right)$$
$$EV\ per\ ton = 1{,}983{,}034$$

Where:

EV per ton : Enterprise Value per ton of factory production capacity (Ton)
EV : Enterprise Value (Rupiah)
Overall Plant Capacity : The total amount of factory production capacity (Ton)

- Formulation of conversion Fund requirements to factory Capacity that can be cooperated
  After knowing the enterprise value per ton (EV per ton), a conversion formula can be set for Enterprise Value per ton with funding requirements.

$$Convertion\ to\ Capacity = \left(\frac{Fund\ Required}{EV\ per\ ton}\right) \quad (3)$$
$$Convertion\ to\ Capacity = \left(\frac{3{,}870{,}775{,}000{,}000}{1{,}983{,}034}\right)$$
$$Convertion\ to\ Capacity = 1{,}952{,}946$$

IOP Conf. Series: Materials Science and Engineering 1072 (2021) 012038 doi:10.1088/1757-899X/1072/1/012038

Where:

Conversition to Capasity	: The amount of factory capacity that can be cooperated (Ton)
EV per ton	: Enterprise Value per ton of factory production capacity (Ton)
Fund Required	: Funds needed in an investment plan (Rupiah)

3.2 Development of mathematical models

At this stage the formulation of variables in the research object is formulated into a linear programming mathematical model and then developed based on existing constraints so that the model can describe the conditions of production allocation to meet demand or sales targets in all provinces. Then the total supply is equal to the total demand, and this problem is called a balanced transportation problem.

The formula model in linear programming in this study takes into account the production capacity of each factory, the target volume in 2020, and the distribution costs from each factory to each province. An explanation of the mathematical formula that has been made according to the basic mathematical model is explained as follows.

Index

Indexes on the mathematical model of this study include :

i : Cement Factory i, $i = 1,2 ..., 8$

j : Market j, $j = 1,2 ..., 34$

Parameters

The parameters in the mathematical model of this study include :

C_{ij} : Cost of cement distribution from factory i to market j

d_j : Demand volume in market j

s_i : Factory production capacity i

Decision Variables

Decision variables in the mathematical model of this study are :

x_{ij} : The volume of cement sent from factory i to market j

Purpose Function

The objective function of this research is to minimize the cost of cement distribution costs at each supply facility and is expressed in equation 4 as follows.

$$Min \sum_{i=1}^{8} \sum_{j=1}^{34} c_{ij} x_{ij} \quad (4)$$

Where,

c_{ij} : Cost of cement distribution from factory i to market j

x_{ij} : The volume of cement sent from factory i to market j

Constraints

Constraints in this problem are the production capacity of each factory, demand or target sales volume, company policies, and non negative constraints. The constraint function is as follows,

(1) Obstacles to cement demand

Constraint equation for the demand for cement which explains that the volume of cement allocated from plant i must be equal to the volume of demand in the market j.

$$\sum_{i=1}^{8} x_{ij} = d_j , \forall_j \quad (5)$$

(2) Capacity constraints for cement production

Constraint equation for the capacity of cement production which explains that the volume of cement allocated from plant i must be equal or lower than the plant volume capacity i.

$$\sum_{j=1}^{34} x_{ij} \leq s_i\,,\ \forall_i \tag{6}$$

(3) Minimal factory allocation volume
Constraint equation in the form of company policy in determining the minimum production capacity to be offered to strategic partners, which is no more than 49%, or if it is converted into factory capacity that the company must maintain at least 51% of total factory capacity i.

$$\sum_{j=1}^{34} x_{ij} \geq 0{,}51\ s_i\,,\ \forall_i \tag{7}$$

(4) The i_1 factory and the i_2 factory only serve j_1-j_{10}
Equality constraints in the form of company policy in supplying to the market, namely the volume of cement supplied from the i_1 factory and the i_2 factory does not serve other areas besides the j_1-j_{10} market.

$$\sum_{j=11}^{34} c_{ij} = \infty\ ,\ \forall_{i=1,2} \tag{8}$$

(5) The i_8 Factory only serves other than the j_1-j_{10} and j_{11}-j_{16} market
Constraint equation in the form of company policy to supply to the market, namely the volume of cement supplied from the i_8 factory does not serve j_1-j_{10} and j_{11}-j_{16} market.

$$\sum_{j=1}^{16} c_{8j} = \infty,\ \forall_{i=8} \tag{9}$$

(6) Non-Negativity Constraint
Non-negativity constraint equation which explains that the volume of cement allocated from plant i to market j cannot be negative.

$$x_{ij} \geq 0 \tag{10}$$

4. Results and Discussions

4.1 Enterprise value calculation results

In accordance with the formulation of Schmidlin [4], it can be calculated that the PT A's Enterprise Value is Rp92,133,750,538,889,- and the Enterprise Value per ton is Rp1,983,034 per ton, it can be easily determined that the total capacity of factories that can work together with strategic partners is 1,952,946 tons.

4.2 Optimization results

An optimal volume allocation has been generated with a total demand volume of 38582921 tons for one year. Markets j_1 to j_{10} as a whole are supplied by i_1 and i_2 factories with support from i_3 factory which supply j_8 and j_9 markets. The j_{11}, j_{12} and j_{13} markets are supplied by i_3 factories supported by i_4 factories supplying almost half of the total volume in j_{13}. The j_{14} market is supplied by factory i_5 with full allocation supported by factories i_4 and i_7. The j_{15} market is fully supplied by factory i_4.

The $i6$ factory is still focused on supplying the j_{16} market, while the i_7 factory besides supplying the j_{16} and j_{14} markets, it also supplies in the j_{17} Market as a single source, and supplies in the j_{18}, j_{19}, j_{20}, j_{31}, j_{32} and j_{32} markets as shown in table 1.

The distribution range between factories has quite large differences, this is due to the size of the target sales volume or demand of each different market area and factory capacity, and this is also reflected in the constraints that create 3 cluster coverage where the size is different, for example cluster 1 has the ratio between demand and capacity is 94.8% (9661000 tons capacity and 9160787 tons demand), cluster 2 is 82.7% (capacity 29400400 tons and demand 24324559 tons) and cluster 3 has 75% ratio (capacity 7400000 tons and demand 5552211 tons).

Table 1. The results of optimization of the allocation of the supply volume of 8 factories to 34 marketing areas can be seen from the allocation of each factory in meeting sales targets of 38582921 tons.

Market = j	Factory Supply = i								Target Sales / Demand
	i_1 Lhonga	i_2 Indarung	i_3 Narogong	i_4 Cilacap	i_5 Rembang	i_6 Tuban SBI	i_7 Tuban SI	i_8 Pangkep	
j_1	1184648	-	-	-	-	-	-	-	1184648
j_2	476352	1540837	-	-	-	-	-	-	2017189
j_3	-	1318637	-	-	-	-	-	-	1318637
j_4	-	1567146	-	-	-	-	-	-	1567146
j_5	-	499906	-	-	-	-	-	-	499906
j_6	-	370127	-	-	-	-	-	-	370127
j_7	-	734897	-	-	-	-	-	-	734897
j_8	-	-	344293	-	-	-	-	-	344293
j_9	-	-	110644	-	-	-	-	-	110644
j_{10}	-	1013299	-	-	-	-	-	-	1013299
j_{11}	-	-	1350085	-	-	-	-	-	1350085
j_{12}	-	-	1861778	-	-	-	-	-	1861778
j_{13}	-	-	2033199	2077361	-	-	-	-	4110560
j_{14}	-	-	-	172848	3000000	-	2446436	-	5619284
j_{15}	-	-	-	949791	-	-	-	-	949791
j_{16}	-	-	-	-	-	2040000	5625862	-	7665862
j_{17}	-	-	-	-	-	-	708686	-	708686
j_{18}	-	-	-	-	-	-	526280	-	526280
j_{19}	-	-	-	-	-	-	363705	-	363705
j_{20}	-	-	-	-	-	-	320889	-	320889
j_{21}	-	-	-	-	-	-	-	630391	630391
j_{22}	-	-	-	-	-	-	-	107379	107379
j_{23}	-	-	-	-	-	-	-	1823403	1823403
j_{24}	-	-	-	-	-	-	-	284604	284604
j_{25}	-	-	-	-	-	-	-	560342	560342
j_{26}	-	-	-	-	-	-	-	99038	99038
j_{27}	-	-	-	-	-	-	-	325744	325744
j_{28}	-	-	-	-	-	-	-	579099	579099
j_{29}	-	-	-	-	-	-	-	228411	228411
j_{30}	-	-	-	-	-	-	-	122318	122318
j_{31}	-	-	-	-	-	-	144949	-	144949
j_{32}	-	-	-	-	-	-	247752	-	247752
j_{33}	-	-	-	-	-	-	-	273612	273612
j_{34}	-	-	-	-	-	-	-	518171	518171
Total Supply Volume	1661000	7044850	5700000	3200000	3000000	2040000	10384559	5552512	38582921

The remaining total capacity resulting from optimization is 7878079 tons, with four factories that have utilization below 100%, namely i_2, i_6, i_7, and i_8 Factories, detailed data are in table 2.

Table 2. Optimal factory capacity and unused maximally. Optimal factory capacity and not used maximally. This table explains that there are variations in the utility of each factory, from 100% to as low as 51%.

Factory = i	Capacity Used	Factory Capacity	Capacity Usage	Unused capacity
i_1	1661000	1661000	100%	(0)
i_2	7044850	8000000	88%	955150
i_3	5700000	5700000	100%	(0)

i_4	3200000	3200000	100%	0
i_5	3000000	3000000	100%	-
i_6	2040000	4000000	51%	1960000
i_7	10384559	13500000	77%	3115441
i_8	5552512	7400000	75%	1847488
Total	38582921	46461000	83%	7878079

4.3 Analysis of Factory Selection

The factory capacity needed for cooperation with strategic partners is only 1952946 tons, while the remaining capacity of the i_2 plant is 955150 tons and the i_8 plant is 1,847,488 tons so that candidate 4 factories have shrunk to 2 factories, i_6 and i_7 factories. Subsequent elections require further analysis to determine which of the two factories has remaining capacity. Then the selection is conducted through an assessment with criteria and linkert scale.

Table 3. Final determination of factories that can be used as a basis for joint venture schemes. Factory i_6 was chosen because it has the highest score.

Num	Criteria	Weight	Factory i_6		Factory i_7		Rating Scale				
			Rating	Score	Rating	Score	1	2	3	4	5
1	Port Ownership	30%	5	1.5	5	1.5	The factory does not have a port and has no agreement on the use of the port with other parties	The factory does not have a port but has a port use agreement with another party	The factory has a port use agreement under 10 years	The factory has a lease agreement for using the port for more than 10 years	The factory has its own port
2	Dock usage capacity (tons per month)	20%	5	1	5	1	> 0 <25000	$\geq$25000 < 50000	$\geq$50000 < 75000	>75000 - <100000	$\geq$ 100000
3	Distance to port (km)	15%	5	0.75	5	0.75	>40 km	>30 km - 40 km	>20 km - 30 km	>10 km - 20 km	< 10 km
4	Projection of share ownership (%)	35%	5	1.75	1	0.35	< 15%	$\geq$15% <25%	$\geq$25% <35%	$\geq$35% <45%	$\geq$45%
	Total Score	100%		5		3.6					

Determination of rating and weighting is based on internal discussions at the top management level which have also met with several potential investors. PT A wants to remain as the majority shareholder of at least 51% and the investor wants to be a shareholder of at least 30% and in addition is also very concerned about infrastructure equipment such as port ownership, dock capacity and access distance from port factories that can ensure supply to their markets overseas can be fulfilled on an ongoing basis. Based on the discussion, the highest weight is in the criteria of share ownership and port ownership.

Table 3 shows result of the assessment is that the i_6 factory was chosen because it has a higher score than the i_7 factory, even i_6 dominates i_7. The main criterion that most influenced the setting up of the i_6 factory was the projected share size of 1.75. With almost the same ownership, making potential investors have higher control in controlling operations and providing certainty of long-term sustainable market allocation abroad while for PT A is in accordance with the minimum shareholding of 51% and remains as the controlling shareholder.

5. Conclusion

The Transportation Problem Model with the associated Enterprise Value can be used as a solution in solving the problem of divesting assets in the context of funding to support business expansion without

IOP Conf. Series: Materials Science and Engineering 1072 (2021) 012038 doi:10.1088/1757-899X/1072/1/012038

reducing operational effectiveness and sustainability of the core business. The optimization results state that there are 4 factories that still have remaining capacity that cannot be absorbed by the Indonesian domestic market, namely i_2, i_6, i_7 and i_8 Factory with a total remaining capacity of 7878079 tons or 17% of capacity overall plant with middle utility. The Enterprise value of PT A by the end of 2019 is Rp92 T with a conversion of Rp1,983,034 per ton of factory capacity. The factory chosen to cooperate with strategic partners is the i_6 factory with a share ownership that is relatively as large as PT A.

References

[1] Noche B, Elhasia T. Approach to innovative supply chain strategies in cement industry; analysis and model simulation. *Procedia Soc. Behav. Sci.* Elsevier BV; 2013 Apr;**75**:359–69. doi.org/10.1016/j.sbspro.2013.04.041

[2] Dubey R, Singh T, Tiwari S. Supply chain innovation is a key to superior firm performance an insight from indian cement manufacturing. *Int. J. Innov. Sci.* Emerald; 2012 Dec;**4**(4):217–30. doi.org/10.1260/1757-2223.4.4.217

[3] Naso D, Surico M, Turchiano B, Kaymak U. Genetic algorithms for supply-chain scheduling: A case study in the distribution of ready-mixed concrete. *Eur. J. Oper. Res.* Elsevier BV; 2007 Mar;**177**(3):2069–99. doi.org/10.1016/j.ejor.2005.12.019

[4] Schmidlin N, editor. *The Art of Company Valuation and Financial Statement Analysis*. John Wiley & Sons, Inc.; 2012 Jan 2. doi.org/10.1002/9781119208020

IOP Conf. Series: Materials Science and Engineering 1072 (2021) 012039 doi:10.1088/1757-899X/1072/1/012039

Feasibility study of waste utilization facilities in cement factory for spent bleaching earth

F F Nugroho[1,2] and M L Singgih[1,3]

[1]Department of Industrial and Systems Engineering, Institut Teknologi Sepuluh Nopember, Surabaya, Indonesia

E-mail: [2]febri.fahrudin@gmail.com, [3]moseslsinggih@ie.its.ac.id

Abstract. In this paper, the problem to be solved is a feasibility study for waste utilization facilities development, especially for spent bleaching by considering the market, technical, and economic aspects. The amount of spent bleaching produced by the palm oil industry in Indonesia reaches 750000 tons/year. Government regulation number 101 2014 categorized spent bleaching earth as hazardous waste with a code B 413. Therefore, the companies must carry out environmental protection and management for spent bleaching earth. This facility built to capture the market of spent bleaching earth waste disposal service, raw material substitution, and alternative fuel (coal). From the market perspective, the facility will capture about 7% market shares. Based on the technical aspect, the calcination process, kiln operational, and quality of clinker won't be disrupted if the capacity of waste facilities is 7-8 tons per hour. Based on the Payback Period calculation, it needs 5.4 years to return the investment. From the Net Present Value and Internal Rate of Return, it can be concluded that the development of waste utilization facility for spent bleaching is feasible. Based on sensitivity analysis the facilities must run above 60% capacity to make sure the facilities still feasible. Further research is expected to analyze the potential for other B3 waste which can managed by the facility and potential to enter end to end waste management business so it can have a greater impact to the company and society.

1. Introduction

Spent bleaching earth (SBE) is a waste product of palm oil bleaching process, which contains 20-30% weight of palm oil. The amount of SBE produced by the palm oil industry in Indonesia reaches 750000 tons/year. Government regulation number 101 2014 categorize SBE as hazardous waste with a code B 413. Therefore, the companies must carry out environmental protection and management for SBE. According to the national vegetable oil industry organization, 17 units of integrated SBE utilization are needed to be Recovered Oil (R-oil) and De-Oiled Bleaching Earth (OBE). However, due to high investment, there are only three utilization units with a production capacity of 300 tons per day. According to research [1], the percentage of residual CPO oil in spent bleaching earth can reach 20-30% and is difficult to separate without special handling.

Waste utilization at PT A, which is a cement factory in East Java, has been carried out since 2017. But, its utilization is limited due to constrained facilities. Since 2017 until now the waste processing has been carried out by putting waste in mixed pile facilities that enter the raw mill. The process limited due to dust collector performance (electrostatic precipitator): if the waste volume increased above standard, it effects dust emission. In 2017 the total waste that successfully utilized was 4348 tons. In 2018, it increased by 76% to 7675 tons. PT D, which is a subsidiary of PT A that newly acquired in 2019, has better waste utilization facilities. The waste facility put waste directly into the kiln inlet. According to the company's annual report in 2018 [2], the total waste treated is 537161 tons with contributing to increase profit (Gross Added Value) of IDR 180 billion.

To solve the problems in disposing of SBE as well as the potential revenue obtained from SBE waste disposal services, the authors intend to analyze the feasibility of SBE waste utilization development facilities in Tuban Plant based on the market, technical, and economic aspects. The result of the feasibility analysis expected to be considerations for decision making by the management of PT A.

2. Literature Review

Indonesia remains the largest palm oil producer in the world, with total production in 2019 is 47.2 million tons. With increasing domestic CPO production, demand for bentonite (bleaching earth) will automatically increase. The composition of bleaching earth consists of SiO_2, Al_2O_3, Fe_2O_3, CaO, and MgO. The amount of bentonite needed as an oil purifier will be the same as 1% - 3% of the amount of palm oil.

Spent bleaching earth (SBE) is industrial waste, mainly produced from vegetable oil processing. It was noted that SBE could present a fire (e.g., spontaneous combustion) because it contains 20-30% by weight of oil. So far there have been several methods to utilize SBE such as 1) recycling as landfills; 2) recycling as palm oil fertilizer; 3) reuse spent bleaching earth as an adsorbent again in the process of bleaching CPO in the cooking oil industry; 4) recovery of CPO residues with porous metal filters into low-quality palm oil and its derivatives; 5) recovery of CPO residues with solvent extraction and supercritical fluid extraction into methyl esters as biofuel feedstocks; 6) reuse SBE to obtain its caloric value and used as briquettes.

2.1. Utilization of waste in the cement industry

Research conducted to utilize biomass waste (rice husks, sawdust, and peanut shells) used as co-processing included in Separated Line Calciner (SLC) [3]. The purpose of the research is to utilize the caloric value of biomass waste as coal substituted.

2.2. Demand aspect

Demand analysis conducted PT A as an SBE waste disposal service from CPO producers in Indonesia. The purpose of the analysis is:

a. To find SBE volume projection from 2020 – 2030 using correlation of population projection and CPO demand projection.
b. Competitor mapping (Company that has business in waste disposal in Indonesia).

2.3. Technical aspect

Technical analysis is needed to determine the capacity of SBE's waste management facility by considering technical process analysis, design analysis, and project schedule. Technical Process analysis is carried out to find out how much capacity of the kiln could burn waste, especially SBE without disrupting calcination process, kiln operational, and clinker quality by simulation of process calculation and operasional. Design analysis is conducted to determine process flow and identify the equipment needed to build SBE waste utilization facilities.

2.4. Economical aspect

Economical aspect is carried out to determine the feasibility of waste utilization facilities development by conducted analysis of cost, revenue, and economic analysis.

2.4.1. Cost analysis (investment & operational cost)

2.4.1.1. Investment cost

Total investment cost to build waste utilization facilities until the commercial. It consists of: preparation and engineering, heavy equipment, civil works, leading equipment, mechanical and electrical work, erection, and commissioning.

2.4.1.2. Operational cost

Total operational costs also called the Cost of Good Sold (COGS). COGS consists of the Cost of Good Manufactured (COGM) and Operating Expense.

2.4.2. Weight average cost of capital

To discount the company's net cash flow annually, the relevant discount rate is the Weighted Average Cost of Capital (WACC), which reflects the weighted average cost of the overall capital used within the company to generate the net cash flow [4]. WACC is determined by using the equation as follows:

$$WACC = (k_e \times W_e) + (k_d[1-T] \times W_d) \quad (1)$$

with:
ke : Cost of equity capital
kd : Cost of debt capital
We : Weight of equity in capital structure
Wd : Weight of debt in capital structure
T : Corporate income tax

Discount rate for equity is obtained by applying Capital Asset Pricing Model (CAPM). This model states that the cost of equity is risk-free interest plus premium to cover the systematic risk of stock securities [4], with a formula as follows:

$$k_e = R_f + \beta(R_m - R_f) \quad (2)$$

with:
Rf : Risk-free investment rate of return
ß : Systematic risk
Rm : Rate of return expected by equity market

Rf (Risk-free rate), is interest rates for instruments considered not to have possible defaults. In Indonesia's case, a risk-free instrument that can be chosen is interest rates on government bonds for long term. Beta is a measurement of the sensitivity rate of stock return towards overall stock market return.

2.4.3. Economic analysis

Investments that are projected to obtain sustainable income for a certain period, the calculation is carried out with a discounted cash flow.

2.4.3.1. Payback period

The payback period describes the period needed to get back the funds that have been invested in a project. Payback Period is determined by using equation as follows:

$$Payback\ Period = \frac{IO}{CFt} \quad (3)$$

with:
IO : Initial Outlay / Investment Cost
CFt : Generated annual cash flow

2.4.3.2. *Net present value /present worth*

NPV is based on the concept of equivalence in which all cash inflows and cash outflows are calculated towards the present time. If NPV is positive, it means the investment can be accepted, or the investment will increase the value of the company, conversely, if the NPV obtained is negative then the investment proposal for the project is rejected, or it can be said that the investment will reduce the value of the company. NPV is determined by using equation as follows:

$$Net\ Present\ Value = \sum_{t=1}^{N} \frac{CFt}{(1+K)^t} - IO \quad (4)$$

with:
IO : Initial Outlay / Investment Cost
CFt : Generated annual cash flow in year t

K : Required rate of return
N : Investment life
t : time period

2.4.3.3. Internal rate of return

The Internal Rate of Return (IRR) is the discount rate that makes the net present value (NPV) of a project zero. In other words, it is the expected compound annual rate of return that will be earned on a project or investment.

$$IO = \sum_{t=1}^{N} \frac{CFt}{(1+IRR)^t} - IO \quad (5)$$

with:
IO : Initial Outlay / Investment Cost
CFt : Generated annual cash flow in year t
N : Investment life
t : period

2.4.4. Sensitivity analysis

Sensitivity analysis determines how different values of an independent variable affect a particular dependent variable under a given set of assumptions. Sensitivity analysis will focus on the minimum utilization of facility and revenue compare to investment cost.

3. Research Methodology

The following are the stages of the research which is implemented.

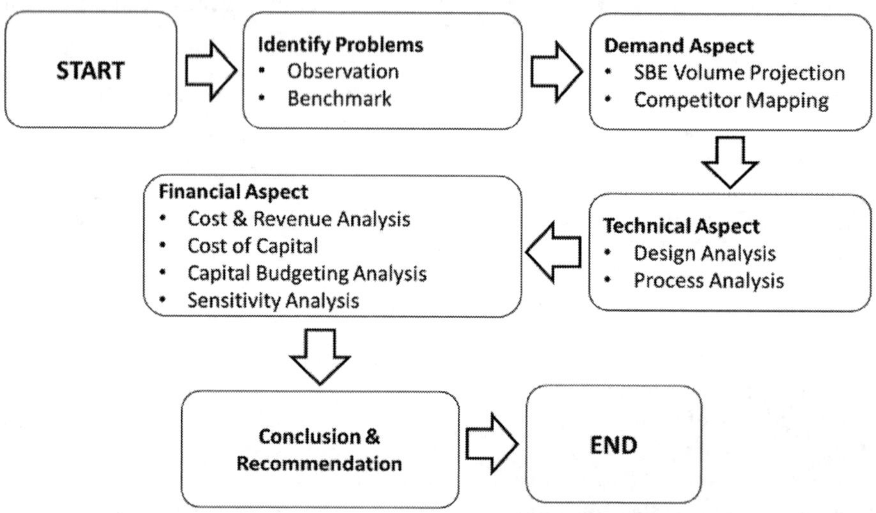

Figure 1. Flowprocess of research method implementation.

4. Results and Discussions

4.1. Demand Analysis

4.1.1. CPO Demand Projection

In projecting CPO demand in 2020 - 2030, this research uses correlation regression analysis. Regression analysis using the form of the relationship between one or more variables / independent variables (X) with one dependent variable (Y) has the equation:

$$Y = a + bX \quad (6)$$

with:
Y : Demand CPO (Dependent Variable)
X : Population (Independent Variable)
a,b : Coefficient

BPS data [5] for Indonesian population and projection from 2010-2030 shown in table 1 below.

Table 1. Indonesian population 2010 – 2030 (based on BPS report 2013).

Year	Population ('000)	Year	Population ('000)
2010	238519	2020	238519
2011	241810	2021	273777
2012	245147	2022	276515
2013	248530	2023	279280
2014	251960	2024	282073
2015	255462	2025	284829
2016	258502	2026	287108
2017	261578	2027	289404
2018	264691	2028	291720
2019	267840	2029	294053
		2030	296405

Historical CPO demand data Indonesian Palm Oil Association [6] shown in table 2 below.

Table 2. Indonesian CPO demand 2010 – 2019 (based on GAPKI report 2019).

Year	CPO Demand ('000)
2010	21800
2011	23500
2012	26500
2013	30000
2014	31500
2015	32500
2016	33637
2017	38166
2018	43108
2019	47180

The results of the regression correlation in table 1 and table 2 show that the population has a positive relationship with CPO demand in Indonesia. This is shown in table 3 where the value of Multiple R and R square is valued above 0.95, while the test result is explained in table 4.

Table 3. Indonesian CPO demand 2010 – 2019.

Regression Statistic	
Multiple R	0.97683
R Square	0.95421
Adjusted R Square	0.94848
Standard Error	1852.11
Observations	10

Table 4. Statistical test result.

	Coefficients	*Standard Error*	*t Stat*	*P-value*	*Lower 95%*	*Upper 95%*	*Lower 95,0%*	*Upper 95,0%*
Intercept	-171165.11	15806.7394	-10.8286	4.6708E-06	-20761	-134715	-20761	-13471
X Variable 1	0.80485811	0.062334805	12.9118	1.2242E-06	0.6611	0.9486	0.6611	0.9486

The correlation formula between population and CPO demand become:

$$Demand\ CPO\ = -171{,}165.11 + 0.805\ Population \quad (7)$$

From the equation 7 above, Indonesia CPO projection demand shown in table 5.

Table 5. Indonesian projection CPO demand 2020 – 2030.

Year	**CPO Demand ('000)**
2020	47005
2021	49187
2022	51390
2023	53616
2024	55863
2025	58082
2026	59916
2027	61764
2028	63628
2029	65506
2030	67399

4.1.2. SBE volume projection

Base on research, the amount of bentonite needed as an oil purifier will be the same as 1.5% - 4.0% of the amount of palm oil [2]. And it was noted that SBE could present a fire because it contains 20 - 40% by weight of oil [3]. From the research, the formula to calculate SBE volume is:

$$Volume\ SBE\ = (Demand\ CPO\ .\ 1{,}5\%)\ x\ 120\% \quad (8)$$

The result showed that in 2020 total volume SBE is 846000 ton/year, which means there are 110 ton/hour SBE in Indonesia. The SBE projection volume for 2020 – 2030 shown in table 6.

Table 6. SBE volume projection 2020 – 2030.

Year	**CPO Demand ('000)**
2020	846
2021	885
2022	925
2023	965
2024	1006
2025	1045
2026	1078
2027	1112
2028	1145
2029	1179
2030	1213

It can be concluded that if the waste utilization facility has a capacity of 8 tons/hour, it will consume 7% of SBE waste in 2020. In other words, the market share target for SBE utilization facilities development in the Tuban plant is 7%.

4.1.3. *Disposal cost*

Disposal costs are costs incurred by companies that produce B3 waste to companies that manage or dispose of the waste. Some of SBE producer near Tuban Plant such as PT Wilmar, PT Barata Elok, PT Bina Karya Prima (Gresik Area), PT Salim Invomas, PT Sinar Mas, PT Ikan Dorang, PT Damai Sentosa (Surabaya & Sidoarjo Area) etc, which has potential SBE waste 500 ton per day. In this research, the assumption of disposal cost is only the disposal cost of SBE based on a contract that has been received previously from PT D. The contract of disposal cost between Rp350,000 – Rp425,000 /ton and assumed that transportation cost from SBE producer to SBE storage is handled by third parties transporter which has waste transport permit. This research assumes that disposal cost is Rp350,000 /ton for the beginning year and increase 1% per three years. The disposal cost assumption shown in Table 7. below.

Table 7. Assumption of SBE disposal cost.

Year	Disposal Cost (Rp/ton)
1	350,000
2	350,000
3	350,000
4	353,500
5	357,035
6	367,746
7	378,778
8	397,717
9	417,603
10	438,483

4.1.4. *Competitor* mapping

The mapping is focused on well-known companies which have B3 waste processing business, both cement companies (exclude PT D) and non-cement companies. Due to the limitations of data to find a specific type of waste by each company, the waste data presented is all of the B3 waste that includes: medical waste and industrial waste.

Table 8. Competitor mapping.

Company	Type of Waste	Location	Capacity / Strenght
PT. PPLI	Medic & Industry	West Java Batam	600,000 ton/year (20 ton/day)
PT Semen Baturaja	Fly ash and Bottom Ash	South Sumatera	Fokus on fly ash and bottom ash, 3 ton/hour
PT. Waste International	Medic & Industry	Banten Centra Java	3 x 36 ton/day
PT. Tenang Jaya Sejahtera	Medic & Industry	West Java	7,2 ton/day
PT. Putera Restu Ibu Abadi	Medic & Industry	Mojokerto	16,8 ton/day
PT. Pengelola Limbah Kutai Kartanegara	Medic & Industry	East Borneo	12 ton/day
PT Arah Environmental Indonesia	Medic & Industry	Solo	12 ton/day
PT Indocement Tunggal Perkasa	Industry	West Java	1 line kiln for industrial waste, 5 ton/hour
PT Cemindo Gemilang	Industry	West Java	1 line kiln for industrial waste, 3 ton/hour

From Table 8 above, there aren't a company that focused on managed and utilized SBE. The total volume of well-known companies above had total capacity about 500 ton per day or 21 ton per hour. So, it can be concluded that PT A still could enter the market, especially for SBE.

4.2. Technical analysis

4.2.1. Process analysis

Technical Process analysis is carried out to find out how much capacity of the kiln could burn waste, especially SBE without disrupting calcination process, kiln operational, and clinker quality. Process analysis determined by simulation of process calculation and operational simulation based on historical data of PT D waste facility.

4.2.1.1. Process simulation

Simulations are carried out to determine the operational impact in the kiln based on the amount of SBE. From the simulation are obtained:

a. SBE input must be maintained in a pre-processing facility to prevent Sulfur (S) level above 1%
b. SBE input of 6 tph impact on decreased of LSF, but it still met the requirement.
c. SBE input of 8 tph requires an increase in raw meal LSF settings so that clinker quality standards are met.

4.2.1.2. Operational simulation

Operational simulations are carried out with reference to the PT D waste facility operational. The parameters of the kiln taken at the time of operation with a feed rate: 7 tph (limitation CO ILC ~ 0.9% fan ID 86% at 500tph kiln feed), it obtained:

a. CO ILC increased from 0.1% to 0.9% at seven tph feed rate (with the same coal rate).
b. There is no impact on the SLF clinker for the short term (for the long term it is necessary to pay attention to the raw mix design).
c. The temperature in ILC Calciner has increased.
d. There was no increase in amps Kilns.

Based on the results of operational and process simulation, the volume of SBE that can be burned into kiln inlet without disrupting the operation and quality of the clinker is between 7-8 tph, on condition that SBE must through a pre-processing process and meet the Service Level Agreement (SLA). The SLA parameter is shown in Table 9 below.

Table 9. Service level agreement of SBE.

Parameter	Unit		Value
GHV (AR)	Kcal/kg		2000±500
Sulfur (S)	%	Max	1
Total Chlor (Cl)	%	Max	0,6
Ash Content	%		40-70
Size	mm	Max	40

4.2.2. Design Analysis

The purpose of design analysis is to determine process flow and identify the equipment needed to build SBE waste utilization facilities. The process is devided into 2 major activities pre-processing and co-processing. Pre-processing activities is all activity to prepare raw waste became processing waste. The process mostly conducted in waste storage facility. Co-processing is all activity or equipment needed to make sure the processing waste is transported and burned into kiln. Figure 2. Shown the flow process of SBE waste utilization facilities.

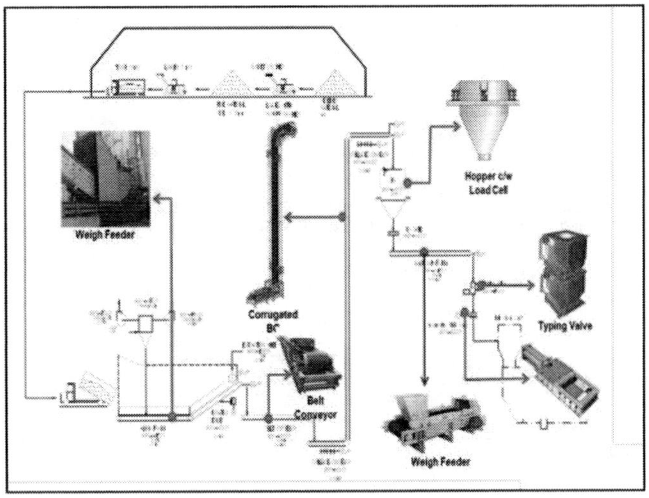

Figure 2. Flow process of SBE waste utilization facility.

Based on Figure 2 flow process diagram, the identification of the work and equipment needed is shown in the Table 10.

Table 10. Work and equipment list.

Item	Amount	Unit
Engineering	1	lot
Pre - Processing		
Waste Storage 8000 Ton (70 m x 36 m)	1	lot
Truck 15 Ton	1	ea
Loader 2 m3	1	ea
Loader with Bucket Crusher	1	ea
Co - Processing		
Civil Work	1	lot
Samson Feeder	1	ea
Corrugate BC	1	ea
Weigh Feeder	1	ea
Double Flap	1	ea
Safety, Fire suppression & Environment	1	lot
Electrical Equipment	1	lot
Commissioning	1	lot

4.2.3. Project schedule

Schedule of the project is last for 18 months from June, 2020 until December, 2021 and devided into 4 milestone. First milestone is completion of the business study, engineering study and feasibility study approval which targeted at Januari, 2021. Second milestone is start construction of the project that targeted in April, 2021. Third milestone is commissioning phase of the project which targeted on December, 2011 and the last milestone is commercial phase which targeted in January 2022.

4.3. Cost analysis

4.3.1. Investment cost

Investment costs are calculated based on the reference price of facilities that have been built by PT D, offers from vendors, company catalogs and other relevant sources by considering of foreign exchange and inflation. Based on Table 10. total investment cost can be seen in Table 11 below.

Table 11. Investment cost.

Item	Amount	Unit	Total Price (Rp)
Engineering	1	lot	300,000,000
Pre - Processing			
Waste Storage 8000 Ton (70 m x 36 m)	1	lot	5,500,000,000
Truck 15 Ton	1	ea	595,000,000
Loader 2 m3	1	ea	423,000,000
Loader with Bucket Crusher	1	ea	752,500,000
Co - Processing			
Civil WorkInvestment	1	lot	11,053,687,000
Samson Feeder	1	ea	5,685,897,000
Corrugate BC	1	ea	6,660,016,000
Weight Feeder	1	ea	2,438,875,000
Double Flap	1	ea	5,799,320,000
Safety, Fire suppression & Environment	1	lot	2,010,602,000
Electrical Equipment	1	lot	5,065,275,000
Commissioning	1	lot	500,000,000
Investment Cost			46,784,172,000
Contingency (5%)			2,339,208,600
Total Investment Cost			**49,123,380,600**

4.3.2. Operational Cost

Operational Cost or Cost of Good Sold (COGS) consists of Cost of Good Manufactured (COGM) and Operating Expense.

4.3.2.1 Cost of Good Manufactured (COGM)

Assumptions for Cost of Good Manufactured can be seen in Table 12 below.

Table 12. Assumptions for cost of good manufactured.

Cost Component	Description				
Labor	7 people/shift	2 Shift/day	Total 17 people; 14x	4 categorized labor	5% increased/y
Electricity	160 kwh total	996 rp/kwh	300 days/year		1% increased/y
Maintenance	6% of investment cost				5% increased/y
Fuel	25 l/hour	12 hour/day	300 days/year	5040 hours/year	1% increased/y

Based on an assumption on table 12 the result of calculation Cost of Good Manufactured can be seen in Table 13 below.

Table 13. Cost of good manufactured.

Cost Component	Year (Rp. Million)									
	Y1	**Y2**	**Y3**	**Y4**	**Y5**	**Y6**	**Y7**	**Y8**	**Y9**	**Y10**
Total Direct Labor	2,226	2,337	2,454	2,577	2,706	2,841	2,983	3,132	3,289	3,453
Electricity	1,147	1,159	1,172	1,184	1,196	1,209	1,221	1,234	1,247	1,260
Fuel	986	995	1,005	1,015	1,026	1,036	1,046	1,057	1,067	1,078
Maintenance	2,947	2,977	3,007	3,037	3,067	3,098	3,129	3,160	3,192	3,224
Total Man. Ovrhd	5,080	5,132	5,183	5,236	5,289	5,342	5,396	5,451	5,506	5,562
Total COGM	**7,306**	**7,469**	**7,638**	**7,813**	**7,995**	**8,183**	**8,379**	**8,583**	**8,795**	**9,015**

4.3.2.2 Operating expense

Assumptions for operating expenses can be seen in Table 14 below.

Table 14. Assumptions for operating expense.

Cost Component			Description		
Labor	2 people/shift	2 Shift/day	Total 4 people ; 14x	Marketing & Sales	5% increased/y
Transportation	Accommodation Cost	Lot	5 mill /month		5% increased/y
Office Supply	1	Lot			5% increased/y
Car Rental	1 ea		3 mill/month		5% increased/y
Marketing Expenses	1	Lot			5% increased/y
Depreciation					
Heavy Equpmt	8 years				
Building & Civ. Work	20 years				
Mechanical	15 years				
Electrical	10 years				

Based on the assumption in Table 14. the result of operating expenses can be seen in Table 15. below.

Table 15. Operating expense and cost of good sold.

Cost Component	Year (Rp. Million)									
	Y1	Y2	Y3	Y4	Y5	Y6	Y7	Y8	Y9	Y10
Salaries	560	588	617	648	681	715	750	788	827	869
Transportation	60	63	66	69	73	77	80	84	89	93
Marketing Expnse	200	210	221	232	243	255	268	281	295	310
G&A Expense	80	84	88	93	97	102	107	113	118	124
Depreciation	5,080	5,132	5,183	5,236	5,289	5,342	5,396	5,451	5,506	5,562
Heavy Equpmt	232	232	232	232	232	232	232	232	232	232
Building & Civ.	869	869	869	869	869	869	869	869	869	869
Mechanical	1,441	1,441	1,441	1,441	1,441	1,441	1,441	1,441	1,441	1,441
Electrical	827	827	827	827	827	827	827	827	827	827
Total Oprt. Expense	4,269	4,314	4,362	4,411	4,463	4,518	4,575	4,636	4,699	4,765
Cost Of Good Sold	**11,576**	**11,783**	**11,999**	**12,224**	**12,458**	**12,701**	**12,955**	**13,219**	**13,494**	**13,780**

4.3.3. Weighted Average Cost of Capital (WACC)

4.3.3.1 Cost of debt

Cost of Debt is calculated using highest interest between short term, long term loan, and bonds payable [8]. The highest interest / coupon is obligation shelf II 2019 B with 9.1% interest rate. Therefore:

$$\boldsymbol{k_d = 7.1\%}$$

4.3.3.2 Cost of equity

For Cost of Equity (k_e) calculation, the data is as follows:

a. Beta = 1.499, in accordance to levered business beta PT A.
b. R_f value = 8.2771%, comes from government bond yield FR0083 series with 20.02-year time to maturity as of April 14, 2020 from http://www.ibpa.co.id/DataPasarSuratUtang/ HargadanYield Harian/tabid/84/Default.aspx.
c. R_m value = 14.5532%, comes from Compound Annual Growth Rate (CAGR) IHSG from January 1, 2009 until January 1, 2020.

According to point 1,2 and 3, equation 2.11 is used and thus **k_e value =17.687%** is obtained as shown in table 16 below.

Table 16. Calculation of k_e value.

R_f	Beta	R_m	k_e
8.2771%	1.499	14.5532%	**17.687%**

4.3.3.3 Weighted Average Cost of Capital (WACC)

For WACC calculation, the data is as follows:

a. k_d value = 7.1%, k_e 17.68% according to the calculation above.
b. Tax (T) = 22%
c. Debt (D) = 45,915 (billion); Equity (E) = 23,892 (billion).

According to point a, b and c, so WACC value is obtained as shown in Table 17 below.

Table 17. Calculation of WACC value.

Debt (D)	Equity (E)	D+E	k_e	W_e	k_d	Tax	W_d	WACC
45.915	33.892	79.807	17,687%	0,425	7,10%	22%	0,58	11,59%

4.4. Revenue Analysis

Revenue analysis from SBE waste utilization facilities calculated by direct revenue and indirect added value or saving. Direct revenue calculated from waste disposal services / disposal cost payed by SBE produced. Indirect added value or savings calculated from raw material substitution, and alternative fuel (coal substitution). Assumptions for revenue analysis can be seen in Table 18. below.

Table 18. Assumptions for revenue component.

Revenue Component	Description		
Revenue			
Disposal Capacity	8 ton/hour		
Working Hour	24 hour/day	300 days/year	
Utilization	70% for y1		5% increased/2y
Disposal Cost	IDR 350,000/ton		2% increased/3y
Saving			
Iron Sand	IDR 286,808/ton		1% increased/y
Silica Sand	IDR 60,627/ton		1% increased/y
Limestone	IDR 37,771/ton		1% increased/y
Clay	IDR 37,233/ton		1% increased/y
Coal	IDR 454,000	4000 Kcal	-2 – 2% / y

4.4.1. Revenue

Based on the assumption in Table 18, the result of the revenue calculation shown in Table 19 below.

Table 19. Revenue projection for Y1-Y10.

Cost Component	Year (Rp. Million)									
	Y1	Y2	Y3	Y4	Y5	Y6	Y7	Y8	Y9	Y10
Disposal Cost	0.35	0.35	0.35	0.35	0.36	0.36	0.37	0.39	0.40	0.42
SBE Vol ('000 ton/year)	57.6	57.6	57.6	57.6	57.6	57.6	57.6	57.6	57.6	57.6
Utilization (%)	70	70	75	75	80	80	85	85	90	90
SBE Revenue	**14,112**	**14,112**	**15,120**	**15,120**	**16,451**	**16,780**	**18,185**	**18,912**	**20,826**	**21,659**
Saving (Indirect Added Value)										
Iron Sand	4,290	4,290	4,643	4,689	5,052	5,102	5,475	5,530	5,914	5,973
Silica Sand	543	543	587	593	639	645	693	700	748	756
Limestone	367	367	397	401	432	436	468	473	506	511
Clay	66	66	71	72	78	79	84	85	91	92
Total Raw Mat Subs	**5,266**	**5,266**	**5,698**	**5,755**	**6,200**	**6,262**	**6,720**	**6,787**	**7,259**	**7,331**
Alternative Fuel	**2,396**	**2,396**	**2,516**	**2,466**	**2,630**	**2,656**	**2,879**	**2,936**	**3,140**	**3,172**
Total Revenue	**21,774**	**21,774**	**23,334**	**23,341**	**25,281**	**25,698**	**27,784**	**28,636**	**31,225**	**32.162**

4.4.2. Income Statement

Based on Table 15 and Table 19, an income statement can be shown as follows:

Table 20. Income statement projection for Y1-Y10.

Cost Component	Year (Rp. Million)									
	Y1	Y2	Y3	Y4	Y5	Y6	Y7	Y8	Y9	Y10
Net Sales	**21,774**	**21,774**	**23,334**	**23,341**	**25,281**	**25,698**	**27,784**	**28,636**	**31,225**	**32,162**
Cost of Goods Manfc,	7,306	7,469	7,638	7,813	7,995	8,183	8,379	8,583	8,795	9,015
Gross Profit	**14,468**	**14,305**	**15,697**	**15,528**	**17,286**	**17,515**	**19,405**	**20,053**	**22,430**	**23,147**
Gross Profit Margin (%)	*66.44*	*65.70*	*67.27*	*66.53*	*68.38*	*68.16*	*69.84*	*70.03*	*71.83*	*71.97*
Operational Expenses										
Marketing Expenses	820	861	904	949	997	1,047	1,099	1,154	1,212	1,272
GA Expenses	80	84	88	93	97	102	107	113	118	124
Depreciation	3,369	3,369	3,369	3,369	3,369	3,369	3,369	3,369	3,369	3,369
Total Opr. Exps.	**4,269**	**4,314**	**4,362**	**4,411**	**4,463**	**4,518**	**4,575**	**4,636**	**4,699**	**4,765**
Operating Profit	**10,198**	**9,991**	**11,335**	**11,117**	**12,823**	**12,997**	**14,829**	**15,417**	**17,731**	**18,381**
Operating Margin (%)	*46.84*	*45.88*	*48.58*	*47.63*	*50.72*	*50.58*	*53.37*	*53.84*	*56.78*	*57.15*
EBIT	**10,198**	**9,991**	**11,335**	**11,117**	**12,823**	**12,997**	**14,829**	**15,417**	**17,731**	**18,381**
Tax Rate (%)	22	22	22	22	22	22	22	22	22	22
Income Tax	2,244	2,198	2,494	2,446	2,821	2,859	3,262	3,392	3,901	4,044
Net Income	**7,955**	**7,793**	**8,841**	**8,671**	**10,002**	**10,138**	**11,567**	**12,026**	**13,830**	**14,337**
Net Income Margin (%)	*36.53*	*35.79*	*37.89*	*37.15*	*39.56*	*39.45*	*41.63*	*41.99*	*44.29*	*44.58*
EBITDA	**13,568**	**13,360**	**14,704**	**14,486**	**16,192**	**16,366**	**18,199**	**18,787**	**21,100**	**21,751**
EBITDA Margin (%)	*62.31*	*61.36*	*63.02*	*62.06*	*64.05*	*63.69*	*65.50*	*65.61*	*67.58*	*67.63*

4.5. Calculation of Net Present Value

According to Table 20 and Table 11, Net Present Value and Internal Rate of Return can be calculated as shown in table 21 below.

Table 21. Net Present Value (NPV).

Cost Component	Year (Rp. Million)									
	Y0	Y1	Y2	Y3	Y4	Y5	Y6	Y7	Y8	Y9
Net Income	-	7,955	7,793	8,841	8,671	10,002	10,138	11,567	12,026	13,830
Depreciation	-	3,369	3,369	3,369	3,369	3,369	3,369	3,369	3,369	3,369
Change in Net Work. Cap.	-	-	-	-	-	-	-	-	-	-
Capital Expenditure	49,123	-	-	-	-	-	-	-	-	-
Free Cash Flow to Firm	**(49,123)**	**11,324**	**11,162**	**12,211**	**12,041**	**13,371**	**13,507**	**14,936**	**15,395**	**17,199**
Cumulative of FCFF	**(49,123)**	**(37,799)**	**(26,637)**	**(14,427)**	**(2,386)**	**10,985**	**24,492**	**39,428**	**54,823**	**72,022**
Payback Period	**4.2 Year**									
WACC	**11.60 %**									
NPV	**19,746**									
IRR	**22.58 %**									

From the calculation in the table above, it is concluded that the NPV is Positive and IRR > WACC so the development of waste utilization facilities in Tuban Plant is **feasible**.

4.6. Sensitivity Analysis

Sensitivity analysis conducted to determine the minimum utilization of facility, change of disposal and investment cost.

4.6.1. Utilization changes

From table 22 below, it can be seen that on 60% utility, the project won't be feasible with negative NPV and IRR < WACC.

Table 22. Sensitivity to Utility Changes

Utility	Net Present Value (Rp. Million)	Internal Rate of Return (%)	Conclusion
70%	19,746	22.6	Feasible
67%	6,697	16.7	Feasible
63%	1,899	14.1	Feasible
60%	**(1,698)**	**12.0**	**Not Feasible**

4.6.2. Disposal cost changes

Table 23 below shows that if disposal cost reduced by 65% from initial cost or became Rp 245.000,- the waste facilities won't be feasible with negative NPV and IRR < WACC.

Table 23. Sensitivity to Annual Cost Changes

Disposal Cost	Net Present Value (Rp. Million)	Internal Rate of Return (%)	Conclusion
100%	19,746	22.6	Feasible
90%	13,641	19.8	Feasible
80%	7,537	16.8	Feasible
75%	4,484	15.3	Feasible
70%	1,432	13.7	Feasible
65%	**(1,620)**	**12.1**	**Not Feasible**

4.6.3. Investment changes

Table 24 shows that the project won't be feasible if investment cost increases to 40 % from the initial project cost.

Table 24. Sensitivity to Annual Cost Changes

Δ Investment Cost	Net Present Value (Rp. Million)	Internal Rate of Return (%)	Conclusion
0 %	19,746	22.6	Feasible
10 %	14,594	19.6	Feasible
20 %	8,764	16.7	Feasible
25 %	6,018	15.5	Feasible
30 %	3,273	14.3	Feasible
35 %	527	13.3	Feasible
40 %	**(2,217)**	**12.2**	**Not Feasible**

5. Conclusion

This study has analyzed feasibility study of waste utilization facilities based on demand, technical and financial aspect. The result is we recommended to management of PT. A to invest and built the facilities in Tuban Plant with Service Level of Agreement (SLA) for the facilities is to run with a minimum 60% utilization. The analysis in this study is limited to spent bleaching earth waste management and only for waste processing facility (transportation assumed by third party transporter) further research is expected to analyze the potential for other B3 waste which can be managed by the facility and potential to enter end to end waste management business with applying for waste transport permit so it can have a greater impact to the company and society.

References

[1] Kheang LS, Foon CS, May CY, Ngan MA. A Study of Residual Oils Recovered from Spent Bleaching Earth: Their Characteristics and Applications. *American Journal of Applied Sciences*. 2006 Oct 1;**3**(10):2063–7. doi.org/10.3844/ajassp.2006.2063.2067
[2] PT. D 2019 *Annual Report of PT. D Year 2018*
[3] Mardiana G, Mahardika R. Pemanfaatan Limbah Biomass Sebagai Bahan Bakar Alternatif Dalam Kegiatan Co-Processing Di Semen Gresik. *Jurnal Rekayasa Kimia Dan Proses*. 2010;1411-4216
[4] Ross SA, Jaffe JF. *Corporate Finance.* Singapore: Irwin/McGraw-Hill; 1999.
[5] Badan Pusat Statistik Indonesia. *Indonesian Population Projection 2010 – 2035*. Indonesia: Catalog: 2101018; 2013.
[6] Indonesia Palm Oil Association. *Kinerja Industri Sawit Indonesia 2019*. Jakarta: GAPKI; 2019.
[7] Newnan DG, Lavelle JP. *Engineering Economic Analysis.* New York: Oxford University Pers; 2004.
[8] PT. A 2020 *Annual Report of PT. A Year 2019*

IOP Conf. Series: Materials Science and Engineering 1072 (2021) 012040 doi:10.1088/1757-899X/1072/1/012040

Regression analysis in predicting coating application lifetime on asset integrity project

R Setiawan[1] and B Santosa[2]

[1,2]Department of Industrial and Systems Engineering, Institut Teknologi Sepuluh Nopember, Surabaya, Indonesia

E-mail: [1]ridwan.setiawanofficial@gmail.com, [2]budi_s@ie.its.ac.id

Abstract. The Asset Integrity project is one of the strategic priorities in maintenance to ensure the functional lifetime of an asset is well maintained to support the nickel processing. The coating protection application is one of the asset integrity strategy to maintain the structural lifetime, however the corrosive environment of nickel processing plant influences the lifetime of the coating design differently on each facility due to different process in the processing line. The objective of this study is to predict the coating design lifetime on Dryer, KILN, Furnace, and Converter facilities. The analysis used regression method for coating lifetime on each facility by considering the independent variables, i.e. SO_2 Intensity, stack emission, NOx emission, rainfall intensity, and underfilm condition. By using regression model, coating design lifetime in Dryer area is 4.95 years, KILN is 5.73 years, Furnace is 6.23 years, and Converter is 6.73 years. Process plant area which having the most significant impact of corrosive environment is Dryer area, causing the coating lifetime in this area is the lowest than other area in the Process Plant area. This result can be used to set maintenance strategy for coating design lifetime lifecycle for the next Asset Integrity project.

1. Introduction

The nickel processing plant has already stood for 50 years, causing a high risk in structural failure since it already reaches the design lifetime limit, any structural failure will cause a huge impact to the operational line. There are 365 assets in the processing plant and to maintain these assets will require proper planning and strategies to set the priority on high risk asset and to execute it. Asset Integrity project is part of the Asset Management Framework (AMF) to focus on planning, and asset controlling based on technical assessment which has been done in the feasibility study. With 7% asset classified as high risk level, 19% classified as medium high risk, and 29% as medium risk, proper planning is required to manage all facilities can be maintained to ensure the production lines is not interrupted. The coating process is part of the action plan to ensure the structure design meets with design lifetime. However, the range of coating durability influenced by many factors. There are a variety of environmental stresses that combine to degrade coatings exposed in a service environment. By identifying these factors and how it correlated with the coating durability will give sufficient information to the maintenance team to predict the coating design lifetime. For this reason, multiple correlation and multiple regression are mathematical tools which allow to identify the linear simple correlation between the variables as well as the multiple correlation coefficients which may reflect the influence of one or the combined influence of two factors on the 3rd one [1].

The regression model can be applied to understand the relation between coating lifetime with SO2 intensity, rainfall intensity, NOx emission, undercoating condition and stack emission as part of the industrial atmosphere in the nickel processing plant focusing on Dryer, KILN, Furnace, and Converter area since this is the main operational line for the nickel processing plant. This research goal is to create a prediction model of coating design lifetime on each facility on the nickel processing plant using attributes of multiple independent variables.

2. Materials and Method

2.1. Corrosion and coating

Corrosion is generally taken to be the wastage of a metal by the action of corrosive agents. However, a wider definition is the degradation of a material through contact with its environment. Thus, corrosion can include non-metallic materials such as concrete and plastics and mechanisms such as cracking in addition to wastage (i.e. loss of material) [2]. Mechanical properties of the materials decrease by corrosion process whereas the corrosion products are released in different forms that may cause a more extreme corrosive environment or harmful side effects in different applications [7]. To overcome this issue and to enhance material properties for specific applications, there have been different methods offered, such as heat treatment, alloying processes, and coatings. Among these solutions, coating processes have the highest portion of material enhancement since coating layers can reduce the cost and neglect scarcity of materials as the thickness of coating layers rarely pass micrometres [7].

Acids consist of inorganic mineral acids such as hydrochloric, sulfuric, and nitric acids, which disassociate completely in water. Acid gases such as sulphur dioxide (SO2), sulphur trioxide (SO3), hydrogen sulphide (H2S), and nitrogen oxide (NOX) react with moisture in the air in the form of precipitation or condensation to form sulfuric and nitric acids [8][9].

Based on table 1 Nickel processing plant environment categories as C5-I for the corrosion category and this environment influence the coating design lifetime very aggressively [10]. Some of the parameters which influence the corrosion based on the feasibility study in the Asset Integrity project are SO2 intensity, surface preparation, rainfall intensity, and Process Plant environment.

Table 1. ISO 12944 (1998) - Atmosphere corrosivity categories and examples of typical environments.

Corrosivity Category	Exterior Environment	Interior Environment
C1 Very low	---	Heated buildings with clean atmospheres, e.g., office, shop, schools, hotels
C2 Low	Atmospheres with low level pollution, mostly rural area	Unheated building when condensation may occur e.g., depots, sports hall
C3 Medium	Urban and industrial atmospheres, moderate sulphur dioxide pollution, coastal area with low salinity	Production rooms with high humidity and some air pollution, e.g., food processing plants, laundries, breweries, dairies
C4 High	Industrial area and coastal area with moderate salinity	Chemical plants, swimming pools, coastal ship-, and boatyards
C5-I Very High (Industrial)	Industrial area with high humidity and aggressive atmosphere	Building area with almost permanent condensation and with high pollution

2.2. Data mining

Data mining is the process of discovering interesting patterns from massive amounts of data. As a knowledge discovery process, the process typically involves data cleaning, data integration, data selection, data transformation, pattern discovery, pattern evaluation, and knowledge presentation. The major dimensions of data mining are data, knowledge, technologies, and applications [3]. In general data mining can be classified in to two categories:

- Descriptive Mining: To characterize property of data inside the target data.
- Predictive Mining: To conduct data interference to create prediction, predictive mining involve variables or field in set of data to predict unknown value or future value from another variable [3].

2.3. ANOVA

ANOVA evaluate the effect of two or more factors as well as the interaction between them. This technique compares the average of one response variable (quantitative) between different groups (different level of one or more factors), informing whether the difference between the sample averages of the experiment is significant to state that, on repeating the experiment, the effects would continue to be verified. For each factor, ANOVA evaluates whether the variability between groups formed by their different level is significantly greater than the variability within them [4]. This is confirmed by the p-value, when the result is lower than or equal to 5% indicating the variable have significant effect to the dependent variable.

2.4. Regression

The linear regression analysis technique is applied in situations which assumes a cause–effect relationship between two or more quantitative variables and which aims to express mathematically this relationship. It is used when there is a data set with two averages for each element and the problem is to find an equation which is able to relate the two averages [4]. Regression analysis provides an objective and systematic way to analyse data. As a result, decisions based on regression are less likely to be subject to bias, they are consistent, the basis for the decisions can be fully explained and they are generally useful [5]. The modelling will be using method as follows:

1. Linear regression with multiple variables
   The regression model with more than one independent variable to explain the relation with dependent variable, the formula for regression with multiple variables as follows [6]:
   $$Y_i = \beta_0 + \beta_1 . X_{1i} + \beta_2 . X_{2i} + e \quad (1)$$
2. Coefficient of determination
   The method is using statistic approach to measure the "*goodness of fit*" of the data in linear regression which represent is proportion or percentage of total variance for dependent variables explained by the independent variables. R-square value range from 0-1, closer to 1 means the observed variation can be clearly explained by the model inputs. in opposite; closer to 0 means the observed variation cannot be explained by the model inputs [6]. The formula for coefficient of determination as follows [6]:
   $$R = \sqrt{\frac{\alpha_1 \sum X_1 Y + \alpha_2 \sum X_2 Y + \cdots + \alpha_i \sum X_i Y}{\sum Y^2}} \quad (2)$$
3. Significance test
   Significance test or F-Test is a test for regression model to measure the linear relation of independent variables with dependent variable by comparing the F-value with F-table with α = 5%. Formula for F- value as follows [6]:
   $$F_{value} = \frac{R^2(N-k-1)}{k(1-|R|^2)} \quad (3)$$
4. T-test
   The t-test is a test to analyse the contribution of independent variables related to dependent variables partially [6]. The t-test takes a sample from each of the two sets and establishes the problem statement of null hypothesis and alternative hypothesis. t-test result will be compared with p-value to identify whether the result is lower than or equal to 0.05 to understand the significance between dependent and independent variables. In t-test the null hypothesis and alternative hypothesis formulate as follows [6]:
   1. H_0 : In partial; there is no significant influence between dependent variable and independent variables
   2. H_1 : In partial; there is significant influence between dependent variable and independent variables

3. Result and Discussion

3.1. Data pre-processing

To evaluate the coating lifetime in processing plant area, data collection was collected from three departments and was used to create regression model. Each of independent variable data was collected from 2011 to 2019. However, it is necessary to ensure the data did not have missing values and noise. To this aim, pre-processing data is conducted to cleanse the collected data, data input for each variable was collected from Dryer, KILN, Furnace, and Converter area. Some of the value in the data was found not recorded in the 2011, therefore the variables will not be considered into the independent variables. Since considering the nickel processing plant layout, the independent variables for each area will be varies therefore some of the variables will not be considered into the modelling.

3.2. CZSQ Result

The data were processed to create regression with multiple variables modelling. The modelling was created by using five independent variables which are SO2 intensity, rainfall intensity, stack emission, NOx emission, and under film condition on four main areas Dryer, KILN, Furnace, and Converter. The variables for each area as follow:

- Dryer: Underfilm condition, SO2 Intensity, Stack Emission
- KILN: Underfilm condition, SO2 Intensity, Stack Emission, and Rainfall Intensity
- Furnace: Underfilm condition, SO2 Intensity, Rainfall Intensity, and NOx emission
- Converter: Underfilm Condition, SO2 Intensity, Stack Emission, and NOx emission

The regression model using software will produce information, i.e. coefficient determination, standard errors, F-value, and a p-value of t-test and regression equation with result as follows:

3.2.1. Coefficient determination result

The observation was conducted on Dryer, KILN, Furnace and Converter area with 50 samples on each area. The result for Coefficient of Determination as shown in Table 2.

Table 2. Coefficient of determination result.

Location	Regression Statistic				
	Multiple R	R^2	Adjusted R^2	Standard Error	Observation
Dryer	0.9887	0.9775	0.9761	0.228	50
KILN	0.9586	0.9189	0.9117	0.382	50
Furnace	0.9115	0.8309	0.8158	0.525	50
Converter	0.9784	0.9572	0.9534	0.282	50

3.2.2. Significance test

In conducting F-test, the null hypothesis and alternative hypothesis as follows:

1. H_0 : there is no significant influence for all independent variables to the coating lifetime
2. H_1 : there is significant influence for all independent variables to the coating lifetime

The F-test will use the assumption to identify whether there is correlation between dependent and independent variables by comparing the p-value with $\alpha = 0.05$ with condition as follows:

1. If p-value is < 0.05 then H0 is rejected and H1 is accepted
2. If p-value is > 0.05 then H0 is accepted and H1 is rejected

Table 3. Significance test result.

Location	F-test				
	DoF	**F**	**Critical Value**	**$F_{significance}$**	**P-Value**
Dryer	46	668,907	2.806	6.243×10^{-38}	0.05
KILN	45	127.492	2.578	6.126×10^{-24}	0.05
Furnace	45	55.282	2.578	8.454×10^{-17}	0.05
Converter	45	252.038	2.578	3.500×10^{-30}	0.05

3.2.3. T-test

T-test will be conducted on this case to identify whether each of the independent variables have a significant impact to the coating lifetime. The null hypothesis and alternative hypothesis as follow:

1. H_0 : There is no significant impact from each of independent variables to coating lifetime
2. H_1 : There is significant impact from each of independent variables to coating lifetime

Table 4. T-test result.

Location	β1		β2		β3		β4		β5	
	t-value	**p-value**	**t-value**	**p-value**	**t-value**	**p-value**	**t-value**	**p-value**	**t-value**	**p-value**
Dryer	-0.431	0.6778	39.099	$6.048\text{x}10^{-37}$	-14.427	$2.262\text{x}10^{-16}$	-	-	-	-
KILN	-0.0086	0.9931	19.750	$8.412\text{x}10^{-24}$	-8.768	$2.694\text{x}10^{-11}$	1.376	0.1754		
Furnace	-0.7377	0.4644	10.274	$2.214\text{x}10^{-13}$	-	-	2.4099	0.0201	-1.5307	0.1328
Converter	0.0603	0.9521	28.334	$2.480\text{x}10^{-30}$	4.6116	$3.310\text{x}10^{-5}$	-	-	0.0068	0.00051

Note: β1 = Under film condition (%), β2 = SO2 intensity (t SO2/t Ni), β3 = Stack emission (mg/m^3),
β4 = Rain intensity (mm/year), β5 = NOx emission (mg/m^3)

3.2.4. Regression equation

The regression result as shown on Table 5 for all areas in process plant.

Table 5. Regression result.

Area	Intercept	Independent Variables				
		β1	**β2**	**β3**	**β4**	**β5**
Dryer	-3.769	$-8.0\text{x}10^{-04}$	14.14	$-2.1\text{x}10^{-02}$	-	-
KILN	-3.042	$-3.8\text{x}10^{-05}$	11.06	$-7.6\text{x}10^{-03}$	$3.2\text{x}10^{-03}$	-
Furnace	-5.098	$-3.4\text{x}10^{-03}$	11.55	-	$8.7\text{x}10^{-03}$	$-1.0\text{x}10^{-02}$
Converter	-4.889	$1.2\text{x}10^{-04}$	12.79	$5.8\text{x}10^{-03}$	-	$2.5\text{x}10^{-02}$

Note: β1 = Under film condition (%), β2 = SO2 intensity (t SO2/t Ni), β3 = Stack emission (mg/m^3),
β4 = Rain intensity (mm/year), β5 = NOx emission (mg/m^3)

3.3. Model Analysis

Regression model analysis for coating design lifetime at process plant area involving five independent variables as follows:

3.3.1. Coefficient of determination

The regression model in Dryer area involves three independent variables with observation of 50 samples (SO2, stack emission, and underfilm condition) with coefficient of determination of 97.75% reflects the variation in the combination of SO2 intensity, stack emission, and underfilm condition around it means with standard error of 0.228. The regression model in KILN area with coefficient of determination of 91.89%, variation explained by the relation of SO2 intensity, rain intensity, stack emission, underfilm condition, and stack emission with standard error of 0.382. The regression model in Furnace area with coefficient of determination of 83,09% indicates variance of the data explained by independent variables of underfilm condition, SO2 intensity, rain intensity, and NOx emission with standard error of 0.525. The regression model in Converter area with coefficient of determination was 95.72% indicates variance of the data explained by independent variables of underfilm condition, SO2 intensity, stack emission, and NOx emission around it means with standard error of 0.282.

3.3.2. Significance test
The significance test for regression model in Dryer area shown that Fsignificance 6.243x10-38 < 0.05, and F value 668,907 > Critical value of 2,806 with degree of freedom of 46. The significance test for regression model in KILN area shown that Fsignificance 6.126x10-24 < 0.05, and F value 127.492 > Critical value of 2,578 with degree of freedom of 45. The significance test for regression model in Furnace area shown that Fsignificance 8.4547 x 10-17 < 0.05, and F value 55.282 > Critical value of 2,578 with degree of freedom of 45. The significance test for regression model in Converter area shown that Fsignificance 3.5004 x 10-30 < 0.05, and F value 252.038 > Critical value of 2,578 with degree of freedom of 45. By this results H0 is rejected, independent variables have significance influence on the dependent variable for Dryer, KILN, Furnace and Converter area.

3.3.3. T-test
Based on result in Table 4 the analysis of t-test for p-value result as follows:

3.3.3.1 Dryer t-test
Underfilm condition result is 0.6778 > 0.05 resulting the H0 is accepted. The underfilm condition in dryer area is not significant influencing the coating lifetime. For any additional underfilm condition causing less significant influence to the coating lifetime. SO2 result is 6.0485x10-37 < 0.05 resulting H0 is rejected. SO2 emission in dryer area is significant to influence the coating lifetime. Stack emission result is 2.2629x10-16 < 0.05 resulting H0 is rejected. Stack emission in dryer area is significant to influence the coating lifetime.

3.3.3.2 KILN t-test
Underfilm condition result is 0.9931 > 0.05 resulting the H0 is accepted. The underfilm condition in KILN area is not significant influencing the coating lifetime. For any additional underfilm increase causing less significant influence to the coating lifetime. SO2 result is 8.4125x10-24 < 0.05 resulting H0 is rejected. SO2 emission in KILN area is significant to influence the coating lifetime. Rain intensity is 0.1754 > 0.05 resulting H0 is rejected. Rain Intensity in KILN area is not significant influencing the coating lifetime. Stack emission result is 2.2629x10-16 < 0.05 resulting H0 is rejected. Stack emission in KILN area is significant to influence the coating lifetime.

3.3.3.3 Furnance t-test
Underfilm condition result is 0.4644 > 0.05 resulting the H0 is accepted. The underfilm condition in Furnace area is not significant influencing the coating lifetime. For any additional underfilm increase causing less significant influence to the coating lifetime. SO2 result is 2.2140x10-13 < 0.05 resulting H0 is rejected. SO2 emission in Furnace area is significant to influence the coating lifetime. Rain intensity is 0.0201 < 0.05 resulting H0 is rejected. Rain Intensity in Furnace area is not significant influencing the coating lifetime. NOx emission result is 0.1328 > 0.05 resulting H0 is accepted. NOx emission in Furnace area is not significant to influence the coating lifetime.

3.3.3.4 Converter t-test
Underfilm condition result is 0.4644 > 0.05 resulting the H0 is accepted. The underfilm condition in converter area is not significant influencing the coating lifetime. For any additional underfilm increase

causing less significant influence to the coating lifetime. SO2 result is 2.4804x10-30 < 0.05 resulting H0 is rejected. SO2 emission in converter area is significant to influence the coating lifetime. Stack emission is 3.3109x10-5 < 0.05 resulting H0 is rejected. Stack emission in converter area is not significant influencing the coating lifetime. NOx emission result is 0.00051 < 0.05 resulting H0 is rejected. NOx emission in converter area is significant to influence the coating lifetime.

3.3.4. Regression Equation

3.3.4.1 Dryer

Based on model analysis the regression showing a good model and can be used for analysis result from regression in Dryer area based on the equation as follows:

$$Y = -3.769 - 0.0008\,(\beta 1) + 14.14(\beta 2) - 0.021(\beta 3) \quad (4)$$

The equation shown a negative relationship between dependent variable and independent variables in Dryer area. Underfilm condition coefficient (β1) contains negative value of 0.0008 means for every additional increase of underfilm condition, the expected coating lifetime decrease by 0.0008 on average holding all other variable constant. However, for every additional increase of SO2 emission (β2), the expected lifetime increases by 14.14 on average holding all variable constant. For every additional increase of stack emission (β3), the expected lifetime is decreasing by 0.021 on average holding all variable constant. The estimated average value of coating lifetime was 3.769 years when all independent variables = 0 (assuming all βi = 0 is within the range of observed values).

3.3.4.2 KILN

Analysis result from regression in KILN area based on the equation as follows:

$$Y = -3.042 - 0.000038\,(\beta 1) + 11.06(\beta 2) - 0.0076\,(\beta 3) + 0.00321\,(\beta 4) \quad (5)$$

The regression equation also shown a negative relationship between dependent variable and independent variables in KILN area. Under film condition coefficient (β1) contains negative value of 0.000038 means for every additional increase of underfilm condition, the expected coating lifetime decrease by 0.000038 on average holding all other variable constant. Also, for every additional increase of stack emission (β3), the expected lifetime is decreasing by 0.0076 on average holding all variable constant. However, for every additional increase of SO2 emission (β2), the expected lifetime increases by 11.06 on average holding all variable constant. It also happens for every additional increase of rain intensity (β4), the expected lifetime is increasing by 0.00321 on average holding all variable constant. The estimated average value of coating lifetime was 3.042 years when all independent variables = 0 (assuming all βi = 0 is within the range of observed values).

3.3.4.3 Furnance

Analysis result from regression in KILN area based on the equation as follows

$$Y = -5.098 - 0.0034\,(\beta 1) + 11.55\,(\beta 2) + 0.0087\,(\beta 4) - 0.010\,(\beta 5) \quad (6)$$

The regression model in Furnace area shown a negative relationship where underfilm condition coefficient (β1) contains negative value of 0.0034 means for every additional increase of underfilm condition, the expected coating lifetime decrease by 0.0034 on average holding all other variable constant. Also, for every additional increase of NOx (β5), the expected lifetime is decreasing by 0.010 on average holding all variable constant. However, for every additional increase of SO2 emission (β2), the expected lifetime increases by 11.55 on average holding all variable constant. It also happened for every additional increase of rain intensity (β4), the expected lifetime is increasing by 0.0087 on average holding all variable constant. The estimated average value of coating lifetime was 5.098 years when all independent variables = 0 (assuming all βi = 0 is within the range of observed values).

3.3.4.4 Converter

Analysis result from regression in Converter area based on the equation as follows:

$$Y = -4.889 + 0.00012\ (\beta1) + 12.79\ (\beta2) + 0.0058\ (\beta3) - 0.0255\ (\beta5) \quad (7)$$

The regression model in Converter area also shown a negative relationship between dependent variable and independent variables. Under film condition coefficient (β1) contains positive value of 0.0012 means for every additional increase of underfilm condition, the expected coating lifetime increase by 0.0012 on average holding all other variable constant. Also, for every additional increase of NOx emission (β5), the expected lifetime is decreasing by 0.0255 on average holding all variable constant. However, for every additional increase of SO2 emission (β2), the expected lifetime increases by 12.79 on average holding all variable constant. It also happens for every additional increase of stack emission (β3), the expected lifetime is increasing by 0.0058 on average holding all variable constant. The estimated average value of coating lifetime was 4.889 years when all independent variables = 0 (assuming all βi = 0 is within the range of observed values).

3.3.5. Testing the regression models

By using the regression equation on each area, we can predict the coating lifetime by giving numbers for each variable. Assuming we are conducting a new Asset integrity project in 2020 and the undercoating variable (β1) will be equal to zero due to there is no undercoating condition for new installed structure. Therefore, only four variables will be used for this regression model. From Table 6, it was shown that Converter has the longest coating lifetime duration followed by Furnace, KILN, and Dryer. From this result, the corrosive environment influences the facilities coating condition, very significant in the Dryer and it's causing the coating lifetime deteriorated faster in this area.

Table 6. Regression equation testing.

		Independent Variables					
Area	**Intercept**	**β1**	**β2**	**β3**	**β4**	**β5**	
		0	**0.710**	**130**	**250**	**20**	**(Yrs)**
Dryer	-3.769	$-8.0\text{x}10^{-04}$	14.14	$-2.1\text{x}10^{-02}$	-	-	4.95
KILN	-3.042	$-3.8\text{x}10^{-05}$	11.06	$-7.6\text{x}10^{-03}$	$3.2\text{x}10^{-03}$	-	5.73
Furnace	-5.098	$-3.4\text{x}10^{-03}$	11.55	-	$8.7\text{x}10^{-03}$	$-1.0\text{x}10^{-02}$	6.23
Converter	-4.889	$+1.2\text{x}10^{-04}$	12.79	$5.8\text{x}10^{-03}$	-	$2.5\text{x}10^{-02}$	6.73

Note: β1 = Under film condition (%), β2 = SO2 intensity (t SO2/t Ni), β3 = Stack emission (mg/m^3),
β4 = Rain intensity (mm/year), β5 = NOx emission (mg/m^3)

4. Conclusion

In summary, the regression analysis shows in the state that SO2 intensity, Stack emission, NOx emission, rainfall intensity may have significant effect for the coating lifetime. The prediction results also can give a relevant information to the maintenance and project team to develop action plan and strategy. Regression model can still be applied to evaluate the effect on each area with additional independent variables. Determine coating specification according to the industrial environment is very crucial to define the coating lifetime. Asset integrity project life cycle can be improved after understanding the impact of the corrosive environment to the coating lifetime.

References

[1] Popescu A. Multiple Correlation and Regression in Predicting Milk Price. *Scientific Papers Series: Management Economic Engineering in Agriculture and Rural Development*; 2015;**15** (4):231-8
[2] Schofield MJ. *Corrosion. Plant Engineer's Reference Book.* Elsevier; 2002;33–1–33–25. doi.org/10.1016/b978-075064452-5/50088-2
[3] Han J, Kamber M, Pei J. *Data mining: Concepts and Techniques.*

[4] Ferreira CA, Serrano CLR, Kuyven PS. Use of analysis of variance and linear regression as prediction tool for mechanical performance of SBR. *Plast. Rubber Compos.* Informa UK Limited; 2011 Feb;**40**(1):40–5. doi.org/10.1179/174328911x12940139029329

[5] Armstrong JS. Illusions in regression analysis. *Int. J. Forecast.* Elsevier BV; 2012 Jul;**28**(3):689–94. doi.org/10.1016/j.ijforecast.2012.02.001

[6] Muhajir AR, Sutoyo E, Darmawan I. Forecasting Model Penyakit Demam Berdarah Dengue Di Provinsi DKI Jakarta Menggunakan Algoritma Regresi Linier Untuk Mengetahui Kecenderungan Nilai Variabel Prediktor Terhadap Peningkatan Kasus. *Fountain Inform. J.* Universitas Darussalam Gontor; 2019 Nov 1;**4**(2):33. doi.org/10.21111/fij.v4i2.3199

[7] Fotovvati B, Namdari N, Dehghanghadikolaei A. On Coating Techniques for Surface Protection: A Review. *J. Manuf. Mater. Process.* MDPI AG; 2019 Mar 25;**3**(1):28. doi.org/10.3390/jmmp3010028

[8] Tator KB, editor. *Organic Coatings and Linings. Corrosion: Fundamentals, Testing, and Protection.* ASM International; 2003; doi.org/10.31399/asm.hb.v13a.a0003692

[9] Tator KB, editor. *Protective Organic Coatings.* ASM International; 2015; doi.org/10.31399/asm.hb.v05b.9781627081726

[10] International Organization for Standardization. *ISO 129444-2 Corrosion Protection of Steel Structures by Protective Paint Systems-Part 2: Classification of Environments*; 1998.

IOP Conf. Series: Materials Science and Engineering 1072 (2021) 012041 doi:10.1088/1757-899X/1072/1/012041

Reducing unplanned downtime using Predictive Maintenance (PdM)

I Yasin[1,2], N Kurniati[1,3] and B Syairudin[1,4]

[1]Department of Industrial & System Engineering, Institut Teknologi Sepuluh November, Surabaya, Indonesia

E-mail: [2]yasin.inman@gmail.com, [3]nanikur@ie.its.ac.id, [4]bsy@ie.its.ac.id

Abstract. Maintenance is one of the essential aspects that contribute to keep production process continue to operate and reduce unplanned downtime. In the cement industry, especially kiln area, unplanned downtime can cause not only low availability but also high maintenance costs due to unplanned demand for spare parts and the need for labour to complete unplanned downtime, hence the right method of maintenance is needed. One of the strategic maintenances could be applied is Predictive Maintenance (PdM). Among others maintenance strategy, PdM provides more accurate task due to real time condition monitoring of equipment based on sensors obtained from server data. In this paper, Mahalanobis-Taguchi System (MTS) is applied to provide reliable multisensory analysis and real-time decision making. MTS includes 1) fuses data from multiple sensors into a single system level performance metric; 2) extends MTS by providing a single tool for fault detection/diagnostic, isolation, and prognostic; and 3) offers a systematic way to determine the key parameters following with fault reducing analysis. Multi-sensor data in the kiln used in this research are speed motor main drive, Pfister Coal feed, kiln feed, speed motor IDFAN1, speed motor IDFAN2, speed motor EPFAN. Based on the MTS calculation, the normal data values are obtained from each sensor data. This method is also used to detect, identify, and classify failures based on tests from scheme data. The proposed diagnostic and prognostics scheme from this paper are useful for guiding the company to predict the failure as well as the maintenance schedule then it will increasing the availability from 80% to 90% by reduce the unplanned downtime.

1. Introduction

Maintenance is a combination of technical and administrative actions, including supervisory activity which is aimed to maintain the tools or repair the tools to its good condition so that it will be functioning in the desired rate. In industry world, the effective maintenance activity requires solid maintenance planning. This is essential because the solid planning allows maintenance action takes place on time, in right quantity, with optimal cost, and safety factor will be satisfied as well. Besides, maintenance planning is able to help to coordinate several stakeholders, both inside and outside of industry organization [1].

The importance of maintenance function in industry also becomes special attention on PT X, which is the biggest company focuses in cement industry in Eastern Indonesia Area. PT X has 5 main plants which are located in South Sulawesi. Plant 1 uses wet method in producing cement and it is very uneconomical, therefore this plant has been shut down since 1984. Currently, there are 4 plants left operating under PT X, including the newly established plant which is Plant 5. This plant is established on 19th February 2014 with installed capacity of 2.5 million tons per year (MTPY).

Even though Plant 5 can be said as new plant, however the electrical and instrumentation (ELINS) equipment at kiln area had availability of 80% in 2018 [2], which is far from World Class Availability standard of 90%. Actual availability of ELINS equipment at kiln field as shown in Figure 1.

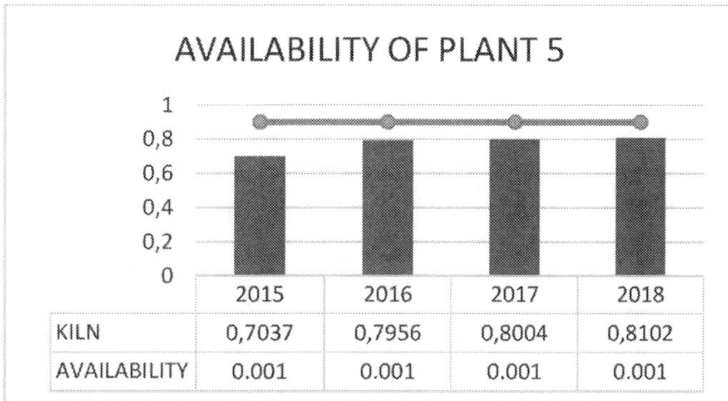

Figure 1. Actual availability of ELINS equipment at kiln field.

One of the causes of unachieved availability is unplanned downtime of ELINS equipment at the kiln. If unplanned downtime happens, therefore, the action taken to compromise the problem is corrective action. Keep performing corrective actions becomes one of the factors causing the low availability of the equipment at said area. This can be carried out as the good opportunity for having improvement in maintenance system of PT X.

To reduce unplanned downtime and corrective action, preventive maintenance is implemented by creating a maintenance schedule through an integrated system [2]. It's shown in Table 1.

Table 1. Schedule maintenance year 2018 PT X.

Order Type	Order	Descryption	2018												TOTAL
			JAN	FEB	MAR	APR	MAY	JUN	JULY	AUG	SEP	OCT	NOV	DEC	INSPECTION
PM03	13608166	**Main drive Motor Inspection**	**1**	**1**	**1**	**1**	**1**	**1**	**1**	**1**	**1**	**1**	**1**	**1**	**12**
		- Torque	√	√	√	√	√	√	√	√	√	√	√	√	
		- Termination	√	√	√	√	√	√	√	√	√	√	√	√	
		- Temperature	√	√	√	√	√	√	√	√	√	√	√	√	
		- Cleanliness	√	√	√	√	√	√	√	√	√	√	√	√	
PM03	13608167	**Pfister Inspection**	**2**	**1**	**2**	**1**	**2**	**1**	**2**	**1**	**2**	**1**	**2**	**1**	**18**
		- Torque	√	√	√	√	√	√	√	√	√	√	√	√	
		- Termination	√	√	√	√	√	√	√	√	√	√	√	√	
		- Temperature	-	-	-	-	-	-	-	-	-	-	-	-	
		- Cleanliness	√	√	√	√	√	√	√	√	√	√	√	√	
PM03	13608168	**Kiln Feed Inspection**	**1**	**1**	**1**	**1**	**1**	**1**	**1**	**1**	**1**	**1**	**1**	**1**	**12**
		- Torque	-	-	-	-	-	-	-	-	-	-	-	-	
		- Termination	√	√	√	√	√	√	√	√	√	√	√	√	
		- Temperature	√	√	√	√	√	√	√	√	√	√	√	√	
		- Cleanliness	√	√	√	√	√	√	√	√	√	√	√	√	
PM03	13608246	**Motor IDFAN 1 Inspection**	**1**	**1**	**1**	**1**	**1**	**1**	**1**	**1**	**1**	**1**	**1**	**1**	**12**
		- Torque	√	√	√	√	√	√	√	√	√	√	√	√	
		- Termination	√	√	√	√	√	√	√	√	√	√	√	√	
		- Temperature	√	√	√	√	√	√	√	√	√	√	√	√	
		- Cleanliness	√	√	√	√	√	√	√	√	√	√	√	√	
PM03	13608247	**Motor IDFAN 2 Inspection**	**1**	**1**	**1**	**1**	**1**	**1**	**1**	**1**	**1**	**1**	**1**	**1**	**12**
		- Torque	√	√	√	√	√	√	√	√	√	√	√	√	
		- Termination	√	√	√	√	√	√	√	√	√	√	√	√	
		- Temperature	√	√	√	√	√	√	√	√	√	√	√	√	
		- Cleanliness	√	√	√	√	√	√	√	√	√	√	√	√	
PM03	13608248	**Motor EPFAN Inspection**	**1**	**1**	**1**	**1**	**1**	**1**	**1**	**1**	**1**	**1**	**1**	**1**	**12**
		- Torque	√	√	√	√	√	√	√	√	√	√	√	√	
		- Termination	√	√	√	√	√	√	√	√	√	√	√	√	
		- Temperature	√	√	√	√	√	√	√	√	√	√	√	√	
		- Cleanliness	√	√	√	√	√	√	√	√	√	√	√	√	

Based on the schedule, data collection is done using a measuring instrument, but because it is not done every week and is not real-time there is still an unplanned downtime due to errors in analyzing the reading of the measuring instrument. Hence other maintenance methods need to be carried out.

Basically, every maintenance action is corrective, retrieval of sensor data for analysis is preventive, predictive is done to determine management policies [16]. Therefore, the existing maintenance system must be improved not only by carrying out corrective maintenance but also by carrying out preventive maintenance which leads to predictive maintenance by carrying out diagnosis and prognosis of failures that occur before.

2. Research Methodology

Focus of this research is to enhance availability of equipment through predictive maintenance system. Predictive Maintenance (PdM) is a maintenance activity based on Condition-Based Monitoring (CBM) and a special method to analyse the failure on equipment [4]. In order to perform PdM, required a set of data collected from various sensors. The more data is being collected, the more accurate the estimation performed towards unplanned downtime. The processed data later becomes prognostic data which influences the policy of which maintenance action will be taken.

Data collection process is including the collection of the most unplanned downtime data as well as sensor data of 2018 in kiln area such as speed motor main drive, Pfister Coal feed, kiln feed, speed motor IDFAN1, speed motor IDFAN2, and speed motor EPFAN. The name of the sensor used in this study is shown in Table 2 below.

Table 2. Sensor information.

Code	Sensor	Unit	Description
X1	Speed Motor Main Drive	RPM	Part of kiln to rotate main drive motor
X2	Pfister Coal Feed	Ton/Hours	Part of kiln for coal feed
X3	Kiln Feed	Ton/Hours	Part of kiln to feed material
X4	Speed Motor IDFAN1	RPM	Part of kiln to rotate fan number 1
X5	Speed Motor IDFAN2	RPM	Part of kiln to rotate fan number 2
X6	Speed Motor EPFAN	RPM	Part of kiln to rotate fan EP

Sensor data of equipment which experiences the most unplanned downtime later will be analysed in term of the failure with application of predictive maintenance. The method that is going to be used to have prediction of failure on the same equipment is Mahalanobis-Taguchi System (MTS) method. In MTS approach, large-sized multivariant data group will be partitioned into two groups which are normal data group and abnormal data group. Normal data group of maintenance diagnosis consists of data from equipment that never experiences failure. In the other hand, abnormal data consists of equipment that experiences failure. The grouping is calculated based on basis scale on all variables to measure the abnormality degree.

Steps to implement MTS consist of as following [5]:

1. Constructing Mahalanobis Space (MS) using mathematical approach. The steps of MTS construction are as follows:
   - Calculate the average of every variable of normal data

$$\bar{x_i} = \frac{\sum_{j=1}^{n} x_{ij}}{n} \tag{1}$$

   - Calculate standard deviation

$$s_i = \sqrt{\frac{\sum_{j=1}^{n}(x_{ij}-\bar{x}_i)^2}{n-1}} \tag{2}$$

   - Transform data by creating matrices

$$Z_{ij} = \frac{(x_{ij}-\bar{x}_i)}{s_i} \tag{3}$$

- Verify that average of normal data is 0

$$\bar{z}_i = \frac{\sum_{j=1}^{n} z_{ij}}{n} = 0 \tag{4}$$

- Verify that the standard deviation of normal data is 1

$$s_z = \sqrt{\frac{\sum_{j=1}^{n} (z_{ij} - \bar{z}_i)^2}{n-1}} = 1 \tag{5}$$

- Construct matrices of normal data correlation

$$c_{ij} = \frac{\sum_{m=1}^{n} (z_{im} z_{jm})}{n-1} \tag{6}$$

- Calculate inverse of correlation matrices
- Calculate Mahalanobis Distance

$$MD_j = \frac{1}{k} Z_{ij}^T C^{-1} \tag{7}$$

2. Validating Mahalanobis Space
MS scale can be stated as good if MD from data of abnormal condition is greater than MD from data of normal condition. If MD scale fails to obtain value greater than 1, therefore that thing is able to identify that selected variable is unable to differentiate between normal and abnormal condition. In this case, adding new variable is necessary to perform [6].
3. Decision Making
For predictive case, failure detection happens once MD is no longer in normal cluster. Predictive action involves [5] operational failure detection, and failure cause identification.

3. Result and Discussions

3.1. MD Calculation from Normal Data at Kiln Area

The Mahalanobis Distance (MD) is calculated from normal data, as shown in Table 3.

Table 3. MD calculation from normal data.

OBJECT	DATA SENSOR						TRANSFORMATION DATA						MD
	X1	X2	X3	X4	X5	X6	Z1	Z2	Z3	Z4	Z5	Z6	NORMAL
1	3.4	20.4	526	594	593	818	0.488	0.654	0.073	-0.08	-1.408	-0.44	0.44520218
2	3.4	21.2	549	595	616	834	0.488	1.189	1.65	0.029	1.152	0.652	1.08761471
3	3.4	18.5	520	589	606	848	0.488	-0.615	-0.339	-0.627	0.039	1.607	0.61444044
4	3.5	18.4	546	591	605	817	0.983	-0.682	1.444	-0.408	-0.072	-0.508	0.65349057
5	3	19.8	508	586	614	835	-1.492	0.254	-1.161	-0.956	0.93	0.72	0.93789764
.....	......	.....	......	.....	......	.....	......	.....	......	.....	......	.....	......
900	3.6	22	504	607	617	815	1.477	1.723	-1.436	1.343	1.264	-0.645	1.8333793
XBAR	3.301	19.42	524.9	594.7	605.6	824.5	5E-14	8E-15	1E-15	-6E-15	-6E-15	-3E-15	
SD	0.202	1.497	14.58	9.137	8.985	14.66	1	1	1	1	1	1	

3.2. MD Calculation from Abnormal Data at Kiln Area

By using mean and standard deviation from normal conditions, abnormal data from each sensor will be obtained. Shown in the Table 4.

Table 4. MD calculation from abnormal data.

DATA	MD NORMAL	MD SPEED MAINDRIVE	MD PFISTER	MD KILN FEED	MD IDFAN1	MD IDFAN2	MD EPFAN
1	0.44520218	6.680519973	21.66747421	14.88588627	10.33778	8.852081	18.6321
2	1.08761471	4.26793149	21.31167981	13.30501533	10.3634	8.928366	18.2037
3	0.61444044	4.506849714	21.64933102	13.8152252	10.56155	8.99712	18.1511
4	0.65349057	6.417338275	21.2921518	13.65659812	10.52512	8.873081	18.1013
5	0.93789764	4.506805092	21.33220743	14.61581641	10.52198	8.9701	17.9896
⋮	......	.....	......	.....	......	.....	......
900	1.8333793	5.269657254	20.67894702	14.15649062	10.69792	8.856295	18.1116
STD DEV	0.35823203	0.914051957	0.283052417	0.606907126	0.135224	0.143826	0.327
MEAN	0.99888889	5.375495885	21.16893554	14.16677373	10.49488	8.848685	18.4482

By comparing normal data with abnormal data, we obtain the threshold shown in the following Table 5.

Table 5. Threshold.

Condition	MEAN	STD DEVIATION	THRESHOLD MINIMUM	THRESHOLD MAXIMUM
Normal	0.998889	0.35823203	-0.075807	2.073585
Abnormal_Speed Main Drive	5.375496	0.91405196	2.63334	8.1176518
Abnormal_IDFAN1	8.848685	0.14382607	8.4172069	9.2801633
Abnormal_IDFAN2	10.49488	0.13522379	10.089213	10.900556
Abnormal_KILN Feed	14.16677	0.60690713	12.346052	15.987495
Abnormal_EPFAN	18.4482	0.32699893	17.467202	19.429196
Abnormal_Pfister	21.16894	0.28305242	20.319778	22.018093

3.3. Failure Diagnostic Prognostic

This research also shows that MD calculation is used as a prognostic diagnostic scheme by entering 2019 sensor data. The 2019 sensor data that is calculated can be said to have a prognostic scheme when MD 2019 is above the normal 2018 threshold group and not included in the threshold group of one of the failures. Prognostic grouping images are shown in Figure 2.

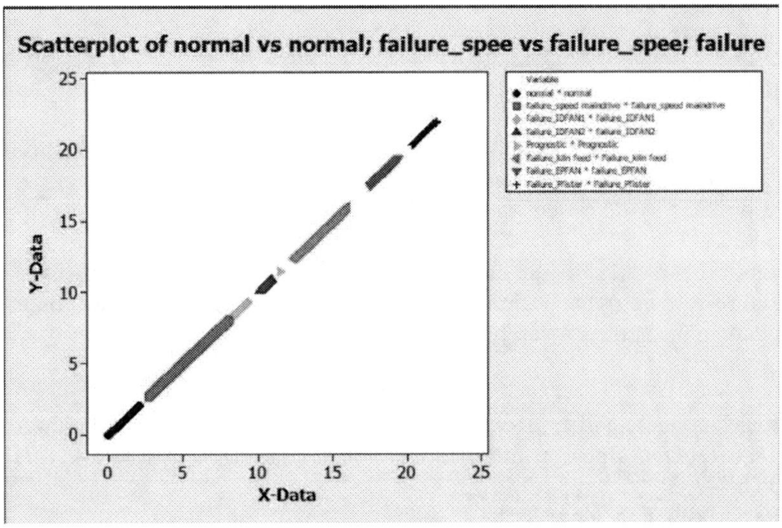

Figure 2. Prognostic grouping.

The prognostic that is carried out will give an idea of when an equipment failure can cause unplanned downtime. This situation becomes a reference from the management to take a new maintenance schedule policy. Maintenance's schedule policy can be in the form of an increase in the frequency of inspections which were once a time in 1 period to 3 times in 1 period as shown in the Table 6 [3].

The inspection improvement activities carried out can reduce the level of unplanned downtime. This reduction directly impacts the increase in equipment availability. Based on PT X's Annual report in 2019, it is known that the availability of kiln equipment rose to 89% from the previous year which was 80% [3].

Table 6. Schedule maintenance year 2019 PT X.

Order Type	Order	Descryption	2019												TOTAL
			JAN	FEB	MAR	APR	MAY	JUN	JULY	AUG	SEP	OCT	NOV	DEC	INSPECTION
PM03	13678856	**Main drive Motor Inspection**	**4**	**4**	**4**	**4**	**4**	**4**	**4**	**4**	**4**	**4**	**4**	**4**	**48**
		- Torque	√	√	√	√	√	√	√	√	√	√	√	√	
		- Termination	√	√	√	√	√	√	√	√	√	√	√	√	
		- Temperature	√	√	√	√	√	√	√	√	√	√	√	√	
		- Cleanliness	√	√	√	√	√	√	√	√	√	√	√	√	
PM03	13681187	**Pfister Inspection**	**2**	**2**	**2**	**2**	**2**	**2**	**2**	**2**	**2**	**2**	**2**	**2**	**24**
		- Torque	√	√	√	√	√	√	√	√	√	√	√	√	
		- Termination	√	√	√	√	√	√	√	√	√	√	√	√	
		- Temperature	√	-	√	-	√	-	√	-	√	-	√	-	
		- Cleanliness	√	√	√	√	√	√	√	√	√	√	√	√	
PM03	13681841	**Kiln Feed Inspection**	**2**	**2**	**2**	**2**	**2**	**2**	**2**	**2**	**2**	**2**	**2**	**2**	**24**
		- Torque	-	-	√	-	-	√	-	-	√	-	-	√	
		- Termination	√	√	√	√	√	√	√	√	√	√	√	√	
		- Temperature	√	√	√	√	√	√	√	√	√	√	√	√	
		- Cleanliness	√	√	√	√	√	√	√	√	√	√	√	√	
PM03	13681859	**Motor IDFAN 1 Inspection**	**4**	**4**	**4**	**4**	**4**	**4**	**4**	**4**	**4**	**4**	**4**	**4**	**48**
		- Torque	√	√	√	√	√	√	√	√	√	√	√	√	
		- Termination	√	√	√	√	√	√	√	√	√	√	√	√	
		- Temperature	√	√	√	√	√	√	√	√	√	√	√	√	
		- Cleanliness	√	√	√	√	√	√	√	√	√	√	√	√	
PM03	13681983	**Motor IDFAN 2 Inspection**	**4**	**4**	**4**	**4**	**4**	**4**	**4**	**4**	**4**	**4**	**4**	**4**	**48**
		- Torque	√	√	√	√	√	√	√	√	√	√	√	√	
		- Termination	√	√	√	√	√	√	√	√	√	√	√	√	
		- Temperature	√	√	√	√	√	√	√	√	√	√	√	√	
		- Cleanliness	√	√	√	√	√	√	√	√	√	√	√	√	
PM03	13682774	**Motor EPFAN Inspection**	**4**	**4**	**4**	**4**	**4**	**4**	**4**	**4**	**4**	**4**	**4**	**4**	**48**
		- Torque	√	√	√	√	√	√	√	√	√	√	√	√	
		- Termination	√	√	√	√	√	√	√	√	√	√	√	√	
		- Temperature	√	√	√	√	√	√	√	√	√	√	√	√	
		- Cleanliness	√	√	√	√	√	√	√	√	√	√	√	√	

4. Conclusion

The conclusions of the research can be summarized as follows:

1. The Mahalanobis-Taguchi System method can be used to determine the type of failure based on a threshold value.
2. Predictive Maintenance has been successfully carried out with prognostic based on MTS calculations by comparing the new sensor data to the threshold.
3. Maintenance policies made based on prognostic calculations can increase equipment availability from 80% to 89%.
4. The limitation of this research is the data used is discrete so that it is not real-time.
5. Therefore, future research can be made a platform that contains real-time data so that decision making on maintenance policies can be more precise.

References

[1] Al-Turki UM, Ayar T, Yilbas BS, Sahin AZ. Integrated Maintenance Planning. *SpringerBriefs Appl. Sci. Technol.* Springer International Publishing; 2014;25–57. doi.org/10.1007/978-3-319-06290-7_3

[2] Annual Report of PT. X Year 2018

[3] Annual Report of PT. X Year 2019

[4] Gouriveau R, Javed K, HisselD.Data Driven Prognostic of Proton Exchange Membrane Fuel Cell Stack with Constrain Based Summation-wafelet Extreme Learning Machine. *6th International Conference on Fundamentals and Development of Fuel cell.* 2015 Feb; 1-8.

[5] Soylemezoglu A, Jagannathan S, Saygin C. Mahalanobis-Taguchi System as a Multi-Sensor Based Decision Making Prognostics Tool for Centrifugal Pump Failures. *IEEE Trans Reliab.* Institute of Electrical and Electronics Engineers (IEEE); 2011 Dec;**60**(4):864–78. doi.org/10.1109/tr.2011.2170255

[6] Anisa. Perbandingan Metode Analisis Diskriminan dan Mahalanobis Taguchi (MT) untuk Data Penderita DM RS. Wahidin Sudirohusodo Makassar.'*Jurnal Matematika, Statistika, dan Komputasi*. 2013 Dec;**10**(1)35-47.

[7] Sahal R, Breslin JG, Ali MI. Big data and stream processing platforms for Industry 4.0 requirements mapping for a predictive maintenance use case. *J Manuf Syst.* Elsevier BV; 2020 Jan;**54**:138–51. doi.org/10.1016/j.jmsy.2019.11.004

[8] Zhao X, Jiang N, Liu J, Yu D, Chang J. Short-term average wind speed and turbulent standard deviation forecasts based on one-dimensional convolutional neural network and the integrate method for probabilistic framework. *Energy Convers. Manag.* Elsevier BV; 2020 Jan;**203**:112239. doi.org/10.1016/j.enconman.2019.112239

[9] Alrabghi A, Tiwari A. State of the art in simulation-based optimisation for maintenance systems. *Comput. Ind. Eng.* Elsevier BV; 2015 Apr;**82**:167–82. doi.org/10.1016/j.cie.2014.12.022

[10] Huynh KT, Grall A, Bérenguer C. Assessment of diagnostic and prognostic condition indices for efficient and robust maintenance decision-making of systems subject to stress corrosion cracking. *Reliab Eng Syst Saf.* Elsevier BV; 2017 Mar;**159**:237–54. doi.org/10.1016/j.ress.2016.11.022

[11] Jardine AKS, Lin D, Banjevic D. A review on machinery diagnostics and prognostics implementing condition-based maintenance. *Mech. Syst. Signal. Process.* Elsevier BV; 2006 Oct;**20**(7):1483–510. doi.org/10.1016/j.ymssp.2005.09.012

[12] Susanto BSP, Kurniati N. Multi Sensor-Based Failure Diagnosis using the Mahalanobis Taguchi System. IOP Conference Series: *Mater. Sci. Eng.* IOP Publishing; 2020 May 28;**847**:012036. doi.org/10.1088/1757-899x/847/1/012036

[13] Jardine AKS, Lin D, Banjevic D. A review on machinery diagnostics and prognostics implementing condition-based maintenance. *Mech. Syst. Signal. Process.* Elsevier BV; 2006 Oct;**20**(7):1483–510. doi.org/10.1016/j.ymssp.2005.09.012

[14] Tiddens WW, Braaksma AJJ, Tinga T. The Adoption of Prognostic Technologies in Maintenance Decision Making: A Multiple Case Study. *Procedia CIRP*. Elsevier BV; 2015;**38**:171–6. doi.org/10.1016/j.procir.2015.08.028

[15] Heng A, Zhang S, Tan ACC, Mathew J. Rotating machinery prognostics: State of the art, challenges and opportunities. *Mech Syst Signal Process*. Elsevier BV; 2009 Apr;**23**(3):724–39. doi.org/10.1016/j.ymssp.2008.06.009

[16] Swanson L. Linking maintenance strategies to performance. *Int J Prod Econ*. Elsevier BV; 2001 Apr;**70**(3):237–44. doi.org/10.1016/s0925-5273(00)00067-0

An integrated analysis of port and coal storage development plan in a mining company with a simulation approach

E Kusuma [1], N I Arvitrida[1*]

[1]Department of Industrial and Systems Engineering, Institut Teknologi Sepuluh Nopember, Surabaya, Indonesia

E-mail: *niniet@ie.its.ac.id

Abstract. Managing inbound logistics activities requires an integrated and comprehensive perspective to understand the operations complexity that affects inbound performance. This study proposes an integrated analysis in port and coal storage development plan in a mining company. The main goal of the analysis is creating a cost-effective plan. A discrete event simulation (DES) approach is used to model the complexity, and the model is applied to analyse the number of barge slip, port loading and unloading equipment, the number and the type of material transport equipment. It also analyses how storage capacity can achieve particular service levels. The model considers the actual restriction and uncertainties, such as tidal time, barge sailing and berthing time, loading and unloading time, material transportation time, availability of barge slip, material transportation, loading and unloading equipment. The simulation using 2020 data as the baseline of demand and 2027 data as the highest demand until 2030, this demand led to high queues at barge slip, making stockpile capacity insufficient, and high utility on trucks followed by long queues. Based on the simulation results, increasing the number of resources can be able to solve the overcome problems.

1. Introduction

Management of inbound logistics activities requires an integrated and comprehensive perspective between one process and another. This is caused by the complexity in understanding the effect of facility changes or process changes on the performance of inbound logistics. One issue that requires a holistic perspective in the scope of inbound logistics is port and warehouse development planning, where port performance and warehouse performance are interrelated. This case-based report (CBR) will discuss the importance of integrated analysis for port development plans and storage capacity (warehouses) by raising a case at PT. X which supporting the project requires the development of the company's port facilities and infrastructures to support operational needs. Therefore, integrated planning is needed. Planning to develop port facilities and infrastructures that must be carried out include coal storage in the port; barge slip area used to transport coal, sulfur, silica sand from suppliers, nickel products produced by the company, and material handling equipment used for loading and unloading at the ports.

The business process flow at the port starting from shipping, loading & unloading, and storage at the port must be integrated with one another so that it can support the sustainability of the company's production processes at the most efficient, reliable, and safe costs. If there is a disruption in the business process at the port, it will be able to stop the nickel processing production process. In actual conditions, there are internal and external factors that greatly affect the sustainability of port operations. Internal factors include productivity and availability of port facilities and infrastructure, while external factors include tides, weather conditions, width, and depth of river channels. These factors add complexity to the operation of the port.

The objectives of this research are to develop a model that integrates all shipping in and out of barges, loading & unloading, transportation to storage, and bulk material storage capacity, especially coal at the company's port.

The problem constraints used in this research are the width of the river channel is 50 meters with a depth of 3 meters LAT (low astronomical tide) so that the size of the barges that can be used is 240 feet and 270 feet with a capacity of 4,100 tons and 5,000 tons respectively. Barges can only cross the river channel when the river reaches a minimum of 4.9 meters, which is the average in the range of 09:00 a.m. - 12:00 p.m. and 18:00 p.m. - 22:00 p.m., the river channel can only be crossed by 1 (one) barge at one time. Material handling in the form of e-cranes has a bucket capacity of 10 tons and trucks carrying it from barge slip to bulk storage have a maximum capacity of 20 tons and cannot use a larger capacity due to safety reasons during operation. Increasing coal storage capacity can only be done in the port area and cannot be planted site areas due to limited space available.

The benefit of this research is supporting the company in optimally developing the port in terms of cost efficiency spent on developing and operating ports. Furthermore, this research could provide an additional reference regarding the need for integrated analysis for investment decisions in companies that have similar characteristics of inbound logistics activities.

The remaining of this paper is organized as follows. Section 2 discusses the literature used in this research. Sections 3 show the model development which includes the conceptual model and computer model. Section 4 presents the model results and analysis. Finally, the conclusion of this research is presented in section 5.

2. Literature Review

2.1. Logistics system

Logistics is a collection of goods, materials, and tools and facilities used by an organization in achieving its objectives. Logistics management is management related to the planning, implementation, and supervision of the distribution of goods, services, and information from the place of origin and ends at the point of consumption to meet the needs of consumers effectively and efficiently. The purpose of managing logistics is to move the right finished goods or materials, in the right amount at the right time and at the right location where it is needed at the lowest total cost. The logistical role has now expanded not only to move finished products and raw materials but also to create a competitive advantage by providing the best service to meet consumer demand. Logistics management functions according to Prihantono, C.R. (2012) is a series of processes consisting of planning and determining functions, budgeting functions, procurement functions, storage and distribution functions, maintenance functions, elimination functions, and control functions.

Christopher (1998) logistics is the process of managing the procurement, movement, and storage of materials, parts, and finished inventory and related information flow through the organization and its marketing channels. Stock and Lambert (2011) stated that logistics no longer solely implemented handling or transportation but do such activities, such as communication, customer service, site localization, reverse logistics, and planning, that were related to trade and production. A logistics system includes supply, transport, servicing, production, warehousing, distribution, and the relationships between them.

2.2. Simulation

Simulation is modeling the process of a system using computer software that is used to evaluate and improve the performance of the system. The reason for using simulations is that the system to be evaluated and upgraded is complex and the trial and error approach requires a large cost, a lot of time, and interferes with the operation of the system. The conceptual model of the simulation contains 5 key activities. These activities are understanding the problem of the situation/system, setting goals, defining inputs, defining outputs, and identifying assumptions and simulations. A system has 4 components, that is entities, activities, queues, and resources (Robinson et al., 2011).

IOP Conf. Series: Materials Science and Engineering 1072 (2021) 012042 doi:10.1088/1757-899X/1072/1/012042

Based on the purpose of the simulation, the simulation can be divided into continuous and discrete events. A system can be represented as a continuous and discrete event. One of the determinants is the discrete-event looks at a certain time unit while continuously considering the changing time or system dynamics (Harrel et al., 2004). A dynamic system can be described as an agent-based simulation (ABM). The agent-based simulation considers all individuals (agents) in a complex and dynamic system. ABM is heterogeneous and autonomous so that they can interact between agents and interact with one another. ABM adds individual components in the system besides the system variables. While in discrete event simulation, one of the software to calculate is Arena. Arena simulation software is a flexible tool in analysis to create animated simulation models that accurately represent virtually all systems.

3. Model development

This section explains the research methodology & model development process. It includes a conceptual model and a computer model to represent the simulation model. Several preliminary results are also presented here. The research methodology in this paper is conducted with discrete event simulation and using Arena Software, the software is used to model a system that takes into account a certain time unit. The variables used in this study are divided into 3 parts, namely the decision variable, the control variable, and the response variable. After determining the variables used, the next step is to collect data. The data collected consists of 2 types, namely primary data and secondary data. The decision variables in the simulation model use the number of trucks 2 units according to the existing number of trucks, the simulation also uses 5 units of trucks to realize the average waiting times of trucks from barge slips to coal, sulfur, and silica stockpiles. The coal stockpiles capacity used in this simulation is the current stockpile capacity of 61,000 tons and the stockpiles capacity of 160,000 tons in 2027 following the demand for coal in that year. The decision variables and data collected in this study is shown in Table 1.

Table 1. The decision variables.

No	Decision variables	Quantity	Units
1	Barge capacity	4,100 and 5,000	ton
2	Trucks from barge slip to stock pile	2 and 5	unit
3	Stock pile capacity	61,000 and 160,000	ton

The control variables used in the simulation model are coal, sulfur, and silica consumption and nickel production in 2020 and 2027. The number of barges that can pass through the river channel is only one (1) unit of a barge at the same time and can only be passed by barges at a minimum depth of 4.9 meters, where the depth of the river is only in a few hours in 1 (one) day according to tides. Shipments of nickel products are carried out every 2 weeks to the buyer and this shipment is the main priority over coal, sulfur, and silica. The flow process of loading and unloading in barge slip until the stockpile uses the existing sequence of processes. The maximum loading and unloading time for coal, coal, and silica is 3 days, meanwhile for nickel product is 4 days, the loading and unloading time is determined by contractual agreement with the suppliers of coal, sulfur, silica, and nickel buyers. The maximum coal inventory used is 4-6 weeks according to the company's policy. The control variables and data collected in this study is shown in Table 2.

Table 2. The control variables.

No	Control Variables	Units
1	Consumption rate of coal, sulphur & silica	ton
2	Nickel production rate	ton
3	The number of barges could be passing thru the river at the same time	unit
4	The river depth of at least 4.9 m	meter
5	Nickel is the first or main priority commodity	types of commodity

6	Nickel shipment every 2 weeks	time
7	Flow process of loading and unloading in barge slip until stockpile	sequence of processes
8	Maximum loading and unloading time of coal, sulfur, and silica	3 days
9	Maximum loading and unloading time of nickel	4 days
10	Maximum inventory level for coal	4-6 weeks

The response variables in this simulation model are the loading and unloading queues of barges at anchorage points both for coal, silica and sulfur barges, by increasing the capacity of the barge, it is expected to reduce barge waiting time and the number of barges waiting in barge slip. Barge slip, e-crane, and truck utilization are also used as variable responses in this simulation model. The response variables and data collected in this study is shown in Table 3.

Table 3. The response variables.

No	Response variables	Units
1	Queues in loading and unloading barge system	Days
2	Barge slip utility	%
3	E-Crane utility	%
4	Truck utility	%

The coal, sulfur and silica barges queue at the anchorage point before berthing at the barge slip and unloading processes is carried out, the anchorage point is approximately 9 nautical miles from the company's barge slip. After ensuring that barge slip, unloading equipment, hauling equipment and stock piles are available, barges are allowed to enter the barge slip through the river channel, this river channel can only be passed by one (1) barge at the same time and at a minimum river depth of 4.9 meters, after the barges are berthed at the company's barge slip, then the unloading process will be carried out using e-crane and transported to the stock pile using trucks.

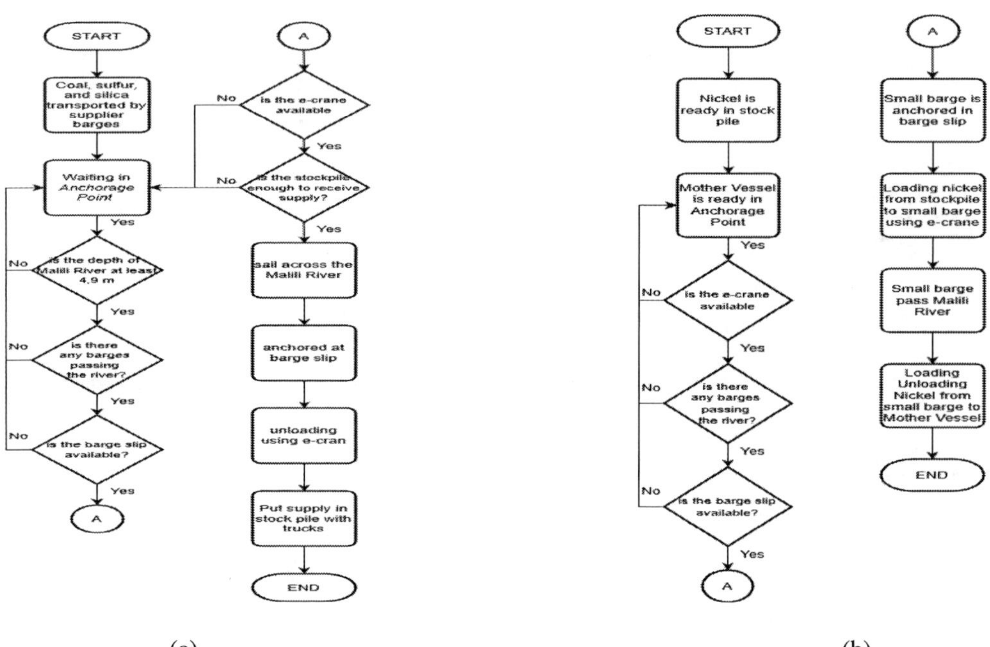

Figure 1. The conceptual model for the process of inbound coal, sulfur, and silica & outbound nickel.

The nickel shipment process is carried out when a mother vessel is available at the anchorage point, the nickel will be loaded onto the barge using an e-crane from a storage warehouse next the barge slip, after the nickel has been loaded into the barge then the barge sails to the anchorage point and the transshipment process is carried out to the mother vessel, the nickel shipments are carried out every 2 weeks. The conceptual model for the inbound and outbound logistics process is used for existing and improvement conditions in 2027. A conceptual model used for the process of inbound coal, sulfur, and silica and the process of outbound nickel is shown in Figure 1.

The existing conditions using 4,100 tons of barges capacity for coal, stockpile capacity is 61,000 tons, and the number of trucks 2 units, meanwhile the condition of improvement using 5,000 tons of barges capacity for coal, stockpile capacity is 160,000 tons and the number of trucks is 5 units. The number and capacity of barge slip and e-crane are remaining the same between existing and improve conditions in 2027. The capacity of barges and stockpile for sulfur, silica, and nickel are using the same capacity between the existing and improvements condition in 2027. Flow process including the existing and improvement conditions is shown in Figure 2.

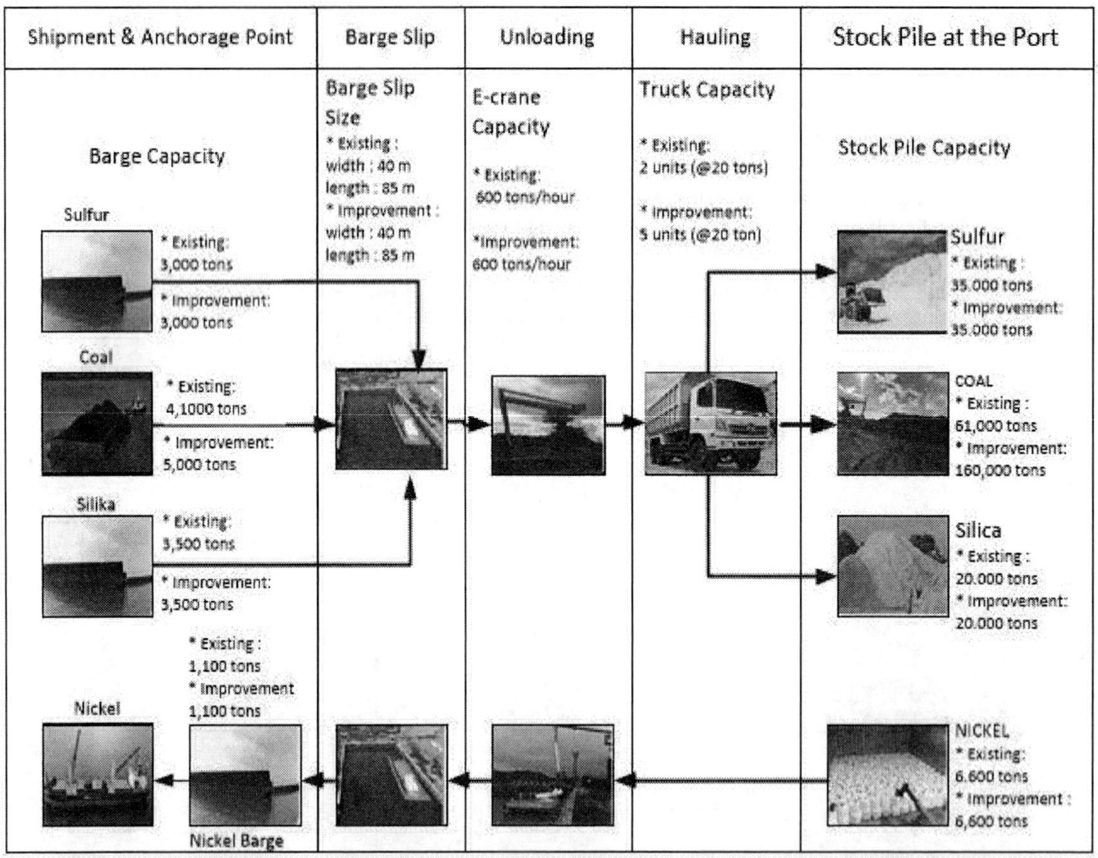

Figure 2. The flow process including the existing and improvement conditions for coal, sulfur, silica and nickel barges, equipment and stock pile.

The average number of waiting in barge slip is the average waiting time of each barge waits in the barge slip for one year. The average waiting time in barge slip is the total average waiting time of all barges waiting at barge slip for one year. The average waiting time in trucks processing is the total average waiting time of trucks during the loading process of barges using e-cranes in one year. The number of barges waiting in barge slips is the total number of barges waiting in a queue at barge slip due to unpreparedness or unavailability of unloading equipment, coal storage, or barge slip itself within one year. The number of supplies that cannot be stored in the stockpile is the amount of coal that cannot be stored in the coal storage area. This result is a one-year simulation process. The simulation process is carried out every day with a replication length of 365 days. The following Table 4 and Table 5 are the simulation results obtained with existing and improved conditions of coal storage capacity, barge capacity and number of trucks in 2027.

Table 4. The simulation results obtained with existing conditions of coal storage capacity, barge capacity and number of trucks in 2027.

No	Year	2027
1	Average number of waiting in barge slip	19.879 hours
2	Average waiting time in barge slip	84.93 hours
3	Average waiting time in trucks processing	12.08 hours
4	Number of barges waiting in barge slip	44 units
5	Number of supplies that cannot be stored in stock pile	99,000 tons (coal)

Table 5. The simulation results obtained with improvement conditions of coal storage capacity, barge capacity and number of trucks in 2027.

No	Year	2027
1	Average number of waiting in barge Slip	0.47 hours
2	Average waiting time in Barge Slip	56.88 hours
3	Average waiting time in trucks processing	3.85 hours
4	Number of barges waiting in barge slip	16 units
5	Number of supplies that cannot be stored in stock pile	0 tons (coal)

4. Analysis

The reason for using the year 2027 is that in that year the highest demand occurred between 2020 and 2030. This demand led to high queues at barge slip, making stockpile capacity insufficient, and high utility on trucks followed by long queues. The long queue will put this company at risk of getting a penalty because it is not following the contract and will not be able to meet the supply of coal to the processing plant to produce nickel because of the unavailability of coal and inadequate coal storage.

Based on the simulation results, the amount of coal that cannot be stored in coal storage in 2027 is 99,000 tons, therefore, the company must increase its coal storage capacity to 160,000 tons from the existing capacity of 61,000 tons. Increasing coal storage capacity from 61,000 tons to 160,000 tons, coal barging capacity from 4,100 tons to 5,000 tons and also increasing the number of trucks in the process of unloading from 2 units to 5 units will reduce the average number of waiting in barge slips from 19,879 hours in 2020 to 0.47 hours in 2027, it will also reduce the average waiting time in Barge Slips from 84.93 hours to 56.88 hours. The improvement also reduces the average waiting time in trucks processing from 12.08 hours in 2020 to 3.85 hours in 2027 and will reduce the number of barges waiting in barge slip from 44 units in 2020 to 16 units in 2027.

Testing results from this arena cannot be directly applied in existing conditions. Implementation in the existing conditions will directly disrupt the ongoing operation process. This application also requires other analyses such as sensitivity analysis and financial analysis and more importantly, it requires considerable effort, time, and cost. However, the results of this paper will greatly assist the company in determining the amount of additional stockpile capacity, barge capacity, and the optimal number of trucks. The optimal amount referred to in this paper is the amount that can produce the most significant queue reduction times, can meet the highest demand in the next 10 years, and can eliminate the risk of companies getting penalized for delays.

5. Conclusions & future research

The existing condition of the company cannot meet future demand and is at risk of being penalized because it cannot fulfill the agreed of maximum loading and unloading time. Improvement conditions by increasing the capacity of barges from 4,100 tons to 5,000 tons, adding stockpile capacity from 61,000 tons to 160,000 tons, and increasing the number of trucks from 2 units to 5 units can overcome this problem.

The contribution of this paper for the company is to help the company develop port optimally following the production plan, generate objective and optimal decisions by considering the decrease in queue time. The contribution of this paper to science is to be a reference material for other companies that have a similar business process flow to determine the optimal addition of resources.

The recommendation for further research is to add cost and financial analysis calculations to get the most optimal and economic addition, and add other scenarios such as increasing the number of e-cranes, the number of barge slip. Sensitivity analysis is also a useful tool to react inconsistency from price. Financial analysis, sensitivity analysis, and adding more scenarios from increasing resources will make this research better.

References

[1] Daellenbach HG, McNickle DC. *J Classif.* Macmillan Education UK; 2005;10–20. doi.org/10.1007/978-0-230-80203-2_2
[2] Daellenbach HG, McNickle DC. Systems thinking. *Manag. Sci.* Macmillan Education UK; 2005;10–20. doi.org/10.1007/978-0-230-80203-2_2
[3] Law, A. M., & Kelton, W. D. 2000. Simulation Modeling and Analysis. New York: Mcgraw Hill.
[4] Husnan dan Suad. 1982. *Teori Antrian*. Yogyakarta: BPFE.
[5] Rosetti. 2016. *Simulation modeling & Arena 2nd edition*, Wiley.
[6] Taha, H.A. 1997. *Riset Operasi : Suatu Pengantar.* Terjemahan Daniel Wirajaya, Jakarta: Binarupa Angkasa.
[7] Hasan dan M.Iqba. 2002. *Teori Pengambilan Keputusan*. Jakarta: Ghalia Indonesia.
[8] Dundovic, Cedomir and Hess, Svjetlana. 2005. *Exploitability of the Port Container Terminal Stacking Area Capacity in the Circumstances of Increased Turnover.* ISEP 2005.
[9] Gurning, Raja Oloan Saut, Budiyanto, dan Eko Hariyadi. 2007. *Manajemen Bisnis Pelabuhan.* Jakarta: PT Andhika Prasetya Ekawahana.
[10] Jinca dan Yamin N. 2011. *Transportasi Laut Indonesia, Analisis Sistem dan Studi Kasus.* Surabaya: Brilian Internasional.
[11] Musso E, Ferrari C, Benacchio M. Port Investment: Profitability, Economic Impact and Financing. *Res. Transp. Econ.* Elsevier BV; 2006 Jan;**16**:171–218. doi.org/10.1016/s0739-8859(06)16008-4
[12] Triatmodjo, B. 2009. *Perencanaan Pelabuhan.*Yogyakarta : Beta Offset.

IOP Conf. Series: Materials Science and Engineering 1072 (2021) 012043 doi:10.1088/1757-899X/1072/1/012043

Determining 150kv transmission tower route using FAHP method

A Muis[1,2], B Santosa[1]

[1,2]Department of Industrial and Systems Engineering, Institut Teknologi Sepuluh Nopember, Surabaya, Kampus ITS Sukolilo, Surabaya, Jawa Timur 60111 Indonesia.

E-mail: [1]muisabeng@gmail.com , [2]budi_s@ie.its.ac.id

Abstract. PT. Vale Indonesia Tbk (PTVI) is a mineral mining company that produces nickel as its main product. PTVI requires high reliability of power supply so that it can produce nickel to the specified target. One of the important equipment to maintain the availability of electricity is the transmission tower. Failure of the transmission tower will have a disastrous effect on nickel production operations. PTVI will lose 165 MW in case of failure on the Larona transmission tower line to plant site and 200 MW on the Balambano transmission tower line to the plant site. Therefore, PTVI will build a new 150kV transmission tower line from Larona switchyard to Balambano switchyard to support the availability of sustainable electricity. The method to be used to determine the best alternative transmission tower route by considering multiple criteria is Fuzzy Analytical Hierarchy Process (FAHP). The results of each alternative are 0.477 for the North route, 0.364 for the Middle route, and 0.159 for the South route. The authors recommend the management of PT. Vale Indonesia, as the owner and operator of the hydroelectric power station, chose the north route for the planned construction of the 150kV transmission tower.

1. Introduction

PT. Vale Indonesia Indonesia, Tbk (PTVI), formerly known as PT INCO, is a mineral mining company that produces nickel as its main product. PT. Vale Indonesia operates in East Luwu Regency, South Sulawesi. In the area around PT. Vale Indonesia, three interconnected lakes that flow downstream of the Larona river which empties into the Ussu bay. With a total area of around 2,800 km2 which is a very large area of rainwater catchment. The Larona river has water in lake Towuti and empties into the gulf of Bone. Lake Towuti (El. +320 m) is a lake system in its upstream, namely lake Mahalona (El. +325 m) and lake Matano (El. +390 m). The three lakes are natural lakes that support the availability of water that functions as a Larona Hydroelectric Power Plant (PLTA) as a supplier for electricity needs at PT. Vale Indonesia's nickel smelting plant.

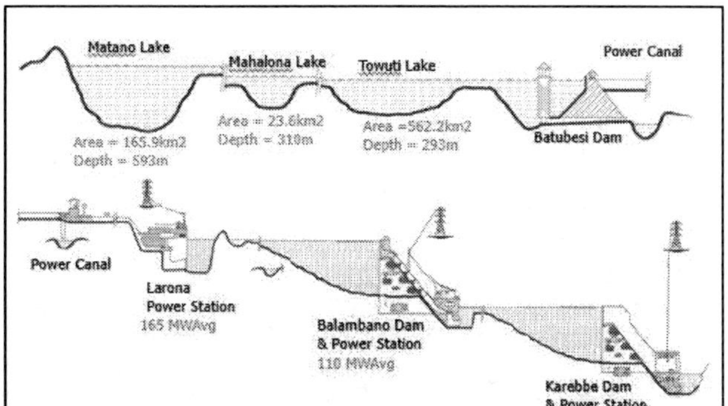

Figure 1. Cascade PLTA PT Vale Indonesia.

The utilization of reservoirs in the Larona river that has stratified characteristics is depicted as shown in Figure 1. Starting from the Matano lake as upstream to the Karebbe dam as downstream. With an average capacity for the Larona hydropower plant at 165 MW, the 110 MW Balambano hydropower plant, and finally the Karebbe hydropower plant at 90 MW. So that the total average supplied reached 365 MW.

PT. Vale Indonesia (PTVI) requires high power supply reliability from hydroelectric power plants (PLTA) to thermal power plants to be distributed to all production facilities so that it can produce nickel as expected. For this reason, the Utilities Department, as the unit responsible for the availability of electricity, must try to ensure that the electricity supply for nickel production must be met continuously. One of the important equipment to maintain electricity availability can be met is the transmission line tower. Currently, PTVI has 153 towers consisting of 57 Larona transmission line towers which are commonly referred to as Line 1 and 2, 76 Balambano transmission line towers and 20 Karebbe transmission line towers. Balambano and Karebbe transmission lines are commonly referred to as Lines 3 and 4. Failure of the transmission line tower will have a disastrous effect on nickel production operations. PT. Vale Indonesia will lose 165 MW if one of the tower lines of the Larona transmission line fails and the other side will lose 200 MW if one of the Balambano tower transmission lines fails. To support the continued availability of electricity, the construction of a 150kV transmission line 5 (Line 5) tower from the Balambano substation to the Larona substation.

To determine the construction of the transmission tower line from Larona switchyard to the Balambano switchyard has many considerations that must be considered because each alternative route has advantages and disadvantages of various aspects. Thus, one of the approaches that correspond to this problem is by the method of Analytical Hierarchy Process (AHP). AHP is a multi-criterion decision making tool with experts' opinions as a defining criterion. However, the preference assessment of the experts is often experienced ambiguity, the confusion in describing a proportion in the form of linguistic information (Dozic et al, 2017). So it is necessary to use Fuzzy to solve this problem. A Fuzzy AHP is a refinement of the classic Multiple Criteria Decision Making (MCDM) problem that overcomes uncertainty because it is caused by uncountable information, incomplete information and unclear information.

2. Methodology

Decision-making models need to tolerate ambiguity because uncertainty is a common feature in many decision-making issues (Yu, 2002). Because decision-makers often provide uncertain answers rather than exact values, the transformation of a qualitative preference to an approximate point may not make sense.

This decision-making is obtained in several stages. By using the hierarchical structure then obtained goals to be achieved, the criteria used and alternatives. The hierarchical structure can be seen in figure 2.

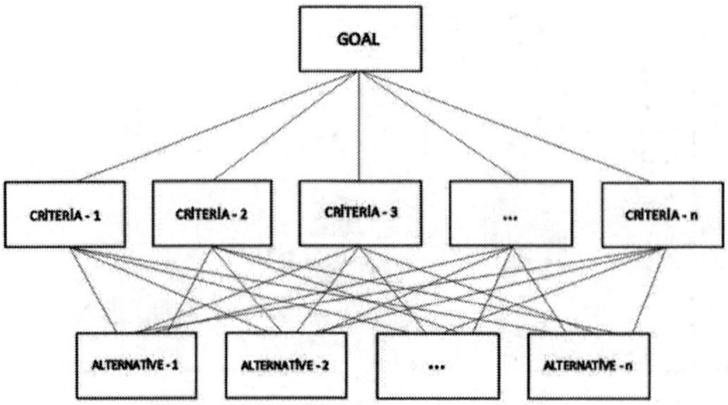

Figure 2. The hierarchical structure.

Pairwise comparisons on the Fuzzy AHP method are illustrated by a ratio scale that is related to the fuzzy scale value. Fuzzy numbers are complete on all scales describe in Table 1.

Table 1. Triangular fuzzy numbers.

Variables	Likert Scale	Positive TFN	Positive reciprocal TFN
Extremely Strong	1	(1,1,3)	(1/1,1/1,1/3)
Intermediate	2	(1,2,4)	(1/4,1/2,1/1)
Very Strong	3	(1,3,5)	(1/5,1/3,1/1)
Intermediate	4	(2,4,6)	(1/6,1/4,1/2)
Strong	5	(3,5,7)	(1/7,1/5,1/3)
Intermediate	6	(4,6,8)	(1/8,1/6,1/4)
Moderately Strong	7	(5,7,9)	(1/9,1/7,1/5)
Intermediate	8	(6,8,9)	(1/9,1/8,1/6)
Equally Strong	9	(7,9,9)	(1/9,1/9,1/7)

Fuzzy AHP procedure follows the steps of the rule by:

1. Create a hierarchical structure of the problem to be solved and determine the pairwise matrix comparison between the criteria and the triangle fuzzy number (TFN) scale.
2. Determine the fuzzy geometric mean value ($\tilde{r}$) of each criterion, sub-criteria, and alternative using the Buckley Method.
3. Determine fuzzy weights ($\tilde{w}$) for each criterion, sub-criteria, and alternatives.
4. Defuzzify using the Center of Area (COA) method.
5. Normalization of vector weight or priority value criteria that have been obtained.

3. Results

3.1 Alternative Route

The following are 3 alternatives transmission tower lines from the Larona Hydroelectric Power Plant to the Balambano Hydroelectric Power Plant which are planned to be built are shown in Figure 3. the three alternative routes are:

1. North route, this route is the longest route, which is 7.505 km. This route follows the existing transmission line route.
2. Middle route, this route has a length of 5.931 km. This route is a new route that has never been cleared before. This route is the shortest route among other alternative routes.
3. South route, this route has a length of 6.292 km. This route is also a new route that has never been cleared before.

Figure 3. Alternative transmission tower routes.

3.2 Expert Selection

Experts in this study are competent expert respondents and can inherit in the decision-making process to evaluate the selection of alternative route 150kV transmission tower lines at PT. Vale Indonesia. In Table 2 below is a data of highly experienced expert respondents selected by the author to assist in this research.

Table 2. Identity of experts.

Expert	Position	Experience
1	Project Manager	12 years
2	Manager of Civil and Hydro Surveillance	12 years
3	Manager of Energy	16 years
4	Senior Energy Engineer	15 years
5	Manager of Utilities Engineering	16 years

3.3 Criteria Selection

Selection criteria are obtained from a combination of the journal references [4] and interviews from expert respondents. All of the results of interview criteria and journal reference selected. There are only a few criteria that are almost the same as one criterion. The summary of the criteria selection can be seen in Table 3 below.

Table 3. Criteria selection.

No.	Alternative	Code
1	Construction costs	T-01
2	Maintenance Access	T-02
3	Soil Condition/Type	T-03
4	Topography	T-04
5	Permit	T-05
6	Land acquisition	T-06
7	Ease of installation	T-07
8	Community Activities around the location	T-08
9	Proximity to other buildings	T-09
10	Proximity to the river	T-10
11	Habitat of Flora and fauna	T-11

3.4 Comparison Matrix

In this phase, all 5 experts were given questionnaires in the form of a comparison matrix of each criterion. Table 4 describes the results of the comparison of expert-1 as shown below.

3.5 Consistency Ratio (CR)

A summary of the results of calculating the consistency ratio (CR) is shown in Table 5. The consistency ratio matrix of the criterion value of each expert indicates good or acceptable consistency because all values are smaller than 0.1 (10%).

Table 4. Comparison matrix expert-1.

Criteria	T-01	T-02	T-03	T-04	T-05	T-06	T-07	T-08	T-09	T-10	T-11
T-01	1	7	3	1/3	1/5	7	2	1/3	3	4	5
T-02	1/7	1	1/5	1/7	1/9	2	1/4	1/7	1/4	1/4	1/3
T-03	1/3	5	1	1/3	1/5	5	2	1/5	3	3	3
T-04	3	8	3	1	1/2	8	5	1	4	5	6
T-05	5	9	5	2	1	9	5	3	6	7	7
T-06	1/7	1/2	1/5	1/8	1/9	1	1/6	1/4	1/5	1/5	1/3
T-07	1/2	4	1/2	1/5	1/5	6	1	3	3	3	4
T-08	3	7	5	1	1/3	4	1/3	1	5	5	5
T-09	1/3	4	1/3	1/4	1/6	5	1/3	1/5	1	1	1/2
T-10	1/4	4	1/3	1/5	1/7	5	1/3	1/5	1	1	1
T-11	1/5	3	1/3	1/6	1/7	3	1/4	1/5	2	1	1

Table 5. Summary of consistency ratio.

	Expert-1	Expert-2	Expert-3	Expert-4	Expert-5
CR	0.0938	0.0989	0.0957	0.0910	0.0899

3.6 Weighting of Criteria with Fuzzy AHP Method

At this stage, the conversion from the comparison matrix to the TFN scale. Because there are 5 experts who do the matrix assessment in pairs, then the geometric mean is used to represent the assessment results of all experts. Then by following steps of the Fuzzy AHP method as mentioned in the sub-chapter of the methodology, obtained the weights of each criterion as shown in Table 6.

Table 6. Summary of weighting criteria using fuzzy AHP method.

Criteria		Weighting	Norm Weighting
T-01	**Construction Costs**	0.130	**0.104**
T-02	**Maintenance Access**	0.046	**0.037**
T-03	**Soil Condition/Type**	0.100	**0.080**
T-04	**Topography**	0.127	**0.101**
T-05	**Permit**	0.370	**0.296**
T-06	**Land acquisition**	0.174	**0.139**
T-07	**Ease of installation**	0.080	**0.064**
T-08	**Community Activities around the location**	0.098	**0.078**
T-09	**Proximity to other buildings**	0.045	**0.036**
T-10	**Proximity to the river**	0.047	**0.038**
T-11	**Habitat of Flora and fauna**	0.034	**0.027**
	Total	1.251	**1**

3.7 *Weighting of Alternative Route with Fuzzy AHP Method*
At this phase, the questionnaire was distributed back to each expert to provide an assessment of the comparison between alternatives of the selected criteria. A summary of the total value of each alternative is shown in Table 7.

Table 7. Summary of total value alternative route.

Criteria	**Weighting**	*North Route*		*Midlle Route*		*Soute Route*	
		Weighting	Total	Weighting	Weighting	Weighting	Total
Construction Costs	0.104	0.080	0.008	0.681	0.071	0.239	0.025
Maintenance Access	0.037	0.553	0.020	0.312	0.012	0.135	0.005
Soil Condition/Type	0.080	0.625	0.050	0.114	0.009	0.260	0.021
Topography	0.101	0.237	0.024	0.650	0.066	0.113	0.011
Permit	0.296	0.594	0.176	0.372	0.110	0.034	0.010
Land acquisition	0.139	0.657	0.092	0.189	0.026	0.154	0.021
Ease of installation	0.064	0.429	0.027	0.480	0.031	0.091	0.006
Community Activities around the location	0.078	0.682	0.053	0.153	0.012	0.165	0.013
Proximity to other buildings	0.036	0.020	0.001	0.335	0.012	0.645	0.023
Proximity to the river	0.038	0.236	0.009	0.138	0.005	0.627	0.024
Habitat of Flora and fauna	0.027	0.612	0.017	0.381	0.010	0.007	0.000
Total			**0.477**		**0.364**		**0.159**

3.8 *Sensitivity Analysis*
One method used to test the vulnerability of results to ranking changes is the weight adjustment method. In the sensitivity analysis phase is carried out by adding criteria weights from 5% to 50% to each criterion. This is done to see the level of sensitivity of the criteria that have been determined. The results of the sensitivity analysis of expert assessment results is shown in Table 8.

Based on the results of the sensitivity analysis of the model based on expert judgment it can be concluded that all of those criteria are robust criteria because the alternative ranking does not change after the simulation is done from 5% to 50%.

Table 8. Sensitivity analysis results.

Alternative		North Route (N)	Middle Route (M)	South Route (S)	Alternative Ranking
Existing		0.477	0.364	0.159	N>M>S
Construction Costs	+5%	0.47728	0.36728	0.16062	N>M>S
	+20%	0.47852	0.37790	0.16434	N>M>S
	+50%	0.48100	0.39912	0.17178	N>M>S
Maintenance Access	+5%	0.47789	0.36432	0.15963	N>M>S
	+20%	0.48095	0.36605	0.16038	N>M>S
	+50%	0.48708	0.36951	0.16188	N>M>S
Soil Condition/Type	+5%	0.47936	0.36420	0.16042	N>M>S
	+20%	0.47936	0.36557	0.16354	N>M>S
	+50%	0.47936	0.36831	0.16977	N>M>S
Topography	+5%	0.47807	0.36704	0.15996	N>M>S
	+20%	0.48166	0.37690	0.16167	N>M>S
	+50%	0.48885	0.39664	0.16510	N>M>S
Permit	+5%	0.48565	0.36924	0.15989	N>M>S
	+20%	0.51200	0.38574	0.16141	N>M>S
	+50%	0.56469	0.41872	0.16444	N>M>S
Land acquisition	+5%	0.48145	0.36506	0.16046	N>M>S
	+20%	0.49520	0.36901	0.16368	N>M>S
	+50%	0.52269	0.37692	0.17013	N>M>S
Ease of installation	+5%	0.47824	0.36528	0.15968	N>M>S
	+20%	0.48235	0.36989	0.16055	N>M>S
	+50%	0.49058	0.37910	0.16231	N>M>S
Community Activities around the location	+5%	0.47954	0.36434	0.16003	N>M>S
	+20%	0.48754	0.36614	0.16197	N>M>S
	+50%	0.50354	0.36972	0.16585	N>M>S
Proximity to other buildings	+5%	0.47690	0.36435	0.16054	N>M>S
	+20%	0.47701	0.36615	0.16401	N>M>S
	+50%	0.47722	0.36976	0.17096	N>M>S
Proximity to the river	+5%	0.47731	0.36401	0.16057	N>M>S
	+20%	0.47865	0.36479	0.16412	N>M>S
	+50%	0.48132	0.36635	0.17121	N>M>S
Habitat of Flora and fauna	+5%	0.47770	0.36427	0.15939	N>M>S
	+20%	0.48020	0.36582	0.15942	N>M>S
	+50%	0.48520	0.36894	0.15948	N>M>S

4. Conclusion

There are some conclusions obtained in the study of determining the 150kV transmission tower route in PT. Vale Indonesia is based on the Fuzzy AHP method. There are eleven criteria selected to make

IOP Conf. Series: Materials Science and Engineering 1072 (2021) 012043 doi:10.1088/1757-899X/1072/1/012043

decision in determining the transmission tower route. The eleven criteria are construction costs, maintenance access, soil condition or type, topography, permit, land acquisition, ease of installation, community activities around the location, proximity to other buildings, proximity to the river, and habitat of flora/fauna. These criteria are then used to weight the comparison matrices of each alternative of the 150kV transmission tower route. the results of value each alternative is 0.477 for the North route, 0.364 for the Middle route, and 0.159 for the South route. From this result, it can also be concluded that the transmission tower route length does not guarantee the best route choice because many other criteria considerations influence the implementation of the transmission tower construction. Therefore, the authors recommend the management of PT. Vale Indonesia, as the owner and operator of the hydroelectric power station, chose the north route for the planned construction of the 150kV transmission tower.

References

[1] Iftikhar and Musher and Anwar 2010 *A Study on Fuzzy AHP method and its applications in a "tie-breaking procedure"* Global Journal of Pure and Applied Mathematics, ISSN 0973-1768 13 pp. 1619-1630.

[2] PT. PLN (Persero) 2014 Buku Pedoman Pemeliharaan Saluran Udara Tegangan Tinggi Dan Ekstra Tinggi (Sutt/Sutet) Dokumen nomor: PDM/STT/10:2014.

[3] Hasan EROGLU and Musa AYDIN 2015 *Optimization of electrical power transmission lines' routing using AHP, fuzzy AHP, and GIS* Turkish Journal of Electrical Engineering & Computer Sciences.

[4] Manitoba Hydro 2018 *Transmission Planning & Design Division Licensing & Environmental Assessment* Birtle Transmission Project Appendix A Transmission Line Routing.

[5] Juho Yli-Hannuksela 2011 *The transmission line cost calculation* vaasan ammattikorkeakoulu university of applied sciences.

[6] Kusumadewi Sri and Hari Purnomo 2006 *Aplikasi Logika Fuzzy Untuk Pendukung Keputusan*, (Yogyakarta: Penerbit Graha Ilmu).

[7] Saaty TL, Vargas LG. Models, Methods, Concepts & Applications of the Analytic Hierarchy Process. *International Series in Operations Research & Management Science*. Springer US; 2001; doi.org/10.1007/978-1-4615-1665-1

[8] Heri Septiadi Alius and Nasrun Hariyanto and Syahrial. (2014) *Perancangan Sistem Transmisi Daya Listrik Bertegangan 150 KV dan Berkapasitas 35 MVA di Kabupaten Bulungan Kalimantan Timur* Jurnal Reka Elkomika Jurusan Teknik Elektro Institut Teknologi Nasional Bandung.

[9] Holland H. Farr 1980 *Transmission Line Design Manual* Water and Power Resources Service Denver.

[10] Serkan Şahin 2016 *Cantegrated Size And Shape Optimization Of Steel Lattice Energy Transmission Line Towers Using Simulated Annealing* Middle East Technical University.

[11] Indar Nurtrihansyah, Udisubakti Ciptomulyono 2011 *Penentuan Jalur Transmisi Sutet (Saluran Udara Tegangan Ekstra Tinggi) Menggunakan Metode Anp-Promethee (Study Kasus: Gardu Induk Pemalang-Gardu Induk Madirancang).*

Optimization operation strategy of electric boilers by mixed integer programming method in nickel production process

W S Djati[1,2] and N Siswanto[1,3]

[1]Department of Industrial and Systems Engineering, Institut Teknologi Sepuluh Nopember, Surabaya, Indonesia

E-mail: [2]wahyu.setiadjati@vale.com, [3]siswanto@ie.its.ac.id

Abstract. Nickel smelting process requires steam in the process of drying and calcining. Utilities Department is responsible for supplying steam either use electric boiler or fossil fuel. With fluctuations in fuel prices, limited electricity supply, boiler reliability, and government emission limits, it is a challenge for department utilities to produce steam. Planning and selection of boiler operations to produce steam to meet the needs of the smelting process are necessary in order to minimize operating costs and maintain boiler reliability in distributing steam. This condition makes mixed integer programming become the best method to solve the model. Mathematical models are used to determine boiler operations in the wet season and dry season by using mixed integer programming methods. In the wet season, boilers that use electricity can operate because electricity from hydropower is sufficient, although not fully the boiler can meet the steam needed. Whereas in the dry season boilers that use electricity cannot operate due to limited electricity supply generated from hydropower.

1. Introduction

Nickel processing in smelters requires supporting facilities to produce SWAP (Steam, Water, Air, Pressure). One of the products that become an important thing on the nickel processing plant is steam. Steam supplied to nickel processing is a product of utilities plant that consists of hydroelectric plant and thermal plant. Most of the steam used on process are produced by thermal plant. Steam are distributed to nickel process plant to fulfill fluctuate demand of steam as per shown at Figure 1.

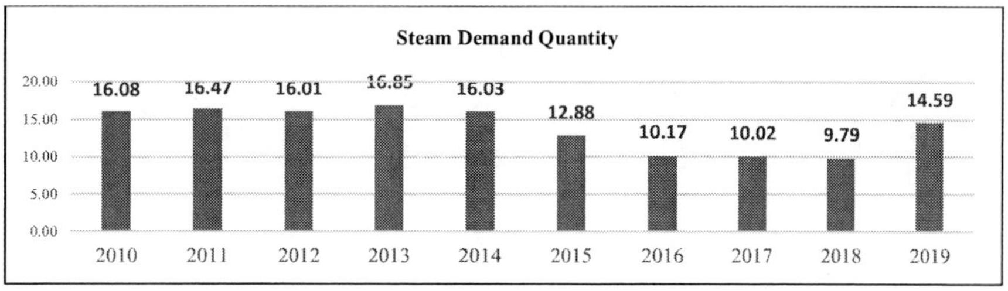

Figure 1. Annual steam demand quantity.

Figure 1 shows that steam demand from process plant has decrease in year 2016 to 2018 due to change fuel in the kiln and dryer which initially used HSFO switched to use coal. In year 2019, steam demand has increase into 14.5 tons/hour due to the nickel processing plant company has the initiative to contribute green environment by reduce the coal consumption. The other alternatives to support the green environment program is using electric boiler. Another consideration that makes electric boilers an option is low operating cost, but on the other side electric boiler required high power consumption. Comparison of steam generation cost and power required shown in Table 1.

Table 1. Steam generation cost comparison.

Steam Generation	Consumption Index	Unit - Fuel	2015	2016	2017	2018	2019
PB#1[a]	98.49	Liter/Ton - HSD	54.17	41.37	47.28	63.03	64.63
PB#2	0.67	Barrel/Ton - HSFO	39.06	24.86	35.51	41.27	47.27
	102.2	Liter/Ton - HSD	55.11	42.08	48.10	64.13	65.75
BW Boiler[b]	99.50	Liter/Ton - HSD	54.73	41.79	47.76	63.68	65.29
WHRB[c]	0	Liter/MWh - HSD	0.00	0.00	0.00	0.00	0.00
EB[d]	0.66	MW/Ton - Hydro Electricity					9.21

[a] PB is Package Boiler
[b] BW Boiler is Babcock & Wilcox Boiler
[c] WHRB is Waste Heat Recovery Boiler
[d] EB is Electric Boiler

The average steam demand for nickel processing plant is 15 tons/hour with required 10 MW of electric power if the steam supplied by electric boiler. This condition will be creating 2 MW lack of electric power condition and become the main constraint. Other constraints to be considered in this paper, to operate the electric boiler is source of water for hydroelectric generation. Hydroelectric water level is depending on the rainfall, during the dry season from October to December can barely operate because of the reduced ability of hydropower plants to generate electricity. It caused by the height of the lake's surface which is supplying water to hydropower plants also decrease.

Each type of boiler operated by company refer to Table 1 has each of its problem. On the other side, increasing nickel production capacity requires higher electric power, that's why electric boiler becomes a bottleneck issue due to limited electric power sources.

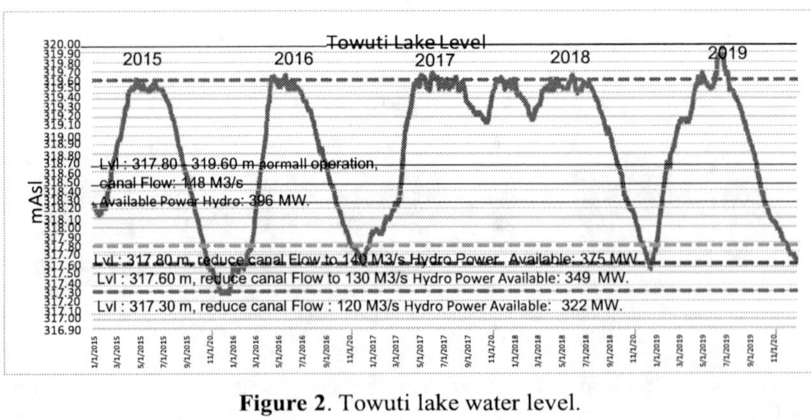

Figure 2. Towuti lake water level.

Company fulfilled the electric power required by using hydropower plant that consist of three hydropower plant which is divided into 3 area, namely Larona, Balambano and Karebbe hydropower plants. These three hydropower plants water source is derived from Towuti lakes and using a cascade system for the type of water flow. The generating capacity of those three hydropower plants is highly dependent on the water level of Towuti lake that fluctuates as shown in Figure 2. Not only water level of Towuti lake, but also HSD and HSFO price that fluctuates and government emission threshold to fulfilled steam demand requirement for nickel processing become a challenge to company to do the best operation strategy that can optimize the operating cost which will be discussed in this study.

The ultimate goal of this paper is to get the optimum boiler operating cost that considers fixed and variable costs. Fixed cost will generate when the boiler operated, that's why fixed cost decision variables will be defined by integer number 1 or 0. On the other side, variable cost define by the steam produced by the boiler, so the number of variable costs allowed not an integer number. According to those case problems, modelling by MIP approach become the best suitable method to solve the problems.

2. Mixed Integer Programming (MILP)

Integer programming is a linear programming in which some or all the variables are required to be non-negative integers. There are two categories for integer programming, the first one is called pure integer programming problem in which all variables are required to be integers, and the second is called mixed integer programming in which only some the variables required to be integers [1]. Linear Programming (LP) models remain much more attractive from a computational point of view for several reasons. When an LP problem is formulated, the software finds an optimal solution. When the formulation requires variables that are naturally integer or binary, the problem becomes a MILP model [2]. The maximum size of instances of MILP models for portfolio optimization that can be solved to optimality strongly depends on the specific model. In MILP models, an increasing number of assets makes the model harder to solve then an increasing number of scenarios, in general, it is the number of binary or integer variables that determine the computational time required. Most models with fixed transaction costs may be solved to optimality within a reasonable amount of time (on the order of hours) on instances with up to several hundreds of variables [2].

The MILP model has been developed from Method for analysis of industrial energy systems (MIND) that has been developed to solve the optimization problem for the iron and steel industry. The model is implemented in LINGO software. LINGO is a software for solving optimization problems developed by Lindo System, Inc., USA. It is a comprehensive tool designed to make building and solving linear, nonlinear and integer optimization models faster, easier and moreefficient [3].

The model in this study is similar to the previous study, the difference from the previous study located on the operating strategy to be implemented [4]. On the previous study, the simulation model is modelling the blackout strategy due to large scale cascading blackout in a power transmission network, in this study will be calculated for strategy operation to get optimum operating cost. The optimal solution by building a model. There are two categories, quantitative and qualitative model. Quantitative model includes linear programming, MILP, MINLP, dynamic programming, and multi-objective programming [5].

0/1 mixed integer linear programming, as the method develop in this document, is an extension of integer programming, due to the fact that new complicating variables (0/1) are introduced. Thus, this is a new approach and the main advantage is that allows coping with a more realistic model, taking into account other important variables [6].

3. Model Development

This study is used to examine cost analysis to determine the boiler operation strategy by developing a mixed-integer programming model using lingo software through data approaches:

1. Production demand
2. Availability of electricity

3. Reliability of the boiler
4. Environmental or emissions factors,
5. Production costs (fixed and variable costs).

In making mathematical models determine the boiler operation strategy in delivering steam to customers is divided into two conditions, namely the wet season and the dry season. In the wet season, which is for January - September or the 1st day to 273th, while the dry season from October to December or the 274th to 365th day. For the mathematical model formulation uses the following notation:

Input

Index:

i	: Boiler
t	: Day

Parameter:

FC_i	= Fixed Cost of the ith boiler.
VC_i	= Variable Cost of the ith boiler.
CP_i	= Capacity of of the ith boiler.
Re_i	= Realibility of of the ith boiler.
EC_i	= Electric Consumption of the ith boiler.
MP_i	= Minimum Production of the ith boiler.

Variables:

BB_{it}	= Value 1 if the ith boiler on the t-day is operating, otherwise value 0.
ST_{it}	= The amount of steam produced from the ith boiler on the t-day
AB_i	= Value 1 if the ith boiler operates, otherwise value 0.

Objective Function:

Minimize Z = (Total fixed cost x Operating boilers) + (Total variable cost x Steam production)

$$\text{Min}\, Z = \sum_{i=1}^{9}\sum_{t=1}^{365} FC_i BB_{it} + \sum_{i=1}^{9}\sum_{t=1}^{365} VC_i ST_{it} \tag{1}$$

There is a constraint also to get an optimum operating cost by modelling the operation strategy boiler. The constraint of this condition is consisting of:

1. Rating Capacity of each boiler PB#1 = 6 Ton/hours, PB#2 = 18 Ton/hours, BW Boiler = 22 Ton/hours, Electric Boiler = 20 Ton/hours, WHRB#1 = 1 Ton/hour, WHRB#2 = 1 Ton/hour, WHRB#3 = 1 Ton/hour, WHRB#4 = 3 Ton/hours, WHRB#5 = 2 Ton/hours. Capacity is calculated daily by multiplying capacity by hours in one day (24 hours). The total steam produced by each plant every day must not exceed its capacity.
$$ST_{it} \le CP_i \ \forall\, i,t \tag{2}$$

2. Steam that supply from the various boiler operated should be fulfil steam demand requirement which is 15 ton/hour. Total demand of steam for 1 day is 15 ton/hour x 24 hours = 360 Ton. Formulation of constraint that describe the steam demand is:
$$\sum_{i=1}^{9} ST_{it} \ge 360 \qquad \forall t \tag{3}$$

3. Electricity power can be provided to produce a steam is only 9 MW and put as an RHS value in equation number (3). Mathematic formula that describe this constraint is
$$\sum_{i=1}^{9} EC_i + ST_{it} \le 9 * 24 \qquad t = 1,2,\ldots,273 \tag{4}$$

4. Total electricity required to produce steam is less than 1MW that refer to formula as below:

$$\sum_{i=1}^{9} EC_i + ST_{it} \leq 1 * 24 \qquad t = 274,275,, \ldots ,365 \tag{5}$$

5. Physical availability of steam supply system when boiler operate during Wet Season the customer gives time for the total boiler shutdown maintenance for 2 days per-year or the Physical Availability target of the steam production system is 92.0%, combination configuration describe in mathematic formula as below:

$$\sum_{i=1}^{273} \sum_{t=1}^{9} ST_{it} \geq 0.992 * 360 \tag{6}$$

6. Physical availability of steam supply system when boiler operate in combination configuration on dry season describe in mathematic formula as below:

$$\sum_{i=274}^{365} \sum_{t=1}^{9} ST_{it} \geq 1 * 360 \tag{7}$$

7. Reliability PB # 1, PB # 2, BW Boiler, and EB are expected to be 100% or 365 days in one year

$$\sum_{t=1}^{365} ST_{it} \leq Re_i * 365 * 360 \qquad i = 1,2, \ldots ,4 \tag{8}$$

8. Reliability WHRB#1 - 5 are expected to be 100% or 365 days in one year

$$\sum_{t=1}^{365} ST_{it} \leq Re_i * 41 * 360 \qquad i = 5,6, \ldots 9 \tag{9}$$

9. Minimum quantity of steam produce from PB#1 is 0.6 ton/hour, PB#2 is 1.8 ton/hour, and BW Boiler is 2.2 ton/hour, and electric boiler is 2 ton/hour. The value of minimum steam produce from each type of boiler is 10% of their rating capacity.

$$ST_{it} \leq M * BB_{it} \qquad \forall t,i \tag{10}$$

$$MP_i - ST_{it} \leq M * (1 - BB_{it}) \qquad \forall t,i \tag{11}$$

10. If EB is the main generator, PB # 2 is the second boiler

$$BB_{2t} \geq BB_{4t} \qquad \forall t \tag{12}$$

11. In the wet season the availability of electricity from hydropower plants can be sufficient from the demand of electric boilers, so that in the wet season the use of electric boilers is optimized. Whereas in the dry season there is no electricity available for electric boilers and during the dry season the PLTD is operated to increase the availability of electricity but the PLTD is only allowed to operate 1000 hours per year because the NOX emissions of the PLTD exceeds the government threshold. Because WHRB utilizes the exhaust gas heat from the PLTD, at that time WHRB could operate for 1000 hours or 41 days

$$\sum_{i=5}^{9} ST_{it} \leq 0 \qquad t = 1,2, \ldots ,273 \tag{13}$$

$$\sum_{i=5}^{9} ST_{it} \leq 0 \qquad t = 314,315, \ldots ,365 \tag{14}$$

12. If a boiler operates at a time, a fixed cost will arise

$$\sum_{t=1}^{365} BB_{it} \leq M * AB_{it} \qquad \forall i \tag{15}$$

Expected output from the mathematic model of mixed linear programming in this study is the decision variable that describes a steam quantity produce and the decision to operate the boiler either should be operate or not. The resulting decision variable will be used in the calculation of total operating cost which is the total of the component cost of the fixed cost and variable cost. Annual total operating cost will be the sum of total operating cost on wet season and dry season. Annual total actual operating cost from model will be compare with the annual existing total operating cost.

4. Result

The results of this analysis can be divided into two mathematical models, namely boiler operation during the wet season for 9 months (January - September) and during the dry season for 3 months (October - December). During the wet season, electricity from hydropower plants can meet the electricity needs for the smelting process of nickel and electric boilers to obtain the cost of producing steam by maximizing the use of electric boilers. Table 2 shows the results of calculations from the mathematical model.

Table 2. Steam production in wet season.

No	Variable	Boiler	Steam production ton per-day
1	X_1	Packed Boiler#1	-
2	X_2	Packed Boiler#2	43.2
3	X_3	BW Boiler	-
4	X_4	Electric Boiler	316

Based on the calculation of steam production from a mathematical model in 9 months, the most optimal for the scenario of boiler operation is by operating an electric boiler and producing steam of 316 ton per-day using 8 MW of electricity. Because the demand from customers is 360 ton per-day of steam for 9 months, based on the results of calculations using lingo the shortage of steam production is met by packed boiler# 2 with a total production of 43.2 ton per-day. With a production scenario like the results of the above calculation, it will be known the cost of producing pressurized steam of $ 1,689,012 in the wet season.

In September to December, the availability of electricity for electric boilers was not available, so that during the season steam can only be supplied from boilers with fossil fuels. Table 3 shows the results of calculations from the mathematical model.

Based on the calculation of pressurized steam production from a mathematical model, in the most optimal 3 months is to maximize the operation of WHRB for 1000 hours and produce steam of 216 ton per-day. Because the demand from customers is 360 ton per-day and the shortage of steam production when WHRB operates is met by packed boiler#2 with a total production of 144 ton per-day tons. Because WHRB cannot operate for more than 1000 hours, packed boiler#2 becomes the main boiler and produces 360 ton per-day. With a production scenario like the result of the calculation above, it will be known that the production cost of pressurized steam is $ 1,615,073 in the dry season.

Three months boiler operating cost in dry season almost same with nine months boiler operating cost in wet season. It caused by in dry season electric boiler cannot operate due to lack of power generated by hydropower plant. Fossil fuel boiler will run in dry season to provide additional power and require more boiler operating cost. Three months boiler operating cost in dry season almost same with nine months boiler operating cost in wet season. It caused by in dry season electric boiler cannot operate due to lack of power generated by hydropower plant. Fossil fuel boiler will run in dry season to provide additional power and require more boiler operating cost.

Table 3. Steam production in dry season.

No	Variable	Boiler	Steam production (Ton)	Remarks
1	X_1	Packed Boiler#1	-	
2	X_2	Packed Boiler#2	144 360	- When backing up WHRB 41 days - After WHRB does not Operate
3	X_3	BW Boiler	-	
4	X_4	Electric Boiler	-	
5	X_5	WHRB#1	24	
6	X_6	WHRB#2	24	
7	X_7	WHRB#3	24	
8	X_8	WHRB#4	72	
9	X_9	WHRB#5	72	

After getting the value of the decision variable and entering that value into the objective function equation of minimizing boiler operating costs using mixed-integer programming, a cost-minimization analysis is performed by comparing the results of optimization with actual costs. In comparing the operating costs of this boiler with the sum of costs in two conditions, namely the wet or dry season or the operating costs of the boiler in 1 year. Operating costs obtained from the mathematical model using the mixed integer programming method during the wet season are $ 1,689,012 and boiler operating costs during the dry season are $ 1,615,073 so the total minimum annual boiler operating costs are $ 3,304,085. If we compare with the actual boiler operating costs 2019 with the results of mathematical model calculations using a mathematical model that calculations with mathematical models, variance only is 3.4% and can be show in Figure 3. During this time the difference between the budget and the actual is greater than 10% refer cost report 2017 – 2019.

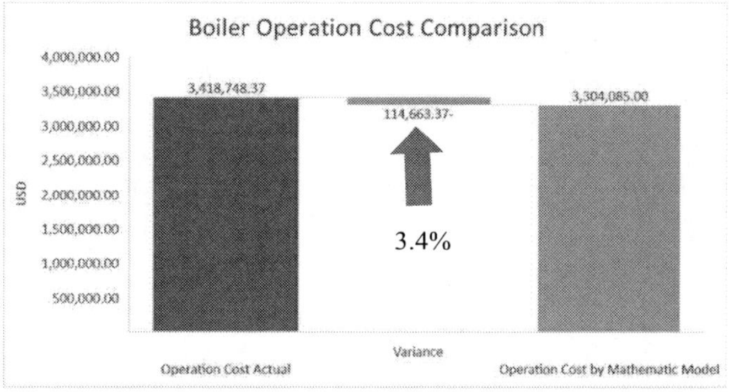

Figure 3. Variance optimization model result with actual operation cost.

5. Conclusion

This optimization study of boiler strategy operations was conducted to obtain optimal boiler operating costs. There are several conclusions based on the results obtained from modelling. The first where the operating costs of the mathematical model is the sum of the operating costs in the rainy and dry seasons which is $ 3,304,085.

The second is achieving optimal boiler operating results with a mathematical model approaching 3.4% of actual boiler operations. When compared with the determination of the budget from 2017 to 2019, the budget difference from 2017 to 2019 has a very significant difference when compared to the actual production costs of boiler operations, which is above 10%.

6. References

[1] Winston WL. *Operation Research Application and Algorithms* 4th edition. Belmont: Thomson Learning Inc; 2004.

[2] Renata M, Wlodzimierz O, Grazia S. *Linear and Mixed Integer Programming for Portofolio Optimization.* London: Springer International Publishing; 2015.

[3] Kong H, Qi E, Li H, Li G, Zhang X. An MILP model for optimization of byproduct gases in the integrated iron and steel plant. *Appl. Energy*. Elsevier BV; 2010 Jul;**87**(7):2156–63. doi.org/10.1016/j.apenergy.2009.11.031

[4] Bienstock D, Mattia S. Using mixed-integer programming to solve power grid blackout problems. *Discrete Optim.* Elsevier BV; 2007 Mar;**4**(1):115–41. doi.org/10.1016/j.disopt.2006.10.007

[5] Wicaksono PA, Pujawan IN, Widodo E. A mixed integer linear programming model for dynamic supplier and carrier selection problems. *Int. J. Procure. Manag*. Inderscience Publishers; 2019;**12**(3):276. doi.org/10.1504/ijpm.2019.099550

[6] Perez Canto S. Using 0/1 mixed integer linear programming to solve a reliability-centered problem of power plant preventive maintenance scheduling. *Optim. Eng*. Springer Science and Business Media LLC; 2011 May 5;**12**(3):333–47. doi.org/10.1007/s11081-011-9146-2

Modeling opportunistic maintenance using discrete event simulation

A L Budiono[1,2,3], N Siswanto[1,4] and N Kurniati[1,5]

[1]Department of Industrial and Systems Engineering, Institut Teknologi Sepuluh Nopember, Surabaya, Indonesia
[2]Department of Maintenance, PT. Semen Indonesia, Tuban, Indonesia

Email: [3]anggozip87@gmail.com, [4]siswanto@ie.its.ac.id, [5]nanikur@ie.its.ac.id

Abstract. This study aims to create a simulation model of Opportunistic Maintenance (OM). In continuous industries and with a large variety of equipment functions, options for maintenance are very small. Determination of the maintenance schedule is based on the time opportunity created by the inventory level conditions, are blocking by the next process or lack of material from the previous process. The research was conducted by using discrete event simulation to generate arrival time of equipment failure and opportunistic maintenance time from scheduled maintenance and arrival time for maintenance due to inventory conditions. The arrival time of failure and the arrival time of opportunistic maintenance will probably intersect. Such conditions will increase the availability of equipment. Then the replication process is carried out to compare the availability of equipment under real conditions and the expected value of availability from the simulation. Finally, scheduled maintenance can be a decision variable to choose the best maintenance time when faced with fluctuations in demand.

1. Introduction

Generally, maintenance can be categorized into two types, namely preventive and corrective maintenance. Preventive maintenance is a part replacement activity, oil change, and adjustments are made before a failure occurs. The purpose of preventive maintenance activities is to increase the reliability of the system from the effects of aging such as wear, corrosion, fatigue, and other related phenomena [1]. While corrective maintenance is an action taken after a failure occurs, it needs to be returned as soon as possible so that the production process can run again.

The application of preventive maintenance in different industries or environments has its uniqueness. In one particular industry has many options for conducting maintenance activities. This kind of industry is an industry that has a lot of equipment units where each equipment has a production capability. This industry is an industry that does not require large variations of equipment function. Every equipment unit that needs maintenance does not have much effect on the amount of production output. Because of these large options, such industries will tend to implement preventive maintenance based on time intervals. Maintenance time intervals depend on the threshold value and the quality of preventive maintenance. In the real world, the quality of preventive maintenance is difficult to be determined precisely. This uncertainty makes it problematic to optimize maintenance policy [2]. This paper discusses optimizing maintenance on different sides, which is related to the inventory.

The other one, implementation of preventive maintenance is more based on opportunistic time. As much as possible, the abnormalities that have founded in the field, will be postponed until there is a chance of time to repair. This opportunity can be obtained due to lack of raw material supply, low market demand or others. In the preventive maintenance analysis, the reliability value threshold may change every time because it follows the available time. The increasing of the reliability value after maintenance activity, is difficult to determine because maintenance work is not repetitive, as mentioned in other

typical preventive maintenance. In other words, industries that implement opportunistic maintenance strategies are those that have the following characteristics:
1. The continuous system, with a large variety of equipment functions, with a limited number of units for each function. Many machines are dedicated to certain functions only.
2. High machine setup costs, so it is necessary to avoid turning off the equipment at any time.

2. Literature Review

2.1 Opportunistic Maintenance

Chang and Banddyopadhayay [3], developed a systematic method for determining maintenance time, which is called maintenance opportunity. A simulation algorithm was developed by utilizing the buffer level to obtain maintenance opportunity time during the production process. The mathematical model of maintenance opportunity time is as follows:

$$\Delta T = \frac{(\dot{m}_1 - \dot{m}_2)T - (N_T - N_0)}{\dot{m}_1} \tag{1}$$

ΔT is the maintenance time opportunity at T.N_0, N_T is the buffer level at the start and T. $\dot{m}_1, \dot{m}_2$ is the production rate of input and output. With this model, it can be predicted that there is an opportunity for maintenance time. If the abnormality can be postponed, an estimated maintenance time ΔT is needed, then set the N_0level, if necessary, set $\dot{m}_2$ and Tc time, so that shortage can be avoided. Such a model is effective at predicting maintenance time opportunities if both the input and output have a deterministic production rate. But in most places, both input and output are stochastic processes.

Wu, et al. 2019, [4] design a two-phase opportunistic maintenance framework based on defect information and integrate it with the conditions of production waiting time into the decision-making process. The waiting time can occur due to low demand or a lack of production material. This model produces probability in three scenarios, namely Corrective replacement, Opportunity replacement and Preventive replacement. Corrective replacement scenario occurs if there is no Preventive maintenance plan, but there is a failure that needs immediate repair or replacement. While Opportunity replacement is an incident when maintenance opportunities come because of the production wait. The use of mathematical analysis is too complicated so it is necessary to assume the presence of the distribution of the arrival time of production. Also in the analysis, there are functions that may need further research to get it because of the limited data available, for example survival function. Besides that, the imperfect value of preventive maintenance activity is also needed, where this variable is varied and interdependent.

2.2 Complex Repairable System

It is important to discuss repairable systems. Real applications are often faced by complex systems that require assumptions so that the model can be derived. The Renewal Process Model (Maximum Repair) assumes that repairs are considered as new equipment. Process renewal is a recorded counting process in sequence. The assumptions underlying the NHPP (Minimal Repair) model imply that, when repairs are made, the latest system conditions are the same as those that occurred right before the failure. The non-homogeneous Poisson process (NHPP) differs from the homogeneous Poisson process only in that the failure rate varies with time rather than being constant. NHPP is the most appropriate model for the reliability of complex systems consisting of infinite components. However, for a limited number of components, this model can only function as an estimate, often poorly, because of the intensity function changes following each improvement [5]. Maintenance activities are often in the two extreme assumptions above.

It is clear that analytical and mathematical approaches are limited in solving complex treatments [6]. By developing analytic and simulation models to solve the same problem Rezg, Chelbi and Xie [7] find that complex analytic models with unrealistic assumptions compared to simulation models provide more flexibility and simpler estimates.

2.3 Optimization Maintenance Based on Simulation

The complexity of the maintenance system has significantly increased. This is partly due to the modern manufacturing system which involves many interactions and dependencies between components [8]. SL.Ho and M.Xie, [9] in their paper, discussing reliability forecasting with the ARIMA approach. The paper states that there is a rational matter if the time between failures (TBF-Time Between Failure) of an upcoming system is closely related to the latest TBF. Or even extends to past values. This is because future TBF depends on the level of success of the present and previous improvement. Here the simulation is presented to generate future behavior based on past behavior.

Even though research on maintenance optimization began a few decades ago [10], simulation-based maintenance became an emerging trend [11]. Roger [12] defines optimization based on simulation as an approach where the optimization engine provides factor input to a simulation program. The simulation will run and give the results of the optimization function objectives. This process will continue repeatedly between the simulation program and the optimization engine until it produces a satisfactory solution or termination due to the specified conditions. The application of simulation to maintenance does not only aim at optimization. Simulation application without optimization is used for comparison, evaluation, and validation purposes.

Simulation applications to optimize equipment availability with an inventory level approach have been demonstrated in a research conducted by Salsabila et al. This paper focuses on experimenting with buffer inventory levels and the capacity of a multi-state manufacturing network to increase production throughput on a company [13]. Increased equipment availability can be achieved through the availability of spare parts needed. Spare parts availability during downtime will reduce repair time (MTTR). Another study from Salsabila et al. develops a simulation model for failure and the whole system to find out the needs of critical parts. The development scenario is made by changing the ROP of the critical part. Availability of critical part when it is needed, will increase the equipment availability itself [14].

3. Research Methodology

Figure 1 shows the validation steps of the model that will be built. The opportunistic maintenance model for the simulation is made as explained in section 4. Reliability data from previous years were collected to predict the current operational condition of the year if it is met with current demand.

We already have performance data from kiln (one of the Cement industrial equipment) in 2019 and demand realization in the same year, from one of the cement companies in Indonesia. Failure data in 2019 is used as a real performance data of equipment. Data equipment failure 4 years earlier (in 2015 until 2018) are needed also. This failure and demand data, used for input simulation data to obtain the performance of equipment from the simulation. Results of the simulation output then compared with real equipment performance data in 2019. The verification and validation process will determine whether this model is suitable for predicting operational in the next years.

4. Proposed Opportunistic Maintenance Model

Kiln operates all the time except when overhaul. Overhaul is planned to conduct annual preventive maintenance activities. During operation there may be an unplanned failure that must be shutting down the kiln system.

For Kiln the conditions can be explained as follows, and can be seen in Figure 2:

1. Opportunity maintenance only during Overhaul, besides that it means corrective maintenance and read as downtime data.
2. Redundancy make increase equipment physical availability, but requires switching time so that it causes shutdown and is read as data downtime.

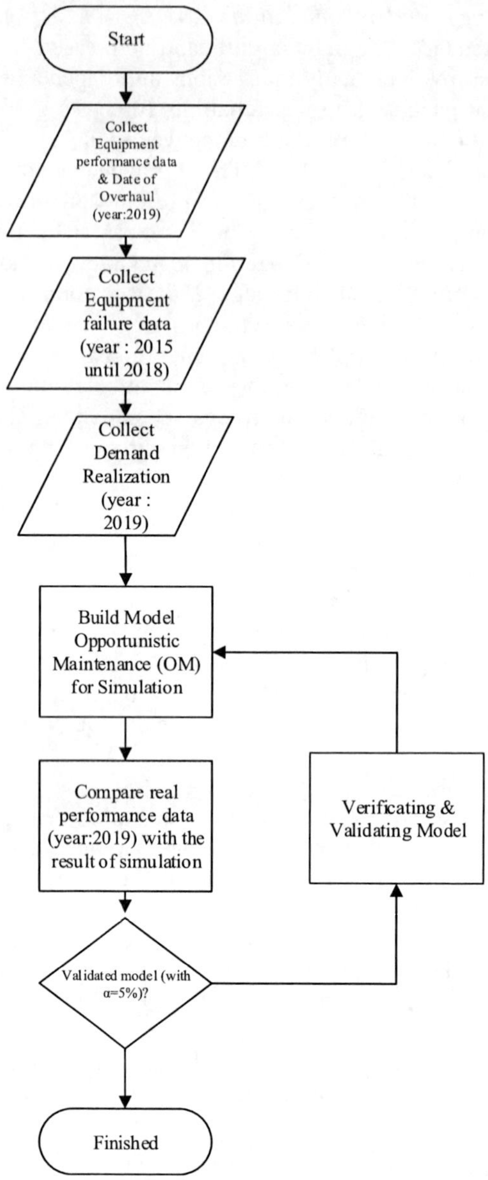

Figure 1. Research methodology.

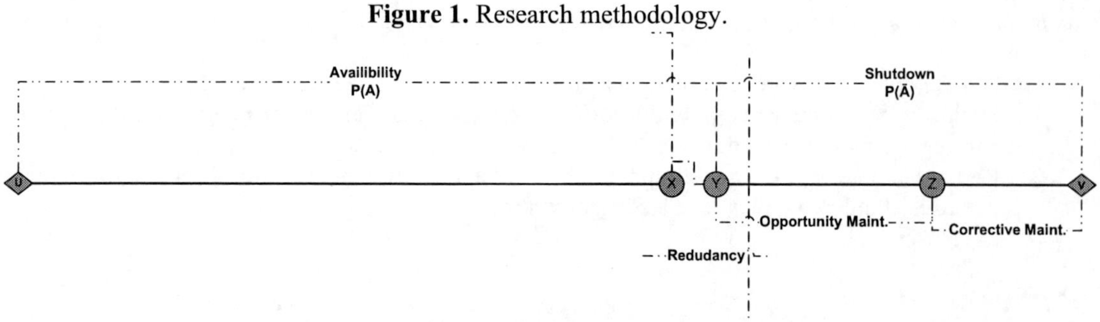

Figure 2. Preventive model for kiln.

In this Kiln model, up (kiln run time) and down (kiln off time) data have been suitable for data failure. We need to capture time horizon, as Figure 2, because sometimes maybe not all down data are carried out by maintenance activity. Down data may appear due to inventory limitations. Kiln, up and down operational data are needed on the model to be generated. These data use to determine the arrival of failure and how long repair finish in simulation. Thus, up data represent time to failure (TTF) and down data represent time to repair (TTR). Table 6 shows up and down data for kilns 1 from 2015 to 2018, the unit is in hours. For kiln 2, kiln 3 and kiln 4 are not shown here, but the results of distribution fittings for all equipment are displayed in Table 3.

T_n is the equipment interval running time
F_n is the frequency of the equipment running during T_n
X is a random variable value between 0 to $\sum_1^n F_n$

Table 1. Equivalence generating TTF in montecarlo simulation and reliability.

EVENT	*CUMULATIVE FREQ.*	
T_1	F_1	$RND(X)$
T_2	$F_1 + F_2$	$RND(X)$
⋮	⋮	⋮
T_n	$\sum_1^n F_n$	$RND(X)$

R (t), reliability at time t is a cumulative function, this is equivalent to the process of generating time T from the TTF (Time To Failure) distribution in Monte Carlo simulations. This also implies that the value of the threshold is a random value based on past history. Another important assumption in this research is related to the increase in reliability after repair activity. After repair, the TTF distribution did not change. TTF distribution still follows the initial distribution that has been input to the simulation. Thus, this model is an NHPP (Non-Homogeneous Poisson Process) or minimal repair model.

Opportunity Maintenance (OM), presence under the following conditions,

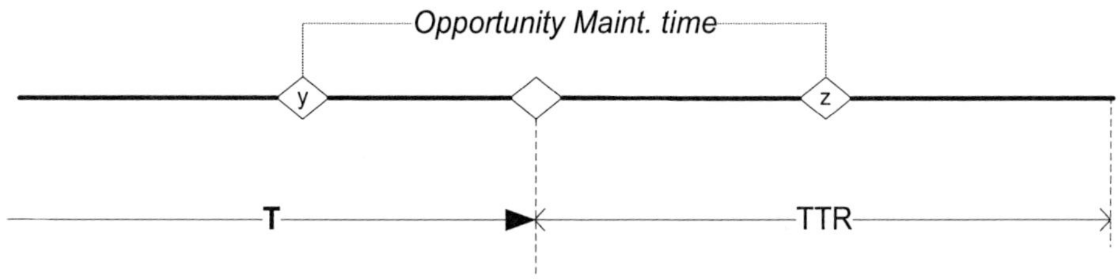

Figure 3. Opportunity maintenance condition.

y is the arrival time for Opportunity time (schedule overhaul)
z is the end of Opportunity time
T is the threshold time for preventive maintenance

For analysis of the preventive maintenance (PM) model, there is an assumption that the repair time can be neglected. This is because repair time is much less than the total running time of the equipment. In this simulation, the duration of maintenance activities, both PM and CM, will be calculated. Thus, it will influence the amount of equipment availability. Preventive maintenance will occur if reliability threshold has not been achieved but repair activity has been taken before. The threshold in this simulation is the end time of each arrival of the TTF (Time To Failure) distribution.

Overhaul Schedule is carried out according to plan (both, duration and time begin), because overhaul schedule is a decision variable in this simulation. The assumption is, overhaul schedule is really well planned, so there is no delay in its execution. If the threshold has not been achieved but the arrival of overhaul has begun before PM, then repair time will be adjusted as long as the length of time for overhaul repair. This happens because the damage is not too severe and at this time it is very possible to control the duration of the repair.

Below, Table 2 is the possibility of the arrival of the failure and overhaul schedule so take effect to the duration of repair. Solid line represent time for overhaul and dash line represent failure time.

Table 2. Repair time in several condition.

No	Condition	Repair time	Explain
1	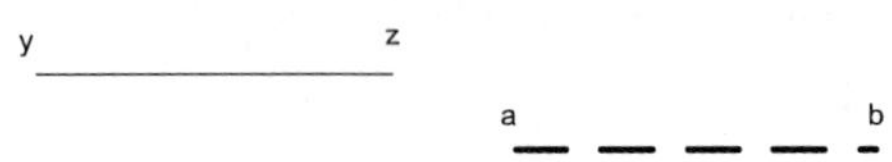	t(y) until t(z) and t(a) until t(b)	There is no intersection. Equipment off for two times, each according to the length of repair.
2	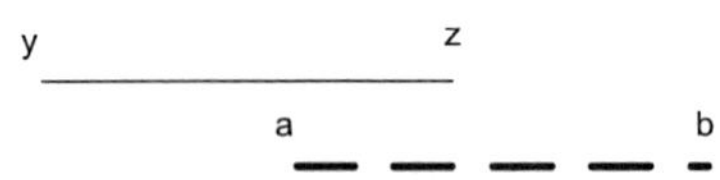	t(y) until t(z)	Failure arrival time is between of schedule overhaul. PM is occur.
3	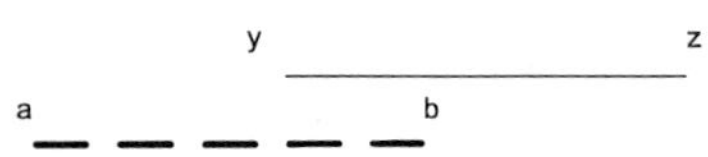	t(a) until t(z)	Failure arrival time before overhaul schedule.CM is occur. Overhaul execution remain as planned.
4	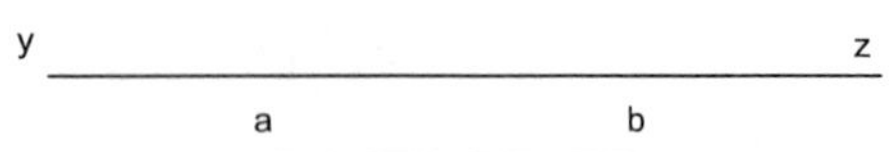	t(y) until t(z)	Failure arrival time between overhaul. Overhaul execute as planned.
5		t(a) until t(b)	Failure arrive before overhaul begin execute. Although CM is occur, overhaul activities remain as planned.

Table 3. Distribution probability TTF and TTR.

Equipment	TTF	TTR
Kiln 1(K1)	1+GAMM(349,0.793)	WEIB(12.2, 0.508)
Kiln 2 (K2)	WEIB(193, 0.798)	WEIB(8.02, 0.517)
Kiln 3 (K3)	EXPO(243)	WEIB(14.4, 0.522)
Kiln 4 (K4)	WEIB(176, 0.641)	WEIB(8.04, 0.478)

The entity in the simulation is running time. The time each hour is given an attachment attribute status of the machine condition. 0 for status off and 1 for status running. For failure arrival time, interval of two entities arrival follow as the distribution of TTF, then entity is hold, duration of hold module follow distribution of TTR. The arrival of overhaul also follows the same rule. When it is time for overhaul, the status attribute is set to 0.

Furthermore, the overhaul status and fail status are combined following the repair duration rules that have been defined in Table 2. Using the numerical algorithm in Table 4 , we will get information of when the system starts and finish runs, how long it runs, how long it shutdown. Finally, performance of the equipment can be calculated based on simulations.

Table 4. Combination fail and overhaul status result system status.

No	Fail status	Overhaul status	Additional condition	System status
1	0	0		0
2	1	1		1
3	1	0		0
4	0	1	$t(y) \leq t(a) \leq t(z)$	1
			else	0

5. Result and Discussion

Below, Table 6 is the result of simulation with 10 replications. The response variable observed are MTBF, Availability, and Production. Table 4 is the t test from the simulation result in Table 6, according to the t test with confidence interval α=5%, we can conclude that the model is valid.

Table 5. Hypothesis testing.

Hypothesis Testing $t\alpha/2=0.025$,df=42	MTBF		Availability		Production	
	REAL	SIM	REAL	SIM	REAL	SIM
Mean	272.4	248.9	7594.7	7897.7	2636409	2599029
Std. deviation	32.2	68.2	205.7	299.7	123281	124459
n	4	40	4	40	4	40
Sp	66.2955		293.9932		124376	
$t_{\alpha/2=0.025}$,df=42	2.0181		2.0181		2.0181	
t	-1.2887		1.9648		-0.5731	

Table 6. Up and down time data for kiln 1, 2015 until 2018.

No	UP	DOWN	No	UP	DOWN	No	UP	DOWN	No	UP	DOWN	No	UP	DOWN	No	UP	DOWN
1	43.8	0.5	21	62.0	415.2	41	411.5	4.9	61	319.7	5.7	81	6.7	210.3	101	100.3	163.2
2	1.1	3.8	22	48.9	0.1	42	389.7	15.6	62	107.0	1.2	82	73.5	9.5	102	239.3	7.5
3	79.6	0.5	23	44.9	1.7	43	40.0	119.6	63	357.3	20.2	83	129.7	2.8	103	617.2	0.7
4	157.8	0.6	24	514.5	0.8	44	175.7	9.3	64	660.7	1.5	84	625.9	1.1	104	2.9	6.0
5	287.3	2.4	25	75.8	4.0	45	295.1	2.8	65	285.9	0.9	85	63.4	48.7	105	409.6	31.1
6	210.7	0.4	26	769.8	0.8	46	189.2	37.1	66	163.9	9.5	86	566.6	2.2	106	3.4	53.6
7	271.8	1.1	27	523.7	22.6	47	321.7	0.3	67	616.5	0.9	87	1354.8	1.0	107	44.1	0.5
8	574.6	0.9	28	119.0	3.4	48	160.7	0.7	68	36.8	26.6	88	263.9	1.5	108	430.1	20.3
9	427.6	0.5	29	699.5	0.3	49	59.9	1.3	69	39.9	6.2	89	344.4	9.2	109	425.3	0.8
10	142.1	1.0	30	150.7	0.6	50	1110.1	13.7	70	436.4	9.4	90	343.6	17.3	110	2.3	24.0
11	153.4	15.1	31	27.9	4.5	51	132.5	38.0	71	184.1	0.9	91	58.8	616.7	111	44.0	2.8
12	366.2	0.9	32	98.4	25.0	52	2.9	16.4	72	1533.7	53.3	92	18.0	131.4	112	715.1	7.5
13	457.9	4.3	33	369.9	15.3	53	118.8	0.3	73	74.7	1.1	93	5.1	6.0	113	34.3	0.8
14	186.7	7.4	34	344.9	1.0	54	648.3	410.9	74	252.4	111.0	94	155.4	13.9			
15	85.3	9.7	35	89.9	1.9	55	11.3	1.9	75	105.9	2.1	95	23.3	0.8			
16	24.8	0.5	36	223.2	0.9	56	366.0	3.1	76	47.8	1.3	96	89.4	6.4			
17	152.0	2.3	37	179.6	9.1	57	392.6	2.9	77	423.0	5.0	97	1113.2	10.8			
18	558.3	2.1	38	339.8	215.8	58	25.4	120.4	78	50.8	15.9	98	1160.1	1.0			
19	218.1	1.1	39	185.5	6.0	59	298.1	1.4	79	81.3	0.5	99	3.0	38.5			
20	432.2	0.9	40	227.6	14.1	60	811.0	15.9	80	122.4	2.3	100	155.9	39.9			

Table 7. Comparison real data and 10 replication simulation result.

MTBF	REAL (Year : 2019)	SIMULATION (n=10 replication)									
		1	2	3	4	5	6	7	8	9	10
KILN 1	284.87	235.7	387.3	264.03	190.84	335.04	395.31	259.95	401.57	297.91	309.38
KILN 2	272.92	172.1	322.08	313.53	182.92	186.83	213.33	273.04	197.43	231.03	234.18
KILN 3	303.90	185.75	275.32	341.59	151.92	268.12	210.82	317.83	231.18	223.96	282.37
KILN 4	228.00	160.81	246.95	217.96	195.3	166.41	226.13	305	137.37	174.75	234.1

Availability	REAL (Year : 2019)	SIMULATION (n=10 replication)									
		1	2	3	4	5	6	7	8	9	10
Total Run Hours K1	7691.42	7365	7909	7996	7885	7882	7819	8008	7697	7885	7702
Total Run Hours K2	7641.90	8010	8443	8235	8399	8474	8055	8442	8238	8229	8485
Total Run Hours K3	7293.60	7727	7951	7621	7573	7976	7431	7870	7659	7797	7824
Total Run Hours K4	7752.02	7812	8083	7504	7852	7413	7749	8006	7530	7618	7752

Production	REAL (Year : 2019)	SIMULATION (n=10 replication)									
		1	2	3	4	5	6	7	8	9	10
Total Prod K1	2472548	2451977	2630726	2658854	2621861	2620879	2601447	2662069	2557852	2621573	2559685
Total Prod K2	2689274	2665484	2810152	2736581	2791391	2816472	2680428	2804833	2738498	2733910	2819735
Total Prod K3	2621946	2577470	2651705	2540984	2527119	2659580	2479083	2624690	2554519	2601076	2608725
Total Prod K4	2761869	2476982	2564775	2382286	2491053	2354124	2463317	2543504	2394529	2419626	2461603

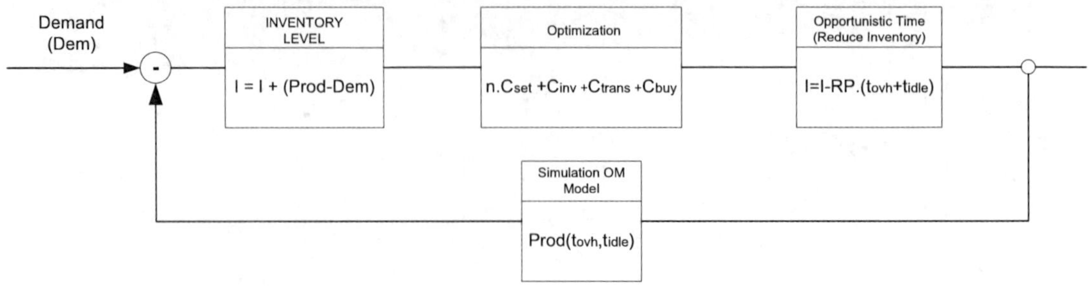

I	Inventory
Prod	Production
Dem	Demand
tovh	Time Overhaul
tidle	Time Idle
n	Frequency of setup
Cset	Cost of setup
Cinv	Cost of Inventory
Cbuy	Cost of buy from other company
Ctrans	Cost of transport product from other plant
RP	Rate of Production

Figure 4. Optimizing inventory level and production.

The simulation results show the OM model that was built is valid. It fit with the real conditions of the 2019 data. In 2019, the opportunity for maintenance time occurs because of the overhaul time interval. Production output is absorbed by the market and does not cause significant problems with the number of inventories. At the present time where demand can be low, it is necessary to pay attention to inventory so that production costs are not high. Inventory reduction is done by equipment shutdown (idle). Idle equipment condition is opportunistic time for maintenance. Equipment shutdown means that the setup costs are higher. Thus, there is a pay off between inventory costs and set up costs that must be minimized. Our proposed model combines the OM model simulation with a minimizing inventory level so that the overall cost of production can be minimized, as can be seen in Figure 4.

Under certain conditions, idle time is held in series with overhauls to reduce setup frequency. The optimization is to minimize existing inventory by regarding the amount of demand only. Other conditions that may occur are low set up costs but limited inventory by certain capacities. This happens as in equipment finish mill (other equipment systems in the cement industry). The optimization model is replaced with inventory limits according to inventory capacity. Thus, the idle time can be determined as opportunistic maintenance time.

A factory that has several production lines, if there is a shortage in a production line, it can be supplied from another line. However, this will drive transportation costs. If the buy option is made for reducing shortage, a purchase cost will appear. In low demand conditions, several lines must be turned off so that inventory is not too much. This means there is idle time. So, there are several optimization models that follow simulation OM model, as can be seen in Figure 4.

6. Conclusion

OM Simulation Model has been proven valid as a model that can be used. In very simple conditions, the opportunistic time is the equipment overhaul time interval only. With this situation, we can localize the validity of this model. However, in a real system, complexity increases when the opportunistic time does not only come from overhaul time but may appear due to minimization of inventory costs. Inventory is closely related to the production scheme. Production scheme strategies can be easily attached to simulations as compare to mathematical analysis. Figure 4 is an alternative combination of simulation and production scheme optimization models that produce low production costs. This model is not carried out in this paper because, as in the original goal, it only focused on Opportunistic Maintenance.

References

[1] E Lewis, *Introduction to Reliability Engineering*, Illinois: John Wiley & Sons, Inc., 1994.
[2] Wu S, Clements-Croome D. Preventive maintenance models with random maintenance quality. *Reliability Engineering & System Safety*. Elsevier BV; 2005 Oct;**90**(1):99–105. doi.org/10.1016/j.ress.2005.03.012
[3] Chang Q, Ni J, Bandyopadhyay P, Biller S, Xiao G. Maintenance Opportunity Planning System. *J Manuf Sci Eng*. ASME International; 2007 Feb 7;**129**(3):661–8. doi.org/10.1115/1.2716713
[4] Wu T, Ma X, Yang L, Zhao Y. Proactive maintenance scheduling in consideration of imperfect repairs and production wait time. *Journal of Manufacturing Systems*. Elsevier BV; 2019 Oct;**53**:183–94. doi.org/10.1016/j.jmsy.2019.09.011.
[5] Percy DF. Preventive Maintenance Models for Complex Systems. *Complex System Maintenance Handbook*. Springer London; 179–207. doi.org/10.1007/978-1-84800-011-7_8
[6] Alrabghi A, Tiwari A. A Review of Simulation-Based Optimisation in Maintenance Operations. 2013 UKSim 15th International Conference on Computer Modelling and Simulation. IEEE; 2013 Apr; doi.org/10.1109/uksim.2013.27
[7] Rezg N, Chelbi A, Xie X. Modeling and optimizing a joint inventory control and preventive maintenance strategy for a randomly failing production unit: Analytical and simulation approaches. *International Journal of Computer Integrated Manufacturing*. Informa UK Limited; 2005 Mar;**18**(2-3):225–35. doi.org/10.1080/0951192052000288152
[8] Duffuaa SO, Ben-Daya M, Al-Sultan KS, Andijani AA. A generic conceptual simulation model for maintenance systems. *J Qual Maint Eng*. Emerald; 2001 Sep;**7**(3):207–19. doi.org/10.1108/13552510110404512
[9] Ho SL, Xie M. The use of ARIMA models for reliability forecasting and analysis. *Computers & Industrial Engineering*. Elsevier BV; 1998 Oct;**35**(1-2):213–6doi.org/10.1016/s0360-8352(98)00066-7
[10] Dekker R. Applications of maintenance optimization models: a review and analysis. *Reliability Engineering & System Safety*. Elsevier BV; 1996 Mar;**51**(3):229–40. doi.org/10.1016/0951-8320(95)00076-3
[11] Garg A, Deshmukh SG. Maintenance management: literature review and directions. *J Qual Maint Eng*. Emerald; 2006 Jul;**12**(3):205–38. doi.org/10.1108/13552510610685075
[12] Rogers P. Optimum-seeking simulation in the design and control of manufacturing systems: experience with OptQuest for Arena. *Proceedings of the Winter Simulation Conference*. IEEE. doi.org/10.1109/wsc.2002.1166371
[13] Salsabila NY, Siswanto N, Widodo E, Rochmadhan OA. Throughput Analysis on a Multi-state Manufacturing System by Considering Availability. *Jurnal Teknik Industri*. Petra Christian University; 2019 Dec 16;**21**(2):69–78. doi.org/10.9744/jti.21.2.69-78
[14] Salsabila NY, Siswanto N. A Simulation study of availability analysis on a chemical process industry considering spare part inventory. *Mater. Sci. Eng. R Rep*. IOP Publishing; 2019 Dec 10;**673**:012079. doi.org/10.1088/1757-899x/673/1/012079

IOP Conf. Series: Materials Science and Engineering 1072 (2021) 012046 doi:10.1088/1757-899X/1072/1/012046

A safety-first portfolio selection framework: Estimating returns of exchange traded funds through regression analysis

M N Young[1]
School of Industrial Engineering and Engineering Management, Mapu'a University, Manila, Phillippines

E-mail: [1]mnyoung@mapua.edu.ph

Abstract. This research provides a possible variant of the basic framework of portfolio selection. First, for the return estimation, considering Asian exchange-traded funds as the selection pool for the investment, regression analysis is applied to generate future return scenarios. Second, for the assignment of weights, the likelihood of these scenarios are assumed to be equally likely. Third, for the portfolio selection model, the safety-first model is utilized with consideration of 5 different diversification factors. The proposed procedure is modeled as a linear programming model into AIMMS optimization software where the resulting portfolios show that safety-first portfolios can outperform the benchmark. These observations can pave the way to a new generic portfolio selection framework that can possibly help investors and may lead to alternative investment instruments.

1. Introduction

the basic framework of portfolio selection is composed of 3 parts namely return estimation, assignment of weights, and portfolio selection model. In the first part, this is where investment pools are identified and future returns are estimated based on the individual perception of investors on future performances of the market. Typically, one can use the historical returns as rough estimates for future returns or they can use sophisticated models to generate return scenarios e.g. regression equations and other technical forecasting techniques. After having these return estimates, the second part of portfolio selection is to provide appropriate weights to these future returns or return scenarios. This is still based on the individual perception of the investor where he/she provides the appropriate weights to the corresponding return scenarios. Commonly, equal weights can be given to returning scenarios or the SP/A (security, potential, and aspiration) theory of [9] can be applied wherein hopeful (fearful) investors will give more weights to favorable (unfavorable) scenarios. There are also other weighting schemes like prospect theory [10], cumulative prospect theory [11], etc. Now that we have the return estimates and corresponding probability weights, the chosen portfolio selection model can now be applied. Normally, the 2 most frequent models utilized are the mean-variance (MV) model [12] and safety-first (SF) model [13]. The basic idea in the MV model is that investors are assumed to be rational such that they will always choose the investment with the highest return (lowest risk) among choices with similar risk (return). In the MV model portfolio return is maximized and risk is minimized considering a risk-return trade-off parameter of the investor. SF model on the other hand is based on the logic that individuals will have their own risk level for corresponding mental frames of the investments [14] as such for each investment investor will have a different loss threshold and the loss probability threshold depending on the source of the invested capital. Thus, in the SF model portfolio return is maximized while limiting the probability of losing a certain percentage of the investment to a specified probability threshold. Thus, this research provides a simple variation of 3 parts of portfolio selection.

In today's investment world, an exchange-traded fund (ETF) is a relatively new investment instrument that aims to track a specific index. It is similar to a mutual fund (MF) in that
regard but ETF can be traded multiple times within a trading day like stocks while MF can only be traded once a day. Moreover, the transaction costs for ETF is relatively cheaper than stocks and MFs.

From 2008-2018, there is a steady increase in cash invested in ETFs. Looking at the world economic outlook, Asia is projected to overtake both US and UK in the year 2022 in terms of GDP. The probable reason according to [15] is a combination of political, regional & global integration, management of renewable sources, and other economic factors. Thus, [8] considered Asian ETFs as the investment pool. They collected market indices of 9 Asian countries and identified the benchmark index and the investment pool. China was considered as the benchmark because it has the largest market share while Malaysia, Singapore, Hong Kong, Indonesia, Korea, Japan, India, and Taiwan are considered into the investment pool. They used equally weighted historical returns as return scenarios and applied a safety-first portfolio selection model to get the optimal portfolios. Then, they compare portfolio performance to the benchmark. They show that their framework can outperform the benchmark and can possibly be a profitable alternative investment option for investors. Thus, this study will expound on their work by focusing more on the return estimation part.

In the work of [8], they consider only the historical returns, but the better estimation of returns may lead to a superior portfolio, so in this research, the work of [8] will be replicated but with the addition of the estimating and generating future return scenarios of the investment pool utilizing regression analysis. The assignment of weights will remain the same by using equiprobable return scenarios. As for the portfolio selection models, the generic safety-first portfolio selection model will be applied. The resulting portfolios will then be compared to the benchmark and also to the corresponding portfolios in [8].

This study flow is as follows. Section 1 provides the basic background of the study. Section 2 discusses the methodologies considered in the proposed portfolio selection framework. Section 3 shows the results and analysis of portfolio performance. Section 4 states the concluding remarks.

2. Methodology

This section discusses the regression equations used to generate the return scenarios, the portfolio selection model, and how the portfolios are evaluated.

2.1. Regression Analysis

As mentioned, similar to [7] and [8], the investment pool considered are 8 Asian ETFs namely Malaysia, Singapore, Hong Kong, Indonesia, Korea, Japan, India, and Taiwan. The benchmark is China ETF. The market index of these countries is collected and assumed that there are corresponding ETFs that perfectly tracts their performance. Thus, in estimating the returns of these ETFs, similar to [5] for each trading day regression analysis is applied to generate return scenarios. Let RM,t, and Ri,t respectively be the returns of the benchmark and ETF i at time t. Then through a single index model, we can calculate for the return of ETF i on the next period by the regression equation:

$$R_{i,t+1} = \beta_0 + \beta_1 R_{M,t+1} + \sigma_{ei} \quad (1)$$

Note that $R_{M,t+1}$ is estimated using EWMA or exponential moving average. σ_{ei} is the error term for ETF i. It is also ensured that residuals are normally distributed before using the corresponding regression equation to generate 500 return scenarios to be used in each trading day. These scenarios are assumed to have equal likelihood of occurrence. Most importantly, the scenarios are anchored based on 1 estimated return of the benchmark. This only mean that for every scenario there is a corresponding return estimate for the benchmark.

2.2. Portofolio Selection Model

To determine the optimal portfolios, the generic safety-first portfolio selection model is applied to the generated and equally weighted return scenarios of the 8 Asian ETFs. The resulting SF portfolios will then be compared to the benchmark (China Market Index) and the corresponding portfolios in [8]. SF model focuses on limiting the probability to lose a certain amount (depending on the investor) to a certain probability (also depending on the investor). Thus, assuming that there are n stocks and m scenarios, let portfolio Y be denoted as $Y = (y_1, y_2, ..., y_n)$, and let r_Y be the return of portfolio Y such that $\sum_{i=1}^{m} y_i = 1$. Let R_L and γ represent the loss threshold and loss probability threshold respectively, the generic SF model is represented as:

$$\max E[r_Y] \quad (2)$$

$$s.t.\quad P(r_Y \leq R_L) \leq \gamma, \quad (3)$$

Equation (2) maximizes the expected return of portfolio Y. $P(r_Y \leq R_L) \leq \gamma$ in Equation (3) is the portfolio's downside risk which will be limited to at most the value of γ.

Consider scenario j as a row vector of returns such that $(r_{1j}, r_{2j}, ..., r_{nj})$, R_{ij} is the return of ETF i on scenario j, Pj as the nominal probability weight on scenario j. Then, for scenario j, denote r_{Y_j} as the return of portfolio Y on scenario j such that $r_{Y_j} = \sum_{i=1}^{m} y_i R_{ij}$. In addition let I denote the threshold portfolio weight that can be allocated to ETF i, then the scenario-based SF model is formulated as:

$$\max E[r_Y] = \sum_{j=1}^{l} r_{Y_j} P_j \quad (4)$$

$$s.t.\quad r_{Y_j} = \sum_{i=1}^{m} y_i R_{ij}, j = 1,2, \dots, m \quad (5)$$

$$R_L - r_{Y_j} \leq M\omega_j, j = 1,2,...,m \quad (6)$$

$$\sum_{j=1}^{m} P_j \omega_j \leq \gamma \quad (7)$$

$$y_i \leq I \quad (8)$$

$$0 \leq y_i \leq 1,\ \omega_j \text{ is binary} \quad (9)$$

In summary, equation (4) reflects equation (2) as it maximizes the expected return of portfolio Y . Equation (5) shows how portfolio return is calculated at scenario j. Equations (6) & (6) reflect equation (3) as it ensures that the probability of losing R_L is limited to at most γ. Equation (8) makes sure that the SF portfolio only I of the total budget where $I = 0.2, 0.4, 0.6, 0.8, 1$.

2.3. Performance Evaluation

Following the comparison format of [1, 2, 3, 4, 5, 6, 7, 8] on evaluating portfolio performance, this study examines the return statistics like mean return, standard deviation, and cumulative returns. For a more rigorous testing pair-return difference t-tests are also applied. The null and alternative hypothesis are:

$$H_0 \text{ - paired portfolio returns} = 0 \quad (10)$$

$$H_1 \text{ - paired portfolio returns} > 0 \quad (11)$$

3. Back-test Results and Analysis

This section contains the data description and the back-test results.

3.1. Data Description

Data are from yahoo finance, where market indices of the 9 Asian countries are collected. These indices are then used to calculate for the corresponding ETFs return. The period of data collected is from 2015-September-1 to 2018-August-20 wherein the back-test period is from 2017-November-11 to 2018-August-20. In total 11 portfolios, 5 SF portfolios or *Pi* portfolios where *Pi* = {*P1, P2, P3, P4, P5*}, 5 counterpart SF portfolios from [8] or SF – I portfolios where *SF – I* = {*SF – 0.2, SF – 0.4, SF – 0.6, SF – 0.8, SF – 1*}, and 1 benchmark (China ETF) or M . Note that the counterpart portfolios are *P1* & *SF – 0.2*, *P2* & *SF – 0.4*, *P3* & *SF – 0.6*, *P4* & *SF – 0.6*, and *P5* & *SF – 1*. As i or I increases the portfolio becomes less diversified. *Pi* portfolios used 500 generated return scenarios and *SF – 1* portfolios used 400 historical returns as the return estimate for the next period estimates for each trading day. The safety-first parameters considered are RL = -5% and γ = 5%.

3.2. Back-test Results

Tables 1 and 2 show the respective performance of *Pi* and *SF – 1* portfolio. Let's first compare the

i

returns of *Pi* with the benchmark M. Looking at the mean returns ($\overline{R_{P_i}}$), it is noticeable that *Pi* portfolios have superior mean returns than M. Moreover, the corresponding risks ($\sigma_{R_{P_i}}$) are also smaller than M. These also translate with the cumulative return wherein *Pi* portfolios have higher CR_{Pi} and $\overline{CRP_i}$ than M. This is also clearly evident in Figure 1. Moreover, through pair-return difference with M, t-tests show that the returns of *P2* and *P4* are significantly larger than M with P-values of 0.1000 and 0.0900 respectively. In addition, *P1*, *P5*, and especially *P3* are very close to being significant. With these observations, it can be said that *Pi* portfolios can outperform the benchmark.

Table 1. Return statistics of P_i portfolios over 100 days test period.

	P1	P2	P3	P4	P5	M
$\overline{RPi}$	-0.0003	0.0001	0.0002	0.0004	0.0005	-0.0018
σR_{Pi}	0.0053	0.0069	0.0078	0.0089	0.0101	0.0149
CR_{Pi}	-0.0296	0.0052	0.0180	0.0380	0.0464	-0.1766
$\overline{CRPi}$	-0.0051	0.0083	0.0271	0.0351	0.0357	-0.0866
P-value with M	0.1480	0.1000*	0.1010	0.0900*	0.1500	
P-value with SF-I	0.9480	0.5770	0.5980	0.6820	0.7370	

$\overline{RP_i}$ denotes the mean return of portfolio Pi
σR_{Pi} denotes the standard deviation of the mean return of portfolio Pi
CR_{Pi} denotes the cumulative return of portfolio Pi
$\overline{CRP_i}$ denotes the mean cumulative return of portfolio Pi
P-value denotes the P-value of the pair-return difference T-test with M
* denotes significance at 0.1 level

Now, let's compare the *Pi* portfolios with their $SF - I$ counterpart. From the return statistics of respective portfolios from Tables 1 and 2, insufficient evidence is available to claim that the *Pi* portfolios are superior than the $SF - I$ portfolios. This is even supported by the resulting P-values of the pair-return t-test between *Pi* and $SF - I$ portfolios.

Table 2. Return statistics of SF – I portfolios over 100 days test period.

	SF-0.2	SF-0.4	SF-0.6	SF-0.8	SF-1	M
$\overline{R_{SF-I}}$	0.0002	0.0000	0.0000	0.0004	0.0008	-0.0018
σR_{SF-I}	0.0052	0.0066	0.0077	0.0087	0.0107	0.0149
CR_{SF-I}	0.0227	-0.0055	0.0020	0.0389	0.0745	-0.1766
$\overline{CR_{SF-I}}$	0.0312	0.0250	0.0261	0.0515	0.0755	-0.0866
P-value with M	0.0880*	0.1310	0.1360	0.1000*	0.0800*	

$\overline{R_{SF-I}}$ denotes the mean return of portfolio *SF – I*
σR_{SF-I} denotes the standard deviation of the mean return of portfolio *SF – I*
CR_{SF-I} denotes the cumulative return of portfolio *SF – I*
$\overline{CR_{SF-I}}$ denotes the mean cumulative return of portfolio *SF – I*
P-value denotes the P-value of the pair-return difference T-test with M
* denotes significance at 0.1 level

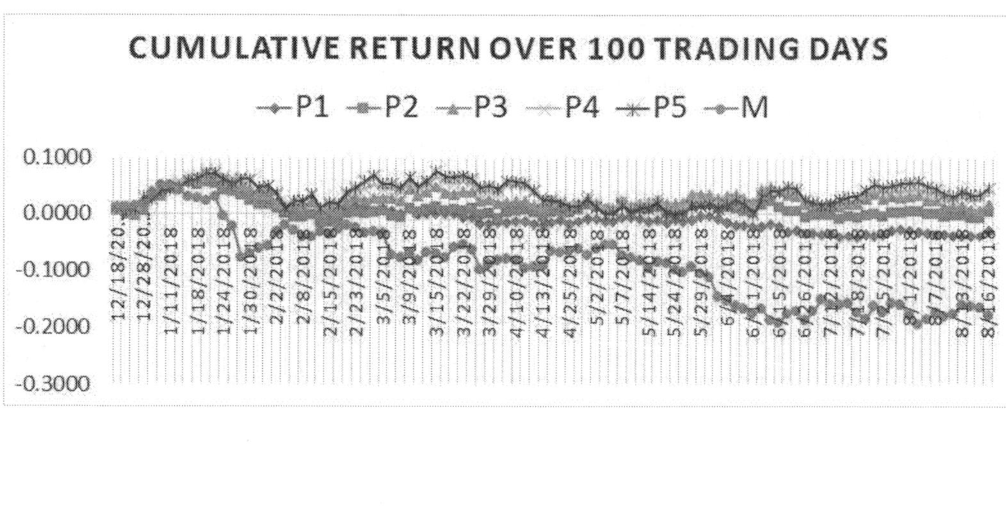

Figure 1. Cumulative returns of P_i portfolios.

Nonetheless, since both Pi and SF − I portfolios can both beat the benchmark M. Then Pi portfolios are still a good investment option. This proposed portfolio selection framework can still be a profitable investment for investors.

4. Conclusion

This research provides a variant of the basic framework of portfolio selection. The proposed framework utilizes Asian ETFs as the investment pool and then applies regression analysis to estimate and generate future return scenarios. On each trading day, the return scenarios are assumed to be equally likely which are then used in the safety-first portfolio selection model to obtain optimal portfolios. The resulting P_i portfolios are compared to the benchmark (M) and to corresponding SF portfolios $SF - I$ that uses historical return as return scenarios. Back-tests reveal that both P_i and $SF - I$ portfolios can outperform the benchmark M, but there is insufficient evidence to claim that P_i portfolios are better than $SF - I$ portfolios. Although this is the case, since P_i can outperform the benchmark, it can still be a profitable investment procedure for some investors.

This study focuses on finding superior methods in the return estimation part of the basic framework of portfolio selection. Although back-test results that the scenario generator technique employed does not show significant superiority over using historical data the research is very optimistic that more sophisticated methods can provide superior portfolios. Thus, this can be the future studies for this research. Other extensions can focus more on the assignment of probability weights to consider the fear and hope levels of the investors. Others can also develop unique models for the selection model. Lastly, finding the best possible combination of techniques for the 3 parts (return estimation, assignment of weights, selection model) of portfolio selection can be the ultimate goal of this research topic.

References

[1] Chang KH, Young MN 2019 Behavioral stock portfolio optimization considering holding periods of B-stocks with short-selling. *Comput Oper Res*. **1** p 112:104773.

[2] Chang KH, Young MN 2019 Portfolios Optimizations of Behavioral Stocks with Perception Probability Weightings. *Ann Econ Financ*. **20**(2).

[3] Chang KH, Young MN, Liu CC, Chung HP 2018 Behavioral Stock Portfolio Optimization through Short-Selling. *Int J Model Optim*. **8**(2) p 125-30.

[4] Chang KH, Young MN, Hildawa MI, Santos IJ, Pan CH 2015 Portfolio selection problem considering behavioral stocks. *In Proceedings of the World Congress on Engineering* 2015 (Vol. 2, pp. 1-3).

[5] Chang KH, Young MN, Diaz JF 2015 Portfolio Optimization Utilizing the Framework of Behavioral Portfolio Theory.. *Int J Model Optim*. **15**(1) p 1-3.

[6] Mercado MJ, Garcia MA, Ilagan JB, Reveche RR, Young MN 2019 An Empirical Analysis of Financial Ratio Trends of Several Companies Listed in the Philippine Stock Exchange. *In Conference Proceeding 10th Annual International Conference on Industrial Engineering and Operations Managemetn* (IEOM) 2019.

[7] Young MN, Chuahay TJTN 2019 Mean-Variance Portfolio Selection Utilizing Exchange Traded Funds in Asia. 2019 IEEE 6th *International Conference on Engineering Technologies and Applied Sciences* (ICETAS). IEEE; doi.org/10.1109/icetas48360.2019.9117564

[8] Young MN, Chuahay TT, Diaz JF 2019 Portfolio Selection Utilizing Safety-First Optimization Model on Exchange Traded Funds in Asia. *In Conference Proceeding 10th Annual International Conference on Industrial Engineering and Operations Management* (IEOM) 2019.

[9] Lopes LL 1987 Between Hope and Fear The Psychology of Risk. *Adv Exp Soc Psychol*. Elsevier; p 255–95. doi.org/10.1016/s0065-2601(08)60416-5

[10] Kahneman D, Tversky A. 1979Prospect Theory: An Analysis of Decision under Risk. *Econometrica*. JSTOR; **47**(2) p 263. doi.org/10.2307/1914185
[11] Tversky A, Kahneman D 1992 Advances in prospect theory: Cumulative representation of uncertainty. *J Risk Uncertain*. Springer Science and Business Media LLC; (4) p 297–323. doi.org/10.1007/bf00122574
[12] Markowitz H 1952 PORTFOLIO SELECTION*. *J Finance*. Wiley; **7**(1) p 77–91. doi.org/10.1111/j.1540-6261.1952.tb01525.x
[13] Roy AD 1952 Safety First and the Holding of Assets. *Econometrica*. JSTOR; **20**(3) p 431. doi.org/10.2307/1907413
[14] Shefrin H 2000 Statman M. Behavioral Portfolio Theory. *J. Financial Quant. Anal*. JSTOR; **35**(2 p 127. doi.org/10.2307/2676187
[15] Kim Y 2018 The Southeast Asian Economic Miracle. Routledge.

IOP Conf. Series: Materials Science and Engineering 1072 (2021) 012047 doi:10.1088/1757-899X/1072/1/012047

Economic evaluation of mobile screening station waterless system operation of nickel laterite ore

A Yustian[1], E Widodo[2] and M L Singgih[3]

[1,2,3]Department of Industrial and System Engineering, Institut Teknologi Sepuluh Nopember. Surabaya, Indonesia.

E-mail: [1]andi.bessi@vale.com

Abstract. The study investigates the economic level of operating the Mobile Screening Station Waterless System (MSSWS) for the first-time application in the nickel laterite mining industry. Three investment alternatives are compared in this study to determine the most economical option, i.e.: maintaining existing Fixed Screening Station/FSS (alternative one), purchasing MSSWS (alternative two), and purchasing new FSS (alternative three). Evaluation is made by performing cost savings analysis, increment cash flow projections, NPV, and IRR calculation. Furthermore, sensitivity analysis of the most economic option will be done as well. The results show that MSSWS gives the highest NPV and IRR. The decreasing of ore hauling distance and water content that are contributed by MSSWS yields significant cost savings in terms of reducing hauled trucks' investment and its operating cost. In accordance with this analysis, the economics of MSSWS is very sensitive to the changed of ore hauling distance. Therefore, this study has been successfully confirmed that purchasing of MSSWS is the most economical option. In addition to this result, a sensitivity analysis will be undertaken based on the situation that the moving of MSSWS over a certain period once the minimum change in the average ore hauling distance occurs.

1. Introduction

A common problem faced in lateritic nickel mining operations is the mining area that moves annually to a new area as the nickel reserves are depleted in the area. When mining moves to a new area causes ore hauling distance will increase farther from the existing fixed screening station. In this case of study, a mining area with ore hauling distance to the fixed screening station has been increased by an average hauling distance from 2 km to 7 km in the last 15 years. Besides, current fixed screening station technology uses a wet screening process (with the addition of water) to separate fractions of the size of -6 inches (containing higher nickel grade) and size of +6 inches (containing lower nickel grade). Thus, it causes water content increased in the screened ore. The increase in hauling distances results decreasing in the overall productivity of mining equipment and an increase in mining operating unit costs each year. Whereas the wet screening process of nickel laterite ore through the use of water causes increasing in carbon combustion energy of the nickel smelting process.

A study of a mobile screening station waterless system (MSSWS) is undertaken to replace the current conventional method of fixed screening station (FSS) with the use of water. MSSWS will move periodically for certain years according to the mine plan but not move perpetually or continuously at any time. MSSWS will increase mining operations flexibility and its productivity by reducing total material hauling distance and water content in nickel ore. So that it will affect on increasing mining costs efficiency over the life of mine.

This study is intended on how to formulate the investment cost and cost savings component resulted from MSSWS operation, and to figure tolerably risk that allows for investment decisions can be made. Three investment alternatives are compared in this study to determine the most economical option, i.e.: maintaining existing FSS (alternative 1), purchasing an MSSWS (alternative 2), and purchasing a new FSS (alternative 3). Evaluation is made by performing cost savings analysis, increment cash flow projections, NPV, and IRR calculation. Furthermore, sensitivity analysis of the most economic option will be done as well.

2. Mining and Screening Process

2.1 Mining Process

After the ore layer is exposed, the ore is mined by excavators and hauled by trucks to the screening station. The screening station will separate the size of +6 "(oversized) fraction which has a nickel grade below 1.5% and a fraction size of -6" (undersized) which has a nickel grade above 1.5%. Whenever fraction size above +6-inch contents nickel grade above 1.5% or a requirement for ore blending to have on specification ore chemistry, it is fed to the crushing system to produce up to size -1 inch. Then, the crushed material will be mixed with the screen products -6 inches. The screening station product (SSP) is loaded with a loader and hauled by truck to the Wet Ore Stockpile (WOS) area. While rejected +6 materials will be hauled to use for mine road construction. MSSWS operations to replace FSS yields overall hauling distance saving by 5 km ore hauling distance. The mining process and the change of overall material hauling distance by MSSWS operation are shown in Figure 1.

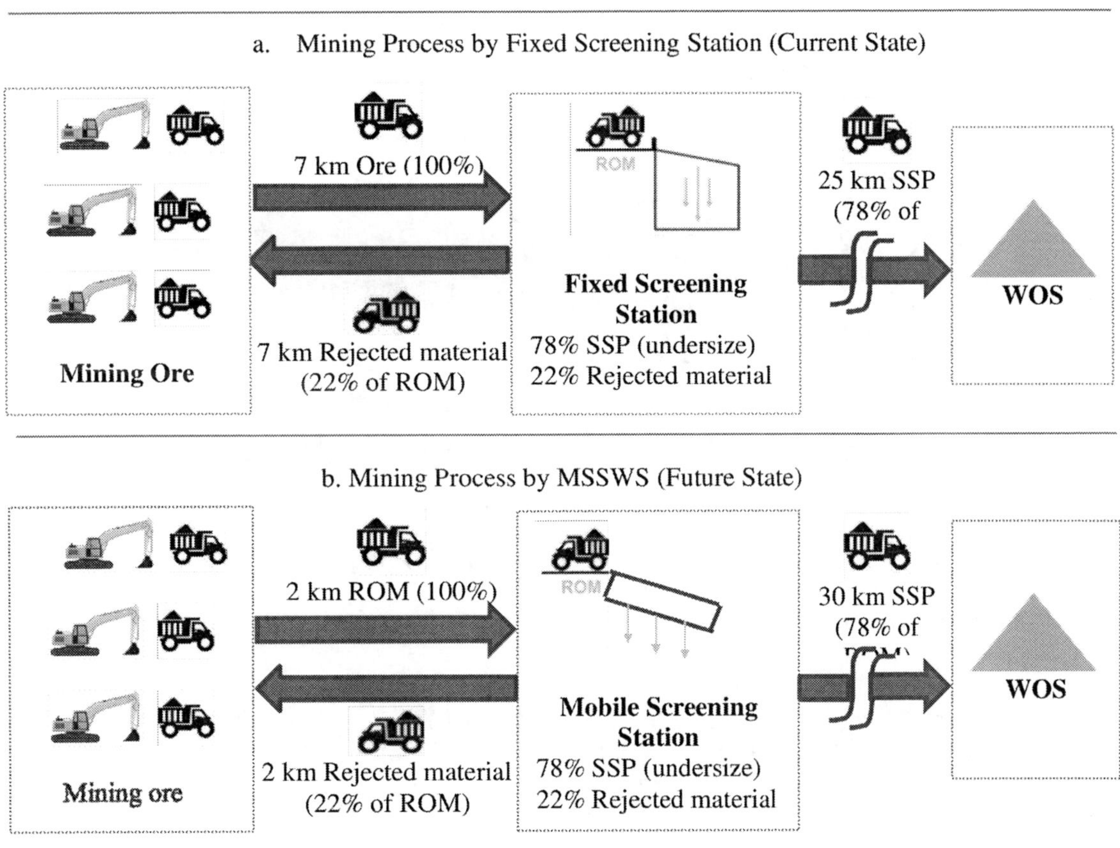

Figure 1. Mining process and material hauling distance.

2.2 Mobile Screening Station Waterless System

The main component parts and material flow of the mobile screening station is shown in Figure 2 [2] and summarized references as follows [1]:

• Bench retaining panels - Fitted to the rear of the heavy-duty main chassis frame, mounted below the feed hopper section and designed for an easy solution to assist with loading, ramp stability, site installation, and reliability. Fully braced and bolted to the main skid – mobile chassis frame.
• Dump truck reversing stop bar - Heavy-Duty and located at the tipping point.
• Main feed hopper with a holding a capacity of 90 m^3 (180mT), designed to accept ROM of -2,000mm material from a dump truck.
• Reciprocating feeder system - Designed so that the feeder table never fully empties or ejects all the material. The feeder provides a deep bed of material, cushioning the impact and load of any large material tipped into the main feed hopper, protecting the feeder table. Any sticky material is consistently pushed into the barrel with no surging or barrel overloading effect.
• Primary Screening Section - Fully welded and robust 2.5m diameter x 11.5m long barrel screening section. Designed for efficiently screening out the -6" ore natural feed material. Will perform either with a dry feed and/or a high moisture content feed.
• Secondary Screening section - Fully welded and robust 2.5m diameter x 3m long barrel screening section for screening out from +6" to - 18" ore material. The flow of material screened at each fraction is ejected gravity from the barrel screen through the respective output channel (chute).
• Self-Cleaning Barrel System - A single free-running rotating self-cleaning system is mounted above the first 3m screening section to assist with any potential pegging within the barrel apertures. The rotating cleaner pushes any sticky material back into the barrel for processing on its return cycle.

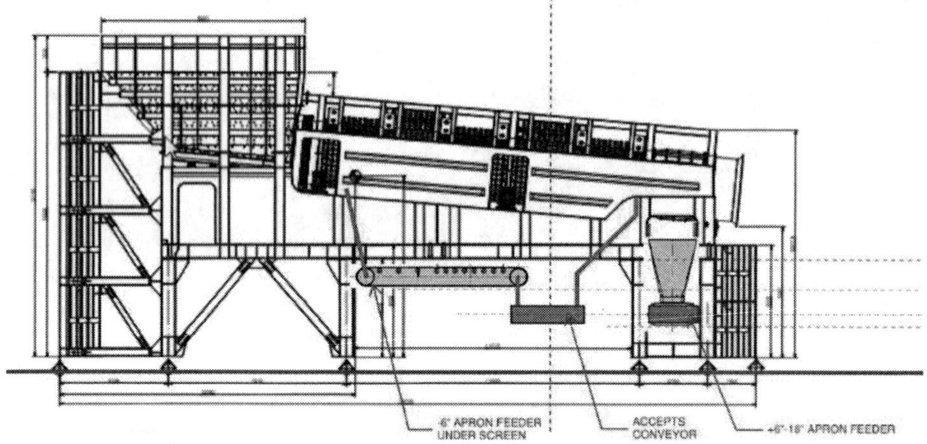

Figure 2. Modular screening station design.

A comparison of screening technology between FSS and MSSWS is shown in Table 1 and Figure 3

Table 1. Screening Technology Comparison.

Description	Fixed Screening Station	MSSWS
Screening Technology	Static grizzly and vibrating steps bar screen (Primary and Secondary Screen)	A barrel of rotary barrel screen (Primary and Secondary Screen)
Static grizzly screening	Yes	No
Rotary barrel screen	No	Yes
Mobile	No	Yes
Feed hopper	Yes	Yes
Transfer conveyors to screen	Yes	No
Stacker belt Conveyor	Yes	Yes
Water added	Yes	No
Feeding capacity	750 Ton/hour	750 Ton/hour

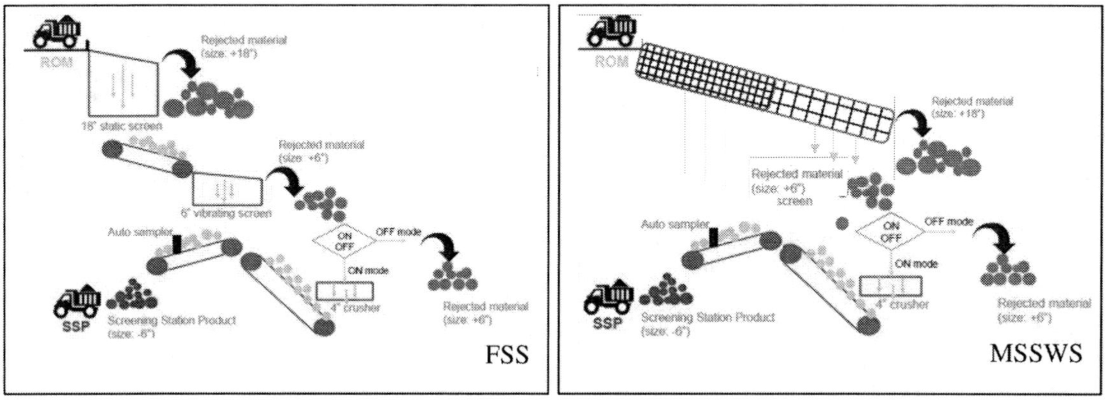

Figure 3. Screening process of FSS and MSSWS.

2.3 Preliminary study MSSWS

MSSWS is a technology that has been applied in mineral rock mining that operates a crushing system to produce aggregate products. However, the screening waterless system has not been applied to screen alluvial or lateritic ore. So that, this MSSWS operation is the first application to the nickel laterite mining industry.

Scoping study (FEL 2) conducted a mini MSSWS trial study with 150 tons/hour feeding rate capacity to examine critical technical and operational issues to screen 6,000 WMT laterite ore samples shows:

- The rotary barrel screen system at mini MSSWS works well to screen laterite nickel ore until -1-inch grain size without added water.
- Rotary barrel screen system at mini MSSWS performs better-screened recovery of -6 inches fraction to compare with FSS. Whereas it requires a longer dimension of rotary barrel screening to produce a fraction until -1 inches.
- The rotary barrel screen system at mini MSSWS produces insignificantly ore dilution both size fraction -6 inches and -1 inches.
- The water content of final products (SSP) by mini MSSWS is lower than FSS. However, the water content is influenced by weather and ore section profile.

3. Economic Evaluation

3.1 Parameter Assumptions

- Depreciation. Based on a statistical analysis of maintenance costs during the operating period of the screening station shows a sharp increase in the cost of screening station maintenance (on average more than doubled) from the 8th year to the 12th year. It occurs due to the span of years had been updated on the main components and building structure reinforcement which causes high maintenance costs in that period. Based on these conditions, the estimated economic life of MSSWS operating that will be used in the analysis of cash flow projections is for an operating of 8 years. While the method and amount of depreciation used in this research are the 25% declined balance of depreciation method [3].
- Income Tax. General rates apply to domestic taxpayers, as well as foreign taxpayers who conduct business or activities in Indonesia through a permanent establishment in Indonesia, are based on the taxable income layer. The amount of the tariff is 25% according to Indonesian government regulation [4].
- Discount rate. Uses the weighted average capital cost (WACC) value which is the cost of capital of a company consisting of shares, use of debt, and retained earnings. The indicator used to measure the variable capital costs is to calculate the WACC using the formula [5]:

$$WACC_{nominal} = D/V.\ K_d.\ (1\text{-}T_c) + E/V.\ K_e \quad (3.1)$$

Where,
D = Market value of the firm's debt
E = Market value of the firm's equity
V = D + E = Total market value of the firm's financing
D/V = Percentage of financing that is debt (30%, internal corporate data)
E/V = Percentage of financing that is equity (70%, internal corporate data)
T_c = Corporate tax rate (34% in Brazil)
K_d = Cost of debt = 3.9% [6]
K_e = Cost of equity

While the cost of equity (K_e) is calculated by the formula [5]:

$$K_e = R_f + \text{CDS} + \beta_s\,(R_m - R_f) \quad (3.2)$$
$$K_e = 2.92\% + 1.59\% + 1.31 \times 4.66\% = 10.6\%$$

Where:
- R_f = Risk-free returns expected on the market = 2.92% [7].
- β_s = Sensitivity to market risks from the average beta of the mining sector = 1.31 [8]
- CDS = Country Default Spread of Indonesia = 1.59% [9]
- (Rm – Rf) = The historical risk premium from 1928-2018 = 4.66% [10]

Then the nominal WACC calculation obtained parameter data from the company in 2019 as follows:

$$WACC_{nominal} = D/V.\ Kd.\ (1\text{-}Tc) + E/V.\ Ke$$
$$WACC_{nominal} = 30\%.\ 3.9\%.\ (1\text{-}34\%) + 70\%.\ 10.6\% = 8.2\%$$

With inflation of 2.13% [11], we get:
$WACC_{real}$ = WACCnominal – inflation
$WACC_{real}$ = 8.20% - 2.13% = 6.07% or round to 6.00%

Based on these calculations, the discount rate to be used in this economic analysis is the same as the $WACC_{real}$ value of 6.00%.

3.2 Cost Analysis

All unit cost components are based on 2019 actual data. MSSWS investment cost is based on the 2019/2020 actual data. Cost savings identified in several activities include:

• Ore hauling cost to the screening station by truck is summarized in Table 2. MSSWS results in a reduction of the average ore hauling distance to the screening station from 7 km to 2 km over an 8-years period. Whereas the new FSS results in an average hauling distance reduction of 4 km.

Table 2. Ore hauling cost summary.

Alternative	Distance (a)	Velocity (b)	Ore Tonnage (c)	Truck Load (d)	Unit Cost (e)	Cost $(\frac{2.a}{b} \cdot \frac{c}{d} \cdot e)$
1 (existing FSS)	7 Km	15 Km/hr	3,416,400 T	87 T	$102/hr	$ 3,738,402
2 (New MSSWS)	2 Km					$ 1,068,144
3 (New FSS)	4 Km					$ 2,136,186

• The hauling cost of screened ore (SSP) to stockpile (WOS) is summarized in Table 3. MSSWS results in an increase of the average SSP hauling distance from 25 km to 30 km over an 8-years period and a 4% reduction of water content in SSP tonnage. Whereas the new FSS results in an average hauling distance reduction of 28 km.

Table 3. SSP hauling cost summary.

Alternative	Distance (a)	Velocity (b)	SSP Tonnage (c)	Truck Load (d)	Unit Cost (e)	Cost $(2 \cdot \frac{a}{b} \cdot \frac{c}{d} \cdot e)$
1 (existing FSS)	25 Km		2,454,223 T			$6,527,749
2 (New MSSWS)	30 Km	26 Km/hr	2,356,054 T	84 T	$107/hr	$ 7,519,960
3 (New FSS)	28 Km		2,454,223 T			$ 7,311,096

• The hauling cost of rejected material +6 " is summarized in table 4. MSSWS results in a reduction of the average rejected material hauling distance from 7 km to 2 km over an 8-years period. Whereas the new FSS results in an average hauling distance reduction of 4 km.

Table 4. Rejected material +6" hauling cost summary.

Alternative	Distance (a)	Velocity (b)	Ore Tonnage (c)	Truck Load (d)	Unit Cost (e)	Cost $(2 \cdot \frac{a}{b} \cdot \frac{c}{d} \cdot e)$
1 (existing FSS)	7 Km					$753,168
2 (New MSSWS)	2 Km	15 Km/hr	751,608 T	95 T	$102/hr	$ 215,220
3 (New FSS)	4 Km					$ 430,440

• Ore screening cost. Due to the time limitations, the MSSWS screening operational costs is assumed to be the same as the FSS cost.
• Relocation and preparation costs of the MSSWS area are conducted every 3 years or twice during the lifetime of the equipment for 8 years with an estimated cost per relocation of $ 1,500,000
• The reduction in overall material hauling distance results in a reduction of a total truck working hours which has the potential to save on investment cost for replacing equivalent with 3 units of trucks for MSSWS operation and 2 units of trucks for new FSS.

Table 5. Cost analysis summary.

Cost Component	Alternative 1	Alternative 2	Alternative 3
Investment cost	None	$10,249,700	$11,442,734
Ore hauling cost	$ 3,738,402	$ 1,068,144	$ 2,136,186
Rejected material hauling cost	$753,168	$ 215,220	$ 430,440
SSP hauling cost	$6,527,749	$ 7,519,960	$ 7,311,096
Screening cost	$546,524	$546,524	$546,524
Relocation cost	0	$3,000,000	0
Truck Investment cost saving	0	$3,000,000	$2,000,000

Where: maintaining existing FSS (alternative 1), purchasing a MSSWS (alternative 2), and purchasing a new FSS (alternative 3).

3.3 Cash Flow Projection and Analysis

Pro Projection and analysis of MSSWS cash flow results in positive NPV of $ 3,269,700 and IRR of 17.4% with a payback period of 4.0 years. Thus, the purchasing and operation of MSSWS are feasible to run because it brings profits to the company where NPV value above zero and IRR value above the required criteria of 6%.

Otherwise, cash flow analysis of a new FSS results in a negative NPV value of - $1,960,600 and an IRR of 0.0%. So that the purchasing and operation of new FSS are not feasible to run because it brings harm to the company where NPV value < 0 and IRR value below the required criteria of 6%. It happens because the new fixed screening station is not designed to move periodically over 8-years period of a lifetime as MSSWS. So that the cost savings by reducing the average ore hauling distance that can be optimized to 4 km over the lifetime of new FSS are inadequate to overcome the investment costs required. Whenever a structure of the new FSS can be modified by a modular system so that it is possible to move periodically over 8-years period, then it results in a positive NPV of $ 845,900 and IRR of 8.7% with a payback period of 5.5 years. However, the NPV and IRR are still lower than MSSWS NPV of $3,269,700 and IRR of 17.4%. Caused the investment cost of MSSWS is lower than the investment cost of new FSS and as well as a contribution to savings from reducing the water content in the SSP by MSSWS. MSSWS screening process is quietly simple where the primary and secondary screening are tided in one rotary barrel screen that provides a relatively smaller investment cost savings compared to the stratified screening system of FSS from the grizzly station, primary, and secondary decks screening which are connected to each other's by transfer belt conveyors system.

3.4 Sensitivity Analysis

The risk of MSSWS investment cost changes can be caused by increasing commodity prices, various modifications needed to fit operational requirements, and the addition of earthwork material due to inaccuracy in topographic conditions.

Sensitivity analysis is undertaken by changing mining operational unit costs show breakeven point (NPV =0) when mining unit costs decreased around 30%. However, the possibility of decreasing the mining unit cost up to 30% for 8 years is unlikely in regard to inflation and commodity price escalation so that it may be neglected. Sensitivity analysis is undertaken by changing investment cost shows a breakeven point when investment cost increased around 40%. Likewise, the possibility is unlikely unless otherwise there is a major changed to meet operational required.

Sensitivity analysis is undertaken with a fixed discount rate, investment, and mining costs by changing in the ore hauling distance due to changes in the mining plan results in breakeven points when hauling distance increase of 62.5% or ore hauling distance becomes about 3.25 km. Sensitivity analysis by changing water content in SSP shows a relatively small affected on NPV and mining costs. Whenever MSSWS is assumed unsuccessful in reducing the water content in SSP yields NPV value is still positive of $1,810,100 with IRR of 12.6%. Likewise, the sensitivity analysis of the discount rate shows a relatively small affected on the economic value of MSSWS. Whereas the breakeven point occurs at the same discount rate as the IRR yielded in alternative 2, which is about 17.4%.

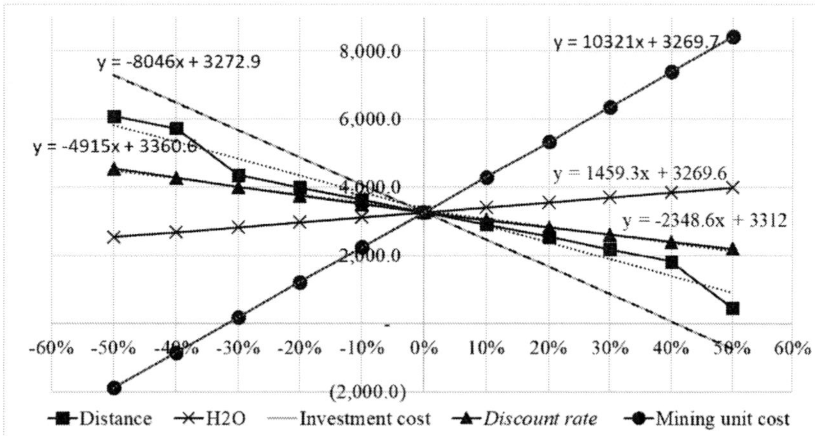

Figure 4. MSSWS sensitivity analysis.

4. Conclusion

An economic evaluation of some selected alternatives shows that purchasing a mobile screening station waterless system (alternative 2) is the most economical option that yields the highest NPV of $3,269,700 and IRR of 17.4% with a payback period of 4 years. The decreasing of ore hauling distance is contributed by MSSWS yields to a significant cost saving in terms of reducing hauled trucks' investment and its operating cost. In accordance with sensitivity analysis, the economics of MSSWS is very sensitive to the changed of ore hauling distance that is likely to occur because of change in the mine plan. In addition, the major changes in investment cost and mine operational cost may be unlikely, but it is still possible to occur that affects the economical of MSSWS. To overcome these associated risks, it is proposed to develop an accurate mine plan, detail MSSWS engineering design, and well-maturing level on the assessment of MSSWS designed capacity to meet with the mine production rate plan. Furthermore, MSSWS yields on reducing the water content of ore that has potential cost saving in energy efficiency for the next ore drying process with High-Speed Diesel (HSD) in kiln drying has not been included in this study. It is recommended to develop a further study to assess the additional economic value of MSSWS that is implied energy combustion cost-saving at drying kiln. The framework presented in this paper is appropriated for the nickel laterite ore at this particular area. Therefore, further study should be required whenever it will be applied in a different geological environment of nickel ore laterite in others area.

References

[1] Jebsen & Jessen 2017 *Technical Proposals for the Supply of Mobile Screening and Stockpiling Equipment*, page 9-11.

[2] PT Bimatekno Karyatama consultant 2019 *Modular Screening Station Design Basis*, Appendix B– Concept Sketches, page 7.

[3] Law (UU) 36 2008 *the Fourth Amendment of Law No.7 of 1983 about Income Tax*, Chapter 11, Article 6, page 18.

[4] Government Regulation (PP) 36 2017 *Income Tax Taken on Certain Income in the Form of Net Assets or Considered as Income*, Chapter 4, Article (1a).

[5] Marshall Hargrave 2019 *Weighted Average Cost of Capital – WACC*, https://www.investopedia.com/terms/w/wacc.asp.

[6] Internal Corporate Data. *Average Yield of the Last Two Years from Global Bonds of Large Mining Companies with Maturing of 2025-2026.*

[7] Trading Economics. *United States 30 Years Bond Yield 1977- 2019 Data*, https://tradingeconomics.com/united-states/30-year-bond-yield

[8] Damodaran, Aswath. January 2020, *Betas by Sector (US)*,

http://pages.stern.nyu.edu/~adamodar/New_Home_Page/datafile/Betas.html
[9] Damodaran, Aswath. January 2020, *Country Default Spreads and Risk Premiums*, http://pages.stern.nyu.edu/~adamodar/New_Home_Page/datafile/ctryprem.html
[10] Damodaran, Aswath. April 2019, *Equity Risk Premiums (ERP): Determinants, Estimation and Implications*, The 2019 Edition, page 34-35, https://papers.ssrn.com/sol3/papers.cfm?abstract_id=3378246
[11] Trading Economics. *United States Inflation Rate 1914-2020 Data*, https://tradingeconomics.com/united-states/inflation-cpi

Empty container reposition using max-min review system: Simulation approach

A Budipriyanto[1], M D Novianti[2] and T Susanto[3]

[1, 2, 3]Industrial Engineering Department, Bakrie University, Jakarta, Indonesia

E-mail: [1]adi.budipriyanto@bakrie.ac.id

Abstract. Global economic conditions and differences in the ability and economic competitiveness of each country cause an imbalance between exports and imports. Shipping of goods is carried out mostly by sea transportation through the use of containers and involves the port as the exit and entry of goods. An imbalance in export-import causes the number of containers in some ports to have a surplus, while others have a deficit. To maintain container availability, shippinglines must adjust the balance by repositioning from a surplus port to a deficit port. Shippinglines must determine the inventory to meet the needs at the port and the number of containers that can be repositioned to the other ports that need it. If the inventory at the port reaches a minimum level, then the order must be immediately carried out, on the contrary, it will be repositioned to another port that needs it. By adopting the concept of a review system (U, D), when the inventory is less than U, an order of Q is immediately placed. But if the inventory is greater or equal to D, repositioning is carried out. Repositioning is done by I-D. In this study, the maximum inventory value is determined based on the value of the inventory service level of 95%. From the simulation results it was concluded that the min-max review system policy with a service level of 95% could reduce inventory by 35.64% and reduce shortage by 8.82%.

1. Introduction

Global economic conditions and differences in the ability and economic competitiveness of each country cause an imbalance between exports and imports [1], [2]. Currently shipments of goods between countries are mostly carried out using sea transportation and shipped using containers [1], [3]–[6]. Based on UNCTAD data in 2011, containers sent from Asia to North America amounted to 12.4 million TEUs, while from North America to Asia only 6.6 million TEUs. In 2012 containers shipped from Asia to North America increased to 13.3 million TEUs, while containers shipped from North America to Asia were only 6.9 million TEUs. [7], [8]. If the port is considered as the origin and destination point, the imbalance of export-import causes a port to become a surplus, while the other port has a deficit.

Shippinglines fill containers from inventory in the form of empty containers owned by shippinglines at the port. The empty container came from a laden container that was sent from another port in the previous period and has been returned to shippinglines in the form of an empty container. The laden container needs time before returning to shippinglines. The location of the return of the container is usually determined by shippinglines, at the depot or at the port. If an empty container is not available shippinglines will rent or request it to be sent from another port. And vice versa, if the inventory at a port is greater than the demand, then the excess empty containers will be sent to the other ports that need them. Shipping empty containers to other ports is known as repositioning. Repositioning the container from the origin port to the destination port requires a lot of time and money.

The cost of repositioning is almost the same as the cost of sending a laden container [2], [9]. The cost of repositioning is high [10], The cost of repositioning is quite high, consisting of storage costs and transportation costs, and other costs that are not different from the shipping costs of laden containers. [2], [9], [11]. Repositioning an empty container requires the same slot as the laden container, so that it will reduce the capacity to carry the laden container. All repositioning costs are borne by shippinglines. [12]. Therefore to meet the demand for containers, shipping lines must determine the right strategy in implementing the repositioning of empty containers and determining the appropriate inventory.

2. Literature Review

Empty container reposition is a crucial problem for shipping lines [13]. According to [2], 20% of the containers sent were empty containers [2]. Empty container repositioning costs almost the same as sending a laden container but it does not provide added value [14], [15]. The ability to efficiently reposition empty containers can create competitive advantages for shippinglines [2]. The right repositioning strategy can increase service levels and increase productivity in container management [16].

Research on repositioning empty containers by considering laden containers simultaneously was developed by [17]. They formulate it as a two-stage problem. Researcher [18] developed a mixed fuzzy decision making model. They solve this problem with two stages, first using fuzzy backorder quantity inventory decision making, the second stage using optimizing mathematical programming network.

Some researchers develop a repositioning model by considering transportation topology. The single threshold policy model using a hub-and-spoke system was developed by [19], while [20] developed a two-threshold policy for a single port, then expanded to a two-treshold policy for a multy port system [20]. The threshold system policy model was expanded to become a hub-and-spoke transport system through a dynamic approach [2].

The empty ontainer repositioning with pricing strategy was developed by [21]–[23]. Acording to [23], pricing strategy can be applied if demand reaches a certain threshold, whereas [21], [22] uses the assumption of the effect of pricing on demand.

Empty container reposition multi commodity, develoved by [24]. The repositioned of empty container consists of good containers and damaged containers, so they need to be repaired before being reused by the next customer. This study considers the probability of a damaged container and the time needed to repair it.

The issue of green supply chain in empty containers was initiated by [25], [26]. To support the implementation of the green supply chain, repositioning can be done with collaboration between shippinglines. For collaboration to be effective, it must consider empty container inventory at each port. Repositioning by means of collaboration between shippinglines was developed by [27] by determining the perceived values of empty containers at each port. This model is solved using a two-stage optimization method. A study of the effect of repositioning on CO_2 emissions was carried out by [28]. The study results show that an efficient empty container reposition can reduce CO_2 emissions significantly.

The combination of cargo routing model with ECR multi services routes, multiple deployed vessels and multiple regular voyages was developed by [29]. The model was solved using two approaches. The first uses the shortest-path based integer programming method, the second uses the two-stage heuristic-rules based integer programming method. A study conducted by [30] was carried out to investigate intermodal repositioning with the aim of minimizing the distribution costs of empty containers. In this model two conditions are considered, first the demand for empty containers is fulfilled from inventory at the port, the second the demands are fulfilled from repositioning.

Returning empty containers can be done directly to the port or returned to the depot. Determining the exact location of returns and the number of depots can minimize costs and reduce the distance of container moving. Research in this area seeks to determine the location of empty container storage, minimize total costs and reduce the distance of container movement. Some models develop flexible depot [31], multi depot [16], and reposition with policy review [20], while [32] develop street-turn models. The model developed by [32] was criticized by [31] because it was difficult to apply. The repositioning of empty containers by considering the route, capacity and frequency of services developed by [33], [34]. They built two models in hinterland transportation, in the first model the ship was allowed to transport containers from different shipping lines, while the second model only transported containers belonging to the shippinglines themselves.

The repositioning of the multi-service route, multi-deployed vessel, and multi-voyage models was developed by [29], [35]. The model is divided into three groups, direct shipping, direct hub and spoke with one transhipment, and direct hub and spoke with two transshipment. The model was completed using two methods, namely the two-stage shortest-path base integer programming methods, and the two-stage heuristic-rules base integer programming methods. To determine the route and number of repositioned containers, researcher [29] uses the feedback control policy method. Researchers [13] developed a model by distinguishing ship sizes, while [14] developed a combination of hub-and-spoke and multi-port-calling models. According to [36] the location of the shipper is not always close to the port, therefore to meet the demand one must consider the location of the nearest depot [36].

Empty container reposition model developed by [2] is different from other models. In this model, the ship functions as an inventory warehouse. Empty container is loaded without specifying the destination port and container is unloaded at the port that needs it. They concluded that flexible destination ports provide better results, but the model can only be applied to medium size vessels, for small and large size vessels, the application of flexible destination port is not significant.

The empty container repositioning uses the inventory review policy approach developed by [20], [37]. Review policy (s,S) developed by [16], while [37] developed an inventory review policy (U, D). Both models use the zero lead time assumption for container leasing.

Repositioning by considering uncertainty was developed by [38]. They use uncertainty parameters, namely demand, supply, residual ship space capacity, and residual ship weight capacity. The sample average approximation method is used to solve small-scale stochastic problems and for large-scale use progressive hedging based algorithms. A model to determine the location of empty container depots by considering inventory uncertainty, supply uncertainty, and demand uncertainty was developed by [10]. The solution uses a two-stage stochastic modeling approach. Researchers [39] developed a multi scenario scenario model for demand uncertainty and supply uncertainty. They developed a hub and feeder model where containers can reach all ports. A cut-and-runs policy is applied to the loading-unloading process.

Other researchers have developed a DSS model for repositioning empty containers where demand and supply are calculated under two conditions, namely deterministic and dynamic conditions [1]. Whereas [9] developed a repositioning model for refrigerator containers.

Technological advances have succeeded in creating containers that can be folded so that repositioning does not require a large place. Some studies related to foldable include [40]–[43]. They developed a mathematical model to minimize the cost of repositioning both foldable and unfoldable containers. Several scenarios were developed and numerical experiments were made to compare the total cost between foldable and unfoldable. Sensitivity analysis is carried out to test changes in purchasing costs and transportation costs for foldable containers. To determine the perception of leasing prices, researcher [44] developed two-stage optimization methods. The first step is to develop a repositioning model by shippinglines. Based on these problems, then used to examine the use of foldable containers. The second stage, based on the solution of the first stage, inverse optimization techniques are used to determine prices at various ports. A mixed-integer non-linear programming model was developed based on the candidate route obtained from the first stage. Then numerical experiments are used to test the effectiveness of the two-stage optimization methods model.

The container planning model with a multi-port multi-period approach to foldable / unfoldable containers was developed by [42]. They build models with the aim of minimizing purchasing costs, repositioning costs and storage costs. Whereas [43], developed a network flow model method for repositioning empty containers by considering foldable containers. They implement two levels of decision, tactical and operational simultaneously. Decisions at the tactical level are needed to determine the route and capacity, while at the operational level it is used to determine the repositioning of containers between terminals.

3. Problem Description

Figure 1 shows the shipping route of the shipping lines from port A to port D and back to port A. The ship takes the route from port A to port D through port B and port C (route: A-B-C-D). From port D, the ship returns to port A goes through port E-F-B and ends at port A (route: D-E-F-B-A).

The ship voyage through the same route so that each port (node) can be function as an initial port (origin) and at the same time as a destination port (destination). For example, if port A is considered as the origin port, then the destination port is port B, C, D, E, and F. Port B is the origin port, while port C, D, E, F, and A will be function as the port of destination, and so on.

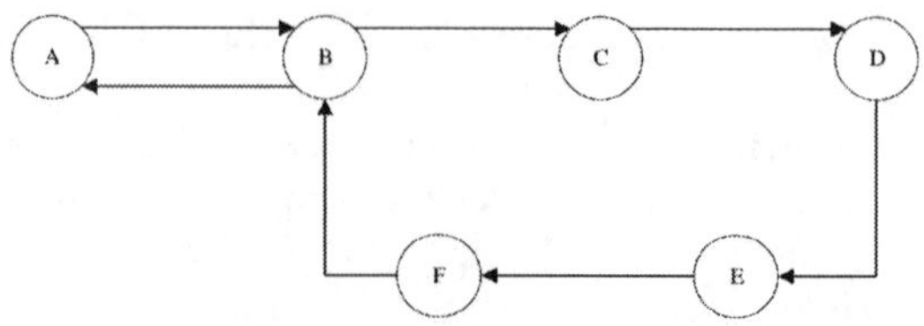

Figure 1. Sailing routes.

Shipping lines only serve shipping containers leased from the shipping lines or containers originating from cooperating forwarders (joining one consortium). Shipping lines must anticipate by making a demand container projection, both to meet the demands at the port itself (ie demand containers that will be sent to other ports) and the demand containers needed at other ports (empty containers sent to other ports). For example, container demand at port A is a container demand to meet customers who will send goods using container from port A to ports B, C, D, E, F, and also empty container demands to meet at ports B, C, D, E , F.

Inventory of empty containers in a port or depot comes from 3 sources: begining inventory, empty containers sent from other ports, empty containers that have returned to port (laden containers that have been empty and returned to shippinglines). Empty containers sent from other ports will become inventory when the containers arrive at the port. Empty container requires delivery time from the original port to the destination, while the laden container that arrives at the destination port must be sent to the consignee, and then the consignee returns the empty container to the depot / port. The time from laden container to empty container cannot be determined. Next, the shipper does stuffing, then sends the laden container to the port.

If the travel time between ports is assumed to be the same, which takes 1 (one) week, then shipping from port A to port B takes 1 week, from port A to port C takes 2 weeks, from port A to port D takes 3 week, and so on. Ships from port D returning to A take 4 weeks, because ships from port D to port C must go through ports E, F, B and C. Ships carrying containers from port D to C, or ships from port E to C and D, as well as F to C, D and E must unload the container at port B. Containers unloaded at port B are transferred to the ship from port A to port C.

The uncertainty of the time required from laden container to empty container affects the inventory of containers at a port. Because inventory depends on the time of returning empty containers, inventory is uncertain. The number of import and export containers (inbound and outbound containers) is not always the same (unbalance). At one port the amount of inbound is greater than outbound, and at other ports the number of outbound is greater than inbound. If repositioning is not manage properly, there will be shortages at one port, and there will always be excess at other ports. Therefore, shipping lines must be able to manage and control inventory at each port to avoid shortages and surpluses at a port.

To control the balance of inventory in each port, shipping lines must set the number of empty containers repositioned to each port. Container repositioning does not add value to shipping lines. All repositioning costs are borne by shipping lines, even shipping lines have to sacrifice the capacity available to carry repositioned empty containers. Therefore, repositioning must be arranged in such a way as to minimize costs and maximize demand fulfillment. Unavailability of inventory (shortage) can cause loss of revenue for shipping lines, because customers will switch to shipping lines or other forwarders, where they are competitors.

4. Methodology

The initial stage of this research is literature review related to repositioning empty containers to get a state of the art about empty container repositioning. The next step is to arrange preliminary studies to several shippinglines companies. Preliminary study aims to obtain sufficient initial information related to routing decisions, forecasting (method) decisions, inventory policy, and customer behavior. Based on the existing conditions made a simulation model using ARENA®. The steps for constructing a simulation model are explained as follows:

1. Analysis of problems and information gathering, including identification of input parameters, performance measures, the relationship between parameters and variables. This information is then represented in a logic flow diagram, or presented in an easy and representative way.
2. Data collection, is needed to estimate the input parameters of the model. Analysts are carried out to formulate the distribution of random variables in the model.
3. Model construction, after the problem is clear and the necessary data is collected, the analyst can continue to build the model and apply it as a computer program or as appropriate software (Software Arena).
4. Verification of the model, to ensure the model is made according to specifications. Model verification is done by inspection, comparing models with specifications. If there are differences corrected by modifying.
5. Model validation, carried out to test the suitability of the model with empirical data (measurement of the real system being modeled).
6. Design and conduct simulation / experiment, after the model is valid, proceed with designing a set of experiments to estimate the performance of the model. The analyst can compile a number of scenarios and run a simulation. To achieve statistical reliability of the performance of each scenario, each scenario is replicated (run several times randomly), so that the average yield variability is small (low).
7. Analysis of output, conducting statistical analysis to determine statistical inference in determining whether one alternative scenario has the best performance so it must be chosen as the best scenario.
8. Recommendations, formulate final recommendations based on the results of output analysis.

Schematically, this research methodology can be explained as in Figure 2.

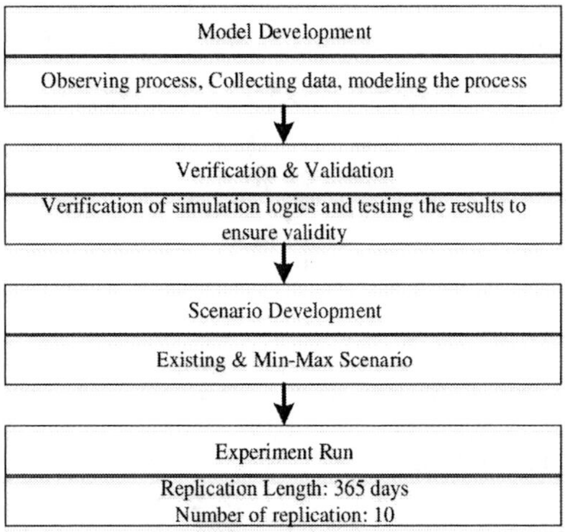

Figure 2. Research methodology.

5. Modeling

In this simulation model, repositioning and empty container inventory are considered simultaneously. Supply and demand for empty containers are both uncertain. Shipping routes are fixed.

5.1. Assumption

The assumptions used in this study are as follows:

1. The number of vessels is assumed to be the same and does not change during the simulation time period.
2. Containers are considered only one type (the same size and type and not container refrigerator);
3. The number of containers owned by shipping lines is fixed, has not been added or reduced;
4. The time taken from a laden container to an empty container to the port / depot and ready to be used follows empirical distribution, where 85% of the container returns until the 7th day, and 15% returns after 7 days.
5. Wharfs and loading / unloading equipment at each port, assumed to be able to serve each ship;
6. Container yards at each port are assumed to be able to accommodate all loading / unloading containers and transhipment containers.
7. Ships and container transshipment (at port B) are not specifically considered (scheduling or berth allocation)
8. Ship service time (pillotage service time, mooring, etc.) is ignored.
9. Dwelling time includes the time from the laden container to the empty container.

5.2. Input Parameters

The input parameters used in this study are as follows:

1. Delivery Time, sailing time, ship turnaround time.
   The arrival time of the ship at the beginning of the period is generated using a uniform distribution (UNIFORM (5,7)) days inter-arrival time. For the next period, the time between arrivals at each port depends on the travel time of the ship (voyage) from the origin port to the destination port. Travel time from one port to another port is considered to be the same and follows a uniform distribution (UNIFORM (5,7)) days.
2. Number of container loading and unloading
   The number of containers loading and unloading at each port varies. The minimum, average, and maximum number of container loading and unloading in each port is presented in Table 1.

Table 1. Number of Container Loading and Unloading

Port	Number of Container Loading			Number of Container Unloading		
	Min	Ave	Max	Min	Ave	Max
A	123	977	1,770	106	1,075	1,889
B	95	427	1,524	74	462	1,692
C	93	1,039	2,116	110	975	2,065
D	194	1,874	2,853	507	2,272	3,128
E	151	1,278	1,831	369	1,367	2,016
F	83	783	1,695	180	1,033	1,834

3. *Demand Generator* and *Demand Fulfillment*
   Container demand is defined as the number of empty container requirements for a certain period. Empty container needs are met from an inventory of empty containers in the port/depot. As stated earlier, inventory originates from begining inventory, empty containers sent from other ports, and laden containers that have returned to port in an empty condition.
   Container demand can be met if the amount of inventory in one port is greater than demand. If inventory is smaller than demand, shortage will occur. Demands are generated using a uniform distribution based on the data in Table 2.

Table 2. Distribution of Demand

Port	Demand Distribution
A	UNIFORM(123, 977, 1770)
B	UNIFORM(95, 427, 1524)
C	UNIFORM(93, 1039, 2116)
D	UNIFORM(194, 1874, 2853)
E	UNIFORM(151, 1278, 1831)
F	UNIFORM(83, 783, 1695)

The proportion of loading-unloading to and from the port can be seen in Table 3.

Table 3. The proportion of inbound and outbound in each port.

From/To	A	B	C	D	E	F
A		0.2	0.1	0.35	0.2	0.15
B	0.1		0.15	0.3	0.25	0.2
C	0.2	0.1		0.35	0.15	0.2
D	0.3	0.15	0.25		0.1	0.2
E	0.2	0.35	0.1	0.1		0.25
F	0.2	0.2	0.2	0.25	0.15	

5.3. Verification and Validation

To ensure the model that is built works in accordance with the characteristics of the original system and the results released from the simulation model are within acceptable levels of accuracy, verification and validation of the model must be carried out [45]. Verification assesses the correctness of the model representation by checking coding and test runs, and checking for consistency statistically. Validation assesses how realistic the assumptions on the model are, by comparing the performance of the model (predictions) obtained from test runs with the system under study [46]. Verification consists of three main activities, namely (1) checking simulation logic, (2) simulating test runs and checking printouts and graphs to check that the simulation logic is correct, (3) simple checking of consistency, and more complex checks, between theoretical and statistical simulation results.

Verification of test runs, carried out in several ways, namely verification of input parameters and statistics from the model output, verification using the debugger, verification with animation, verification using statistics or formulation to ensure the model matches to the logic. For Arena simulation, it is necessary to verify more than just models that are built according to the logic or not, but it need to be verified through performance analysis. Verification of animation is used to observe the operating behavior of the model during the running simulation. Several entities are observed to move from entering the system to exiting the system, thus it can be seen whether the entity moves correctly in accordance with the predetermined modeling concept. From the observations of the animation it appears that the entity (ship) moves according to the specified step. The ship came to the arrival station, the ship carried out the process of unloading and loading. From the animation, when the ship is at the berth (the ship is not moving), it is followed by unloading activities for inbound containers. After all outbound containers have finished loading, the ship moves to leave the port. The whole movement is well visualized in animation and in accordance with real conditions. The transfer of laden containers to consignees and from consignees back to the port can be visualized well. Thus the model is verified.

Validation is very important to develop a credible model. The standard approach to model validation is to collect data from the system under study, and compare it with simulation results. In this study validation was carried out on the interarrival time, the number of container loading and unloading, and the amount of inventory. Inventory is adopted from the periodic review system (R, S), where the status of the inventory is reviewed periodically, and when the inventory level reaches point R (reorder point), the order to fulfill a number of S is immediately carried out. In the min-max review system, when the inventory is greater or equal I_{max} (maximum inventory), then repositioning is R ($I-I_{max}$). The maximum inventory at each port is determined based on the desired inventory service level, in this study the service level is set at 95%. For example, I_{max} at port A is 547, if in a period the amount of inventory at port A is less than or equal to 547, then no repositioning is made, but if it is greater than 547, excess inventory will be repositioned to other ports. The maximum inventory in each port is shown in Table 4. Based on the simulation results, it can meet the specified conditions and provisions, thus the model can be said to be valid.

Table 4. Maximum inventory at each port.

Port	Inventory Max
Port A	547
Port B	319
Port C	811
Port D	895
Port E	676
Port F	327

5.4. Run Set Up and Variable Respons

Before the simulation is run, run set up is done first. In this simulation the set up running is done as follows:

Number of replication : 10 replication
Replication Length : 8760 hours (24 hour x 365 days)
Hour per day : 24 hours

Response variables used to assess the performance of the system in this model are the amount of inventory and the amount of demand that cannot be met (shortage).

6. Results and Disscussions

6.1. Existing Condition

Table 5 shows the average inventory under existing conditions at each port. At port A the average inventory in replication 1 is 1860 containers, replication 2 is 549 containers, replication 3 is 1161, and so on. The average inventory of replication 1-10 is 1205 containers. At port B the average inventory in replication 1 is 799 containers, replication 2 is 2265 containers, replication 3 is 2547, and so on. The average inventory of replication 1-10 is 2317 containers. At port C the average inventory in replication 1 is 328 containers, replication 2 is 790 containers, replication 3 is 772, and so on.

Table 5. Average inventory at existing conditions.

	A	B	C	D	E	F
Replication 1	1860	799	328	766	937	1077
Replication 2	549	2265	790	1671	579	2046
Replication 3	1161	2547	772	2260	650	1461
Replication 4	936	1158	474	875	679	1345
Replication 5	1268	1487	311	1105	502	1252
Replication 6	627	5037	671	1169	444	1006
Replication 7	1589	4364	530	1569	491	839
Replication 8	1225	3754	345	754	363	1886
Replication 9	1863	902	587	963	759	1306
Replication 10	975	860	728	3216	546	707
Average	1205	2317	554	1435	595	1293

The average inventory of replications 1-10 is 554 containers. At port D the average inventory in replication 1 is 766 containers, replication 2 is 1671 containers, replication 3 is 2260, and so on. The average inventory of replication 1-10 is 1435 containers. At port E, the average inventory in replication 1 is 937 containers, replication 2 is 579 containers, replication 3 is 650, and so on. The average inventory of replication 1-10 is 595 containers. At port F the average inventory in replication 1 is 1077 containers, replication 2 is 2046 containers, replication 3 is 1461, and so on. The average inventory of replication 1-10 is 1293 containers.

Table 6 shows the average shortage in the existing conditions at each port. At port A, the average shortage of replication 1 is 409, replication 2 is 528 containers, replication 3 is 498, and so on. The average shortage of replication 1-10 is 401 containers. At port B, the average shortage of replication 1 is 307 containers, replication 2 is 529, replication 3 is 294 containers, and so on. The average shortage at port B of replication 1-10 is 401 containers. At port C, the average shortage of replication 1 is 644 containers, replication 2 is 843, replication 3 is 705 containers, and so on. The average shortage at port C from replication 1-10 is 703 containers. At port D, the average shortage of replication 1 is 843 containers, replication 2 is 862, replication 3 is 555 containers, and so on. The average shortage at port D of replication 1-10 is 719 containers. At port E, the average shortage of replication 1 is 593 containers, replication 2 is 553 containers, replication 3 is 452 containers, and so on. The average shortage at port E from replication 1-10 is 562 containers. At port F, the average shortage of replication 1 is 502 containers, replication 2 is 514, replication 3 is 553 containers, and so on. The average shortage at port F from replication 1-10 is 488 containers.

Table 6. Average shortage of existing conditions.

	A	B	C	D	E	F
Replication 1	509	307	644	843	593	502
Replication 2	528	529	843	862	553	514
Replication 3	498	294	705	555	452	553
Replication 4	541	397	699	833	473	436
Replication 5	424	380	589	580	609	432
Replication 6	558	0	636	604	530	515
Replication 7	437	574	652	710	518	329
Replication 8	564	675	690	863	641	663
Replication 9	441	417	789	811	543	519
Replication 10	449	432	778	526	707	419
Average	495	401	703	719	562	488

6.2. Min-Max Review

Table 7 shows the average inventory in the min-max review system for each replication at each port. At port A, the average inventory in replication 1 is 623 containers, replication 2 is 599 containers, replication 3 is 499 containers, and so on. The average inventory of replication 1-10 is 573 containers.

Table 7. Average inventory scenarios for min-max review system.

	A	B	C	D	E	F
Replication 1	623	419	495	695	681	577
Replication 2	599	544	578	881	671	475
Replication 3	499	665	648	1056	555	691
Replication 4	455	562	338	843	588	510
Replication 5	789	554	702	646	575	558
Replication 6	491	614	576	1087	618	540
Replication 7	499	547	592	843	416	633
Replication 8	603	556	765	948	578	605
Replication 9	681	640	733	972	415	603
Replication 10	495	414	655	1052	699	593
Average	573	552	608	902	580	578

At port B the average inventory in replication 1 is 419 containers, replication 2 is 544 containers, replication 3 is 665 containers, and so on. The average inventory of replications 1-10 is 552 containers. At port C the average inventory in replication 1 is 495 containers, replication 2 is 578 containers, replication 3 is 648 containers, and so on. The average inventory of replication 1-10 is 608 containers. At port D the average inventory in replication 1 is 695 containers, replication 2 is 881 containers, replication 3 is 1056, and so on. The average inventory of replication 1-10 is 902 containers. At port E the average inventory in replication 1 is 681 containers, replication 2 is 671 containers, replication 3 is 555 containers, and so on. The average inventory of replications 1-10 is 580 containers. At port F the average inventory in replication 1 is 577 containers, replication 2 is 475 containers, replication 3 is 691, and so on. The average inventory of replication 1-10 is 579 containers.

Table 8 shows the average container shortage for each replication at each port. At port A, the average number of container shortage in replication 1 is 509 containers, replication 2 is 427 containers, replication 3 is 395 containers, and so on. The average inventory of replication 1-10 is 470 containers.

Table 8. Average number of container shortage for min-max review system.

	A	B	C	D	E	F
Replication 1	509	382	578	705	665	427
Replication 2	427	387	609	837	498	355
Replication 3	395	270	818	623	513	434
Replication 4	517	393	549	683	538	435
Replication 5	388	399	608	601	487	444
Replication 6	519	283	504	795	401	376
Replication 7	325	283	654	686	420	419
Replication 8	821	321	602	827	651	346
Replication 9	439	401	679	786	494	386
Replication 10	363	270	603	906	586	393
Average	470	339	620	745	525	402

At port B, the average shortage of replication 1 is 382 containers, replication 2 is 387, replication 3 is 270 containers, and so on. The average shortage at port B from replication 1-10 is 339 containers. At port C, the average shortage of replication 1 is 578 containers, replication 2 is 609, replication 3 is 818 containers, and so on. The average shortage at port C from replication 1-10 is 620 containers. At port D, the average shortage of replication 1 is 705 containers, replication 2 is 837, replication 3 is 623 containers, and so on. The average shortage at port D of replication 1-10 is 745 containers. At port E, the average shortage of replication 1 is 665 containers, replication 2 is 498, replication 3 is 513 containers, and so on. The average shortage at port E from replication 1-10 is 525 containers. At port F, the average shortage of replication 1 is 427 containers, replication 2 is 355, replication 3 is 434 containers, and so on. The average shortage at port F from replication 1-10 is 402 containers.

6.3. Existing Vs Min-Max Review System

The comparison of inventory and shortage at port A under existing conditions and the min-max review system. In the existing condition, the average inventory is 1205 containers, while in the mi-max review system, the average inventory is 573 containers. The min-max review system can reduced inventory by 632 containers. The average inventory reduction at port A was very significant, with a decrease of 52.45%.

In the existing condition, the average number of container shortage of 495 containers, while in the min-max review system by 470. Min-max review system can reduce the average container shortage by 25 containers or decrease to 470 containers (5,05%). Although the percentage of reduction in shortage is not as large as the decrease in inventory, overall the min-max review system can reduce the average inventory and the average of shortages simultaneously. Comparison of average inventory between existing conditions and min-max review system can be seen in Figure 3, while the comparison of the average shortage between existing conditions and min-max review system can be seen in Figure 4.

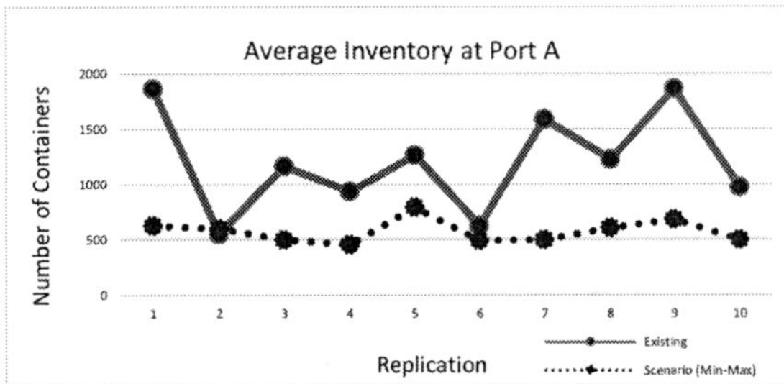

Figure 3. Average inventory at port A.

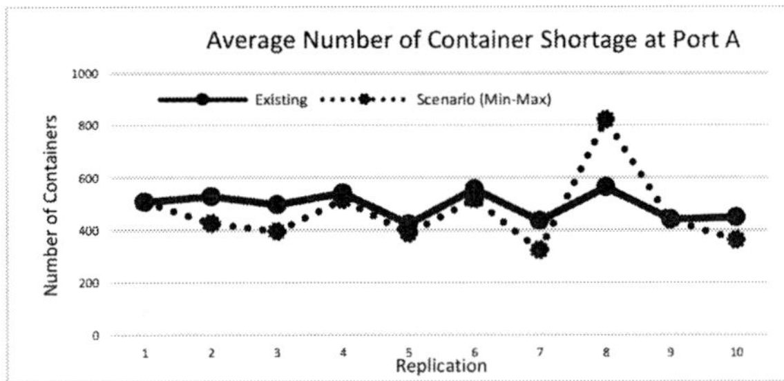

Figure 4. Number of container shortage at port A.

Average inventory at port B under existing conditions existing condition is 2317 containers, while the min-max review system is 552 containers. The min-max review system can reduce inventory significantly by reducing the number to 1765 containers. In the existing condition, the average number of shortages is 401 containers and in the min-max review system is 339 containers. The min-max review system can reduce the number of shortages by 62 containers. Although the amount is not too large, the min-max review system can reduce the average amount of inventory and shortage simultaneously. Figure 5 and Figure 6 respectively show the comparison of average inventory and average shortage at port B under existing conditions and min-max review system.

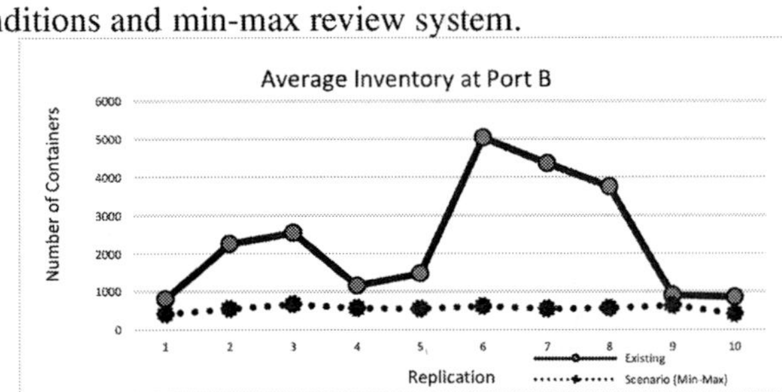

Figure 5. Average inventory at port B.

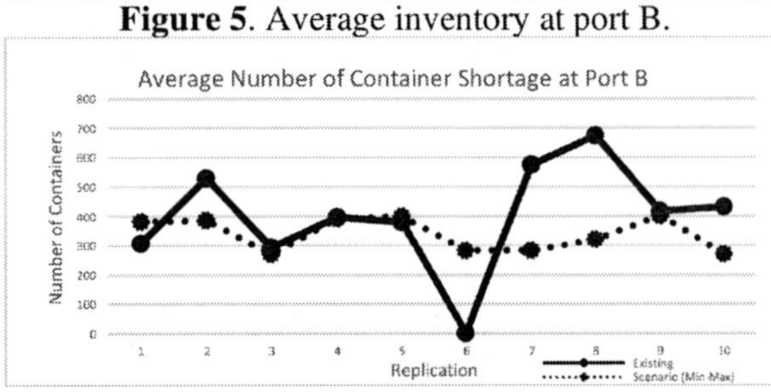

Figure 6. Number of container shortage at port B.

Figure 7 shows the average inventory and shortage that occurred at port C based on existing conditions and min-max review system. In the existing condition, the average inventory is 554 containers, whereas in the min-max review system the average amount of inventory is 608 containers. In the min-max review system there was an increase in the average amount of inventory by 54 containers or an increase in the average amount of inventory by 9.75% compared to existing conditions. Figure 8 shows the average shortage that occurred at port C in the existing condition was 703 containers, whereas in the min-max review system the average shortage was 620 containers. The min-max review system reduced the number of shortages by 83 containers.

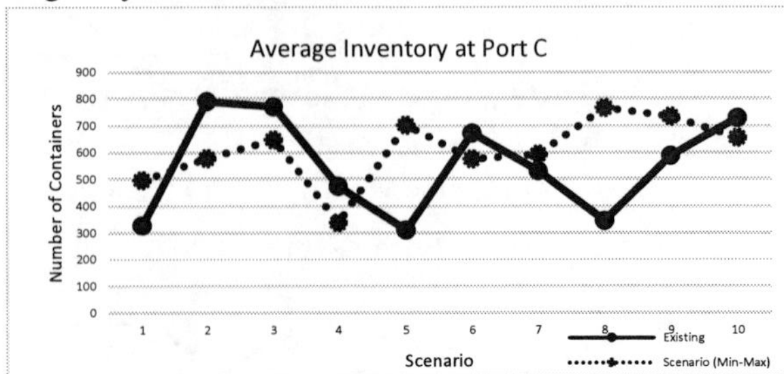

Figure 7. Average inventory at port C.

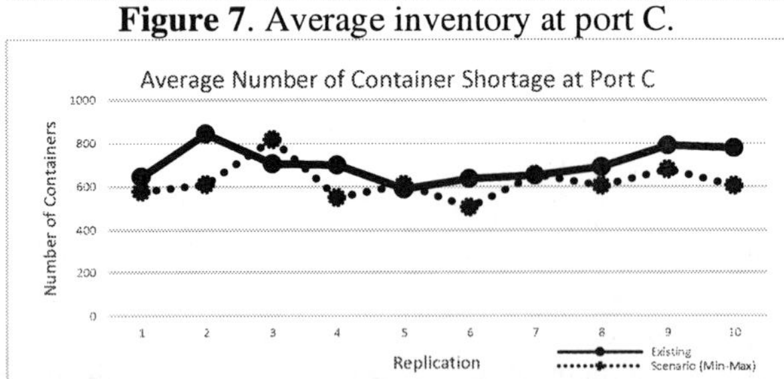

Figure 8. Number of container shortage at port C.

Figure 9 shows the average inventory and shortage at port D in the existing conditions and in the min-max review system. In the existing condition, the average amount of inventory is 1435 containers, whereas in the min-max review system the average inventory is 902. Min-max review system can significantly reduce the average inventory amount, that is, by 553 containers. In the existing conditions, the average shortage of 719 containers, while in the min-max review system the average shortage increased to reach 745 or an increase of 3.62%.

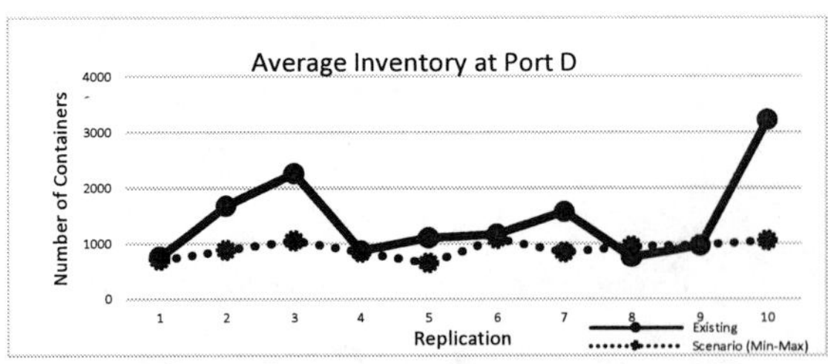

Figure 9. Average inventory at port D.

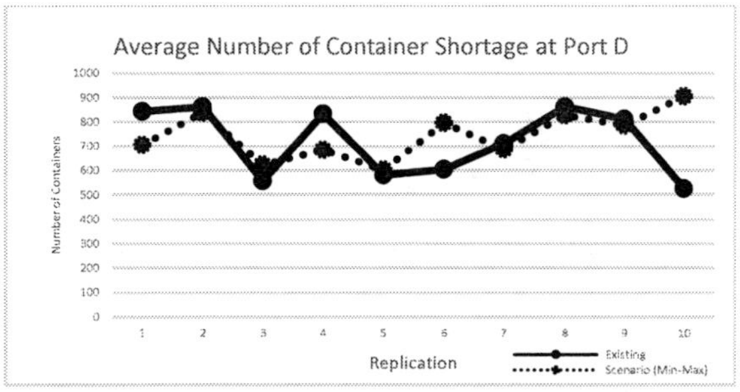

Figure 10. Number of container shortage at port D.

Figure 11 shows the average inventory and average port shortage in the existing conditions and min-max review system. In the existing condition, the average inventory is 595 containers, while in the min-max review system the average inventory is 580. The average shortage in the existing condition is 562 containers and 525 containers for the min-max review system. Although not significant, a min-max review system can simultaneously reduce inventory and shortage. Figure 12 show the average number of container shortage in the existing conditions and min-mac review system at port E.

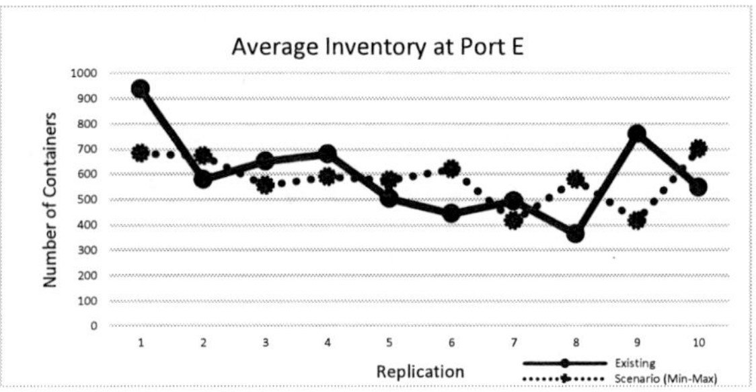

Figure11. Average inventory at port E.

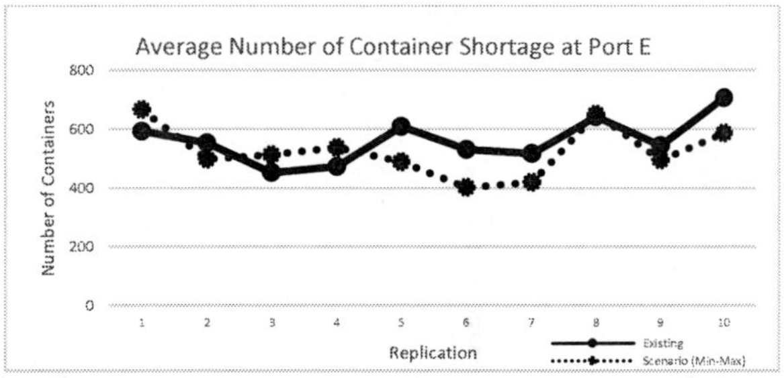

Figure 12. Number of container shortage at port E.

Figure 13 shows the average inventory for existing conditions and the average inventory for min-max review system at port F. In the existing condition, the average inventory is 1293 containers, while the average number of inventory in the min-max review system are 578 containers. Min-max review system can significantly reduce the average inventory amount, which is 715 containers. Figure 14 shows the average number of shortages in the existing condition is 488 cotainers, whereas in the min-max review system the average number of container shortage is 402. Min-max review system can reduce the average inventory and average shortage significantly and simultaneous.

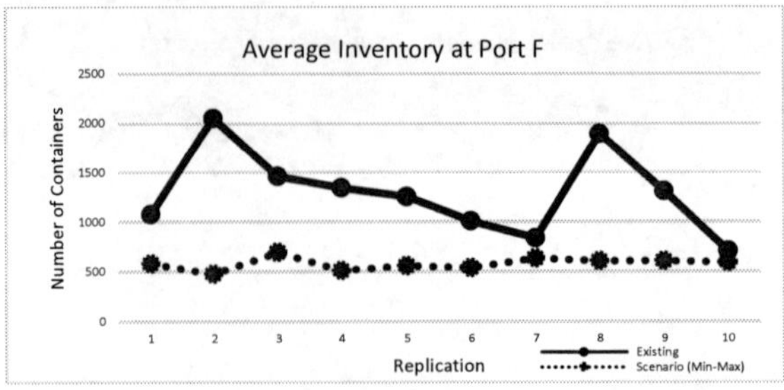

Figure 13. Average inventory at port F.

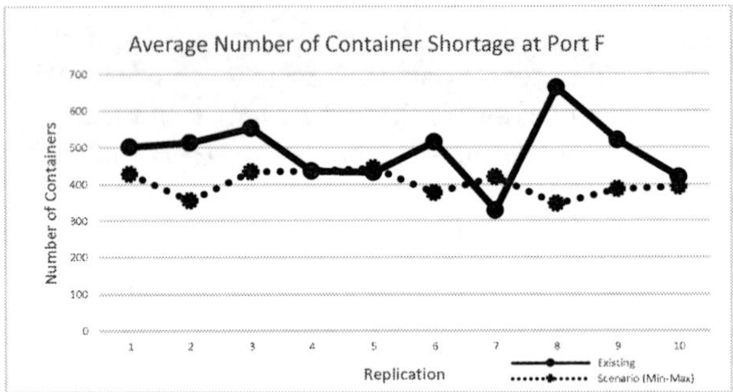

Figure 14. Number of container shortage at port F.

7. Conclussion

From this research, it can be summarized that in the min-max review system, repositioning is carried out by R (difference between current inventory and max inventory) when the inventory is greater or equal to the maximum inventory with 95% of service level. Overall, the min-max review system policy can reduce inventory surplus by 35.64% and shortage of inventory by 8.82%. The reduction in the amount of inventory that occurs with this system in detail is as follows a much as 52,45% at Port A, 76,17% at Port B, and 37,17% at Port D, where there is an increase at Port C by 9.75%. Another advantage by implementing this system through reducing the number of shortages shortages are as follows namely as much 5.05% at Port A, 15.46% at Port B, 11.81% at Port C, 6.58% at Port E, and 17.62% at Port F, while there is an increasing number of shortage by 3.62% at Port D.

8. Future Research

The next research that can be developed from this research include the following:

1. Research to determine the limits (minimum inventory and maximum inventory) that can be used as a basis for determining ordering and repositioning limits.
2. Research to determine the optimum repositioning value by considering the minimum-maximum inventory at each port.
3. Research to compare the combination of periodic inventory policy and continuous inventory policy, as well as the simultaneous use of policies.

References

[1] Bandeira DL, Becker JL, Borenstein D. A DSS for integrated distribution of empty and full containers. *Decis Support Syst.* Elsevier BV; 2009 Nov;47(4):383–97. doi.org/10.1016/j.dss.2009.04.003

[2] Song D-P, Dong J-X. Effectiveness of an empty container repositioning policy with flexible

destination ports. *Transp Policy*. Elsevier BV; 2011 Jan;18(1):92–101. doi.org/10.1016/j.tranpol.2010.06.004
[3] Mason R, Nair R. Supply-side strategic flexibility capabilities in container liner shipping. Xu J, editor. *J Logist Manag*. Emerald; 2013 May 17;24(1):22–48. doi.org/10.1108/ijlm-05-2013-0053
[4] Feng M, Mangan J, Lalwani C. Comparing port performance: Western European versus Eastern Asian ports. *J Phys Distrib & Logist Manag*. Emerald; 2012 Jun 8;42(5):490–512. doi.org/10.1108/09600031211246537
[5] Asgari N, Farahani RZ, Goh M. Network design approach for hub ports-shipping companies competition and cooperation. *Transp Res Part A: Policy and Pract*. Elsevier BV; 2013 Feb;48:1–18. doi.org/10.1016/j.tra.2012.10.020
[6] Parola F, Sciomachen A. Intermodal container flows in a port system network. *J Prod Econ*. Elsevier BV; 2005 Jul;97(1):75–88. doi.org/10.1016/j.ijpe.2004.06.051
[7] UNCTAD, "Review of Maritime Transport 2013," 2013.
[8] Hsu WK. Improving the service operations of container terminals. Xu J, editor. *J Logist Manag*. Emerald; 2013 May 17;24(1):101–16. doi.org/10.1108/ijlm-05-2013-0057
[9] Chao S-L, Chen C-C. Applying a time–space network to reposition reefer containers among major Asian ports. *Res Transp Bus Manag*. Elsevier BV; 2015 Dec;17:65–72. doi.org/10.1016/j.rtbm.2015.10.006
[10] Mittal N, Boile M, Baveja A, Theofanis S. Determining optimal inland-empty-container depot locations under stochastic demand. *Res Transp Econ*. Elsevier BV; 2013 Jun;42(1):50–60. doi.org/10.1016/j.retrec.2012.11.007
[11] Song D-P. Characterizing optimal empty container reposition policy in periodic-review shuttle service systems. *J Oper Res Soc*. Informa UK Limited; 2007 Jan;58(1):122–33. doi.org/10.1057/palgrave.jors.2602150
[12] Di Francesco M, Crainic TG, Zuddas P. The effect of multi-scenario policies on empty container repositioning. *Transp Res Part E: Logist and Transp Rev*. Elsevier BV; 2009 Sep;45(5):758–70. doi.org/10.1016/j.tre.2009.03.001
[13] Dong J-X, Song D-P. Container fleet sizing and empty repositioning in liner shipping systems. *Transp Res Part E: Logist and Transp Rev*. Elsevier BV; 2009 Nov;45(6):860–77. doi.org/10.1016/j.tre.2009.05.001
[14] Meng Q, Wang S. Liner shipping service network design with empty container repositioning. *Transp Res Part E: Logist and Transp Rev*. Elsevier BV; 2011 Sep;47(5):695–708. doi.org/10.1016/j.tre.2011.02.004
[15] Kuzmicz KA, Pesch E. Approaches to empty container repositioning problems in the context of Eurasian intermodal transportation. *Omega*. Elsevier BV; 2019 Jun;85:194–213. doi.org/10.1016/j.omega.2018.06.004
[16] Dang Q-V, Yun W-Y, Kopfer H. Positioning empty containers under dependent demand process. J *Comput Ind Eng*. Elsevier BV; 2012 Apr;62(3):708–15. doi.org/10.1016/j.cie.2011.11.021
[17] Shintani K, Imai A, Nishimura E, Papadimitriou S. The container shipping network design problem with empty container repositioning. *Transp Res Part E: Logist and Transp Rev*. Elsevier BV; 2007 Jan;43(1):39–59. doi.org/10.1016/j.tre.2005.05.003
[18] Chou C-C, Gou R-H, Tsai C-L, Tsou M-C, Wong C-P, Yu H-L. Application of a mixed fuzzy decision making and optimization programming model to the empty container allocation. *Appl Soft Comput*. Elsevier BV; 2010 Sep;10(4):1071–9. doi.org/10.1016/j.asoc.2010.05.008
[19] Du Y, Hall R. Fleet Sizing and Empty Equipment Redistribution for Center-Terminal Transportation Networks. *Manag Sci*. Institute for Operations Research and the Management Sciences (INFORMS); 1997 Feb;43(2):145–57. doi.org/10.1287/mnsc.43.2.145

[20] Li J-A, Liu K, Leung SCH, Lai KK. Empty container management in a port with long-run average criterion. *Math Comput Mod*el. Elsevier BV; 2004 Jul;40(1-2):85–100. doi.org/10.1016/j.mcm.2003.12.005

[21] Zhou W-H, Lee C-Y. Pricing and competition in a transportation market with empty equipment repositioning. *Transp Res Part B: Method*. Elsevier BV; 2009 Jul;43(6):677–91. doi.org/10.1016/j.trb.2008.12.001

[22] Chen R, Dong J-X, Lee C-Y. Pricing and competition in a shipping market with waste shipments and empty container repositioning. *Transp Res Part B: Method*. Elsevier BV; 2016 Mar;85:32–55. doi.org/10.1016/j.trb.2015.12.012

[23] Xu L, Govindan K, Bu X, Yin Y. Pricing and balancing of the sea–cargo service chain with empty equipment repositioning. *Comput Oper Res*. Elsevier BV; 2015 Feb;54:286–94. doi.org/10.1016/j.cor.2014.03.001

[24] Hjortnaes T, Wiegmans B, Negenborn RR, Zuidwijk RA, Klijnhout R. Minimizing cost of empty container repositioning in port hinterlands, while taking repair operations into account. *J Transp Geog*. Elsevier BV; 2017 Jan;58:209–19. Available from: http://dx.doi.org/10.1016/j.jtrangeo.2016.12.015

[25] Li L, Wang B, Cook DP. Reprint of "Enhancing green supply chain initiatives via empty container reuse. *Transp Res Part E: Logist and Transp Rev*. Elsevier BV; 2015 Feb;74:109–23. doi.org/10.1016/j.tre.2014.12.007

[26] Li L, Wang B, Cook DP. Enhancing green supply chain initiatives via empty container reuse. *Transp Res Part E: Logist and Transp Rev*. Elsevier BV; 2014 Oct;70:190–204. doi.org/10.1016/j.tre.2014.06.018

[27] Zheng J, Sun Z, Gao Z. Empty container exchange among liner carriers. *Transp Res Part E: Logist and Transp Rev*. Elsevier BV; 2015 Nov;83:158–69. doi.org/10.1016/j.tre.2015.09.007

[28] Song D-P, Xu J. An operational activity-based method to estimate CO2 emissions from container shipping considering empty container repositioning. *Transp Res Part D: Transp and Environ*. Elsevier BV; 2012 Jan;17(1):91–6. Available from: http://dx.doi.org/10.1016/j.trd.2011.06.007

[29] Song D-P, Dong J-X. Cargo routing and empty container repositioning in multiple shipping service routes. *Transp Res Part B: Method*. Elsevier BV; 2012 Dec;46(10):1556–75. doi.org/10.1016/j.trb.2012.08.003

[30] WANG B, WANG Z. Research on the Optimization of Intermodal Empty Container Reposition of Land-carriage. *J Transp SysT Eng and Inf Technol*. Elsevier BV; 2007 Jun;7(3):29–33. doi.org/10.1016/s1570-6672(07)60020-8

[31] Lei TL, Church RL. Locating short-term empty-container storage facilities to support port operations: A user optimal approach. *Transp Res Part E: Logist and Transp Rev*. Elsevier BV; 2011 Sep;47(5):738–54. doi.org/10.1016/j.tre.2011.01.004

[32] Furió S, Andrés C, Adenso-Díaz B, Lozano S. Optimization of empty container movements using street-turn: Application to Valencia hinterland. *Comput Ind Eng*. Elsevier BV; 2013 Dec;66(4):909–17. doi.org/10.1016/j.cie.2013.09.003

[33] Braekers K, Caris A, Janssens GK. Optimal shipping routes and vessel size for intermodal barge transport with empty container repositioning. *Comput Ind*. Elsevier BV; 2013 Feb;64(2):155–64. doi.org/10.1016/j.compind.2012.06.003

[34] Alfandari L, Davidović T, Furini F, Ljubić I, Maraš V, Martin S. Tighter MIP models for Barge Container Ship Routing. *Omega*. Elsevier BV; 2019 Jan;82:38–54. doi.org/10.1016/j.omega.2017.12.002

[35] Song D-P, Dong J-X. Long-haul liner service route design with ship deployment and empty container repositioning. *Transp Res Part B: Method*. Elsevier BV; 2013 Sep;55:188–211. doi.org/10.1016/j.trb.2013.06.012

[36] Crainic TG, Gendreau M, Dejax P. Dynamic and Stochastic Models for the Allocation of

Empty Containers. *Oper Res*. Institute for Operations Research and the Management Sciences (INFORMS); 1993 Feb;41(1):102–26. doi.org/10.1287/opre.41.1.102
[37] Young Yun W, Mi Lee Y, Seok Choi Y. Optimal inventory control of empty containers in inland transportation system. *J Prod Econ*. Elsevier BV; 2011 Sep;133(1):451–7. doi.org/10.1016/j.ijpe.2010.06.016
[38] Long Y, Lee LH, Chew EP. The sample average approximation method for empty container repositioning with uncertainties. *J Oper Res*. Elsevier BV; 2012 Oct;222(1):65–75. doi.org/10.1016/j.ejor.2012.04.018
[39] Di Francesco M, Crainic TG, Zuddas P. The effect of multi-scenario policies on empty container repositioning. *Transp Res Part E: Logist and Transp Rev*. Elsevier BV; 2009 Sep;45(5):758–70. doi.org/10.1016/j.tre.2009.03.001
[40] Shintani K, Konings R, Imai A. The impact of foldable containers on container fleet management costs in hinterland transport. *Transp Res Part E: Logist and Transp Rev*. Elsevier BV; 2010 Sep;46(5):750–63. doi.org/10.1016/j.tre.2009.12.008
[41] Moon I, Do Ngoc A-D, Konings R. Foldable and standard containers in empty container repositioning. *Transp Res Part E: Logist and Transp Rev*. Elsevier BV; 2013 Jan;49(1):107–24. doi.org/10.1016/j.tre.2012.07.005
[42] Myung Y-S, Moon I. A network flow model for the optimal allocation of both foldable and standard containers. *Oper Res Lett*. Elsevier BV; 2014 Sep;42(6-7):484–8. doi.org/10.1016/j.orl.2014.08.004
[43] Wang K, Wang S, Zhen L, Qu X. Ship type decision considering empty container repositioning and foldable containers. *Transp Res Part E: Logist and Transp Rev*. Elsevier BV; 2017 Dec;108:97–121. doi.org/10.1016/j.tre.2017.10.003
[44] Zheng J, Sun Z, Zhang F. Measuring the perceived container leasing prices in liner shipping network design with empty container repositioning. *Transp Res Part E: Logist and Transp Rev*. Elsevier BV; 2016 Oct;94:123–40. doi.org/10.1016/j.tre.2016.08.001
[45] N. N. Huynh, C. M. Walton, and R. River, "Methodologies for reducing truck turn time at marine container terminals," *Cent. Transp. Res. Univ. Texas Austin*, vol. 7, no. 2, pp. 1–144, 2005.
[46] T. Altiok and B. Melamed, *Simulation modeling and analysis with Arena*. 2007.

Utilization of facial expression analysis in the pricing strategy formulation

H Sari[1,2], M S Rahadian[1] and A Rizkiana[1]

[1]Bandung Institute of Technology, Bandung, Indonesia

E-mail: [2]hasrini@mail.ti.itb.ac.id

Abstract. Value-based pricing is one of the pricing strategies to determine the appropriate price plan based on the product's value. This article examines how the facial expression of the customer can be used to measure intention to buy for the price plan developed by value-based pricing. The results of this study indicate that participants' positive emotions were significantly higher than negative emotions for all the products displayed formulated using value-based pricing, consistent with the self-reported questionnaire result. In conclusion, facial expression measured by FaceReader can be used to validate the price plan formulated using value-based pricing.

1. Introduction

Price is one of the marketing tools that have a significant impact on the profitability of business organizations [1]. Price has been an important factor that has historically influenced consumer preferences. Therefore, a business organization must carefully choose the right pricing strategy for its products. One of these pricing strategies is value-based pricing. Value-based pricing is a pricing strategy based on the value inherent in the product in the eyes of consumers [2]. Value-based pricing is done after analyzing consumer needs and perceived value. For customers, the perceived value is the ratio between the benefits obtained when consuming a product or service (perceived quality) and the costs incurred when using a product or service (perceived cost) [3]. High perceived value means that the customer is more willing to buy the product. The consumer prefers value-based pricing because it can help them satisfy their needs and wants [4].

Although value-based pricing theoretically can be used to determine the price plan for the products based on its value, there is still a need to test the validity of the price plan to the potential customer. The validity test is used to determine the customer's intention to buy the product with the price plan prepared using value-based pricing. Then, based on the test result, the price plan can be further refined. One of the disciplines that can help the price validity test process is neuromarketing.

Neuromarketing is a discipline of economics that uses clinical information about nerve function and mechanisms to help explain consumer behavior [5]. Neuromarketing can help understand how unconscious mind processing influences the decision to purchase, providing a better understanding of the consumers' thoughts, emotions, feelings, needs, and motivation [6]. Emotional signals especially have a contribution in the pre-decision phase [7]. Reference [7] proposed two approaches in neuromarketing, namely psychological and neurological. The psychological approach, also called biometrical, measures body reactions on specific marketing stimulus. While the neurological approach, or neuromeric, measures brain reactions on specific marketing stimulus. Each approach uses different types of signals so that each has weaknesses and strengths. Biometric measurement in neuromarketing includes measuring facial expressions, eye tracking, electrodermal activities, heartbeat and perspiration, and response time.

The biometric measurement used in this research to test the validity of price plan is facial expression. Facial expression is a series of muscle contractions that are semi-universal and associated with emotions [8]. Facial expressions are crucial for understanding humans. Emotions, pain, and intentions are among human feelings that can be deduced from the cues displayed on facial expressions [9]. In marketing, facial expression can help the marketer develop a marketing strategy. For example, [10] uses facial expressions of the visitors of an online or offline store to differentiate between window shoppers, potential customers, and buyers. The result is then used to determine customer handling strategies. Facial expression in this research will be used to measure customer's intention to buy based on the price plan to develop using value-based pricing.

2. Research Methodology

Price plan validity test is done using the neuromarketing experimental method that is equipped with a self-reporting questionnaire. This test was conducted to determine the intention of respondents to buy products with a price plan that has been developed using a value-based pricing strategy. Reference [11] said that there is a strong correlation between emotions felt by customers and customer purchase intentions. Customers who display positive emotions are considered to have purchase intentions on the products offered [12].

The biometric measurement instrument used in the experiment is FaceReader developed by Noldus Information Technology. FaceReader is an online platform for face expression analysis through the web. FaceReader was chosen because FaceReader is a reliable indicator of financial expressions of basic emotions and does not need human correction [13].

To illustrate the testing framework clearly, this research uses Swifter (SF) product, a woman's clothing brand, as the research object. The respondents are customers who have ever purchased an SF product and meets SF's target market criteria. The criteria are women aged between 16-30 years. The sampling technique used in this study is a non-probability sampling technique, i.e., purposive sampling. The sample size required for simple experimental studies with strict experimental controls is between 10 and 20 [14]. Researchers contacted people who met the sample criteria and asked for their willingness to be the object of the experiment. If they agreed, then a link that will connect the participant with the FaceReader website is shared with prospective participants. Participants must meet several conditions when opening a link, as follows [15]:

a. Open the link using a handphone, a laptop, or a computer that has a camera or webcam.
b. In a room or location having enough lighting.
c. The position of the camera is right in front of the participant's face.
d. Participants should focus on watching the video displayed and not doing other activities.

When the link was opened for the first time, the face of participants was scanned. After that, videos showing SF products were displayed. While the participants were watching the video, the camera will record the facial expressions of the participants.

Then, emotion analysis is carried out by detecting the micro expression of participants' faces when they watch videos showing products at the proposed price. FaceReader classifies facial expressions into six basic emotions: happiness, sadness, anger, surprise, fear, and disgust.

Emotion is a good predictor of the intention to buy. Reference [16] states that customers tend to buy when their positive emotion is higher than negative emotion toward a product. Therefore, the t-test between positive and negative emotion is used as a predictor of intention to buy. According to [16], positive emotion that influences buying behavior is happiness, while sadness, anger, surprise, fear, and disgust show negative emotions.

The procedure of the experiment is as follows:

1. Participants are checked whether they meet the research criteria.
2. The researcher explained the research protocol to the participants.
3. Video exposure consists of:
   a. A product picture with the proposed price.
   b. A fixation cross image at every product transition, to let the brain enter a resting phase so the reactions to the following product image will not overlap and are easier to analyze [17].
4. Participants were asked to answer a self-reported questionnaire to measure intention to buy. The self-reported questionnaire contains several questions about the intention to buy and the opinion

about the appropriate prices. Intention to buy is measured using a Likert scale (1-5). Every time participants watch a product image along with its price tag, they were asked to answer two questions: a). I will buy the product or b). The product price is worthy.

If the price offered is not perceived as worthy, the respondent is asked to provide an appropriate price recommendation for the product. The combination of the self-reported questionnaire and the expression analysis is done to get a better understanding of consumer perception towards the product and its price. The questionnaire was used as a complement and comparison for the analysis of the facial expression.

3. Result

Tests were carried out on 16 respondents for all SF products in the 2019 SF's product catalog. FaceReader analyzes respondents' expressions from videos that are exposed every 0.1 seconds. The emotions analyzed in this study are emotions of happiness, sadness, anger, and disgust. The four emotions were chosen because they are included in the category of positive emotions and negative emotions that can be used as predictors of purchase intentions [16]. Mean and standard deviation of data collected from FaceReader can be seen in Table 1.

Table 1. Data mean and standard deviation.

Product Name	Type of Emotion							
	Happiness		Sadness		Anger		Disgust	
	μ	σ	μ	σ	μ	σ	μ	σ
Kaleena	0.30	0.17	0.06	0.11	0.03	0.03	0.02	0.03
Naima	0.32	0.12	0.04	0.03	0.06	0.06	0.05	0.05
Calyta	0.31	0.17	0.04	0.14	0.02	0.01	0.04	0.04
Malone	0.29	0.25	0.04	0.04	0.03	0.04	0.05	0.09
Frila	0.27	0.23	0.03	0.04	0.03	0.03	0.04	0.03
Lyah	0.18	0.12	0.04	0.06	0.04	0.04	0.04	0.04
Adys	0.27	0.23	0.06	0.05	0.06	0.06	0.06	0.08

Table 2. T-test results.

Product Line Name	T-test				
	Happiness vs Sadness	Happiness vs Disgust	Happiness vs Anger	Positive Emotion vs Negative Emotions	Difference Between Means
Kaleena	6.827	9.225	9.092	8.443	Significant
Naima	13.109	12.139	11.888	12.223	Significant
Calyta	7.192	8.865	9.736	8.921	Significant
Malone	5.633	5.067	5.813	5.494	Significant
Frila	6.021	5.897	6.093	6.859	Significant
Lyah	6.192	6.427	6.534	3.988	Significant
Adys	5.286	5.140	5.244	3.719	Significant

A normality test was carried out to find out whether the data can be well-modelled with normal distributions. The result showed that each data set is having p-value of more than 0.05. Therefore, it can be concluded that the data were normally distributed.

Then, dependent samples t-tests were carried out to determine if there is a significant difference between the means of positive and negative emotions with α=5% and degree of freedom 14 (tα=1.761). The result can be seen in Table 2.

Table 2 shows that all the positive emotions is statistically different from negative emotions for each product. Recapitulation of respondents' answers to the self-reported questionnaire is in Table 3:

Table 3. Willingness to buy.

Product Line Name	Item of Question		Frequency (*n*=16)	%	Price Recommendation (IDR)
Kaleena (IDR 244.000)	Intention to buy	Willing	15	94%	NA
		Unwilling	1	6%	220,000-235,000
	Price worthiness	Unworthy	2	12%	NA
		Worthy	14	88%	NA
Naima (IDR 194,000)	Intention to buy	Willing	14	88%	NA
		Unwilling	2	12%	170,000
	Price worthiness	Unworthy	1	6%	NA
		Worthy	15	94%	NA
Calyta (IDR 164,000)	Intention to buy	Willing	12	75%	NA
		Unwilling	4	25%	150,000
	Price worthiness	Unworthy	3	19%	NA
		Worthy	13	81%	NA
Malone (IDR 204,000)	Intention to buy	Willing	13	81%	NA
		Unwilling	3	19%	200,000
	Price worthiness	Unworthy	2	12%	NA
		Worthy	14	88%	NA
Frila (IDR 244,000)	Intention to buy	Willing	15	94%	NA
		Unwilling	1	6%	195,000
	Price worthiness	Unworthy	1	6%	NA
		Worthy	15	94%	NA
Lyah (IDR 144,000)	Intention to buy	Willing	13	88%	NA
		Unwilling	3	12%	130,000-135,000
	Price worthiness	Unworthy	3	19%	NA
		Worthy	13	88%	NA
Adys (IDR 194,000)	Intention to buy	Willing	16	100 %	NA
		Unwilling	0	0%	NA
	Price worthiness	Unworthy	1	6%	170,000
		Worthy	15	94%	NA

Table 3 shows that all the price formulated using value-based pricing are acceptable for the respondent. Majority of the respondent willing to buy the product on the proposed pricing.

4. Discussion

Reference [18] stated that nonverbal messages were sometimes more important than verbal messages, one of them was facial expressions. Reference [9] argued that "The face is one of the most powerful channels of nonverbal communication." Previous researches have investigated the relationship between facial expression with certain aspects of marketing [10][19][20], but none on pricing strategy.

The results of this study indicate that participants' positive emotions were significantly higher than negative emotions for all the products displayed. Therefore, it can be concluded that all the prices formulated using value-based pricing method are acceptable. The results of a self-reported questionnaire to measure purchase intention also show a relatively high level of willingness to buy for all products, which are more than 75%. The same results were also found for price worthiness.

However, the result of this study should be taken carefully. According to [13] the accuracy of FaceReader is depend on the quality of the posed photograph used, which is difficult to achieve in the spontaneous (and more ecologically valid) facial expressions. Also, the lack of comparative measures, such as facial EMG, heartbeat, or skin conductance, make this research results limited.

5. Conclusion

This research uses the facial expression to measure the intention to buy for the customer based on the price plan developed using value-based pricing. The results of this study indicate that the significance of participants' positive emotions measured by facial expression can be used to measure intention of buy. This is aligned with the self-reported questionnaire result. So, it can be concluded that all the prices formulated using a value-based pricing method are acceptable for the customer. However, the result should be taken carefully because of the limitations of FaceReader and the lack of comparative measures, such as Facial EMG, heartbeat, or skin conductance, to validate the FaceReader reported emotions.

Acknowledgments
This research is funded by ITB, through P3MI Research Funding scheme.

References

[1] De Toni D, Milan G S, Saciloto E B and Larentis F 2017 Pricing strategies and levels and their impact on corporate profitability. *Rev. de Admin.* **52**(2) 120–33. doi.org/10.1016/j.rausp.2016.12.004
[2] Kotler P and Keller K L 2016 *Marketing Management 16th Ed* (New Jersey: Pearson Education Inc)
[3] Monroe 2003 *Pricing Making Profitable Decision* (New York: The Mc GrawHill Companies, Inc)
[4] Ferrel, O C, and Hartline D 2011 *Marketing Strategy (5th Edition)* (South Western: Thomson Corporation)
[5] De Toni D, Milan G S, Saciloto E B, and Larentis F 2017 Pricing strategies and levels and their impact on corporate profitability. *Rev. de Admin.* **52**(2) 120–33. doi.org/10.1016/j.rausp.2016.12.004
[6] Sebastian V 2014 Neuromarketing and Evaluation of Cognitive and Emotional Responses of Consumers to Marketing Stimuli. *Proc. Soc. and Beh. Sci.* **127** 753–7. doi.org/10.1016/j.sbspro.2014.03.349
[7] Palokangas L, Jyrki S, Jarmo H, Sini, M and Jussi N 2012 Measuring the Willingness To Purchase using methods of neuromarketing. *Bus. Mngmt. Prgrm.*.
[8] Lewinski P, Fransen, M L and Tan E S 2016 Embodied resistance to persuasion in advertising *Front. in Psych*, **7**(AUG), 1–12. doi.org/10.3389/fpsyg.2016.01202

[9] Torre F, De and Cohn J F 2011 Visual Analysis of Humans. *In Vis. Anlys of Hum.* doi.org/10.1007/978-0-85729-997-0

[10] Shergill G S, Zealand N, Diegel O, Sarrafzadeh A, Sciences M and Shekar A 2003 Computerized Sales Assistants: The Application of Computer Technology To Measure Consumer Interest – a Conceptual. **January**, 176–91

[11] Poels K, and Dewitte S 2011 How to Capture the Heart? Reviewing 20 Years of Emotion Measurement in Advertising. *SSRN Elect. Jrnl.* doi.org/10.2139/ssrn.944401

[12] Morris J D, Woo C, Geason J A and Kim J 2002 The power of affect: Predicting intention *Jrnl of Adv. Rsrch* **42**(3) 7–17 https://doi.org/10.2501/JAR-42-3-7-17

[13] Lewinski P, Den Uyl T M and Butler C 2014 Automated facial coding: Validation of basic emotions and FACS AUs in facereader *Jrnl of Neuro, Psych, and Eco* **7**(4) 227–36. doi.org/10.1037/npe0000028

[14] Sekaran and Bougie 2016 *Research Methods for Business: A Skill-Building Approach 4th Edition* (New York: John Wiley and Sons)

[15] Uyl T, Den 2018 Benefits and best practices of online facial expression analysis what is facereader online ? *White Paper by Noldus Information Technology*.

[16] Consoli D 2009 Emotions That Influence Purchase Decisions And Their Electronic Processing. *Ann. Univ. Apl. Srs Oecon* **2**(11) 1–45.

[17] Dong Y, Nakamura K, Okada T, Hanakawa, T, Fukuyama H, Mazziotta J C, and Shibasaki H 2005 Neural mechanisms underlying the processing of Chinese words: An fMRI study *Neuro. Rsrch* **52**(2) 139–45. doi.org/10.1016/j.neures.2005.02.005

[18] Steward D W, Hecker S and Graham J L 1987 It' More Than What You Say; Assessing the influence of Nonverbal Communication in Marketing. *Psyc. and Mrkt* **4**(4), 303–22

[19] Howard D J and Gengler C 2001 Emotional Contagion Effects on Product Attitudes *Jrnl of Cnsmr Rsrch* **28**(2) 189–201. doi.org/10.1086/322897

[20] Ausin-azofra J M, Guixeres J, Bigné E and Alcañiz M 2017 Facial Expressions to Evaluate Advertising: A Laboratory versus Living Room Study *Adv. in Adv. Rsrch* **VIII** 109–22. doi.org/10.1007/978-3-658-18731-6

Conceptual modeling of resilience measurement during natural disasters for SMEs

I D Utami[1], I Santosa[2] and E Rifa'i[3]

[1,3]Department of Industrial Engineering, Universitas Trunojoyo, Madura, Indonesia
[2]Department of Informatics, Universitas Trunojoyo, Madura, Indonesia

E-mail: [1]i.d.utami@trunojoyo.ac.id, [2]iwan@trunojoyo.ac.id,
[3]emon.rifai@trunojoyo.ac.id

Abstract. This study focuses on how to generate resilience information services to natural disasters for SMEs. The information in this study was using software or application developed based on a conceptual model formed as a framework for SMEs' managers in dealing with natural disasters. The conceptual model was shaped based on actual data from SMEs' owners and leaders, as well as stakeholders in four regencies in the Madura Island, Indonesia, namely Bangkalan, Sampang, Pamekasan and Sumenep, by identifying the variables or dimensions of SMEs' resilience to natural disasters. The conceptual model was produced using sequence and use case diagrams to identify the relationship between risk and resilience dimensions to measure the level of resilience of SMEs. The level SMEs' resilience to natural disasters was the basis of information received by SMEs' owners and leaders. This study is expected to help SMEs to determine the level of SMEs' resilience to the risk of natural disasters to minimize the existing risks.

1. Introduction

SMEs are very potential in Indonesia because Indonesia is the largest archipelagic country in the world with 13,466 islands and a sea area of 3,257,483 km^2, as well as a land area of 1,922,570 km^2 [1]. SMEs can help to improve the welfare and economy of the community as well as multiplying the regional economy by providing new jobs, producing goods and services, and paying taxes [2,3]. The SME operational process cannot be separated from the presence of several roles of the elements, including the SME owners and leaders, traders, government, environment, and location [2,4]. Indonesia, a nation with thousands of islands, is prone to natural disasters. The magnitude of the potential natural disasters causes human powerlessness due to the impact of after natural disasters, including the emergence of problems for SME owners and leaders and industries, such as not optimal production and marketing processes in SMEs. Internal and external factors of SMEs consist of six points, which are the quality of human resources and natural resources, production system, financial system, marketing strategy, partnership system, and infrastructure and regulation quality. SMEs can operate and develops if these six points can be formed and implemented well [5]. The occurrence of natural disasters is undeniable that the points above will experience an extraordinary reduction if the level of SMEs' resilience is extremely weak against natural disasters. Thus, the problem that arises is how SMEs can find out the level of resilience of their SMEs to anticipate the impacts of natural disasters and increase the resilience against the impacts. Therefore, a model that can help SME managers in measuring the level of resilience of SMEs in dealing with natural disasters is required [6,7].

This research presents a conceptual model of an information system that can be used as a framework for generating an information system for an online resilience measurement system. The conceptual model is created using social and technical perspectives and will be used as a basis in the establishment of software or applications for measuring SMEs' resilience to deal with natural disasters. The model needs to be formed to analyze the elements of a real system and further build a logic model that contains logical relationships between system elements and variables that affect the system, which are programming algorithm flowcharts that are then developed in a computer program [8].

SMEs are one of the business sectors that provide a significant contribution to the economic growth, which include small businesses with the household scale that have less than 20 employees and medium-sized businesses with less than 100 employees. The main problem in SMEs is in the sector of exposure, where the level of marketing effectiveness is influenced by two factors, including internal and external factors [4]. Internal factors are related to the aspects of production, human resources, natural resources, and SMEs operating practices, while external factors deal with the aspects of marketing, competitor, and environment [2]. Both internal and external factors must be considered carefully by the management of SMEs because these two factors are highly influential in the continuation of SMEs. SMEs' owners and leaders need local governments' attention to support the SME netting process, such as training and public facilities [9].

Natural disasters are events caused by the interaction of nature on environmental conditions that have major impacts on humans, disrupting humans' life patterns and causing them to fall into helplessness and suffer, and in these conditions, humans ultimately need adequate food, clothing and others [6]. Natural disasters certainly have effects, one of which is the decline in the economy of the community or the local government, for example, causing the nets of SMEs and other businesses to become less optimal and even stop [10]. The impacts of natural disasters are extremely detrimental to humans, and therefore, special attention needs to be given and anticipating action is necessary to be made in various ways, one of which is always monitoring natural phenomena that can potentially lead to natural disasters and even impose limits on the public, especially to large industries, to stop activities that cause damage to nature [11].

The conceptual model is a diagram that consists of a set of relationships between factors or elements that are believed to have an impact on a target or goal [7]. A good conceptual model must describe a system of thought that is easily understood and logically bring up the relationship between factors or elements with more effective outputs [8]. Modeling is essential to support research activities, especially in the development of information and analysis systems [12].

The process of building a conceptual model includes two general stages. The first stage involves creating an initial conceptual model that shows what is happening at the project site before starting the project. This model explains the condition of the target location, underlying factors and initial relationship to start the project. The second stage comprises placing the initial conceptual model as a basis and then adding project activities to be carried out to achieve the goals and objectives. After completing the second stage, a complete project conceptual model is successfully created that shows the expectation toward the project to affect the situation at the project site [7,8].

2. Methodology

2.1. Data Collection

Data in this study were collected through both field and literature studies. The field study was carried out by directly visiting the locations of fifty SMEs in four regencies, including Bangkalan, Sampang, Pamekasan and Sumenep, in the island of Madura, to acquire and find out the production and marketing processes of SMEs' products, stakeholders playing a role in the businesses, the risks faced by SMEs during natural disasters and mitigation efforts undertaken by SMEs. Literature studies were useful to support the formation of conceptual models to identify problems and their solutions based on previous research, journals, scientific publications, or related theories. In modeling a complex system, the elements of the system, covering variables or dimensions of resilience, were identified. The conceptual model was then formed using sequences and case diagrams to spot the relationship between risk and resilience to measure the level of resilience of SMEs.

2.2. Sequence Diagrams

A sequence diagram depicts the interaction between objects and indicates the communication between these objects to identify the extent of the relationship between objects [13,14]. In this study, the sequence diagram was used to portray the behaviors in scenarios and describe how entities and systems interacted, including messages used during the interactions in sequential order of execution [15,16]. A sequence diagram is closely linked to the use case diagram, in which for one case, a sequence diagram will be formed [17].

2.3. Use Case Diagram
The use case diagram is a UML diagram model that is utilized to describe the desired functional requirements of a system [18]. A use case diagram is benefitted to briefly define who uses a system and what the system can do [19]. This type of diagram was used in this research to model and express the functional units or services provided by the system to the user. It does not explain its use in detail but gives a brief description of the relationship between the actors and the system [17].

3. Results and Discussion
The conceptual model was formed as an information system that could provide services to measure the level of resilience of an SME to natural disasters.

3.1. Results of Use Case Diagram
This diagram explains how SMEs' actors or users accessed the menus in the SMEs' Resilience Measurement Information System for Natural Disasters.

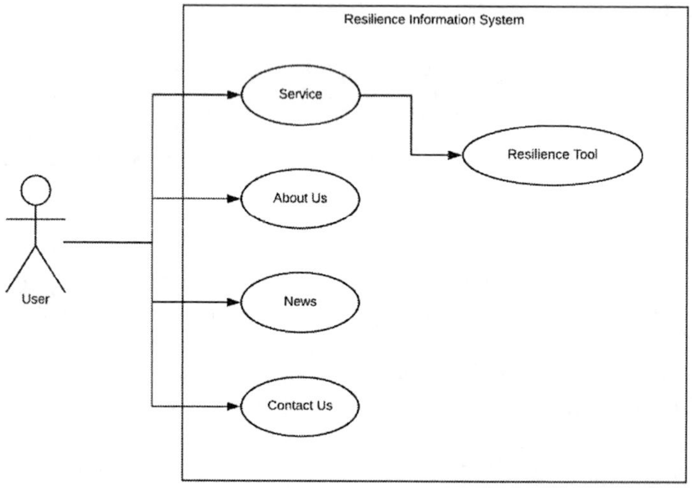

Figure 1. Use case diagram in the SMEs' resilience measurement information system for natural disasters.

Figure 1 communicates that the Service menu has a Resilience Measurement Submenu, which contains a page with survey points and an explanation of how to complete the survey. The About Us menu contains pages that describe researchers and their research on the level of resilience of an SME to natural disasters. The News covers various news about the resilience of SMEs to natural disasters. The Contact Us menu comprises contact information that can be accessed to submit questions, suggestions, feedback, or reports about the resilience of SMEs to natural disasters.

3.2. Results of Sequence Diagram
This diagram explains how the users carried out the resilience measurement process by filling out a survey and the responses were later sent to the users' emails.

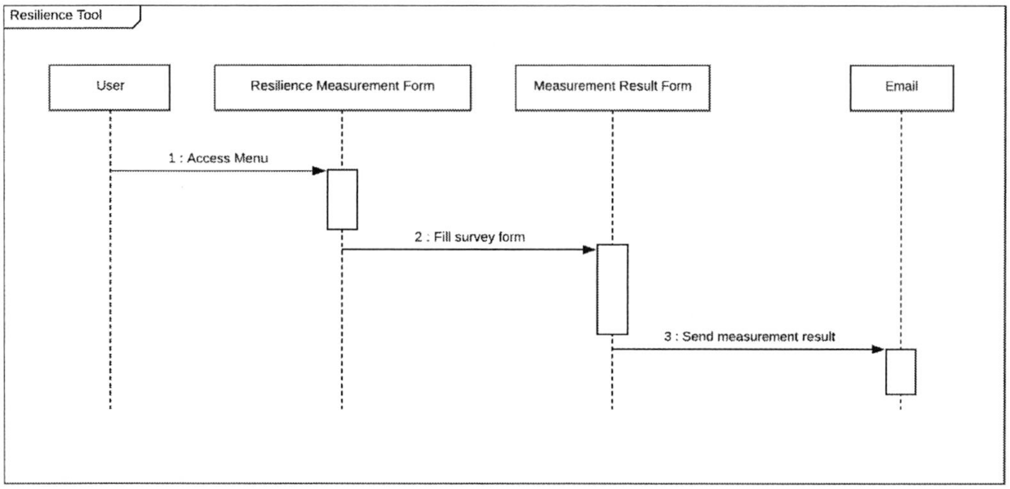

Figure 2. Sequence diagram in the SME resilience measurement information system for natural disasters.

Figure 2 demonstrates that SMEs or Users accessed the Resilience Measurement submenu in the Service menu displaying a screen containing the survey form. The survey form contains 25 points, responded according to the levels of the provided responses. After the survey was submitted, the final step was entering an email address used to receive the result of the measurement in the form of a score of an SME's resilience against natural disasters and some suggestions for improving the resilience of the SME.

3.3. Flowchart Diagram

Figure 3 presents the process flow of measuring the resilience of SMEs to natural disasters in four regencies in Madura Island, Indonesia, in which the initial step was taken in measuring the resilience of SMEs using surveys for data collection. The survey was conducted by making use of software or application that has been developed, with access to survey forms for measuring resilience. The data obtained from the survey were then inputted via email for submission. When the submission process was successful, the responses later proceeded to the calculation stage. However, when the submission process failed, the data inputting process had to be repeated. The results of the calculation were displayed as information by sending via email to the SMEs' manager.

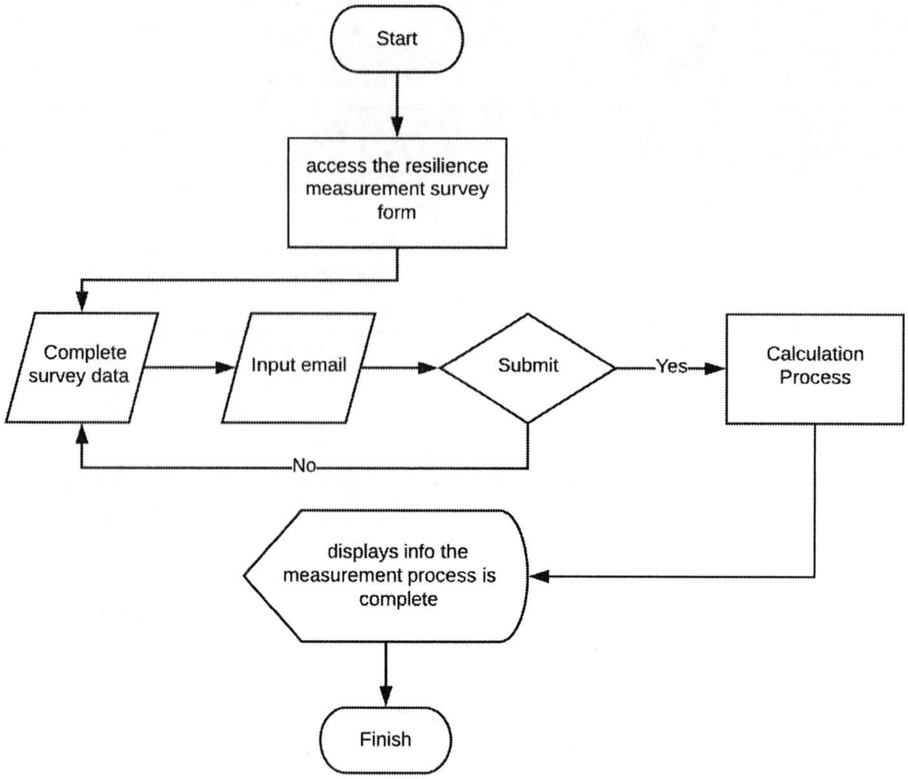

Figure 3. Flowchart diagram in the SMEs' resilience measurement information system for natural disasters.

4. Conclusion
This study concludes that a conceptual model can be used as one approach to help solve problems in a system. The conceptual model in this study was formed through several stages with the use case diagram and sequence diagram. The model formed was later translated into a computer model to produce an information system about measuring SMEs' resilience to natural disasters.

References
[1] National Disaster Management Agency of the Republic of Indonesia. Risiko Bencana Indonesia (Indonesia's Disaster Risk). 2016. http://inarisk.bnpb.go.id/pdf/Buku RBI_Final_low.pdf
[2] Iborra M, Safón V, Dolz C. What explains resilience of SMEs? Ambidexterity capability and strategic consistency. *Long Range Plann.* 2019;101947. doi.org/10.1016/j.lrp.2019.101947
[3] Günerergin M, Penbek Ş, Zaptçıoğlu D. Exploring the Problems and Advantages of Turkish SMEs for Sustainability. Procedia - Soc Behav Sci. 2012;58:244–51.
[4] Albats E, Alexander A, Mahdad M, Miller K, Post G. Stakeholder management in SME open innovation: Interdependences and strategic actions. *J Bus Res.* 2019;(April 2018):1–11. doi.org/10.1016/j.jbusres.2019.07.038
[5] Ansori N, Sutalaksana IZ, Widyanti A. Comparison Between Key Success Factors in Safety Behavior in Small- and Medium-Sized Enterprises (SMEs) and Large Industries, and Development of a Hypothetic Model for Safety Behavior in Indonesian SMEs. *KnE Life Sci.* 2018;4(5):582.
[6] Henneman A, Thornby KA, Rosario N, Latif J. Evaluation of pharmacy resident perceived

impact of natural disaster on stress during pharmacy residency training. *Curr Pharm Teach Learn*. 2020;12(2):147–55. doi.org/10.1016/j.cptl.2019.11.008

[7] Kalkhoven JT, Watsford ML, Impellizzeri FM. A conceptual model and detailed framework for stress-related, strain-related, and overuse athletic injury. *J Sci Med Sport*. 2020. doi.org/10.1016/j.jsams.2020.02.002

[8] Bianchi Janetti E, Guadagnini L, Riva M, Guadagnini A. Global sensitivity analyses of multiple conceptual models with uncertain parameters driving groundwater flow in a regional-scale sedimentary aquifer. *J Hydrol*. 2019;574(April):544–56. doi.org/10.1016/j.jhydrol.2019.04.035

[9] Leckel A, Veilleux S, Dana LP. Local Open Innovation: A means for public policy to increase collaboration for innovation in SMEs. *Technol Forecast Soc Change*. 2020;153(December 2019):119891. doi.org/10.1016/j.techfore.2019.119891

[10] Boudreaux CJ, Escaleras MP, Skidmore M. Natural disasters and entrepreneurship activity. *Econ Lett*. 2019;182:82–5. doi.org/10.1016/j.econlet.2019.06.010

[11] Murray M, Watson PK. Adoption of natural disaster preparedness and risk reduction measures by business organisations in Small Island Developing States - A Caribbean case study. *Int J Disaster Risk Reduct*. 2019;39(November 2018):101115. doi.org/10.1016/j.ijdrr.2019.101115

[12] Verdonck M, Gailly F, Pergl R, Guizzardi G, Martins B, Pastor O. Comparing traditional conceptual modeling with ontology-driven conceptual modeling: An empirical study. *Inf Syst*. 2019;81:92–103. doi.org/10.1016/j.is.2018.11.009

[13] Baidada C, Bouziane EM, Jakimi A. A New Approach for Recovering High-Level Sequence Diagrams from Object-Oriented Applications Using Petri Nets. *Procedia Comput Sci*. 2019;148:323–32. doi.org/10.1016/j.procs.2019.01.040

[14] Campean F, Yildirim U. Enhanced Sequence Diagram for Function Modelling of Complex Systems. *Procedia CIRP*. 2017;60:273–8. doi.org/10.1016/j.procir.2017.01.053

[15] Refsdal A, Runde RK, Stølen K. Stepwise refinement of sequence diagrams with soft real-time constraints. *J Comput Syst Sci*. 2015;81(7):1221–51. doi.org/10.1016/j.jcss.2015.03.003

[16] Sellami A, Hakim H, Abran A, Ben-Abdallah H. A measurement method for sizing the structure of UML sequence diagrams. *Inf Softw Technol*. 2015;59:222–32. doi.org/10.1016/j.infsof.2014.11.002

[17] Arora PK, Bhatia R. Agent-Based Regression Test Case Generation using Class Diagram, Use cases and Activity Diagram. *Procedia Comput Sci*. 2018;125:747–53.

[18] El-Attar M. Evaluating and empirically improving the visual syntax of use case diagrams. *J Syst Softw*. 2019;156:136–63. doi.org/10.1016/j.jss.2019.06.096

[19] Skersys T, Danenas P, Butleris R. Extracting SBVR business vocabularies and business rules from UML use case diagrams. *J Syst Softw*. 2018;141:111–30. doi.org/10.1016/j.jss.2018.03.061

Analysis of usability and user experience on the *batik*'s website design

S Hartomo[1] and T S C Bakal[2]

[1,2]Industrial Engineering Department, Faculty of Industrial Technology Islamic University of Indonesia, Yogyakarta, Indonesia

Email: [1]hartomo@uii.ac.id, [2]cikalsalatoen@gmail.com

Abstract. Batik is one of Indonesia's cultural heritage that has been recognized internationally such that this commodity is in demand by most local and international markets. For that reason, it is significant to promote and make an on offer to those markets in an effective and attractive way. E-marketing is a method of the company to communicate, promote, and sell products and services over the internet. However, some Batik industry in Indonesia has not applied this technique optimally though they have the website. It is indicated by a number of customers interacting with the company are still low which is around 3,735 visitors for 4 months. And they utilize the Batik's website for only the tour, not executing the purchase. Thus, it is beneficial for evaluating the quality of the Batik website. The objective of this study is to analyze the usability and responsiveness level of one of the Batik websites in a certain company in Indonesia. Usability test and user experience method were utilized in this study where about the 30 experienced respondents is participated. Descriptive statistical analysis is conducted to explain the status of the website. Result of this study shows that the usability level of the existing Batik website is not effective, not efficient, not satisfaction. While the level of responsiveness is not responsive (Bad Level). This is indicated a score of attractiveness, perspicuity, efficiency, dependability, stimulation, novelty is 0.12, -0.25, -0.24, 0.12, 0.08, -0.20 respectively. The recommended design in this study is valid to satisfy the user requirements at 5% of significant levels statistically.

1. Introduction

Batik is one of Indonesia's cultural heritage that has been recognized internationally with approved by UNESCO (United Nations of Educational, Scientific, and Cultural Organization) at October 2, 2009 as the artefact cultural heritage of Indonesia. This recognition makes a glory and at the same time a very strategic milestone for Indonesia as the cultural identity as well as bringing to increase the economic value where the fact show that Indonesian's Batik is in demand by most local and international markets. For this reason, it is significant to promote and make an on offer to them in an effective and attractive way. E-marketing is a method of the company to communicate, promote, and sell products and services over the internet [1].

Currently, the internet is a media used mostly by people in the world and likewise in Indonesia. Internet users in Indonesia are increasing from 132.7 million in 2016 to 143.26 million in 2017 from a 262 million populations [2]. It was also noted that everyday people (43.89% of population) in Indonesia operating a minimum duration of internet about 1 to 2 hours. As for 50% more of people uses internet for 4 up to 7 hours per day [2]. This fact raises business opportunities to be able to increase the sales of Batik products, one of which is by utilizing the website as an e-marketing method.

However, some Batik industry in Indonesia has not mostly taken advantage of this opportunity optimally though they have the website. It was indicated that a number of customers who interacts with the company through the website are still low around 3,735 visitors in May to August 2019. And they utilize the Batik's website for only the tour, not executing the transaction. This circumstance emerges a question about how good is the quality of the website system? It may the design of the website is not easy to use and attractive for a customer such as it was stated by [3] in which the difficulties in the use of the system will give an impact to the reputation of a company.

Thus, it is beneficial for evaluating the quality of the Batik website. The objective of this study is to analyze the usability and the responsiveness level of one of the Batik websites in a certain company in Indonesia.

2. Research Method

2.1. Survey
The survey was conducted by interviewing the company's Co-Founders who were business partners and handled the sales and digital marketing systems of the Batik website. This activity aims to explore the problems experienced by the company related to suboptimal function of the website used as an e-marketing media. It was indicated with the data of sales. The survey was conducted in 2 times on May 29 and July 30, 2019.

2.2. Empirical Study
The empirical study was carried out to collect the relevant data on usability and user experience of Batik websites.

2.2.1. Subject
The respondents participated in this study are some customers who familiar with e-marketing and have experienced buying a product via the websites. Their age is in range 15-27 years as well as have 6 to 16 years in utilizing the internet. They are consisted of 15 males and 15 females.

2.2.2. Apparatus
This study used some apparatus as the follows:

- Questionnaires of System Usability Scale (SUS) and User Experience Questionnaire (UEQ) for analyzing usability and user experience aspects. SUS is a questionnaire that refers to the Standard Usability Questionnaires that are used to measure satisfaction aspect. And the main goal of the UEQ is to allow a fast and immediate measurement of user experience which includes aspects of attractiveness, perspicuity, efficiency, dependability, stimulation, and novelty [4].
- Screen recording software is used to record respondent's activities when accessing the website.
- Corel X7 software, Canva application, Adobe XD software, Marvel App application, are used to redesign initial websites and make prototypes.

2.2.3 Experimental Design
The experiment was done in the Ergonomics Laboratory which participants were asked for doing the task by using laptop Intel Core i3 version. A prelimenary test was conducted to identify an experience of respondents in using website. They are in sitting position to operate the computer in comfortable position and complete the task for 30-60 minutes. Subsequently the respondent is asked to fill in the questionnaires to test usability and user experience of the website. Then the interview is conducted to explore some matters in using Batik website to make a deal with the company. Figure 1 describes the layout design of experiment.

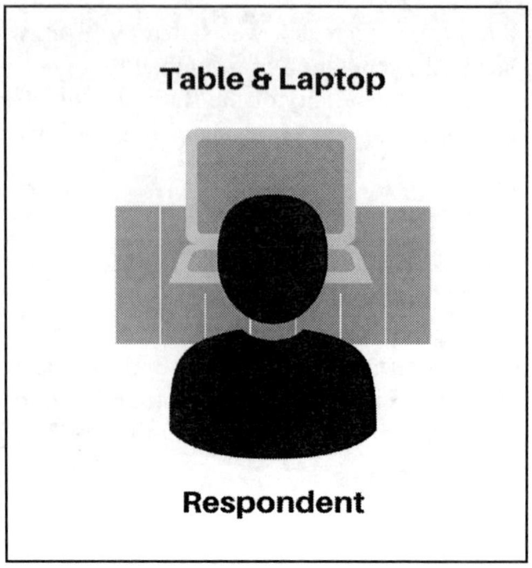

Figure 1. Experimental layout design.

2.2.4. Task Design
A task conducted in this study was composed of three parts which is the result of discussions with the company's Co-Founders who want to find out how good the quality of the 3 main features of the website are. The first task is to buy 1 batik clothes with a price in range 440,000 rupiahs to 545,000 rupiahs for male and 1 batik clothes with a price more than 950,000 rupiahs for female by using full payment. The second task is to confirm the payment done. And last task is to carry out the process of returning and exchanging goods purchased because of there was a discrepancy between the items received and those ordered.

2.2.5. Procedure of Experiment
Experiment study was conducted as following procedure. Respondent was informed about the objective and the rules of this study. Each respondent was asked to do the task by clicking the start button on the screen recording software. After each task is completed, the respondent tells the researcher to press the stop button and save the video. Then, each respondent was asked to fill in the SUS and UEQ questionnaire related to their assessment of the website. After completing the questionnaires, each respondent submits the questionnaire to the researcher. Lastly, each respondent was asked about their perceptions and responses experienced when interacting with the Batik website. The researcher shows the website design that is tested and asks questions about the website based on the answers to the questionnaires. The researcher closes the interview session and thanks the respondent for the time given.

2.3. Statistical Analysis
Descriptive statistical analysis is conducted to test the hypothesis of the website. The data is presented in graphical form of average calculation and percentage calculation of each aspect.

3. Result and Discussion

3.1. Result of Usability Analysis
Figure 2 describes the result of usability test. It explains the effectiveness in use of the Batik website which the average success rate in completing the task is 85.6%. This phenomenon shows that user of the website still experiences some error in access that is 13 respondents from 30 (43.3%). The errors

occurred is resulted from uninformative and unresponsive of interface as well as incomplete command. This condition produces confusion of user in accessing the website to make transaction.

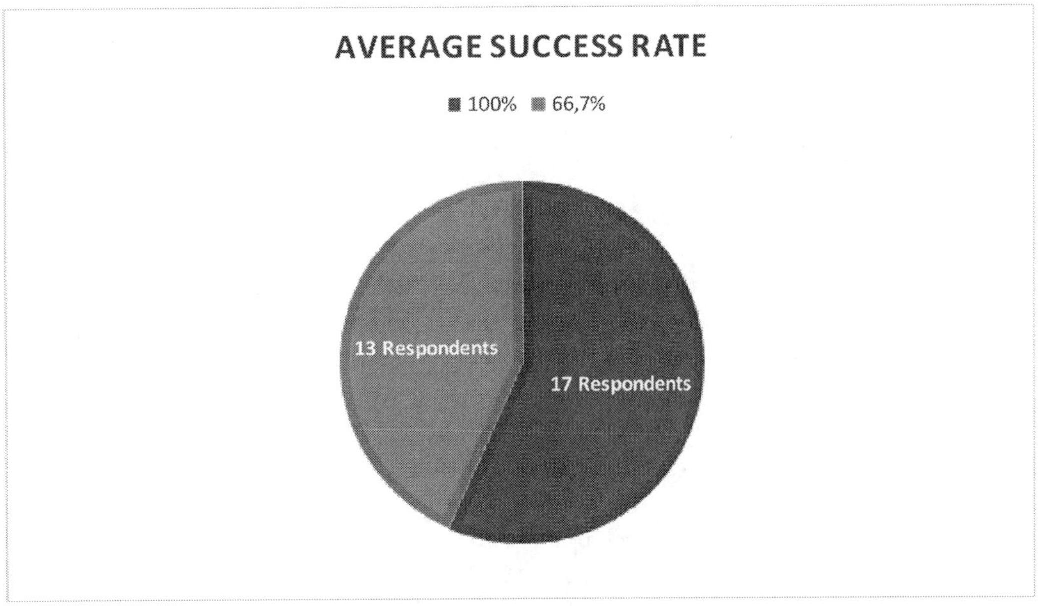

Figure 2. Average success percentage of effectiveness.

As for Figure 3 presents a level of error for each task. This figure explains level of error in completing task 1; buy 1 batik clothes for male and 1 batik clothes for female in which as much 232 errors were occurred consisting of mistake in selecting a size (53 times), mistake in typing an address (10 times), mistake in filling some data such as a misspelt word in typed (25 times), mistake in filling out the address (3 times), mistake in choosing filter and catalogue (12 times), mistake in choosing the clothes (2 times), mistake in choosing the quantity of clothes (10 times), mistake in clicking unrelated features with the given task (85 times), mistake in doing revision activity errors task (19 times), mistake in name input (2 times), mistake in repeating clothes buying activities (11 times). In completing task 2; to confirm the payment done in which as much 77 errors were occurred consisting of mistake in filling account number (7 times), mistake in filling name of account owner (1 time), mistake in filling date (2 times), mistake in filling some data such as a misspelt word in typed (44 times), mistake in doing revision activity errors task (7 times), mistake in filling data name (8 times), mistake in clicking unrelated features with the given task such as opening the basket feature, click payment, click home, click make order (8 times). And in completing task 3; to carry out the process of returning and exchanging goods purchased in which as much 150 errors were occurred consisting of mistake in not selecting the return feature (1 time), mistake in clicking unrelated features with the given task (149 times) such as clicking catalog feature, blog feature, search feature, home and page refresh feature, FAQ feature, privacy policy feature, my account feature, size guide feature, promotions feature, store location feature, order tracking feature, login feature, Instagram feature, confirm payment feature, Facebook feature, chat feature.

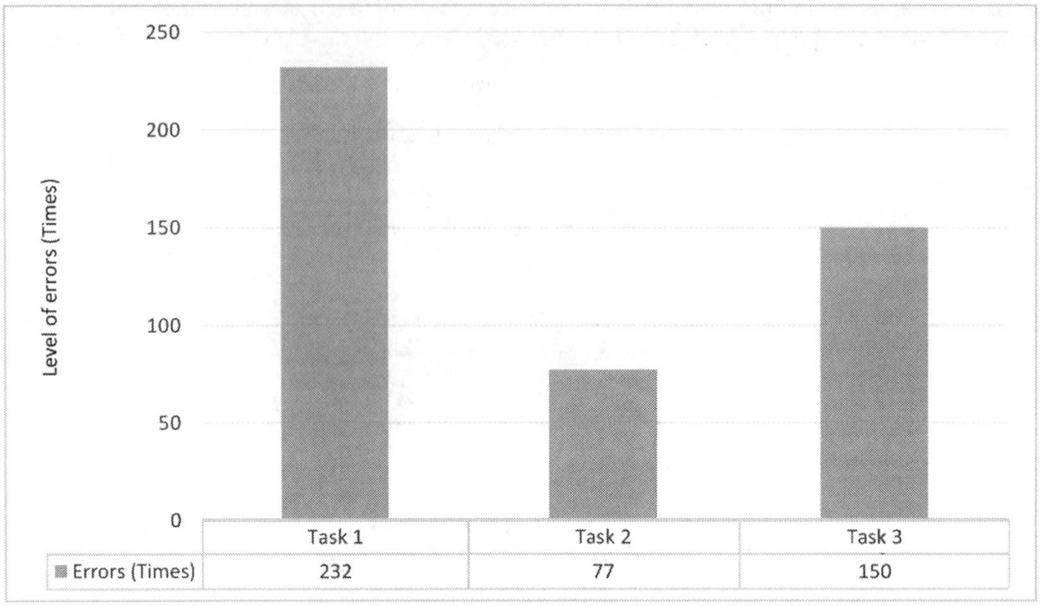

Figure 3. Level of error for each task.

Figure 4 explains the efficiency level of customer in using the Batik website in which the average time in completing the task 1 is 567 second. It because of the customer experienced some error, 232 errors, when ordering 1 batik clothes for male and 1 batik clothes for female. While task 2 is 136 second where this was happened because of the customer experienced some error, 77 errors, when confirm the payment done. And task 3 is 194 second where this was happened because of the customer experienced some error, 150 errors, when carry out the process of returning and exchanging goods purchased.

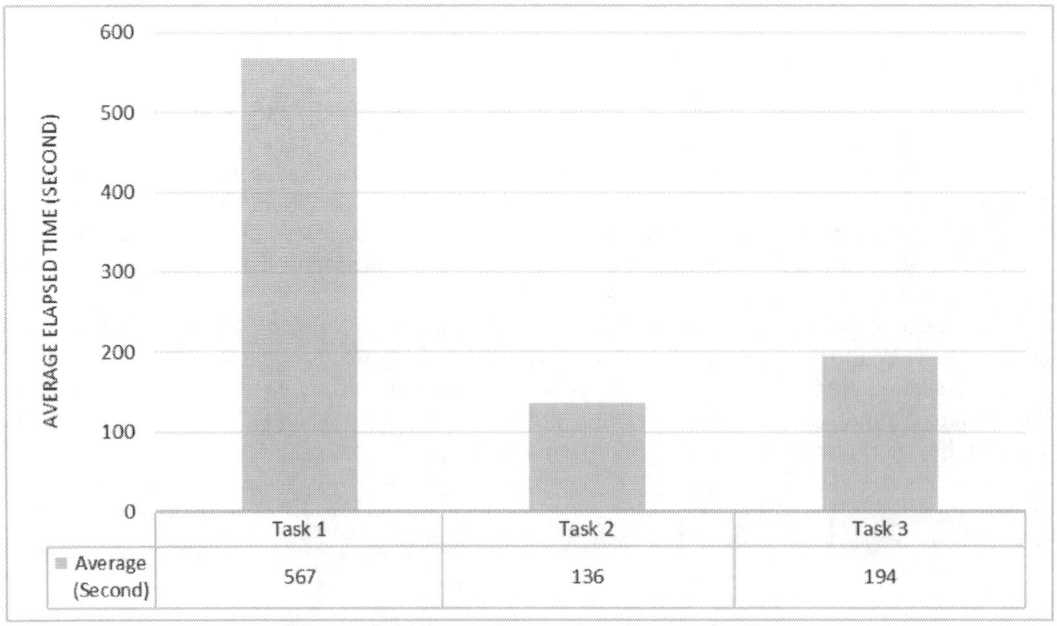

Figure 4. Average elapsed time for each task.

Figure 5 explains the level of customer satisfaction in using the Batik website as many as 12 respondents (40%) assess the average satisfaction level of the website with 62.1 results categorized as Ordinary and Marginally Acceptable. It means the Batik website can still be accepted within marginal limits because the customer feels no significant obstacles and is not too confused. But there are still many features that are inconsistent so that customers make mistakes. While as many as 15 respondents (50%) assess the average website satisfaction level with 34.5 results that are categorized as Poor and Not Acceptable. And as many as 3 respondents (10%) rate the average website satisfaction level with 21.7 results categorized as Worst Imaginable and Not Acceptable. It means the Batik website is not informative in which the content is still difficult to understand so as the most errors occurred. And moreover, the Batik website is also not responsive in which if there is some error, no immediate feedback to the user. Thus, this phenomenon proves that the existing Batik website does not accumulate the requirements of the customer.

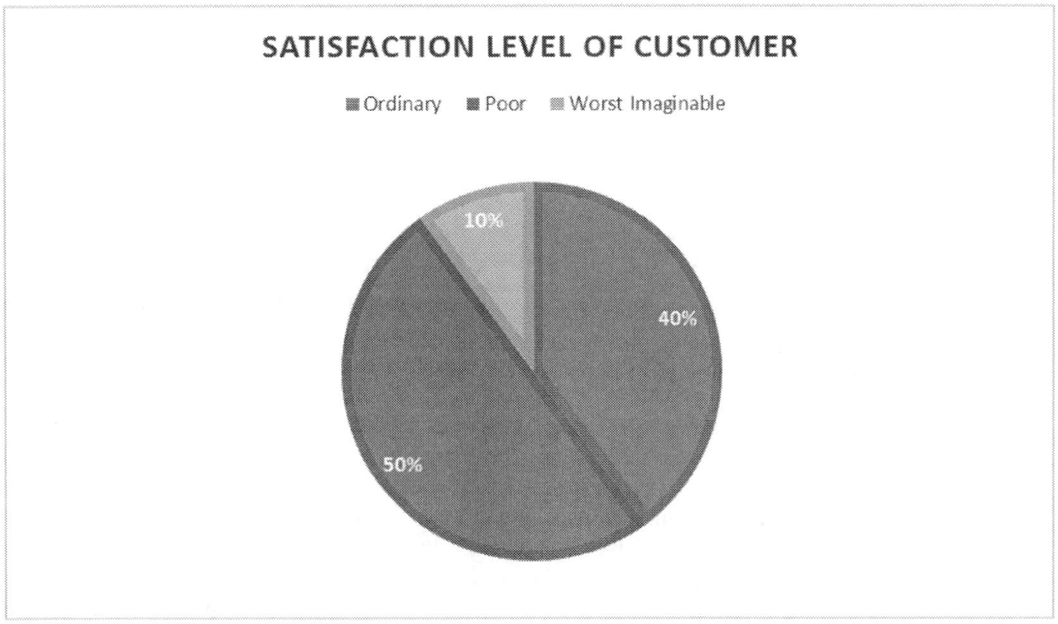

Figure 5. Satisfaction level of customer.

3.2. Result of User Experience

Figure 6 describes the result of user experience (UX) test on bases customer survey using User Experience Questionnaire. This survey involves six criteria. They are attractiveness, perspicuity, efficiency, dependability, stimulation, and novelty. Attractiveness is user's impression of the product including like or dislike, enjoyable or pleasing and bringing out the user to experience. While perspicuity is related to easy to get familiar with the product, easy to learn, and easy to understand. And efficiency is associated with an effort used and speed in completing the task. Dependability involves the user feeling in control of the interaction and safety. As for stimulation is a user's feeling on exciting product and also fun. Finally, novelty is an innovation of product that brings about user's attention [5]. The result of UX test shows that the average score of each criterion is 0.12 for attractiveness, -0.25 for perspicuity, -0.24 for efficiency, 0.12 for dependability, 0.08 for stimulation, and -0.20 for novelty. Subsequently, these scores were benchmarked with the result of previous research by [6] that using 401 products evaluated. Result of this benchmarking is presented in Figure 6 and Table 1.

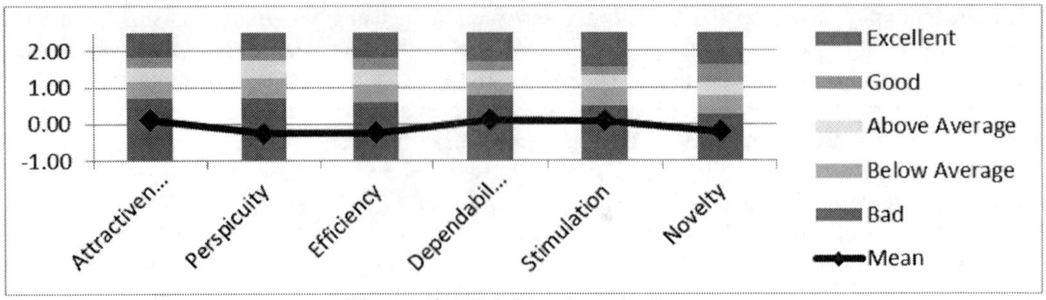

Figure 6. UEQ benchmark chart.

Table 1. Comparison to benchmark.

Criteria	Score	Score References	Interpretation
Attractiveness	0.12	0.70	Bad
Perspicuity	-0.25	0.71	Bad
Efficiency	-0.24	0.57	Bad
Dependability	0.12	0.77	Bad
Stimulation	0.08	0.50	Bad
Novelty	-0.20	0.25	Bad

Figure 6 and Table 1 explained that the Batik website investigated was classified into Bad level. It because of the score obtained for each criterion was below the score reference which this level describes the website does not accommodate the user criterion so as its quality is identified as not attractive, not perspicuity, not efficient, not dependable, not stimulate, and not innovative. This state of affairs demonstrates that the Batik website used currently is not responsive to implement in the user mind. Thus, the existing website should be improved based on UX concept. According to [7], UX will help to understand the users need. The impact is the user will be able to memorize an experience obtained so as it may be repeated and trust to make some transactions at another opportunity. While Mc Knight et. al in [8] explained that if consumers see the good quality of the website, they will be encouraged to willing doing an online transaction and it will grow satisfaction and loyalty on customers. Oliver in [9] states that loyalty is a commitment held firm to repurchase a product or service.

4. Conclusion

The conclusion of this study can be mentioned as follow:

- The usability level of the existing Batik website is not effective, not efficient, not satisfaction in which score of them respectively shows that an average of success rate is 85.6%, the level of error in completing task 1 is 232 errors, in completing task 2 is 77 errors, in completing task 3 is 150 errors for effectiveness criteria. As for the average time in completing the task 1 is 567 second while task 2 is 136 second, and task 3 is 194 second for efficiency criteria. While for satisfaction criteria in using the Batik website is only 40% of user with score 62.1 that is classified as Ordinary and Marginally Acceptable, 50% of user with score 34.5 that is categorized into Poor and Not Acceptable, and 10% of user with score 21.7 that is categorized into Worst Imaginable and Not Acceptable.
- The user experience level of the existing Batik website is not responsive (Bad Level) in which the website is not attractive (0.12), not perspicuity (-0.25), not efficient (-0.24), not dependable (0.12), not stimulate (0.08), and not innovative (-0.20).

5. Recommendation

The recommended design in this study as a prototype of the Batik website for improvement can be seen in the appendix in which the design was developed based on the usability and user experience concept. However, this design also involved a guideline developed by the U.S. Department of Health and Human Services (HHS) and U.S. General Service Administration in [10] on The Research-Based Web Design & Usability. And in the second place, the proposed design is valid to satisfy the user requirements at 5% of significant levels statistically.

Acknowledgement

We wish to acknowledge the Directorate of Student Development Islamic University of Indonesia Yogyakarta-Indonesia for funding this project.

References

[1] Sharma, G. K., and Kumar, P 2006 *Handbook of Management Terms* (Punjagutta: The Icfai University Press)
[2] APJII 2017 *Penetrasi & Perilaku Penguna Internet Indonesia (*Asosiasi Penyelenggara Jasa Internet Indonesia)
[3] Maguire, M 2001 Method to Support Human-Centered Design *J Hum Comput Stud* vol. 55 (4), pp 587-634. doi.org/10.1006/ijhc.2001.0503
[4] Schrepp M., Hinderks A., Thomaschewski J 2017 Construction of a Benchmark for the User Experience Questionnaire (UEQ) *J Interact Multimedia and Artif Intell* vol. 4(4), pp 40-44. doi.org/10.9781/ijimai.2017.445
[5] Schrepp, M.; Hinderks, A. & Thomaschewski, J 2014 Applying the User Experience Questionnaire (UEQ) in Different Evaluation Scenarios. In: Marcus, A. (Ed.): Design, User Experience, and Usability. Theories, Methods, and Tools for Designing the User Experience. Lecture Notes in Computer Science, Volume 8517, 383-392, Springer International Publishing. doi.org/10.1007/978-3-319-07668-3_37
[6] Schrepp, M 2019 User Experience Questionnaire Handbook: All You Need To Know To Apply The Ueq Successfully In Your Projects.
[7] Cheng, N 2014 What is UX Research and Why Should I Care?. http://chaione.com/what-is-ux-research-and-whyshould-i-care/
[8] Chang, H. H. & Chen, S., W 2008 The impact of online store environment cues on purchase intention: trust and perceived risk as a mediator. Online Information Review, vol. 32, pp. 818-841 doi.org/10.1108/14684520810923953
[9] Hurriyati, R 2010 *Bauran Pemasaran dan Loyalitas Konsumen* (Bandung: Alfabeta)
[10] Leavitt, M. O. & Shneiderman, B 2006 *Research-Based Web Design & Usability Guidelines* (s.l.: U.S. Department of Health and Human Services (HHS) and U.S. General Services Administration (GSA))

Appendix A. Recommended design

Table A.1 Proposal design results 1.

Before	After

	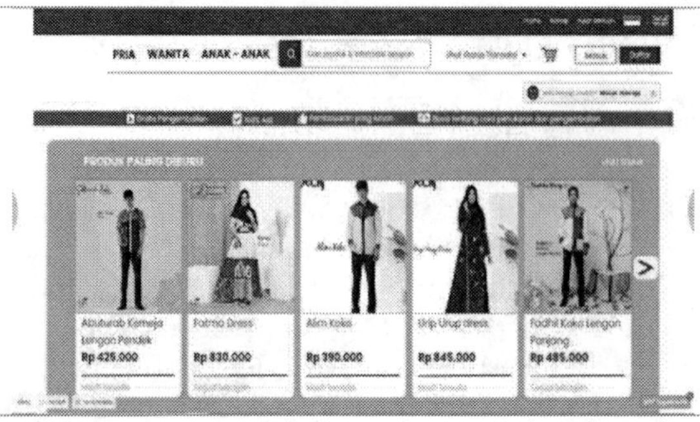
Avoiding cluttered appearance, creating a level of importance of information, minimizing the number of clicks or pages that correspond to related information.	

Table A.2 Proposal design results 2.

Before	After
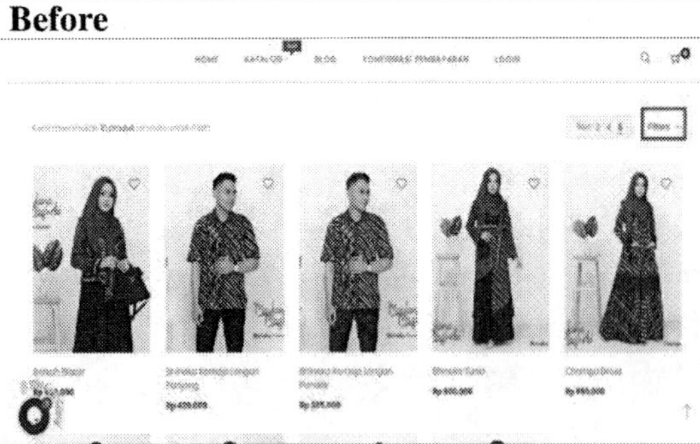	
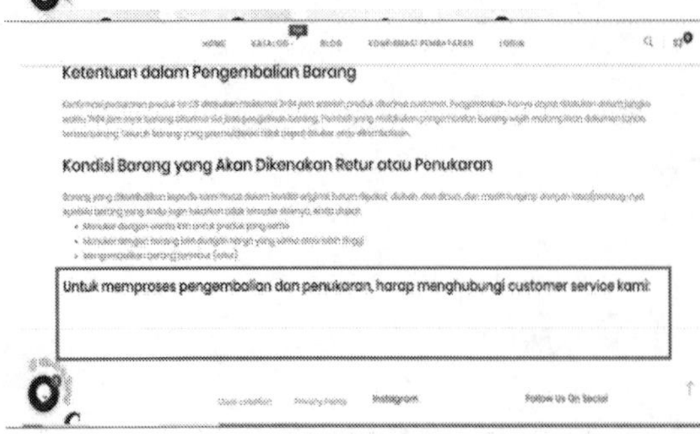	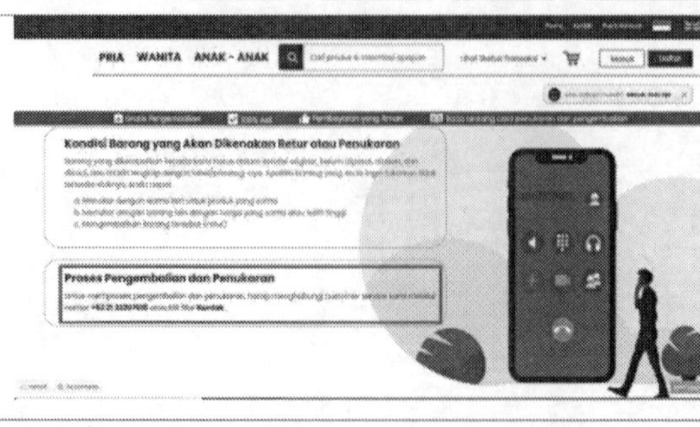
Ensure that the required information is displayed by adding supporting documents.	

Table A.3 Proposal design results 3.

Before	**After**

Place important features that are often accessed at the top center of the page.	

Table A.4 Proposal design results 4.

Before	**After**
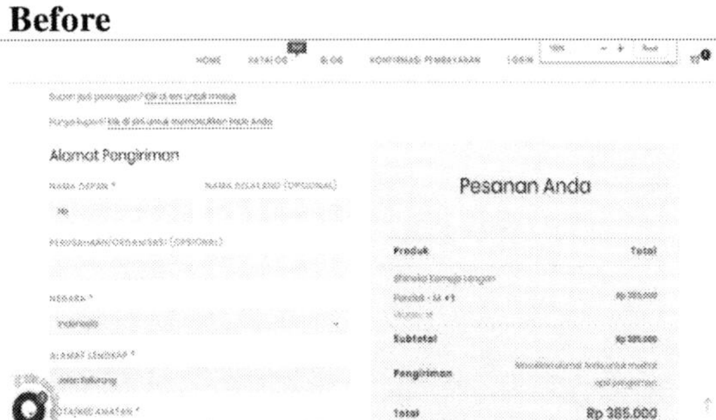	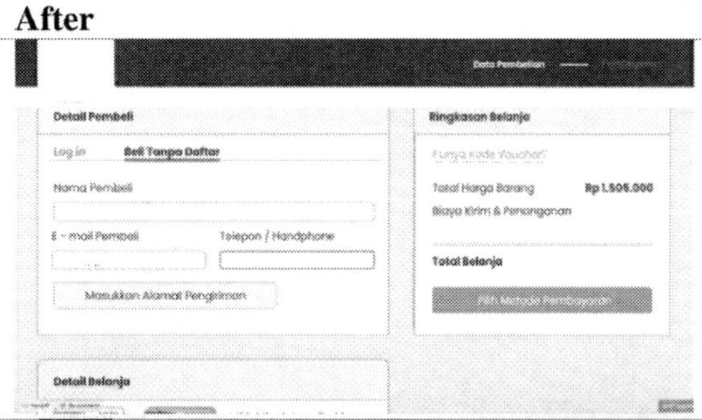
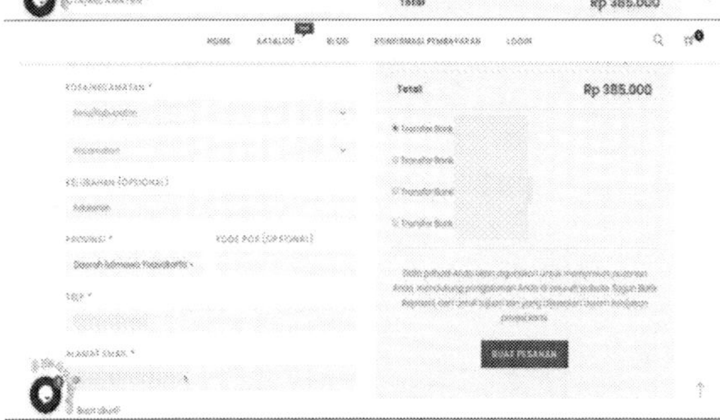	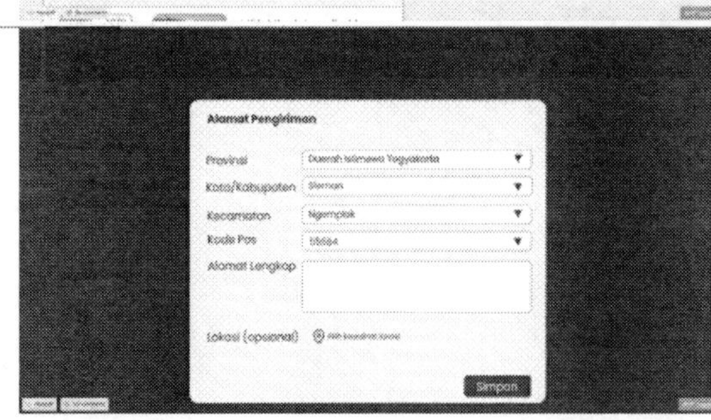
Only display information that is needed by removing information that is less useful or optimizing it to make it more useful.	

IOP Conf. Series: Materials Science and Engineering 1072 (2021) 012051 doi:10.1088/1757-899X/1072/1/012051

Table A.5 Proposal design results 5.

Before	**After**
	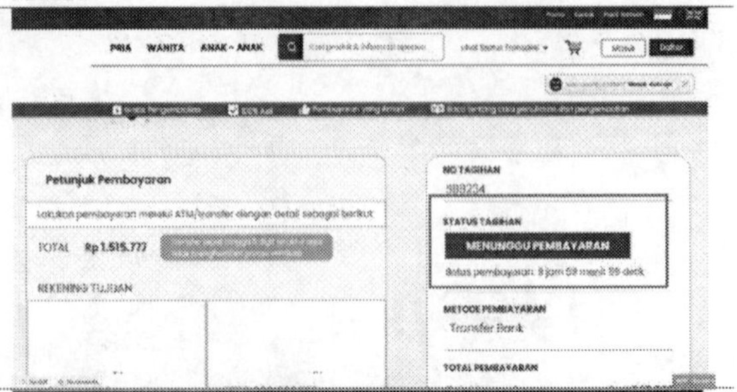 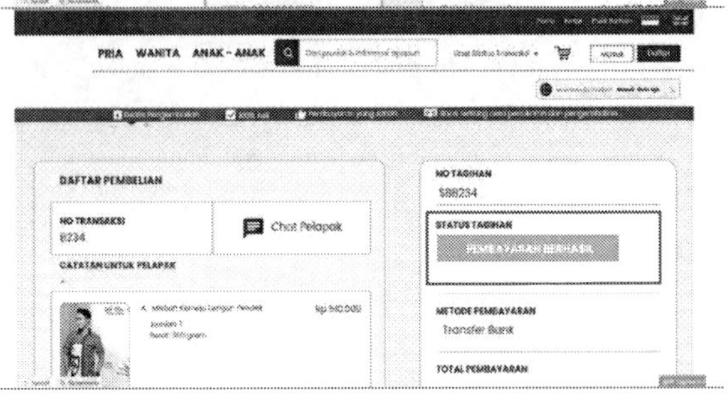
Announce changes by adding notification display and message.	

Quality control of palm oil crude (CPO) using six sigma method in PT Karya Tanah Subur West Aceh regency

Sofiyanurriyanti[1], G Putra[2], A R Arifin[3]

[1,2,3]Industrial Engineering Departement, Faculty of Engineering, University Teuku Umar, Alue Peunyareng, Ujong Tanoh, Darat, Mereubo, Meulaboh, Aceh Barat 23681, Indonesia

E-mail: [1]Sofiyanurriyanti20@gmail.com, [2]Gaustamaputra@utu.ac.id, [3]Refands02@gmail.com

Abstract. The rapid world demand for palm oil shows an increasing tendency towards product consumption in palm oil raw materials. PT. Karya Tanah Subur is a company engaged in processing fresh fruit bunches that produce palm oil and palm kernel. Six Sigma is a method used to provide solutions and find the cause of product defects by providing a proposed improvement in several stages. The results showed that the causes of CPO oil quality discrepancies were obtained in the cause-and-effect diagram analysis, namely factors of workers, production machinery, work methods, materials. While the defect per million opportunities in free fatty acid levels is 35,528, impurities content is 18,899 and water content is 14,719. Based on the value of the average sigma level for free fatty acid levels that is equal to 3.38, the level of impurities sigma value of 3.57 and the water content of the sigma value of 3.68 sigma. Proposed improvement of product defects mismatch with the 5W + 1H method, namely in free fatty acid levels, work system improvements in the sorting of fresh fruit bunches, provide training to operators in the selection of appropriate fruits while impurities by monitoring oil purification operators, machine maintenance cleaning periodically at the sorting station, periodic checking every hour, while the moisture content of the corrective action that must be taken is to supervise the sterilizer operator.

1. Introduction

Indonesia is one of the largest producers and exporters of palm oil. The rapid world demand for palm oil shows an increase in line with the world population from year to year increasing due to the increasing consumption of products with basic raw materials crude palm oil [5].

According to [3] of the rapid world demand for palm oil consumption at this time, Indonesia is the largest producer of palm oil in the world. According to data from the United States Department of Agriculture (USDA) that domestic consumption of palm oil in 2019 is estimated to reach 12.75 million tons or around 17% of total world consumption which reaches 74.48 million tons. While the second largest country is India, reaching 10.19 million tons or 13.68% of total world consumption. Furthermore, the third largest was China, with 7.22 million tons or around 9.7%.

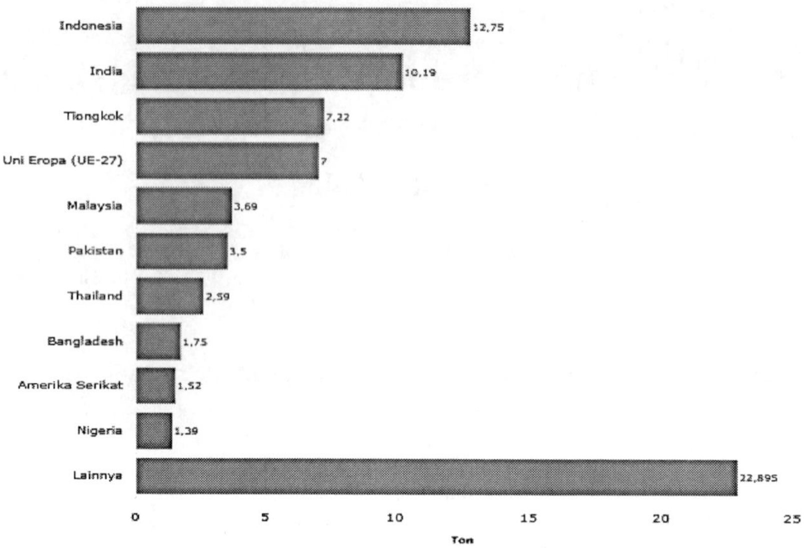

Figure 1. Ten countries with the world largest crude palm oil consumption.

2. Literature Review

Therefore, the high demand in the palm oil industry for the types of processed raw materials for palm oil is increasingly rapid and competition is also increasingly competitive. Companies also have to compete to survive by showing the superiority of the products produced. Quality control is a company's tactics and strategy to achieve market excellence, competitive advantage by having the character to be able to satisfy the needs of consumers [5]. Therefore, if a product produced does not meet a need in accordance with the wishes of consumers, there will be problems with quality. An issue can be controlled by measuring the performance of the industrial system according to the strategy [7].

Palm oil parameter standards in the company based on company operational standard specifications that have been determined through the government BSN (National Standard Agency) have set the standardization of CPO palm oil contained in SNI-01-2901-2006 is the level of ALB (Free Fatty Acid) (2.5-3.0%), impurities content (0.01-0.02%) and water content (0.1-0.2%) [9].

PT. Karya Tanah Subur is engaged in processing fresh fruit bunches that produce palm oil and crude palm oil. The parameters that determine the quality of palm oil are free fatty acid content, water content, and impurities. The current situation shows that in processing palm oil, the quality produced turns out that some still do not meet the standard specifications set by the company. Efforts to reduce defects in palm oil products are by controlling the quality of palm oil.

Based on the types of parameters of palm oil shows that there are still types of defects in the three parameters, namely the levels of ALB in July-December 2018 shows that there are still many data that are above the standard limits that have been set and in 2019 ALB levels that exceed the standard limit is at in June, while for the level of impurities that is data that exceeds the disability limit occurs in September-November 2018 and for 2019 occurred from March-June which exceeds the established standard limits, while the water content that is not according to the standards set by the company namely in May-June 2019.

In their research [12] with the title "Analysis of Quality Control of CPO (Crude Palm Oil) Using Six Sigma in the CPO Processing Plant of PT. Gunajaya Karya Gumilang, Kedawangan District, Ketapang Regency. The results of his research showed that free fatty acid quality remained at the level of 1.3 sigma with a DPMO value of 317,500 products that did not meet the requirements specified by SNI. Water content is at level 3 and impurities are approaching level 4 sigma.

Based on the description in the background of this research, it is necessary to analyze the efforts to control the quality of palm oil in PT. Karya Tanah Subur with the aim to determine the factors that cause

CPO palm oil quality mismatch, determine the sigma level in the process of controlling palm oil quality and provide suggestions for improvement of product defects with 5W+1H method.

3. Methods

3.1 Quality Control

According to [8] Quality is a key success factor for overall business growth characteristics of a product and service from marketing, maintenance, design, price and durability of product availability. If a product produced by a company is not in accordance with the standards to be set then the product is said to be a defective product group [7].

Quality in the world of manufacturing can be interpreted as an item or result to determine or decide on consumers who have bought the product/goods will not return to buy it [1].

Quality control is one of the techniques or methods that will be carried out before the end of the production process by producing the final product in accordance with established standards [10].

Steps in determining quality standards according to [6] there are four standards including:

1. Considering competition and competitor product quality
2. Considering the use of the final product
3. The selling price of the product is in accordance with the expected quality
4. Consist of several teams in the field carried out

3.2 Quality Control Aids

Quality control has seven statistical tools that are used to control quality, including Check Sheet, Stratification or Histogram, Histogram, Control Chart, Pareto Diagram, Scatter Diagram, Cause Chart (Fishbone Chart) [13].

3.3 Six Sigma

Six Sigma [4] generally consist of two words namely six meaning six, sigma means symbol or symbolic standard deviation which can be interpreted as a unit size by symbolizing a process or measure of sigma value. The Six Sigma method approach is also needed to approach science, equipment, training and measurement. The approach used in the Six Sigma method is define, measure, analyze, improve and control.

3.4 Steps of the Six Sigma Method

The steps for the Six Sigma method include:

1. Define
   Is a stage of improving quality by determining the proportion of defects to damage to production failures so you can use the SIPOC diagram to determine the process model"SIPOC (Suppliers, Inputs, Processes, Outputs, Customers).
2. Measure
   Measure is an improvement in the quality of Six Sigma, there are three things that need to be done, among others: determining the quality characteristics by determining CTQ (Critical to Quality), determining the control variable map and measuring sigma values.
3. Analyze
   Quality improvement programs that can be carried out include grouping types of disabilities, making pareto diagrams, making cause and effect.
4. Improve
   Improve is the cause of quality problems done by establishing an action plan to improve the quality of Six Sigma. In this improve phase there are stages that need to be done including determining the corrective action with the 5W+1H method, determining the action using the tree diagram method.

5. Control
   Supervision is needed to be able to know the effect of the achievement process on the quality of product produced by documenting the procedures documented in the standard work guidelines.

3.5 Six Sigma Method as a Measurement Tool
Six Sigma as a measurement system uses defect per million opportunities as a unit of measurement. DPMO is a good measure for product or process quality, because it correlates directly with defects, costs and time wasted, to be able to see more clearly about sigma levels can be seen in Table 1. The relationship between Six Sigma and DPMO [1].

Table 1. Relationship between sigma and DPMO.

Sigma	Parts per Million
6 Sigma	3.4 defects per million
5 Sigma	233 defects per million
4 Sigma	6,210 defects per million
3 Sigma	66,807 defects per million
2 Sigma	308,537 defects per million
1 Sigma	690,000 defects per million

3.6. SIPOC Diagram (Supplier, Input, Process, Output, Customer)
The SIPOC diagram is one of the Six Sigma tools used by the process improvement team to identify each element in the process improvement project before the process is run [4].

4. Results
4.1 Data Collection and Data Processing
Data collection was obtained by interviewing, documenting the company and fielding directly in the field at PT. Karya Tanah Subur located in Padang Sikabu Village, West Aceh Regency, Aceh Province. Data collection which includes data on the amount of production, causes of disability and conditions that exist in the production department. Whereas in data processing using the DMAIC method (Define, Measure, Analyze, Improve and Control).

4.2 Palm Oil Production Data (CPO)
Palm oil Production Data (CPO) in 2018-2019 can be seen in Table 2. as follows.

Table 2. Total palm oil production (CPO) in july 2018-june 2019.

Month	Amount of Production CPO (Ton)
July 2018	5503
August 2018	5696
September 2018	5564
October 2018	5293
November 2018	5084
December 2018	5403
January 2019	6715
February 2019	6688
March 2019	6370
April 2019	6669
May 2019	6685
June 2019	6983

4.3 Data on the Type of Palm Oil Damage (CPO)
Data on the type of damage to palm oil (CPO) products of PT. Karya Tanah Subur, there are three types of damage including free fatty acid (ALB), impurities content and water content, while data on the types of damage to palm oil (CPO) products in 2018-2019 can be seen in Table 3. as follows.

Table 3. Data on damage types of palm oil products (CPO).

Month	Type Parameter Product CPO					
	Free fatty acid %	Total Reject	Dirt %	Total Reject	Moisture %	Total Reject
July 2018	3.18	537	0.02	312	0.17	243
August 2018	3.97	521	0.02	326	0.22	226
September 2018	3.17	540	0.03	372	0.20	252
October 2018	3.13	541	0.03	385	0.25	265
November 2018	3.30	555	0.03	314	0.22	243
December 2018	3.31	552	0.02	339	0.19	248
January 2019	2.84	506	0.02	348	0,20	283
February 2019	2.52	576	0.02	361	0.22	261
March 2019	2.43	568	0.03	389	0.22	259
April 2019	2.60	592	0.03	339	0.29	245
May 2019	2.71	506	0.03	336	0.32	267
June 2019	3.48	581	0.03	345	0.30	259

4.4 Data processing
Data processing in this study uses the DMAIC (Define, Measure, Analyze, Improve and Control) approach which is a holistic approach to solve problems and improve six sigma processes. The approach will be described in stages as follows

4.4.1 Define Stage
Making the sequence of processes that occur can be done using SIPOC diagrams (Suppliers, Inputs, Processes, Outputs, Customers). At this stage, the process of processing fresh fruit bunches into CPO oil will be explained. The SIPOC diagram for the CPO oil processing is as follows:

1. Suppliers
   Suppliers of raw materials in the form of palm fruit or known as fresh fruit bunches used for the process brought in from the community and PT. Karya Tanah Subur itself.
2. Input
   Input is a raw material in the form of fresh fruit bunches that will be processed for processing palm oil imported from the community and the Oil Palm Plantations (PKS) unit of PT. Karya Tanah Subur itself.
3. Process
   It is a step or process of production activities ranging from receiving fresh fruit bunches to becoming CPO oil.
4. Results (Output)
   Output is the result obtained from the processing of fresh fruit bunches which are processed into Crude Palm Oil (CPO).
5. Customers
   Is a person or group that receives Crude Palm Oil (CPO) oil products produced by PT. Karya Tanah Subur, that sold to PT. Belawan Tangki Indonesia North Sumatra.

4.4.2 Measure Stage

Measure is the measurement phase is the second operational step in the Six Sigma method. The steps are as follows:

a. Determination of Quality Characteristics Critical to Quality (CTQ)
   CPO product standards indicated the type of disability called CTQ (Critical to Quality), to further clarify the definition of the factors that affect the disability of CPO oil products.

Table 4. CTQ on *crude palm oil* (CPO).

No	CTQ *(Critical to Quality)*	Type of CPO Oil	Specification Quality Standard CPO	Description
1.	CTQ-1	Free Fatty Acid (ALB)	Grading Value ALB Between 2.5-3.0%	High concentrations of free fatty acids present in palm oil are very detrimental. This high free fatty acid results in lower oil yields and low oil quality, in food ingredients, fatty acids with levels greater than the weight of fat will result in unwanted and smell bad.
2.	CTQ-2	Dirt	Dirt Between 0.01-0.02%	To get a better oil can be done by removing dirt, so if one is palm oil processing companies can reduce levels of dirt by level the smallest, then the oil already has the requirements to be a good oil.
3.	CTQ-3	Moisture	Moisture Between 0.1-0.2%	The quality of palm oil must be maintained by removing volatile substances. Moisture in this case is one of the volatile substances when it is at temperatures above 100°C. High levels of water can cause odorous oils and reduce the quality of the palm kernel oil.

b. P Full Chart (P-chart)
   P control chart (P-chart) is used to determine the production process that occurs within the limits of control or not. If the process is within the control limits then it can be said that the process is stable, on the contrary if the process is outside the control limits then the process is said unstable. For making a control chart the first steps in making a control chart is as follows:
   1. Check Sheet
      Check Sheet method is useful to simplify the process of controlling CPO oil quality to determine the problem area based on the frequency of the type or cause and make a decision to make improvements or not. Check Sheet total oil production and type of damage CPO show in Table 5.
   2. Charts of Free Fatty Acid (ALB)
      The control chart of Free Fatty Acid (ALB) can be seen in the Full Control Chart Figure which is adjusted to the standards set as Figure 2:
      Based on Figure 2, it can be explained that there are still points that are outside the upper control limit and the lower control limit. There are 8 points that are outside the control limit, namely at points 1, 2, 3, 4, 5, 6, 9 and 12 or 12 samples while points that are inside the control limit are 4 points including 7, 8, 10, and 11. Based on that, so that it can be said that the process is not controlled because of the point of fluctuation and irregularity this shows that the levels of Free Fatty Acid (ALB) in CPO oil products.

Table 5. Check sheet total oil production and type of damage CPO.

Month	Type Parameter Product CPO					
	Levels ALB %	**Total Reject**	**Levels of Dirt%**	**Total Reject**	**Levels Moisture %**	**Total Reject**
July 2018	3.18	537	0.02	312	0.17	243
August 2018	3.97	521	0.02	326	0.22	226
September 2018	3.17	540	0.03	372	0.20	252
October 2018	3.13	541	0.03	385	0.25	265
November 2018	3.30	555	0.03	314	0.22	243
December 2018	3.31	552	0.02	339	0.19	248
January 2019	2.84	506	0.02	348	0,20	283
February 2019	2.52	576	0.02	361	0.22	261
March 2019	2.43	568	0.03	389	0.22	259
April 2019	2.60	592	0.03	339	0.29	245
May 2019	2.71	506	0.03	336	0.32	267
June 2019	3.48	581	0.03	345	0.30	259
Total Reject		**6575**		**4166**		**3051**

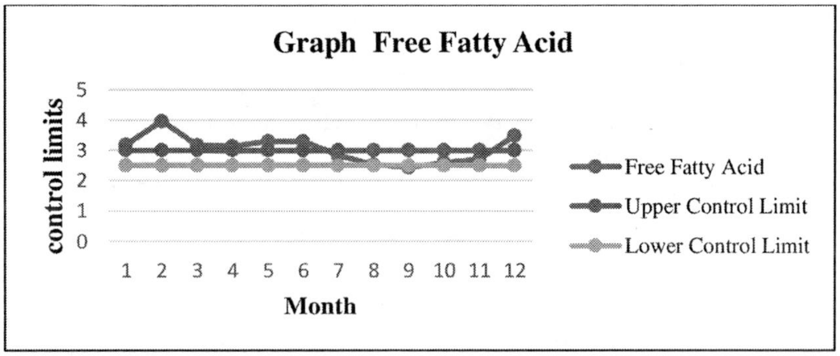

Figure 2. Graph of free fatty acid (ALB).

3. Moisture
   The control chart for impurities from the CPO oil processing is based on the Table 4 adjusted to the standards set can be seen in Figure 3 set as follows:

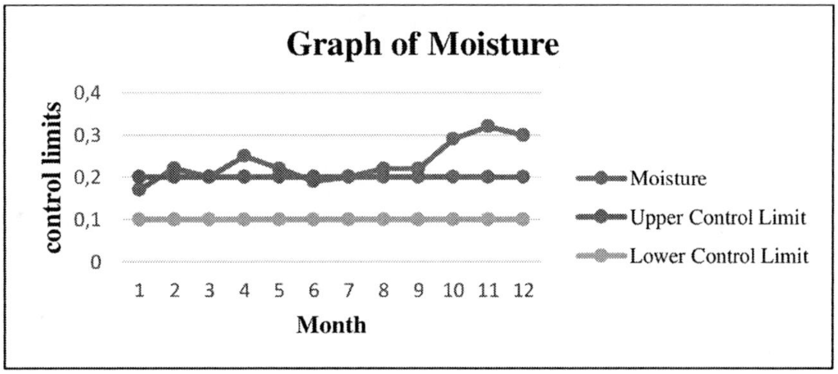

Figure 3. Graph of moisture.

Based on Figure 3 above showed that there are still points that are outside the upper and lower control limits. There are 7 points that are outside the control limit including at the point or sample to 3, 4, 5, 9, 10, 11 and 12 while points that are inside the control limit are 5 of them at points 1, 2, 6, 7 and 8.

4. Dirt

Control chart for the water content processing is based on the Table 4 adjusted to the standards set can be seen in Figure 4 set as follow:

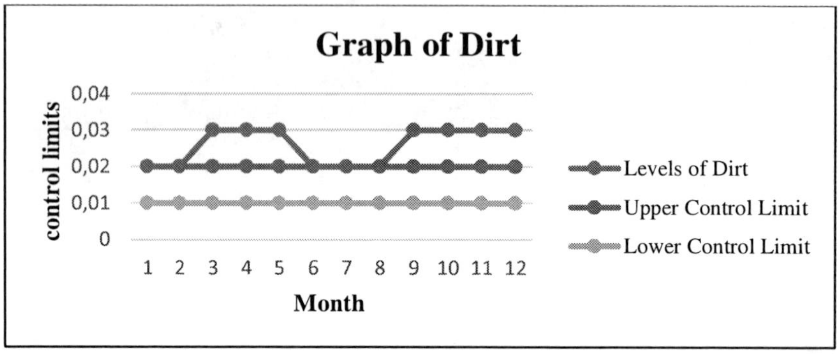

Figure 4. Graph of dirt.

Based on Figure 4 above, showed that there are still points that are outside the control limits in accordance with the standards set by the company. There are 8 points that are outside the control limit including those at point 2, 4, 5, 8, 9, 10,11 and 12 while points that are inside the control limit are 4 of them at point 1, 3, 6 and point 7. Based on that, so that it can be said that the process is not controlled.

4.4.3 Sigma Level

As for calculating the sigma level is used based on these calculations, then for the monthly calculations on the CPO oil processing produced by PT. Karya Tanah Subur works in the same way from each type of quality parameters can be seen in Figure 5. Level Sigma Crude Palm Oil set as follows:

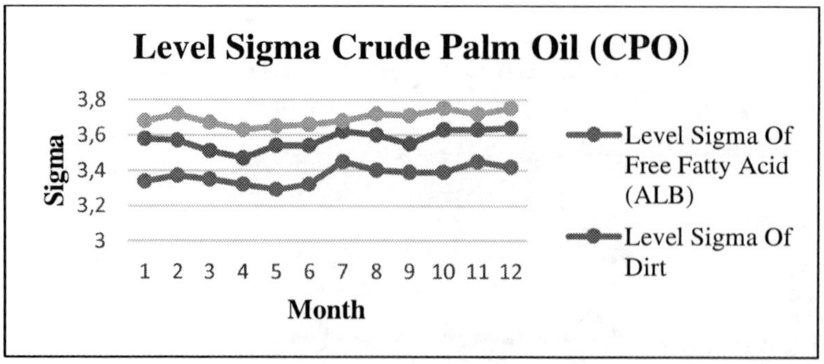

Figure 5. Graphs of sigma from the CPO oil processing process in PT. Karya Tanah Subur.

4.4.4 Analysis Phase

In the analysis phase, there are three stages carried out, among others, stratification, making pareto diagrams, and making causation.

4.4.5 Histogram

Based on Figure 6, showed that the highest total defect is in the level of Free Fatty Acid (ALB) with a total disability of 6,575 with a total percentage of 47.67%, the second highest disability is in the total dirt level with 4,166 with a total percentage of 30.21% and the smallest is moisture content of 3,051 with a percentage of 22.12%.

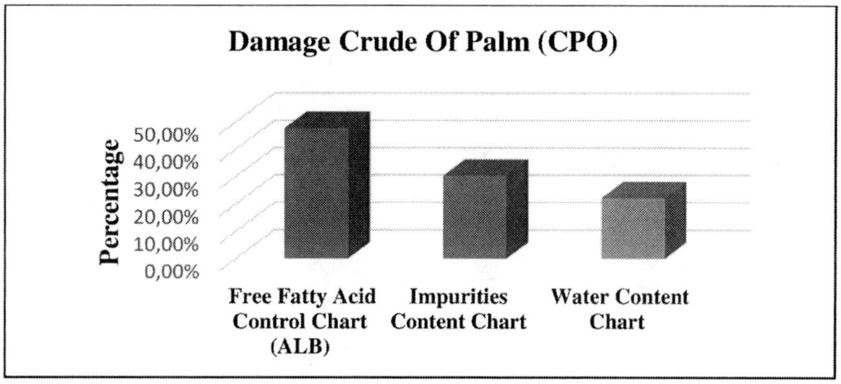

Figure 6. Histogram percentage of CPO oil damage parameter types.

4.4.6 Diagram Pareto

Based on the Figure 7, the highest total percentage is Free Fatty Acid (ALB) with a total percentage of 47.46%, for the second largest percentage, namely the amount of impurities with a percentage of 30.21% and the smallest percentage, namely the level of impurities with a total percentage of 22.12% of total production.

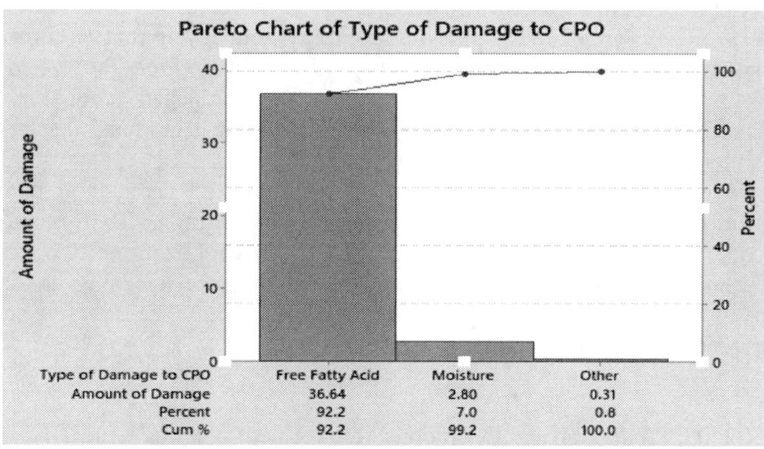

Figure 7. Pareto chart of type damage to CPO.

4.4.7 Cause and Effect Diagrams

Analyze what factors are the causes of product damage. Factors that influence and cause product damage in general can be classified as follows:

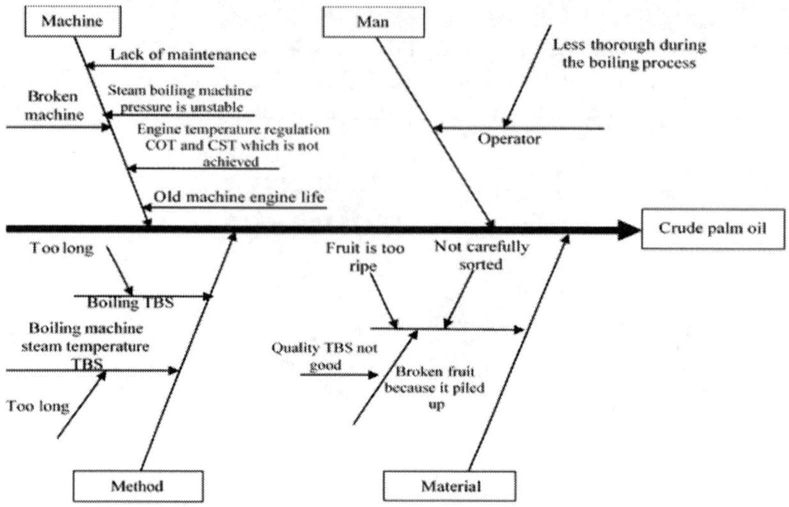

Figure 8. Palm oil cause and effect diagram.

4.4.8 Control Stage

Efforts to suppress the factors that can cause a decrease in the quality of crude palm oil (CPO) Table 6. The mechanism of controlling the processing of Crude Palm Oil can be seen in the appendix.

5 Conclusion

Based on the analysis several conclusions are obtained including:

1. Factors causing CPO oil quality discrepancies are obtained in the cause-and-effect analysis diagram, known factors causing damage in production are derived from human/worker factors, production machines, work methods, materials/raw materials.
2. Based on the average Sigma Level value for free fatty acid levels that is equal to 3.38 and the Indonesian national standard that is equal to the levels of ALB (Free Fatty Acid) (2.5-3.0%), for impurities levels of sigma value of 3.57 and standards the Indonesian national level for impurities (0.01-0.02%) and the water content of the sigma value of 3.68 sigma and the Indonesian national standard for water content (0.1-0.2%).
3. Proposed improvement of product defects mismatch with the 5W+1H method, namely the levels of Free Fatty Acid that must be improved work systems in sorting FFB fruit, provide training to operators in the selection of appropriate fruit, while for the level of impurities, the improvement is to supervise operators oil refining, periodic engine maintenance, cleaning at the sorting station, conducting periodic checks every hour, while for moisture content the corrective action that must be taken is to supervise sterilizer operators.

References

[1] Assauri, S. 2008. Production and Operations Management. Jakarta: LPFEUI
[2] National Standardization Agency, (2006), SNI 01-2901-2006, items 5.4. Palm Oil, Jakarta.
[3] Databoks, “10 Countries with the World's Largest Palm Oil Consumption”. 18 March 2020, from https://databoks.saiddata.co.id/datapublish/2019/12/16/10- the world's biggest palm oil-rich country.
[4] Gaspert. V. 2002. *Analysis Method for Quality Improvement*. Jakarta: Gramedia Pustaka Utama.
[5] Ginting, E. I. 2017. Analysis of Production Quality Control Using Six Sigma Method. Semarang: Universitas Diponogoro.
[6] Gitosudarmo, I. 2012. Marketing Management. second edition, second printing. Yogyakarta: BPFE.

[7] Gusti I. Ayu and Gede Mertha. 2019. *Quality Control of Arabica Coffee Production Process at UD. Cipta Lestari in Pujungan Village. J Manag.* 8(4). Pp 2495-2523. doi.org/10.24843/EJMUNUD.2019.v08.i04.p22
[8] Hardjosoedarmo, S. 2004. *Total Quality Management.* Revised Edition. Yogyakarta: Andi Offset
[9] Indonesian National Standards Agency, 2019, http://sispk. bsn.go.id/ SNI/ Detail SNI/7123
[10] Ilham, N. 2012. Analysis of Product Quality Control Using Statistical Processing Control (SPC) at PT. Bosowa Graphic Media. Thesis. Faculty of Economics and Business. Makassar: Hasanuddin University.
[11] National Standardization Agency, (2006), SNI 01-2901-2006, item 5.4 Oil Palm Oil, Jakarta.
[12] Prawiro I. S., Kusrini N, Nurliza N 2016 Analisis Pengendalian Mutu CPO (Crude Palm Oil) Menggunakan Six Sigma di Pabrik Pengolahan CPO PT. Gunajaya Karya Gumilang Kecamatan Kendawangan Kabupaten Ketapang *J Soc Econ Agric* 5(2):28. http://dx.doi.org/10.26418/j.sea.v5i2.17911
[13] Render, B. 2006. Operations Management, Issue seven, Jakarta: Salemba Empat.

Appendices A. The Mechanism of Controlling the Processing of Crude Palm Oil

PT. KARYA TANAH SUBUR	Palm Oil Processing Procedure	
	CPO Production Process	
Purpose of the Procedure: To minimize the increase and variation in Crude Palm Oil levels (CPO) to conform to specified specifications		
Aspect	**Maintenance Plan**	**Control**
Equipment Use/ Boiling Machine	Tool Checking and Machine Before Process	Provide special time to ensure that the engine is in condition good to avoid Process errors that can result in damage to the product or work accident. Provide training to each operator about usage the right tools / machines, good for new operators, as well as old operator. Arrange the rules of work usage steps every tool / machine. Provide written information in the form of instructions regarding the steps to use the tool and placed in a location that makes it easy for the operator to know and carry out the steps and the procedure in particular in the engine temperature setting Make instructions regarding stages of maintenance and schedule implementation of the maintenance process
Material Utilization Raw and Sorting	Selection of Raw Materials	The quality of raw materials is certainly very influential in quality products, therefore the selection of the right raw materials should be done starting at the time of procurement, and when sorting TBS fruits.
		Knowledge about material management needs to be increased, especially for operator, this can be done by providing special training on the supply of materials at the beginning of the production phase.
	Material Handling	Material Handling Raw material should be placed in a closed area, away from influence
Management of Operators / Factors Human	Coaching and Training Upgrade HR Quality	weather conditions (rain and heat). High water content in fruit, can cause oxidation processes

PT. KARYA TANAH SUBUR	Palm Oil Processing Procedure	
CPO Production Process		
Purpose of the Procedure: To minimize the increase and variation in Crude Palm Oil levels (CPO) to conform to specified specifications		
Aspect	**Maintenance Plan**	**Control**
	Enhancement Work motivation	Work routines can sometimes cause a decrease in concentration and the spirit of workers in carrying out activities. Therefore, the management needs to spend a little time so that the motivation of workers increases again. One of them is through togetherness activities, recreation and meetings which are more relaxed, and full of harmony.
		Providing compensation in accordance with the workload provided
		There is a specialization in work, so that mastery of the task becomes more leverage.
Procedure / Method	Enhancement inspection stage	Perform periodic sample checks once every hour
		Perform control chart calculations and calculate sigma values every period periodically.

Designing sweet biscuits packaging by considering the level of attractiveness based on eye tracking data

I M Ronyastra[1], A H Kusumo[2], M Hartono[3], and E S Tantoisworo[4]

[1, 2, 3, 4]Department of Industrial Engineering, University of Surabaya, Raya Kalirungkut, Surabaya, 60293, Indonesia.

E-mail: [1]imade.ronyastra@yahoo.com

Abstract. The diversity of types of packaging of the product is not entirely desirable by consumers. This study aims to conduct a ranking analysis on packaging, an analysis of the correlation between fixation to the level of attractiveness, and packaging design that is of interest to consumers. Data collection was obtained through eye tracking and post exposure questionnaire for 49 respondents. The analysis shown that the correlation between fixation to level of attractiveness is not correlated, due to external factors such as brand image, brand familiarity, and others. Then, the design of the product packaging is done where the basic orange color packaging concept and its packaging elements are obtained based analogous theory. A concept testing was done afterwards and revealed that the overall packaging design can meet the needs of consumers with a percentage of fulfillment above 70%.

1. Introduction

Product packaging is one important element that needs to be present in a product, because one of the functions of product packaging aims to attract consumer buying interest. In addition, one factor of product quality is the product packaging itself. Apart from the packaging aspect, the elements on the packaging also need to be considered in order to avoid mistakes in delivering information to consumers. Therefore, this research conducted analysis of ranking based on eye fixation on biscuit products packagings, analysis of the correlation between fixation and attractiveness, and designing packaging design concepts that are of interest to consumers. Through this research, it is expected to be a benefit for companies to make attractive packaging design recommendations so as to increase company sales, find out the correlation between fixation and attractiveness, and determine the placement of packaging elements that are of interest to consumers.

Visual attention from human vision is divided into several stages, namely the stage of giving a stimulus, the position of the eye fovea looking for a stimulus that attracts attention, and the eye fovea focus on a stimulus because it is considered attractive [1]. The model of eye movement is divided into 2, namely saccades which are reflex eye movements and fixation of eye movements that are centered and stable [1]. Then, eye tracking is an eye tracking method used to study visual attention in the human eye. Eye tracking can determine 3 important points, namely in which part of the eye's eye view, how long the eye sees at a point, and the cycle of eye sight movement from one point to another.

Eye tracking methods have been widely used to analyze user experience. User experience measurements performed using an eye tracker will produce eye movement data that can be tracked and recorded. This is closely related to the cognitive processing activities of users over time [2]. This method is used to get consumers' visual attention which is measured and analyzed based on gaze plot and heat map analysis [3]. Santos et al. (2018) measured areas of interest (AOI), namely areas on posters and videos that had been marked to obtain more detailed information. AOI is marked in the form of a product brand from a company and the type of product advertised so that for each AOI obtained time fixation (time viewed), time of first fixation (time to first view) and number of fixation (fixation) [4].

According to Estiri et. al. [5] product packaging is a layout that contains visual attributes of information that consumers want to receive. Some functions of product packaging are as follows namely as a media that protects the product from external influences, as a product identity, and serves to improve product efficiency. In addition, the elements contained in product packaging include company names, brands, raw materials, net weight, nutritional value information, barcodes, and product images [6]. Often, product packaging is the initial interaction between the product and its users. There are six factors forming packaging visual appeal, namely the layout factor, brand factor, illustration factor, typographic factor, characteristic factor, and color factor. Therefore, product packaging must be designed by considering the visual appeal of consumers because the packaging will directly deal with consumers so that consumers will have their own image of the product that reflects the desired brand image of the company. Product packaging also influences purchasing decisions which are classified into labels (some clear information), shapes, materials, colors and images. Overall influence consumer purchasing decisions by 53.40% [7]

In this study, there are 3 important variables that considered as dependent variables, namely recall, likeability, and suitability. Recall is "to remember something", which has the meaning to remember a thing that has been seen or felt. Then for likeability comes from the word likeable, which is "easy to like", which means to like something easily. Then for suitability that is "the degree to which someone or something is suitable for a particular job or purpose" which means an assessment of the level of suitability of a thing. These terms was used in the research done by Husić-Mehmedović et.al [8] where recall variable is used to measure something that is remembered, then for the likeability variable is used to measure the level of interest in a thing, and for the suitability variable is used to measure the level of suitability of a thing.

The statistical test used in this study consisted of several methods, namely the Pearson Product Moment validity test, the Alfa Cronbach reliability test, and the Bivariate Pearson correlation test. Pearson Product Moment validity testing according to Sugiyono [9] is considered valid when the data can be used to measure what you want to know or measure. Bivariate Pearson correlation test is hypothesis testing which aims to find out the relationship / relationship between the independent variable with the dependent variable [9].

After statistical testing, the product design and development process are carried out. According to Ulrich [11], product design and development is an activity that aims to add or renew the use value of an existing product or a completely new product. The steps undertaken to carry out the process of product design and development include the following, namely identifying consumer needs, establishing new product specifications, composing new product concepts, selecting product concepts, testing the concept of product selection results, and making new product prototypes.

2. Research Methodology

The research process is carried out in a laboratory that is soundproof and free of all content that interferes with the running of the instrument. Respondents sit in front of a 32-inch Full HD LED monitor and eye tracker position which is located below the monitor screen. The eye tracker device used is Gazepoint (https://www.gazept.com/) with a GP3 model that captures eye sight at 60Hz frequency. In this eye tracking process, 2 types of eye movement models can be obtained, namely fixation & saccades [1]

In this study, the results of eye tracking in the form of many fixations are independent variables, while the dependent variables come from the results of the post exposure questionnaire. The process of this research was conducted on 49 students of the University of Surabaya from various faculties as a sample target. Then for the research instrument divided into 4 stages, namely the calibration of the eye tracker tool to the pupil area of each respondent's eyes, the appearance of images of 20 types of biscuit products displayed in a row, the individual performances of the 20 types of biscuit products available and the size of the enlarged image (zoom in), and questionnaire after the data collection process with an eye tracker to measure 3 variables, namely recall, likeability, and suitability.

After the eye tracking instrument process is completed, the data is analyzed statistically related to the results of eye tracking and the questionnaire. The first is to test the validity of the recall, likeability, and suitability questionnaire variables. The second is testing the reliability of the questionnaire variables recall, likeability, and suitability. Third, testing the correlation between variables from eye tracking to the questionnaire variables. Then the results of correlation analysis of eye tracking data with a post exposure questionnaire are associated with the process of product design and development to produce a new product packaging design based on the level of attractiveness of consumers.

3. Results and discussions

Data processing was carried out on the results of eye tracking instruments, namely the results of the questionnaire along with the data on the results of eye tracking. Based on the results of the instrument, the following demographic data of research respondents:

Table 1. Demographic data of respondents.

Sample Structure		N	%
Gender	Male	33	67
	Female	16	33
Age	<18 years old	1	2
	18 years old	1	2
	19 years old	12	24
	20 years old	14	29
	21 years old	15	31
	>21 years old	6	12

Based on the results of statistical tests regarding the validity testing on recall, likeability, and suitability questionnaire variables indicate that all variables are valid, because the Pearson Correlation value of each brand of biscuits above the r table value of 0.2377 ($\alpha = 0.1$; N = 49). Then, for reliability testing shows that the variable recall (0.951), likeability (0.805), and suitability (0.860) have Cronbach's Alpha coefficient above the standard, which is 0.70, so that the three variables are classified as reliable. After testing the validity and reliability of product ranking analysis, the ranking is based on the longest fixation time, because the results of eye tracking variables are more quantitative in nature when compared to the questionnaire results variables. The following table is the result of eye tracking and also a questionnaire sorted by the longest fixation time:

Table 2. Ranking sequence based on average percentage of fixation.

	Average Fixation [%]	Recall [%]	Likeability [%]	Suitability [%]
Gery Malkist Salut Coklat	7,56	68,57	62,86	70,20
Roma Sari Gandum Sandwich Coklat	7,29	69,39	62,45	63,67
Slai O'lai	6,93	76,33	64,08	71,02
Wonderland Biskuit Kelapa	6,55	47,35	52,24	60,82
Oreo	5,97	72,24	77,96	71,84

Based on the Pearson correlation test in Table 3, it can be seen that there is no significant correlation between the average percentage of fixation to the percentage of recall, likeability, and suitability. This is evidenced by the Pearson Correlation between the average fixation of recall, the average fixation of likeability, and also the average fixation of the suitability is negative and also the value for p-value for each variable. Based on the interval scale for Pearson correlation figures between the average fixation of recall and suitability have a weak correlation, while for the average fixation of suitability has a very weak correlation.

Table 3. Pearson correlation results eye tracking variables to questionnaire variables.

		Average Fixation [%]	Recall [%]	Likeability [%]	Suitability [%]
Average Fixation [%]	Pearson Correlation	1	-0.298	-0.231	-0.073
	Sig. (2-tailed)		0.202	0.327	0.761
	N	20	20	20	20
Recall [%]	Pearson Correlation	-0.298	1	0.671**	0.572**
	Sig. (2-tailed)	0.202		0.001	0.008
	N	20	20	20	20
Likeability [%]	Pearson Correlation	-0.231	0.671**	1	0.726**
	Sig. (2-tailed)	0.327	0.001		0.000
	N	20	20	20	20
Suitability [%]	Pearson Correlation	-0.073	0.572**	0.726**	1
	Sig. (2-tailed)	0.761	0.008	0.000	
	N	20	20	20	20

The relationship between these variables is not significant due to many factors, one of which is the brand familiarity factor [4]. This causes a bias when the respondent fills out the questionnaire, so the results of the questionnaire are inconsistent and do not match the results of eye tracking. When the respondent evaluates the questionnaire on the likeability variable, the brand familiarity factor plays an important role in the following evaluation process, which is that the respondent often gives high ratings of biscuit products to the research instrument because of the level of product popularity, the taste of the product, and many other factors.

However, for recall and likeability variables, based on correlation testing as in [8] there was a significant correlation and the level of correlation was strong. This was also proven by several products that were successfully recalled by respondents, namely Ovaltine (81.22%) and Better (77.14%) were the products with the highest percentage of likeability. Therefore, it can be concluded that the products in the memory of the respondents are also the preferred products in terms of packaging. Then the recall variable with suitability is proven to have a significant correlation and has a moderate level of correlation. This is evidenced by several products such as Ovaltine (79.59%), Slai O'lai (71.02%), and also Better (70.61%), which are some of the products rated by respondents as having a high level of conformity for product categories biscuits in terms of packaging and also managed to be recalled by respondents. Therefore, it can be concluded that the respondent's assessment of the product in terms of packaging is appropriate, bearing in mind that the product which is in the consumer's memory is a product with a high suitability rating in terms of packaging.

In addition, for the variable likeability to suitability is proven to have a strong correlation. This is evidenced in some biscuit products favored by respondents having a high level of conformity assessment, namely Ovaltine (79.59%), Oreo (71.84%), Biskuat Coklat (71.43%), and also Better (70, 61%). Therefore, it can be concluded that the respondents' appraisal of the biscuit product in terms of packaging is also a product that is preferred by respondents. Thus, products that are given a high suitability value in terms of packaging are also products that are preferred by respondents in terms of packaging.

After that the product design and development process is carried out, while the needs of consumers usually obtained through direct interviews, but in this study the needs of consumers are obtained from the interpretation of the results of eye tracking along with the questionnaire. Table 4 summarizes the consumer needs and their metrics. Consumer needs related to an attractive packaging concept is found in the results of a post exposure questionnaire about likeability which shows that a particular product is an attractive product design concept.

Table 4. Customer needs, engineering metrics, & units.

Customer Needs	Metrics	Units
Attractive packaging concept	Color selection	-
	Packaging size	Cm
Brands are easily seen & contrast	Font size	Pt
	Font color	-
	Font type	-
Attractive product image	Image size	Cm
	Image color	-
Packaging form adjusts the shape of the biscuits	Packaging form	-
	Basic packaging material	-

Consumer needs related to product brands are easily seen & contrast is obtained and also attractive product images are obtained from eye tracking results. Consumer needs regarding packaging form adjusts the shape of the biscuits obtained from the suitability assessment of the results of the post exposure questionnaire.

After obtaining the list of consumers' needs, a questionnaire was distributed to 10 University of Surabaya students to measure the importance of existing consumer needs and the results were obtained that the most important requirement was an attractive packaging concept and followed by a product brand that was easily seen and contrasted, images of products, and the shape of the packaging matches the shape of the biscuit. After that the weighted score is calculated as one of the requirements for the preparation of House of Quality (HOQ), as shown in Figure 1. The product concept is shown in Figure 2.

Based on the HOQ design, the most important metric (having the highest relative weight percentage) is the packaging size. It's just that in this study other metrics such as color selection, font size, font color, font type, image size, packaging form, and basic packaging materials are also included and used as considerations for designing attractive packaging designs for consumers. According to the results of the questionnaire, products that have a conformity assessment in terms of packaging according to respondents are Ovaltine products, so the basic metrics related to the shape and basic ingredients of the packaging are obtained from the recommendations on Ovaltine products and the form of packaging design used is the shape of the tube with the basic ingredients of the packaging being in the form of metallize paper.

In the selection of basic colors of packaging and packaging size obtained from the results of the post exposure questionnaire on the likeability variable, the results taken are regarding the level of interest of respondents in the product in terms of packaging. Similar to the results of the questionnaire on suitability variables, Ovaltine is a product that is preferred by 11 out of 50 respondents when compared to 19 other products, so the basic color of the packaging is orange and can hold 13 biscuits with a net weight of 130 grams.

Font color refers to the analogous color harmony theory, if the packaging is orange then the contrasting color is light orange to yellow. Then for the type of font using fonts that have no legs and separate letters to make it easier for consumers to see and read the elements on the packaging. Then the font size for the product brand is calculated according to what was formulated by Kusumo, et. al. [11] and the reading distance is 5 meters, namely letters height of 3.8 cm, width of letters 2.533 cm, distance between 2 letters 0.76 cm, distance between 2 words 2.533 cm, and distance between 2 lines of sentence 3.8 cm. While the metrics regarding the color of the image are selected based on the results of eye tracking and also the results of the questionnaire variable likeability, namely chocolate biscuits with chocolate jam. Detailed image size is 4.5 cm high image, because the shape of the image is square, the width is 4.5 cm equal to the length.

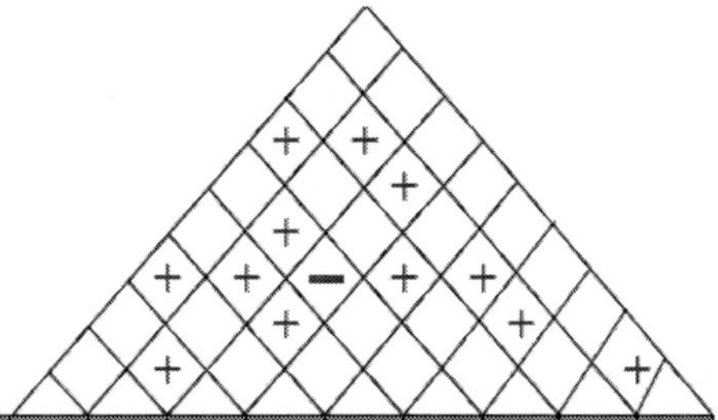

				Column #	1	2	3	4	5	6	7	8	9						
				Direction of Improvement: Minimize (▼), Maximize (▲), or Target (x)	X	X	X	▲	▲	X	▲	X	X	Competitive Analysis (0=Worst, 5=Best)					
Row #	Max Relationship Value in Row	Relative Weight	Weight / Importance	Quality Characteristics (a.k.a. "Functional Requirements" or "Hows") / Demanded Quality (a.k.a. "Customer Requirements" or "Whats")	Color Selection	Packaging dimension (cm)	Font size (mm)	Font Color	Font facetype	Picture size (mm)	Color of the picture	Shape of the packaging	Material for the packaging	XYZ	Gery Malkist Salut Coklat	Roma Sari Gandum Sandwich Coklat	Slai O'lai	Wonderland Biskuit Kelapa	Oreo
1	9	28,5	4,9	Interesting Packaging Concept	Θ	Θ	O	O	O	O	O	O	O	4	5	4	5	4	5
2	9	27,3	4,7	Product's Brand is easily visible and contrast	O	Θ	Θ	Θ	Θ	O		O		5	5	4	5	4	5
3	9	25,0	4,3	Interesting picture of the product	O					Θ	Θ	O		4	5	5	5	4	4
4	9	19,2	3,3	The shape of the packaging adjusts the shape of the biscuits	O	Θ						Θ	Θ	4	5	5	5	5	5
				Difficulty (0=Easy to Accomplish, 10=Extremely	5	5	6	6	4	4	5	3	5						
				Max Relationship Value in Column	9	9	9	9	9	9	9	9	9						
				Weight / Importance	470,9	675,0	331,4	331,4	331,4	392,4	310,5	415,1	258,1						
				Relative Weight	13,4	19,2	9,4	9,4	9,4	11,2	8,8	11,8	7,3						

Figure 1. House of quality product packaging design.

Figure 2. Product packaging looks overall blanket tubes.

After obtaining the product packaging concept in Figure 2, a concept test was carried out by distributing questionnaires to 10 respondents regarding the level of packaging fulfillment to consumer needs. The results state that the level of fulfillment of the needs of the packaging is above average, namely the level of fulfillment of attractive packaging concepts (80%), brands are easily seen and contrast (88%), attractive product image (78%), and packaging form adjusts the shape of the biscuits (88%). Figure 3 displays the appearance of the front and back packaging design.

Figure 3. Product packaging front view (left) & rear view (right).

Table 5. Detail types and font sizes on the package.

Viewpoint	Element	Font Type	Font Size [pt]
Front	XYZ (brand)	Supercraft	108.9
	Net weight	Helvetica Neue	14.2
	EST (company logo)	Arcon	28.4
Rear	XYZ (brand)	Supercraft	56.7
	Nutritional value information	Helvetica Neue	Title : 8.5 Content : 8.5
	Ingredients	Helvetica Neue	Title : 15 Content : 8
	Number below the barcode	Courier	8

Table 6. Detail image size on the package.

Viewpoint	Element	Image	Size [cm]	
			Length	Width
Front	Image of a biscuit		4.5	4.5
Rear	Image of a biscuit		3.5	3.5
	Barcode		4	2.5

In addition, after the packaging design display is obtained, the details of the image size and font type and size used in the packaging design are shown in Table 5 and Table 6. However, there are differences in the size of the biscuit image located on the front and back, this is because the front is a priority area so it is made with a larger size and has a reading distance of about 5 meters according to calculations according to Kusumo, et. al. [11]. In addition to the size of the barcode adjusted to the size of the existing packaging, so it can meet the available space.

4. Conclusion

This research exposed respondents to a visual of 20 different biscuits packaging and the visual response were recorded through eye tracking device. It is found that among the 20 brands, the most visually sought were Gery Malkist Salut Coklat, Roma Sari Gandum Sandwich Coklat, Slai O'Lai, Wonderland Biskuit Kelapa, and Oreo. It is also found that the orange color is a better attention grabber compared to other color. Then for some packaging elements according to respondents such as company name, brand, and product image, it is categorized as important and must be on the packaging.

The eye tracking data reveals that there is no significant relationship between duration of sight at a point (fixation) and brand recall. The brand recall seems to be influenced by other external factors such as brand image and brand familiarity from the respondents. There is a significant correlation observed between likeability and recall which suggest if the respondents liked the brand or product, s/he would likely able to recall the brand during the post exposure questionnaire. There was no correlation found between the variables of eye tracking and the questionnaire variables. The biscuit packaging concept designed based on the eye tracking result is tested to be interesting in terms of visuals, fulfillment of consumer needs is above 70% overall. The drawbacks of this research are that we used familiar products to be exposed to respondents and the respondents were focused on university students which may have influence in the data gathered. To improve the data, it is suggested for further research the selection of products on instruments using foreign products or totally mock-up products so that the respondents do not have prior knowledge of the brands. Demographic expansion can also be done to better understand the customers perception in a wider perspective.

References

[1] Duchowski, A. (2007). Eye tracking methodology: Theory and practice. In Eye Tracking Methodology: Theory and Practice.

[2] Rayner K. Eye movements in reading and information processing: 20 years of research. *Psychol Bull*. American Psychological Association (APA); 1998;124(3):372–422. doi.org/10.1037/0033-2909.124.3.372

[3] Nisiforou EA, Michailidou E, Laghos A. Using Eye Tracking to Understand the Impact of Cognitive Abilities on Search Tasks. *Lect Notes Comput Sci*. Springer International Publishing; 2014;46–57. doi.org/10.1007/978-3-319-07509-9_5

[4] Alonso Dos Santos M, Calabuig Moreno F, Sánchez Franco M. Congruence and placement in sponsorship: An eye-tracking application. *Physiol Behav*. Elsevier BV; 2019 Mar;200:159–65. doi.org/10.1016/j.physbeh.2018.05.032

[5] Estiri M, Hasangholi T, Yazdani H, Nejad HJ, Rayej H. Food Products Consumer Behaviors: The Role of Packaging Elements. *J Appl Sci*. Science Alert; 2010 Jul 1;10(7):535–43. doi.org/10.3923/jas.2010.535.543

[6] Mukhtar S, Nurif M. PERANAN PACKAGING DALAM MENINGKATKAN HASIL PRODUKSI TERHADAP KONSUMEN. Jurnal Sosial Humaniora. Lembaga Penelitian dan Pengabdian kepada Masyarakat ITS; 2015 Nov 1;8(2):181. doi.org/10.12962/j24433527.v8i2.1251

[7] Nuryanti BL, Rahman AY. PENGARUH VARIASI DAN KEMASAN PRODUK TERHADAP KEPUTUSAN PEMBELIAN TEH KOTAK ULTRAJAYA (Survei pada Mahasiswa FPIPS Universitas Pendidikan Indonesia). Strategic : Jurnal Pendidikan Manajemen Bisnis. Universitas Pendidikan Indonesia (UPI); 2008 Sep 13;8(2):31. doi.org/10.17509/strategic.v8i2.1022

[8] Husić-Mehmedović M, Omeragić I, Batagelj Z, Kolar T. Seeing is not necessarily liking: Advancing research on package design with eye-tracking. *J Bus Res*. Elsevier BV; 2017 Nov;80:145–54. doi.org/10.1016/j.jbusres.2017.04.019

[9] Sugiyono, P. D. Metode penelitian kuantitatif, kualitatif,dan R&D. , Alfabeta, cv. (2016).

[10] Ulrich, K. (2001). Perancangan dan Pengembangan Produk. Irwin McGraw-Hill.

[11] Kusumo AH, Hartono M, Wahyudi RD. Product design with integration of Kansei engineering and TRIZ to promote sustainable tourism. *Explor Resour, Process Des For Sustain Urban Dev*: Proceedings of the 5th International Conference on Engineering, Technology, and Industrial Application (ICETIA) 2018. AIP Publishing; 2019; doi.org/10.1063/1.5112489

The Reliability-Centered Maintenance (RCM) effect on plant availability and downtime loss in the process industry

R Khasanah[1]*, I Sodikin[2], A P Penirewod[3], B Rachmad[4], N A Pratama[5]

[1,2]Department of Industrial Engineering, AKPRIND Institute of Science and Technology, Yogyakarta, Indonesia
[3,4,5]PT. Pupuk Kaltim, Bontang, Indonesia

E-mail: *rahayu.khasanah@akprind.ac.id

Abstract. The process industry needs an optimal preventive maintenance strategy because the even breakdown of one machine can cause an unexpected shutdown for the whole production process. The case study company (CSC), one of the fertilizer companies in Indonesia implements Reliability-centered Maintenance (RCM) as their preventive maintenance strategy to decrease downtime and increase the availability of the production plants. Eversince the RCM implementation started in 2007, there is no study to evaluate the RCM implementation. Relations between the two parameters (downtime loss and availability) are analyzed using the Pearson correlation test as a statistical method to evaluate the RCM implementation. The results of the analysis show that the R-value of downtime loss and availability of K-3 Ammonia and Urea plants are increased, while the R-value of K-4 Ammonia and Urea Plants are decreased. The RCM implemented in K-4 plants should be evaluated respectively, because the decreasing of the correlation means the RCM implementation on the plant still has not met its propose and there are stills hidden factors that affect the availability and downtime loss of the plant, regardless the RCM implementation.

1. Introduction

In this era of the ever-growing industry, companies compete in giving satisfaction to the consumer to meet the needs demanded by consumers. To maintain company productivity, the high availability of industrial facilities is required. One of the major facilities that strongly supports the productivity of a company is the condition of the equipment or machine. Damage (breakdown) on one machine can cause unexpected shutdowns for the whole process in continuous production, such as the process industry [1]. In these conditions, a series of an optimal preventive maintenance strategy is required on the machine because it is well-known that maintenance, production, and quality are strongly linked to each other [2].

Effective maintenance is critical to many operations. It extends equipment life, improves equipment availability and retains equipment in proper condition. Conversely, poorly maintained equipment may lead to more frequent equipment failures, poor utilization of equipment and delayed production schedules [3]. Maintenance is defined as the combination of all technical and administrative actions, including supervision, which ensures that a system is in its required functioning state [4]. A well-implemented integrated asset management and maintenance system can impact every section of an organization, increasing asset uptimes, reducing maintenance costs, increasing benefits, and enhancing the reputation of the business with its customers [5].

A lot of maintenance strategies have been developed during the last few years. RCM has been one of the most recent strategies in maintenance around the world. Originally, RCM is defined as a systematic process for development and optimization of the maintenance requirements of a physical resource in its operating context to realize its inherent reliability by logically incorporating the maintenance strategies like reactive, preventive, condition-based and proactive maintenance. Around the world, it is an imperative technology in the industrial maintenance field that can be functional to improve the equipment availability and reliability and reduce operational and maintenance costs. The theory of RCM is the function of the operating system to recognize the consequences of the failure by the failure analysis and system function [6].

RCM implementation has to be evaluated by comparing the value of some variables before and after RCM has been applied as a measurement of the effectiveness of RCM implementation in the company. Nabhan (2010) stated that after the measurement of RCM effectiveness was carried out, some industries are not applying RCM effectively. Regarding Nabhan (2010) statement, the evaluation of the effectiveness of RCM implementation should be done.

In the previous researches, certain parameters are used to evaluate RCM. Unexpected failures rate, plant availability and maintenance cost are the typical global measurements of the RCM program constructed. Fore and Mudavanhu evaluate RCM using gaps in production output, total downtime, and availability and machine utilization. Despite total downtime, other related parameters should be considered are the mean time between failures (MTBF) and mean time to repair (MTTR) [5].

The case study company (CSC) has been implementing Reliability-Centered Maintenance (RCM) as its preventive maintenance strategy in the various plants. The company began to implement RCM in 2007, starting from K-3 urea & ammonia plant and K-4 urea & ammonia plant, and then spread to other plants. Earlier the company applying Root Cause Analysis (RCA) in the execution of its maintenance. Since its implementation in 2007, there is no study to evaluate the RCM implementation whether it meets its purpose to increase the availability and reliability of the plant or assets. The study analyzing the relation of downtime loss and availability of the plant after RCM is implemented then proposed as this study, to show the effect of RCM implementation on the availability and reliability of the plant and whether the purpose of RCM has met the expectation or not. Both parameters (availability and downtime loss) are used because high availability is one of the objectives of maintenance in the company. Despite the availability, downtime cost is another indicator highlighted in key literature [8]. In this study, downtime cost represented by downtime loss is one of the indicators used to determine the successful implementation of RCM. The purpose of this research is to analyze reliability-centered maintenance (RCM) practice relation with availability and downtime loss in CSC plants. The study is conducted on K-3 Urea and Ammonia plant, and K-4 Urea and Ammonia plant, because that plant has a level of the highest average production.

2. Research Methodology

As mentioned before, the analysis of RCM implementation is done by using the parameters of availability and downtime loss. Both parameters are analyzed using a statistical method. The statistical method used is the Pearson correlation. The Pearson correlation test has been conducted to understand the relationship between maintenance measurement factors [1]. Correlation tests performed on the availability and downtime loss value at the time before and after the implementation of RCM. Before performing the correlation and calculating the value of availability and downtime loss, we need to identify the amount of downtime that occurs in each plant. Downtime is used to calculate the value of availability and downtime loss. Definition of downtime, availability, downtime loss and correlation tests used are as follows:

2.1. Downtime

Downtime is a time when the machine/equipment is not in a condition to be able to perform its function. Downtime is calculated from when the machine does not function until the machine back in a position to function as it should, after repair. The downtime machine is idle or length of time in which the unit can no longer perform its function as expected. This occurs when a unit experiencing problems such as damage to the machine which can interfere with the performance of the machine as a whole including the quality of the product or its production speed so it takes a certain time to restore the function of these units in the original condition [9] [10].

Goyal and Nand (2012), in their research, covering 29 ammonia plants, stated that downtime in ammonia plants is due to plant-related and external problems. The plant-related problems consist of problems due to mechanical, electrical, instrumentation, process and miscellaneous failures of plant and equipment. The external problems cause downtime such as shortage of raw materials, failure of external power, shortage of water and labour problems [11].

2.2. Availability

Availability is a ratio that illustrates the use of the time available for the operations of machinery and equipment. The availability rate is influenced by two components, namely equipment failure and setup and adjustment losses. The availability is the ratio of the operation time, by eliminating unplanned downtime of equipment to the loading time. Operation time is a time of plant operation include the total downtime. While the loading time is the only time plant operation does not include planned downtime. The formulations used are as follows [12]:

$$Loading\ Time = Base\ Production\ Time - Planned\ Downtime \quad (1)$$

$$Operation\ Time = Loading\ Time - Unplanned\ Downtime \quad (2)$$

$$Availability = \frac{Operation\_Time}{Loading\_Time} x100\% \quad (3)$$

2.3. Downtime Loss

Khasanah (2015) state that the downtime loss is the loss of production caused by the downtime. Maintenance-related downtime is the biggest contributor to low production performance [14]. Total downtime comprised of unplanned downtime and planned downtime. Downtime can be defined in a loss of money, the amount of production (tonnes), or time. Variables used to calculate are the actual production (AP), operating time and total downtime. The formula is as follows:

$$Actual\_Production\_Rate = \frac{Actual\_Production}{Operation\_Time} \quad (4)$$

$$Downtime\ Loss = \text{Total Downtime x Actual Production Rate} \quad (5)$$

2.4. Correlation Test (Pearson Correlation)

Correlation is a statistical term that states the degree of a linear relationship (unidirectional not reciprocal) between two or more variables, which was discovered by Karl Pearson at the beginning of 1900. Therefore, known as the Pearson Correlation Product Moment (PPM). Correlation is a statistical analysis technique most used by researchers. The magnitude of the correlation numbers is called the correlation coefficient is expressed by the symbol (R). Formulations for R as follows [15].

$$R_{yx} = \frac{n.\sum X.Y - (\sum X)(\sum Y)}{\sqrt{\left[n.\sum X^2 - (\sum X)^2\right]\left[n.\sum Y^2 - (\sum Y)^2\right]}} \quad (6)$$

The R-value generated is not more than 1 and less than -1 (-1 <r <1). The resulting R-values can show the correlation pattern between variables used (x and y). The correlation patterns among others walking the same direction if the R-value is positive (+), running in the opposite direction if the R-value is negative (-), and not correlated if R-value = 0. The coefficient R if close to 1 / -1 then increasingly strong correlation exists, otherwise if closer to 0, then the correlation is lower. The size of the correlation can be observed interpretation in Table 1 [15] [16].

Based on the theory, the relation of availability and downtime loss should increase after the implementation of RCM. That is, R-value after implementation of RCM will be higher than before the implementation of RCM. Because after implementation of RCM the plant could prevent breakdown and unexpected shutdown.

Table 1. Interpretation of correlation coefficient (R).

R (+/-)	Interpretation
0	Not Correlate
0.01 - 0.20	Very low
0.21 - 0.40	Low
0.41 - 0.60	Rather Low
0.61 - 0.80	Enough
0.81 - 0.99	High
1	Very high

2.5. Mathematic Relation between Availability and Downtime Loss

A mathematical relation of availability and downtime loss is performed by using the substitution of equations (3) and (5). The comparison is done as follows [12]:

Based on the equation (3), then

$$Operation\ Time\ (OT) = Availability\ x\ Loading\ Time\ (LT) \tag{7}$$

Based on the equation (5), then

$$Downtime\ Loss = Downtime\ x\ \frac{Actual\ Production\ (AP)}{Operation\ Time\ (OP)} \tag{8}$$

Substitute equation (7) and (8), then

$$\mathbf{Downtime\ Loss} = \frac{Actual\ Production\ (AP)}{Availability\ x\ Loading\ Time}\ x\ Downtime \tag{9}$$

$$\mathbf{Downtime\ Loss} = \frac{\mathbf{1}}{\mathbf{Availability\,x\ Loading\ Time}}\ x\ \frac{\mathbf{AP\,x\ Downtime}}{\mathbf{1}} \tag{10}$$

Based on the equation (10), it is known that the downtime loss (DL) is inversely proportional to the value of availability, which means the value of downtime loss and availability will run in the opposite direction. That is, if downtime loss is little, then availability value is high, and vice versa.

2.6. Data Collection

2.6.1. Location and Time Research. The study was conducted at the factory K-3 and K-4 plant of ammonia and urea in CSC. The data used in this research is secondary data of the company that started in 2001 to 2018.

2.6.2. Data Collection and Processing Methods. Data needed in this research is secondary data of companies that include data on the amount of production, production time data, and the data of downtime occurrence. Other tasks include supporting data relating to research as previous research literature, and theoretical foundations derived from books and journals. Then the data processing is done to calculate the value of availability and downtime loss using the formula equation above. Then, test the correlation between the availability and downtime loss at the time before and after the implementation of RCM to determine differences RCM relationship after implementing at the plant. A correlation test was done manually using the above formula equation and performed using the software as a tool.

3. Data Result, Analysis and Discussion

3.1. Data on Loading Time and Operation Time

Data on loading time and operation time found using equation (1) and (2). Loading time found by using data of base production time and planned downtime. While operation time found by using data of loading time and unplanned downtime. The data of loading time four plants as the following Table 2.

Table 2. Data on loading time and operation time.

RCM Implementation	Year	Loading Time (Hour)				Operation Time (Hour)			
		Ammonia K-3	Urea K-3	Ammonia K-4	Urea K-4	Ammonia K-3	Urea K-3	Ammonia K-4	Urea K-4
BEFORE RCM	2001	8160.00	8160.00	-	-	7120.80	6683.52	-	-
	2002	8760.00	8760.00	1200.00	5856.00	8457.60	8319.60	1164.00	2123.04
	2003	7786.32	7776.00	8760.00	8760.00	7663.68	7166.16	6417.36	6459.36
	2004	8784.00	8784.00	8418.96	8403.36	8387.76	8310.48	8108.40	6258.96
	2005	8760.00	8760.00	8456.16	8435.76	8760.00	8760.00	8308.56	7597.68
	2006	8256.00	8256.00	8760.00	8760.00	7648.80	7534.80	7408.08	6492.96
AFTER RCM	2007	8760.00	8760.00	8424.00	8424.00	8565.84	8463.84	8075.28	7664.64
	2008	8373.60	8245.20	8784.00	8784.00	8145.36	7799.04	8685.36	8516.16
	2009	8760.00	8760.00	8376.00	8376.00	8752.56	8712.00	8193.12	8027.76
	2010	8269.20	8254.80	8760.00	8760.00	8109.84	8044.32	8541.36	8367.84
	2011	8760.00	8760.00	8760.00	8760.00	8760.00	8666.64	8483.04	8079.60
	2012	8784.00	8784.00	8348.64	8342.88	8717.04	8695.92	8234.40	8147.76
	2013	8507.28	8487.12	8760.00	8760.00	8377.47	8303.52	8668.80	8691.60
	2014	8760.00	8760.00	8760.00	8760.00	8611.20	8421.84	8668.56	8230.08
	2015	8256.00	8265.12	8760.00	8760.00	7592.64	7818.72	8725.20	8085.25
	2016	8784.00	8784.00	8223.84	8115.84	8590.08	8480.16	7816.08	7071.60
	2017	8256.00	8256.00	8760.00	8760.00	6979.92	6440.64	8650.08	8491.68
	2018	8760.00	8760.00	8211.12	8268.96	8731.20	8462.64	8211.12	8227.44

In 2001 the factory K-4 has not been put into operation so that loading time and operation time cannot be calculated. Based on Table 2, it is known that the implementation of RCM began in 2007 onward. The data from 2001 to 2006 represents the condition before RCM implemented on the plant. The highest to the lowest average of loading time and operation time is start from K-3 ammonia plant, K-3 urea plant, K-4 urea plant, and K-4 ammonia plant in 18 years.

3.2. Data on Availability and Downtime Loss

The availability is calculated using equation (3). The downtime loss is calculated using equation (4) and (5). Downtime loss is found by using data of actual production, operation time, and total downtime. The availability and downtime loss value for each plant per year is shown in Table 3.

Table 3. Data of availability and downtime loss, before and after RCM implementation.

RCM Implementation	Year	*Availability (%)*				*Downtime Loss (ton)*			
		Ammonia K-3	Urea K-3	Ammonia K-4	Urea K-4	Ammonia K-3	Urea K-3	Ammonia K-4	Urea K-4
BEFORE RCM	2001	87.26	81.91	-	-	71044.9	146286.7	-	-
	2002	96.55	94.97	97.00	36.25	13332.0	31058.6	879.2	212586.5
	2003	98.42	92.16	73 26	73.74	49343.6	112763.1	88273.9	133962.5
	2004	95.49	94.61	96.31	74.48	18759.3	35646.8	30156.0	149503.1
	2005	100.00	100.00	98.25	90.07	0.0	0.0	20254.3	73389.4
	2006	92.65	91.26	84.57	74.12	49081.8	87929.4	58580.0	136252.9
	Mean	**95.06**	**92.49**	**89.88**	**69.73**	**33593.6**	**68947.4**	**39628.6**	**141138.9**
AFTER RCM	2007	97.78	96.62	95.86	90.99	8446.8	22056.3	29501.8	69028.4
	2008	97.27	94.59	98.88	96.95	27999.1	73265.4	4268.8	17776.7
	2009	99.92	99.45	97.82	95.84	327.6	3544.7	24379.6	51381.5
	2010	98.07	97.45	97.50	95.52	28742.8	51600.1	9473.8	26994.2
	2011	100.00	98.93	96.84	92.23	0.0	6955.1	12121.5	48568.4
	2012	99.24	99.00	98.63	97.66	3039.6	6417.6	24186.6	45938.9
	2013	98.47	97.84	98.96	99.22	17425.1	32692.0	4110.8	4839.7
	2014	98.30	96.14	98.96	93.95	6702.5	24676.0	4086.3	39149.8
	2015	91.97	94.60	99.60	92.30	51735.4	67528.5	1553.7	47780.1
	2016	97.79	96.54	95.04	87.13	8752.0	21513.8	39579.7	104754.8
	2017	84.54	78.01	98.75	96.94	77899.8	176522.8	4845.6	17558.3
	2018	99.67	96.61	100.00	99.50	1284.8	21560.7	24086.3	38342.8
	Mean	**96.92**	**95.48**	**98.07**	**94.85**	**19363.0**	**42361.1**	**15182.9**	**42676.1**

Correlation test is used to determine the correlation R-value between availability and downtime loss at the time before and after the implementation of RCM. The variables that used are the availability value as X and downtime loss as Y. The availability value is used as an independent variable of the correlation test because it is one of the purposes of maintenance, which ensures the optimum level of the facility. While downtime loss is used as the dependent variable of correlation test and as one of the parameters to determine the success of the RCM implementation. Downtime loss is also used to find out the amount of loss generated based on the availability value. This correlation test calculation using the SPSS software program. It can be seen the R-value that will demonstrate the significance of the correlation between availability and downtime loss. The correlation test was conducted with the two treatments, that are before the implementation of RCM in 2001-2006 and after implementation of RCM in 2007-2018. The correlation value and mean value of availability and downtime loss before and after the implementation of RCM are summarized in Table 4.

Table 4. Summary of availability and downtime loss correlation

Plant	RCM Implementation	R-value comparation			Mean value comparation			
		Correlation value	Correlation pattern	Interpretation	Availability (%)	Δ Availability	Downtime Loss (ton)	Δ Downtime Loss
Ammonia K-3	Before RCM	-0.7745	Opposite direction	Enough	95.06	1.86	33,593.6	-14,230.6
	After RCM	-0.9504	Opposite direction	High	96.92		19,363.0	
Urea K-3	Before RCM	-0.9188	Opposite direction	High	92.49	3.00	68,947.4	-26,586.4
	After RCM	-0.9526	Opposite direction	High	95.48		42,361.1	
Ammonia K-4	Before RCM	-0.9514	Opposite direction	High	89.88	8.19	39,628.6	-24,445.8
	After RCM	-0.6438	Opposite direction	Enough	98.07		15,182.9	
Urea K-4	Before RCM	-0.9551	Opposite direction	High	69.73	25.12	141,138.9	-98,462.8
	After RCM	-0.8383	Opposite direction	High	94.85		42,676.1	

Even though the mean value of availability and downtime loss before and after RCM shows that the availability increase and downtime loss decrease after RCM has been implemented to the Ammonia and Urea plants, but the correlation shows that for some plant, the correlation decrease after RCM has implemented to the plant.

Based on the equation (10), it is known that the downtime loss (DL) is inversely proportional to the availability value, which means the value of downtime loss and availability will run in the opposite direction, that is, if downtime loss is little then availability value is high, and vice versa. While affecting the level of relations between the two parameters is Loading Time (LT), which means Planned Downtime at the mill is very influential on the level of relationship of downtime loss and availability. Then the actual production and total downtime can also affect the relationship of downtime loss and availability of the plant. In theoretical, the relation of availability and downtime loss will increase after the implementation of RCM. That is R-value correlation after implementation of RCM will be higher than before the implementation of RCM. Because after implementation of RCM the plant could prevent production failure. As in theory, the correlation value will increase if the relation of availability and downtime loss are inversely proportional (the value of availability increases, so the downtime loss decreases). So, comparing with the theory, it's true that the relation of availability and downtime loss in the K-3 plant occurs in the opposite direction and will increase after the implementation of RCM. But, in the K-4 plant does not seem following the theory.

By analysing the correlation and the mean value of availability and downtime loss before and after RCM implementation, it shows that the implementation of RCM in the K-3 plant can improve the relationship between availability and downtime loss. It is proven that the high availability value can reduce the level of downtime loss, and the correlation of those two variables increases after RCM has been implemented to the plant. Meanwhile, the implementation of RCM at the K-4 plant has not been able to improve the relationship between availability and downtime loss because, after the implementation of RCM, the correlation value K-4 plant has not increased. Even though RCM does increase the availability and reliability (shows from the increase of the mean value of availability and the decrease of the mean value of downtime loss), but there are hidden factors that affect the correlation between those variables to be different from the theory, despite the RCM implementation. Therefore, the company must evaluate further the overall implementation of RCM in K-4 plant, because the statistics show that the implementation of RCM in K-4 plant does not increase the correlation between availability and downtime loss even though the mean value of each variable proves that RCM able to improves the availability and reliability of the plant.

4. Conclusion and Recommendation

Based on the analysis and discussion of the results of research conducted at the four plants in the CSC, it can be concluded that after the implementation of RCM, the availability and reliability of the plants proven to be improved by comparing the mean value of each variable before and after RCM has been implemented. The results of the analysis show that the R-value of downtime loss and availability of K-3 Ammonia and Urea plants are increased, while the R-value of K-4 Ammonia and Urea Plants are decreased. The RCM implemented in K-4 plants should be evaluated respectively, because the decreasing of the correlation means the RCM implementation on the plant still has not met its propose and there are stills hidden factors that affect the availability and downtime loss of the plant, regardless the RCM implementation.

The CSC should conduct a comprehensive evaluation related to the implementation of RCM at K-4 plant so that the correlation between availability and downtime loss becomes more significant. For future research, the implementation of RCM at the K-4 plant can be evaluated thoroughly to improve the effectiveness of the RCM implementation and the hidden factors affecting the correlation of availability and downtime loss could be inspected further.

References

[1] Shahaji GB, Patil AMMPP. Maintenance performance measurement – a case of the sugar industry. *J Qual Maint Eng* 2018;24:79–100. doi:10.1108/13552510910961110
[2] Nourelfath M, Nahas N, Ben-Daya M. Integrated preventive maintenance and production decisions for imperfect processes. *Reliab Eng Syst Saf* 2016;148:21–31. doi:10.1016/j.ress.2015.11.015
[3] Swanson L. Linking maintenance strategies to performance. *Int J Prod Econ* 2001;70:237–44. doi:10.1016/S0925-5273(00)00067-0
[4] Basri EI, Razak IHA, Ab-Samat H, Kamaruddin S. Preventive maintenance (PM) planning: A review. *J Qual Maint Eng* 2017;23:114–43. doi:10.1108/JQME-04-2016-0014
[5] Pourjavad E, Shirouyehzad H, Shahin A. Analyzing RCM Indicators in Continuous Production Lines: A Case Study. *Int Bus Res* 2011;4:115–24. doi:10.5539/ibr.v4n4p115
[6] Gupta G, Mishra RP. A SWOT analysis of reliability centered maintenance framework. *J Qual Maint Eng* 2016;22:130–45. doi:10.1108/13552510910961110
[7] Nabhan MB. Effective implementation of reliability centered maintenance. *AIP Conf. Proc.*, vol. 1239, 2010, p. 88–95. doi:10.1063/1.3459791
[8] Wijesinghe D, Mallawarachchi H. A systematic approach for maintenance performance measurement: Apparel industry in Sri Lanka. *J Qual Maint Eng* 2019;25:41–53
[9] Ben-Daya M, Duffuaa SO, Knezevic J, Ait-Kadi D, Raouf A. Handbook of maintenance management and engineering. 2009. doi:10.1007/978-1-84882-472-0
[10] Pritchard G. Preventive and Predictive Maintenance. World Cem 2007
[11] Goswami M, Goyal VK, Nand S. Ammonia-Urea Plants. Indian J Ferilizers 2012
[12] Khasanah R. RELIABILITY-CENTERED MAINTENANCE (RCM) EVALUATION IN THE INDUSTRY APPLICATION, CASE STUDY: FERTILIZER COMPANY,

INDONESIA. 2012

[13] Khasanah R. Reliability-Centered Maintenance (RCM) Evaluation Method in the Industry Application, Case Study: Fertilizer Company, Indonesia. *J Teknol TECHNOSCIENTIA* 2015;8:56–60

[14] Fore S, Mudavanhu T. Application of RCM for a chipping and sawing mill. *J Eng Des Technol* 2011;9:204–26. doi:10.1108/17260531111151078

[15] Pearson Correlation. *A Pract. Approach to Using Stat. Heal. Res.*, 2018. doi:10.1002/9781119383628.ch17

[16] Singh YK. Fundamental of Research Methodology and Statistics. 2006. doi:10.1017/CBO9781107415324.004

IOP Conf. Series: Materials Science and Engineering 1072 (2021) 012055 doi:10.1088/1757-899X/1072/1/012055

Supplementary quality control features for the production department in Odoo ERP

A S Ahmadiyah[1,2], Y Y Ratna[1,3], N N Yotifa[1,4], and I Dinillah[1,5]

[1]Informatics Department, Sepuluh Nopember Institute of Technology, Surabaya, Indonesia

E-mail: [2]adhatus@if.its.ac.id, [3]yukiratna.17051@mhs.its.ac.id, [4]nitama.17051@mhs.its.ac.id, [5]izzahdinillah.17051@mhs.its.ac.id

Abstract. Odoo is one of the top Enterprise Resource Planning (ERP) applications. In particular, it is supported by a quality control module with the ability to control, trigger an alert, and check its purpose. This research strives to supplement the existing Odoo quality control module with parameters from past inspection data obtained before quality control migrates to Odoo. To support the study, we used a machine learning method to discover the intrinsic pattern from the dataset of quality control in the production process of baby biscuit products. The experiment shows that the additional quality control feature can provide the product measurement and tolerance threshold fed into Odoo quality control module. The additional feature is helpful for decision making and error minimization in setting quality control parameters and tolerance threshold. Furthermore, a high accuracy rate of 95.71% is obtained from the employed Decision Tree algorithm.

1. Introduction

Quality control (QC) is vital to ensure that the final product or service meets the requirements and a set of quality standards. It supports companies to maintain and protect the quality of products or services, which leads to achieving the target market opportunity through customer satisfaction [1]. Specifically, in the manufacturing industry, quality control appears as quality control of raw materials, quality control in the production process, and quality control of the final product.

In a large-scale business, quality control is recorded into an integrated management system such as an Enterprise Resource Planning (ERP). Odoo is one of the popular open-source ERP tools. Odoo is equipped with quality inspection in which one can create an inspection plan based on business processes and company needs. The role of ERP in quality inspection is to speed up and simplify the inspection process.

Odoo quality control consists of three functions, i.e., control, check, and alert functions. In defining the quality control point using take measure type, one needs to input product measurement along with the threshold value during the production process. This feature can only be used for the currently inspected data. However, this feature does not accommodate past data before the company migrates quality control to Odoo, which potentially holds essential information about measurement and threshold.

This paper aims to tackle this problem by suggesting additional quality control features by extracting inherent information from the past data. We utilized a machine learning method to enable the learning process. Machine learning is a powerful tool to conduct quality inspection [2][3][4][5][6][7]. Our proposed solution can automatically find the product measurement and threshold values fed to the Odoo quality control module by using the baby biscuit production process in our experiment.

2. Methodology

Our concern is to incorporate essential information from past inspection data generated before the QC migration to Odoo as a recommendation fed to define quality control points. As seen in Figure 1, our proposed process is illustrated in the bottom left side. It starts with the preparation of the past inspection dataset. Then, a Decision Tree algorithm is utilized to discover intrinsic patterns. Finally, the recommended product measurement and threshold from the past is obtained. The term past refers to quality control conducted without the Odoo quality control module.

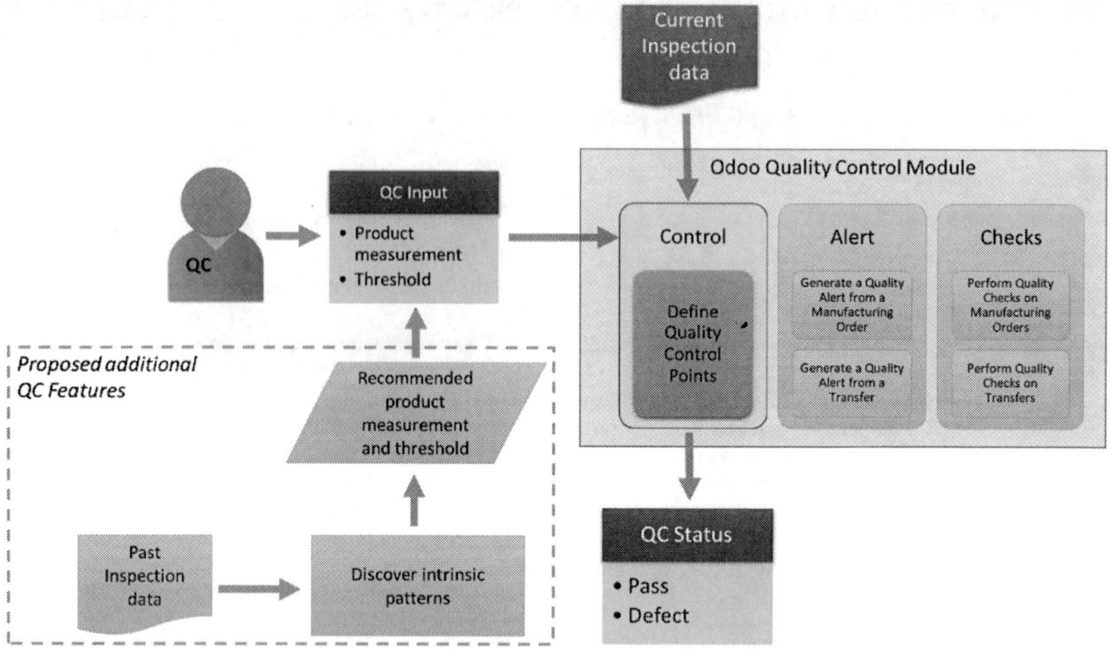

Figure 1. The position of our proposed process against the existing Odoo quality control module.

As seen in Table 1, the past inspection dataset from baby biscuit is prepared. From 233 collected records, the quality control dataset has several dimensional measurements or attributes. They are shape, diameter in centimeters, weight in grams, and color in RGB, separated into three-color channels ranging from zero to 255. The products are divided into five codes; there are M01, M02, M03, M04, and M05; each represents the flavor of the baby biscuits.

Table 1. Samples of baby biscuit quality control in production process.

Product Code	Circular Shape	Diameter (cm)	Weight (gm)	Color Channel		
				Red	Green	Blue
M03	Yes	5.9	9	191	173	109
M02	Yes	5.8	8.9	189	179	98
M04	Yes	6	10.83	192	161	101
M01	Yes	4.9	10.83	193	173	111
M05	No	5.6	11	213	182	101
M05	Yes	5.7	12.1	190	162	101
M04	Circle	6	8	189	175	110
M04	Circle	5.8	12.5	201	170	104

Next, the discovering intrinsic patterns step is performed using the Iterative Dichotomiser 3 (ID3) algorithm [8], one type of the Decision Tree, to train each data in the dataset. The ID3 algorithm generates a tree structure. At first, The ID3 selects the root node, which has the highest information gain value. It is followed by setting the next attribute with the highest information gain value among the rest of the attributes as the branch node. This process is repeated until all attributes are counted.

From this step, we have a model to classify each data into pass or defect type. This model's performance is then tested to ensure the accuracy of the model, leading to trusted results fed into the Odoo quality control module. Lastly, we calculate product measurement tolerance from the model built, which is also needed by the Odoo quality control module.

3. Results and Discussions

From 233 records used in the training phase, the classifier can predict 95.71% of the quality control status correctly, leaving the 4.29% misclassified. Meanwhile, Table 2 shows the result of the testing dataset using the model built using the ID3 algorithm. The column labeled Expected Output is the result of the past inspection. In contrast, the column labeled Real Output represents the result obtained using ID3. In this case, The ID3 algorithm is proved to perform well in revealing the intricate pattern of the dataset.

Table 2. Quality control test results.

Test #	Product Code	Circular Shape	Diameters (cm)	Weight (gm)	Color Channel			Expected Output	Real Output
					Red	Green	Blue		
1	M03	Yes	5.6	10	203	177	101	Pass	Pass
2	M02	Yes	5.8	9	210	172	101	Pass	Pass
3	M04	Yes	5.7	11	205	181	109	Pass	Pass
4	M01	Yes	4.9	10	201	170	109	Defect	Defect
5	M05	No	5.8	9	202	180	109	Defect	Defect
6	M05	Yes	5.6	10	216	170	105	Defect	Defect
7	M04	Yes	5.7	8	201	171	102	Defect	Defect
8	M01	No	5.8	9	191	179	109	Defect	Defect

In Table 3, we organized product measurement and tolerance threshold similar to the Odoo quality control module, particularly defining the quality control point feature. The product measurement name, type, and unit of measurement (UoM) are obtained directly from the dataset attributes. In contrast, the minimum, maximum, and tolerance value of the quantitative type are obtained from machine learning, in our case, the ID3 algorithm. Holding that information, the quality control team has the option to directly use the recommendation from our supplementary quality control feature or make adjustments. This can lead to minimizing the effort of setting quality control points and human error.

Table 3. Product measurement and tolerance threshold.

Name	Type	Minimum tolerance	Minimum	Maximum	Maximum tolerance	UoM	Qualitativ e value
Circular shape	qualitative	0	0	0	0	-	2 records (Yes, No)
Diameters	quantitative	0.2	5.8	5.8	0.2	centimeters	-
Weight	quantitative	2	10.83	10.83	2	grams	-
Red channel	quantitative	0	185	220	0	-	-
Green channel	quantitative	0	140	165	0	-	-
Blue channel	quantitative	0	0	80	0	-	-

To be used by other companies, this solution can be adopted directly. However, the high accuracy of machine learning may vary due to the inspection of the dataset's characteristics. The accuracy of the model built under different inspection dataset needs to be checked before continuing to be filled into the Odoo quality control module.

4. Conclusion

When the quality control is established long before migrating to Odoo, the past quality control data can be extracted automatically using a machine learning method, specifically the ID3 algorithm. Usually, the accuracy rate is the one sought. Furthermore, the splitting criteria from the ID3 algorithm are beneficial in providing the tolerance threshold.

References

[1] Tata J, Motwani J and Prasad S. Benchmarking Quality Management Practices: US Versus Costa Rica. *Multinatl. Bus. Rev.* 2000;**8**(2):37-42. Retrieved from: https://www.researchgate.net/publication/292017239_Benchmarking_quality_management_practices_US_versus_Costa_Rica

[2] Anand S, Priya L. *A Guide for Machine Vision in Quality Control*. Chapman and Hall/CRC; 2019 Dec 23; doi.org/10.1201/9781003002826

[3] Mohammadi P, Wang ZJ. Machine learning for quality prediction in abrasion-resistant material manufacturing process. *2016 IEEE Can Conf Electr. Comput. Eng. (CCECE).* 2016 May; doi.org/10.1109/ccece.2016.7726783

[4] Ordukaya E, Karlik B. Quality Control of Olive Oils Using Machine Learning and Electronic Nose. *J. Food Qual*. 2017;**2017**:1–7. doi.org/10.1155/2017/9272404

[5] Alonzo LMB, Chioson FB, Co HS, Bugtai NT, Baldovino RG. A Machine Learning Approach for Coconut Sugar Quality Assessment and Prediction. *2018 IEEE 10th International Conference on Humanoid. Nanotechnology. Inf. Technol, Commun Control. Environ. Manag (HNICEM).* 2018 Nov; doi.org/10.1109/hnicem.2018.8666315

[6] Li F, Wu J, Dong F, Lin J, Sun G, Chen H, et al. Ensemble Machine Learning Systems for the Estimation of Steel Quality Control. *2018 IEEE Int Conf on Big Data* (Big Data). 2018 Dec; doi.org/10.1109/bigdata.2018.8622583

[7] Bordekar L, Velingkar H, Fernandes E, Bandekar HH, Harmalkar AG, Antonio Pinto BJ. Cashew Nut Grade Identification and Quality Testing Using Machine Learning. *2018 Second Int. Conf. Inventive Commun. Comput. Technol (ICICCT).* 2018 Apr; doi.org/10.1109/icicct.2018.8473058

[8] Quinlan, J.R. Induction of decision trees. *Mach Learn 1.* 1986:81–106. doi.org/10.1007/BF00116251. Retrieved from: https://link.springer.com/article/10.1007/BF00116251

Design improvement of pallet in automotive industry

A Bakhtiar[1] and S A Puspitasari[1,2*]

[1]Department of Industrial Engineering, Diponegoro University, Semarang, Indonesia

E-mail: [2*]steffanyaudina@gmail.com

Abstract. In automotive manufacturing industry, supplies of different parts are required to construct products. Products are manufactured as per a company's production targets, and in the effort to do so, some that are sent from suppliers must arrive on time and accordingly. The process of delivery incorporates the use of shuttles where parts are packed into pallets and then sent to warehouses. High demand yields high production that implies high trucking costs. The more pallets used in the transport of goods, the more shuttles are used and the higher the trucking costs. Based on observations of problems, the low effectiveness of the transporting process of goods, among others, is factored by the less than maximum pallet design giving rise to unused spaces within shuttles upon shipment.

1. Introduction

In a supply chain, the pallet becomes a very vital supporting equipment. The role of pallet is much related to warehousing, inventory management and trucking. By definition pallet is a means to transport products or goods, in a large enough amount and in a manner that the products can be clearly classified into types of product. The product or goods are arranged vertically or horizontally depending on the warehousing system used. The pallet has several sizes and types that can be adjusted based on the trucking transportation system and product capacity.

The need for pallets increases as production capacity increases. There are several challenges or course of actions that must be taken when a company possesses its own stocks of pallets, namely (i) the need to supply wood and other supporting materials to maintain the existing pallets, (ii) overseeing of pallets damage that may give rise to claims from vendors, (iii) provision of a warehouse space to store damaged pallets and excess stock, and (iv) human resources that are not specifically trained in handling wood pallets or pallet racks.

An automotive company normally carries very complex process flow, where each workstation requires specific goods or parts compatible for that segment of work, implying the need for supplied goods from a variety of suppliers. In the present case, suppliers do not directly send goods or products to the corresponding company, but rather goods are first delivered into a warehouse. Suppliers are sectored according to geographical area into a Distribution Centre (DC) not far from the company. In the shipment process to the DC, cross docking is carried out where goods or products are stored from the suppliers and then sent to the automobile company. Suppliers transports products or parts in pallets, either in the form of racks, wood pallets, or boxes, according to the specified category of request. The pallets are then loaded into shuttles or trucks that are assigned to several supplier locations.

Cross docking, to some degree, allows savings in logistics and moving costs of products. Nonetheless, avoidable cases of improvidence practices in the transport process as goods are shipped in low or non-optimal quantities, in addition to imbalances that may occur within the pallet loads. In practice, there are only a small number of pallets sent because pallet stacking cannot be carried out, so the shipping process is not optimal with large logistics costs. From this, the researcher will choose the best method to redesign the pallet in order to maximize the quantity and great efficiency in order to reduce logistics costs. Design itself is an activity that results in planning from designing, building and engineering for better innovation. The method that will be discussed and then chosen for product design is the House of Quality, TRIZ, Kano, and DFMEA methods.

2. Literature Review

Based on the introduction above regarding product design, here are some methods available for pallet product design.

2.1. Quality Function Deployment (QFD)

How to prioritize needs and desires as desired by consumers is the understanding of Quality Function Deployment (QFD). The purpose of the QFD principle is to ensure that the needs and desires of customers can be met in the process of reducing a product. Therefore, QFD stems from the voice of the customer (VOC = Voice of Customer) so that it is often referred to as customer-driven product development or customer focused design. The first goal of QFD is to always avoid marketing missed products that fall into the market due to losing competition. And the next goal is to increase the rate and efficiency and the product development process.

Overall benefits of QFD according to [1], namely (i) product design prioritizes the needs and desires of consumers so that it is more easily understood, (ii) analysing the company's performance against its competitors, (iii) concentrate the overall design effort so that it will reduce the time of the product planning process, (iv) reduce the frequency of changes in the design so as to reduce costs, (v) encourage teamwork in making decisions.

There are several deficiencies in the implementation of QFD, which are difficult to distinguish between the different and conflicting needs of consumers, it is difficult to meet the needs of different consumers groups or segments, so it usually takes a long time.

2.1.1. House of Quality (HOQ)

House of Quality (HOQ) is part of the Quality Function Deployment (QFD) which defines consumer desires through consumer opinions, which are then given priority and interpret them into measurable specifications (engineering requirements) and set targets according to existing needs. QFD is also a tool used to determine problems that need to be resolved appropriately, where HOQ is used to build information and documentation acquisition.

There are two aspects that must be clear in HOQ according to [2]. First, specification of dimensions that can support the product. Size is a necessity in conceptualizing a product, where the size requires a goal value that helps in measuring performance, the value can be specific. Second, the point of view of the functional needs of users or consumers.

This aspect is one of the supporting aspects of the consolidation of a product. This is because during the process of determining product specifications, it must be more comfortable to use than the previous concept and fulfill the fundamental needs of its use: (i) Identifying user needs into product attributes to find out how the user's or consumer's opinion of a product is generally the consumer states his opinion the most important thing is to identify the consumer's desire to avoid misinterpretation. (ii) Determine the relative importance of attributes by giving an average weight to each attribute using a priority scale based on questionnaire and interview data for (HOQ). (iii) Evaluate the attributes of existing products to analyze satisfaction ratings on existing products and respondents have used to assess each statement of needs on a scale of 1 to 5. (iv) Making a matrix of relationships between attributes and product characteristics. (v) Identifying the relationship between product attributes and technical characteristics to provide a score making it easier for the design team to identify the characteristics that most influence customer satisfaction. (vi) Identifying relevant interactions from the results of this stage will be placed on HOQ. (vii) Determine the picture to be achieved satisfying the desires of consumers and improving their products beyond existing competing products. Tools of HOQ are presented in Figure 1.

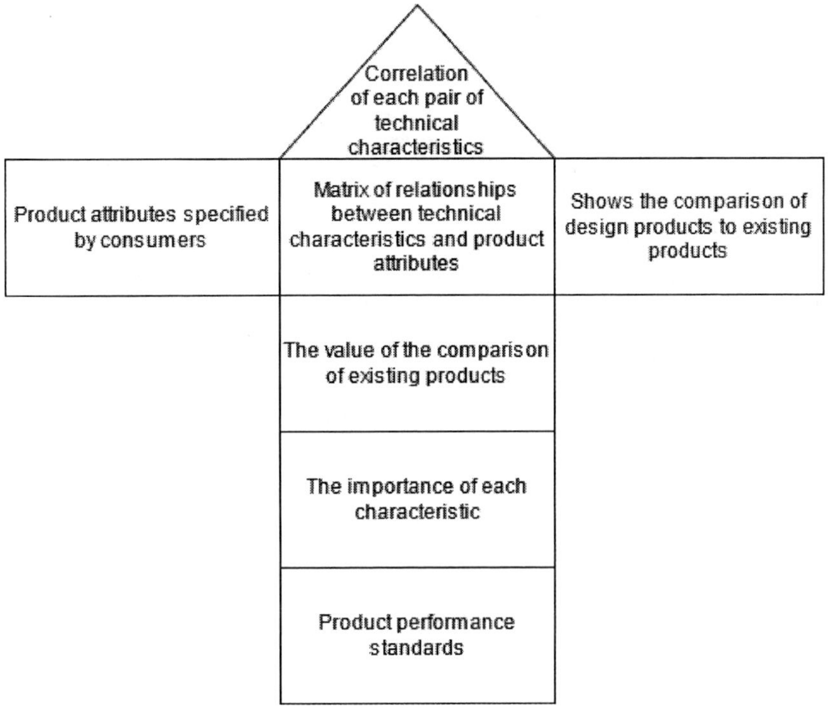

Figure 1. House of Quality tools [3].

2.2. Kano model

The model that can be used to classify product or service attributes based on the level of customer needs is also called the Kano model. Where the canoe model helps in increasing the level of expectation and satisfaction of its customers.

Mapping consumer choices is a way to understand the service attributes that are expected and chosen by customers to get satisfaction. The canoe method can do reference mapping so that an acceptable understanding and influence consumers will be able to save a lot of expenditure and energy for the product provider so that it can determine a good strategy for maximum results [4].

There are several advantages of classifying customer voices with canoe diagrams, namely (i) prioritizing needs for future development, (ii) can identify critical customer requirements that have a large impact on customer satisfaction, (iii) can get exchange value if customers have. There are different needs that cannot be met at the same time it can minimize the impact on customer satisfaction.

According to [5] there are several classifications of product and service attributes in the canoe model, namely there are six categories: One dimensional, then the second must be, then attractive, indifferent, reverse, and questionable. Where asking this clarification is done by paired questionnaires with questions that are functional and dysfunctional. Then the results are reclassified using the Kano evaluation table is presented in Table 1.

2.3. Theory of Inventive Problem Solving (TRIZ)

One method to help solve problems based on data and logic, thus accelerating the team to solve problems creatively is the definition of the Theory of Inventive Problem Solving (TRIZ) method. Where the purpose of TRIZ is asked to be able to solve problems creatively in which there is an ideal design, contradiction and innovation process.

This contradiction is a conflict where if one parameter causes other parameters to change or decrease. Where the main purpose of TRIZ is to find the ideal or perfect solution, in which there are 4 aspects, namely perfection, function, contradiction and the use of resources.

In conducting problem solving, TRIZ has 4 steps and this can be applied in the design of pallet products, namely (i) Specific Problems, which define and analyze problems in the innovation process. (ii) General Problem, analyzing the cause and effect that existed at the beginning of the problem which will then be translated into 39 parameters. (iii) General Solution, here improving and worsing features which are then incorporated into the TRIZ contradiction matrix with 40 principles to find solutions. (iv) Specific Solution, an analysis of the application of principles chosen based on priority. Section of contradiction matrix is presented in Figure 2.

The advantages of TRIZ in contradiction can be resolved by the application of innovative solutions. There are 3 basic principles, namely according to [3] (i) ideal design is the goal, (ii) contradiction helps to solve problems, (iii) the innovation process can be configured systematic.

As the definition, the aim of Triz is to solve problems creatively. The basic concept itself consists of contradictions, ideal designs, and innovation processes. Contradiction means opposition. This usually occurs when we make an increase in one parameter, but cause the other parameters to go down. The contradiction itself is divided into 2, namely technical and physical contradictions.

Table 1. Kano evaluation table [5].

Customer needs		Dysfunctional form of the question				
		1.I like this feature omitted	**2. I need this feature omitted**	**3. I am neutral about this feature**	**4. I can live with omitting this feature**	**5. I dislike omitting this feature**
Functional form of the question	1. I like this feature included	Q	A	A	A	O
	2. I need this feature included	R	I	I	I	M
	3. I am neutral about this feature	R	I	I	I	M
	4. I can live with including this feature	R	I	I	I	M
	5. I dislike including this feature	R	R	R	R	Q

Notes: A – Attractive need; O – One dimensional need; M- Must be need; I – Indifferent; R – Reverse; Q - Questionable

Worsening Feature → / Improving Feature ↓		Weight of Moving Object	Weight of Stationary Object	Length of Moving Object	Length of Stationary Object	Area of Moving Object	Area of Stationary Object	Volume of Moving Object	Volume of Stationary Object	Speed	Speed
		1	2	3	4	5	6	7	8	9	39
1	Weight of Moving Object	–	–	15,8, 29,34	–	29,17, 38,34	–	29,2, 40,28	–	2,8, 15,38	35,3, 24,37
2	Weight of Stationary Object	–	–	–	10,1, 29,35	–	35,30, 13,2	–	5,35, 14,2	–	1,28, 15,35
3	Length of Moving Object	8,15, 29,34	–	–	–	15,17, 4	–	7,17, 4,35	–	13,4, 8	14,4, 28,29
4	Length of Stationary Object	–	35,28, 40,29	–	–	–	17,7, 10,40	–	35,8, 2,14	–	30,14, 7,26
5	Area of Moving Object	2,17, 29,4	Area of	14,15, 18,4	–	–	–	7,14, 17,4	–	29,30, 4,34	10,26, 34,2
6	Area of Stationary Object	–	30,2, 14,18	–	26,7, 9,39	–	–	–	–	–	10,15, 17,7
7	Volume of Moving Object	2,26 29,40	–	1,7,4, 35	–	1,7,4, 17	–	–	–	29,4, 38,34	10,6, 2,34
8	Volume of Stationary Object	–	35,10 19,14	19,14	35,8, 2,14	–	–	–	–	–	35,37, 10,2
9	Speed	2,28, 13,38	–	13,14 8	–	29,30, 34	–	7,29, 34	–	–	–
10											
11											
12											
13											
14											
15											
16											
17											
18											
19											
20											
21											
22											
23											
24											
25											
26											
27											
28											
29											
30											
31											
32											
33											
34											
35											
36											
37											
38											
39	Productivity										

Figure 2. Section of contradiction matrix [3].

2.4. Design Failure Mode Effects Analysis (DFMEA)

DFMEA stands for Design Failure Mode Effects Analysis. This methodology helps identify potential failures in product design at the development stage. In most cases, engineers use this as a procedure to explore possible design failures in real-world situations or uses. Matrix DFMEA is presented in Figure 3.

DFMEA is an analysis of failure models that can occur due to product design. DFMEA is almost the same as PFMEA (Process Design Failure Mode Effects Analysis). The difference is that PFMEA is for process analysis, whereas DFMEA is more focused on product design. The output from DFMEA is used for PFMEA analysis. The relationship between these two documents is very important to be able to produce products according to the initial design. For example, failure mode in DFMEA and failure mode in PFMEA can produce the same potential effects. In this case, the effect of design failure mode must be shown on the potential effect and severity value of DFMEA and PFMEA.

Indeed, not all elements in DFMEA will be seen directly related to elements in PFMEA, because the focus is different. Information from each column is not always the same. For example, the potential design failure mode is not the same as the potential process failure mode, the potential village cause is not the same as the potential process cause. But by comparing the entire contents, we can get the connection. For example the relationship in terms of special characteristics in DFMEA and PFMEA.

The steps to design a product design using DFMEA's Design Failure Mode Effects Analysis are: (i) Design Review. To help ensure all team members are familiar with the product and its design, identifying each of the main components of the design. (ii) Review Potential Failure. Review existing documentation and data for instructions. Consider potential failure modes for each component and interface. (iii) List of Potential Failure Effects. There is more than one effect for each failure. The effect is directly related to the ability of certain components to perform the intended function. (iv) Assign Severity. Determine the severity rating for each effect that has been identified. Severity is an estimate of how serious the effect will be. Severity rating is based on a relative scale ranging from 1 to 10. Rating "10" means the effect has a very high severity level that leads to danger without warning and vice versa. (v) Assign Occurrence. Event rating is based on how often the cause of failure is likely to occur. So it is necessary to know the potential causes to rank the events because, as the severity ranking is driven by effects, the ranking of events is a function of the causes. Severity rating is on a relative scale from 1 to 10. Occurrence rating of "10" means the appearance of failure mode is very high, it happens all the time and vice versa. (vi) Assign Detection. Detection rating is based on the possibility of failure will be detected before the customer finds it. To assign a detection rating consider the design or control associated with an existing product for each failure mode and then assign a detection rating to each control. The detection rating of "1" means the possibility of detecting failure is almost certain and vice versa. (vii) Calculating RPN. RPN gives us a relative risk rating. The higher the RPN, the higher the potential risk. Calculate RPN for each mode and failure effect. The RPN formula is RPN = Severity x Occurrence x Detection.

Each of the three relative ranking scales ranging from 1 to 10, the RPN will always be between 1 and 1000. (viii) Developing an Action Plan. Taking action means reducing the RPN. RPN can be reduced by decreasing one of the three ranks (severity, occurrence, or detection) individually or in combination with each other. (ix) Taking Action. Implementing identified improvements such as outlining what steps are needed to implement the solution, then responsibilities and target completion dates for the specific actions to be taken are identified. (x) Recalculate the Resulting RPN. Re-evaluate each potential failure once repairs are made and determine the impact on the RPN.

Prototype Pre-launch Production FMEA Number:		Key Contact / Phone	Date (Orig.) / Revision Date / Rev. Level
Part Number / Revision Level:		Core Team:	Plant Approval / Date
Part Name / Description:			Customer Engineering Approval (if required)
Device Code:	Plant Code:	Supplier / Plant:	Customer Quality Approval (if required)

Part Number	Number and Description of Operation	Process Function	Potential Failure Mode	Potential Effect of Failure	Y/N	SEV	Potential Cause(s Mechanism(s) o Failure	OCC	Current Controls Detection / Prevention	DET	RPN	Recommended Action	Responsible & Completion Date	Action Results: Action Take	SEV	OCC	DET	PRN

Figure 3. Matrix DFMEA [6].

3. Method Analysis

Based on the background, the automotive company needs to design a pallet product so that it can increase the effectiveness and efficiency of cross docking activities, namely by using a shuttle to carry pallets containing parts needed by automotive companies, in order to reduce the cost down the company wants.

Here are the advantages and disadvantages of each method are presented in Table 2.

Table 2. Advantages and disadvantages of the method.

Methods	Advantages	Disadvantages
QFD	- According to the wants and needs of consumers, - Quality improvement, Consumer satisfaction, - Reduction of costs, profits, - Time to market - Comparison of existing products	- It's hard to distinguish between conflicting needs, - Difficult to meet all segments, - Long time - Large questionnaire
TRIZ	- Knowing contradictions, ideal design, systematic, technological principles - Includes many categories	- Not focus on one segment - Too many variables
DFMEA	- Knowing the product failure, - Know the level of risk of severity, incidence and detection, - Priority of action, - Customer satisfaction, - Reducing time and costs, - Failure control	- Different focus, - Long time - Design shoes
KANO	- Easy to understand product specifications according to needs - Easy to determine the priority product attributes into specifications - Easily differentiate and compete with other products	- Only presents classification results not quantitative values - Only the results of the selection of the level of customer satisfaction without explanation.

The QFD method cannot be chosen because QFD uses a questionnaire from the customer, thereby increasing the effectiveness and time efficiency of a product, but in the case of pallets, the company does not need the customer's voice to get the desired pallet design according to the customer's voice, so what is needed is between suppliers and companies only.

The TRIZ method has advantages because it covers a broad scope with many basic principles and has contradictions, which all have in designing product designs but in the case of pallet to 40 these basic principles are too broad and not specific. While the canoe method has the advantage of knowing how well the attributes are to satisfy consumers, but also the disadvantages that canoes have because they are less able to capture all the voices of consumers.

The chosen method is the FMEA method which is then continued with DFMEA. So that in this FMEA, the product can be analyzed for the failure rate and the level of risk that exists in the product, then it will be continued with DFMEA which is to find failures in product design after finding a failure of the process.

The main purpose of DFMEA is to uncover potential product design failures. A basic review of DFMEA is carried out using a product scheme, DFMEA helps analyze potential product design failures at the design and development level. In addition, this methodology will reduce production costs and also the possibility of failure in product launches.

Steps that can be taken in designing the pallet using the DFMEA method are: (i) design review, (ii) reviewing the potential failure, (iii) list of potential failure effects, (iv) setting the severity, (v) setting the occurrence, (vi) setting the detection, (vii) calculate RPN, (viii) develop an action plan, (ix) take action, (x) recalculate the resulting RPN.

4. Conclusion

Based on the discussion in this review literature, the researcher concludes several research results, that: The method chosen in designing the pallet design uses the DFMEA method. DFMEA is a document that continues to grow. All updates and changes to the product development cycle are made for the product or process. This change can and often is used to recognize new failure modes.

DFMEA indicates the likelihood of failure, the severity of the risk, existing controls, recommendations and improvements after the recommended action. The main objective of both is to reduce or avoid major product and production deviations. This methodology can reduce design or operating costs too. Furthermore, some failures might not be reduced at FMEA. This failure can be transferred to DFMEA to take the control needed for pallet product upgrades.

References

[1] Daetz, D.B. Barnard, R. Norman. 1995. Customer Integration : The Quality function Deployment (QFD) Laders Guide for Decision Making. (New York: John Wiley & Sons,Inc)

[2] Otto, K.N. Wood, K.L. 2001. *Product Design: Techniques in Reverse Engineering and New Product Development.* Retrieved from: https://books.google.com/books?hl=id&lr=&id=0X54fSKq7bkC&oi=fnd&pg=PR13&ots=53vaeBMtWY&sig=IgWWbZtzEsTg7wsC71g56-7LWPA

[3] Ginting, R. 2013. *Perancangan Produk.* (Yogyakarta: Graha Ilmu).

[4] Berger, C. Blauth, R. Bolster, C. Burchill, G. DuMouchel, W. Pouliot, F. Richter, R. Rubinoff, A. Shen, D. Timko, M. and Walden, D. Kano's methods for understanding customer-defined quality". *Center Qual Manag J.* 1993;**2**(4):3 – 36.

[5] Tan, K. C., Xie, M., & Shen, X. X. Development of Innovative Products Using Kano's Model and Quality Function Deployment. *Int. J. Innov. Manag..* 1999 Sep;**3**(3):271-286. doi.org/10.1142/S1363919699000141

[6] Dr. Gregory, K. 2020. *Fmea and Reliability.* Ohio University. Retrieved from: https://www.ohio.edu/mechanical/design/snrdesign07_08/me471/fmea&reliability.pdf

The development of Kansei-based mining model for robust service design

M Hartono[1,2]

[1]Department of Industrial Engineering, University of Surabaya, Surabaya, Indonesia

E-mail: [2]markus@staff.ubaya.ac.id

Abstract. This study proposes a model which shows the importance of Kansei Engineering (KE) methodology in supporting the design of robust services. The KE methodology is enhanced by the Kansei-based mining process, SERVQUAL, and Kano categorization in order to conceptualize robust service design and development. Due to complexity and contextual based, Kansei words as emotion representative are usually formed in fuzzy and abstract terms. It may lead to ambiguous and unclear meaning. Mapping and structuring more representative Kansei is needed. Hence, through Kansei mining system, historical data includes customer Kansei feedbacks are critical. KE model incorporating Kansei mining process followed by expected contribution which captures more organized and captured Kansei of product and service experience is proposed and discussed.

1. Introduction

Affective design or also known as Kansei Engineering-based (KE-based) product design and development has received much attention from both researchers and practitioners. Incorporating customer affective needs into product design elements will fulfill and increase customer satisfaction, especially the emotional satisfaction. Not only in product design, KE also applies to service design and development [1]. Kansei Engineering (KE) has shown its ability and superiority in capturing, identifying, and mapping customer affective/emotional needs into design characteristics [2][3]. This method has been used ranging from physical products to service attributes, e.g., shampoo bottle, audio system, virtual kitchen, hotel services, logistics services [1][2][4][5]. Related to services, this study shows the milestones of KE application in services [5], and current research direction (see Table 1). This current research is, inherently, to enhance the KE methodology in representing an entire service experience.

Table 1. Milestones of KE application.

Study	Contexts/Tools/Methods								
	General	Servqual	Kano	TRIZ	Culture	Sustainability	Logistics	Robust	Mining*
[1]	√	√	√						
[4]	√	√	√	√			√	√	
[6]	√		√						
[7]	√	√	√						
[8]	√								
[9]	√	√	√		√	√			
[10]	√	√	√		√				
[11]	√						√		
[12]	√	√	√	√		√			
Current	√	√	√						√

**new direction*

Referring to KE research mapping (see Table 1), there is an opportunity to extend KE studies through service contexts and methods. It is quiet challenging to map customer emotional needs/expressions/emotions, namely Kansei, to perceptual design elements. One of critics is that how to get representative Kansei, though many studies are still using wordings [2][3][4][5]. Kansei words have been widely used to represent the emotional needs of customer and user. It is still relevant. In addressing correct and perfect Kansei is not easy. It is a challenging part in the entire KE methodology. To collect Kansei, at present, majority of research utilizes observation, survey, literature review, in-depth interview, and ethnography study [1]. Kansei collection, called as spanning the semantic space, is the first step in KE methodology. Due to limited budget, sample size, and time, the validity of Kansei is questioned. How can we obtain the truly representative and valid Kansei? Hence, to understand the customer affective needs (also called Kansei words) accurately and then to match them with design characteristics is really challenging.

In some cases, it is very difficult to grasp and interpret the customer Kansei due to linguistic factors and cultures. The impact is that the Kansei will be relatively short-lasting, imprecise, and ambiguous. Moreover, the Kansei applied to certain industry or product in a country will not be compatible to that of other countries, even though the Kansei is the same. Thus, in this study, in matching and refining Kansei structure, Kansei mining system is introduced. Reuse of knowledge from past sales records, product specification, and customer reviews will bring potential benefit for this mining process. This is linked to KE methodology for robust design and development. Some methods related to service quality and robust design are embedded to refine the KE methodology. It is expected that this study will provide a practical guideline for service provider to promote more in-depth understanding of customer emotional satisfaction and ways to satisfy them.

2. Kansei Engineering, Kansei-based mining, and Robust Service Design

Kansei is referred to emotional or affective needs. It is translated to be emotional state or psychological feelings of customer due to product experience or service encounter. Kansei is one of literatures to describe emotions and feelings of customers related to product design, marketing, and psychological aspects for service encounters. According to Nagamachi [3], affective design is the inclusion of representation of affect and psychological state (e.g., subjective impression, visual perception, and emotional response) in the product and service design.

Kansei is a function of perceived product and service experience in the KE methodology. Kansei can be captured by human five senses [3]. KE uses Kansei words to express customer affective needs and emotional states. Then, these Kansei words will be translated into perceptual design elements. Next, it will be moved to designer's side. It is quite often the designers have no underlying understanding of the relationship between customer Kansei and design elements. Hence, mapping out and highlighting the difference between customer Kansei and design elements are critical.

Due to complexity and contextual based, Kansei words are usually formed in fuzzy and abstract terms. It may lead to ambiguous and unclear meaning. It is also influenced by the psychological condition of respondents. A condition when they are happy or unhappy will give different Kansei regardless the service experience. The question is asked again. What Kansei is appropriate? How to obtain Kansei correctly? Reusing historical data is deemed to be sufficient to facilitate the handling of Kansei information. Also, it gives information regarding the design elements. In other words, this process known as data mining is recognized as a decision supporting tool through efficient knowledge discovery of historical data. Related to Kansei, it is called as Kansei mining system. This Kansei mining system will be proposing to utilize valuable and reliable historical information of customer affect or impression on existing service experiences. Some methods can be used such as Kawakita Jiro (KJ) or affinity diagram, multi-pickup method (MPM), and Kano model and categorization. More direct methods used by marketing such as conjoint analysis with fuzzy systems, focus group discussion, in-depth interview, and similarity-dissimilarity attribute rankings will bring benefit in finding latent customer needs. Those methods are quite related to marketing and customer, which are not directly to engineering characteristics. Thus, there is a gap between marketing and engineering sides. These approaches are usually used for any product or service design at the very beginning of stage, a new and clean sheet of paper. In facts, most products or even called as new product or service evolves from the current one. Hence, historical data includes customer feedbacks are critical. KE methodology which captures more organized and captured Kansei of product and service experience is required.

In designing service, ideally, it should be ensuring that all service attributes will have significant impacts on Kansei. On the other hand, there will be service traits or noise which are not intended by both customer and designer. Unfortunately, they are sometimes blended in the package of service attributes offered [4]. Robust Kansei service design has been proposed [4] using Taguchi method to determine the optimal service design settings by analyzing the complex relationships between the controllable factors (i.e., service attributes), the uncontrollable factors (i.e., noise factors), and the service quality performance.

3. Framework Development

This framework is modified from the previous works [4] by incorporating the refined step of span the semantic space/Kansei. The refinement of spanning the semantic space/Kansei adopts the concept of Kansei mining system. The entire framework or model of Kansei-based mining for robust service design is provided in Figure 1.

It starts with the choice of service domain, followed by spanning the semantic space/Kansei and the service attribute space. In refining Kansei, some methods of mining system are utilized, such as expert knowledge, online content text mining, or KJ method. According to historical data of Kansei, mapping, identification, refinement, and measurement of Kansei will be conducted. According to Hartono & Tan [1], Kano categorization and SERVQUAL model are included to map the prioritization of service attributes. The satisfaction score is calculated to give the understanding of which service attributes will be followed up for improvement purpose.

Once there is negative satisfaction, it will be the focus for improvement. Assuming that Kansei is a function of perceived service attributes with negative satisfaction and Kano's A & O categories, some potential service attributes will be identified and regarded as WHATs at the House of Quality (HoQ), then translated into engineering characteristics of services (known as HOWs). We apply Taguchi method to find factors and levels which can minimize the noise or error. The final objective is that to achieve robust design-based for service that increases customer emotional satisfaction (Kansei).

4. Discussion and Implication

This study mainly focuses on short/brief literature on Kansei mining system and robust service design. This topic of study on Kansei Engineering (KE) methodology is still interesting due to the dynamics of customer need and satisfaction. Satisfaction today will not be the same with that of tomorrow. More specifically, in terms of customer satisfaction dimension, Kansei or affect will bring more influence on the success of service offer than that of cognition. Model of Kansei-based mining for robust service design has been introduced. It has been modified from the study by Hartono & Tan [1] and Hartono & Santoso [4]. This model should be tested through the real case study to see its applicability. Several innovative service settings can be considered such as e-commerce, logistic, internet provider, online taxi, and online courses.

Kansei related through mining process might be "fast, excited, responsive, happy, and et cetera". Each Kansei will be linked to all perceived service attributes and they are provided in linear models. Through Kano's A & O category, negative satisfaction score and Kansei impact, the critical service attributes will be collected. They are put on WHATs and continued to find HOWs through Taguchi method. The expected result is robust design elements which are sensitive and critical to representative Kansei [4].

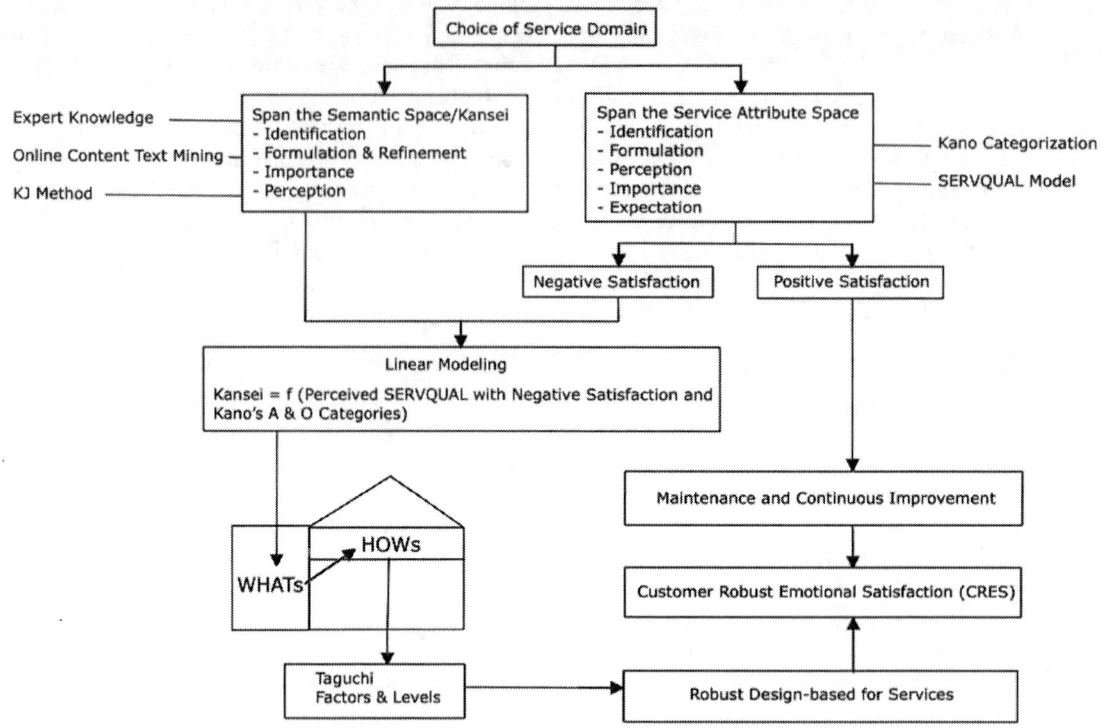

Figure 1. Model of Kansei-based mining for robust service design.

In order to make this model work well, appropriate and sufficient number of samples should be considered. In addition, sample plan whether it is probability or non-probability should be decided. For those who experience certain service well will be targeted as respondent. Some criteria such as duration of use and frequency of use should be considered.

5. Conclusion and Further Research

This study proposes an integrative model of Kansei Engineering (KE) methodology incorporating Kansei-based mining methodology for robust service design and development. The objective is that to refine the Kansei identification and structure as Kansei is the most critical component in any KE studies. The utilization of service quality and management tools combined with statistical tools will provide much benefits both theoretically and practically. Theoretically, this study provides a combined refine Kansei for KE methodology in robust service design. Practically, service provider will be equipped with a guidance to prioritize service attributes for improvement and find the optimal alternative of solutions. Surely, this study is of limited due to validation and verification. For future work, thus, it is required to test the proposed framework into real service sector. A new generation of service industry such as co-working space or e-commerce service will be of high interest.

References

[1] Hartono M, Chuan TK. How the Kano model contributes to Kansei engineering in services. *Ergon*. 2011 Oct 25;**54**(11):987–1004. doi.org/10.1080/00140139.2011.616229

[2] Nagamachi M. Kansei Engineering: A new ergonomic consumer-oriented technology for product development. *Int. J. Ind. Ergon*. 1995 Jan;**15**(1):3–11. doi.org/10.1016/0169-8141(94)00052-5

[3] Nagamachi M. Kansei Engineering in Consumer Product Design. *Ergon Des: The Quarterly of Human Factors Appl*. 2002 Apr;**10**(2):5–9. doi.org/10.1177/106480460201000203

[4] Hartono M, Santoso A. Kansei engineering-based robust design model for logistics services. Puspitasari P, Suryanto AH, Lenggo Ginta T, Srikun I, Ayu Permanasari A, Niaz Akhtar M, et al., ed. *MATEC Web of Conf.* 2018;**204**:03009. doi.org/10.1051/matecconf/201820403009

[5] Hartono M, Santoso A, Prayogo DN. How Kansei Engineering, Kano And QFD can Improve Logistics Services. *Int. J. Technol*. 2017 Dec 26;**8**(6):1070. doi.org/10.14716/ijtech.v8i6.689

[6] Llinares C, Page AF. Kano's model in Kansei Engineering to evaluate subjective real estate consumer preferences. *Int. J. Ind. Ergon.* 2011 May;**41**(3):233–46. doi.org/10.1016/j.ergon.2011.01.011

[7] Hartono M. Incorporating Service Quality Tools into Kansei Engineering in Services: A Case Study of Indonesian Tourists. *Procedia Econ Finance.* 2012;**4**:201–12. doi.org/10.1016/s2212-5671(12)00335-8

[8] Rasamoelina F, Bouchard C, Aoussat A. Towards a Kansei-based User Modeling Methodology for Eco-design. *Int. J. Affect. Eng.* 2013;**12**(2):337–48. doi.org/10.5057/ijae.12.337

[9] Hartono M, Chuan TK, Peacock JB. Applying Kansei Engineering, the Kano model and QFD to services. *Int. J. Serv. Econ. Manag.*. 2013;**5**(3):256. doi.org/10.1504/ijsem.2013.054958

[10] Hartono M, Raharjo H. Exploring the mediating role of affective and cognitive satisfaction on the effect of service quality on loyalty. *Total. Qual. Manag. Bus. Excell.* 2015 Jul 31;**26**(9-10):971–85. doi.org/10.1080/14783363.2015.1068595

[11] Chen M-C, Hsu C-L, Chang K-C, Chou M-C. Applying Kansei engineering to design logistics services – A case of home delivery service. *Int. J. Ind. Ergon.*. 2015 Jul;**48**:46–59. doi.org/10.1016/j.ergon.2015.03.009

[12] Hartono M. The Extended Integrated Model of Kansei Engineering, Kano, and TRIZ Incorporating Cultural Differences into Services. *Int. J. Technol.*. 2016 Jan 30;**7**(1):97. doi.org/10.14716/ijtech.v7i1.1789

IOP Conf. Series: Materials Science and Engineering 1072 (2021) 012058 doi:10.1088/1757-899X/1072/1/012058

Analysis on acquisition of Philippine Civil Registry Documents and inclination towards paperless e-government

L E A Villena[1,4], R A C Robielos[2,5] and Wei-Jung Shiang[3,6]

[1] School of Graduate Studies, Mapua University, 658 Muralla St., Intramuros, Manila, 1002 Metro Manila, Philippines
[2] School of Industrial Engineering & Engineering Management, Mapua University, 658 Muralla St., Intramuros, Manila, 1002 Metro Manila, Philippines
[3] Department of Industrial and Systems Engineering, Chung Yuan Christian University, No. 200, Zhongbei Road, Zhongli District, Taoyuan City, Taiwan

E-mail: [4]lizetteerinvillena@gmail.com, [5]racrobielos@mapua.edu.ph, [6]wjs001@cycu.edu.tw

Abstract. In the Philippines, the Civil Registry Documents (CRD) acquired from Philippine Statistics Authority (PSA) such as Birth Certificate, Marriage Certificate, Certificate of No Marriage and Death Certificate are being used as one of the requirements in different transactions in academics, government, travel, employment, religious matters and other transactions (e.g. proof of identification, Title/Career advancement, etc.). The current acquisition ways of these Civil Registry Documents are through online, PSA Office, Travel Agency/Courier Company and Satellite/Mobile PSA Office. In this study, the different acquisition ways of Civil Registry Documents are analyzed through service blueprint. Furthermore, a survey was conducted to determine the satisfaction rate and preferences of the CRD acquirers and was analyzed through crosstabulations. With the help of the analyzed Service Blueprints, the possible problems that arises on each process of acquisition were identified. This paper proposed an instant, electronic (involves bar coding/QR code/Reference number), paperless and less time-consuming substitute to the existing ways of acquiring Civil Registry Documents.

1. Introduction

1.1. Background

The power of the internet and web technologies has been clearly established in business, as epitomized by the enormous success of electronic commerce [1]. Technology lets people reap beneficial things interdependently. E-government as one of the subcategory of information technology, made it possible for the governments to provide the information and services efficiently as quickly and as less costly as possible through the use of modern information technology [2]. E-government is a strong and strategic tool for governance policy, as well as for improving the efficiency and effectiveness of government functions and to propel regularizations. All agencies must manage the challenges and tensions they encounter [3]. E-government also offers improved quality of information supply and fewer administrative burdens [4].

In Taiwan, every administrative agency or unit has its own website. Taiwanese people's lives currently appear to have been more convenient than before due to these well-developed web services. While, in South Korea the government is improving to a partnership-based framework whereas the government together with its citizens are working beyond bilateral participation. This was deemed necessary since demands are also increasing for e-government services to comply with the constant influences in the interactive services on social networks [5].

The utilization of Information and Communication Technology (ICT) by the government units has generated valuable improvements in the offer of public services being provided by the different levels of governments to their citizens [6]. However, not all countries have the same success stories since the success adoption of e-government contingent upon citizens' trust and their willingness to use it [7].

The entire business world has been revolutionized by the Information Communication Technologies (ICTs). These ICTs could bestow potent strategic and tactical aids which, if used properly could pave ways to great gains in upholding organizations competitiveness. The airline industry in particular has adapted an ICTs dependency since most of airline activities are reliant of them [8]. Airline industries are gradually resorting to the internet because of its ability to acquire information, for the provision of electronic/paperless, for showing the pricing and its values incorporated with it, self-booking functions, and the extensive marketing advantages it provides [9].

However, the increasing developments, growth and consumption of goods and services in the Asia region, where economies are combination of developed and developing ones, is one of the results of the factors to blame for the upsurge of municipal solid waste generation [10]. One of the major fractions of municipal solid wastes produced all across the world is paper waste, Though, most of the paper wastes could be reprocessed in pulp and paper mills but the quality of remanufactured paper is less than the virgin pulp and the incur of transportation costs make the recycling of paper alternative uneconomic [11]. According to an article on waste management, in the year 2009 it was recorded that among the 15 Asian countries studied, Philippines is the fifth country with the most paper waste composition based on percent wet weight and that the Philippines has 0.3-0.7 kg/capita/day waste generation [10].

Every citizen interacts with his or her government numerous times to avail services. The registration of births, requesting of child support, asking for identity documents, and reliance on the government for education and health, to open businesses, buy houses, and assistance during unemployment application for employments [12]. Furthermore, it was also indicated on the same study that there are three steps involved whenever a citizen requests for service. First, the data are being collected through interfaces between the government and the citizen. It is emphasized that the typical data being required are the data being used to identify the citizen and the service details. After then, the government keeps the data on databases and the data undergo processing and analyses if needed. Lastly, the data then are being used to provide services to citizens. These data are being examined to make decisions on the applications and convey the decisions to the citizens [12].

1.2. Scope

This study is within the following measures: Filipino survey respondents who can access the online survey; Civil Registry Documents acquisition from the Philippines Statistic Authority only and do not include acquisition from the Local Registries nor NSO (National Statistics Office); the process-mapping using Service Blueprint was based on the data given by the interviewed respondents (customer perspective); the creation of the Civil Registry Document database for the paperless electronic system is not included in this study; and the actual implementation of the paperless electronic system for acquiring Civil Registry Documents is beyond the scope of this study.

1.3. Objectives and motivation of the study

In this study, it is being aimed to: Identify the different ways of acquiring Civil Registry Document; Determine if the respondents are satisfied on the identified ways of acquiring Civil Registry Documents; Propose ways to improve the current system of acquiring Civil Registry Documents; Propose a paperless electronic acquisition workflow of Civil Registry Documents that does not have waiting time and does not involve queueing in line; and Identify if the respondents are inclined towards paperless and reference-code led CRD acquisition.

This study aimed to be beneficial for the Filipino citizens who are acquiring Civil Registry Documents from Philippine Statistics Authority and for the organizations and companies requiring Civil Registry Documents as one of their requirements on their transactions. It aimed to propose an easier, efficient and more environmentally inclined way of acquiring and submitting Civil Registry Documents.

1.4. Civil Registry Documents (CRD)
The Civil Registry Documents are the documents that serves as records for different data such as but not limited to birth, marriage or death of a person.

1.4.1. Birth certificate
It is a Civil Registry document containing particular birth details of a person including his/her birthplace, birthdate and time, parents and parents' occupation.

1.4.2. Marriage certificate
It is a document that shows social union or a legal contract between people that creates kinship. Such a union, often formalized via a wedding ceremony, may also be called matrimony. A general definition of marriage is that it is a social contract between two individuals that unites their lives legally, economically and emotionally [13].

1.4.3. Death certificate
It is the official document entailing details relating to a dead person, including the name of the individual, the date of birth and the date of death is called a Death Certificate [13].

1.4.4. CENOMAR. a.k.a. certificate of no marriage
As described in the Philippine Statistics Authority site, a CENOMAR is a certification issued by the PSA that states that a person has not contracted any marriage. It is also called as a Certificate of No Record of Marriage or Certificate of Singleness [13].

1.5. Organizations and companies requiring Civil Registry Documents on their transactions and/or applications
In the Philippines, there are numerous organizations and companies using the Civil Registry Documents as one of the requirements in their transactions/applications [14]. In this study, these organizations/companies were categorized into the following groups: Government; Academic Employment; Travel Agencies & Embassies; Religious; and other Organizations such as involving Title/Career Advancement.

1.6. Factors in service satisfaction
Five factors are defined in this study with regards to service/product satisfaction and in relation to previous studies. Also, these factors will be addressed through paperless electronic system of Civil Registry Documents.

1.6.1. Convenience
The concept of convenience has become progressively more significant to consumers, especially in a context of adoption of Information and Communication Technology (ICT). [13] In the Civil Registry document acquirement, convenience may include the following: easy acquirement, less effort to exhaust in acquiring CRD, good facilities, availability & accessibility of locations.

1.6.2. Dependability
Dependability is to ensure that systems will deliver the expected services to its users. In order to ensure the dependability of large safety-critical systems. In the past several decades, a significant amount of attention has been devoted to the dependability assessment of safety critical control systems from some perspectives such as reliability, availability, safety, and security [16].

1.6.3. Environmental Sustainability (En)
There was a study conducted that showed that the best practices which have the most positive impact on Environmental Sustainability (En). From that it was found out that the practices such as material

saving, energy saving, use of cleaner technologies, and less resources consumption have a direct and positive impact on the improvement of environmental performance [17].

1.6.4. Price
The price for Civil Registry Documents varies depending on what way of acquirement the Civil Registry Documents are acquired. Service quality illuminates the mechanisms through which service quality is associated with price perception of service [18].

1.6.5. Timeliness
In this study, the service provided in the acquirement of Civil Registry Documents involve speed of service, queueing time, waiting time and delivery Lead time.

2. Methodology

2.1. Survey & IBM® SPSS® Statistics v.20
An online survey form was created through Google Docs. A link for the said survey was distributed to 246 respondents to answer. Four-point scale was used, from 1 'strongly disagree/dissatisfied' to 4 'strongly agree/satisfied'. The created survey aimed to get information regarding frequency, purposes, insights, experiences, preferences and suggestions of the respondents with regards to the acquirement of Civil Registry Documents. It is also aimed in the survey to measure the likelihood of interest towards paperless and reference-code led acquisition of Civil Registry Documents. The acquired data from the survey were analyzed through IBM® SPSS® Statistics v.20.

2.2. Service Blueprint (SB)
The succeeding service blueprints shows the mapped process of different ways on acquiring Civil Registry Documents.

2.2.1. Online process
As shown on the Online CRD Acquirement SB, the acquirement of Civil Registry Documents through Online involves accessing the website which is the www.psaserbilis.com.ph, filling-up of the online form, paying for the requested Civil Registry Documents and waiting for the delivery which takes 3-7 days.

When, visiting www.psaserbilis.com.ph, a window would pop-up once you enter the site that says the details regarding payment. First to be filled out is the contact and delivery information form, followed by choosing which CRD/s to purchase. Then, questions will be asked pertaining to previous transactions as well as corrections done on the Civil Registry Document being requested. Personal details are then asked to be filled out. Then after submitting, summary of details window will pop up and for confirmation of the indicated details. Then, an acknowledgement window will open. This entails the batch request number and the payment and delivery instructions. On the right portion of the window, options will be provided on how the transaction will be paid. The transaction can be paid through partner payment centers and banks.

There is one inconvenient waiting point W1 as shown in Figure 1 whereas the CRD acquirer needs to wait for the lead time of the delivery and two failure points as shown on F1 and F2 whereas an error might occur if the CRD acquirer filled out the form incorrectly as well as for the payment of the requested CRD through payment centers/online payment channels.

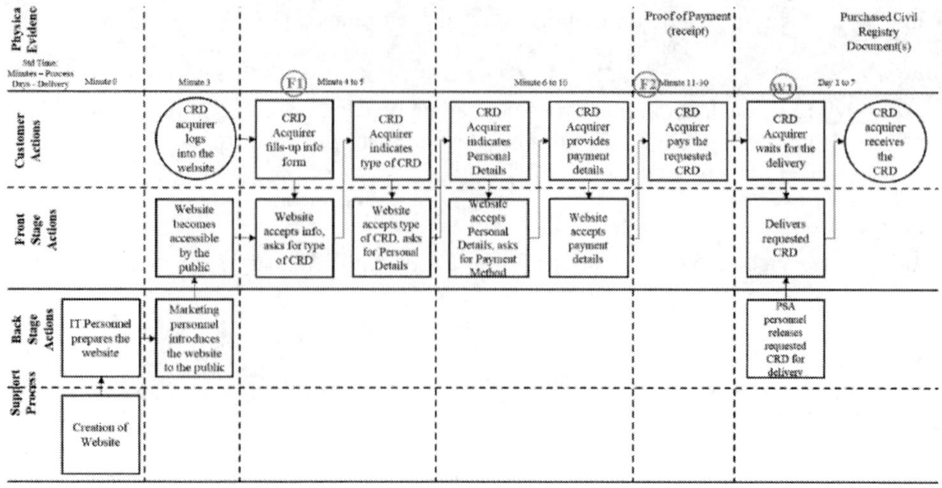

Figure 1. Current online process of acquiring CRD.

2.2.2. *Walk-in on PSA office process*

As shown in the Walk-in on PSA Office service blueprint, the acquirement of the Civil Registry documents through walking-in at Philippine Statistics Authority (PSA) Office involves visiting the PSA Office, queuing in line, filling up form, submitting the form, paying for the requested Civil Registry Document to the Cashier and claiming the requested Civil Registry Document from the Releasing section or the requested document will be delivered to the specified address of the requestor if the requestor does not wish to wait for 4-5 hours to get the document on the same day it was requested. The delivery would take 3-7 days.

Furthermore, some of the respondents/interviewees have indicated that some of PSA authority/personnel verifies the valid ID, asks the relationship of the owner of the Civil Registry Document with the requestor and asks authorization letter and additional documents if not directly related to the owner of the Civil Registry Document.

There are two inconvenient waiting points W2 & W3 as shown in Figure 2, whereas the CRD acquirer needs to wait in queue until its number is called out and to wait for the lead time of the delivery and failure points F3, F4 and F5 whereas the CRD acquirer needs to utilize a paper from the queue machine, an error might occur in filling out the form as well as the payment for the requested CRD through payment centers/online payment channels.

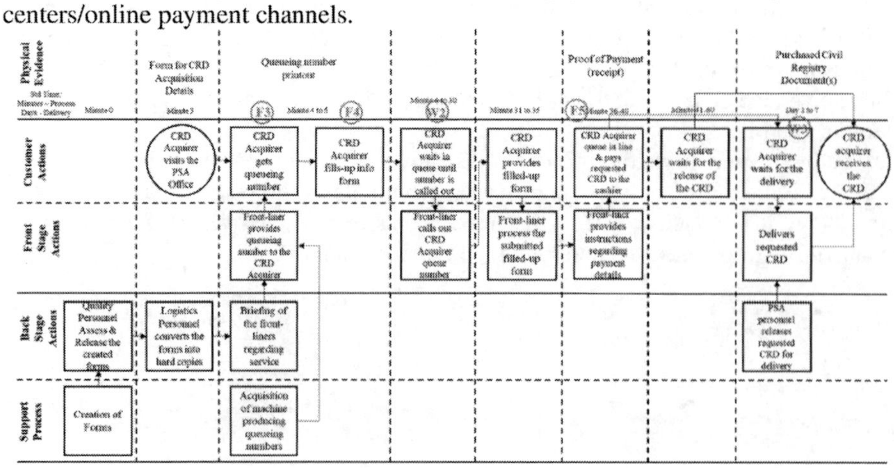

Figure 2. Current walk-in on PSA office process in acquiring CRD.

2.2.3. Travel agency/courier company process.

As shown on the Travel Agency/Courier Company service blueprint, the acquirement of Civil Registry Documents through Travel Agency/Courier Company involves visiting the travel agency office, providing details by filling-up form, paying the fee to the office or partner payment centers and delivery of the document to the address provided within 3-5days.

There are two inconvenient waiting points W4 & W5 as shown in Figure 3, whereas the CRD Acquirer needs to wait in line until its queueing number is called out and CRD acquirer needs to wait for the lead time of the delivery and failure points F6, F7 and F8 whereas the CRD acquirer needs to utilize a paper from the queue machine, an error might occur error in filling out of the form as well as for the payment of the requested CRD through payment centers/online payment channels.

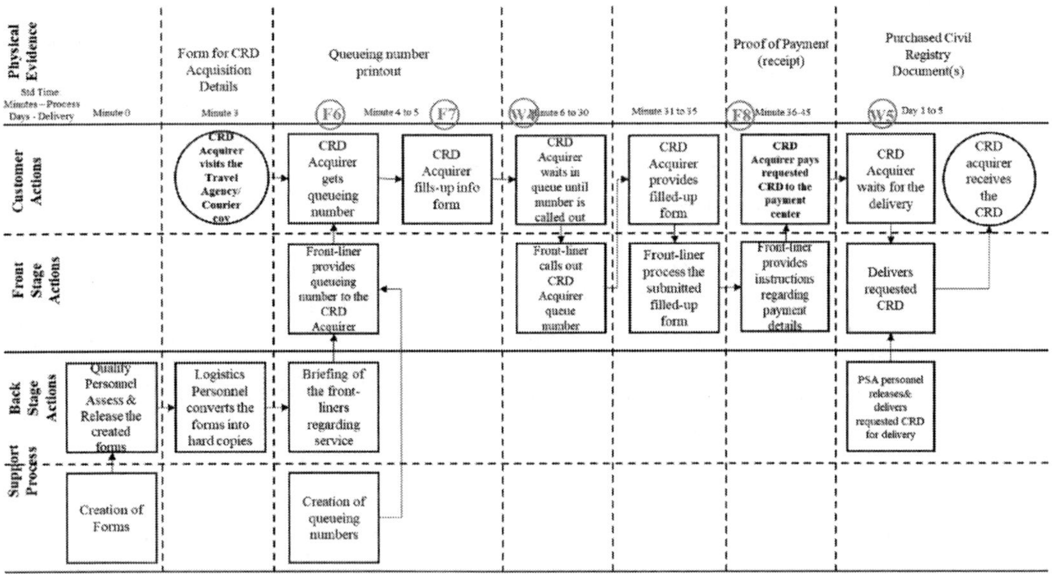

Figure 3. Current travel agency/courier company process in acquiring CRD.

2.2.4. Satellite/mobile/roaming PSA office.

As shown on the Satellite/Mobile PSA Office service blueprint, the acquirement of Civil Registry Documents through Satellite/Mobile PSA Office involves visiting Satellite/Mobile PSA Office, provide details through filling up the form, paying the fee through the Satellite/Mobile PSA Office or through online or partner payment centers, and picking up of the document on the date specified by the Satellite/Mobile PSA Office or the delivery of the document to the address provided within 3-7 days. Note: Documents requested from Satellite/Mobile PSA Office can't be claimed on the same day these documents were requested.

There are three inconvenient waiting points W6, W7 & W8 as shown in Figure 4, whereas the CRD Acquirer needs to wait in line until its queueing number is called out and CRD acquirer needs to wait for the lead time of the delivery/pickup availability of the CRD and failure points F9, F10 and F11 whereas the CRD acquirer needs to utilize a paper from the queue machine, an error might occur in filling out the form as well as the paying of the requested CRD through payment centers/online payment channels.

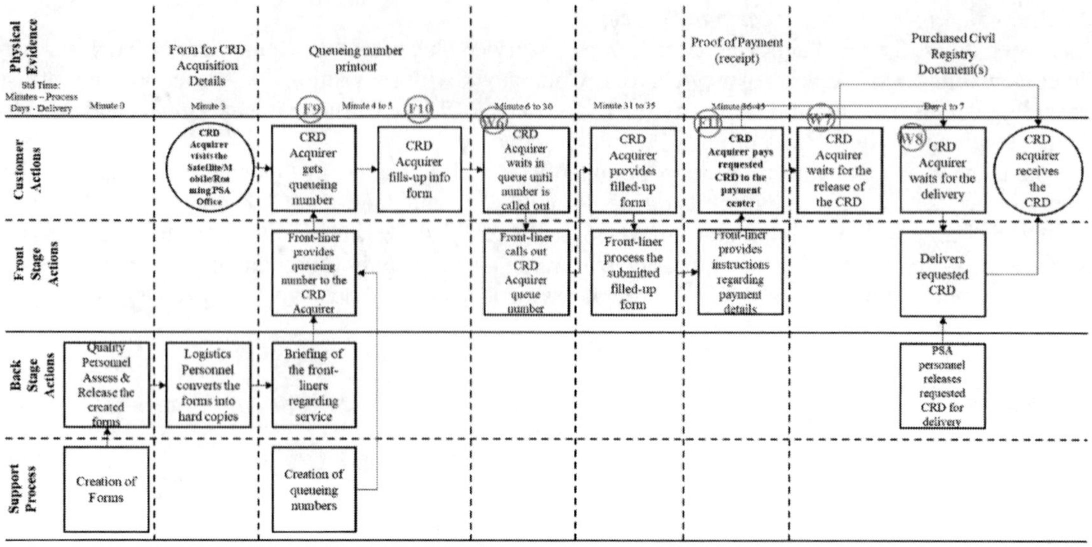

Figure 4. Current satellite/roaming/mobile process of acquiring CRD.

2.2.5. Summary of service blueprints possible failure and waiting points.
As shown in Table 1, in total, there are 11 failure points and 8 waiting points identified on the current ways of acquiring CRD.

Table 1. Summary of service blueprints possible failure and waiting points.

	Possible Failure Pt	Category	Waiting Pt	*Category*
Online	F1	Error Form Fill-up	W1	Delivery Leadtime
	F2	Error Payment	-	-
Walk-in on PSA Office	F3	Usage of Paper Queue No.	W2	Queue Line
	F4	Error Form Fill-up	W3	Delivery Leadtime
	F5	Error Payment	-	-
Travel Agency/Courier Company	F6	Usage of Paper Queue No.	W4	Queue Line
	F7	Error Form Fill-up	W5	Delivery Leadtime
	F8	Error Payment	-	-
Satellite/Mobile/Roaming PSA	F9	Usage of Paper Queue No.	W6	Queue Line
	F10	Error Form Fill-up	W7	Release of CRD
	F11	Error payment	W8	Delivery Leadtime

3. Results

3.1. Proposed process flow for acquiring CRD

In line with the identified failure points from the Service Blueprints of each process, in this section acquiring of Civil Registry Documents is plotted into a process flow that the process would become paperless and would utilize reference codes such as bar codes, QR codes and Reference codes. The process was patterned on how a boarding pass is being acquired nowadays. This aims to eliminate the identified failure points namely: Error Fill-up Form and Usage of Paper Queue. Also, it is being highlighted on the service blueprint that the waiting points involving queueing lines, pick-availability and delivery lead time will be eliminated as well. This process involves three entities namely: *Civil Registry Document Database* (storage that will keep copies of all the CRDs), *User* (individual that acquires the CRD/s from the CRD Database and/or submits its acquired CRD/s to the receiver for various purposes, and *Receiver* (organization/company whom the User submits its acquired CRD/s). The proposed service blueprint is divided into two: Acquirement and Submission Downloading Process as shown on Figures 5 and 6.

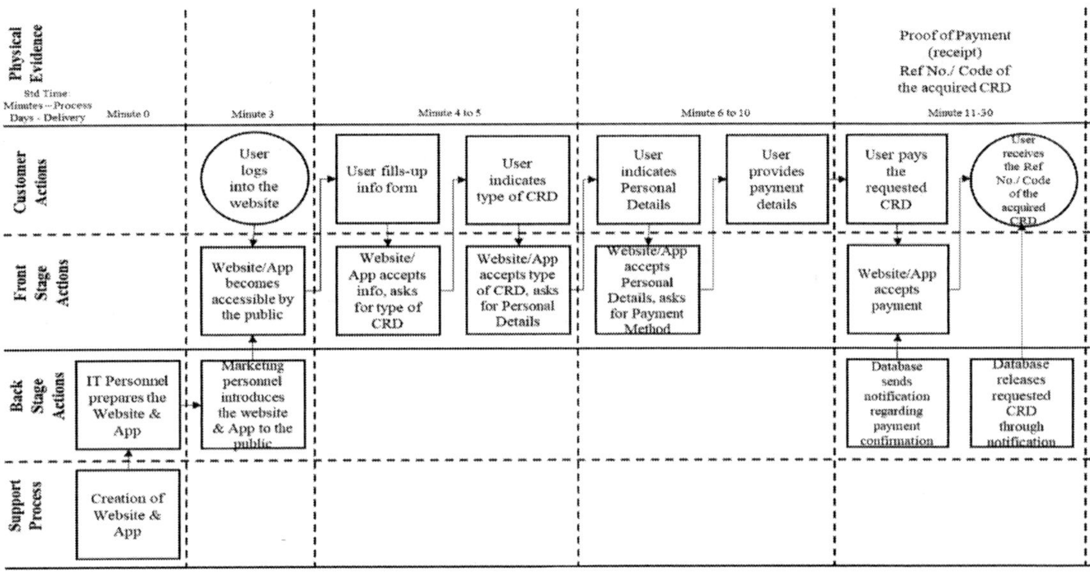

Figure 5. Proposed acquirement process of CRD.

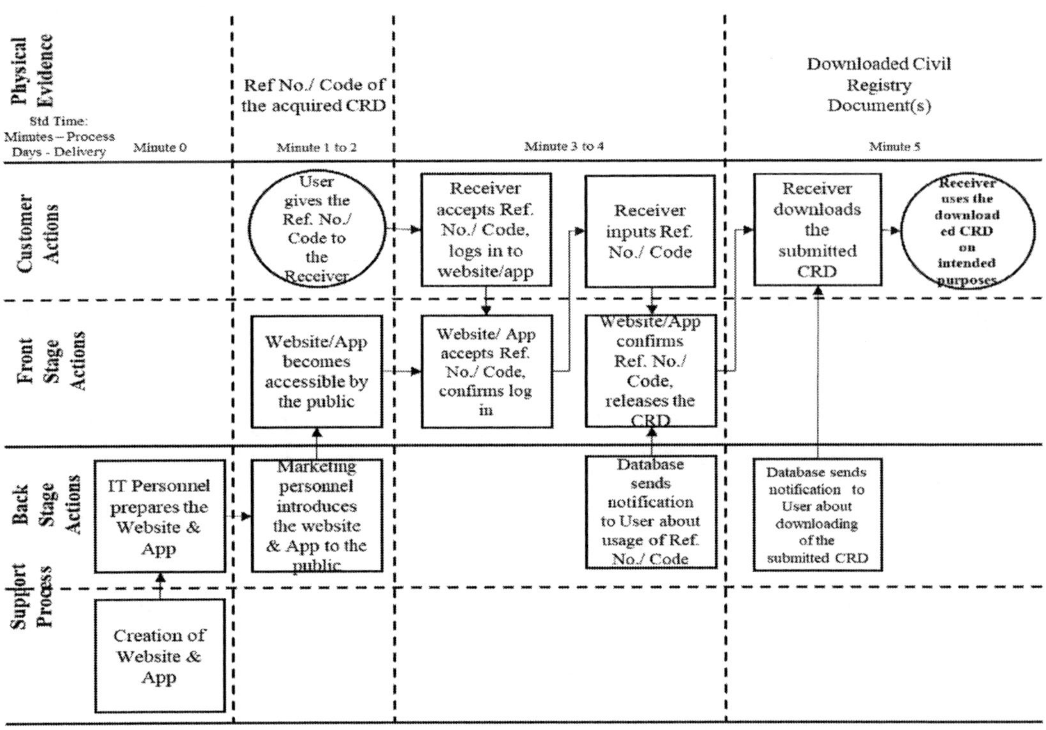

Figure 6. Proposed submission and downloading process of CRD.

IOP Conf. Series: Materials Science and Engineering 1072 (2021) 012058 doi:10.1088/1757-899X/1072/1/012058

3.2. Respondents demographics and technology dependency effect on PSA satisfaction rating and preference for paperless and reference code system of acquiring CRD

In order to evaluate the relationship between the Age group, location, employment status and technology dependency effect on PSA Satisfaction Rating and Preference for paperless and reference code system of acquiring CRD, ANOVA test has been performed. The results are as shown on Table 2. The following were assessed based on respective ranges and descriptive scales: age range (21-40 and 41-60 years old), location (urban, rural and international), employment status (employed, unemployed and student) and technology dependency (neutral, somehow dependent and highly dependent).

Table 2. ANOVA results to examine effects of respondent's demographics towards satisfaction on current process & preferences on paperless and reference code system.

Survey Respondents Demographics	Service Satisfaction and Preference for Change					
	Current Process Satisfaction Rating		Change to Paperless System Preference		Change to Ref. Code System Preference	
	Mean	P-value	Mean	P-value	Mean	P-value
Age						
21-40	2.08	0.13**	3.36	0.11**	3.45	0.51**
41-60	1.94		3.3		3.39	
Location						
Urban Philippines	2.05	0.04*	3.37	0.27**	3.53	0.00*
Rural Philippines	2.04		3.24		3.21	
International	2.27		3.67		3.47	
Employment Status						
Employed	2.03	0.00*	3.36	0.00*	3.44	0.75**
Unemployed	2.29		3.39		3.39	
Student	2.04		3.12		3.56	
Technology Dependency						
Neutral	2.18	0.00*	3.05	0.00*	3.24	0.00*
Somehow Dependent	2.12		3.45		3.46	
Highly Dependent	1.8		3.4		3.61	

*Significant ($p<0.05$)
**Not significant ($p >0.05$)

3.3. Respondents experience effect on PSA satisfaction rating and preference for paperless and reference code system of acquiring CRD

In order to evaluate the relationship between the Acquirement way, purpose of acquirement and factors in service satisfaction effect on PSA Satisfaction Rating and Preference for paperless and reference code system of acquiring CRD, ANOVA test has been performed. The results are as shown on Table 3. The following were assessed based on respective ranges and descriptive scales: acquirement way (online, walk-in on PSA Office, Travel Agency/Courier company and Satellite/Mobile/Roaming PSA Office), purpose of acquirement (academic, employment, government, religious and travel), factors in service satisfaction (convenience, dependability, environmental sustainability, price and timeliness).

Table 3. ANOVA results to examine effects of respondent's experience towards satisfaction on current process & preferences on paperless and reference code system.

Variable	Survey Respondents Experience with					
	Current Process Satisfaction Rating		Change to Paperless System Preference		Change to Ref. Code System Preference	
	Mean	P-value	Mean	P-value	Mean	P-value
Acquirement Way						
Online	2.13	0.00*	3.39	0.10**	3.53	0.00*
Walk-in on PSA Office	2.12	0.14**	3.28	0.06**	3.39	0.00*
Travel Agency/Courier Company	1.52	0.00*	3.28	0.17**	3.56	0.63**
Satellite/Mobile/Roaming PSA Office	2.02	0.00*	2.93	0.00*	3.09	0.00*
Purpose of Acquirement						
Academic	2.31	0.00*	3.51	0.07**	3.57	0.02*
Employment	2.21	0.00*	3.35	0.03*	3.47	0.01*
Government	2.12	0.07**	3.39	0.08**	3.49	0.01*
Religious	2.12	0.00*	3.23	0.00*	3.23	0.03*
Travel	2.33	0.16**	3.28	0.03*	3.4	0.04*
Factors in Service Satisfaction						
Convenience	1.86	0.00*	3.27	0.09**	3.44	0.02*
Dependability	1.85	0.04*	3.52	0.01*	3.55	0.00*
Environmental sustainability	1.99	0.00*	3.46	0.07**	3.59	0.00*
Price	2.16	0.00*	3.19	0.01*	3.41	0.16**
Timeliness	2.14	0.00*	3.39	0.01*	3.49	0.04*

*Significant ($p<0.05$)
**Not significant ($p >0.05$)

The ANOVA tests done have shown that age and location significantly affect the satisfaction rate for the current process as well as for the preference for reference code system. While, employment status significantly affects the satisfaction rate for the current process and the preference for paperless system. But the technology dependency has shown that it significantly affects the satisfaction for the current process and preference for both paperless and reference code system.

While, in terms of respondents' experience on acquiring CRD, the ANOVA test shows that all are statistically significant affecting the current process satisfaction rating except for acquirement of CRD thru walk-in on PSA Office and government & travel as purpose of acquirement. While, for preference for paperless, all are significantly affecting the preference to change into paperless system except acquiring CRD through online, walk-in on PSA Office and Satellite/Roaming PSA Office, having academic and government as purpose of acquirement and convenience as factor in service satisfaction. And lastly for preference to change into reference code system, all variables are significant except acquiring the CRD thru travel agency/courier company and having the factor of price in service satisfaction.

3.4. Current process rating effect on preference for paperless and reference code system

As shown on the Tables 4 and 5, the preference for paperless and reference code system is significantly affected by the satisfaction rating on the current process.

Table 4. Result of preference for paperless system and current process rating.

Factor	Mean	Std. Dev	F	P-value
Strongly Disagree (SD)	3.5	5.20	$F_{3,381}$= 11.35	0.00081*
Disagree (D)	14.25	10.78		
Agree (A)	23.5	12.07		
Strongly Agree (SA)	55	20.12		

*Significant ($p<0.05$)
**Not significant ($p>0.05$)

Table 5. Result of preference for reference code and current process rating.

Factor	Mean	Std. Dev	F	P-value
Strongly Disagree (SD)	1	2	$F_{3,381}$= 9.21	0.00194*
Disagree (D)	3.12	6.24		
Agree (A)	12.26	24.53		
Strongly Agree (SA)	9.68	19.36		

*Significant ($p<0.05$)
**Not significant ($p>0.05$)

4. Discussion

This study assessed the current acquirement processes of Civil Registry Documents, summarized into service blueprints and identified the failure and waiting points in the process. The service blueprints served as instruments in providing structured way of showing the process, provided a base for reference [19], implied that service is a complex relationship between human, product and process [20] and identified opportunities for improvement [21]. From the identified 11 failure points and 8 waiting points, a new process was proposed involving paperless and utilization of reference codes in an electronic system for a more efficient acquisition way [22]. This process involves three entities namely: *Civil Registry Document Database, User* and *Receiver***.** The proposed service blueprint is divided into two: *Acquirement and Submission Downloading Process*. There has been published literature indicating that the Philippine government is brewing its EGMP plan (E-Government Master Plan) that is being crafted by Department of Information and Communications Technology (DICT). The EGMP aims to integrate Information and Communication Technologies on the whole government including institutions, agencies, processes, resources, and policies [22] which makes the proposed new process viable since it is inclined towards e-government implementation of acquiring CRD in the Philippines. Furthermore as perceived in related studies, going paperless is a means of improving operational efficiency [24], whereas the business sector has seen benefits on making use of technological advancements [25].

On previous study, the components of negative customer interaction were explored [26]. With regards to that previous study, it was determined in this research that after evaluation of survey results, the majority of the survey respondents are not satisfied with the current processes of acquiring Civil Registry Documents and have asserted interests towards the paperless and reference code system of acquiring CRD. This was done in line with the related literatures that explores the influencing factors regarding e-government implementation [27] as well as the key stakeholders' roles in driving effective implementation within public sectors [28] and for promoting sustainability in the downstream supply chain practices especially in the developing countries [29].

5. Conclusion

An assessment was conducted to evaluate the current processes of acquiring Civil Registry Documents in the Philippines using mapping of the processes into Service Blueprint. The assessment showed the failure and waiting points of the current processes. From that identification of failure and waiting points, a proposed process was shown that involves paperless way of acquiring Civil Registry Documents through the use of reference codes concept.

Also, done in this study is a survey to 246 respondents that aimed to capture their experience in acquiring Civil Registry Documents, satisfaction on the current process and preferences towards paperless and reference code system of acquiring CRD. From this, it is concluded that generally, the experience of the respondents in acquiring Civil Registry Documents has led to the poor current service satisfaction ratings. Furthermore, the dissatisfaction on the current processes of acquiring CRD has statistically been proven that it has paved the inclination of the respondents towards the paperless and reference-code led acquisition of CRD.

This study can serve as guide for researchers aiming to gain knowledge on the processes of acquiring Civil Registry Documents in the Philippines. Moreover, this can serve as driver towards identification of possibilities of adapting e-government for the improvement of satisfaction as well as efficiency on the government services. Further studies could involve the application of this paperless and reference-code concept on Academic Records (e.g. Transcript of Records, Certificate of Graduation and etc.) and Hospital Records (e.g. Laboratory Results, Medicine Prescription and etc.) to eliminate/minimize the usage of paper on those documents. Thus, it is recommended to include risk analysis on further studies.

References

[1] Huang Z, Benyoucef M. Usability and credibility of e-government websites. *Gov Inf Q*. 2014 Oct;**31**(4):584–95. doi.org/10.1016/j.giq.2014.07.002

[2] Mahmoodi RK, Nojedeh SH. Investigating the Effectiveness of E-government Establishment in Government Organizations. *Procedia Soc Behav Sci*. 2016 Sep;**230**:136–41. doi.org/10.1016/j.sbspro.2016.09.017

[3] Sun P-L, Ku C-Y, Shih D-H. An implementation framework for E-Government 2.0. *Telemat.Informatics*. 2015 Aug;**32**(3):504–20. doi.org/10.1016/j.tele.2014.12.003

[4] Hamid AA, Razak FZA, Bakar AA, Abdullah WSW. The Effects of Perceived Usefulness and Perceived Ease of Use on Continuance Intention to Use E-Government. *Procedia Econ Finance*. 2016;**35**:644–9. doi.org/10.1016/s2212-5671(16)00079-4

[5] Sangki J. Vision of future e-government via new e-government maturity model: Based on Korea's e-government practices. *Telecomm. Policy*. 2018 Nov;**42**(10):860–871. doi.org/10.1016/j.telpol.2017.12.002

[6] Serra LC, Carvalho LP, Ferreira LP, Vaz JBS, Freire AP. Accessibility Evaluation of E-Government Mobile Applications in Brazil. *Procedia Comput. Sci*. 2015;**67**:348–57. doi.org/10.1016/j.procs.2015.09.279

[7] Alzahrani L, Al-Karaghouli W, Weerakkody V. Analysing the critical factors influencing trust in e-government adoption from citizens' perspective: A systematic review and a conceptual framework. *Int. Bus. Rev*. 2017 Feb;**26**(1):164–75. doi.org/10.1016/j.ibusrev.2016.06.004

[8] Elbahnasawy NG. E-Government, Internet Adoption, and Corruption: An Empirical Investigation. *World Dev*. 2014 May;**57**:114–26. doi.org/10.1016/j.worlddev.2013.12.005

[9] Yu S-F. Price perception of online airline ticket shoppers. *J. Air Transp. Manag*. 2008 Mar;**14**(2):66–9. doi.org/10.1016/j.jairtraman.2007.11.002.

[10] Othman SN, Zainon Noor Z, Abba AH, Yusuf RO, Abu Hassan MA. Review on life cycle assessment of integrated solid waste management in some Asian countries. *J. Clean. Prod.*. 2013 Feb;**41**:251–62. doi.org/10.1016/j.jclepro.2012.09.043.

[11] Gupta D, Mahajani SM, Garg A. Effect of hydrothermal carbonization as pretreatment on energy recovery from food and paper wastes. *Bioresour. Technol.*. 2019 Aug;**285**:121329. doi.org/10.1016/j.biortech.2019.121329

[12] Scholta H, Mertens W, Kowalkiewicz M, Becker J. From one-stop shop to no-stop shop: An e-government stage model. *Gov Inf Q.* 2019 Jan;**36**(1):11–26. doi.org/10.1016/j.giq.2018.11.010
[13] Philippine Statistics Authority. Philippine Statistics Authority. *Philipp. Stat. Auth.* 2018.
[14] S. Authority. Strengthening Civil Registration and Vital Statistics A case study of Strengthening Civil Registration and Vital Statistics A case study of. 2014. Retrieved from: https://psa.gov.ph/civilregistration/requesting-civil-registry-document/cenomar
[15] Okazaki S, Mendez F. Exploring convenience in mobile commerce: Moderating effects of gender. *Comput. Hum. Behav.* 2013 May;**29**(3):1234–42. doi.org/10.1016/j.chb.2012.10.019
[16] Kaur R kamal, Pandey B, Singh LK. Dependability analysis of safety critical systems: Issues and challenges. *Ann. Nucl. Energy.* 2018 Oct;**120**:127–54. doi.org/10.1016/j.anucene.2018.05.027
[17] Piyathanavong V, Garza-Reyes JA, Kumar V, Maldonado-Guzmán G, Mangla SK. The adoption of operational environmental sustainability approaches in the Thai manufacturing sector. *J. Clean. Prod.*. 2019 May;**220**:507–28. doi.org/10.1016/j.jclepro.2019.02.093
[18] Liu C-HS, Lee T. Service quality and price perception of service: Influence on word-of-mouth and revisit intention. *J. Air Transp. Manag.* 2016 Apr;**52**:42–54. doi.org/10.1016/j.jairtraman.2015.12.007
[19] Seppelt R, Fath B, Burkhard B, Fisher JL, Grêt-Regamey A, Lautenbach S, et al. Form follows function? Proposing a blueprint for ecosystem service assessments based on reviews and case studies. *Ecol. Indic.* 2012 Oct;**21**:145–54. doi.org/10.1016/j.ecolind.2011.09.003
[20] Lee C-H, Wang Y-H, Trappey AJC. Service design for intelligent parking based on theory of inventive problem solving and service blueprint. *Adv. Eng. Inform.* 2015 Aug;**29**(3):295–306. doi.org/10.1016/j.aei.2014.10.002
[21] Berkley BJ. Analyzing service blueprints using phase distributions. *Eur. J. Oper. Res.*. 1996 Jan;**88**(1):152–64. doi.org/10.1016/0377-2217(94)00157-x
[22] C. Apinan. Implementing a paperless and integrated pharma quality system. *Asian J. Pharm. Sci.* 2016;**11**(1):47
[23] Khalid SA, Lavilles RQ. Maturity Assessment of Local E-government Websites in the Philippines. *Procedia Comput. Sci.* 2019;**161**:99–106. doi.org/10.1016/j.procs.2019.11.104
[24] Caldeira M, Serrano A, Quaresma R, Pedron C, Romão M. Information and communication technology adoption for business benefits: A case analysis of an integrated paperless system. *Int. J. Inform. Manag.*. 2012 Apr;**32**(2):196–202. doi.org/10.1016/j.ijinfomgt.2011.12.005
[25] Gosnell GK. Communicating Resourcefully: A Natural Field Experiment on Environmental Framing and Cognitive Dissonance in Going Paperless. *Ecol Econ.* 2018 Dec;**154**:128–44. doi.org/10.1016/j.ecolecon.2018.07.020
[26] Go M, Kim I. In-flight NCCI management by combining the Kano model with the service blueprint: A comparison of frequent and infrequent flyers. *Tour. Manag.* 2018 Dec;**69**:471–86. doi.org/10.1016/j.tourman.2018.06.034
[27] Nzaramyimana L, Susanto TD. Analysis of Factors Affecting Behavioural Intention to Use E-Government Services in Rwanda. *Procedia Comput. Sci.* 2019;**161**:350–8. doi.org/10.1016/j.procs.2019.11.133
[28] Ashaye OR, Irani Z. The role of stakeholders in the effective use of e-government resources in public services. *Int. J. Inform. Manag.* 2019 Dec;**49**:253–70. doi.org/10.1016/j.ijinfomgt.2019.05.016
[29] Adjei-Bamfo P, Maloreh-Nyamekye T, Ahenkan A. The role of e-government in sustainable public procurement in developing countries: A systematic literature review. *Resour Conserv Recycl.* 2019 Mar;**142**:189–203. doi.org/10.1016/j.resconrec.2018.12.001

Eye-hand movement in 3D displays: A structural equation modeling approach

Y T Prasetyo[1*] and R Widyaningrum[2,3]

[1]School of Industrial Engineering and Engineering Management, Mapúa University, 658 Muralla St., Intramuros, Manila 1002, Philippines
[2]Department of Industrial Engineering, Sepuluh Nopember Institute of Technology, Kampus ITS Sukolilo, Surabaya 60111, Indonesia

Email: *ytprasetyo@mapua.edu.ph , [3]retno.widyaningrum09@gmail.com

Abstract. Eye-hand movements are two important parameters to evaluate the User Experience (UX) in the stereoscopic display (3D). The current study utilized structural equation modeling (SEM) to derive the causal relationship between parallax, index of difficulty (ID), eye movement, hand movement, eye index performance (IP eye), and hand index performance (IP hand) simultaneously. 10 participants voluntarily participated in the current study and they were asked to perform a multi-directional Fitts' Law task in the projection-based stereoscopic display. The results indicated that parallax had significant effects on eye movement time and hand movement time. In addition, ID was found to have significant effects on eye movement time and hand movement time. Interestingly, parallax and ID were found to have significant effects on IP eye, however, no significant effects were found on IP hand. These findings could be very beneficial for UX researchers and projection-based stereoscopic display developers especially in terms of evaluating UX of an interface.

1. Introduction

Today, User Experience (UX) plays an important role in modern technology. It is an attempt to encompasses all aspects of the experiences (perception, reaction, behavior) of a user during the interaction with a system [1]. The term UX is concerned less with technology, product design, or an interface and more with the aid of a digital device, creating a sensible and meaningful experience for the user [1]. UX has been a major concern in one of the hottest topics in modern technology: virtual reality.

Virtual reality has been developed significantly over the past 10 years [2]. It is designed to make possible a human sensorimotor and cognitive activity in a digitally created artificial world, which can be imaginary, symbolic, or a simulation of certain aspects of the real world [3]. Big companies, such as Alibaba, have started to utilize virtual reality in their commercial business. One of the most commonly used platforms to create virtual reality is a projection-based stereoscopic display.

The projection-based stereoscopic display has been utilized in many VR platforms [4]-[6]. It generates stereoscopic display by creating binocular disparity or widely known as parallax. Parallax is a lateral shift or difference between the spatial positions of corresponding right and left eye images [5]-[7]. One of the most widely known devices to measure the performance in the projection-based stereoscopic display is the eye tracker [4]-[8] and it is very valuable to analyze the effect of parallax and index of difficulty in the projection-based stereoscopic display.

The purpose of the current study was to analyze the causal relationship between parallax, index of difficulty (ID), eye movement time, hand movement time, index of eye performance (IP eye), and index of hand performance (IP eye) by utilizing structural equation modeling. Structural equation modeling (SEM) is a very powerful statistical analysis to derive the interrelationship among selected variables [8]-[14]. The finding of this study could be very beneficial for UX researchers and projection-based stereoscopic display developers especially in terms of evaluating UX of an interface.

Figure 1 represents the proposed 9 hypotheses in the current study. Parallax was hypothesized had significant effects on eye movement time (H1) [6], IP eye (H2), IP hand (H3), and hand movement time (H4). In addition, ID was hypothesized had significant effects on eye movement time (H5) [6], IP eye (H6), IP hand (H7), and hand movement time (H8). Finally, IP eye was hypothesized had a significant effect on IP hand. H1 and H5 were proven and the details had been published in our previous study [6].

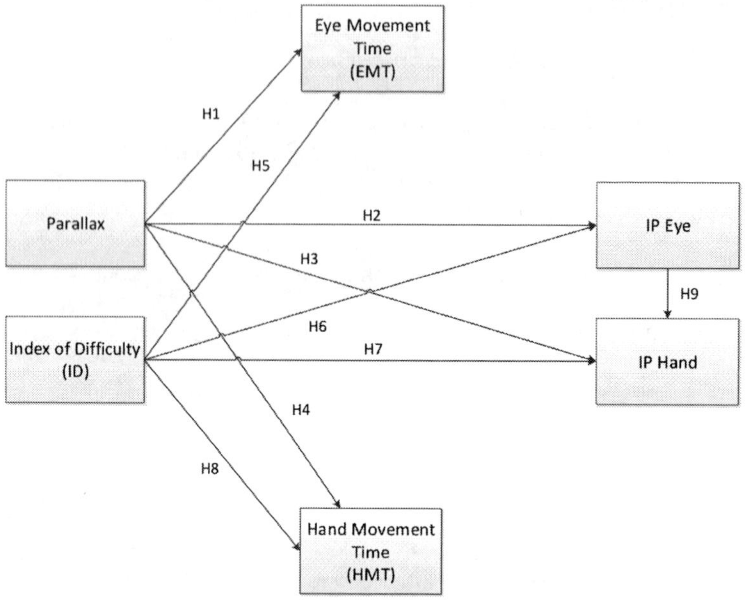

Figure 1. The SEM hypothesis constructs for eye-hand movements in the projection-based stereoscopic display.

2. Methodology

2.1. Participants
10 voluntary participants (seven males and three females) participated in the current study. All of them were graduate students at the National Taiwan University of Science and Technology (NTUST). All participants had normal or corrected to normal visual acuity with 1.0 in decimal units. In addition, all participants were right-handed. They were required to fill the consent form and pass the capability to perceive the object in the projection-based stereoscopic display.

2.2. Apparatus
Tobii X2-60 remote eye-tracking system was utilized to record the eye movements. The sampling rate was set to 60 Hz. As recommended by Tobii, the fixation filter Tobii Studio version 3.3.2 was utilized for calibration, testing, and data analysis. Raw eye fixation data collected from Tobii Studio was filtered using an I-VT fixation filter with 30 degrees/second velocity threshold.

Participants were instructed to wear a pair of View Sonic 3D glasses PDF-250 to perceive the 3D in the projection-based stereoscopic display. The 3D glasses PDF-250 were integrated with a 3D View Sonic (PJD 6251) projector and a 3D vision IR Emitter NVIDIA. The IR Emitter NVIDIA was placed in a certain distance from Tobii X2-60 to eliminate the shuttering effect so the signal of 3D emitter NVIDIA and the infrared from Tobii X2-60 would not affect the eye movement registration [6-8]. The width and length of the screen were 108 x 143 cm respectively. Finally, a Logitech C-920 webcam was utilized to record the eye movement data from the screen display.

The distance between the screen and the participants was 181 cm. The 3D View Sonic (PJD 6251) projector was placed 89 in front of the screen and the Tobii eye tracker was placed 60cm in front of the participants. All task was performed in a dark room (3.6x3.2x2.5m) covered by black curtains to prevent the light and create a sophisticated 3D environment.

2.3. Independent variables
Following our previous studies [4]-[8], parallax and ID were selected as independent variables. Six levels of ID and three levels of parallax [4]-[8] were utilized in the current study. Table 1 shows the details of the ID and Figure 2 shows the details of the parallax [4]-[8].

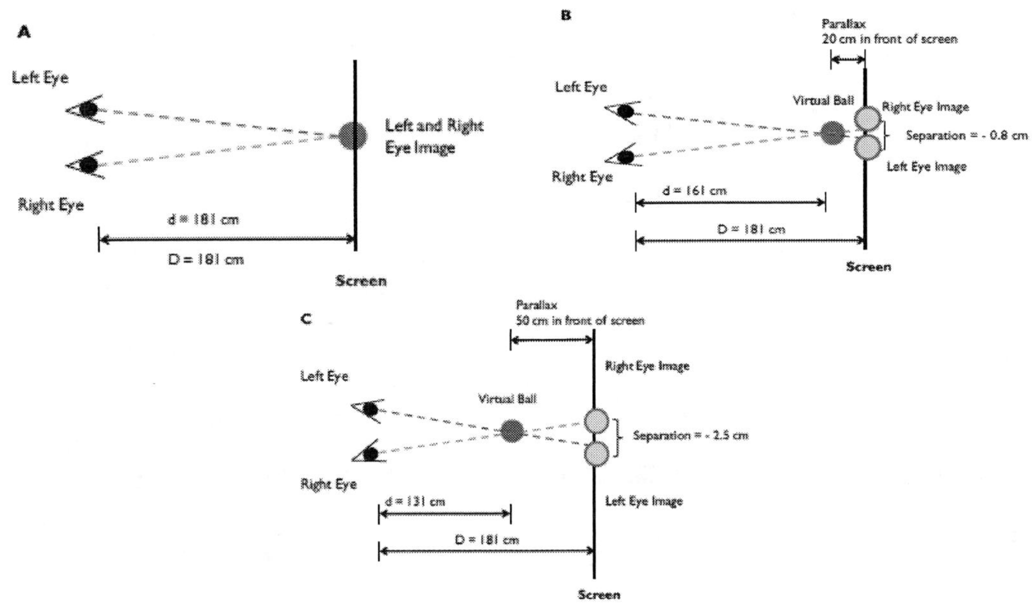

Figure 2. Three different levels of parallax in the current study [4]-[8].

Table 1. ID and task precision level [4]-[8].

Distance (unity unit)	Width (unity unit)	ID (bits)	Task Precision Level
40	3.3	3.7	Low
40	2.3	4.2	Medium
40	0.6	6.1	High
20	3.3	2.8	Low
20	2.3	3.3	Low
20	0.6	5.1	Medium

2.4. Dependent variables
There were four dependent variables in the current study: (1) eye movement time, (2) hand movement time, (3) IP eye, and (4) IP hand. Eye movement time was defined as an elapsed time from the eye fixation point on the origin to the fixation point of the next target [4]-[8]. Hand movement time was defined as an elapsed time from the hand point on the origin to the target [4]. IP eye and IP hand were defined as the ID/MT. The details of each calculation had been published in our previous study [4].

2.5. Experimental procedures
Participants were required to complete the calibration process to ensure Tobii X2-60 could detect the participant's eye gaze movement. The regular Tobii calibration setting with five red dots was utilized to capture the participant's eye gaze binocularly. Participants needed to pass this test in order to continue

the next test. In addition, they were also instructed to fill the consent form which conducted according to the ethical guidelines of the National Taiwan University Research Ethics Committee.

Multi-directional Fitts' Law task was used as a task in this study [15]-[17]. Participants were instructed to click the virtual red ball as accurately as fast as possible [4]-[8]. Unity 3D platform version 4.3.4 was utilized to create the virtual 3D ball. Figure 3 shows the sequence of multi-directional Fitts' Law task based on ISO 9241-9 [4]-[8]. Finally, the data were analyzed using our previous algorithm and the detailed of the algorithm has been presented in our previous study [4].

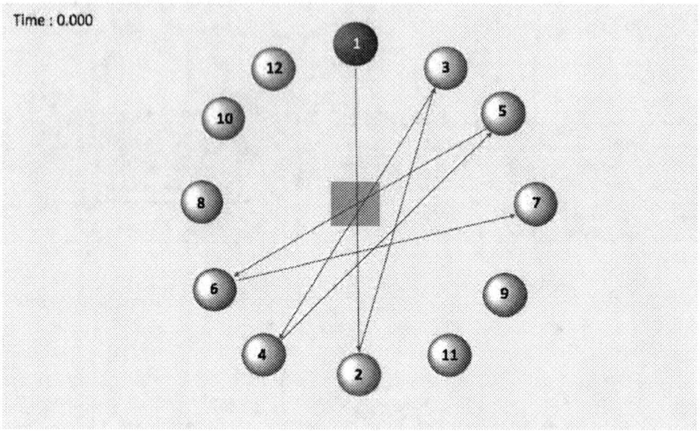

Figure 3. The sequence of multi-directional Fitts' Law task based on ISO 9241-9 [4]-[8].

2.6. *Structural equation modeling*

Figure 1 shows that our SEM had six variables, including two exogenous variables (parallax and ID) and four endogenous variables (eye movement time, hand movement time, IP eye, and IP hand).

SEM was derived by utilizing Amos 22 with the Maximum Likelihood estimation approach. Four sets of tests were used to measure the difference between the observed data and the hypothesized model: full model test, incremental fit indices, the goodness of fit index, and badness of fit index [8]. Since there were three different levels of parallax, six different levels of ID, and tested on 10 participants, a total of 180 data was analyzed. The bootstrapping technique with the bias-corrected 95% confidence interval percentiles was conducted since the sample sized was under 250 [8]. Finally, a two-tailed Pearson-correlation with an alpha value of 0.05 was also conducted to double-check the beta coefficient of each path in the proposed model (Table 2).

Table 2. Descriptive statistics and correlations among variables.

		Mean	SD	Range	(1)	(2)	(3)	(4)	(5)	(6)
(1)	Parallax				1					
(2)	ID				0	1				
(3)	HMT	1.587	0.561	3.92	0.172*	0.394**	1			
(4)	EMT	0.165	0.075	0.51	0.156*	0.371**	0.812**	1		
(5)	IPH	2.805	0.785	4.78	-0.141	0.324**	-0.412**	-0.352**	1	
(6)	IPE	27.9666	9.521	49.27	-0.167*	0.320**	-0.589**	-0.598*	0.751**	1

3. Results & Discussion Conclusions

The initial SEM of eye-hand movements in the projection-based stereoscopic display is presented in Figure 4. Based on this figure, two paths were found not significant. Therefore, a revised model was derived by removing these two paths: Parallax-IP Hand and ID-IP Hand. The revised SEM of eye-hand movements in the projection-based stereoscopic display is presented in Figure 5 and the parameter estimates and goodness of fit are presented in Table 3.

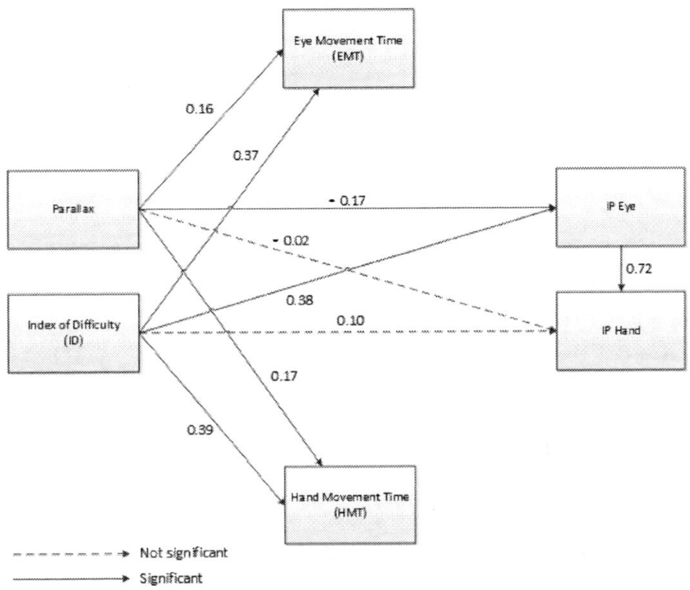

Figure 4. The initial SEM of eye-hand movements in the projection-based stereoscopic display.

The SEM approach indicated that parallax had significant effects on eye movement time (β: 0.16), hand movement time (β: 0.17), IP eye (β: -0.17), however, no significant effect was found on IP hand (β: -0.02). Higher parallax was found to lead to longer eye movement time, hand movement time, and lower IP eye. Similarly, ID was found had a significant effect on eye movement time (β: 0.36), hand movement time (β: 0.39), IP eye (β: 0.32), however, no significant effect was found on IP hand (β: 0.10). The higher ID was found to lead to longer eye movement time and hand movement time. Interestingly, the effect of parallax on hand movement time was found slightly higher than on eye movement time, however, the effect of ID on hand movement time was found higher than on eye movement time. Future research to investigate further about this phenomenon through a more comprehensive eye-hand coordination study would be very valuable.

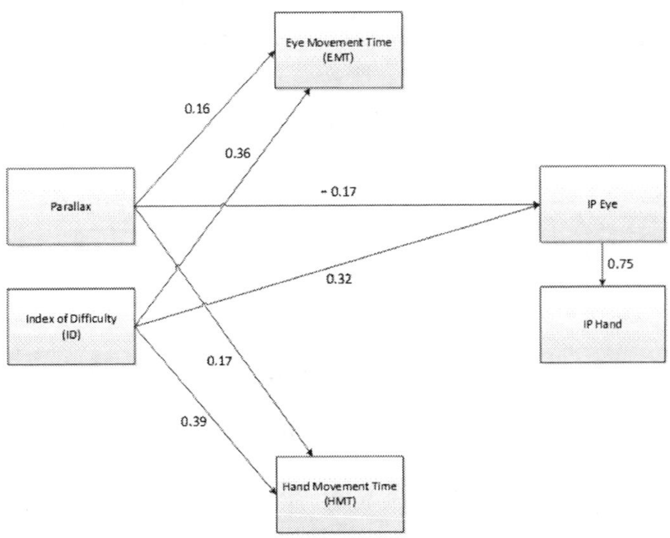

Figure 5. The revised SEM of eye-hand movements in the projection-based stereoscopic display.

Table 3. The goodness of fit measures of the SEM.

The goodness of fit measures of the SEM	Parameter Estimates	Suggested cut-off	Recommended by
p-value for Chi-square (x^2)	0.238	> 0.05	[8]-[14]
Chi-square statistic (x^2)	4.229		
Degree of freedom	3		
Normed chi-square	1.410	< 2.0	[8]-[14]
Incremental Fit Indicates			
Normed Fit Index (NFI)	0.994	> 0.95	[8]-[14]
Tucker Lewis Index (TLI)	0.990	> 0.95	[8]-[14]
Comparative Fit Index (CFI)	0.998	> 0.96	[8]-[14]
Goodness-of-fit index			
Goodness of Fit Index (GFI)	0.992	> 0.95	[8]-[14]
Adjusted Goodness of Fit Index (AGFI)	0.946	> 0.90	[8]-[14]
Badness-of-fit index			
Root Mean Square Error of Approximation (RMSEA)	0.048	< 0.07	[8]-[14]

Interestingly, we found that the effect of ID on eye movement time (β: 0.36) was higher compared with the effect of parallax on eye movement time (β: 0.16). Similarly, the effect of ID was found to be higher on hand movement time (β: 0.39) compare with the effect of parallax (β: 0.17). Future research would be very worth to investigate this phenomenon.

4. Conclusions

Virtual reality has been a key parameter in today's UX [18]-[24]. The current study utilized structural equation modeling (SEM) to derive the causal relationship between parallax, index of difficulty (ID), eye-hand movements, eye index performance (IP eye), and hand index performance (IP hand) simultaneously. Seven males and three females voluntarily participated in the current study and they were asked to perform multi-directional Fitts' Law task in the projection-based stereoscopic display. The results indicated that parallax had significant effects on eye movement time and hand movement time. In addition, ID was found to have significant effects on eye movement time and hand movement time. Interestingly, parallax and ID were found to have significant effects on IP eye, however, no significant effects were found on IP hand. In addition, the effect of parallax on hand movement time was found slightly higher than on eye movement time, however, the effect of ID on hand movement time was found higher than on eye movement time. These findings could be very beneficial for UX researchers [25] and projection-based stereoscopic display's developers especially in terms of evaluating UX of an interface.

Acknowledgments

We would like to thank Ministry of Science and Technology of Taiwan for supporting this study (MOST 103-2221-E-011-100-MY3).

References

[1] W. Nagel 2016 *Multiscreen UX design: developing for a multitude of devices*. (Waltham, MA: Morgan Kaufmann, an imprint of Elsevier)

[2] Bojko 2013 *Eye tracking the user experience: a practical guide to research*. (Brooklyn, NY: Rosenfeld Media).

[3] Patterson RE 2014 Introduction to Human Factors of Stereoscopic 3D Displays. *Human Factors of Stereoscopic 3D Displays*. **29** p 1–5. doi.org/10.1007/978-1-4471-6651-1_1

[4] Lin, C. J., & Widyaningrum, R 2016 Pointing in Stereoscopic Displays. *J Eye Mov Res*. **9**(5).

doi.org/10.16910/jemr.9.5.4
[5] Lin CJ, Widyaningrum R 2018 The effect of parallax on eye fixation parameter in projection-based stereoscopic displays. *Appl Ergon.* **69** p 10–6. doi.org/10.1016/j.apergo.2017.12.020
[6] Lin, C. J., Prasetyo, Y. T., & Widyaningrum, R 2018 Eye movement parameters for performance evaluation in projection-based stereoscopic display. *J Eye Mov Res.* **11**(6). doi.org/10.16910/jemr.11.6.3
[7] Prasetyo YT, Widyaningrum R, Lin CJ 2019 Eye Gaze Accuracy in the Projection-based Stereoscopic Display as a Function of Number of Fixation, Eye Movement Time, and Parallax. *2019 IEEE International Conference Industrial Engineering and Engineering Management (IEEM).* doi.org/10.1109/ieem44572.2019.8978502
[8] Lin CJ, Prasetyo YT, Widyaningrum R 2019 Eye movement measures for predicting eye gaze accuracy and symptoms in 2D and 3D displays. *Disp.* **60** p 1–8. doi.org/10.1016/j.displa.2019.08.002
[9] J. F. Hair 2014 *Multivariate data analysis.* (Edinburg: Pearson).
[10] Pedram S, Palmisano S, Richard S, Perez P, Farrelly M 2020 Investigating the process of mine rescuers' safety training with immersive virtual reality: A structural equation modelling approach. *Comput Edu.* **153** 103891. doi.org/10.1016/j.compedu.2020.103891
[11] Makransky G, Petersen GB 2019 Investigating the process of learning with desktop virtual reality: A structural equation modeling approach. *Comput Edu.* **134** p 15–30. doi.org/10.1016/j.compedu.2019.02.002
[12] Ai-Lim Lee E, Wong KW, Fung CC 2010 How does desktop virtual reality enhance learning outcomes? A structural equation modeling approach. *Comput Edu.* **55**(4) p 1424–42. doi.org/10.1016/j.compedu.2010.06.006
[13] Prasetyo YT 2020 Factors Affecting Gross Manual Dexterity: A Structural Equation Modeling Approach. *2020 IEEE 7th Int. Conf Ind Eng Appl (ICIEA).* doi.org/10.1109/iciea49774.2020.9102096
[14] Prasetyo YT 2020 Factors Affecting Fine Dexterity. *Proceed 2020 2nd Int Conf on Manag Sci Ind Eng.* doi.org/10.1145/3396743.3396776
[15] Lin CJ, Chen H-J, Choi J 2015 The postural and control-display gain effects of distal pointing on upper extremity fatigue. *Ergon.* **59**(1) p 73–84. doi.org/10.1080/00140139.2015.1055824
[16] Lin CJ, Ho S-H, Chen Y-J 2015 An investigation of pointing postures in a 3D stereoscopic environment. *Appl Ergon.* **48** p 154–63. doi.org/10.1016/j.apergo.2014.12.001
[17] Chen H-J, Lin CJ, Lin P-H 2019 Effects of control-display gain and postural control method on distal pointing performance. *In. J. Ind. Ergon.* **72** p 45–53. doi.org/10.1016/j.ergon.2019.04.004
[18] Kim J-E, Nembhard DA 2019 Eye movement as a mediator of the relationships among time pressure, feedback, and learning performance. *Int. J. Ind. Ergon.* **70** p 116–23. doi.org/10.1016/j.ergon.2018.12.006
[19] Li X, Schroeter R, Rakotonirainy A, Kuo J, Lenné MG 2020 Effects of different non-driving-related-task display modes on drivers' eye-movement patterns during take-over in an automated vehicle. *Transp Res Part F Traffic Psychol Behav.* **70** p 135–48. doi.org/10.1016/j.trf.2020.03.001
[20] Chen S-C, Hsiao M-S, She H-C 2015 The effects of static versus dynamic 3D representations on 10th grade students' atomic orbital mental model construction: Evidence from eye movement behaviors. *Comput. Hum. Behav.* **53** p 169–80. doi.org/10.1016/j.chb.2015.07.003
[21] Chen G, Huang T, Fan Z, Zhang X, Liao H 2019 A naked eye 3D display and interaction system for medical education and training. *J Biomed Inform.* **100** 103319. doi.org/10.1016/j.jbi.2019.103319
[22] Zhou J, Wang L, Yin H, Bovik AC 2019 Eye movements and visual discomfort when viewing stereoscopic 3D content. *Digit Signal Process.* **91** p 41–53. doi.org/10.1016/j.dsp.2018.12.008
[23] Yaramothu, C., Jaswal, R., & Alvarez, T. L 2009 Target eccentricity and form influences disparity vergence eye movements responses: A temporal and dynamic analysis. *J Eye Mov Res.* **12**(4).

doi.org/10.16910/jemr.12.4.7
[24] Iskander J, Hossny M, Nahavandi S 2019 Using biomechanics to investigate the effect of VR on eye vergence system. *Appl Ergon.* **81** 102883. doi.org/10.1016/j.apergo.2019.102883
[25] Prasetyo YT 2020 Standardizing Human Factors and Ergonomics Education for the Undergraduate Programs in Industrial Engineering: A Comparative Analysis between Indonesia, Philippines, and Taiwan. *2020 IEEE 7th Int. Conf Ind Eng Appl (ICIEA).* doi.org/10.1109/iciea49774.2020.9101950

Kansei engineering application in redesigning *carica* packaging to support local-small industry in Central Java

H Prastawa[1,2], M Mahachandra[1,3], and D A H Donida[1,4]

[1]Industrial Engineering Department, Faculty of Engineering, Diponegoro University, Jl. Prof. Soedarto, SH, Kampus Undip Tembalang, Semarang, Indonesia 50275

Email: [2]heru.prastawa@gmail.com, [2]manik.mahachandra@gmail.com, [3]deandyahd4@gmail.com

Abstract. Packaging is one of the important elements of product sales. The packaging is one form of branding business of a business party. This study was conducted to apply the principles of Kansei Engineering to packaging products of *Carica*, one of the MSMEs in Wonosobo. This research was conducted to find out what kind of design desired by consumers of *Carica* products to encourage them to make purchases of these products. The method used is the Kansei Engineering method which will then be collaborated with the Kano model method and conjoint analysis. Kansei Engineering will help this research in determining what aspects consumers want to feel when they buy a *Carica* product. After that, using the Kano model principle will be determined categories from each aspect that consumers feel into several parts, namely basic attributes, linear attributes, attractive attributes, and non-significant attributes. Aspects included in the non-significant attribute category are considered not suitable to be used as a new design concept of *Carica* packaging. Finally, the packaging aspects that consumers want to feel when buying these products will be linked to the new design using conjoint analysis. This research concludes that the new packaging will be square and made of glass and has a label that has a fruit picture and has an orange theme.

1. Introduction

Carica sweets are one of the most famous from Wonosobo, Central Java, Indonesia. The fruit *Vasconcellea Cundinamarcencis* generally grows around the hills which have a fairly cold climate. Nowadays, *Carica* sweets are even sold online, showing that the businesses are in the promising development phase. MSME Embun Pagi is one of more than 200 *carica* sweets confectionery industries in Wonosobo is a business that is fostered by us, which has been producing since 2016. To compete with other similar businesses, MSME Embun Pagi requires a good and distinctive image of the product. Currently, Embun Pagi packages *Carica* sweets in a 7.5 cm diameter and 4 cm height plastic cup. It is also equipped with a yellow color label, which was designed around two years ago. The packaging costs IDR1.750 today. Figure 1 shows the existing packaging.

Figure 1. Existing packaging of *carica* sweets of MSME Embun Pagi.

The packaging is a science, art, and technology that aims to protect a product when it will be sent, stored, or sold or it can also be a production process that aims to package. In the past people said, "packaging protects what is sold", but now it turns into "packaging selling what is protected" [1]. Therefore, packaging must also have an attraction aspect for consumers [2]. Because in the end, the purpose of packaging is actually so that consumers buy the product [3]. Packaging also functions as the first communication medium between products produced by producers and consumers to display the image of the product itself. The current packaging of *carica* sweets in MSME Embun Pagi does not yet represent the company's identity and urgently need new packaging so that it looks more attractive. As common products in the market, the design process was only based on a designer's intuition and perception, not involving any scientific technique [4]. Therefore, attractive packaging design is needed by consumers effective and can protect the product itself.

Kansei Engineering aims to gain consumers' desires through feelings felt by consumers. When consumers want a product or service, their sensory needs will appear elegant, cheap, unique, etc., which is called Kansei. Therefore, Kansei can be interpreted as the psychological and physiological senses of consumers of the desired product or service [5]. Through Kansei's words obtained from various sources, the Kansei concept collects consumer feelings and turns them into product design specifications. Concerning customer satisfaction, Kano Model can also help product designers to get products that are following the customer's desires.

Therefore, this study aims to design *carica* sweets packaging that can show the identity of the product, increase consumer buying interest in the product, and protect the product so that the quality reaches consumers. The packaging design was carried out based on the assessment of the aspects needed by a consumer. As well, this research combined Kansei engineering with the Kano model concept to produce aspects that are truly important and needed by packaging.

2. Methodology

The research was conducted in an observatory method using two types of questionnaires, each of them for container and label of the packaging. Many 105 people were voluntarily involved as participants. Data were collected via online questionnaires.

The preliminary stage was an open question to gain Kansei's words. Related Kansei words were also determined from previous studies in product design. Then the first questionnaire was distributed to assess any aspects that participants want when interacting with *carica* sweets products. Participants were asked to rate packaging design according to each Kansei words by selecting one of five points of Likert-like scale. This scale was applied to quantify their impressions and feelings towards the design of *carica* packaging in general. This process was used as input to the second questionnaire.

In the second questionnaire, the packaging was divided into two parts, namely the container section and the label section. In the container section, the idea of packaging design elements was divided into ideas of shape, size, and material. While in the label section, it was divided into colors and illustrated images. Furthermore, Kansei word mapping was conducted with the Kano model diagram. At this stage, each aspect of packaging will be obtained using the Kano concept so that it divides the packaging aspects into four categories, i.e. basic attributes, linear attribute, attractive attributes, and non-significant attributes. The non-significant category was discarded because it was considered to not affect the changes in the packaging. The final step was to design a packaging design proposal. It was carried out using the analysis to determine the relationship between Kansei words and packaging design elements, so that the combination of packaging design elements desired by consumers was well explained.

3. Result and Discussion

3.1. Respondents profile

The selected respondents consisted of 49 males and 56 females, aged between 15 and 40 years, with a total of 105 people. Respondents are students of University X in Indonesia. Participants are recruited based on whether they are familiar with e-learning and e-commerce sites. This consideration encourages the selection of respondents from students, and the sampling plan used is non-probability sampling, more precisely purposive sampling. The mean age is 17.6 years (SD = 3.45); 46.6% are male, and 53.4 % female. Residence, 36.1% are Central Java and 63.9 % are others. Demographic data in this group is shown in Table 1.

Table 1. Profile of respondent.

Variable	Frequency	Percentage
Gender		
Male	49	46.6
Female	56	53.4
Age		
15 – 25	21	20
26 – 36	52	49.5
37 – 47	32	30.5
Educated		
Under graduate	63	60
School	40	38.1
Others	2	1.9
Province		
Central Java	38	36.1
Others	72	63.9

3.2. *Consumer identification*
In product design, it is necessary to know who the users for mapping the needs. Unfortunately, in the current condition, there was no focus on the market for the sweet *carica* products of the MSMEs. However, by knowing *carica* sweets products are a typical souvenir product, it can be assumed that the product will be purchased by consumers who are likely not domiciled in the area, which was travelers from outside Wonosobo. Therefore, especially for consumers who are new to the product, *carica* sweet product was expected to have an interesting point and convince consumers to make purchases. Moreover, the product was expected to have simple, easy to carry, easy to store, practical, and lightweight for traveling needs.

3.3. *Kansei words collection*
The first step in this research is data collection in the form of Kansei words. At this stage, the data collected comes from two sources. The first source was taken from references on packaging product development and the second source was taken through a questionnaire distributed to participants regarding their feelings about *carica* packaging products. Twenty-eight Kansei words had been collected as follows: unique, portable, easy to store, have product image, reusable, luxurious, single-use, interesting, easy to remember, easy to open, informative, convincing, tight, tough, transparent, simple, safe, limited edition, eco-friendly, hygienic, ergonomic, practical, durable, up-to-date, brightly colored, plain, lightweight, and economic.

3.4. *First questionnaire*
After collecting the data of each Kansei words on a Likert-like scale, validity tests were conducted. With 4 times repetitions, 15 Kansei words were passed the validity tests. In a reliability test with Cronbach's Alpha, all data were concluded as reliable and can be further analyzed.

3.5. Generate design idea

By paying attention to consumer needs for products that have been described before, this research chose products that can translate that needs. By benchmarking, the design elements of the container were divided into three parts, namely material, shape, and size. In the materials section, the sample design elements were divided into glass and plastic, concerning consumer needs for safe packaging, plastic materials are chosen rather than glass-based packaging. Glass materials, on the other hand, referred to consumer needs for unique packaging. In the shape section, the sample design elements were divided into balls, boxes, and tubes. The tubes and boxes were based on consumer needs for portable and easily stored packaging. As an alternative, ball shape can fill consumers' needs for uniqueness. The size section was divided into small and large, to complete all consumers' needs. Figure 2 shows the container samples.

The design elements of the label were divided into color and image. In the color section, there were yellow and orange, which generalized the selection from male and female perspectives. These colors can also symbolize the color of the *carica* fruit itself as the basic ingredient of the product. Whereas the image section was categorized into fruit, mountain, and plain elements. These selections were considered simple and directly related to the product. Figure 3 comprises the samples of labels.

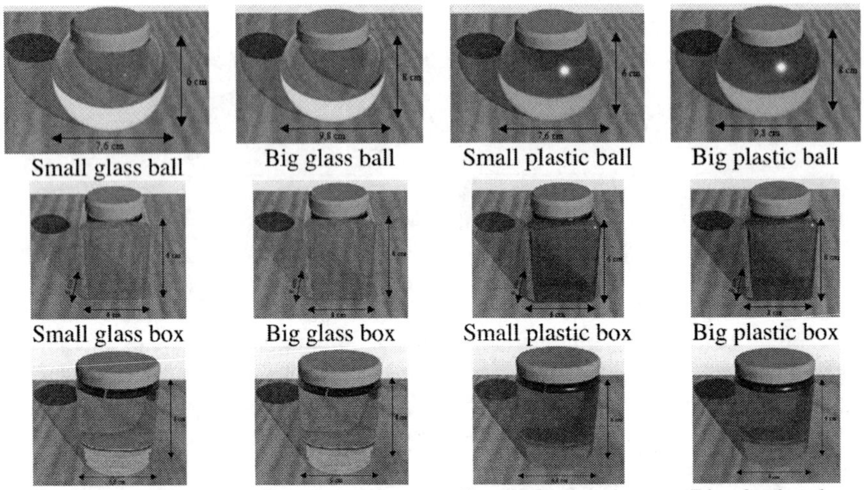

Figure 2. Research samples of container.

Yellow fruit | Yellow mountain | Yellow plain | Orange fruit | Orange mountain | Orange plain

Figure 3. Research samples of label.

3.6. Second questionnaire

In the container, there are eleven Kansei words, i.e. unique, portable, easy to store, easy to open, single-use, safe, practical, durable, easy to remember, tight and lightweight. For the label section, there were six Kansei words, i.e. unique, have product images, easy to remember, informative, bright, and plain colors. Besides, there are additional questions in this second questionnaire, whether the participants saw the container or label as good or not.

3.7. *Kano model mapping*

The data needed in mapping the Kano model in both the container section and label section are, among others, the average value of each Kansei words in the first questionnaire and the second questionnaire data, as proposed in [6]. Two categories indicate the tendency of an attribute to appear according to the participants called the positive category. Vice versa, the negative category was the category that shows the tendency of an attribute to be eliminated according to the participants.

To find out the Kansei words section of the container, the average Kansei words section of the container was obtained by summing all the values given by participants to Kansei words and dividing by the number of participants. This value will determine which data from the second questionnaire was included in the positive or negative category. After the previous stage was complete, then the next step is to find the correlation value of each Kansei words to its global evaluation. Each correlation value for all Kansei words either positive or negative will be the source of making the Kano model in the cartesian graph in the final stage. The ordinate was derived from the Kansei words correlation value for the positive global evaluation, and the abscissa was derived from the Kansei words correlation value for evaluation the global category was negative, and the center of the graph was sourced from both averages. Finally, Kano display models can be made in the form of diagrams for parts of containers and labels, respectively in Figure 4 and 5. The results of the mapping Kano Model were Kansei words which were categorized as insignificant attributes, that were portable and easy to remember for the container, and easy to remember for the label. For Kansei words that are not included in the category of attributes that are not significant, they will still be used at the next stage in this research because Kansei words were stated to have a value that consumers consider in choosing this product.

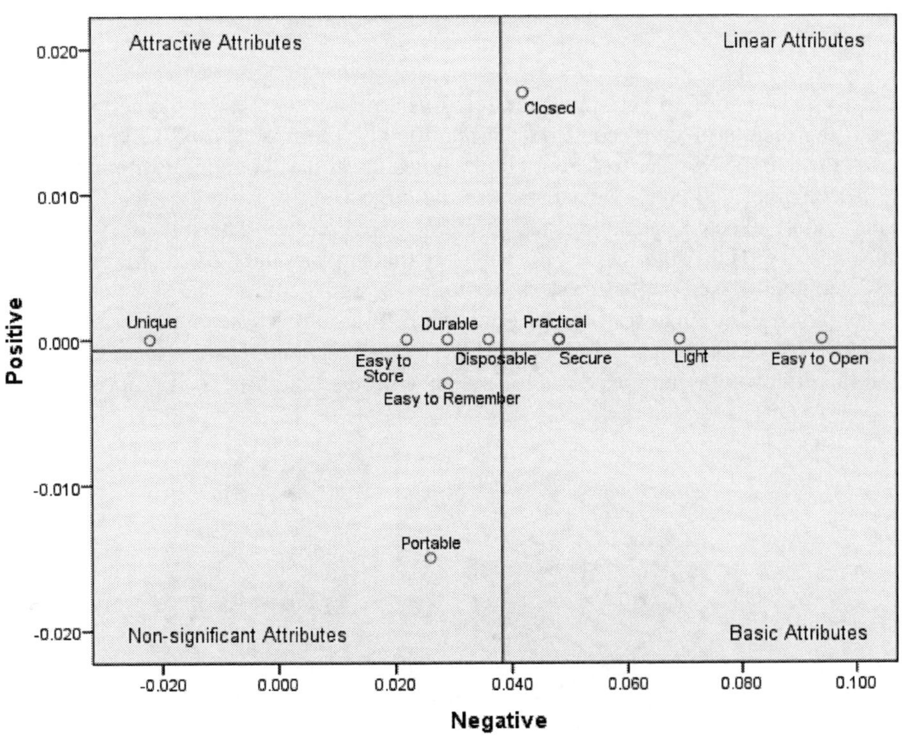

Figure 4. Kano diagram of a container.

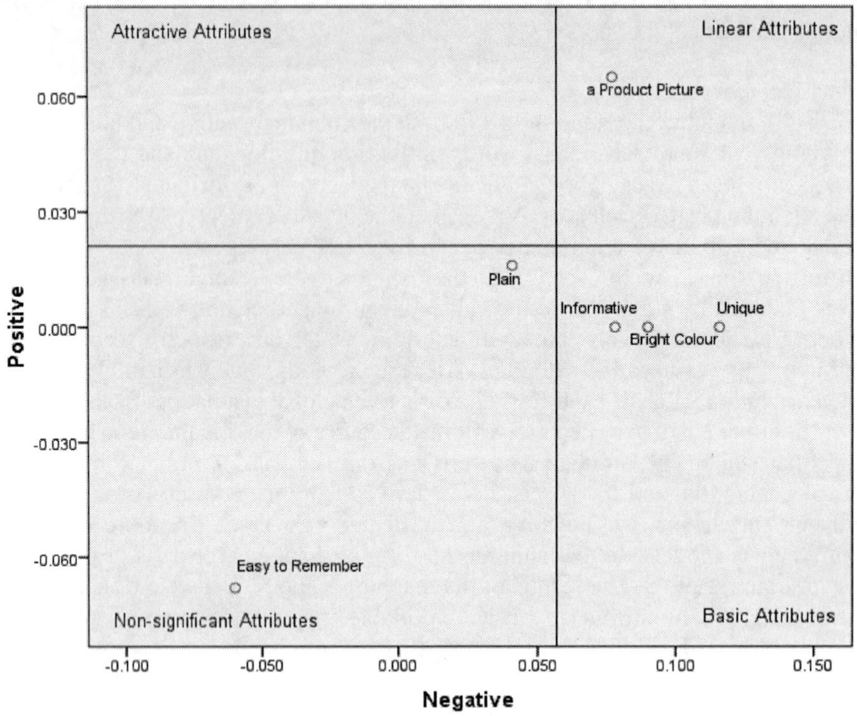

Figure 5. Kano diagram of a label.

3.8. *Conjoint analysis*

Conjoint analysis was used to determine the relationship between design elements and Kansei words according to the results of the second questionnaire. The input data used to carry out this conjoint analysis was the average Kansei word data in each design in the second questionnaire. Later on, from the utility of container and label, we can choose the largest utility value of each item for each Kansei words. The overall selected packaging design elements were then determined by the category mode chosen in Kansei words. In other words, the category that appears most often will be the category in the overall selected item. Based on that process, container design elements and selected labels was a small glass box container and orange fruit label, as shown in Figure 6. These characteristics were correlated with Kansei words easy to open similar to Rahmayani research result [7], safe, easy to store [8], practical, tight, lightweight, unique, durable, and have product pictures on it propose this study, single-use [9].

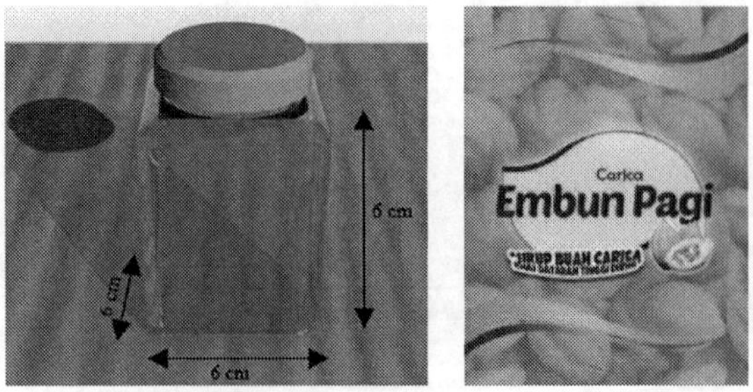

Figure 6. Selected design combination.

MSME Embun Pagi was then recommended to change the packaging of the product into a small glass box and labeled with an image of *carica* fruit in it with an orange theme color. Additional costs were needed to change the product packaging now. However, MSMEs should make this research recommendation as to their business investment through engineering their products in the packaging field. This is supported by various studies listed in the book on consumer behavior [10], where it is said that packaging is an important aspect that is directly felt by consumers when interacting with a product, not to mention *carica* products in these MSMEs. This study also supports the opinion that affective factors increasingly show a significant trend in influencing the selection of a product [11].

4. Conclusion

A new design of *carica* sweets packaging was proposed to involve the emotional aspects of consumers: a small glass box container and an orange label with pictures of *carica* fruits on it. As the packaging is an important aspect of a product, the new design was hoping to gain more attention of the prospective customers, as it is easy to open, safe, practical, tight, lightweight, unique, easy to store, single-use, durable, and have product pictures on it.

Acknowledgment

The authors thank MSME Embun Pagi as the study object. This research was financially supported by The Faculty of Engineering, Diponegoro University, Indonesia through Strategic Research Grant 2019 number 3161/3/UN7.3.3/PG/2019.

References

[1] Kertajaya, H 1996 *Business Strategy: Win and Survive in the Asia Pacific Century*. (Jakarta: Gramedia Pustaka).

[2] Kotler, P 2003 *Marketing Insight from A to Z*. (New Jersey: Wiley).

[3] Cenadi, C. S 2000 The role of packaging design in the marketing world. *Nirmana*, **2**(1) p 92-103. doi.org/10.9744/nirmana.2.2.

[4] Azrifirwan, D.T 2017 D*evelopment of soft drink packaging design models based on Kansei Engineering*. *Dissertation*. (Bogor: Bogor Agricultural University Repository)

[5] Nagamachi M, Lokman AM 2016 *Innovations of Kansei Engineering*. CRC Press; doi.org/10.1201/ebk1439818664.

[6] Llinares C, Page AF 2011 Kano's model in Kansei Engineering to evaluate subjective real estate consumer preferences. *Int. J. Ind. Ergon.* **41**(3) p 233–46. doi.org/10.1016/j.ergon.2011.01.011

[7] Rahmayani, N., Yuniar, & Desrianty, A 2015 The design of loose powder packs using the Kansei Engineering method. **3**(4) 2338-5081.

[8] Geraldine, A. 2017 *Marco Macaroni's packaging redesign using Kansei Engineering*. Undergraduate *Thesis*. (Bandung: Parahiyangan Catholic University Repository).

[9] Hidayat, R 2015 Re-design the packaging with the Kansei Engineering method. *Jurnal Al-Azhar Indonesia Seri Sains dan Teknologi*. **2**(4) p 215-223. doi.org/10.36722/sst.v2i4.156

[10] Dickson, P. R. & Sawyer, A. G (1986) *Point-of-purchase Behavior and Price Perceptions of Supermarket Shoppers*. (Massachusetts: Marketing Science Institute).

[11] Prastawa H, Ciptomulyono U, Laksono-Singgih M, Hartono M 2019 The effect of cognitive and affective aspects on usability. *Theor Issues Ergon Sci.* **20**(4) p 507–31. doi.org/10.1080/1463922x.2018.1547458

IOP Conf. Series: Materials Science and Engineering 1072 (2021) 012061 doi:10.1088/1757-899X/1072/1/012061

Trends and future studies on policies to improve renewable energy share

J Sulistio[1,2,3], B Wirjodirdjo[2] and P D Karningsih[2]

[1]Industrial Engineering Department, Islam Indonesia University, Yogyakarta, Indonesia
[2]Industrial Engineering Department, Sepuluh Nopember Institute of Technology, Surabaya, Indonesia

Email: [3]jokosulistio@uii.ac.id

Abstract. Every country needs to set up robust policies to reach its future targets in renewable energy development. The objective of this article is to explore the trends and research opportunities in this subject area. Systematic Literature Review (SLR) is applied together with bibliometric analysis to satisfy the objective. The database used in this study is Scopus considering its size and wide acceptance by major publishers. This study applies structural literature review and bibliometric analysis within one of its stages. A simple meta-analysis is cooperated prior to other analyses in this study. Various models are exercised to analyse important aspects in renewable energy policies, namely Input-output models, CGE models, Theory-based evaluation, multi-criteria analysis, and hybrid approach. Nonetheless, renewable energy is considered as a complex system. Therefore, advanced simulation techniques such as the discrete event simulation model, the agent-based model, system dynamics, and mixed methods are considered. Due to its characteristics, the mixed method is the most comprehensive approach. However, because mix methods have not been supported by proper tools or simulation software, agent-based model and system dynamics are proposed individually. Future research also should consider other than a market-based approach to develop robust policies to reach the ideal renewable energy share.

1. Introduction

World leaders are now more concerned about energy due to its growing demand and importance for economic development. It is projected that in 2035 energy demand will reach 778 Etta Joule and will continue to multiply in line with population growth and the world economic progress in 2050. Unfortunately, it is attained by environmentally unfriendly sources. In 2015, most energy needs were met by petroleum, followed by coal and natural gas [1–3].

The United Nations Framework Convention on Climate Change (UNFCCC) established the Conference for Parties (COP) as the highest forum in dealing with global environmental issues. The third meeting (COP-3) was held in 2005 and was attended by 37 countries. All members of the country agreed and committed to reduce the adverse effects of development by running the Greenhouse Gas Protocol (GHG). This protocol is known as the Kyoto Protocol which was later amended with the Doha Agreement. The Kyoto Protocol ended in 2012 and was replaced by a new agreement at the twenty-first COP meeting (COP-21) known as the Paris Agreement. In accordance, the United Nations submitted a Sustainable Development Goals (SDG) document. The document is a continuation of the Millennium Development Goals (MDG) which ended in 2015 [4].

The countries representative agreed to increase the proportion of energy sources that are more environmentally friendly. A series of policies are needed to endeavor countries' success in developing energy supply from renewable sources [5]. The purpose of this article is to investigate the trends of policies to develop a renewable energy mix.

2. Method

A literature review was applied to meet the objective of the study. Although there are several methods available to carry out the literature review, this research uses Systematic Literature Review (SLR) together with Bibliometric Analysis (BA). SLR explicitly contains information needed to be processed into some relevant information. The SLR is used to produce a comprehensive summary of the literature that is closely related to the research question. SLR identifies gaps in the literature [6] and results in classifications based on key themes in the area studied [7]. The literature review results are presented in two parts of the analysis, which are metadata analysis and renewable energy policy insights.

The SLR method is divided into four parts namely, (1) Identification; (2) Screening; (3) Eligibility; and (4) Inclusion [8]. BA is introduced at the identification stage to determine the keywords that will be used. BA is able to describe features and trends in energy policy research by combining information visualization technology and traditional citation analysis [9]. The tool used in BA is the VOS viewer version 1.6.13. The software displays the network and cluster of documents in visual form. The type of analysis used is co-occurrence with keywords as the unit of analysis. That is, relationships are determined based on the number of documents in which the keywords used appear. While the counting method used is full counting. That is, each relationship has the same weight. In addition, at least 5 keywords are also determined to form one cluster. The following are the detailed stages of SLR:

1. Identification. This step uses the Scopus database because large publishers such as Elsevier, Emerald, Taylor and Francis, Springer, and Willey can be found in it. In accordance with the research area, the first keyword used is "energy mix". These keywords produce 3.263 documents. Furthermore, BA is conducted to sharpen the research area by exploring topics related to the energy mix. Figure 1 shows that the topic of energy mix forms a cluster of subject areas. The first cluster consists of the energy mix and energy policy. The second cluster is the energy mix and renewable energy. The third cluster is the energy mix and sustainable development. The fourth cluster is the energy mix and the environmental impact (environmental impact) and carbon dioxide (carbon dioxide). From BA, it can be determined that search keywords with better relevance are "energy mix", "renewable energy", "sustainable development", "energy share" and "energy policy". Next, search for these keywords in the Scopus database with the format: (TITLE ("energy mix" OR "energy share" OR "renewable energy" OR "sustainable development") AND TITLE ("policy" OR "policies")).
2. Screening. Base to the previous search involved 160 keywords and then proceed to the screening process. This step considered only the articles with keywords related to the policy. These: Energy Policy, Renewable Energy Policy, Environmental Policy, Public Policy, Policy-Making, Policy Implementation, Policy Implementation, Policy Approach, Policy Development, Policy Instruments, Policy Strategy, Policies, Social Policy, Climate Policy, Health Care Policy and Renewable Energy Policies. It found 836 documents as result.
3. Eligibility. This step is to maintain the eligibility of the subject restricted by the subject area related to the purpose of the study, namely "energy". The result of this step found 466 documents related to the meta-analysis in the next section in this paper.
4. Inclusion. This stage verified the articles that are relevant to the research questions. In general, quantitative research is divided into three types, namely: correlational, comparative, and experimental research [10]. Correlational research tries to find the relationship between one or more independent variables on the dependent variable [11]. Comparative research aims to explain how a phenomenon occurs in different cases, variables, and relations [12]. Experimental research is intended to analyse the effects of changes in several independent variables on the dependent variable [13].

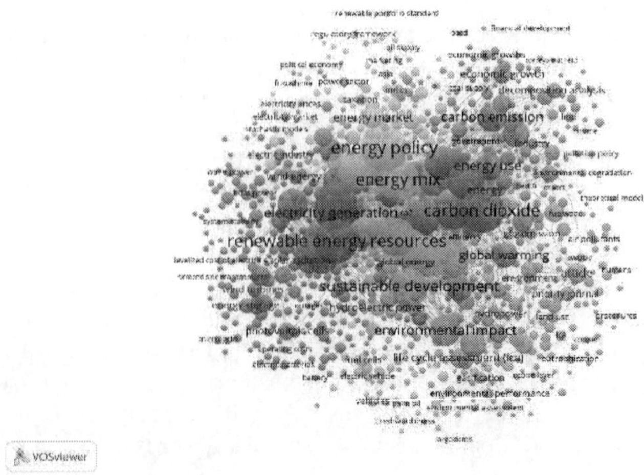

Figure 1. Bibliometric analysis.

3. Metadata analysis

The metadata analysis in this study is limited to simple descriptive statistical analysis. As seen in Figure 2, research on energy mix policies began in 1989 and the number of studies continues to increase. In the 1990s, the number of studies in this area increased and has exponentially grown. A total of 50 articles were published in 2019. It indicates that this area of research is becoming increasingly relevant to ongoing developments and dynamics.

From 117 journals and proceedings, Figure 3 shows 11 most article sources. Energy Policy is a journal included in Q1 and has published 133 articles or about 28% of the total journals and proceedings. This number is the largest in this research area. In 2019, Energy Policy has a cite score of 6.01 and an impact factor of 4.88. Renewable and Sustainable Energy Reviews ranks second in the number of publications. A total of 55 articles or about 11% of the total journals and proceedings in this area. Nevertheless, this journal is included in Q1 and has a cite score of 14.36 in 2019 and an impact factor of 10.55.

Figure 4 shows ten authors who have the most citations. The article entitled "The politics and policy of energy system transformation - Explaining the German diffusion of renewable energy technology" with authors Jacobsson S. and Lauber V. received 537 citations [14]. Jacobsson himself managed to publish 58 articles indexed by Scopus and had an h -index of 27. Lauber publishes 26 articles indexed by Scopus and has an h-index of 10. Both are authors who have high credibility in this area of research. The second article that has the most citations entitled "Prices versus quantities: Choosing policies for promoting the development of renewable energy" published in 2003 by Menanteau P., Finon D., and Lamy M.-L [15].

This area of research is carried out extensively in many countries. Figure 5 shows the 20 countries with the most studies. America is the country with the most amount of research related to energy policy. A total of 68 articles or about 11% discuss energy policy in America. The study covers energy sources, energy policies at the state and national levels. The second country is China with 55 articles or around 9%, followed by Britain with 41 articles or about 7%. Germany as a developed country in the development of renewable energy is discussed in 33 articles or about 5% of the total articles in this research area.

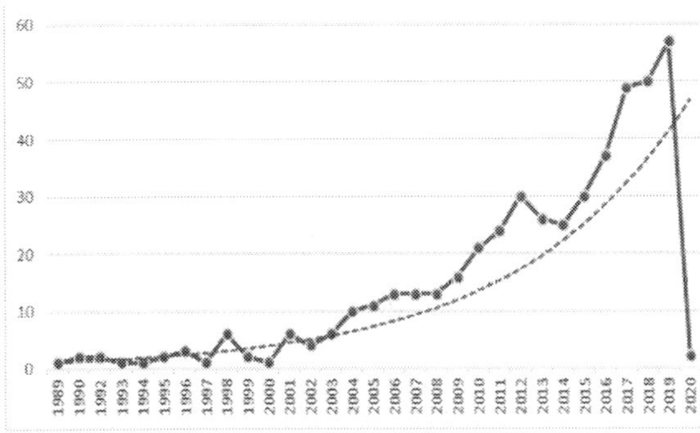

Figure 2. Publication by year.

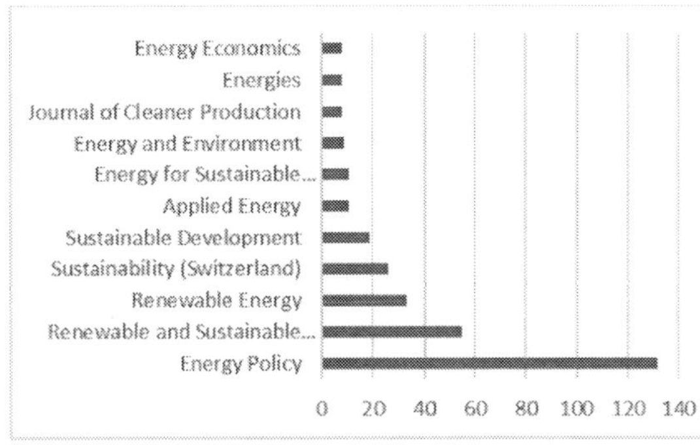

Figure 3. Influential journals on energy policy.

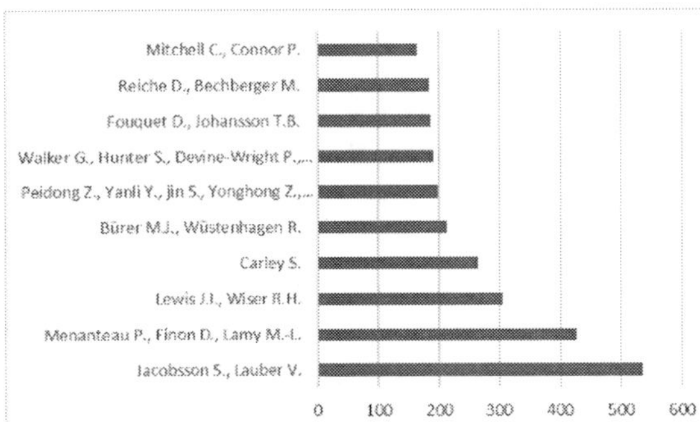

Figure 4. The top ten authors in the energy policy research area.

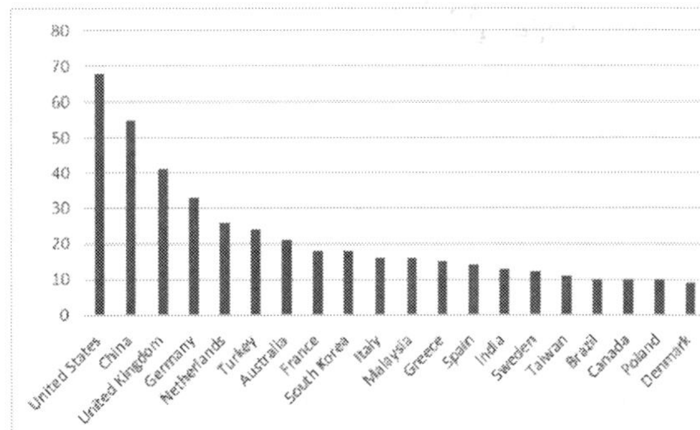

Figure 5. Publications by country.

4. Insight on renewable energy

Renewable energy is energy produced through the utilization of environmentally friendly energy sources. These energy sources are continuously filled by nature, come from the sun directly or indirectly, and which come from natural movements in nature. Fossils and their derivatives are not included in renewable energy because they are non-renewable so that they will eventually be used up. Some of the renewable energy sources are biomass energy, geothermal energy, hydropower energy, marine energy, solar energy, and wind energy [16].

Biomass energy is an energy that is sourced from all biological materials, both living and dead. Biomass can come from trash, organic fertilizer, water disposal, plants, and animals and gas combustion. This energy source has advantages compared to other renewable energy sources. Processing biomass into energy does not require technology that is too sophisticated, so it has good economic value. Biomass is a versatile energy source because it can be converted into solid, liquid, and gas forms. In addition, the abundant availability of biomass makes energy reserves from this source important to meet energy needs in the future. One of the critical concerns about this energy source is its contribution to CO2 emissions [17].

Geothermal energy is the energy obtained from the conversion process from inside the earth. The utilization of geothermal energy has been done long before BC. The first time this energy was used for energy and industrial purposes was in 1904. Since then, this energy source has been developed by many countries such as America, the Philippines, Italy, Mexico, Indonesia, and others. Geothermal can be used in two ways. The first way is to use it directly for purposes such as baths, space heaters, greenhouses, and so forth. This method can use high temperature and low temperature. Characteristics of use in this way are the dependence on the location of geothermal sources. This limits the use of energy for broader interests. The second application is to use geothermal energy which comes from a depth of about 3 km. The use of geothermal in this way requires more intensive technological applications.

Despite this, due to its abundant availability, this energy can be relied upon for the development of renewable energy [18].

The wind is a promising source of energy to be converted into clean and environmentally friendly energy. Wind energy uses wind movement to drive turbines that produce electricity. This energy is formed from differences in pressure on the surface of the earth because of exposure to the sun. It is estimated that 1-3% of solar energy is converted to wind energy [19]. No complicated technology is needed to convert wind power into other energy such as electricity. Nevertheless, the main obstacle to developing wind power is its intermittent nature.

Energy sources originating from the sea are known by several terms such as marine energy, ocean energy, or blue energy. Marine energy can be taken from six sources, namely waves/waves, tidal range, tidal currents, ocean currents, differences in ocean temperature energy conversion, and salinity gradients. This energy potential is huge considering that around 70% of the earth consists of seas. The use of the sea as an energy source must be shared with other potentials as economic drivers in the fields of transportation, tourism, fisheries, and others. This requires integrated management through Marine Spatial Planning (MSP) and Environmental Impact Assessment (EIA) [16].

Energy sourced from solar energy (solar energy) is obtained through two ways, namely solar photovoltaic (PV) systems and concentrating solar power (CSP) systems. PV systems use PV cells, which are semiconductors that convert sunlight directly into electrical energy. PV cells are interconnected through PV modules with batteries, inverters. and other electrification applications forming PV systems. CSP works by concentrating sunlight to heat liquids, solid objects, or gases which are then used to generate electricity [16]. The potential of this energy source can be measured using global horizontal irradiation (GHI) and direct normal irradiation (DNI). GHI is the total solar radiation on the horizontal surface on Earth in units of kWh / m2. DNI is the power received from solar electromagnetic radiation in units of W / m2. There are several countries on earth that have great potential in utilizing this solar power.

5. Energy models for policy evaluation

Models are needed to represent real systems and involve certain simplifications in several aspects. Nevertheless, the credibility of the model needs to be supported by the level of model detail, flexibility, adaptive capacity for new research questions, and transparency. The models used in energy policy evaluation include: Input-output models, CGE models, Theory-based evaluation, multi-criteria analysis, and hybrid approach [20].

1. Input-output models. This model was first developed by Wassily Leontief in 1936. The input-output (I / O model) model is very popular in its use in the economic analysis by reviewing the relationships between production, consumption, labor, and export imports. The application of I / O models in the renewable energy sector is counted less by researchers especially to enlarge the mix of renewable energy. Kung, 2016 uses the Big Data approach to analyse Taiwanese government policy in the bioenergy sector. The data used are the results of I / O modeling of crop production, demand elasticity, international customs and excise, process technology, commodity prices, and energy prices [21].
2. Multi-criteria analysis. Each country has a set of policies in developing renewable energy. Therefore, it is important for the government to determine the policy priorities taken to achieve the renewable energy mix target. Some research is dedicated to sorting policy priorities using established methods such as multi-criteria decision making (MCDM). The popular method used is the Analytic Hierarchy Process (AHP) [22–24]. Several studies apply fuzzy MCDM to accommodate the ambiguity within the criteria of assessment [25–27]. Similar combinations are necessary to address certain problem properties. Some of the combined methods are MULTIMOORA and Topsis [28], and Stochastic Multi-Criteria Acceptability Analysis and Long-range Energy Alternatives Planning [29].
3. CGE model. One of the developing optimization models is the Computable General Equilibrium Model (CGE Model). This model was adopted from the general equilibrium theory in the market

where supply and demand meet. CGE model has been widely applied in the social field especially to explain phenomena in the economic system [30]. In its development, this model is also applied in other fields including the renewable energy policy area. At least 68 energy policy analysis tools have been developed based on this model. A total of 37 of them were evaluated to provide guidance on choosing the right tool according to the needs of the analysis. The field of application for this tool varies from energy analysis for one building to the national energy system. In addition, other considerations such as the energy sector, technology used, and time parameters were included [31,32]. Some tools that are widely used to analyse policies in increasing the renewable energy mix are EnergyPLAN [33], TIMES [34], OseMOsys [35], and Promotheus [36]. These tools intend to minimize the total cost of energy supply for short and also long term activities. This includes technology choices and policy instruments. However, these models cannot describe the behaviour of individuals involved in the energy system.

4. Theory-based evaluation (TBE). Policies are often taken without enough theoretical basis or academic paper. Successful implementation of policies must be seen from the outcome studies that are produced. For this reason, a theory-based evaluation study is needed. TBE has two objectives, namely, to develop new theories and to prove theories empirically [20]. The TBE study methods that are often used are the Payback Framework, Contribution Analysis (CA), and Outcome Evaluation [37]. One study in this area examined the causal relationship between renewable energy consumption and economic growth in South Korea. This study examines the theory of neoclassical production functions from capital, labor, and renewable energy. The test used in this study is an autoregressive distributed lag bounds test. The results show that renewable energy consumption has a negative impact on economic growth. In addition, the causality test did not show a reciprocal relationship between the two variables. The author suggests that the South Korean government concentrates on economic growth rather than extending the use of renewable energy to grow the economy [38]. Other studies in this field examined the correlation of the type of energy consumed on total factor productivity (TFP). The results show that the use of fossil will decreases TFP growth and vice versa renewable energy consumption will increase it. Nonetheless, there are some results showing inconsistencies in different sub-panels [39].
5. Hybrid. This approach, also known as a mixed method of the research. It combines quantitative and qualitative methods to answer research questions. This method can be classified into three types of studies. The first type is called the convergent mixed method. This method merges qualitative and qualitative approaches to get an adequate understanding of the problems. The second type is called the explanatory mixed method. This approach carries out quantitative studies first and then continues with qualitative studies. The third type is the exploratory mixed method. This research is the opposite of explanatory studies. The study will carry out a qualitative study first and then continue with a quantitative study [40]. One of the mixed methods explanatory studies in this research area was conducted to analyse the effectiveness of energy policies to increase the renewable energy mix. In particular, the policy on investor preferences in the renewable energy technology sector. The study involved 60 venture companies in Europe and America. From this study, it is known that the Tariff Feed-in policy is effective in increasing investor preferences to invest in the renewable energy generation sector. So that will ultimately increase the renewable energy mix [41]. These results of this study similar to the research had conducted in Germany [42].

The models mentioned above could not capture the whole complexity appear in the energy system. Policy studies should be able to analyse its effectiveness in the long run as well as the short run. Therefore, a more complex model is needed.

6. Discussion

Many studies mentioned that energy systems have complex systems attributes, due to the interactions between diverse actors in turning energy sources [43]. Energy systems showed the characteristics of complex systems, namely (1) System uncertainties; (2) Nonlinear relationships between system variables; (3) Feedback structures; and (4) Causal relationships [44,45].

Non-linearity is often the cause of complexity in the system. However, the existence of feedback behaviour is more responsible for complexity in the system. That is, changes in the behaviour of one actor or element will cause the other actor or element to change. But at a certain point, it will change the behaviour or the element itself. Numerous actors and elements involved in a complex system make it hard to decide which is the first element causing the change. In some cases, the interactions formed a complex network. The system is considered to have spontaneous cause and effects relations. Consequently, the system will be sensitive to change and have no dominant elements (robustness and lack of central control). In complex systems, macro behaviour is an accumulation of several behaviours at the level of the microsystem (emergence) which indicates a smaller system involved within the larger system (hierarchical organization) [46].

Discrete Events (DE), Agent Base Model (ABM), System Dynamics (SD), and Multi-approach Model (MM) are models that are often used to analyse complex systems [20,47]. The complexity in the DE model expects to appear from the uncertainty of events in the system with discrete changes over time. It is possible to apply DE for systems behaviour projection in the long run. Nevertheless, energy systems involve sources that have continuous flow over time. Thus, DE is not a suitable approach for the purposes of the next study.

ABM capable to model individual interactions between agents and predict behaviours in their environment even though it seldom drives the model dynamics [48]. ABM unable to analyse the dynamics between elements in the long run which is a major drawback because energy policy analysis involves a long period of time as shown by energy policy targets in many countries [49]. ABM views an individual or agent as a unique and independent entity that interacts with other agents and their environment. Thus, ABM will be able to produce patterns at the macro level, only a way of determining properties and interactions at the micro-level. ABM has been a appropriate tool directed to implement in various socio-technical systems, including energy [50]. In the area of renewable energy research, this model is used to analyse agents' behaviour in the energy market [51–53]. ABM can also be used to analyse the selection of renewable energy policies as shown in several studies such as German [54].

SD model is capable to solve problems with dynamic complexity. There is no formal definition for the term dynamics complexity, but it can be recognized by the characteristics. The system will always experience changes in various time units, and it is common for these units to interact with each other. The change that happened in the system is rarely proportional to its causes which adds system complexity called non-linearity. The change in an element will change itself in a certain time unit. It shows that the system is controlled by a feedback causal loop, therefore has adaptive and self-organizing properties. The policy is an important element because it determines the behaviour of the energy system in the long run. However, in certain conditions, the system shows behaviour against the direction of the policy (policy resistant) and which is contrary to conjecture (counter-intuitive) [55]. Although it is very good to model interactions between agents in the system over time, SD cannot show the behaviour of agents individually. Several studies have made a conceptual model to describe the elements and relationships that exist in order to analyse the policy of developing renewable energy in Indonesia [56]. SD is also used to analyse systems in the context of reducing carbon emissions by increasing investment in the solar PV and wind power sectors and eliminating fossil-based energy subsidies [57]. In addition, an analysis of the conversion of fossil-based fuels to hydrogen and biofuel-based fuels [58].

MM saw the weaknesses of the earlier approaches and tried to expect them to use other approaches. Complex system analysis requires a various level of abstraction required so that modelers need more than one approach to formulating models. Although conceptually interesting, MM is still hard to find in studies especially in energy policy. This is mainly due to the availability of software as a tool for formulating models [47].

7. Conclusion

In terms of model, the mixed model shows a promising comprehensive result. It captures the micro as well as macro behaviour as responses to renewable energy applied. Availability and access are the drawbacks of this model. The agent-based model and system dynamics are proofed to be a suitable model to analyse renewable energy policy. In many cases, individuals and several combinatory scenarios show positive outcomes of the renewable energy mix. However, there is still a wide gap to fill because none of those scenarios can reach the national or regional renewable energy mix targets. Future research could consider the demand side or non-market-based approach to accelerate the growth of the renewable energy mix in order to reach the targets.

Acknowledgments
I would like to acknowledge the supports of these following institution in making this research and presentation possible:
1. LPDP (Indonesia Endowment Fund for Education)
2. Institut Teknologi Sepuluh Nopember
3. Universitas Islam Indonesia

Reference

[1] Rafiee A, Khalilpour KR 2019 Renewable Hybridization of Oil and Gas Supply Chains. Polygeneration with Polystorage for Chemical and Energy Hubs. p 331–72. doi.org/10.1016/b978-0-12-813306-4.00011-2
[2] Yoon J, Sun Y, Rogers JA 2010 Flexible Solar Cells Made of Nanowires/Microwires. *Semiconductor Nanomaterials for Flexible Technologies.* **201** p 159–96. doi.org/10.1016/b978-1-4377-7823-6.00006-4
[3] Azad K, Rasul MG, Khan MMK, Sharma SC 2019 Introduction to sustainable and alternative ecofuels. *Advances in Eco-Fuels for a Sustainable Environment.* p 1–14. doi.org/10.1016/b978-0-08-102728-8.00001-2
[4] Transforming Our World: The 2030 Agenda for Sustainable Development. *A New Era in Global Health.* Springer Publishing Company. doi.org/10.1891/9780826190123.ap02
[5] Zemin J 2010 Reflections on Energy Issues in China. *Research on Energy Issues in China.* p 7–52. doi.org/10.1016/b978-0-12-378619-7.00002-6
[6] Tranfield D, Denyer D, Smart P 2003 Towards a Methodology for Developing Evidence-Informed Management Knowledge by Means of Systematic Review. *Br. J. Manag.* **14**(3) p 207–22. doi.org/10.1111/1467-8551.00375
[7] Seuring S, Müller M, Westhaus M, Morana R. Conducting a Literature Review — The Example of Sustainability in Supply Chains. *Research Methodologies in Supply Chain Management.* Physica-Verlag. p 91–106. doi.org/10.1007/3-7908-1636-1_7
[8] Tseng M-L, Islam MS, Karia N, Fauzi FA, Afrin S. A 2019 literature review on green supply chain management: Trends and future challenges. *Resour Conservation Recycling.* **141 p**145–62. doi.org/10.1016/j.resconrec.2018.10.009
[9] Huang L, Zhou M, Lv J, Chen K 2020 Trends in global research in forest carbon sequestration: A bibliometric analysis. *J. Clean. Prod* **252** 119908. doi.org/10.1016/j.jclepro.2019.119908
[10] Moissenko F, Braicu C, Tomuleasa C, Berindan-Neagoe I 2015 Types of Research Designs. *Cancer Research and Clinical Trials in Developing Countries* p 29–39. doi.org/10.1007/978-3-319-18443-2_3
[11] Kampen JK 2010 A methodological note on the making of causal statements in the debate on anthropogenic global warming. *Theor. Appl. Climatol* **104**(3-4) p 423–7. doi.org/10.1007/s00704-010-0355-y
[12] Esser F, Vliegenthart R. 2017 Comparative Research Methods. *The International Encyclopedia of Communication Research Methods.* p 1–22. doi.org/10.1002/9781118901731.iecrm0035
[13] Panciu E, Anca DE, Chisamera M 2016 The Influence of Cooling Rate and Equivalent Carbon on the Chill Tendency and Cooling Curves Parameters in Uninoculated and Inoculated Grey Cast Iron. Solid State Phenom **254** p 255–60. doi.org/10.4028/www.scientific.net/ssp.254.255
[14] Jacobsson S, Lauber V 2006 The politics and policy of energy system transformation—explaining the German diffusion of renewable energy technology. *Energy Policy.* **34**(3) p 256–76.

doi.org/10.1016/j.enpol.2004.08.029

[15] Menanteau P, Finon D, Lamy M-L 2003 Prices versus quantities: choosing policies for promoting the development of renewable energy. *Energy Policy*. **31**(8) p 799–812. doi.org/10.1016/s0301-4215(02)00133-7

[16] Ellabban O, Abu-Rub H, Blaabjerg F 2014 Renewable energy resources: Current status, future prospects and their enabling technology. *Renew. Sust. Energ. Rev.* **39** p 748–64. doi.org/10.1016/j.rser.2014.07.113

[17] Alookandeh AE, Vaez-Zadeh S. A 2019 Comparative Review of Renewable Energy Potential, Policy Targets, and Implementation in Iran. *IEEE International Conference on Environment and Electrical Engineering and 2019 IEEE Industrial and Commercial Power Systems Europe (EEEIC / I&CPS Europe)*. IEEE doi.org/10.1109/eeeic.2019.8783528

[18] Fridleifsson IB. 2001 Geothermal energy for the benefit of the people. *Renew. Sust. Energ. Rev.* **5**(3) p 299–312. doi.org/10.1016/s1364-0321(01)00002-8

[19] Kaygusuz K 2009 Wind Power for a Clean and Sustainable Energy Future. *Energ Source Part B*. **4**(1):122–33. doi.org/10.1080/15567240701620390

[20] Horschig T, Thrän D 2017 Are decisions well supported for the energy transition? A review on modeling approaches for renewable energy policy evaluation. *Energy Sustain. Soc.* **7**(1). doi.org/10.1186/s13705-017-0107-2

[21] Kung C-C, Zhang L, Chang M-S 2017 Promotion policies for renewable energy and their effects in Taiwan. *J. Clean. Prod.* **142** p 965–75. doi.org/10.1016/j.jclepro.2016.05.034

[22] Hasibuan S, Thaheer H, Nugroho RE, Hidayati J 2017 The development strategy of sustainable bioenergy investment in Indonesia based palm biomass feedstock. *Proceedings of the International Conference on Industrial Engineering and Operations Management*. **2017** p 775.

[23] Okioga IT, Wu J, Sireli Y, Hendren H 2018 Renewable energy policy formulation for electricity generation in the United States. *Energy Strategy Rev.* **22** p 365–84. doi.org/10.1016/j.esr.2018.08.008

[24] Malkawi S, Al-Nimr M, Azizi D 2017 A multi-criteria optimization analysis for Jordan's energy mix. *Energy*. **127** p 680–96. doi.org/10.1016/j.energy.2017.04.015

[25] Yuan J, Li C, Li W, Liu D, Li X 2018 Linguistic hesitant fuzzy multi-criterion decision-making for renewable energy: A case study in Jilin. *J. Clean. Prod.* **172** p 3201–14. doi.org/10.1016/j.jclepro.2017.11.038

[26] Wu Y, Xu C, Zhang T 2018 Evaluation of renewable power sources using a fuzzy MCDM based on cumulative prospect theory: A case in China. *Energy*. **147** p 1227–39. doi.org/10.1016/j.energy.2018.01.115

[27] Lee H-C, Chang C-T 2018 Comparative analysis of MCDM methods for ranking renewable energy sources in Taiwan. *Renew. Sust. Energ. Rev.* **92** p 883–96. doi.org/10.1016/j.rser.2018.05.007

[28] Streimikiene D, Balezentis T, Krisciukaitienė I, Balezentis A 2012 Prioritizing sustainable electricity production technologies: MCDM approach. *Renew. Sust. Energ. Rev.* **16**(5) p 3302–11. doi.org/10.1016/j.rser.2012.02.067

[29] Rahman MM, Paatero JV, Lahdelma R, A. Wahid M 2016 Multicriteria-based decision aiding technique for assessing energy policy elements-demonstration to a case in Bangladesh. *Appl. Energy*. **164** p 237–44. doi.org/10.1016/j.apenergy.2015.11.091

[30] Hayati M 2013 Pemahaman Dasar Analisis Model Computable General Equilibrium (CGE). *Agriekonomika*. **2**(1) p 66-75. https://doi.org/10.21107/agriekonomika.v2i1.667

[31] Connolly D, Lund H, Mathiesen BV, Leahy M 2010 A review of computer tools for analysing the integration of renewable energy into various energy systems. *Appl. Energy*. **287**(4) p 1059–82. doi.org/10.1016/j.apenergy.2009.09.026

[32] Weijermars R, Taylor P, Bahn O, Das SR, Wei Y-M 2012 Review of models and actors in energy mix optimization – can leader visions and decisions align with optimum model strategies for

our future energy systems? *Energy Strategy Rev.* **1**(1) p 5–18. doi.org/10.1016/j.esr.2011.10.001

[33] Noorollahi Y, Itoi R, Yousefi H, Mohammadi M, Farhadi A. Modeling for diversifying electricity supply by maximizing renewable energy use in Ebino city southern Japan. *Sustain. Cities Soc.*, 2017 Oct;**34**:371–84. doi.org/10.1016/j.scs.2017.06.022

[34] Mondal MAH, Ringler C, Al-Riffai P, Eldidi H, Breisinger C, Wiebelt M 2019 Long-term optimization of Egypt's power sector: Policy implications. *Energy*. **166** p 1063–73. doi.org/10.1016/j.energy.2018.10.158

[35] Dreier D, Howells M. OSeMOSYS-PuLP 2019 A Stochastic Modeling Framework for Long-Term Energy Systems Modeling. *Energies*. **12**(7) p 1382. doi.org/10.3390/en12071382

[36] Fragkos P, Kouvaritakis N, Capros P 2015 Incorporating Uncertainty into World Energy Modelling: the PROMETHEUS Model. *Environ. Model. Assess.* **20**(5) p 549–69. doi.org/10.1007/s10666-015-9442-x

[37] Belcher BM, Davel R, Claus R 2020 A refined method for theory-based evaluation of the societal impacts of research. *Methodsx*. **7** 100788. doi.org/10.1016/j.mex.2020.100788

[38] Lee S-H, Jung Y 2018 Causal dynamics between renewable energy consumption and economic growth in South Korea: Empirical analysis and policy implications. *Energy Environ.* **29**(7) p 298–315. doi.org/10.1177/0958305x18776546

[39] Rath BN, Akram V, Bal DP, Mahalik MK 2019 Do fossil fuel and renewable energy consumption affect total factor productivity growth? Evidence from cross-country data with policy insights. *Energy Policy*. **127** p 86–99. doi.org/10.1016/j.enpol.2018.12.014

[40] Creswell JW, Creswell JD 2018 *Research Design: Qualitative, Quantitative, and Mixed Methods Approaches.* (Los Angeles: SAGE Publications Inc.).

[41] Bürer MJ, Wüstenhagen R 2009 Which renewable energy policy is a venture capitalist's best friend? Empirical evidence from a survey of international cleantech investors. *Energy Policy*. **37** p 4997–5006. doi.org/10.1016/j.enpol.2009.06.071

[42] Held A, Ragwitz M, Haas R 2006 On the Success of Policy Strategies for the Promotion of Electricity from Renewable Energy Sources in the Eu. *Energy Environ.* **17**(6) p 849–68. doi.org/10.1260/095830506779398849

[43] Basu S, S. E. Bale C, Wehnert T, Topp K 2019 A complexity approach to defining urban energy systems. *Cities*. **95** p 102358. doi.org/10.1016/j.cities.2019.05.027

[44] Mutingi M, Mbohwa C, Kommula VP 2017 System dynamics approaches to energy policy modelling and simulation. *Energy Procedia*. **141** p 532–9. doi.org/10.1016/j.egypro.2017.11.071

[45] Qudrat-Ullah H. 2015Modelling and Simulation in Service of Energy Policy. *Energy Procedia*. **75** p 2819–25. doi.org/10.1016/j.egypro.2015.07.558

[46] Ladyman J, Lambert J, Wiesner K 2012 What is a complex system? *Eur. J. Philos. Sci.* **3**(1) p 33–67. doi.org/10.1007/s13194-012-0056-8

[47] Kremers EA 2012 *Modelling and Simulation of Electrical Energy Systems through a Complex Systems Approach using Agent-Based Models*. (KIT Scientific Publishing).

[48] Bale CSE, Varga L, Foxon TJ 2015 Energy and complexity: New ways forward. *Appl. Energy*. **138** p150–9. doi.org/10.1016/j.apenergy.2014.10.057

[49] Van Dyke Parunak H, Savit R, Riolo RL 1998 Agent-Based Modeling vs. Equation-Based Modeling: A Case Study and Users' Guide. *Lecture Notes in Computer Science*. p 10–25. doi.org/10.1007/10692956_2 5

[50] Iychettira KK, Hakvoort RA, Linares P, de Jeu R 2017 Towards a comprehensive policy for electricity from renewable energy: Designing for social welfare. *Appl. Energy*. **187** p 228–42. doi.org/10.1016/j.apenergy.2016.11.035

[51] Iwagami A, Suzuki H, Aihara K 2012 Analysis of an Agent-based Electricity Market Model with Renewable Energy Power Plants by Wind and Solar Power. *IEEJ Trans. Power Energy*.

132(5) p 468–77. doi.org/10.1541/ieejpes.132.468
[52] Deissenroth M, Klein M, Nienhaus K, Reeg M 2017 Assessing the Plurality of Actors and Policy Interactions: Agent-Based Modelling of Renewable Energy Market Integration. *Complexity*. **2017** p 1–24. doi.org/10.1155/2017/7494313
[53] Reeg M, Hauser W, Wassermann S, Kast T, Klann U, Nienhaus K, et al. 2012 AMIRIS: An Agent-Based Simulation Model for the Analysis of Different Support Schemes and Their Effects on Actors Involved in the Integration of Renewable Energies into Energy Markets. 2012 23rd International Workshop on Database and Expert Systems Applications. IEEE; doi.org/10.1109/dexa.2012.54
[54] Beckenbach F, Daskalakis M, Hofmann D 2017 Agent-Based Analysis of Industrial Dynamics and Paths of Environmental Policy: The Case of Non-renewable Energy Production in Germany. *Comput Econ*. **52**(3) p 953–94.doi.org/10.1007/s10614-017-9773-6
[55] Sterman JD 2000 *Business dynamics: systems thinking and modeling for a complex world*. (Boston: Irwin/McGraw-Hill).
[56] Hidayatno A, Dhamayanti R, Destyanto AR 2018 Model conceptualization for policy analysis in renewable energy development in Indonesia by using system dynamics. *Int. J. Smart Grid Clean Energy*. p 54–8. doi.org/10.12720/sgce.8.1.54-58
[57] Kelly C, Onat NC, Tatari O 2019 Water and carbon footprint reduction potential of renewable energy in the United States: A policy analysis using system dynamics. *J. Clean. Prod.* **228** p 910–26. doi.org/10.1016/j.jclepro.2019.04.268
[58] Shafiei E, Davidsdottir B, Leaver J, Stefansson H, Asgeirsson EI 2015 Simulation of Alternative Fuel Markets using Integrated System Dynamics Model of Energy System. *Procedia Comput. Sci.*, **51** p 513–21. doi.org/10.1016/j.procs.2015.05.277

IoT based kWh Meter Recorder System for Electrical Substation

Moh Khozain[1] and Adhatus S Ahmadiyah[2]

[1]Maintenance Department, State Electricity Company (PLN), Jl. Embong Wungu PLN UP2D Jatim, Surabaya, Indonesia

[2]Informatics Department, Institut Teknologi Sepuluh Nopember, Jl. Raya ITS Kampus ITS Sukolilo, Surabaya, 60111, Indonesia

E-mail: moh.khozain@pln.co.id[1], adhatus@if.its.ac.id[2]

Abstract. The plan to abolish the substation operator raises new problems in reading the electrical energy stand, which at present all State Electricity Company units conduct it manually. This paper aims to provide a cost-effective meter data recording system to replace manual reading. The proposed solution is developed using WI-FI based internet of things (IoT) technology. Our proposed solution is implemented as a meter-recording device installed at the substation. Since deployed in December 2018, this tool works very well. Since it actively supports East Java's performance in the pilot distribution control unit, it can be implemented in other units.

1. Introduction

At present, the reading of the electrical energy stands at feeders in the substations of all Indonesia State Electricity company units is still conducted manually. Officers or operators need to come to the substation to get meter data as a reconciliation of postpaid energy consumption. When the Supervisory, Control, and Data Acquisition (SCADA) system fails, the operator of the substation needs to read the electrical energy parameters at any time. Manual reading is highly vulnerable to human error because the meter position is higher than the eye, and the recording process is not practical.

However, the effort to automate meter reading at substations faces some challenges. First, the existing system with a multi drop system fails to integrate communication of kiloWatt hour (kWh) meter reading in feeders. The second is the high-cost problem of adding modems and using GSM / GPRS card for communication. The feeder meter serves as the main upstream energy reconciliation facility, instead of a direct transaction point with the customer. Therefore, the development of Advanced Metering Infrastructure (AMI) in the proposed system must be pursued to promote efficient operating costs.

Plans for eliminating the operator of the substation operator can still minimize the occurrence of deviations. It is in line with the unit's mission and not disrupt the existing SCADA system by the end of 2018, experiencing constraints due to the two problems earlier. Therefore, developing a kWh meter reading system with the IoT WI-FI system is a highly relevant solution [1] since IoT has been an accessible technology lately [2-4].

2. Methodology

The results of data collection in the field have several different brands and types of kWh meters. In one substation, there are more than five kWh meters, some even 50 kWh meters close together. Each kWh meter uses one reader device, but each substation is only enough to use one router, except for a separate substation, two separate buildings need two routers. Figure 1 displays the required equipments for the proposed solution.

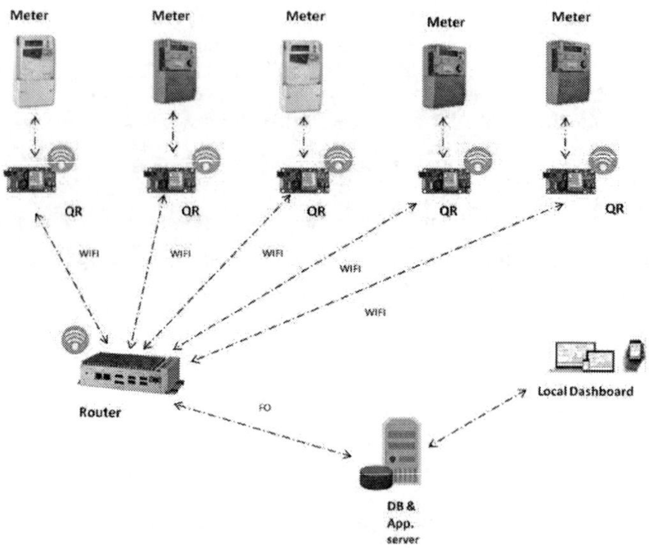

Figure 1. Quality Control Program Flowchart

2.1 Hardware

QR (Queue Reader) is a part that receives data from meters with a variety of protocol meters. The data from the protocol meter convert into a smaller size MQTT data protocol and send to the server with a push system via the Fiber Optic pathway of the existing SCADA system. Since MQTT is a protocol above the TCP / IP level for IoT implementation, data conversion needed to keep sending data to server from interfering with SCADA communication.

ESP32 (shown in Table 1) is a System on Chip that provides Wi-Fi plus Bluetooth 4.2 Solution combo on the 2.4 GHz band using 40 nm technology. Node MCU ESP-32 is a development board kit for internet of things applications. This board consists of a very feature-rich IoT ESP32 module and a USB to-serial CP2102 chip that makes it easy to program this module. One core dedicates to WI-FI and Bluetooth connectivity and the other core dedicates to user applications.

Table 1. ESP32 Specification

Items	Specifications
RF certification	FCC/CE-RED/IC/TELEC/KCC/SRRC/NCC
Wi-Fi certification	Wi-Fi Alliance
Bluetooth certification	BQB
Green certification	RoHS/REACH
Reliablity	HTOL/HTSL/uHAST/TCT/ESD
Protocols	802.11 b/g/n (802.11n up to 150 Mbps)
	A-MPDU and A-MSDU aggregation and 0.4 μs guard interval support
Frequency range	2.4 GHz ~ 2.5 GHz

The QR block diagram as shown in Figure 2 is designed to be small enough to make it possible to place it securely on 20 kV cubicle panel or stamp meter.

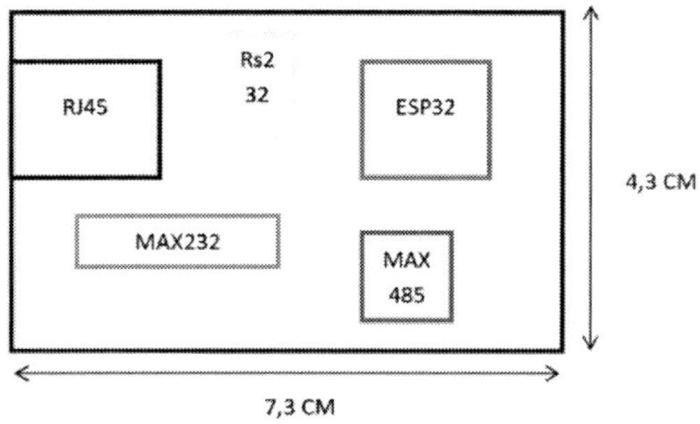

Figure 2. Block diagram of QR

2.2 Web Application

PHP is a scripting language used to create dynamic web pages and is a server-side script language. Dynamic means the displayed web page is processed at the time. The web page requested by the client. This mechanism cause information received by clients is always the latest.

The kWh meter parameter data that has been sent to the server can be displayed informatively via the web application. The master IoT METER web is designed with standard parameters needed for the analysis and evaluation of upstream electricity reading. In the future, it enables easy integration with AMICON, in which a single application monitoring meter data from the base to the ends of the network at the customer. The main parameters of reading data on the web application include the following: (a) Dashboard, (b) Load profile (real-time history data), (c) Detailed readings in real-time (complete with phasors), (d) Flow chart, (e) Voltage graph, and (f) EoB record stand.

Granting access managed by the super administrator in East Java distribution control center - operating unit to be give to certain parts of interest, for example, the UP3 and UP2B energy transaction section, the operation section, and the protection meter of UP2D JATIM. The access restriction is intended to prevent electrical energy transaction data from being misused by unauthorized users.

This innovation is implemented in 105 substations and 711 feeders in the East Java Region with several steps of development. The development plan of this design system integrates the existing AMICON system and balance sheet application as shown in the chart in Figure 3.

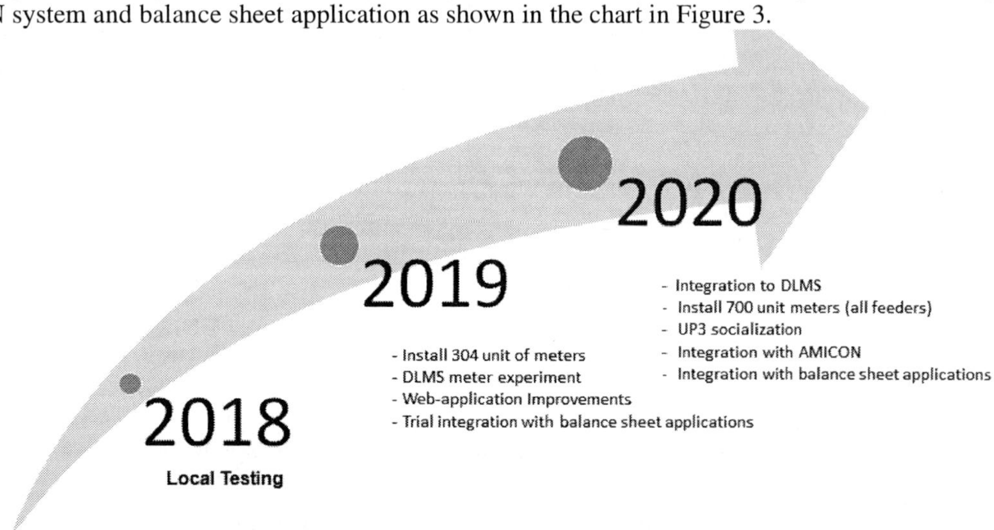

Figure 3. Block Diagram of development system

3. Result and Discussion

In the application of this system uses hardware (tools) and software (web browsers) as shown in Figure 4. The hardware (tool) functions to send the meter data to the Router / Serial Link and send again to the server using the IoT system, which uses the MQTT protocol so as not to interfere with the existing SCADA communication. After the data have sent to the server, the software (web application) processes and displays the data into accurate and easy to understand information.

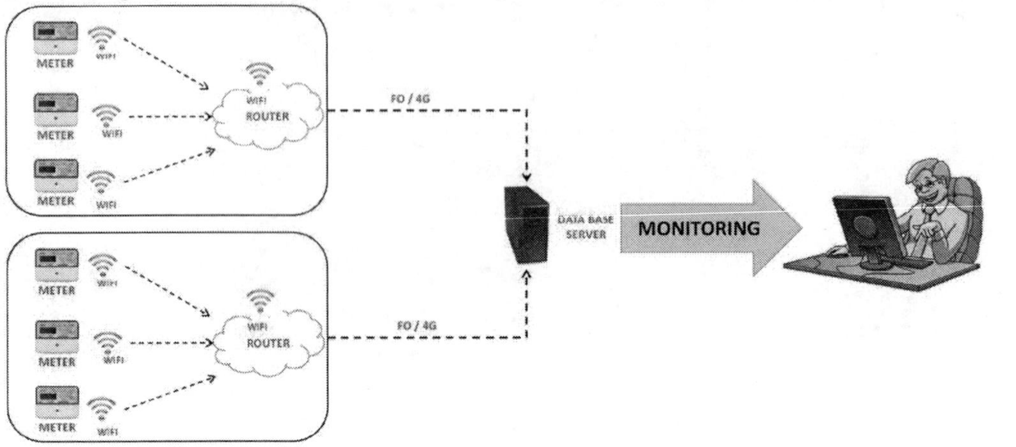

Figure 4. Topology of IoT Meter Main Station for two locations

3.1 Hardware Testing and Installation

Wi-Fi components generally consist of Access Point (AP) and Wireless Client that can be integrated with a user's device or external card. AP can be connected to existing LAN network. The individual test results of each meter can be done by checking the success of the Wi-Fi connection from mobile to QR. Through the browser on our cell phone, we can check the success or failure of the connection based on the default IP address of the QR. Meter data settings and naming ID feeders can be done directly in the field, without having to register via the web.

The results of the data transmission test can also be seen from the physical QR, the active QR power indicator light, and the indication for receiving data from the meter.

This QR is designed without additional voltage supply to minimize the additional wiring that has the potential for anomalous DC Ground in a 20 kV cell installation

Some meters have different internal power supplies, QR requirements can be lit normally, that is, 9VDC voltage and QR power consumed by 2 watts. Ports of communication owned by the brand and meter type are also different, so it needs several programming segments.

3.2 Web Monitoring Test

System testing can be known from the features prepared in the web display. The dashboard shows the number of meters the QR has been installed, successful reading, and total points failed to send data. The number of End of Billing (EoB) data records for reconciliation transactions between units is also displayed.

Data load profile, in this case, the web is not taking data from parameters stored by the meter. However, it is retrieving data sent by QR in the form of instant data collection. Each data load profile history can be checked in detail because the source of the data taken is instantaneous reading. Energy measurement data from the load profile can display voltage, current, frequency, cos phi, phasor, and others. Based on these data, the deviation evaluation during reconciliation can be carried out earlier and more quickly.

Based on the results of the overall system implementation test, repeated communication failures occurred (indicated with red highlight in Figure 5). The failure is detected using a certain hardware. When we analized, most communication failures are dominated by port number 23, while port number

1883 works well. After replacing the communication to port number 1883, a more stable communication with success reading of 91% is achieved.

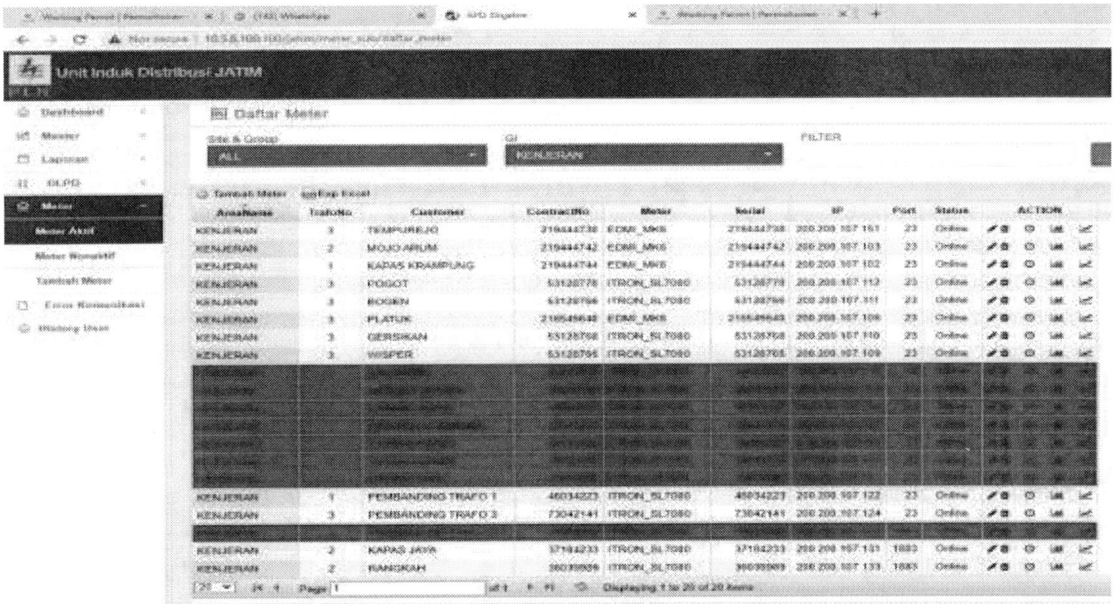

Figure 5. Display of Communication fail

4. Conclusion

The proposed system has proved its easy and safe installation without additional power supply and communication to the router via WI-FI. In addition, it can directly be integrated with existing communication networks without additional routine connection fees because it uses the MQTT protocol that is able to utilize the SCADA line without disrupting the system that is running. The analysis can be done by its monitoring application which displays instant and EoB recording data.

References

[1] Schwartz M-O 2014 Internet of Things with Arduino (England)

[2] S. Gérald. The Internet of Things: Between the Revolution of the Internet and the Metamorphosis of Objects. European Commission Community Research and Development Information Service. Retrieved August 2017.

[3] G.F.Nama, M.Komarudin, H.Priambodo, Mardiana,H.D. Septama, 2014. Electricity, Temperature, and Network Utilization Monitoring at Lampung University Data Centre Using Low Cost Low Power Single Board Mini Computer. In The Regional Conference on Computer and Information Engineering (RCCIE), Yogyakarta Indonesia, pp 184-189.

[4] D.Despa, M.A.Muhammad, H.Gusmedi, N.Amaro. 2017. Sistem Monitoring Besaran Listrik Dengan Teknologi IoT (Internet of Things). Bandar Lampung

IOP Conf. Series: Materials Science and Engineering 1072 (2021) 012063 doi:10.1088/1757-899X/1072/1/012063

Analysing the barriers of reverse logistics implementation: A case study

S N W Pramono[1], M M Ulkhaq[1,2*] and Z Aulia[1]

[1]Department of Industrial Engineering, Diponegoro University, Semarang, Indonesia
[2]Department of Economics and Management, University of Brescia, Brescia, Italy

Email: *ulkhaq@live.undip.ac.id

Abstract. An increase in both consumers' environmental awareness and government regulation related to the environment encourages industries to be more responsible for their products after consumption. As a result, a relatively new concept in logistics called reverse logistics (RL) is introduced. While the forward (or traditional) logistics refers to the process of distributing products from companies (or producers) to consumers, RL refers to the process of moving products from consumers to producers. This practice has attracted not only scholars but also practitioners in the field of supply chain and sustainability. A company could gain benefits from the implementation of RL since it can lead to higher profit. However, some barriers regard its implementation despite the benefits that make the company reluctant to implement it. Therefore, this study aims to investigate the barriers to implementing RL in a bottled drinking water company. Moreover, this study also analyses the interaction among the barriers to looking for the possible "root barrier". The interpretative structural modelling and MICMAC analysis were used to accomplish the objective of the study. It is expected that the study could help the management to better formulate policies about the implementation of RL.

1. Introduction

The concept of reverse logistics (RL) in recent decades has attracted huge attention in industries. Shortly, this concept refers to the process of moving products from the consumers to the producers, i.e., the company. To be more precise, it is related to the process from planning, executing, and monitoring the cost-efficient flow of materials; controlling the inventory; to monitoring the delivery of finished products from the point of consumption to the point of origin [1]. Comparing to the forward or traditional logistics, it is a "reversed" process, i.e., allowing products (and information) to be returned to the company.

There are several reasons why RL has received huge attention, from a pure environmental concern to competition and marketing motives. A company that manages RL practice well can gain several benefits, such as higher revenue, lower operational costs, and higher customer satisfaction [2]. Other advantages that can be mentioned are environmental protection and efficient resource usage [3][4]. In addition, an increase in consumer awareness and company social responsibilities towards the environment also drives the company to implement RL [5].

Nevertheless, regardless advantages, several challenging endeavours might hinder the implementation especially since RL could generate cost that might affect the financial performance of the company. In the beginning, it is obvious that RL implementation is costly; comparing to forward logistics, it can cost to five times greater [6]. Also, the products that are moved back to the company might pile up in the company's warehouse [7].

Most of the research regarding the analysis of RL implementation barriers were conducted in developed countries; only a few researches were conducted in developing countries (e.g. [8][9][10]). This phenomenon is not really surprising because RL has become a mandatory practice in the developed countries, but, in the developing countries, it is still in its infancy state [11][12]. This condition has motivated us to study this issue and its practice in Indonesia, as one of the developing countries.

The object of this research is a bottled drinking water company located in Semarang, Central Java Province, Indonesia. The rate of consumption of bottled drinking water in Central Java Province shows an increasing trend, from 9.2% in 2007 to 68.1% in 2017, meaning that this is a promising sector. In addition, according to ASPADIN, an association of Indonesian bottled drinking water companies, total national revenues generated from this sector is about USD 1.676 million in 2013 and it is projected to grow about 11.1% per year to 2017. However, the bad effect of this increased consumption is that the waste generated is also increasing. The disposable packaging, i.e. the plastic bottle, the straw and its packaging, as well as plastic labels covered the bottle, obviously have a harmful impact on the environment. The materials used to produce those things would take more than 20 years to decompose if those are thrown into the landfills.

Although disposable packaging can be recycled to reduce the environmental effects, the practice remains limited, especially from the manufacturer's perspective. Therefore, a proper RL practice is required to improve the effectiveness of the process. As has been stated previously, the implementation of RL has some barriers. The barriers could influence both the RL operations and also influence one another. It is important to recognize the mutual relationship among the barriers. The investigation of the possible "root barrier" and the barriers that are influenced by others might help the management to better execute the RL practice. In addition, it could be good support for taking appropriate action to deal with the barriers to implementing RL.

The objectives of the study are then threefold. Firstly, this study aims to identify RL implementation barriers in a bottled drinking water company. It used a literature review and interviews. Secondly, this study aims to investigate the interaction of the barriers by using interpretative structural modelling (ISM). Lastly, this study aims to map the barriers into the driver-dependence diagram by employing MICMAC analysis.

2. Research design

To accomplish the objective of the study, the barriers of RL implementation have to be identified first. ISM then was employed to identify the contextual relationship among the barriers. It also can be used to identify the "root" barrier(s), i.e., the barrier(s) in which most influence(s) other barriers. Finally, *Matrice d'impacts croisés multiplication appliquée á un classement* (MICMAC) or in English: cross-impact matrix multiplication applied to the classification analysis was used to investigate the dependence and the driver powers of the barriers.

2.1. Barriers of reverse logistics implementation

To investigate the barriers of RL implementation at the object of the research, a literature review was performed through examining several related studies (i.e. [1][5][10][13]-[16]). The barriers found from the literature review were required to be refined due to several considerations (e.g., different locations and conditions). Some experts also had been interviewed for giving such insight into the barriers which hinder the implementation of RL at the object of the research. Finally, there are thirteen barriers which are described as follows.

The first barrier is the lack of information systems (IS) and technology (BR1). IS is considered as a vital factor for good RL implementation [17]. If a company desires to implement a good RL, it has to establish a proper IS, particularly for the process of tracing and tracking products' status after delivering to the consumers. However, the situation at the bottled drinking water company as the object of the research is that the company manages an internal IS, meaning that the information regarding materials and finished products only available from the suppliers to the company and from the company to the retailers. After the products left from the retailers, the information cannot be traced anymore. The second barrier is some problems related to product condition (BR2). Comparing to the forward logistics as the company could maintain the uniformity of the quality of the product, the condition of the returned products is heterogeneous. The products can be returned because they are indeed defective, guarantee returns, or because of violating the environmental issue [1]. The condition at the object of the research is that the company does not have a good gatekeeping to do a screening of the defective products and unwarranted products. The third barrier relates to firm policies (BR3), particularly in managing product returns. Most returned products are crushed or dumped in the garbage. The returned products are rarely remanufactured, recycled, or reused. Interestingly, a study by [18] identified that more than one-third of more than one hundred and fifty managers who responsible for RL practice believe that firm policies eventually negatively affect RL practice.

The fourth barrier is the resistance of the management to fully implement RL (BR4). An organizational change could generate resistance and scepticism in the company's employees, causing it hard to achieve improvement. RL implementation somehow needs a radical change in the employee as well as employers' mindset and practice especially when the previous system is not really supported. The fifth barrier is financial constraint (BR5). This financial issue is considered as one of the major barriers in implementing RL because it is important to support not only the workforce but also the infrastructure of the RL practice. Good training and IS indeed need the company's financial support. If the company does not have a desire to establish a good RL practice, the company then will not assign the funds for investing in RL. The sixth barrier is lack of commitment by top management (BR6). The top management has to provide a clear vision to the RL implementation. They also have to deliver good strategic plans regarding the RL practice and also action plans to successfully execute the plans.

The seventh barrier is unawareness of the customers (BR7). Demands from consumers nowadays are the most critical type of external pressure. If the customers demand firms to implement RL, it will be a "burden" for such firms if they do not fulfil customers' demands. The eighth barrier is lack of government support (BR8). The regulation from the government can either encourage or eventually discourage the process of innovation adoption since the government is the entity who sets the regulations for the industry [19]. If the government is reluctant or discourage to implement "the new concept" in the logistics practice, i.e., RL, it might be a major barrier. The ninth barrier is lack of training related to RL (BR9). Training of the company's employees is one of the main requirements in order to achieve success in any company; hence, a lack of competent or qualified employees is a barrier in RL implementation. A good RL training has to be performed for all people in the company, not only for top-level management but also for the operator level as well.

The tenth barrier is the fear of failure when adopting RL (BR10). This might lead to monetary losses or losing a competitive advantage. The eleventh barrier is the lack of environmental management system (EMS) certification (BR11), for instance, ISO 14001. A study by [20] examined the multifaceted connection green supply chain management (GSCM) practices and ISO 14001 certification in Japanese companies. The result showed that ISO 14001 could promote GSCM activities. It implies that a lack of knowledge of EMS could represent as a barrier in RL implementation [21]. The twelfth barrier is the lack of corporate social responsibility (CSR) (BR12). CSR recommends that a company is keen to go beyond trivial compliance and consider public consequences of its actions [22]. It suggests the company to do voluntary but strategic practices, for instance, eliminating wastes and informing consumers about the negative environmental effect of the products. Therefore, if the firm is unaware of this responsibility, it will hinder the firm to implement RL practice.

The last barrier is the lack of support from dealers, distributors, and retailers (BR13). In fact, there is no such policy that administers the products to be returned from consumers to the distributors (or dealers or retailers); or in other words, the consumers can return the products with any condition. This condition might harm the distributors because there is no pricing policy. If the distributors decline to do cooperation in RL, the company might meet problems in collecting the products that will be returned to the company.

2.2. Interpretative structural modelling

ISM is a well-established method to spot relations among variables. In the process of this method, a group of various directly- and indirectly- related variables will be structured into a comprehensive systematic model. The model can picture the structure of a posh problem by graphics also as words [23]. ISM has been broadly used in many areas, e.g., [5], [10], [13]–[16].

ISM begins with variables identification and then expands through group problem solving; ISM also can be utilized for individual problem-solving. A relevant secondary or subordinate relation is selected contextually. A structural self-interaction matrix (SSIM) is then generated after determining contextual relation on each element set. A reachability matrix (RM) is then generated by converting the SSIM. The next step is to check the transitivity and then to obtain a matrix model. After a matrix model is found, the partitioning of the elements and extraction of the structural model is then generated.

In this study, the identified thirteen barriers are examined by employing the ISM to identify the relationship among them. A thorough examination of the relationship can provide a clearer portrait of the situation than only considers one barrier alone. ISM can also be employed as a method to examine an individual interaction state of affairs since the idea of variable relationships as well as the overall structure are often extracted from the system [24].

2.3. MICMAC analysis

MICMAC analysis is usually used to study the dependence and driver powers among variables which have relations. The principle is according to the multiplication of matrices to classify the important variables that drive the system in various categories. The MICMAC analysis has been widely used as a supplement of the ISM method, see for example [4], [9], [12], [14], [15], and [15].

Depending on the dependence and driver powers, the variables are categorized into four groups or quadrants, i.e., autonomous, linkage, independent, and dependent quadrant. The autonomous quadrant belongs to the variables that have both weak driver power and dependence power. The linkage quadrant belongs to the variables which have both strong driver power and dependence power. The independent quadrant includes variables which have weak dependence power but strong driver power. Lastly, the dependent quadrant includes variables which have strong dependence power but weak driver power.

3. Results and Discussion

3.1. Interpretative Structural Modelling Hierarchy

After thirteen barriers were recognized, interviews with five employees of the object of the research were conducted. They are a manager of human resource and general affairs department, a manager of marketing department, three employees of marketing, logistics, and production departments. They were asked to answer several questions about the relationships among thirteen barriers which have been identified previously.

The contextual relationship among the barriers—which is then depicted in the SSIM—concerning which sets of barriers are inspected. The relationship between barrier *m* and barrier *n* is labelled by four symbols as follows:

- V: Barrier *m* will facilitate to reduce barrier *n*;
- A: Barrier *m* will be reduced by barrier *n*;
- X: Both barriers will facilitate to attain each other;
- O: Both barriers are unconnected.

If the symbol, let say, for barrier 6 (BR6) to barrier 10 (BR10) is V, meaning that when the commitment of the top management about RL practice is increasing, then the fear of failure to implement RL will decrease. Another example, when A is assigned for barrier 6 (BR6) to barrier 1 (BR1), meaning that elimination of the lack of commitment of the top management will help to reduce the lack of IS and technology. In the other words, when the top level of management has a clear vision regarding RL implementation, the company agrees to assign its funds to be invested in IS and technology for enabling RL implementation. When the relationship between barrier 3 (BR3) to barrier 4 (BR4) is assigned with X, meaning that when firm policies barrier and resistance to change to RL are eliminated, it might facilitate to attain each other. If O is assigned to the relationship between barrier 6 (BR6) and barrier 7 (BR7), meaning that the respondents thought these barriers do not have any relationship.

The next step is establishing an initial RM. The previous symbols O, X, A, and V are converted into a binary matrix:

- If (m,n) entry in SSIM is O, the entry (m,n) in RM is 0 and (n,m) is 0.
- If (m,n) entry in SSIM is X, the entry (m,n) in RM is 1 and (n,m) is 1.
- If (m,n) entry in SSIM is A, the entry (m,n) in RM is 0 and (n,m) is 1.
- If (m,n) entry in SSIM is V, the entry (m,n) in RM is 1 and (n,m) is 0.

Five initial RMs are recapped to establish total RM. A threshold of 3 (three) is set, meaning that if an entry in total RM is equal to or more than 3, the entry is assigned to number 1 (one) in final RM; vice versa. Table 1 shows the final RM. Calculating the dependence power and driver power for each barrier is the next step to be performed. The dependence power is the number of barriers (including itself) that facilitates to attain it; whereas the driver power of a certain barrier is the number of barriers that it might facilitate to attain. These dependence and driver powers are used in designing the MICMAC diagram.

Table 1. Final reachability matrix.

Barrier	1	2	3	4	5	6	7	8	9	10	11	12	13	Driver Power
BR1	1	0	0	0	0	0	0	0	0	1	0	0	0	2
BR2	0	1	0	0	0	0	0	0	0	1	0	0	0	2
BR3	1	1	1	1	1	0	1	1	1	0	0	1	1	10
BR4	0	0	1	1	1	0	0	0	1	0	0	0	1	5
BR5	1	0	1	1	1	1	0	0	0	1	0	0	1	7
BR6	0	1	1	1	0	1	1	0	0	0	0	0	0	5
BR7	1	1	1	0	0	0	1	0	0	0	0	0	0	4
BR8	1	1	1	1	0	1	1	1	1	1	1	0	0	10
BR9	1	1	0	1	0	1	1	0	1	1	0	0	1	8
BR10	0	1	0	1	1	1	0	0	0	1	0	0	0	5
BR11	1	1	1	0	0	1	0	0	0	0	1	1	0	6
BR12	1	1	1	1	0	1	1	0	1	1	0	1	0	9
BR13	1	1	1	1	0	0	0	0	0	1	0	0	1	6
Dependence power	9	10	9	9	4	7	6	2	5	8	2	3	5	

RM is subdivided into several levels. The reachability set of the particular barrier contains the barrier itself and others that might facilitate to attain. The predecessor set contains the barrier itself and others that might facilitate to attain them. Then, the intersection set can be established, it is the intersection between barriers in the reachability set and in the predecessor set. The "top level" barrier is investigated from the intersection set that consists of all barriers from reachability set. After identifying "top level" barrier, the barrier is casted off from other remaining barriers. The iteration is continued till all levels are discovered. The iteration process is depicted in Table 2.

Table 2. Final level partition.

Barrier	Reachability Sets	Predecessor Sets	Intersection Sets	Level Partitions
BR2	2, 10	2, 3, 6, 7, 8, 9, 10, 11, 12, 13	2, 10	I
BR1	1	1, 3, 5, 7, 8, 9, 11, 12, 13	1	II
BR7	3,7	3, 6, 7, 8, 9, 12	3, 7	III
BR13	4,13	4, 5, 9, 13	4,13	IV
BR6	6	5, 6, 8, 9, 11, 12	6	V
BR4	5,9	5, 8, 9, 12	5,9	VI
BR10	5	5, 8, 9, 12	5	VI
BR9	9	8, 9, 12	9	VI
BR5	5	5	5	VI
BR12	12	11, 12	12	VII
BR11	11	8, 11	11	VIII
BR3	8	8	8	VIII
BR8	8	8	8	VIII

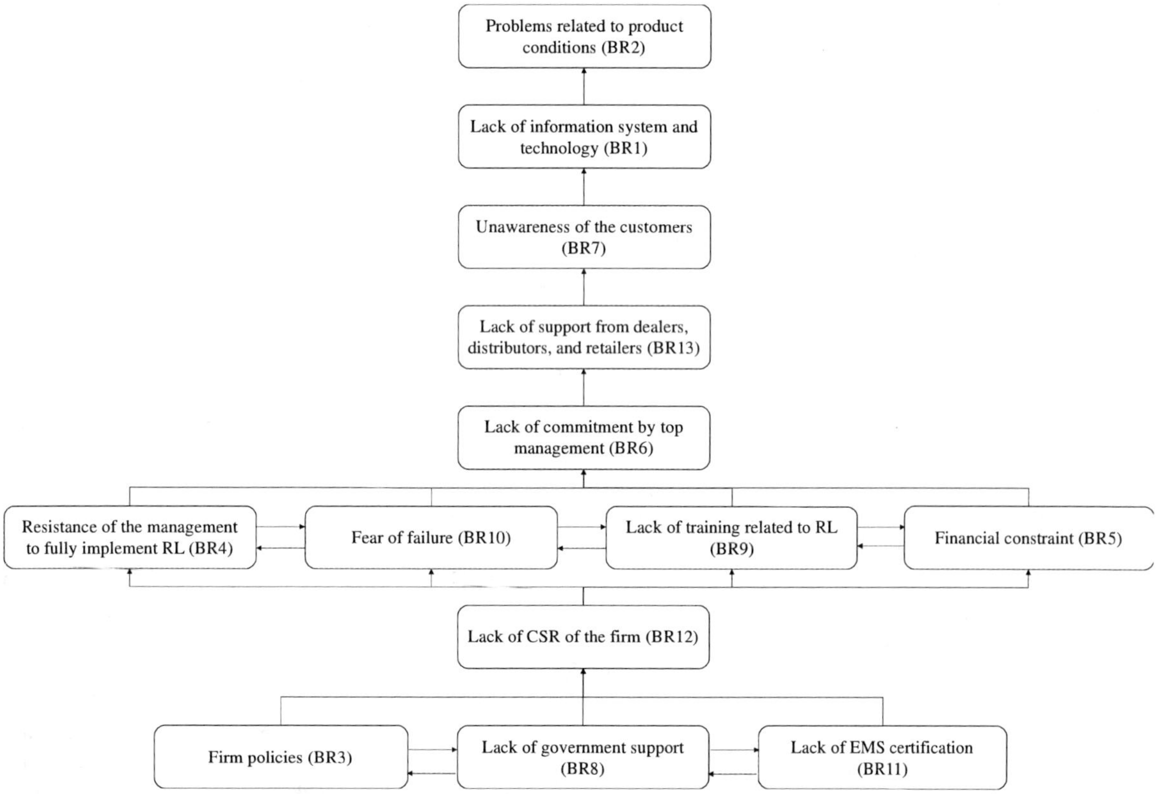

Figure 1. ISM-based model for the barriers of reverse logistics implementation: The final digraph.

According to the iteration, the structural model can be established. If the relation exists between barrier *m* and *n*, an arrow from *m* to *n* is assigned. The resulting model is called a digraph which is shown in Figure 1. Note that the BR3 (firm policies), BR8 (lack of government support), and BR11 (lack of EMS certification) are the most critical barriers because they are located at the bottom of the digraph. They are called the "root barriers". They lead to lack of CSR of the firm (BR12), and so on.

As a consequence of this result, the company must be able to eliminate these root barriers for successfully implementing the RL. To eliminate BR3, the firm has to provide a policy especially related to product stewardship. The company must ensure the process of handling safely, as well as the use of its products during their life cycle. Legislation and regulation are the mechanisms that are required for the proper governance of business organisations containing their environment as well. In particular, the environmental laws and regulations are also crucial because they are intended to initiate sustainable development and clean production in a society. The government of Indonesia could adapt the regulations applied in the developed countries: they implement the extended producer responsibility (EPR). In this regulation, the manufacturers are required to take financial and physical responsibility through a return program to reduce the volume of waste in landfills [24]. The implementation of EPR is not yet legally enforced, so that most product stewardship programs are limited to the voluntary participation of the manufacturers [25]. The EPR's objectives are to decrease waste disposal, raise recycling rates, resource conservation, and promote environmental design products [26]. In sum, support and guidance from authorities, i.e., the government, in the area of green SCM, especially in RL, are necessary for prevention of pollution and reduction of environmental load to achieve sustainable society.

The ISO 14001 standard is the key international standards for an organization which has an objective to apply an EMS as well as to get an environmental certification process. Since the launch of ISO 14001, the certification has acknowledged huge growth and widely implemented in numerous companies across the globe [28]. In sum, it is expected that the company needs to have EMS certification since this certification can be such a catalyst in implementing RL. Minimizing or fully eliminating those barriers are supposed to eliminating other barriers (which have positions "above" the rot barriers) as well.

3.2. MICMAC diagram

MICMAC diagram is employed to study the connection (or the relationship) between dependence and driver powers. The horizontal axis of the diagram shows the dependence power while the vertical axis shows the driver power. The barriers are categorized into four quadrants, i.e., autonomous (I), dependent (II), linkage (III), and independent (IV). For instance, BR4 has dependence power of 9 and driver power of 5, so that this barrier has the vertical axis of 5 and the horizontal axis of 9. The MICMAC diagram is shown in Figure 2.

Barriers belong to autonomous quadrant (I) have both weak driving and dependence powers. These are not really related to the system under consideration. Also, these might not have huge relations with others so that the barriers can be eliminated. According to Figure 2, there are five barriers belong to this cluster (BR5, BR6, BR7, BR11, and BR13), meaning that these barriers are not connected to the system. Barriers belong to dependent quadrant (II) have weak driver power but strong dependence power. They are BR1, BR2, BR4, and BR10. Barriers located in linkage quadrant (III) have both strong dependence power and driver power. Any action against the barriers might affect others and become feedback on the barriers themselves. There is only one barrier in this quadrant, i.e., BR3. Barriers located in independent quadrant (IV) have weak dependence power but strong driver power. They are BR8, BR9, and BR12. The improvements can be focused on these barriers.

As a consequence of this result, strategic issues such as government support, CSR of the firm, and training related to RL implementation must be considered in order to fruitfully applying the RL practice, since those barriers have high value of driving power. In other words, it is the priority of the company to manage those barriers because they can affect other barriers. The company is also recommended not to put a huge endeavour to remove the low driving power barriers; in this case, the company must provide adequate financial funding to support RL practice; the company must integrate RL into its policy; and the company must focus on building planning and strategies to increase employees' awareness about RL by let say organizing an event explaining that RL is important and explaining that implementing the program might benefit to the company.

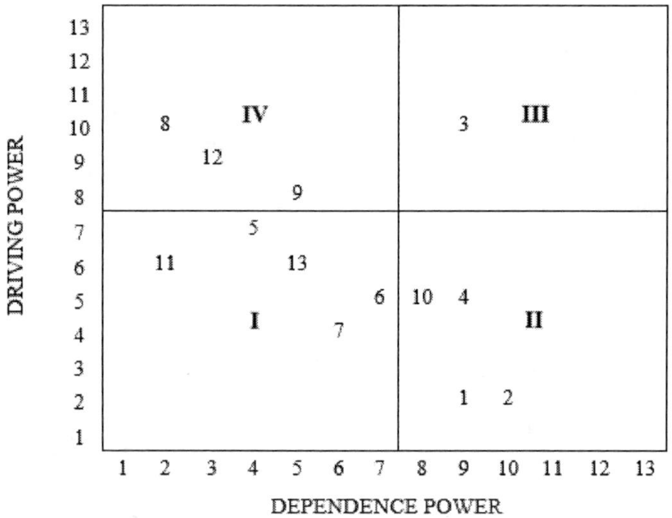

Figure 2. MICMAC diagram.

4. Conclusion

RL implementation barriers in a bottled drinking water company have been investigated by employing a literature review and interviews. Thirteen barriers influence RL implementation, i.e., lack of IS and technology (BR1), problems related to product condition (BR2), firm policies (BR3), resistance of the management to fully implement RL (BR), financial constraint (BR5), lack of commitment by top management (BR6), unawareness of the customers (BR7), lack of government support (BR8), lack of training related to RL (BR9), fear of failure (BR10), lack of EMS certification (BR11), lack of CSR of the firm (BR12), and lack of the support from dealers, distributors, and retailers (BR13). The barrier is expected to influence others; hence, the relationships among the barriers are needed to be understood to investigate root barrier(s) as well as the most influenced barrier(s). Interviews with five employees of the object of the research were conducted to investigate the relations among barriers.

Root barriers and their relationships are shown in the digraph (see Figure 1). It describes that root barriers RL implementation in the object of the research are firm policy, lack of government support, and lack of EMS certification. It means that the firm has to have a willingness to implement RL (fully) so that the policies related to RL implementation will be established. The government also has to fully support the implementation of RL in the industry by providing clear regulations and laws. Finally, the firm needs to have EMS certification, such as ISO 14001, since this certification can be such a catalyst in implementing RL. Minimizing or fully eliminating those barriers is supposed to eliminating other barriers (which have positions "above" the rot barriers) as well.

MICMAC diagram is used to analyse the connection between dependence and driver powers of the barriers. MICMAC diagram is partitioned into four quadrants. The autonomous quadrant is placed by five barriers, i.e., BR5, BR6, BR7, BR11, and BR13. The dependent quadrant consists of four barriers, i.e., BR1, BR2, BR4, and BR10. Only BR3 is located in the linkage quadrant. The independent quadrant consists of three barriers, i.e., BR8, BR9, and BR12. The company is recommended to minimize (or reduce) barriers located in independent quadrant because the barriers have strong driver powers which might affect others.

References

[1] Rogers DS and Tibben-Lemke RS. 1998. *Going Backwards: Reverse Logistics Trends and Practices.* Reverse Logistics Executive Council.
[2] Frota Neto JQ, Bloemhof-Ruwaard JM, van Nunen JAEE, van Heck E. Designing and evaluating

sustainable logistics networks. *Int. J. Prod. Econ.*, 2008 Feb;**111**(2):195–208. http://dx.doi.org/10.1016/j.ijpe.2006.10.014

[3] Fernández E, Kalcsics J, Nickel S, Ríos-Mercado RZ. A novel maximum dispersion territory design model arising in the implementation of the WEEE-directive. *J Oper Res Soc.* 2010 Mar;**61**(3):503–14. http://dx.doi.org/10.1057/jors.2009.70

[4] Gunasekaran A, Spalanzani A. Sustainability of manufacturing and services: Investigations for research and applications. *Int. J. Prod. Econ.*, 2012 Nov;**140**(1):35–47. http://dx.doi.org/10.1016/j.ijpe.2011.05.011

[5] Ravi V, Shankar R. Analysis of interactions among the barriers of reverse logistics. *Technol Forecast Soc Change.* 2005 Oct;**72**(8):1011–29. http://dx.doi.org/10.1016/j.techfore.2004.07.002

[6] Pishvaee MS, Kianfar K, Karimi B. Reverse logistics network design using simulated annealing. *Int. J. Adv. Manuf. Technol.*, 2009 Jul 17;**47**(1-4):269–81. http://dx.doi.org/10.1007/s00170-009-2194-5

[7] Zareinejad M, Javanmard H, Arak I. Evaluation and selection of a third-party reverse logistics provider using ANP and IFG-MCDM methodology. *Life Sci.*, 2013;**10**(6s):350-5.

[8] Miao Z, Cai S, Xu D. Exploring the antecedents of logistics social responsibility: A focus on Chinese firms. *Int. J. Prod. Econ.*, 2012 Nov;**140**(1):18–27. http://dx.doi.org/10.1016/j.ijpe.2011.05.030

[9] Zhu Q, Geng Y. Drivers and barriers of extended supply chain practices for energy saving and emission reduction among Chinese manufacturers. *J. Clean. Prod.*, 2013 Feb;**40**:6–12. http://dx.doi.org/10.1016/j.jclepro.2010.09.017

[10] Sari DP, Ulkhaq MM, Rinawati DI and Rasyida DR. Barriers of reverse logistics implementation: a case study in a car battery industry in Indonesia. *Int. J. Supply Chain Manag.,* 2018;**7**:53-67.

[11] Sarkis J, Zhu Q, Lai K. An organizational theoretic review of green supply chain management literature *Int. J. Prod. Econ.*, 2011 Mar;**130**(1):1–15. http://dx.doi.org/10.1016/j.ijpe.2010.11.010

[12] Zhang T, Chu J, Wang X, Liu X, Cui P. Development pattern and enhancing system of automotive components remanufacturing industry in China. *Resour Conserv Recycl.* Apr;**55**(6):613–22. http://dx.doi.org/10.1016/j.resconrec.2010.09.015

[13] Luthra S, Kumar V, Kumar S, Haleem A. Barriers to implement green supply chain management in automobile industry using interpretive structural modeling technique: An Indian perspective. *J. Ind. Eng. Manag.*, 2011 Jul 14;**4**(2). http://dx.doi.org/10.3926/jiem.2011.v4n2.p231-257

[14] Mudgal RK, Shankar R, Talib P, Raj T. Modelling the barriers of green supply chain practices: an Indian perspective. *Int. J. Logist. Syst. Manag.*, 2010;**7**(1):81. http://dx.doi.org/10.1504/ijlsm.2010.033891

[15] Jayant A, Azhar M. Analysis of the Barriers for Implementing Green Supply Chain Management (GSCM) Practices: An Interpretive Structural Modeling (ISM) Approach. *Procedia Eng.*, 2014;**97**:2157–66. http://dx.doi.org/10.1016/j.proeng.2014.12.459

[16] Mathiyazhagan K, Govindan K, NoorulHaq A, Geng Y. An ISM approach for the barrier analysis in implementing green supply chain management *J. Clean. Prod.*, 2013 May;**47**:283–97. http://dx.doi.org/10.1016/j.jclepro.2012.10.042

[17] Daugherty PJ, Myers MB, Richey RG. INFORMATION SUPPORT FOR REVERSE LOGISTICS: THE INFLUENCE OF RELATIONSHIP COMMITMENT. *J. Bus. Logist.*, 2002 Mar;**23**(1):85–106. http://dx.doi.org/10.1002/j.2158-1592.2002.tb00017.x

[18] Rogers DS, Tibben-Lembke R. AN EXAMINATION OF REVERSE LOGISTICS PRACTICES. *J. Bus. Logist.*, 2001 Sep;**22**(2):129–48. http://dx.doi.org/10.1002/j.2158-1592.2001.tb00007.x

[19] Scupola A. The Adoption of Internet Commerce by SMEs in the South of Italy: An Environmental, Technological and Organizational Perspective. *J. Glob. Inf. Technol. Manag.*,

2003 Jan;**6**(1):52–71. http://dx.doi.org/10.1080/1097198x.2003.10856343
[20] Arimura TH, Darnall N, Katayama H. Is ISO 14001 a gateway to more advanced voluntary action? The case of green supply chain management. *J Environ Econ Manage*. 2011 Mar;**61**(2):170–82. http://dx.doi.org/10.1016/j.jeem.2010.11.003
[21] Stevels A. Green supply chain management much more than questionnaires and ISO 14.001. *Conference Record 2002 IEEE International Symposium on Electronics and the Environment (Cat No02CH37273)*. 2002:96-100. http://dx.doi.org/10.1109/isee.2002.1003247
[22] Sharma S. MANAGERIAL INTERPRETATIONS AND ORGANIZATIONAL CONTEXT AS PREDICTORS OF CORPORATE CHOICE OF ENVIRONMENTAL STRATEGY. *Acad Manage J*. 2000 Aug 1;**43**(4):681–97. http://dx.doi.org/10.2307/1556361
[23] Raj T, Attri R. Identification and modelling of barriers in the implementation of TQM. *Int. J. Product. Qual. Manag.*, 2011;**8**(2):153. http://dx.doi.org/10.1504/ijpqm.2011.041844
[24] Govindan K, Palaniappan M, Zhu Q, Kannan D. Analysis of third party reverse logistics provider using interpretive structural modeling. *Int. J. Prod. Econ.*, 2012 Nov;**140**(1):204–11. http://dx.doi.org/10.1016/j.ijpe.2012.01.043
[25] Lee S, Na S. E-Waste Recycling Systems and Sound Circulative Economies in East Asia: A Comparative Analysis of Systems in Japan, South Korea, China and Taiwan. *Sustainability*. 2010 Jun 4;**2**(6):1632–44. http://dx.doi.org/10.3390/su2061632
[26] Agamuthu P, Victor D. Policy trends of extended producer responsibility in Malaysia. *Waste Manag Res*. 2011 Jul 19;**29**(9):945–53. http://dx.doi.org/10.1177/0734242x11413332
[27] Kaffine D and O'Reilly P. 2015. *A Survey of the Recent EPR Economic Literature*.
[28] Setyorini NA, Ulkhaq MM, Rasyida DR, Setiowati PR, Trianto R. Assessing campus sustainability: An ISO 14001 approach. *Int. J. Adv. Agric. Environ. Eng.,* 2016;**3**:245-48.

Data analytics in asset valuation under uncertainty: A case study of unexplored oilfields

F R Pratikto[1,2,*], S W Indratno[2], K D Suryadi[2] and D Santoso[2]

[1]Parahyangan Catholic University, Jl. Ciumbuleuit 94, Bandung, Indonesia
[2]Bandung Institute of Technology, Jl. Ganesha 10, Bandung, Indonesia

Email: *frianp@unpar.ac.id

Abstract. Valuing unexplored oilfields is challenging since it deals with multiple sources of uncertainty and covers a lengthy period. Previous research suggests that the major source of uncertainty in the valuation of unexplored oilfields is reservoir condition. It is usually represented by a bunch of parameters based on which reserve volume and production rates can be estimated. The values of these parameters are revealed only after exploration well is drilled; we refer them as post-discovery parameters. Prior to exploration drilling, data on some reservoir characteristics, called the pre-discovery parameters, are available. This research aims to develop a model to estimate the probability distribution of post-discovery parameters based on pre-discovery data. The model is data-driven, built using the database of proven reservoirs. We use the Bayesian network to develop the model and apply the k-fold cross validation to test the results. By adding a window parameter to the target variable, the model provides information regarding the trade-off between accuracy and confidence. In general, compared to the initial model that uses a priori clustering based on the reservoir's lithology and depth, our Bayesian network model produces lower variances.

1. Introduction

This work is part of our research to develop a valuation model of unexplored oilfields. The valuation model is to serve situation where an investor is participating in an auction to place a bid on the right to operate an unexplored oilfield. In doing so, the investor needs information about the oilfield value. Valuing unexplored oilfields is very complex since is deals with multiple sources of uncertainty and covers a lengthy period. The uncertainty may be of market-related like oil prices, exchange rates, and investment climate, technical-related like exploration outcome and reservoir condition, and even social and geopolitical factors like war, political upheaval, etc.

Past research on oilfield valuation usually focus on the market and technical-related factors. Oil prices are the most common source of uncertainty and usually represented as a geometric Brownian motion (GBM) as in Pickles and Smith [1], Sunnevag [2], and Dias [3], or mean-reverting stochastic process as in Cortazar and Schwartz [4], Park et al. [5], and Haque et al. [6]. Exchange rates and investment climate are incorporated in the valuation model by Fan and Zhu [7]. Previous research by One the major sources of uncertainty in the valuation of unexplored oilfields is reservoir condition. Reservoir condition affects the oilfield value by determining reserve volume and production rates. Some research in the past represents reservoir condition by directly making assumption about reserve and/or production rates, whether it follows a certain probability distribution as in Zettl [8], Dias [3] and Park et al. [5], or stochastic process as in Smith and McCardle [9]. The only research that represents reservoir condition by its parameters and accordingly estimates the reserve volume and production rates is Hultzsch et al. [10].

Hultzsch et al. [10] uses a bunch of parameters based on which reserve volume and production rates can be estimated. The values of these parameters are revealed only after exploration well is drilled; we refer them as post-discovery parameters. In Hultzsch et al. [10], the probability distributions of post-discovery parameters are estimated from data of proven reservoirs that are considered geologically similar, where similarity is inferred by lithology and depth. In this sense, lithology and depth are the pre-discovery parameters. Pre-discovery parameters are those whose values are known prior to exploration drilling. Based on these two parameters the proven reservoirs are a priori clustered, and the probability distribution in each cluster is estimated. Hultzsch et al. [10] assumes that all post-discovery parameters follow lognormal distribution and allows associations between them which are represented using simple correlation. We adopt the same approach in our initial model but modify the way associations between post-discovery parameters are represented. Instead of using simple correlation, we use the Gaussian copula to represent the associations between parameters.

This research aims to develop a data-driven model to estimate the probability distributions of post-discovery parameters based on pre-discovery data. This empirical approach is chosen because there has not been a solid analytical model representing the relationship between those two groups of parameters. In doing so, the proven reservoirs are posteriori clustered with respect to each post-discovery parameter. The cluster numbers are ordered such that greater number corresponds to reservoirs with greater parameter value. We use the Bayesian network to develop the model to predict the cluster of a reservoir of interest and apply the k-fold cross validation to test the results. The Bayesian network is chosen since we are interested in learning the causal relationship between the pre and post discovery parameters. We use the Tertiary Oil Recovery Information System (TORIS) database which was developed by the US government and contains data of proven reservoirs that have undergone secondary recovery.

2. Methods

Our valuation model is a multistage decision model with stochastic state variables, represented as a dynamic programming problem (see Figure 1).

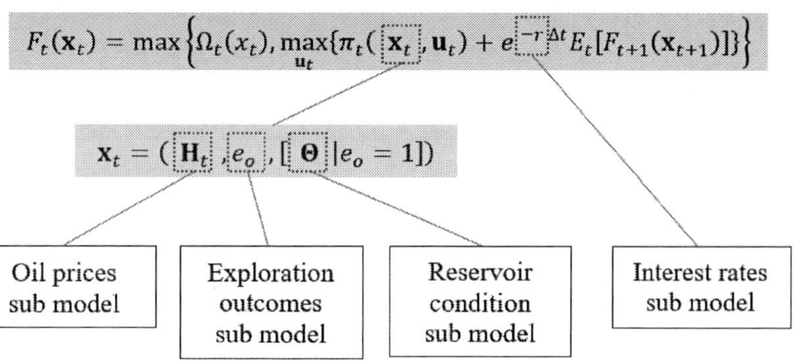

Figure 1. Relationship between reservoir condition and other sub models in our valuation model.

$F_t(\mathbf{x}_t)$ is the oilfield value at time t represented as a function of vector of stochastic state variables, $\mathbf{x}_t$ and optimal choice of vector of decision variables, $\mathbf{u}_t$. At any time t, the investor chooses between abandoning the field which results in the payoff of $\Omega_t(x_t)$, and continuing the operation. The expression of the payoff for continuing the operation is a typical approximate dynamic programming representation which consists of the immediate profit, $\pi_t(\mathbf{x}_t, \mathbf{u}_t)$, and the expected oilfield value in the next period, continuously discounted to time t, $e^{-r\Delta t}\mathrm{E}_t[F_{t+1}(\mathbf{x}_{t+1})]$. The stochastic state variables are oil prices ($\mathbf{H}$), exporation outcome ($\mathbf{e}_o$), and the vector of reservoir parameters ($\mathbf{\Theta}$) given that the exploration is successful. This paper focus on the sub model of the reservoir condition. In our valuation model, reserve volume and production rates are estimated using the compressible-liquid tank model. Table 1 lists the parameters needed as inputs to run the tank model, where the data of the first eight are available in the TORIS database and hence the Bayesian network model of those can be developed.

Table 1. Inputs to the compressible-liquid tank model.

No	Parameters	Remarks
1	Net pay	Estimated empirically
2	Porosity	Estimated empirically
3	Initial oil saturation	Estimated empirically
4	Formation volume factor	Estimated empirically
5	Oil permeability	Estimated empirically
6	Oil viscosity	Estimated empirically
7	Initial pressure	Estimated empirically
8	Bottomhole pressure	Estimated empirically
9	Shape factor	Estimated using expert judgment
10	Skin factor	Estimated using expert judgment
11	Water compressibility	Estimated theoretically
12	Oil compressibility	Estimated theoretically
13	Formation compressibility	Estimated theoretically
14	Well radius	Assumed

The data of the last six parameters are not available and hence the data-driven model of those cannot be built. Instead, we use their theoretical values (for water compressibility, oil compressibility, and formation compressibility), common practice value (for well radius), and expert judgment based (for shape factor and skin factor). The pre-discovery parameters are chosen based on data availability in the TORIS. The list of the parameters and their corresponding measurement scale are presented in Table 2.

Table 2. Pre-discovery parameters and their corresponding values.

No	Parameters	Scales
1	Lithology	Nominal, 4-point scale
2	Depth	Ratio
3	Depositional system and its degree of confidence	Nominal, 165-point scale
4	Digenetic overprint and its degree of confidence	Nominal, 30-point scale
5	Structural compartmentalization and its degree of confidence	Nominal, 36-point scale
6	Predominant element of heterogeneity	Nominal, 3-point scale
7	Trap type	Nominal, 3-point scale

All post-discovery parameters are stochastic in nature. The idea of predicting the values of such parameters does not make sense. Instead, our model aims to predict the cluster of a reservoir with respect to a post-discovery parameter, based on which the probability distribution of that reservoir's parameter will be estimated. We cluster the reservoirs based on their magnitude of the parameter of interest using the k-means clustering algorithm [11]. The clusters are labelled such that greater-numbered cluster corresponds to group of reservoirs with greater parameter value. The Bayesian network model aims to predict the cluster in which a reservoir belongs with respect to each post-discovery parameter. By doing so, we wish to come up with estimated probability distribution for each parameter with lower variance. Let $\boldsymbol{\tau} = (\tau_1, \tau_2, \tau_3, \tau_4, \tau_5, \tau_6, \tau_7)$ be the vector of pre-discovery parameters where each element represents parameter in to the order of that in Table 2, and let $\boldsymbol{\Theta} = (\Theta_1, \Theta_2, \Theta_3, \Theta_4, \Theta_5, \Theta_6, \Theta_7, \Theta_8)$ be the vector of post-discovery parameters where Θ_q corresponds to the q-th parameter in Table 1.

Bayesian network (denoted by B) is a representation of joint probability distributions of n interrelated random variables (denoted by $\mathbf{X} = (X_1, \dots X_n)$) in a directed acyclic graph or DAG (denoted by G) with parameters representing the conditional probability of each variable in G (denoted by $\boldsymbol{\Pi}$). If D_i is the domain of X_i then the joint domain of all random variables is $\mathrm{D} = \prod_{i=1}^{n} D_i$. The DAG G consists of nodes which represent the random variables (denoted by $V = \{X_1, \dots X_n\}$) and arcs which represent relationships between variables (denoted by E). Relationships between any two variables can be determined using the Pearson chi-square or the likelihood ratio test.

We use the Markov blanket algorithm to develop our classification model. Markov blanket of a node representing X_i is the set of nodes comprising its parent nodes (denoted by R_{X_i}), its children nodes (denoted by C_{X_i}), and the nodes sharing a child with this node (denoted by M_{X_i}).
Conditional probability $\mathbf{\Pi}$ in Bayesian network B is represented by $\{\pi_{ijk}\}_{i\in 1\ldots n, j\in D_{P_{X_i}}, k\in D_i}$ where $\pi_{ijk} = P_B(X_i = x_{ik}|R_{X_i} = w_{ij})$, in which $P_B(\cdot)$ denotes the probability in Bayesian network B, x_{ik} is the k-th value of X_i, and w_{ij} is the j-th configuration of R_{X_i}. Then, the joint probability of all elements of $\mathbf{X}$ in B can be calculated as follows

$$P_B(X_1, \ldots X_n) = \prod_{i=1}^{n} P_B(X_i|P_{X_i}) \tag{1}$$

This algorithm aims to obtaining posterior joint probability of target variable by learning from the data. This requires the set of variables V and the data of their corresponding values as training dataset. In this research, $V = \{\tau_1, \tau_2, \tau_3, \tau_4, \tau_5, \tau_6, \tau_7, \Theta_i\}$. There are two learning processes in this method, structural learning to determine nodes and arcs and parameter learning to estimate conditional probability in each node. The objective of the Markov blanket algorithm is to obtain a Bayesian network that can predict the target variable and fit with the dataset. In this research, the fitness is evaluated using the likelihood-equivalence Bayesian Dirichlet (BDe) scoring function [12] which uses the following notation:

$P(B)$	:	the prior probability of network B
n	:	the number of variables in $\mathbf{X}$
r_i	:	the number of possible states of X_i
x_{ik}	:	the k-th value of X_i
$q_i = \prod_{X_j\in R_{X_i}} r_j$	:	the number of possible configurations of R_{X_i}
w_{ij}	:	the j-th configuration of R_{X_i}, $1 \le j \le q_i$
N_{ijk}	:	the number of records in dataset T where X_i takes the value of x_{ik} variables in R_{X_i} takes configuration w_{ij}
$N_{ij} = \sum_{k=1}^{r_i} N_{ijk}$	:	the number of records in dataset T where variables in R_{X_i} takes configuration w_{ij}

For Bayesian network B using dataset T, the BDe score can be calculated as follows:

$$BDe(B,T) = log(P(B)) + \sum_{i=1}^{n}\sum_{j=1}^{q_i}\left(log\left(\frac{\Gamma(N'_{ij})}{\Gamma(N_{ij}+N'_{ij})}\right) + \sum_{k=1}^{r_i} log\left(\frac{\Gamma(N_{ijk}+N'_{ijk})}{\Gamma(N'_{ijk})}\right)\right) \tag{2}$$

In the above equation, $N'_{ijk} = N' \times P(X_i = x_{ik}, R_{X_i} = w_{ij}|G)$ and $N'_{ij} = N' \times P(R_{X_i} = w_{ij}|G)$ where N' is the equivalent sample size indicating the confidence of the prior probability, and Γ is the gamma function. We use the following algorithm to train our Bayesian network model:

Step 1 Test the significance of each arc between any two nodes in V using the Pearson chi-square or the likelihood ratio. For these significant arcs, find the smallest set S such that the relationship between the nodes connected by those arcs conditioned on S, are significant. If S is empty, remove the arc.

Step 2 Determine the direction of each arc.

Step 3 Identify the Markov blanket of target variable Θ_i (denoted by MB_{Θ_i}).

Step 4 Determine the value of $\{\pi_{ijk}\}_{i\in 1\ldots n, j\in D_{P_{X_i}}, k\in D_i}$ that maximize $\mathrm{BDe}(\mathrm{MB}_{\Theta_i}, T)$ using Equation 2.

Once we get the values of $\pi_{ijk} = P_B(X_i = x_{ik}|R_{X_i} = w_{ij})$, the conditional probability for each target variable Θ_i can be determined.

Since the number of records is limited, we use the k-fold cross validation to test our model. Figure 2 depicts the mechanism of the validation process. In this validation process, the data are divided equally into k folds. Subsequently, for $i = 1, \ldots, k$ we take the i-th group as the validation dataset and and the remaining data as the training dataset. The training dataset are then used to train the classification model, and the predicted values of target variable are compared to those of actual to measure the fold's performance. The fold's performance is measured by the proportion of record whose cluster are correctly predicted. The model performance is measured as the average performance of all folds.

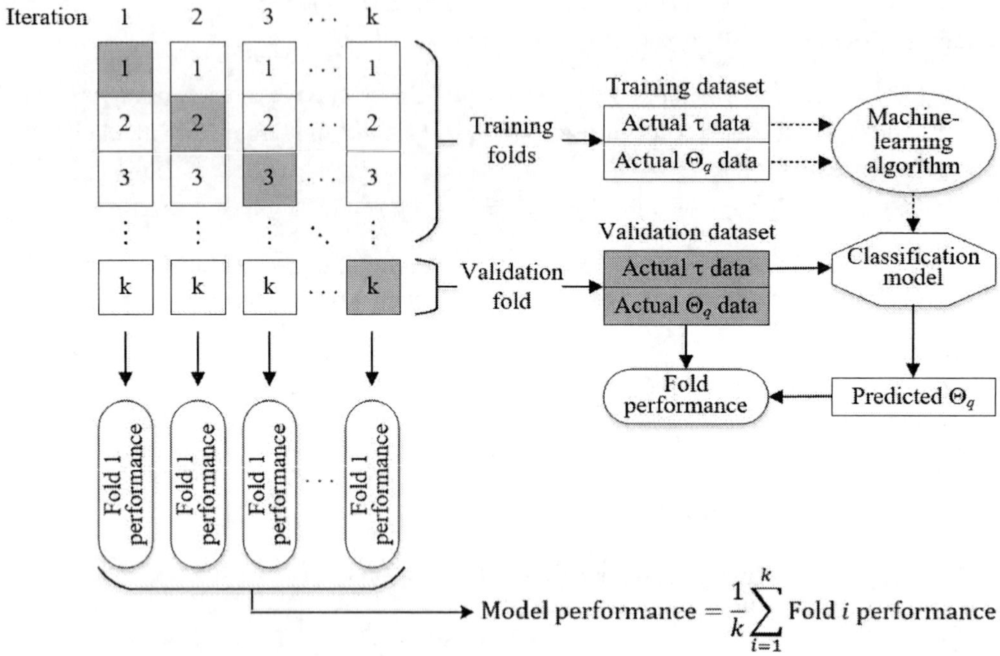

Figure 2. The k-fold cross validation process.

When the classification model assigns a reservoir to cluster i with respect to parameter Θ_q, a nonnegative integer $w \leq 3$ is added such that when all records are assign to cluster $[\max(1, i-w) \cup \ldots \cup \min(i+w, k)]$ the proportion of records whose clusters are correctly predicted is at least 95%, where k represents the number of clusters. For example, if $w = 1$ and $C_I = \{1, 2, 3, 4, 5\}$ denotes the initial cluster resulted from the classification model, then the final cluster would be $C_F = \{1_F, 2_F, 3_F, 4_F, 5_F\} = \{1 \cup 2, 1 \cup 2 \cup 3, 2 \cup 3 \cup 4, 3 \cup 4 \cup 5, 4 \cup 5\}$. The performance of all folds and models will be evaluated by considering the final cluster after the window parameter is added.

3. Result and Discussion

The TORIS database has about 2,500 records with about 1,400 records are available for public, out of which 498 records are complete and used to train our classification model. There are two k's in the methods that we use, the first one refers to the k-means clustering algorithm, and the second refers to the k-fold cross validation. We pick $k = 5$ for the k-means, while for the cross-validation we pick $k = 10$ since it is considered the best practice. The distribution of reservoirs resulted from the k-means clustering is presented in Table 3.

The target variable of the post-discovery parameter in our Bayesian network model is the cluster number. Figure 3 depicts the structure of our Bayesian network model for each target variable resulted from the training using all data in the training dataset.

Table 3. Distributions of reservoirs in each cluster.

Post-discovery parameters		**Number of reservoirs in cluster**				
		1	**2**	**3**	**4**	**5**
Net pay ratio	(Θ_1)	77	117	115	153	36
Porosity	(Θ_2)	126	161	99	110	2
Initial oil saturation	(Θ_3)	7	51	162	229	49
Formation volume factor	(Θ_4)	303	131	53	9	2
Permeability	(Θ_5)	340	100	42	14	2
Viscosity	(Θ_6)	480	11	2	2	3
Initial pressure	(Θ_7)	182	167	106	36	7
Bottomhole pressure ratio	(Θ_8)	51	71	98	130	148

Table 4 shows the result of the k-fold cross validation and the model's performance in term of the proportion of records that are correctly classified.

Table 4. The proportion of records that are correctly classified by the model.

Folds	**Post-discovery parameters**							
	Θ_1	Θ_2	Θ_3	Θ_4	Θ_5	Θ_6	Θ_7	Θ_8
	$w = 3$	$w = 2$	$w = 2$	$w = 2$	$w = 2$	$w = 1$	$w = 1$	$w = 3$
1	0.980	0.900	0.940	0.960	0.980	0.980	0.900	0.980
2	1.000	1.000	0.960	0.960	0.900	0.980	0.960	0.940
3	0.980	1.000	0.900	1.000	0.960	0.980	0.980	1.000
4	0.980	0.980	0.860	0.940	1.000	1.000	0.900	0.860
5	0.980	0.980	0.900	0.960	0.960	0.980	0.920	0.820
6	0.940	0.980	0.840	0.960	1.000	0.980	0.900	0.920
7	0.940	0.960	0.980	0.980	1.000	0.960	0.960	0.900
8	0.960	0.960	0.900	0.960	0.980	0.980	0.980	0.900
9	0.939	1.000	0.816	0.980	0.980	0.980	0.939	1.000
10	0.959	1.000	0.959	0.959	1.000	0.959	0.959	1.000
Folds average	0.966	0.976	0.906	0.966	0.976	0.978	0.940	0.932
All data	0.996	0.994	0.998	1.000	0.990	0.996	1.000	0.988

The window parameters are set such that the proportion of correct classification of both the model and the cross-validation is at least 0.90. It is interesting to see if the classification model produces lower variance compared to our initial model which uses a priori clustering based on lithology and depth. Table 5 shows the comparison between the initial and the Bayesian network model in term of the total weighted variance implied by the reservoir classification suggested by the model.

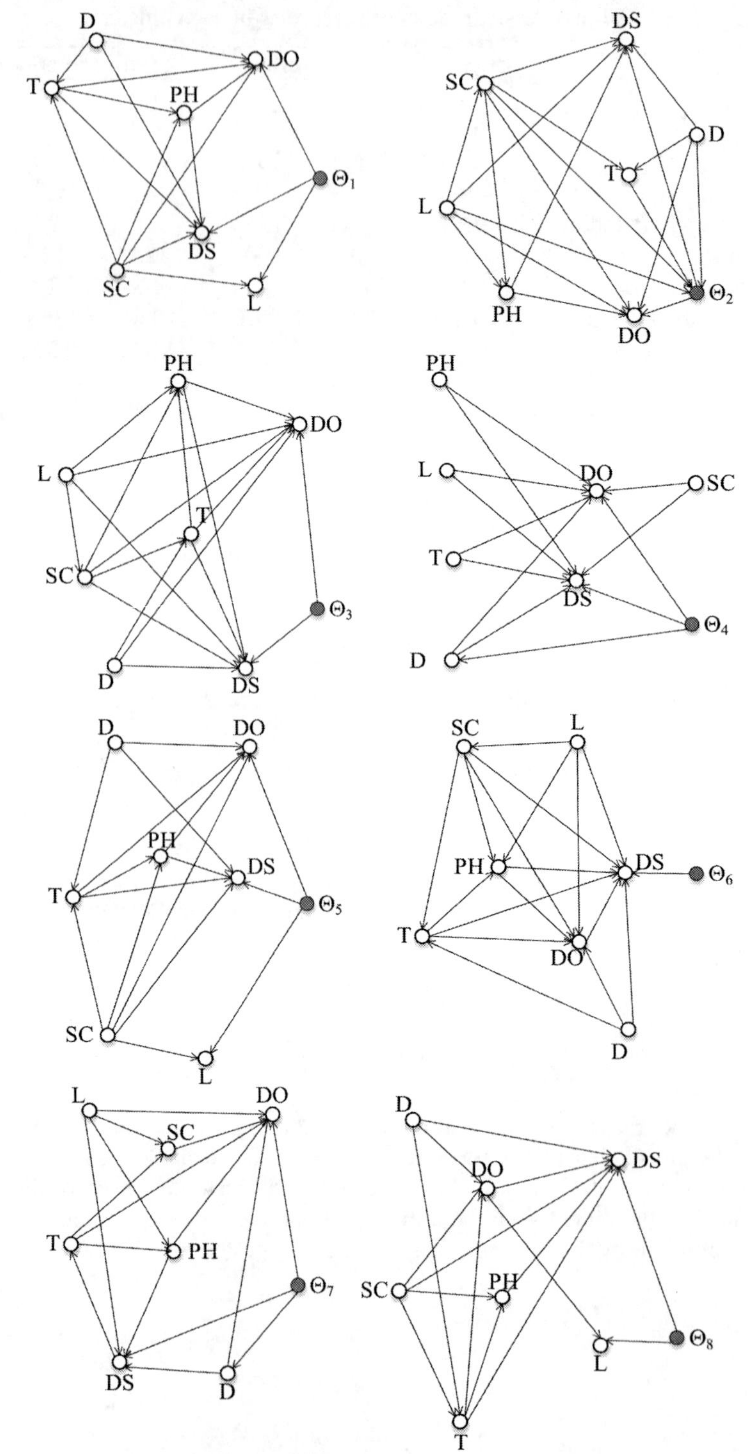

Figure 3. The structure of our Bayesian network models.

Table 5. Total weighted variance of the initial and the Bayesian network model.

Models	Post-discovery parameters							
	Θ_1 $w=3$	Θ_2 $w=2$	Θ_3 $w=2$	Θ_4 $w=2$	Θ_5 $w=2$	Θ_6 $w=1$	Θ_7 $w=1$	Θ_8 $w=3$
Initial	0.0637	0.0051	0.0091	0.0226	343,516	124,436	383,338	0.0717
Bayesian network	0.0456	0.0033	0.0053	0.0212	205,220	2,990	1,088,988	0.0443
Variance change	-28.4%	-35.3%	-41.8%	-6.2%	-40.3%	-97.6%	184.1%	-38.2%

It can be observed from Table 5 that the Bayesian network model produces classification with lower variance except for parameter Θ_7, initial pressure. The result of the k-fold cross validation provides information about the confidence of our Bayesian network model, while the window parameter gives the accuracy of the post-discovery parameter estimates. Greater value of w refers to greater overlap between two consecutive final cluster, which corresponds to greater variance. Our initial model which is based on Hultzsch et al. [10] does not provide this information. In this sense, our Bayesian network model is considered better.

4. Conclusion

We develop a Bayesian network classification model to determine the cluster of a certain reservoir with respect to a post-discovery parameter, based on which the probability distribution of the parameter will be estimated and used in the stochastic valuation model. The classification model aims to specify the cluster based on the value of pre-discovery parameters.

Our Bayesian network classification model is better than our initial one which uses a priori clustering based on lithology and depth. It is not only because the Bayesian network model produces lower variance on almost all parameters, but more than that, it provides information regarding accuracy and confidence. This information would be useful when the model is implemented and the data evolves overtime, and the models are retrained, accordingly.

Acknowledgments

FRP thanks Parahyangan Catholic University for their support in conducting this research.

References

[1] Pickles E, Smith JL. Petroleum Property Valuation: A Binomial Lattice Implementation of Option Pricing Theory. *Energy J.*, 1993 Apr 1;**14**(2). doi.org/10.5547/issn0195-6574-ej-vol14-no2-1

[2] Sunnevåg K. An option pricing approach to exploration licensing strategy. *Resour. Policy*. 1998 Mar;**24**(1):25–38. doi.org/10.1016/s0301-4207(98)00005-1

[3] Dias MAG. Valuation of exploration and production assets: an overview of real options models. *J. Pet. Sci. Eng.*, 2004 Oct;**44**(1-2):93–114. doi.org/10.1016/j.petrol.2004.02.008

[4] Cortazar G, Schwartz ES. Implementing a Real Option Model for Valuing an Undeveloped Oil Field. *Int Trans Oper Res*. 1997 Mar;**4**(2):125–37. doi.org/10.1111/j.1475-3995.1997.tb00069.x

[5] Park C, Kang JM, Min B. Compound Real Options Incorporated With a Stochastic Approach for Evaluating an Uncertainty in Petroleum Exploration. *Energ Sources Part B*. 2013 Jul;**8**(3):252–62. doi.org/10.1080/15567240903117617

[6] Haque MA, Topal E, Lilford E. Evaluation of a mining project under the joint effect of commodity price and exchange rate uncertainties using real options valuation. *Eng Econ*. 2016 Jul 28;**62**(3):231–53. doi.org/10.1080/0013791x.2016.1217366

[7] Fan Y, Zhu L. A real options based model and its application to China's overseas oil. *Energy Economics*. 2010:**32**:627-637. doi.org/10.1016/j.eneco.2009.08.021

[8] Zettl M. Valuing exploration and production projects by means of option pricing theory. *Int J Prod Econ.* 2002 Jul;**78**(1):109–16. doi.org/10.1016/s0925-5273(01)00215-8
[9] Smith JE, McCardle KF. Valuing Oil Properties: Integrating Option Pricing and Decision Analysis Approaches. *Oper Res*. 1998 Apr;**46**(2):198–217. doi.org/10.1287/opre.46.2.198
[10] Hultzsch P, Lake LW, Gilbert RB. Decision And Risk Analysis Through The Life Of The Field. *Hydrocarbon Economics and Evaluation Symposium*. 2007. doi.org/10.2118/107704-ms
[11] Strauss DJ, Hartigan JA. Clustering Algorithms. *Biometrics*. 1975 Sep;**31**(3):793. doi.org/10.2307/2529577
[12] Heckerman D, Geiger D, Chickering DM. Learning Bayesian Networks: The Combination of Knowledge and Statistical Data. *Uncertainty Proceedings 1994*. 1994;293–301. doi.org/10.1016/b978-1-55860-332-5.50042-0

IOP Conf. Series: Materials Science and Engineering 1072 (2021) 012065 doi:10.1088/1757-899X/1072/1/012065

Work system analysis on puncher bar repair process in the production division of Akademi Teknik Soroako (Soroako Technical Academy)

Burhanuddin[1,2] and R S Dewi[1,3]

[1]Department of Industrial Engineering, Sepuluh Nopember Institute of Technology, Surabaya, Indonesia

Email: [2]burhanats007@gmail.com, [3]ratna.sari.dewi80@gmail.com

Abstract. One of the regular services in the Production Division of Akademi Teknik Soroako (Soroako Technical Academy) is repairing puncher bars of PT Vale Indonesia, One of the stages of his work is to straighten each punch bar. For straightening up each puncher bar, the worker needs to hammer each puncher bar with a 5 kg iron hammer manually. This operation requires enormous physical energy and high cognitive effort for controlling body movement, causing fatigue and cumulative musculoskeletal trauma. The research focuses on assessing the risk of musculoskeletal disorders on the straightening employing Ovako Work Posture Analysis System (OWAS) and Rapid Upper limb Assessment (RULA). The result of the study shows that the work postures while straightening the puncher bar has OWAS score = 2 which means it needs corrective action in the near term, and RULA score = 7 which means the work posture improvement should be made right now. Based on the OWAS and RULA score, several improvement programs and techniques are suggested in this study.

1. Introduction

Akademi Teknik Soroako, which was initially named the Inco Sumitomo Memorial Technical Training Center (ISTC) was established as a form of commitment from PT. INCO, now called PT. VALE INDONESIA Tbk, in collaboration with Sumitomo Metal Mining on community empowerment for the people around the company. The Soroako Technical Academy became a tertiary institution in 1993 which organized a Mechanical Maintenance Study Program diploma level 3. Akademi Teknik Soroako has a production unit which is engaged in production services which only acts as a manufacturing unit because the customer has prepared raw materials. Some product design orders are carried out through the Soroako Academy Production Engineering, so that the added value obtained varies. One of the regular services in the Production Division of Akademi Teknik Soroako (Soroako Technical Academy) is repairing puncher bars of PT Vale Indonesia, TBK. Figure 1 shows the shape of the puncher bars before they are repaired. There are several stages in repairing the puncher bars, i.e., measuring a puncher bar dimensions, cutting the puncher bar with an oxy-acetylene cutting machine, straightening up each puncher bar, grinding the tip of each puncher bar and lastly, adding a new puncher bar plate using a welding machine. There are a lot of Musculo Skeletal Disorder (MSD) related complaints from the workers [1] in the straightening up the puncher bar workstation. In this operation, the workers need to straighten the puncher bars by using a 5 kg iron hammer repeatedly. The operation needs large energy and requires worker's concentration in the movement of the body. Therefore, it is important to do the analysis of the work system related to straightening process. The analytical methods used in this study were the Ovako Work Posture Analysis System (OWAS) [2] and Rapid Upper Limb Assessment (RULA) [3] to measure the risk of musculoskeletal disorders due to improper work postures [4]. OWAS and RULA methods are selected because the puncher bar repair is done manually. Since the dominant work postures are standing, bending, and lifting the load, the assessment of the entire body is needed. The analysis conducted by both methods can show the risk level of work postures and indicate the potential of musculoskeletal disorders resulting from the unergonomic work postures.

Figure 1. Puncher bars before they are repaired.

2. Research methodology

This study mainly applied OWAS and RULA methods [5] to evaluate the repair puncher bar workstation in the ATS production unit by observing the postures of 3 operators. The three operators were selected since they have been working for quite long period, experienced, and doing the process of puncher bar repairs more than other workers. The repair processes performed by three sampled workers were captured using a camera in video format. The recordings were then analysed and highlighted on the postures that requires large work energy and unsafe body positions. Preliminary data collection is also conducted by distributing Nordic Body Map questionnaires [6] to 20 operators to measure the significance of their complaints regarding muscle fatigue during the repair process of the puncher bar. This study also collected the repairing time of each puncher bar in each workstation.

3. Result and analysis

3.1. Repair time of the puncher bar

Table 1 shows time spent, and tools used for each activity in the repairing process. The time to straighten up the puncher bars is the longest compared to the other activities. As presented at Figure 1, the straightness of each puncher bar may vary. If they are bent significantly, it will take longer time to straighten up the puncher bars.

Table 1. Puncher bar repairing time.

Activity	Time	Tool Used
Measure	0.2 seconds	Roll meter
Cut	0.5 seconds	Oxygen / Acetylene
Straighten up	2.65 minutes	*Paron* (base) and iron hammer (5 kg)
Grinding up	1.5 minutes	Pedestal Burrs
Weld plate bar	2 minutes	Welding machine
	Total = 6,85 minutes	

3.2. Nordic Body Map questionnaire data

Table 2 presents the recapitulation of data collected by the Nordic Body Map questionnaire among the 20 workers of the puncher bar workstation. For each type of pain, the distributions among the pain degree (i.e., "painless", "pretty sick", "sick", and "very ill") are presented as the percentage of the participant number. As can be seen in Table 2, the waist is the body part which has pain degree "sick" for 75% of the workers surveyed. Around 25% to 30% of the workers also reported having pain on their backs, shoulders (either right or left), and left upper arms.

Table 2. Percentage of types of complaints of workers.

No	Type of pain	Grade of complaints							
		Painless	%	Pretty sick	%	Sick	%	Very ill	%
0	Paint/stiff in the upper neck	8	40	10	50	2	10	-	-
1	Pain in the lower neck	8	40	10	50	2	10	-	-
2	Pain in the left shoulder	3	15	12	60	5	25	-	-
3	Pain in the right shoulder	2	10	13	65	5	25	-	-
4	Pain in the left upper arm	2	10	13	65	5	25	-	-
5	Pain in the back	1	5	13	65	6	30	-	-
6	Pain in the right upper arm	1	5	15	75	4	20	-	-
7	Pain in the waist	3	15	2	10	15	75	-	-
8	Pain in the buttock	17	85	2	10	1	5	-	-
9	Pain in the bottom	18	90	1	5	1	5	-	-
10	Pain in the left elbow	12	60	5	25	3	15	-	-
11	Pain in the right elbow	5	25	12	60	3	15	-	-
12	Pain in the left lower arm	5	25	12	60	3	15	-	-
13	Pain in the right lower arm	4	20	15	75	1	5	-	-
14	Pain in the left wrist	8	40	11	55	1	5	-	-
15	Pain in the right wrist	2	10	17	85	1	5	-	-
16	Pain in the left hand	9	45	10	50	1	5	-	-
17	Pain in the right hand	3	15	16	80	1	5	-	-
18	Pain in the left thigh	18	90		0	2	10	-	-
19	Pain in the right thigh	17	85	1	5	2	10	-	-
20	Pain in the left knee	18	90	1	5	1	5	-	-
21	Pain in the right knee	19	95		0	1	5	-	-
22	Pain in the left calf	19	95		0	1	5	-	-
23	Pain in the right calf	19	95		0	1	5	-	-
24	Pain in the left ankle	19	95	1	5		0	-	-
25	Pain in the right ankle	18	90	2	10		0	-	-
26	Pain in the left foot	19	95		0	1	5	-	-
27	Pain in the right foot	18	90	1	5	1	5	-	-

3.3. OWAS and RULA assessment

Based on the observation to 3 sampled workers, in general, the worker's posture in conducting the straightening puncher bar operation is uniform. Therefore, it can be concluded that by evaluating the working posture of one worker will represent the working postures of the other workers. The whole operation of straightening the puncher bar could be divided into two sub-activities, i.e., lifting the 5kg iron hammers to the air and hammering the puncher bars. Figure 2 and 3 show the worker's posture for each of these two sub-activities, respectively.

Figure 2. Worker is lifting the 5kg iron hammers to the air.

Figure 3. Worker is hammering the puncher bars.

The OWAS and RULA scores for each sub-activity are presented in Table 3 and 4, respectively. The OWAS score [7] for lifting 5 kg iron hammer and hammering puncher bar is 2. This result suggests that both postures are slightly dangerous for the musculoskeletal system, and corrective action is needed. While for the RULA [8], both sub-activities' scores are 7, meaning that the workers are working in the worst posture with an immediate risk of injury from their work posture. Investigation on the reasons is needed and changing the postures the near future is required to prevent an injury.

Table 3. OWAS and RULA scores for lifting 5 kg iron hammer.

Method	Score	Description
OWAS	2	This posture is dangerous for the musculoskeletal system; work posture produces tension; and needs improvement in the future.
RULA	7	The risk is in high-level category; and the current changes are required.

Table 4. OWAS and RULA scores for hammering puncher bar.

Method	Score	Description
OWAS	2	This posture is dangerous for the musculoskeletal system; work posture produces tension; and needs improvement in the future.
RULA	7	The risk is in high-level category; and the current changes are required.

3.4 Work intervention and improvement

OWAS and RULA analyses suggest that work intervention should be conducted to prevent potential injury. Therefore, this study suggests replacing the manual straightening process with a hydraulic press machine. While being straightening with the press machine, the puncher bar is supported by block V as the base. Figure 4 shows the straightening process with the recommended procedure. Since the press machine replaces the manual operation, the high-risk postures of lifting 5 kg iron hammer and hammering the puncher bar are eliminated. Figure 5 shows how to set up the puncher bar on a hydraulic press machine, while Figure 6 shows the working posture when applying pressure to the machine for straightening the puncher bar.

Figure 4. Recommended procedure for straightening a puncher bar.

Figure 5. Setting up a puncher bar in a hydraulic machine.

Figure 6. Work posture when straightening puncher bars on a hydraulic press machine.

The OWAS and RULA scores for each of the sub activities of the straightening process with a hydraulic press machine presented on Table 5 and 6, respectively. OWAS score for positioning iron hydraulic press machine is 2. These results indicate that both postures are slightly harmful to the musculoskeletal system, and corrective action is required. As for the RULA, the second score of the sub activity is 6, meaning that workers work at moderate risk, need for further handling, and a change from their working posture.

Table 5. OWAS and RULA scores for setting the puncher bar in a hydraulic machine.

Method	Score	Description
OWAS	2	This posture is dangerous for the musculoskeletal system; work posture produces tension; and needs improvement in the future.
RULA	6	Medium risk, advanced handling and need changes.

Table 6. OWAS and RULA scores for when straightening puncher bars on a hydraulic press machine.

Method	Score	Description
OWAS	2	This posture is dangerous for the musculoskeletal system; work posture produces tension; and needs improvement in the future.
RULA	6	Medium risk, advanced handling and need changes.

Since the process of alignment with the recommended procedure is replacing the manual operation, the high-risk posture of lifting the iron hammer 5 kg and hammering process were eliminated. In addition, this new recommended procedure eliminates the potential of MSD for workers, it also requires shorter operating time. Table 7 shows the time required for each step in the new method. By using the new procedure, the total operating time for each punch bar is 6.18 minutes or approximately 0.6 minutes faster than the original method. That means with a press machine, there are 9.7 puncher bars that can be straightened every hour. Table 8 shows comparison of capacity improvements for both methods.

Table 7. Process time repair puncher bar with hydraulic press machine.

Activity	Time	Tool Used
Measure	0.2 seconds	Roll meter
Cut	0.5 seconds	Oxygen / Acetylene
Straighten up	1.98 minutes	Hydraulic press machine
Grinding up	1.5 minutes	Pedestal Burrs
Weld plate bar	2 minutes	Welding machine
	Total = 6.18 minutes	

Table 8. Comparison of repairing capacity on the existing and recommended Procedure.

Description	Time (minute)	Puncher bar per hour
Iron hammer 5 kg	6.85	8.3
Hydraulic press machine	6.18	9.7

4. Conclusion

Among all operation for repairing puncher bars, straightening up the puncher bar is the riskiest process due to its high potential causing MSD injuries. Analysis using the OWAS and RULA methods result in score 2 and 7, respectively, suggesting that corrective action is needed soon. Proposed improvement to the unergonomic work posture is conducted by replacing the manual puncher bar straightening process with a hydraulic press machine. Besides that the new procedure will eliminate the potential of MSD injuries, it also saves the operation time for about 0,6 minute for each puncher bar. This improvement will surely be beneficial for the Production Division of Akademi Teknik Soroako for both its occupational health and safety and financial aspects.

References

[1] Stack T, Ostrom LT, Wilhelmsen CA. *Occupational Ergonomics*. John Wiley & Sons, Inc; 2016 May 9;8-9. doi.org/10.1002/9781118814239

[2] Pramestari D. Analisis Postur Tubuh Pekerja Menggunakan Metode Ovako Work Analysis. *Ikraith-Teknologi*. 2017;**1**;19-22.

[3] Sie KMS, Valentino F, Dearosa EY, Rahardjo B. Analisis Resiko Postural Stress Pada Pekerja Di UD.XYZ Dengan Metode Rapid Upper Limb Assessment. *Jurnal Rekayasa Sistem Industri*. 2017 Oct 31;**6**(2):149. doi.org/10.26593/jrsi.v6i2.2498.149-154

[4] Colombini D. *Working Posture Assessment: The TACOs (Time-Based Assessment Computerized Strategy) Method*. 2018 Sep 3;17. doi.org/10.1201/b22442

[5] Indan R, Pratiwi NK, Z MM. Analisis Postur Kerja Karyawan Dibagian Pengangkatan Bahan Baku Dengan Menggunakan Metode OWAS (Ovako Working Posture Analysis) Dan RULA (Rafid Upper Limb Assesment). *Majalah Ilmiah UPI YPTK*. 2019;**26**;36-44.

[6] Rahdiana N. Identifikasi Risiko Ergonomi Operator Mesin Potong Guillotine Dengan Metode Nordic Body Map. *Industry Xplore*. 2017;**2**;1-12.

[7] Zetli S. Analisis Postur Tubuh Pekerja Manual Material Handling dengan Pendekatan OWAS (Ovako Working Posture Analysis). *Rekayasa Sistem Industri*. 2017;**3**;16-25.

[8] Dzikrillah N, Yuliani ENS. ANALISIS POSTUR KERJA MENGGUNAKAN METODE RAPID UPPER LIMB ASSESSMENT (RULA) STUDI KASUS PT TJ FORGE INDONESIA. *Jurnal Ilmiah Teknik Industri*. 2017 May 22;**3**(3). doi.org/10.24912/jitiuntar.v3i3.466

Sustainable-value stream mapping to improve manufacturing sustainability performance: Case study in a natural dye batik SME's

S Hartini[1]*, J Manurung[1] and R Rumita[1]

[1]Department of Industrial Engineering, Diponegoro University, Semarang 50275 Indonesia

Email: *srihartini@lecturer.undip.ac.id

Abstract. The lean manufacturing system has proved its capability to eliminate waste and produce environmental gains. The integration of sustainable manufacturing indicators in lean manufacturing analysis tools is expected to improve manufacturing sustainability performance. This paper proposes sustainable-value stream mapping to evaluate sustainability performance using a lean manufacturing approach. This new tool will analyze all activities in the manufacturing process incorporated sustainability indicator includes economic, environmental, and social dimensions. The integration of sustainability indicator into a lean manufacturing tool is validated through a mini case study in a small-medium enterprise that produce natural dye batik. The key contribution of the study is a practical framework for identifying non-value-added activities at the production process of natural dye batik. This tool can identify the sources of problems in economic, environmental, and social dimensions. This article can recommend the improvement plan to improve the sustainability performance of a company, especially in batik SMEs.

1. Introduction

Value stream mapping (VSM) is a fundamental lean technique to identify waste and value improvement opportunities Its purpose is to identify value-adding and non-value-adding activities in the value stream so that wasteful activities can be eliminated, and production can be aligned with demand [1]. Value stream mapping is an effective system analysis for diagnosing manufacturing system problems and identifying improvement opportunities [2] [3]. Both lean and sustainable manufacturing systems aim to improve organizational performance and provide both operational and sustainable benefits. Integrated lean and sustainable implementation can create value for customers by eliminating waste consistently and adopting environmentally friendly, economical and safe processes for employees [4] [5] [6] [7] [8]. The application of Value Stream Mapping [VSM] as a tool to improve the sustainability performance of a production process is still very limited, especially when compared to the large amount of scientific research that focuses on 'traditional' VSM tools [9]. Traditional VSM framework does not account for environmental and societal performance. A number of studies have addressed the extension of VSM to incorporate additional criteria. A vast majority of these efforts have focused on adding environmental and energy-related metrics to VSM. The Sus-VSM added environmental and social metrics into VSM to enhance sustainability performance [1], [10] [11]. Previous research has reviewed sustainable-VSM that has been studied in several industrial sectors, including the automotive industry, agricultural food, electronic industry, and wood industry. Sustainable-VSM studies in the textile industry are still very limited, especially in the production process of natural dyieng batik.

The batik industry is Indonesia's mainstay commodity. The batik industry is the largest commodity (18.98%) with a target of increasing sales up to 6-8% in 2019. Coloring process is one of the important elements in creating beautiful batik art [12]. Before synthetic dyes became a trend today, the textile dyeing process used natural dyes. However, as the needs and technological advances increase with the discovery of synthetic dyes for textiles, the use of natural dyes is increasingly eroded. Unlike natural dyes, synthetic dyes will be easier to obtain in the market, guaranteed color availability, various kinds of colors, and more practical in their use [13]. Related to that, many batik craftsmen who have not treated their waste properly. Washed water that is mixed with chemicals including batik dyes, is just flowed into the gutter or directly into the river [14]. As a result, the river becomes discolored and creates an unpleasant odor. This waste can kill organisms that live in the river.

Bonfante et al [15] explored the opportunities and difficulties of implementing sustainable-VSM in the small and medium enterprises as part of a greening production plan. Bonfante et al found that SMEs have limitations in understanding social and environmental sustainability, even at the management level. SMEs flexibility is noted as driving the adoption of the Sus-VSM. This study aims to apply sustainable-VSM in Batik SME and examines the ability of this tool in order to improve manufacturing sustainability performance.

2. Methodology

2.1. *Sustainable-VSM as a tool to enhance manufacturing sustainability performance*

The manufacturing sector is a major contributor to the economy, unfortunately, many manufacturing processes have a negative impact on environmental and society. However, the manufacturing process consumes excessive scarce resources and produces hazardous waste and emissions [16]. Efforts to develop sustainability assessment in the manufacturing sector have been carried out before [17] [18]. Simons & Mason [19] proposed Sustainable Value Stream Mapping (SVSM) as a means of increasing sustainability in product manufacturing by analyzing greenhouse gas emissions, CO2, and value-added ratios. Some subsequent studies on the environment-VSM (e-VSM) are still limited by adding environmental indicators to VSM [20]. Sustainable-VSM was developed by Faulkner & Badurdeen [10] to analyzes the economic, environmental, and social dimensions applied to satellite dish manufacturing companies. Indicators analyzed include time, water consumption, materials, and energy, work environment and the physical work of the workforce. Helleno et al [7] propose a conceptual method to integrate a new group of sustainability in dicators into the VSM tool to assess manufacturing processes. The method capability test was conducted in three case studies, and the results showed that the proposed method was able to identify various levels of sustainability of the manufacturing process. Hartini et al [20] designed Sustainable–value stream mapping of the furniture company. Sustainable-VSM is able to identify the potential problems on furniture company. The problems include economic, environmental, and social aspect: high setup time, high defect, high breakdown machine and work in process in almost all processes, high amount of wood waste during sawmill process and energy during oven process and the work posture unergonomic and potential of accident rate still high in sawmill process. If these problems can be reduced the sustainability level will improve [21].

2.2. *Manufacturing Sustainability Performance*

Sustainable company performance should have three measurement elements, namely: (i) economic, (ii) social, and (iii) environmental. A sustainability indicator is a single parameter used to measure the condition of an aspect in sustainability, namely index [22]. An index can be used to effectively compare sustainability performances over different years for the same production process, which can help decision-makers to evaluate improvement in performance and effectiveness of changes made to the company [18]. An index can be approached by multiplying between the indicator score and its weight [23]. Indicator weights can be approximated by equal weights [24].

2.3. *Indicator Scoring*

The score of the indicator is determined by using the concept of efficiency, which is the comparison between the use of the sources with an added value toward the total use [23]. The score of the economic indicator is the reflection toward efficiency of time, quality, inventory, and cost. The score of environment indicator is a reflection toward the efficiency of the use of green material, water, material waste, recovered waste. The score of the social indicator is the reflection toward efficiency of workers in terms of health, satisfaction and the development of human resources. The level of work risk uses the level of risk from OHSAS [25]. The indicators that will be involved in evaluating the performance of sustainability and the scoring indicators are summarized in Table 1.

IOP Conf. Series: Materials Science and Engineering 1072 (2021) 012066 doi:10.1088/1757-899X/1072/1/012066

Table 1. The indicator and the indicators and formulas for determining indicator scores.

	Indicator	Formula	Reference
Economic	Time	$\frac{\text{Value added time}}{\text{Cycle time}}$	[25], [26], [7], [23]
	Quality	$\frac{\text{Number of inputs} - \text{number of rejects}}{\text{Number of inputs}}$	[23]
	Inventory	$\frac{\text{Value added time}}{\text{Value added time} + \text{Inventory time}}$	[23]
	Cost	$\frac{\text{Value added cost}}{\text{Total cost}}$	[26] , [23]
Environment	Green material	$\frac{\text{Quantity of green raw material}}{\text{Quantity of raw material}}$	[7]
	Water consumption	$\frac{\text{Ideal amount of water consumtion}}{\text{Total amount of water consumtion}}$	[25], [26], [7]
	Material waste	$\frac{\text{Total amount of waste}}{\text{Total amount of input}}$	[7],[23]
	Recovered waste	$\frac{\text{Amount of waste recovered}}{\text{Total waste}}$	[18]
Social	Satisfaction	$\frac{\text{Number of employees} - \text{number of resign}}{\text{Number of employees}}$	[25], [23]
	Health	$\frac{\text{Number of employees} - \text{number of absent}}{\text{Number of employees}}$	[25], [23]
	Training	$\frac{\text{The number of employee who get training}}{\text{Number of employees}}$	[22], [23]
	Safety	$\frac{\text{Number of employees} - \text{number of accident}}{\text{Number of employees}}$	[22], [23]

3. Case study

3.1. Production Process of the Natural Dyeing Batik

The colouring process in batik is an important step in making good quality batik [12]. Before synthetic dyes became a trend today, the textile dyeing process used natural dyes. As the development of synthetic dye technology for textiles, the use of natural dyes for batik production began to decrease. Synthetic dyes are easily accepted by producers because they are easily obtained in the market, guaranteed colour availability, various types of colours, and more practical in their use [13].

The problem arises when batik craftsmen do not process their waste properly. Used rinsing water containing chemicals, including batik dyes, is discharged into waterways or rivers [14]. This causes the river to change colour and smell. The liquid waste can kill organisms that live in the river which have a negative impact on the survival of fish and other animals in it.

To minimize the risks posed by batik waste, the use of natural dyes has again become an alternative choice in batik production. Batik natural dyes can be obtained from plant extracts [27]. These materials can come from roots, stems, bark, leaves, flowers and fruit. From these materials various colours will be produced though not as complete as when using chemical batik dyes [28]. For example, reddish purple staining using mangosteen skin and indigo blue using indigofera tinctoria and mangrove waste. Natural dyes are more environmentally friendly and produce natural classic batik colours. Batik that uses natural dyes has high economic value because it has artistic values and distinctive colours, environmentally friendly so it has an ethnic and exclusive impression [27].

The process of making batik with natural dyes begins with the mordanting process, which is the process of removing starch and fats that stick to the fabric so that the colour can stick well using a solution of metal salts such as alum [29]. Then, the stamping process is carried out. The next process is colouring with natural dyes which usually require two types of dyes. Natural dyes such as mangrove or indigo waste must go through an extraction process. The colouring process is carried out by five times dyeing and drying to get the desired colour intensity. After staining, the fixation process using chemicals such as lime, alum and tunjung. The *'nemboki'* process to cover the parts that do not want to follow the next colour using wax that is heated on the stove. The next colouring process uses another colouring material in the same way as the previous colouring and then is fixed. The last process is removing the wax from the fabric called *'nglorod'* (washing). The production process of natural dye batik is shown in Figure 1, supplemented by material consumption and non-product output.

3.2. Sustainable VSM of Natural Dyes Batik Production

This research has observed the producing process of natural dyeing batik. Observations were made on all activities from material preparation to product completion. Economic, environmental and social performance is measured at each stage of the process. Economic performance is measured based on efficiency of processing time, level of quality, work in process and production costs. Environmental performance is measured based on the efficiency of Green material, Water consumption, Material waste and Recovered waste. Social performance is measured based on the level of employee satisfaction, health, job risk and resource development through training.

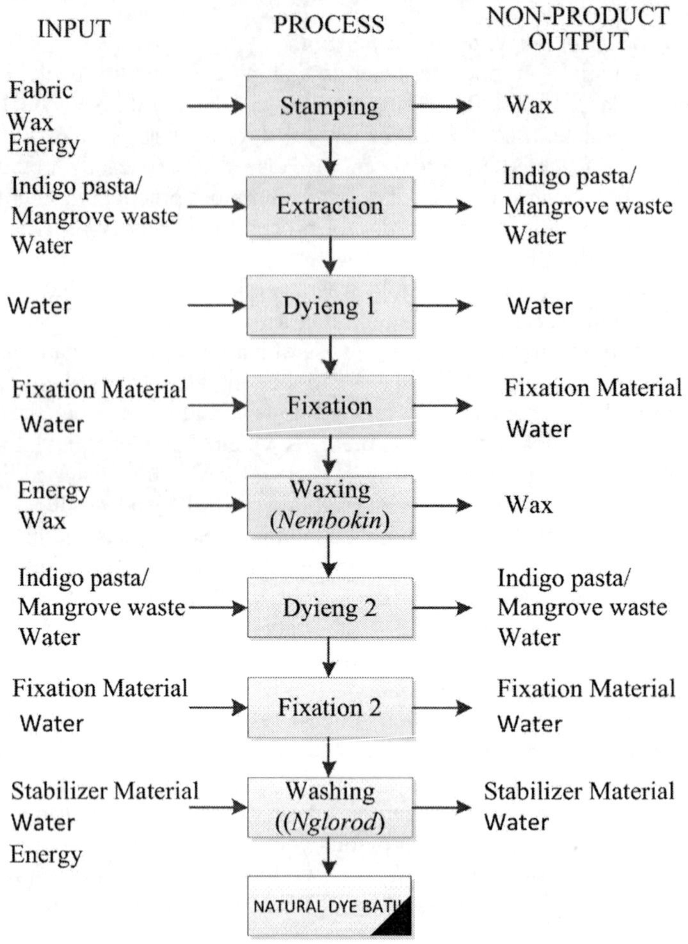

Figure 1. The input-output diagram of Batik production process.

- Economic indicator

The stamping process has the wasting time for equipment preparation and transportation time to the stamping location. The extraction process is the process of boiling natural dyes for the colouring process. The extraction process still uses firewood, so it requires a long setup time. The colouring is done with 5 times to get the desired colour intensity. Thus, the process of moving from the colouring site to the drying occurs repeatedly. Table 2 is the recapitulation of time efficiency in the economic dimension. The natural dye batik production process has a cycle time of 3889 minutes with a value added time of 3337 minutes, non-value added time of 403 minutes and necessary non value added time of 148.9 minutes. The value of the time efficiency of each process is still below 90% except the extraction process, so these processes must be improved to improve time efficiency. The time of production of batik natural dyes in each process is shown in Table 2.

Batik production process has a good level of quality, rarely found defective products. However, there is still a lot of material waiting to be found that triggers a work in process. Waiting time occurs in almost all processes causing low inventory levels. Factors that influence cost efficiency are the consumption of time and material. The amount of wasting time will be directly proportional to the waste of costs. The comparison between value added costs and non-value-added costs due to time is non-value added is 11% and value added is 89. While the amount of material usage costs is obtained from the price and the amount used in each process. Based on the consumption of time and material, the cost efficiency is still below 90%. The stamping process was 86%, the extraction process was 79%, the colouring process 1 and 2 were 85% and 83%, *nemboki* 88%, fixation 1 and 2 were 81% and

82% and *nglorod* was 86%. This value is considered to still need to be increased so that it becomes a priority for improvement.

Table 2. Production process time of natural dyeing batik.

No	Process	Process time (min)						VA	NVA	NNVA	Efficiency
		O	I	T	D	S	Total				
1	Stamping	347.4	19.1	9.1	10.1	0	385.7	317	40.4	28.1	82%
2	Extraction	372.5	0	1.1	31.9	0	405.5	373	18.5	14.5	92%
3	Dyeing 1	754.8	0	67.3	11.6	0	833.7	704	75.6	53.8	84%
4	Fixation 1	166.1	0	21.6	10.2	0	197.9	166	31.8	0	84%
5	Waxing	617.9	29.1	8.8	39.5	0	695.3	618	77.4	10	89%
6	Dyeing 2	693.3	0	64.6	12	0	769.9	641	76.6	52.5	83%
7	Fixation 2	167.6	0	22.4	20.4	0	210.4	180	30.8	0	85%
8	*Nglorod*	338.5	0	29.7	22.2	0	390.4	339	51.9	0	87%
	Total	3458	48.2	225	157.9	0	3889	3337	403	158.9	86%

- Environmental indicator

Raw materials for natural dye batik are grouped into two parts, natural (green) and chemical. Materials used for a batch process are measured and grouped according to this classification. Table 3 is the recapitulation of the green production efficiency. The impact of the material used in the batik production process on the environment shown in Table 4.

Table 3. The recapitulation of the green production efficiency.

No	Process	The number of materials		Efficiency
		Natural (kg)	Chemical (kg)	
1	Stamping	0	7	0%
2	Extraction	4	0	100%
3	Dyeing 1	2	0,5	80%
4	Fixation 1	0	2	0%
5	Waxing	0	14	0%
6	Dyeing 2	2	0	100%
7	Fixation 2	0	2	0%
8	*Nglorod*	0	0.5	0%
		Efficiency rate		35%

Table 4. Environmental impact of material.

No	Materials		Type	Impact
1	Cloth		Chemical	The environment will be difficult to decompose the material directly [12]
2	Colouring:	Indigo pasta	Natural	Safe for the environment [12] [28]
		Mangrove waste		
3	Wax		Chemical	River and soil pollution. Evaporation of a candle can worsen air quality in the work environment [30]
4	Fixation material		Chemical	River and soil pollution [28]
5	Sodium carbonate		Chemical	
6	TRO		Chemical	

The efficiency of water consumption is approached by comparing the amount of water used to ideal water consumption. The ideal amount of water consumption uses references from batik experts. Table 5 is a recapitulation of the water consumption efficiency. Table 6 is the recapitulation of material efficiency and recovered waste efficiency.

Table 5. The recapitulation of the water consumption efficiency.

No	Process	Actual	Ideal	Efficiency
1	Stamping	-	-	-
2	Extraction	40 litres	40 litres	100%
3	Dyeing 1	10 litres	10 litres	100%
4	Fixation 1	120 litres	80 litres	67%
5	Waxing	-	-	-
6	Dyeing 2	10 litres	10 litres	100%
7	Fixation 2	130 litres	80 litres	62%
8	*Nglorod*	270 litres	170 litres	63%
		Efficiency		82%

Table 6. The recapitulation of material efficiency and recovered waste efficiency.

No	Process	Material			Eff. of material	Recovered	Eff. of recovered
		Material	Input	Waste			
1	Stamping	Fabric: mori	20 sheets	0	93%	0	0%
		Wax	7 kg	0.5 kg		0	0%
2	Extraction	Indigo pasta	2 kg	1 kg	50%	0	0%
		Mangrove waste	2 kg	1 kg		0	0%
		Water	40 litres	20 litres		0	0%
3	Dyeing 1	Fabric: mori	20 sheets	0	90%	0	0%
		Indigor pasta	1 kg	0.1 kg		0	0%
		Water	10 litres	1 litre		0	0%
4	Fixation 1	Fabric: mori	20 sheets	0		0	0%
		Fixation material	1 kg	0.75 kg	25%	0	0%
		Water	120 litres	115 litres		0	0%
5	*Nemboki/* Waxing	Fabric: mori	20 sheets	0	89%	0	0%
		Wax	14 kg	1.5 kg		0	0%
6	Dyeing 2	Fabric: mori	20 sheets	0		0	0%
		Mangrove waste	1 kg	0.1 kg	90%	0	0%
		Water	10 litres	1 litre		0	0%
7	Fixation 2	Fabric: mori	20 sheets	0		0	0%
		Fixation material	1 kg	0.75 kg	25%	0	0%
		Water	130 litres	115 litres		0	0%
8	*Nglorod/* Washing	Fabric: mori	20 sheets	0	40%	0	0%
		Hydrogen Peroxide	0.5 kg	0.3 kg		0	0%
		Water	270 litres	255 litres		0	0%
		Efficiency			63%		0%

The amount of recycled waste will determine the level of recovered waste by the company. The recovered waste efficiency is approached by comparing the amount of treated to the total waste produced. The recapitulation of the recovered waste efficiency shown in Table 5. The company has not done the waste treatment properly.

- Social Indicator

The level of employee satisfaction is calculated based on the number of employees who resigned for five years from 2014 - 2019, because it was assumed to be dissatisfied with the work environment, salary, or work system in the company. There is only one employee who resigns the *nemboki* process so that the efficiency value of the *nemboki* process is 93% while for other processes 100%. The level of health indicators was obtained from company absence data for the past three years. The health efficiency in the stamping process and *nglorod* each by 99% while the other processes respectively by 97%. While the employee training efficiency of all process by 0% because the company has never conducted training for its employees.

The level of the work risk indicator is obtained based on the identification of the type of hazard and the probability of the occurrence of the hazard. In the process of stamping, *nemboki* and *nglorod* have H-4 values because this process is in contact with chemicals material that can cause allergies and irritation to the skin. Besides smoke and evaporation at wax can disrupt the respiratory system. The risk of working in the process is classified as moderate but has a high probability. One of the factors causing the high occupational risk is the lack of attention and awareness to use the Self Protection Equipment (SPE). Whereas for other processes it has a low risk because it does not deal with hazardous chemicals so that the probability of occurrence of the hazard is classified as low. However, the level of work risk indicators is a priority for improvement and evaluation because it has a considerable influence on sustainability in batik.

The results of economic, environmental and social performance as pillars in sustainable manufacturing are visualized in sustainable value streams (Figure 2).

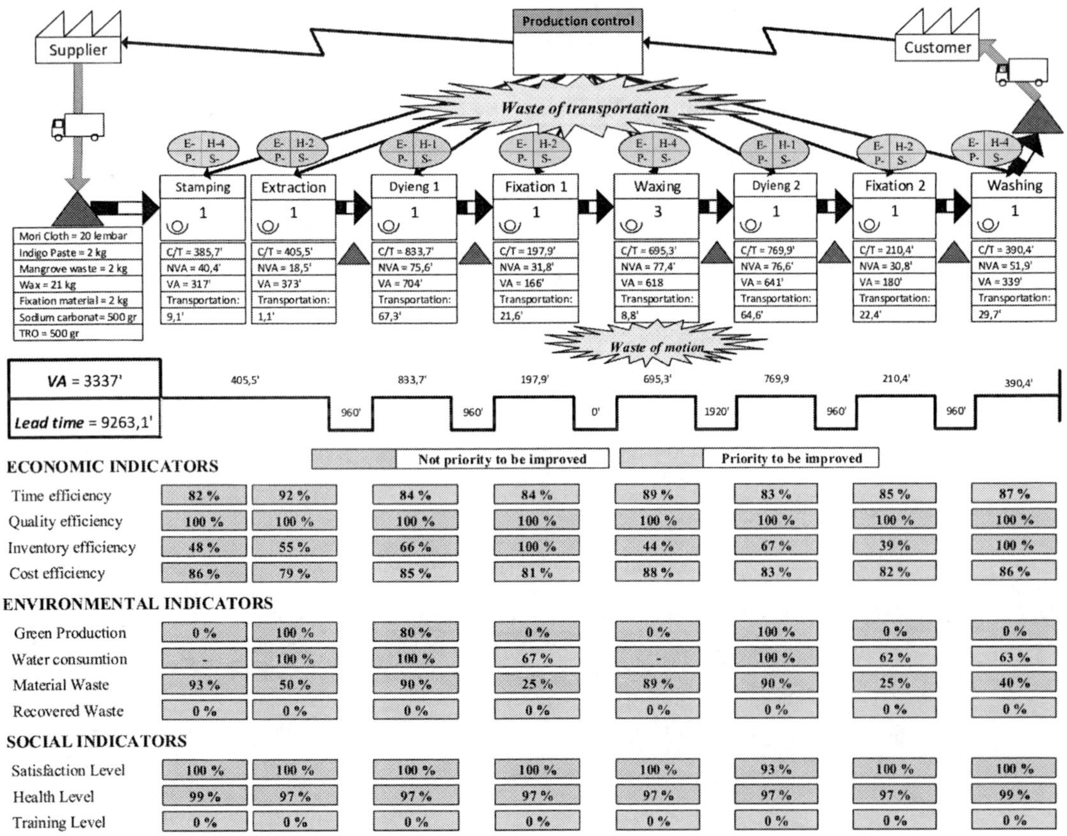

Figure 2. Sustainable-value stream mapping of natural dyeing batik production process.

4. Discussion

4.1. *The Opportunities and Limitation*

Sus-VSM has the opportunity to be applied in SMEs. Sus-VSM is made with clear steps based on the existing production process in the company. Relevant indicators are determined based on factors that have an effect on achieving optimal sustainability performance. That is, companies can achieve economic goals, while being able to protect the environment and be safe and socially comfortable based on their limitations. The efficiency approach used to assess sustainability performance requires simple calculations. Based on case studies at batik SME, the application of the Sus-VSM is able to identify and evaluate sustainability performance based on indicators that are relevant to the process of producing natural dye batik. Based on the Sus-VSM, the company can find out the performance that is not optimal based on indicators and activities.

The obstacle that occurs in implementing Sus-VSM lies in the process of measuring the required metrics. Production batch sizes in SMEs are usually small, making it difficult to get normally distributed data. This can be approached by taking data on several batches of production. In addition, SME companies do not have good data records, so almost all data must be based on direct measurements. However, the Sus-VSM can provide information and recommendations to improve sustainability performance

4.2. Recommendation to Improve Sustainability Performance

Sustainable-value stream mapping is able to identify sources of inefficiency that occur in the production process of natural dyes batik. To improve the company's sustainability performance, this study provides several recommendations as shown in Table 7.

Table 7. The recommendation to improve the sustainability performance.

Indicator	**Process**	**Recommendation**	**Expected improvement**
Time	Dyeing, Fixation, Waxing (*Nemboki*), Washing (*Nglorod*)	Material handling equipment design to help move from the colouring location to the drying location by adding wheels to the *'gawangan'*. *Gawangan* is the designation for a tool used to move cloth after colouring.	Reducing transportation time
		Workplace layout improved using the 5 S principle (*Seiri, Seiton, Seiso, Seiketsu, Shitsuke*)	Reducing setup time
Recovered waste	Waxing	Recovery wax using *gondorukem*. Benchmarks in other companies, recovering wax using *gondorukem* can save wax use by 60%.	Reducing wax and consumption and improving recovered waste
	Dyeing, Extraction *Nglorod*	Wastewater treatment installation	
Water consumption	Extraction and dyeing	Use of stop tap and flow meters	Reducing water consumption
Work risk	All process	Implement standard operation procedures (SOP): use personal protective equipment (PPE) such as filter masks (waxing process) and gloves as well as aprons (all process)	Reducing work risk
	Waxing (*Nemboki*)	The Ergonomic design of work tools such as *gawangan*, chair, stove holder	Reduce the risk of musculoskeletal disorder in the batik body
Training	All process	Employees are included in training programs, for example about natural dyeing and others	Training has a significant effect on Batik productivity [31]

Proposed methods for improving sustainability performance are visualized in the future Sus-VSM (Figure 3). The estimation of sustainability performance obtained when applying the proposed method is also informed in the future Sus-VSM. Comparison between current sustainability performance and expected sustainability performance in Table 8.

Table 8. Comparison between current sustainability performance and expected sustainability performance.

Indicators	Current efficiency	Future efficiency
Time	86.0	90.8
Quality	100.0	10.0
Inventory	64.8	64.8
Cost	83.8	88.4
Green material	35.0	35.0
Water consumption	82.0	90.0
Material waste	63.0	63.0
Recovered waste	0.0	52.5
Satisfaction	99.1	99.1
Health	97.5	97.5
Training	0.0	95.9
Safety	62.5	100.0
Improvement	64.5	81.4

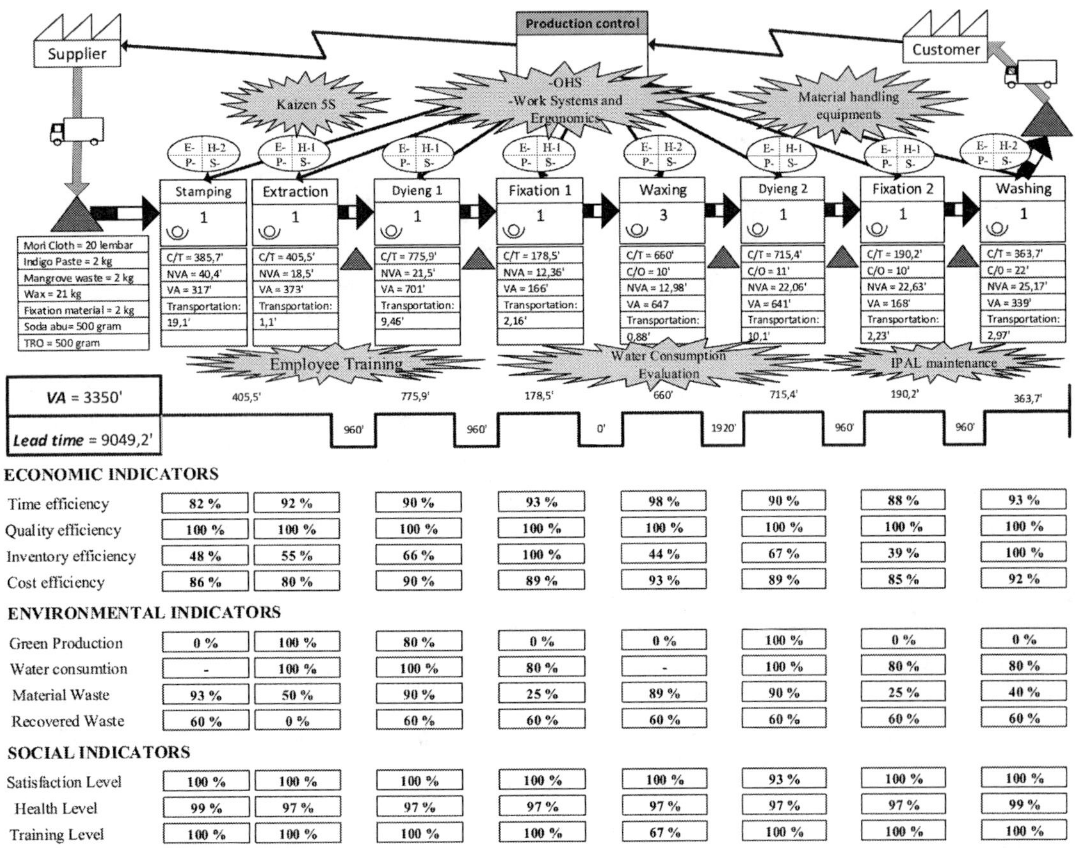

Figure 3 The future Sus-VSM.

4.3. *Managerial and Theoretical Implications*

The basic principle of lean manufacturing is continuous improvement. Value stream mapping is a tool that can identify non-value-added activities that can be used as a basis for making improvements. The design of Sus-VSM in a natural dye batik company has been proven to be able to identify activities that are not yet efficient. Indicators identified include economic, environmental, and social dimensions. With the existence of Sus-VSM, companies can understand the types of activities and indicators that need to be improved. In this case, time indicators, waste management, water consumption, and training are still not optimal in almost all activities. This study provides several recommendations that are expected to improve the sustainability performance of batik production.

In terms of theoretical value, this study complements previous research about evaluating sustainability indicators using lean manufacturing tools at the factory level [7] [23] [25]. By considering the relevant indicators in each sector cannot be the same, then designing the Sus-VSM to evaluate the sustainability performance in batik companies enriches the study for the development of theories in the field of measuring sustainability performance.

Batik production with natural dyes has relevance to the use of natural materials and has a lot of water consumption. Key indicators that differentiate it from other sectors, such as furniture companies, are green materials and water consumption. Given the scarcity of previous research on the use of Sus-VSM for evaluating the sustainability performance and relevant indicators in each sector can be different, then evaluating the performance of sustainability in companies in other sectors is interesting for further study. This study only evaluates the manufacturing process. Sustainable manufacturing should consider the overall scope.

5. Conclusion

Sustainable-value stream mapping is able to identify the source of problems that occur in the production process of natural dyeing batik. Inefficient activities occur almost throughout the process. This study provides recommendations by 3R (reduce, reuse and recycle) to improve the efficiency of the economic and environmental dimensions. The reduction is done by reducing the setup time and transportation time by improving the work environment with 5S and the use of material handling equipment. Reuse is done by reusing wax by using *gondorukem*, thereby reducing material consumption and increasing recovered waste. Recycle is done by treating wastewater so that it increases recovered waste and reduces water consumption. In the end, the reduction of time and material affects the reduction in production costs. The performance of the social dimension is enhanced by improving the work environment, employee awareness in using personal protective equipment and improving training programs.

This research still has not improved all the inefficiencies that have occurred. Further research can be done by designing stamping tools that can reduce processing time and employee risk. The collaboration program with the surrounding community in making natural colouring extracts can reduce the extraction time and add colour variations. The indicator weights used to determine the sustainability index are still assumed to be the same. Further research can be done by evaluating the weight of each indicator by making the expert judgment. Further studies will be more interesting if an evaluation of the sustainability performance of a batik company is carried out by considering a broader scope such as the pre-manufacturing, consumption, and post-consumption stages.

Acknowledgement

This research was financially supported by The Faculty of Engineering, Diponegoro University, Indonesia through Strategic Research Grant 2020.

References

[1] Norton A, Fearne A. *Handbook of Waste Management and Co-Product Recovery in Food Processing*. 2009. doi.org/10.1533/9781845697051.1.3
[2] Ben Fredj-Ben Alaya L. VSM a powerful diagnostic and planning tool for a successful Lean implementation: a Tunisian case study of an auto parts manufacturing firm. *Prod. Plan. Control.*, 2016 Apr 11;1–16. doi.org/10.1080/09537287.2016.1165305

[3] Dadashnejad A-A, Valmohammadi C. Investigating the effect of value stream mapping on overall equipment effectiveness: a case study. *Total. Qual. Manag. Bus. Excell.*, 2017 Apr 18;**30**(3-4):466–82. doi.org/10.1080/14783363.2017.1308821

[4] Hartini S, Ciptomulyono U. The Relationship between Lean and Sustainable Manufacturing on Performance: Literature Review. *Procedia Manuf.*, 2015;**4**:38–45. doi.org/10.1016/j.promfg.2015.11.012

[5] Zhan Y, Tan KH, Ji G, Chung L, Chiu ASF. Green and lean sustainable development path in China: Guanxi , practices and performance. *Resour Conserv Recycl.* 2018 Jan;**128**:240–9. doi.org/10.1016/j.resconrec.2016.02.006

[6] Caldera HTS, Desha C, Dawes L. Exploring the role of lean thinking in sustainable business practice: A systematic literature review. *J. Clean. Prod.*, 2017 Nov;**167**:1546–65. doi.org/10.1016/j.jclepro.2017.05.126

[7] Helleno AL, de Moraes AJI, Simon AT. Integrating sustainability indicators and Lean Manufacturing to assess manufacturing processes: Application case studies in Brazilian industry. *J. Clean. Prod.*, 2017 Jun;**153**:405–16. doi.org/10.1016/j.jclepro.2016.12.072

[8] Ruben RB, Vinodh S, Asokan P. Lean Six Sigma with environmental focus: review and framework. *Int. J. Adv. Manuf. Technol.*, 2017 Sep 27;**94**(9-12):4023–37. doi.org/10.1007/s00170-017-1148-6

[9] Garza-Reyes JA, Torres Romero J, Govindan K, Cherrafi A, Ramanathan U. A PDCA-based approach to Environmental Value Stream Mapping (E-VSM). *J. Clean. Prod.*, 2018 Apr;**180**:335–48. doi.org/10.1016/j.jclepro.2018.01.121

[10] Faulkner W, Templeton W, Gullett D, Badurdeen F. Visualizing Sustainability Performance of Manufacturing Systems using Sustainable Value Stream Mapping (Sus-VSM). *Proceedings of the 2012 International Conference on Industrial Engineering and Operations Management.* 2012 Jul 3;815–24

[11] Hartini S, Ciptomulyono U, Anityasari M. Life cycle – Value stream mapping: Evaluating sustainability using lean manufacturing tools in the life cycle perspective. *EXPLORING RESOURCES, PROCESS AND DESIGN FOR SUSTAINABLE URBAN DEVELOPMENT: Proceedings of the 5th International Conference on Engineering, Technology, and Industrial Application (ICETIA) 2018.* 2019. doi.org/10.1063/1.5112428

[12] Felaza E, Priadi CR. Implementation of Cleaner Production in a Natural Dye Batik Industry SME: A way to Enhance Biodegradability of Batik Wastewater? Park CJ, Morita H, Huang HH, Khan MR, Cheng CK, de Leon RL, editors. *MATEC Web Conf.*, 2016;**62**:05003. doi.org/10.1051/matecconf/20166205003

[13] Indrawati S, Sunaryo, Budi Raharjo A, Clara Shinta R, Budi Khoirani A. Reliability Centered Maintenance on Critical Component in Indonesian Commercial Trains. *Proc. - 2018 4th Int. Conf. Sci. Technol. ICST 2018.*, 2018 Aug. doi.org/10.1109/icstc.2018.8528632

[14] Hasibuan S, Hidayati J. The Integration of Cleaner Production Innovation and Creativity for Supply Chain Sustainability of Bogor Batik SMEs. *Int J Ind Manuf Eng.* 2018;**12**(5). zenodo.org/record/1474275

[15] Bonfante MC, Ferreira JCE, Fernandes S, Back H. Lessons Learned from Sustainable Value Stream Mapping (Sus-VSM) Application in a Small Enterprise. *Springer Proceedings in Business and Economics.* 2019 Oct 15;1045–53. doi.org/10.1007/978-3-030-23816-2_103

[16] Saad MH, Nazzal MA, Darras BM. A general framework for sustainability assessment of manufacturing processes. *Ecol. Indic.*, 2019 Feb;**97**:211–24. doi.org/10.1016/j.ecolind.2018.09.062

[17] Duarte S, Cruz Machado V. Green and lean implementation: an assessment in the automotive industry. *Int. J. Lean Six Sigma.* 2017 Mar 6;**8**(1):65–88. doi.org/10.1108/ijlss-11-2015-0041.

[18] Huang A, Badurdeen F. Metrics-based approach to evaluate sustainable manufacturing performance at the production line and plant levels. *J. Clean. Prod.*, 2018 Aug;**192**:462–76. doi.org/10.1016/j.jclepro.2018.04.234

[19] Simons D, Mason R. Firms are under pressure to prove their environmental credentials . *Now a win-win way of weaving 'green' considerations into business decisions is emerging Lean*

and green: 'doing more with less.'. 2003;84–91. leanuk.org/downloads/general/lean_and_green.pdf

[20] Hartini S, Ciptomulyono U, Anityasari M. Extended value stream mapping to enhance sustainability: A literature review. *AIP Conference Proceedings*. 2017. doi.org/10.1063/1.5010647

[21] Hartini S, Ciptomulyono U, Anityasari M, Sriyanto, Pudjotomo D. Sustainable-value stream mapping to evaluate sustainability performance: case study in an Indonesian furniture company. Ma'mun S, Tamura H, Purnomo MRA, editors. *MATEC Web of Conferences*. 2018;**154**:01055. doi.org/10.1051/matecconf/201815401055

[22] Lee JY, Kang HS, Noh SD. MAS2: an integrated modeling and simulation-based life cycle evaluation approach for sustainable manufacturing. *J. Clean. Prod.*, 2014 Mar;**66**:146–63. doi.org/10.1016/j.jclepro.2013.11.029

[23] Hartini S, Ciptomulyono U, Anityasari M, Sriyanto. Manufacturing sustainability assessment using a lean manufacturing tool. *Int. J. Lean Six Sigma*. 2020 Feb 21;ahead-of-print(ahead-of-print). doi.org/10.1108/ijlss-12-2017-0150

[24] Singh RK, Murty HR, Gupta SK, Dikshit AK. An overview of sustainability assessment methodologies. *Ecol. Indic.*, 2009 Mar;**9**(2):189–212. doi.org/10.1016/j.ecolind.2008.05.011

[25] Faulkner W, Badurdeen F. Sustainable Value Stream Mapping (Sus-VSM): methodology to visualize and assess manufacturing sustainability performance. *J. Clean. Prod.*, 2014 Dec;**85**:8–18. doi.org/10.1016/j.jclepro.2014.05.042

[26] Vinodh S, Ben Ruben R, Asokan P. Life cycle assessment integrated value stream mapping framework to ensure sustainable manufacturing: a case study. *Clean Technol Environ Policy*. 2015 Aug 11;**18**(1):279–95. doi.org/10.1007/s10098-015-1016-8

[27] Yaacob MR, Faizah N, Zain M. Environmental awareness and perception on environmental management of entrepreneurs in batik industry in Kelantan, Malaysia. *J Trop Resour Sustain Sci*. 2016;**4**(1):54–60

[28] Purwanto P. Exploration of natural dyes as alternative substitutes of synthetic dyes on batik making fabrics. *J Phys Conf Ser*. 2019 Nov;**1375**:012023. doi.org/10.1088/1742-6596/1375/1/012023

[29] Susanty A, Hartini S, Puspitasari D, Arsiwi P. Measuring Efficiency of Using Resource in the Production Process of Making Stamped-Batik: A DEA Approach. *Mediterr J Soc Sci*. 2015 Sep 1. doi.org/10.5901/mjss.2015.v6n5s2p318

[30] Munthe EL, Surjanto E, Yunus F. Dampak Pajanan Asap Lilin Batik (Malam) terhadap Fungsi Paru dan Asma Kerja pada Pekerja Industri Batik Tradisional Occupational Asthma of Traditional Batik Industry Workers. *J Respir Indo*. 2014;**34**(3):149–57

[31] Murdiyanto A. Analisis Pengaruh Motivasi dan Pelatihan Terhadap Produktvitas (Pada Industri Kecil Kelompok Wirausaha Kerajinan Batik di Kecamatan Gunungpati Kota Semarang). *Prosiding Call For Paper Fakultas Ekonomika dan Bisnis UNTAG Semarang Tahun 2015 ANALISIS*. 2015;1–20

Integration lean manufacturing and 6R to reduce wood waste in furniture company toward circular economy

S Hartini[1,2], P A Wicaksono[1], M D Rizal A[1] and M Hamdi[1]

[1]Department of Industrial Engineering, Universitas Diponegoro, Semarang, 50275, Indonesia

Email: [2]srihartini@lecturer.undip.ac.id

Abstract. The furniture industry is suspected to have triggered global warming. The need for wood as raw material becomes a necessity for cutting down trees. The concept of a circular economy aims to reduce the use of natural resources by utilizing production waste or post-use product waste into raw materials. One strategy that can be used to support the circular economy is by implementing 6R. The application of 6R in a circular economic perspective on the furniture industry will be interesting to do. This study uses value stream mapping as a lean manufacturing analysis tool to identify furniture company waste. Value stream mapping is developed by integrating green manufacturing indicators such as material consumption and waste management. Practically, the furniture company can find out the map of production waste so that it can utilize waste as raw material for derivative products as an effort to reduce tree felling. Theoretically, this research will enrich the integration of green-value stream mapping toward the circular economy by implementing the 6R strategy in furniture companies. Reuse wood waste as a raw material for a variety of table products can reduce the amount of wood waste while reducing the resources of wood logs.

1. Introduction

Circular economy (CE) was first introduced by Ellen MacArthur Foundation [1]. In recent years, the concept of CE has gained significant momentum throughout the world. This concept improves the traditional concept of linear economy, which produces with the "take-make-dispose" model so that it is difficult to answer the challenges of sustainability which simultaneously requires sustainable economic growth, environmental protection, and social welfare [2]. The change from a linear economic model to a circular economy is beneficial for companies: cost efficiency, increased competitive advantage and reduction of negative environmental and community impacts [3]. The definition of CE itself is still being debated [4], but in general many associate it with a combination of reduce, reuse and recycle (3R) activities [5]. In addition to the 3R combination, CE also creates sustainable value that encourages economic growth, better environmental quality and social justice [4]-[6]. With resources steadily depleting at an alarming rate, the world is in serious trouble with its activities of meeting its never-ending needs through extraction, production, and consumption activities where most of the products end up in landfills (linear economy). The CE message is very strong because it is based on reducing resources through efforts to increase resource efficiency through reuse, reduce, recycle, recovery, redesign and remanufacturing [6R]. CE is no longer an option, but it is inevitable to continue economic prosperity and ecological and social balance [2]. Related to the urgency of implementing the CE concept, to date the study is still conceptual (6R). Only a few try to implement on certain commodities and are limited to the application of 3R [7]. Wood-based furniture is a sector that triggers global warming. Tree felling is considered to trigger environmental damage. This can be reduced by saving wood and utilizing its wood waste. Wood waste is a mass of wood that cannot be utilized at a stage of production. Wood waste can be distinguished based on the location of the occurrence of waste, namely waste from timber harvesting in the forest and wood processing waste that is in the industry. Timber harvesting waste is wood that is not brought from the forest because it is considered to be of no economic value by the company or the one that is cutting down. The types of wood waste in the form of non-commercial wood, decorative wood with certain uses, round wood with a diameter of less than 30 cm without length limit and round wood with a length of less than two meters. Wood processing waste is wood waste that is not utilized in the form of sawdust, slabs, bark, pieces, shavings, off-cut of logs and wood dust. The sawmill waste was 50.8% and wood processing industry waste was 61.0%. Sawmill waste consists of sawdust (19.6%), slabs (25.9%), and cuts (14.3%). Wood processing industry waste consists of log cuts (5.6%), sawdust (0.7%), wet veneer waste (24.8%), dry veneer waste (12.6%), peeled residues (11.0 %) and wood edge cutting (6.3%). This wood waste problem poses unique challenges to eliminate, reduce, or manage effectively. A satisfactory solution to these challenges calls for a systematic integrated approach to minimize wood waste to reduce its concomitant environmental impacts [8]. In the last decade, interest has been growing in utilizing this resource and millions of tons of solid wood waste are available for recycling into a myriad of products. Reuse in the wood waste brings not only important benefits in terms of material and energy savings but also avoidance of GHG emissions in the forest and landfill [9].

Wood waste identification must be carried out throughout the production process to be able to utilize waste using the 6R strategy. Lean manufacturing uses value stream mapping to identify 7 types of waste. Value stream map is a collection of all value-added activities and non-value added ones, ranging from raw materials to end customers, that involves information flows, materials flow, and decision-making flows [10]. The purpose of VSM is to identify value-added and non-value-added activities in a value streams so that non-value-added activities can be eliminated [11]. Integrating VSM and other concepts had been done to improve their capability. For example, VSM was combined with supply chain management [12]-[13], product development methodologies [14]-[15], agent-based systems [16]. Value stream mapping as a mean to evaluate the sustainability indicator was introduced by Simon and Mason. A number of studies have addressed the extension of VSM to incorporate additional criteria [17].

The study of the circular economy in the furniture industry has been carried out by previous researchers. There are investigate how much furniture companies know about Circular Economy practices, barriers and drivers of CE [18], the expansion of the circular economy by identifying the challenges and opportunities for a furniture cluster [19], design a business model for the furniture industry to address the circular economy concept using game theory [20], investigate the relationship between varying ESCC practices on the CE-targeted performances [21]. Empirical studies of the potential waste from the furniture production process and the challenges of utilizing the waste in a circular economy perspective are still rarely conducted. The Circular Economy, a new economic framework, can be combined with the 6R-based elements of sustainable manufacturing in order to relieve these challenges by establishing a closed-loop material flow [22]. This study aims to develop a framework that integrates value stream mapping with indicators that can identify potential waste throughout the production process while also involving the 6R strategy to realize a circular economy. By integrating 6R into VSM, activities that have the potential to reduce, reuse, recycle, recovery, redesign and remanufacture are more easily identified. To achieve this goal, this research will be carried out in stages. In the first stage, this research will develop a VSM that involves indicators of quality and material consumption. Potential defects that still occur in the production process lead to waste. Likewise with material consumption that is not optimal, the remaining material will potentially become waste. The next step is to identify the type of waste as the basis for the 6R design. Furthermore, the 6R design will be the basis for realizing a circular economy. Utilization of waste into products will benefit the company both in economic aspects (saving raw material purchases) and environmental aspects (reducing wood waste). Recycling of wood waste is beneficial in delaying and avoiding carbon emissions stored in trees [23]. To validate the applicability of the framework, a case study was conducted in a furniture firm.

2. Methodology

2.1. Circular Economy

Macarthur introduces the concept of circular economy (CE), derived from the term restorative industrial economy, which enables the effective flow of material, energy, labour, and information. Restorative means the flow of material can be returned again from the end consumer to the initial economic activity (manufacturing). Thus, it is possible to take corrective actions and reuse materials. Kirchherr, Reike and Hekkert [5] examines transparency regarding the latest understanding of the concept of circular economics. For this purpose, 114 definitions of circular economics have been collected. His findings are that circular economics is most often described as a combination of reduce, reuse, and recycle, but often the article does not highlight the importance of systemic changes in circular economy.

Genovese *et al.* [24] examines Sustainable Supply Chain Management and its role in Circular Economy. Case studies are carried out in various process industries (chemical and food), by comparing the performance of traditional and circular production systems across a range of indicators. This paper confirms that the integration of circular economic principles in sustainable supply chain management provides clear advantages from an environmental perspective. Need to study other industries to compare the benefits of using linear and circular economic practices.

Kalmykova, Sadagopan and Rosado provides an overview of the literature on Circular Economy (CE) strategies and analyse implementation cases. The findings are that the 3R strategy stands out in the manufacturing process [25]. Accordingly, Lewandowski examines the principles of waste design for reuse [26]. Manufacturers need to think of methods to make this happen, for example by recycling. Economically, CE practices make it possible to minimize the potential for material extraction from nature. Environmentally, CE practices will reduce waste emissions to nature. This is one way to solve the problem of scarcity of resources and environmental impact. Huang et al conducted an analysis of construction waste management through the 3R principle [27]. Management policies and situations are investigated and analysed based on the principles of reduction, reuse and recycling. Barriers to reuse of waste due to lack of guidelines for effective waste collection and sorting, lack of knowledge and standards for reuse of construction waste. Moktadir *et al* [28] state that knowledge of the circular economy is very important for implementing sustainable manufacturing practices in the Bangladesh leather industry [28]. This lesson will help managers of leather companies to formulate strategies so that the utilization of available resources can be optimal, and be able to do waste reduction in the context of a circular economy.

2.2. *Extended-Value Stream Mapping to identify the Potential Waste*

Faulkner and Badurdeen [29] and Brown et al [30] developed a methodology for Sus-VSM to capture the economic, environmental, and societal sustainability of manufacturing firms. They also identified suitable metrics and visual symbols to develop Sus-VSM and conducted a study at satellite television dishes manufacturing firm. Folinas, Aidonis, & Karayannakidis [31] explores the application of the VSM tool so as to determine the waste that has an environmental impact in a specific agri-food supply chain; the production of the canned peach. The main objective of this study is to propose a four-step approach for measuring the environmental performance of supply chains in the food sector based on Lean thinking techniques so as to identify sources of waste in the selected supply chain.

Hartini, Ciptomulyono, Anityasari, Sriyanto, & Pudjotomo [32] develop sustainable-VSM for wood furniture companies. Sustainable-VSM involves indicators of lead time, defects, inventory, material consumption, energy consumption, and level of risk. The defect and material consumption indicators in sustainable-VSM are able to identify that there is potential waste in the production process both from the material preparation stage to the packaging stage. Potential waste has not been linked to a strategy to create a circular economy.

2.3. *Principles of 6R*

The 6R methodology consists of reduce, reuse, recycle, recovery, redesign, and remanufacturing [22]. Reduce refers to subtraction use of resources in pre-manufacturing, including reduction of energy consumption, material consumption, and other resources during manufacture, and reduction of emissions and waste during the use phase. Reuse refers to the overall reuse of a product, or its components, for the next life cycle, to reduce the use of virgin material to produce new products and components. Recycle refers to the process of material conversion which should be considered waste, become new material or product. The process of collecting products at the end of the use stage, disassembling, sorting, and cleaning for utilization at the next life cycle of this product is called Recovery. Redesign activity involves the next action to redesign generation products, which will use components, materials, and resources recovered from previous life cycles, or previous product generations. Remanufacturing involves reprocessing products that have been used for recovery to their product's original state or form as new through reuse as much as possible parts possible without loss of functionality. The integration of lean and 6R concepts is implemented in a company to analyze its ability to reduce waste and natural resources toward a circular economy. Research Method is shown in Figure 1.

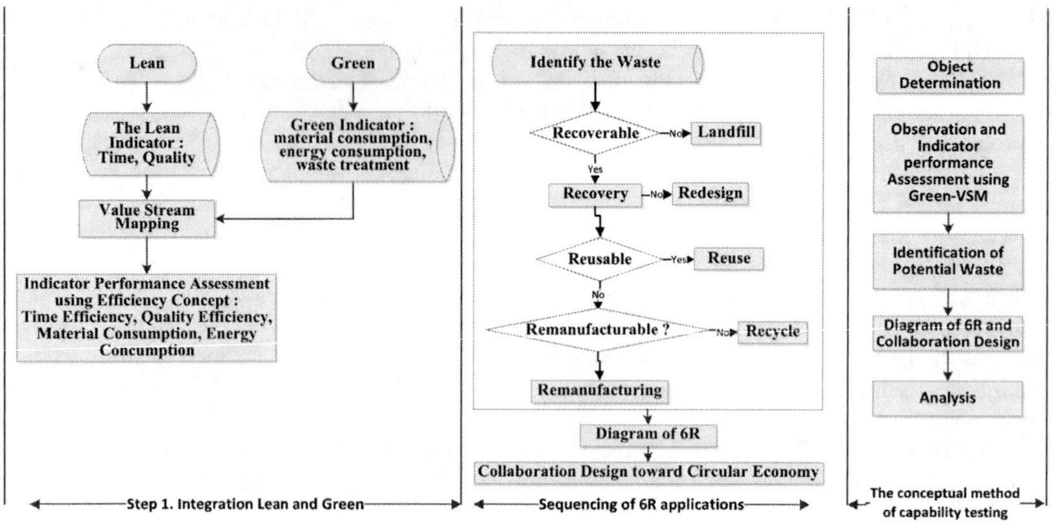

Figure 1. Research method.

Step 1 aims to evaluate the performance of the production process with green-VSM by measuring time efficiency, quality, energy, material consumption, and waste treatment. Table 1 shown some previous studies involving these indicators.

The score of the indicator is determined by using the concept of efficiency, which is the comparison between the use of the sources with added value towards the total use. The score of economic indicator is the reflection towards the efficiency of time and quality. The score of the environment indicator is reflection towards the efficiency of the use of material, energy, and waste management. The common formula of efficiency that will be used ratio between the use of resource and total resource. Detail of formula shown in Table 1.

The next step is to identify waste and the potential to reduce, reuse, recycle, recovery, redesign, and remanufacture. The identification results are used as the basis for making the 6R diagram. Next will be the design of the company collaboration toward a circular economy. To evaluate the integration of value stream mapping and the 6R concept in identifying waste and reducing waste in the perspective of a circular economy, a case study was developed at a furniture company.

Table 1. Lean and green indicators in value stream mapping.

Manufacturing	Related-metric	Measurement Method	Reference
Lead time Efficiency (LTE)		LTE= VAT/TT	[29], [33], [34], [35]
n : process on sawing, construction, assembly, finishing	Value-added time during process (VAT)	VAT = $\sum_{i=1}^{n}(VATi)$	
	Non-value-added time (NVAT)	$NVAT = \sum_{i=1}^{n}(NVATi)$	
	Total time during process (TT)	TT = VAT + NVAT	
Quality Level Efficiency (QL)	Quality level (QL)	QL = 1 – ND/TP	[36]
	Number of defect (ND)		
	Total product/component (TP)		
Material Efficiency (ME)	Material efficiency (ME)	ME =TMU/TM	[34,37], [38], [7]
	Total material used (TMU)	TMU = $\sum_{i=1}^{n}(TMUi)$	
	Total material (TM)		
	Product/material defect (MD)	TMU = TM - MD	
Energy Efficiency (EME)	Energy efficiency during process (EE)	EE = VAE/ET	[34,37], [38], [7]
	Value added energy (VAE)	VAE = $\sum_{i=1}^{n}(VAEi)$	
	Non-value added energy (NVA)	NVAE = $\sum_{i=1}^{n}(NVAEi)$	
	Energy total during process (ET)	ET = VAM + NVAM	
Waste Efficiency (WE)	Number of waste to landfill/river (ML)	WE = 1 – WL/TW	[7]
	Total waste during process (TW)		

2.4. Green-Value Stream Mapping

Case study was done at PT X, wooden furniture company in Semarang City, Indonesia. The manufacturing process can be characterized as Flow Shop, and it presents a high volume and low variety of products. The sequence of the manufacturing process is shown in Figure 2.

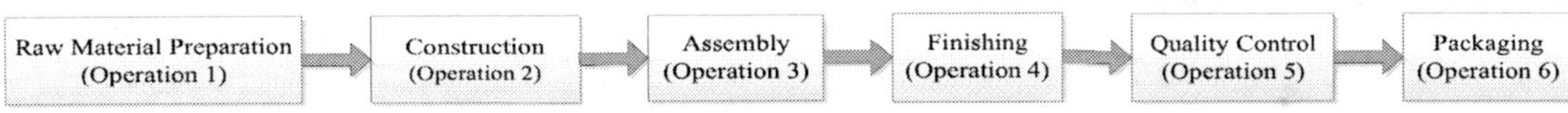

Figure 2. Furniture manufacturing process.

Raw material preparation is turning wood logs into boards. The construction process is forming the boards according to the product design. Assembly is the process of combining the various components. Finishing is the last process to finalize the product, e.g. smoothening of the wood surface. Quality control is done before packaging to ensure that products are in accord with the specifications. Green-VSM is shown in Figure 3.

The top management state that the performance of indicators is safe if the efficiency value has a value of more than 90 and is declared critical if below 65. The efficiency score that is between 60 – 90. A Green-VSM has been able to identify non-value-added activities that occur during the production process so inhibit company performance. The company has very low efficiency in the time indicators of raw material cutting (45.7%), quality indicators in the assembly process (37.5%) and finishing (0%), material consumption indicators in material preparation (46%) and waste recycling indicators (20%) in all activities. Low time efficiency in cutting material is caused by high waiting time, changeover time, and transportation time. Waiting time occurs because the operator does other work and the production area is not neat. Changeover time is high because of the many variations of components in one machine and the setup time is done by trial and error. Transportation time is high because it doesn't use material handling equipment. Un Ergonomic work posture causes the operator to get tired easily so the work completion takes longer.

Low-quality efficiency in the assembly and finishing process caused by hot room temperatures causes shrinkage of wood moisture so that the wood is bent, cutting is not precise, or wood fibers are not straight. Quality control is not carried out in every process so that the quality of materials that are not good can pass to the production department. The efficiency of material consumption is low at the material preparation stage because of inappropriate product size tolerance, type of logwood coming from community forests with low quality, and too much wood cutting depth resulting in large waste and incorrect product size. Meanwhile, waste treatment is still limited.

2.5. Identification of Wood Waste in Furniture Company

The types of wood waste that occur in furniture companies that are case studies shown in Table 2. There are the sawdust, slabs, bark, wooden board, and pieces of wood.

Table 2. The types of wood waste in furniture companies.

No.	Description	Type of waste
1	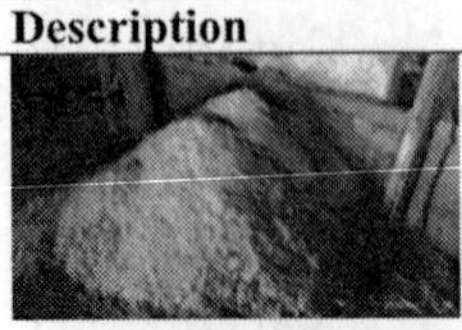	Sawdust is wood waste obtained from sawing wood using both manual and machine sizes of about 0.25 mm - 2.00 mm. Sawdust which is left to rot, stacked, or burned will have a negative impact on the environment. Its lightweight in a dry state is easily blown away by the wind and can damage breathing.
2		Slabs are wood waste that is produced from the logwood breakdown process. A piece of wood has a width of 15-20 cm, thickness 1-3 cm, and a length of 3-4 cm.
3		The bark is wood waste that comes from the outer layer. The bark is very soft so it cannot be material for wood furniture. The bark is usually removed during the sawing process. The layer of wood adjacent to the bark is also discarded because of its soft texture and high water content so it is easy to shrink, brittle, and break easily.
4		The wooden board is damaged wooden boards that occur due to poor sawing processes, the quality of logs with many holes and cracks, or the quality of saw blades. The company should check the raw materials to be processed and check the cutting tools before use.
5		Pieces of wood produced from the process of reducing and cutting materials to get products that match the planned product.

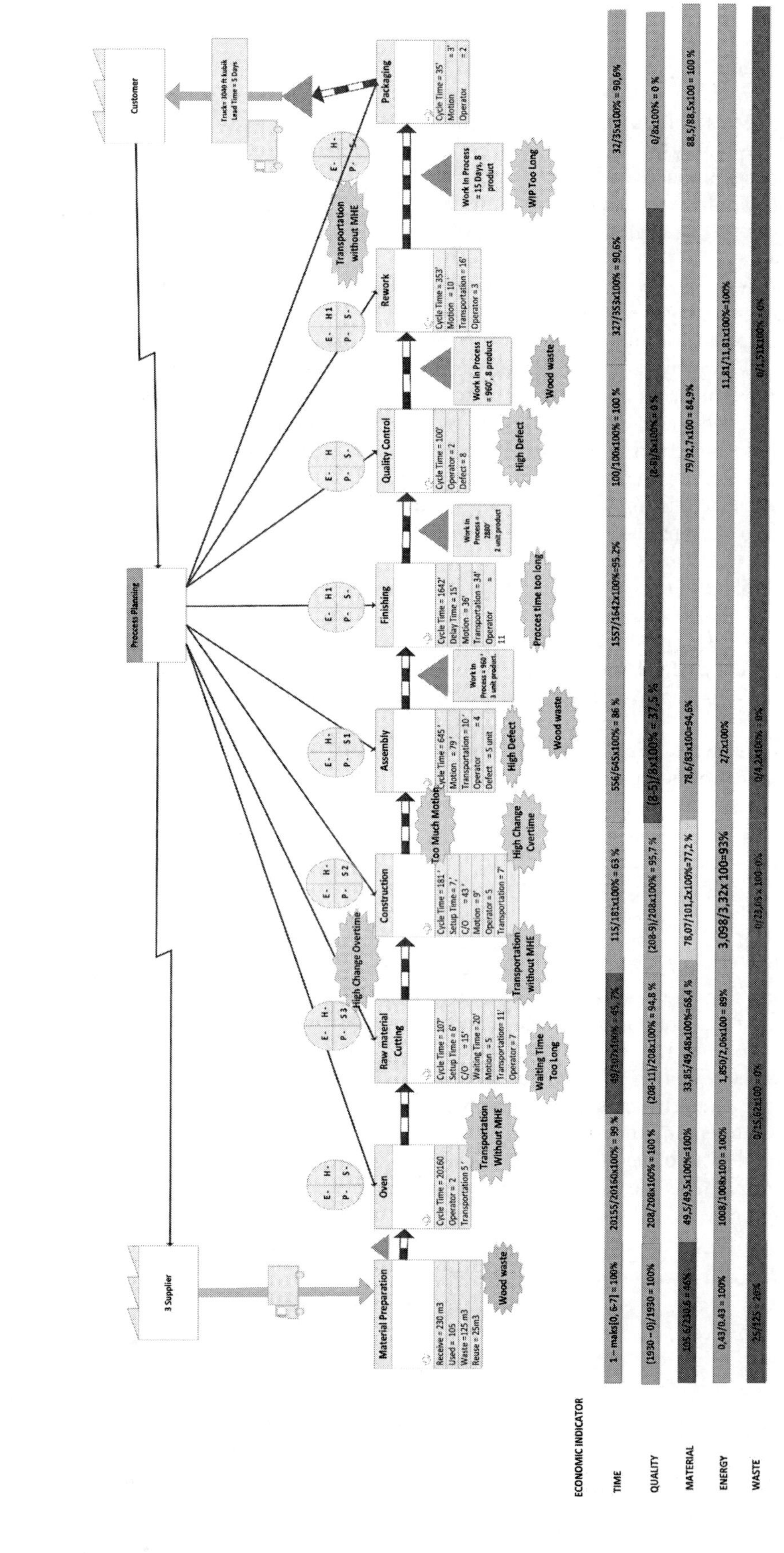

Figure 3. Green-value stream mapping of Furniture company.

IOP Conf. Series: Materials Science and Engineering 1072 (2021) 012067 doi:10.1088/1757-899X/1072/1/012067

3. Analysis and Discussion

3.1. Reducing wood furniture waste with the 6R strategy

The resulting Green-VSM is able to identify non-value added activities that occur in the furniture production process. Low-quality performance results in components/defect products. Low material consumption performance produces a lot of waste. If the company does not have good waste management, this waste will have a negative impact on the environment. To reduce the impact of wood waste on the environment, 6R strategies need to be developed. Some of the strategies implemented are explained in Table 3.

Basically, the 6R strategy can be applied to furniture suppliers, producers, and consumers. The 6R strategy in this furniture company can be grouped based on the 6R principles. The reduction activity is carried out with reducing the use of resources in the pre-manufacturing stage, by implementing a selective cutting system. Only cut down the right trees, the right amount, and reforest. Make saving using energy, materials and other resources during manufacturing, for example using glue and paint as needed, turning off the engine when not in use, turning off the water tap when not in use, reducing the use of fuel for kilns, reducing errors in cutting wood and others. Reducing emissions and waste by consumers by recovering products and not burning post-use products. Reuse is done by reusing waste to reduce the use of raw materials, for example using sawdust that is not utilized into board particles that can be processed into other products. Recycle converts waste into raw material for other products. Some furniture company waste recycling processes include utilizing the remnants of wood pieces into other products such as children's toys, assembly tables, and others.

Utilizing sawdust for briquettes, bricks, organic fertilizer, growing media, paving blocks. Recovery includes the process of collecting products at the end of the use, dismantling, sorting, and cleaning stages for utilization in the next life cycle of the product. Products that have expired in the hands of consumers can be recovered by making improvements either by furniture service services or by working with furniture manufacturers to make recovery. Redesign activities can be done by adding product variants by utilizing existing wood waste. For example using waste boards and pieces of wood into tables. Remanufacture involves reprocessing a product that has been used for recovery to the original state or form as new through reusing as many parts as possible without losing functionality. For example, companies do repair furniture products that are damaged by consumers rather than thrown away.

3.2. Reusing wood waste as raw material: a coffee table product

Waste in the company should be used as a variety of products so as not to have an impact on the environment. The product design tries to vary the table with raw materials from the company's waste. The product design was developed using the TRIZEE method. The TRIZEE design method was introduced by Issac Lim Sing Sheng and Teoh Kok-Soo [39]. The concept analysis takes a minimalist, multifunctional, and unity concept. The minimalist concept is used to get products that are easily made and require a faster process. The minimalist concept eliminates complicated details while still prioritizing space efficiency, quality, and function over quantity. The multifunctional concept is done by adding storage such as drawers, cabinets, and shelves. Furniture with a multifunctional concept is one aspect that must be present in product design. The concept of unity is used to bring together small blocks into a blockboard. These block boards are shaped like ordinary square boards that can be processed into furniture products like primary wood raw materials. The design and prototype of the product is shown in Figure 4.

Table 3. Input-output of furniture company and 6R strategy.

Stage	Process	Input	Output: Product Output	Output: Non-Product Output	6R Strategy
Pre-Manufacturing	Logging	Tree	Log	a. Pieces of wood log edges b. Pieces of wood c. Root d. Branches and twigs e. Foliage f. Sawdust	a. Reuse (firewood) b. Recycle (toys and tables) c. Reuse (firewood) d. Recycle (antique table) e. Reuse (Animal feed) f. Recycle (particle board, crafts, briquettes, organic fertilizer, mushroom growing media, bricks and paving blocks)
	Sawmills	Log	Beam and Board according to specifications	a. Sawdust b. The remaining pieces of wood log edges	a. Recycle (particle board, crafts, briquettes, organic fertilizer, mushroom growing media, bricks and paving blocks b. Reuse (energy for kiln)
	Drying in the kiln	Beam and 3. Board according to specifications	Beams and boards of appropriate size and moisture content	Emission	Reduce (reduce fuel use)
	Construction	Beams and boards of appropriate size and moisture content	Component	a. Sawdust b. Piece of wood	a. Recycle (particle board, crafts, briquettes, organic fertilizer, mushroom growing media, bricks and paving blocks) b. Reuse (firewood, toys, table)
	Assembly	a. Component b. Glue	Product	a. Remaining glue b. Components that cannot be assembled (reject)	a. Reducing glue waste b. Recycle (become children's toys and table assemblies)
	Finishing	Product	Finished Product	a. Sanding dust b. The rest of the painting c. Remaining spraying d. Residual bleaching fluid	a. Reduce with air circulation b. Dust machine c. Reducing paint waste d. Reduce (reduce excessive spraying)
Production	Packaging	a. Finished product b. Glue c. Cardboard d. Staples e. Rope	Product is ready to delivery	a. Remaining glue b. Remaining cardboard c. The remaining Styrofoam d. The remaining staples e. Piece of rope	a. Reuse (reuse for packaging) b. Reduce (reduce the use of excess staples) c. Reuse (reuse for packaging)
Post-Use	End of life	broken furniture	furniture has been recovered	a. Glue b. The rest of the painting c. Remaining spraying d. Piece of wood	a. Reducing glue waste b. Reducing paint waste c. Reduce (reduce excessive spraying) d. Recycle (toys)

The efficiency of material consumption on sawmills is around 46%, with wood waste averaging 19.44 m3/month. Wood processing efficiency of around 49%. the company can make blocks with a size of 1200 x 600 x 30 as many as 89 pieces. It the tables made from the wood waste volume of about

1,116,304 mm3, so the production capacity of about 56 tables. And the waste that has been successfully reused is around 64 m3.

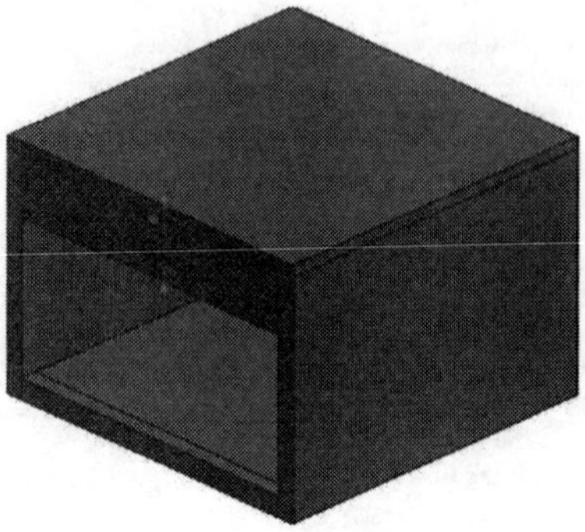

Figure 4. The design and prototype of the product.

4. Conclusion

VSM is able to identify the waste that occurs during the production process. By adding indicators from green manufacturing, Green-VSM is able to identify the efficiency of material consumption and the amount of waste that has been treated. Furniture companies that use logs as raw material have a potential waste of more than 50%. The quality of wood, cutting technology, and operator workmanship affect the amount of waste produced. However, wood waste can be reduced by implementing the 6R strategy. This research proves that wood waste can be reused as raw material for table products. This strategy is environmentally beneficial because it reduces wood waste and is economically beneficial because it saves the purchase of logwood. This confirms the statement

Jawahir et al [2] and Bradley [22] that the application of 6R in the perspective of a circular economy can reduce the use of resources. Although this research has detailed the types of wood waste produced by wood furniture companies and their potential uses, it is still limited in its implementation. This research is limited to using a strategy to reuse and redesign wood waste into coffee table products. Further research can be done by implementing other strategies, such as remanufacturing and recovery at the post-used stage. Research in other sectors will be interesting to study in order to realize the circular economy and efforts to reduce the use of natural resources.

Reference:

[1] Korhonen J, Nuur C, Feldmann A, Birkie SE. Circular economy as an essentially contested concept. *J Clean Prod.* 2018;**175**:544–52. doi.org/10.1016/j.jclepro.2017.12.111

[2] Jawahir IS, Bradley R. Technological Elements of Circular Economy and the Principles of 6R Based Closed-loop Material Flow in Sustainable Manufacturing. *Procedia CIRP.* 2016;**40**:103–8. doi.org/10.1016/j.procir.2016.01.067

[3] Stratan D. Success Factors of Sustainable Social Enterprises Through Circular Economy Perspective. *Visegr. J. Bioeconomy Sustain.* 2017 May 24;**6**(1):17–23. doi.org/10.1515/vjbsd-2017-0003

[4] D'Amato D, Droste N, Allen B, Kettunen M, Lähtinen K, Korhonen J, et al. Green, circular, bio economy: A comparative analysis of sustainability avenues. *J Clean Prod.* 2017;**168**:716–34. doi.org/10.1016/j.jclepro.2017.09.053

[5] Kirchherr J, Reike D, Hekkert M. Conceptualizing the circular economy: An analysis of 114 definitions. *Resour. Conserv. Recycl.*, 2017 Dec;**127**:221–32. doi.org/10.1016/j.resconrec.2017.09.005.

[6] Korhonen J, Honkasalo A, Seppälä J. Circular Economy: The Concept and its Limitations. *Ecol. Econ.*, 2018 Jan;**143**:37–46. doi.org/10.1016/j.ecolecon.2017.06.041
[7] Huang B, Wang X, Kua H, Geng Y, Bleischwitz R, Ren J. Construction and demolition waste management in China through the 3R principle. *Resour. Conserv. Recycl.*, 2018 Feb;**129**:36–44. doi.org/10.1016/j.resconrec.2017.09.029
[8] Frank Eshun J, Potting J, Leemans R. Wood waste minimization in the timber sector of Ghana: a systems approach to reduce environmental impact. *J. Clean. Prod.*, 2012 May;**26**:67–78. doi.org/10.1016/j.jclepro.2011.12.025
[9] Bais-Moleman AL, Sikkema R, Vis M, Reumerman P, Theurl MC, Erb K-H. Assessing wood use efficiency and greenhouse gas emissions of wood product cascading in the European Union. *J. Clean. Prod.*, 2018 Jan;**172**:3942–54. doi.org/10.1016/j.jclepro.2017.04.153
[10] Forno AJD, Pereira FA, Forcellini FA, Kipper LM. Value Stream Mapping: a study about the problems and challenges found in the literature from the past 15 years about application of Lean tools. *Int. J. Adv. Manuf. Technol.*, 2014 Feb 26;**72**(5-8):779–90. doi.org/10.1007/s00170-014-5712-z
[11] Pereira Librelato T, Pacheco Lacerda D, Henrique Rodrigues L, Rafael Veit D. A process improvement approach based on the Value Stream Mapping and the Theory of Constraints Thinking Process. *Bus. Process Manag. J.*, 2014 Oct 28;**20**(6):922–49. doi.org/10.1108/bpmj-07-2013-0098
[12] Boonsthonsatit K, Jungthawan S. Lean supply chain management-based value stream mapping in a case of Thailand automotive industry. *2015 4th International Conference on Advanced Logistics and Transport (ICALT)*. 2015 May. doi.org/10.1109/icadlt.2015.7136593
[13] Suarez-Barraza MF, Miguel-Davila J-Á, Vasquez-García CF. Supply chain value stream mapping: a new tool of operation management. Sampaio, Maria Sameiro Carvalho and P, editor. *Int. J. Qual. Reliab. Manag.*, 2016 Apr 4;**33**(4):518–34. doi.org/10.1108/ijqrm-11-2014-0171
[14] Tyagi S, Choudhary A, Cai X, Yang K. Value stream mapping to reduce the lead-time of a product development process. *Int. J. Prod. Econ.*, 2015 Feb;**160**:202–12. doi.org/10.1016/j.ijpe.2014.11.002.
[15] Ciarapica FE, Bevilacqua M, Mazzuto G. Performance analysis of new product development projects. *Int. J. Product. Perform. Manag.*, 2016 Feb 8;**65**(2):177–206. doi.org/10.1108/ijppm-06-2014-0087
[16] Xie Y, Peng Q. Integration of value stream mapping and agent-based modeling for OR improvement. *Bus. Process Manag. J.*, 2012 Jul 20;**18**(4):585–99. doi.org/10.1108/14637151211253747
[17] Hartini S, Ciptomulyono U, Anityasari M. Extended value stream mapping to enhance sustainability: A literature review. *3rd International Materials, Industrial and Manufacturing Engineering Conference (MIMEC2017)*. 2017. doi.org/10.1063/1.5010647
[18] Barbaritano M, Bravi L, Savelli E. Sustainability and Quality Management in the Italian Luxury Furniture Sector: A Circular Economy Perspective. *Sustain.* 2019 May 31;**11**(11):3089. doi.org/10.3390/su11113089
[19] Oliveira FR de, França SLB, Rangel LAD. Challenges and opportunities in a circular economy for a local productive arrangement of furniture in Brazil. *Resour. Conserv. Recycl.*, 2018 Aug;**135**:202–9. doi.org/10.1016/j.resconrec.2017.10.031
[20] Wicaksono PA, Hartini S, Sutrisno, Nabila TY. Game Theory Application for Circular Economy Model in Furniture Industry. *IOP Conf. Ser. Earth Environ. Sci.*, 2020 Apr 4;**448**:012061. doi.org/10.1088/1755-1315/448/1/012061
[21] Susanty A, Tjahjono B, Sulistyani RE. An investigation into circular economy practices in the traditional wooden furniture industry. *Prod. Plan. Control.*, 2020 Jan 3;1–13. doi.org/10.1080/09537287.2019.1707322
[22] Bradley R, Jawahir IS, Badurdeen F, Rouch K. A total life cycle cost model (TLCCM) for the

circular economy and its application to post-recovery resource allocation. *Resour. Conserv. Recycl.*, 2018 Aug;**135**:141–9. doi.org/10.1016/j.resconrec.2018.01.017
[23] Ng R, Shi CWP, Tan HX, Song B. Avoided impact quantification from recycling of wood waste in Singapore: an assessment of pallet made from technical wood versus virgin softwood. *J. Clean. Prod*, 2014 Feb;**65**:447–57. doi.org/10.1016/j.jclepro.2013.07.053
[24] Genovese A, Acquaye AA, Figueroa A, Koh SCL. Sustainable supply chain management and the transition towards a circular economy: Evidence and some applications. *Omega*. 2017 Jan;**66**:344–57. doi.org/10.1016/j.omega.2015.05.015
[25] Kalmykova Y, Sadagopan M, Rosado L. Circular economy – From review of theories and practices to development of implementation tools. *Resour. Conserv. Recycl.*, 2018 Aug;**135**:190–201. doi.org/10.1016/j.resconrec.2017.10.034
[26] Lewandowski M. Designing the Business Models for Circular Economy—Towards the Conceptual Framework. *Sustainability*. 2016 Jan 18;**8**(1):43. doi.org/10.3390/su8010043
[27] Huang A, Badurdeen F. Metrics-based approach to evaluate sustainable manufacturing performance at the production line and plant levels. *J. Clean. Prod.*, 2018 Aug;**192**:462–76. doi.org/10.1016/j.jclepro.2018.04.234
[28] Moktadir MA, Rahman T, Rahman MH, Ali SM, Paul SK. Drivers to sustainable manufacturing practices and circular economy: A perspective of leather industries in Bangladesh. *J. Clean. Prod.*, 2018 Feb;**174**:1366–80. doi.org/10.1016/j.jclepro.2017.11.063
[29] Faulkner W, Badurdeen F. Sustainable Value Stream Mapping (Sus-VSM): methodology to visualize and assess manufacturing sustainability performance. *J. Clean. Prod.*, 2014 Dec;**85**:8–18. doi.org/10.1016/j.jclepro.2014.05.042
[30] Brown A, Amundson J, Badurdeen F. Sustainable value stream mapping (Sus-VSM) in different manufacturing system configurations: application case studies. *J. Clean. Prod.*, 2014 Dec;**85**:164–79. doi.org/10.1016/j.jclepro.2014.05.101
[31] Folinas DK, Aidonis D, Karayannakidis P. Greening the canned peach production. *J. Agric. Informatics*,. 2015 Jan 13;**6**(1). doi.org/10.17700/jai.2015.6.1.154
[32] Hartini S, Ciptomulyono U, Anityasari M, Sriyanto, Pudjotomo D. Sustainable-value stream mapping to evaluate sustainability performance: case study in an Indonesian furniture company. Ma'mun S, Tamura H, Purnomo MRA, editors. *MATEC Web of Conferences*. 2018;**154**:01055. doi.org/10.1051/matecconf/201815401055.
[33] Ng R, Low JSC, Song B. Integrating and implementing Lean and Green practices based on proposition of Carbon-Value Efficiency metric. *J. Clean. Prod.*, 2015 May;**95**:242–55. doi.org/10.1016/j.jclepro.2015.02.043
[34] Vinodh S, Ben Ruben R, Asokan P. Life cycle assessment integrated value stream mapping framework to ensure sustainable manufacturing: a case study. *Clean Technol. Environ. Policy.*, 2015 Aug 11;**18**(1):279–95. doi.org/10.1007/s10098-015-1016-8
[35] Prasanna M, Vinodh S. Lean Six Sigma in SMEs: an exploration through literature review. *J. Eng. Des. Technol.*, 2013 Oct 7;**11**(3):224–50. doi.org/10.1108/jedt-01-2011-0001
[36] Dissanayake CK, Cross JA. Systematic mechanism for identifying the relative impact of supply chain performance areas on the overall supply chain performance using SCOR model and SEM. Int J Prod Econ. 2018 Jul;**201**:102–15. doi.org/10.1016/j.ijpe.2018.04.027
[37] Vinodh S, Selvaraj T, Chintha SK, K E K V. Development of value stream map for an Indian automotive components manufacturing organization. *J. Eng. Des. Technol.*, 2015 Jul 6;**13**(3):380–99. doi.org/10.1108/jedt-08-2010-0054
[38] Helleno AL, de Moraes AJI, Simon AT. Integrating sustainability indicators and Lean Manufacturing to assess manufacturing processes: Application case studies in Brazilian industry. *J. Clean. Prod.*, 2017 Jun;**153**:405–16. doi.org/10.1016/j.jclepro.2016.12.072
[39] Sheng. Eco-Efficient Product Design Using Theory of Inventive Problem Solving (TRIZ) Principles. *Am J Appl Sci*. 2010 Jun 1;**7**(6):852–8. doi.org/10.3844/ajassp.2010.852.858

Design of therapy equipment for osteoarthritis patients based on static bikes using QFD and human centre design method

H Prastawa[1,2], A A S M Mahachandra[1], R Salsabila[1] and D Sanu[1]

[1]Industrial Engineering Department, Faculty of Engineering, Diponegoro University, Jl. Prof. Soedarto, SH, Kampus Undip Tembalang, Semarang, Indonesia 50275

Email: [2]heru.prastawa@gmail.com

Abstract. The most frequently affected by Osteoarthritis (OA) is the knee joint (genu osteoarthritis). Therapeutic exercises for sufferers of knee OA are divided into 6 (six) programs, including strengthening to increase the strength of muscle contraction. The design to develop is based on the existing tools and input from users so that it is easily accessed widely for OA sufferers or doctors or therapists. The best choices for knee OA therapy are aerobics and strengthening of the thigh muscles by training the quadriceps muscles. There are several methods used to train quadriceps muscles, including straightening the legs when sitting and bending the legs while lying on his stomach. Meanwhile, aerobics training requires technology in the form of a static bicycle. This study was conducted based on the product development process, it has several stages that start with making a mission statement. The stages are: identify customer's needs, establish target specifications, generate product concepts, select product concept(s), test product concept(s), and the last is set the final specifications. This stage of the study resulted in 3 alternative designs and 1 design selective, that have been synthesized from the expectation and desires of the stakeholders and will continue with the design details until the prototype is presented.

1. Introduction

Osteoarthritis (OA) is a degenerative disease in the joints that often occurs in the hands, hips, and knees involving cartilages, joint layers, ligaments, and bones, causing pain and stiffness in the joints [1]. The most frequent part affected by OA in the knee joint (genu osteoarthritis). In this joint, OA classification can be grouped into five levels 0-4, where the knee joint can function properly and there is no pain or can be said to be healthy. Level 4 of experiencing severe osteoarthritis is indicated by extreme pain and discomfort to walk even to move the joints in the knee due to reduced joint space and the cartilage is almost completely gone [2],[3].

As a result of OA for the elderly, especially in the knees, namely the disruption of daily life activities due to limited movement in the capsular pattern of the knee joint, impaired joint stability, and decreased functional knee to walk and as a buffer of body weight. Pain becomes the main symptom in joints that experience OA. Knee pain that is felt by people with OA causes a person to be afraid of making movements so that it reduces the quality of life [4].

Therapeutic exercises for sufferers of knee OA are divided into 6 (six) programs, including strengthening to increase the strength of muscle contraction (lifting), aerobics to increase cardiorespiratory endurance (swimming, running), flexibility exercises to increase range of motion and flexibility of joint muscles (stretching hamstring), neuromotor training to improve balance and coordination (rocking board, walking on foam), mind-body training by integrating awareness/relaxation into physical movements (tai chi, yoga), and mixed exercises (more than one type of exercises) [5]

The best choices for knee OA therapy are aerobics and strengthening of the thigh muscles by training the quadriceps muscles [5]. Based on the results of an interview with Dr. Hari Peni Julianti, Sp.KFR, M. Kes, it was explained that there were several methods used to train quadriceps muscles, including straightening the legs when sitting and bending the legs while lying on his stomach. Meanwhile, aerobics training requires technology in the form of a static bicycle. Manual training in question is strengthening the thigh muscles by training quadriceps muscles. Straightening your legs when sitting is done by giving a weight of 1 kg of sandbags. Then the sandbag is tied to the ankle then the training is done by doing Flexi (from bottom to top) and extensions (from top to bottom) on the foot.

Currently, the tools used to carry out therapy for patients with OA include En-Tree [6] static bicycles [7], sandbags [7], stairs, parallel bars, and others. Based on the results of interviews, a tool that is often used for therapy of knee OA patients is a static bicycle. The tool can be used by people with OA in both mild and severe severity, the difference is in the need for help from nurses or therapists due to the difficulty of patients moving the feet. However, these tools have expensive prices and limited quantities for therapeutic use. Also, static bicycles do not have safe use so that therapy patients can fall while using it. Then, during the process of therapy requires help from nurses, while the number of nurses is limited. So, we need a therapeutic tool that can be used without having to use a nurse's help to operate it. Besides, a static bicycle is needed especially for people with OA with a severity of 3 (three) or osteoarthritis to the severity of 4 (four) or severe osteoarthritis. The safe and independent factor is expected to be able to reach and provide wider access to patients to be able to do healing therapy better.

Based on the above problem, the researcher intends to develop a design of a safe static bicycle therapy device, where the tool can be used independently without the help of nurses and pay attention to the safety of use. The design to develop is based on the existing tools and input from users so that it is easily accessed widely by OA sufferers or doctors or therapists

2. Literature Review

2.1. Knee Osteoarthritis and its severity

The knee is the most common part of osteoarthritis to take place in. Symptoms that occur are joint pain in the knee that is felt after activity and disappears after resting. Also, **crepitations** are found that are increasingly apparent with the increasing weight. This symptom is caused by the friction of both surfaces of the joint when the joint is moved [1]. As a result of knee OA for the elderly is the disruption of daily life activities due to limited movement in the capsular pattern of the knee joint, impaired joint stability, and decreased functional knee for walking and as a burden on the body. Knee pain felt by people with OA causes a person to be afraid of making movements so that it reduces the quality of life [2]. The severity of knee OA seen from the radiographic picture is categorized into 5 (five) groups based on the Kellgren-Lawrence criteria which are outlined in table 1 and illustration is shown in figure 1.

Table 1. Classification of radiological features of OA [2].

Degrees	Classification	Radiological Picture
0	Normal	There is no abnormal radiographic impression.
1	Doubtful	The appearance of small osteophytes due to proteolytic damage to the cartilage matrix, not accompanied by pain.
2	Minimal	Look osteophytes, normal joint gaps. The presence of fibrillation and erosion on the surface of the cartilage.
3	Moderate	Osteophytes, narrowing of joint gaps, synovial inflammation
4	Severe	Narrowing of the joint gap and the presence of sclerosis, the drastic decrease in synovial. Extraordinary pain, stiffness in the joints, and people with difficulty moving

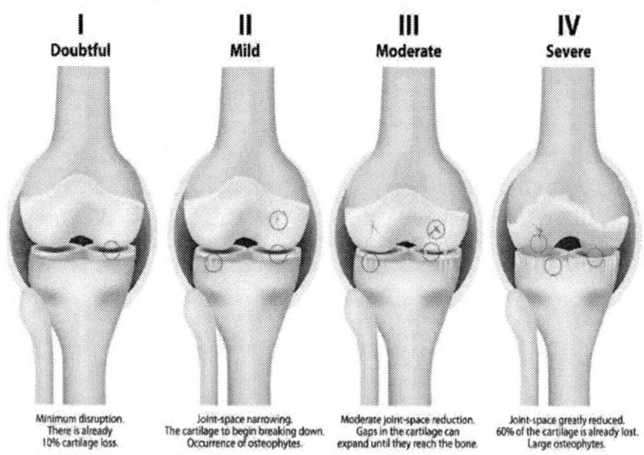

Figure 1 Severity of knee OA.
Source: www.ruperthealth.com

2.2. Product Development Process

This study was conducted based on [8] product development process, it had several stages starting with making a mission statement. The stages were: identify customer's needs, establish target specifications, generate product concepts, select product concept(s), test product concept(s), and the last was set the final specifications.

2.2.1. The House of Quality

The House of Quality is a part of Quality Function Deployment and has a purpose to process the customer's needs through their opinion and statement into specifications and characteristics (engineering requirement) that can be measured and set a target for each of those [9]. Two aspects need to be concerned, they are the dimension of the product and the viewpoint of the needs from the customer's perspective. Based on [10] there are seven steps to build a House of Quality. First of all, is the identification of a customer's needs (Voice of Customer) into a product attribute. Next is to set a priority rating for each of its attributes from the VOC. Then evaluate the existing product using the attribute(s) that have been made. After that, making a relationship matrix that describes the correlation between attributes and technical characteristics of the product. And then identify and analyze both of them based on the designer's viewpoint about their correlation using four levels of scoring. Next is to identify and analyze the relevant interaction of each technical characteristics to another. And the last is to set the target or goal that can be measured to satisfy the customer's needs.

2.2.2. Identification of customer's needs (Voice of Customer)

Identification was conducted by asking several questions through interviews and questionnaires. Several aspects concerned that had been asked [8] there were desired use of the product, things that customers fond of about the existing product and things that customers did not like about it, and the last was about the opinion from them if they had any idea on how the product could be improved.

2.2.3. Established House of Quality

After the Voice of Customer (VOC) had been done and the attributes of the product had been listed, it was still considered as cannot be used directly to generate the concept. Therefore, the VOC must be defined again to the technical characteristics. Then to build the House of Quality, it needed to identify and analyze the attributes and technical characteristics of the product and look for the importance of each attributes to the customer and the satisfaction of existing products based on each attribute [10].

3. Methodology

3.1. Participants

The participants of this study consisted of two groups and both of them consisted of patients, doctors or nurses, lecturers in public health, and practitioners in elder gymnastics. The first group consisted of 30 people, this group was selected as a participant in identifying customer's needs and in making it to several attributes that related to design a therapeutic aid in reducing pain caused by knee Osteoarthritis. The second group consisted of 5 people, this small group was participated in helping to make House of Quality and Selection Concept(s).

3.2. Procedure

Compiling and circulating open questionnaires given to users/prospective users to get user's statements against static bicycle-based therapeutic tools.

Translating it into User's Needs based on user's statements, for example, after a user's statement could be pedaled by sitting, a static bicycle design requirement could be used in a sitting position. Based on the results of an open questionnaire, a closed questionnaire was created that was used to determine the relative importance and evaluate the competitor's products by giving a percentage weight to each attribute.

Conducting reliability testing for products and competitors by calculating the size of Cronbach's Alpha, as a material to consider whether or not the data can be used for the next calculation process.

The next step was to determine the relative importance of the attributes. The ranking of these attributes could be done by giving a percentage weight to each attribute by using a priority scale or level of importance. Evaluating the attributes of a competitor's product. The performance of a competitor was analyzed by giving a percentage weight to each attribute using a scale of priority or importance. Making a resistance matrix and identifying the relationship between product attributes and technical characteristics. Identifying relevant interactions between technical characteristics and determining the description of targets to be achieved for technical characteristics. The final step was to determine the target that must be achieved for each technical characteristic including, the level of difficulty of making the product, the level of importance, and the estimated cost level of each technical characteristic. For example, to determine the level of difficulty for product composition could be calculated with the steps: Difficulty Product Composition = (Weight of Each Technical Characteristics)/ (Weight of Technical Characteristics) x100%

4. Result and discussion

4.1. *Determination of Customer* Requirements

After the user's statement was collected, the list of statements was processed into a list of user needs for a static bicycle-based therapeutic device. This translation was done to facilitate designers in developing products so that the results could meet the desires and needs of users that must be owned by a static bicycle-based therapeutic tool. The results of translating user statements to a static bicycle-based therapeutic device contained 17 user requirements that must be owned by a static bicycle-based therapeutic product.

4.2. Data Validity and Reliability

Questionnaires that had been distributed and collected again were first tested for their validity and reliability before processing the data. From the calculation results, it was known that all variables were declared valid and reliable. The reliability test results for products and competitors could be obtained by Cronbach's Alpha for Products of 0.77 and Competitors of 0.79. It could be concluded that the Cronbach's Alpha value obtained for the product and the competitor had passed the reliability test so that the data could be used for the next calculation process.

4.3. Determination of Characteristics

In determining the characteristics that had the aim to determine the desires and needs of consumers for products based on static bicycle therapy tools used the QFD (Quality Function Deployment) method. Consumer desires obtained from closed questionnaires in the form of attributes for work chairs will be adjusted to the characteristics of the technique. This was done using the House of Quality.

4.4. Designing with QFD

The design of the QFD was carried out to determine the level of difficulty, degree of importance, and estimated costs of the product characteristics seen from the House of Quality (HOQ). Difficulty Level; Level 2 was obtained for the time of painting and the quality of the machine where it was not too difficult to process and provide in the design. Degrees of Interest; A value of 19% is obtained for the type of material, where the type of material was important in the design as a consideration in making the design. Estimated Cost; the value was 11% on the time of painting and the quality of the machine, which is quite moderate/quite expensive because it was used to support the attractiveness of the product being designed. The House of Quality was a combination of all technical characteristics, attributes desired by users, product position and competitors against the same attributes depicted in Figure 2.

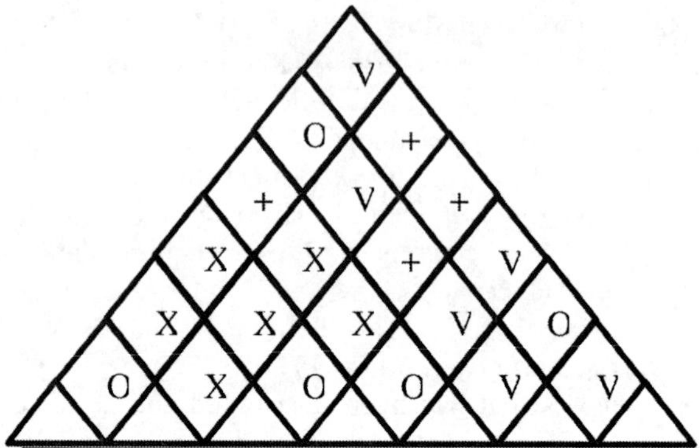

Degree of Relationship

X = Strong positive Relationship = 4
O = Medium positive Relationship = 3
V = Medium negative Relationship = 2
+ = Strong Negative Relationship = 1

	Customer Requirement	Relative Importance	Product composition	Material Type	Material Strength	Tool Weight	Life of use	Duration of Painting	Machine Quality	Existing Product User Perception
Design	The sitting position is used	5	+	x	o	+	v	+	o	4
	Triangular frame shape	4	o	x	x	o	x	+	v	4
	Rectangular pedal shape	4	+	x	o	o	v	+	o	5
	Load weight according to practitioner/doctor's recommendation	5	v	x	x	o	x	+	x	3
	Black colour	3	x	v	+	+	v	x	x	4
Function	The main function of therapeutic devices	5	+	x	x	x	x	+	o	4
Material	Iron frame	4	x	x	x	x	x	o	x	4
	Plastic pedal	4	o	x	x	x	x	+	v	4
	Rubber frame leg linings	4	v	x	x	x	x	+	+	4
	Rubber pedal adhesive	5	x	x	x	o	o	+	v	4
Degree of difficulty			3	3	3	3	3	2	2	
Degrees of Interest			13%	19%	17%	15%	16%	6%	14%	
Cost estimation			16%	16%	16%	16%	16%	11%	11%	

Degree of difficulty
1 = Easy = 0-5%
2 = Easy enough = 6-11%
3 = Difficult = 12-17%
4 = Very Difficult = 18-

Cost estimation
0–10 = Cheap
11–20 = Medium
21–30 = Expensive

Degree of interest
0 – 10 = Less Important
11 – 20 = Important
21 – 30 = Very Important

Figure 2. The house of quality.

4.5. *Alternative Design*

This stage aimed to gather alternatives that could be used to solve problems in the design of static bicycle pedal therapy equipment products using the Morphological Chart is showed in table 2

Table 2. Morphological chart.

Function	How to realize Functions		
	1	2	3
Position of Use	Sit	Lie down	Stand
Frame Shape	Triangle	Rounded	Box
Pedal Shape	Rectangle	Box	Hexagon
Weight Load	recommended by practitioners/doctors	1/2 load of ordinary static bicycle	Same with an ordinary static bicycle
Colour	Black	Gray	White
The main function	Therapeutic	Sports	Play
Frame Material	Iron	Aluminium	Plastic
Pedal material	Plastic	Rubber	Aluminium
Frame Leg Upholstery	Rubber	Plastic	Fabric
Pedal Adhesive	Rubber	Plastic	Fabric
	Alternative I	Alternative II	Alternative III

The Morphological Chart of static bicycle pedal therapy equipment products was displayed in the form of a matrix of 10 x 3. 10 was a function that must be achieved and 3 was an alternative that might be applied. The combination formula used was:

$$ {}_{10}C_3 = \frac{10!}{3!(10-3)!} = \frac{10!}{3!7!} = 120 $$

From the combination calculation results, the total number was very large, then a solution was found to reduce the total number to be more organized, namely the consideration of the interview results by sorting the answers of the most respondents and getting the 3 alternatives.

4.6. *Concept Selection*

The three alternative concepts will be selected using the objective weighted method and making concept 1 as a reference so that the selection criteria obtained from user needs and product attributes are made simpler [8]. Concept selection is done by distributing online questionnaires containing information related to descriptions, illustrations, how to use the product, and some questions regarding the selection of concepts that have been made.

1 2 3

Figure 3. Alternative concept.

Respondents at this stage consisted of 7 experts who had backgrounds as practitioners of elderly gymnastic instructors and or academics who were experts in the field of ergonomics. In conducting concept selection there are 5 assessments where 1 (one) is very bad compared to the reference of up to 5 (five) for very better than the reference shown in Table 4. Concept 1 is a reference so that each criterion is given a value of 3 (the same as a reference) but the team considered that there were 1 criterion which were not appropriate so according to Ulrich and Eppinger, concept 2 was used as a reference for this case and concept 1 would be given a value 2 [8]. Based on the calculation results, concept 3 had the highest final value of 354 so concept 3 selected as the final design of a static bicycle-based therapeutic apparatus.

Table 3. Concept selection.

Selection Criteria	Weight	Concept					
		1 Reference		2		3	
		Value	Value Weight	Value	Value Weight	Value	Value Weight
Functional	25						
Lightweight therapeutic apparatus	12	3	36	3	36	4	48
Suitable for various positions	13	3	39	5	65	3	39
Comfort	20						
Tied to the ends of both sides	10	3	30	4	40	4	40
Pedal comfort in pedaling	10	3	30	4	40	5	50
Flexibility of use	15						
Used at different locations	5	3	15	1	5	4	20
Can change the weight of the load	5	3	15	4	20	5	25
Height can be adjusted	5	2	10	3	15	2	10
Ease of use	20						
Easy to clean	6	3	18	3	18	4	24
Ease of storage when not in use	5	3	15	2	10	3	15
Ease to adjust height/paddle load	9	3	27	4	36	3	27
Ease of manufacture (manufacturing process)	20						
Low cost of raw materials	8	3	24	3	24	4	32
Low component complexity	6	3	18	2	12	1	6
The number of assembly stages is small	6	3	18	3	18	3	18
Total value		**295**		**339**		**354**	
Ranking		**3**		**2**		**1**	
Continue?		**No**		**No**		**Develop**	

Figure 2 shows concept 3 that combines a simple design, strong, and safe material used so that the therapeutic apparatus is designed not to be damaged easily. Therapeutic devices are made to reduce the magnitude of the dimensions of the product that fills the room by eliminating the seat on a static bicycle so that the product is designed as a pedal without a seat. Also, the therapeutic apparatus is designed to be used simply and flexibly so that it can be used in various seats with the same function as a static bicycle.

The design has a size of p x l x t of 39 cm x 35 cm x 28 cm based on anthropometric considerations. Besides, a hole is given in the oval-shaped tool leg with a size of 2.5 cm with a radius of 1.5 cm. The tool is given a hole in the leg to prevent the tool from sliding when in use so that it can tie a rope that will be connected to the seat used. Pedals are also equipped with adhesive so that the foot is more comfortable during the paddle process made of fabric. The tool has two functions, namely therapeutic tools and sports equipment. So, the load on the paddle weights an ordinary static bicycle by using a magnetic load that does not cause noise. Pedal load level adjustment can be done by turning the knob on the tool leg. A monitor on the body of the device that can show the number of turns, time, and calories is added to provide information as well as the round body shape of the tool is designed to add aesthetics.

5. Conclusion

The conclusion from the results of QFD processing obtained 10 attributes of the product design of a static bicycle pedal therapy device. The level of difficulty of the product has a type of technical characteristics that are not too difficult at the time of painting and the quality of the machine with the highest degree of importance is the type of material with a value of 19%. This can be the researchers' priority as a reference to improve the design of a static bicycle pedal therapy device. Then from the QFD results obtained 3 alternative designs for static bicycle pedal therapy products using the Morphological Chart under the wishes and needs of users.

Acknowledgment

"This research was financially supported by The Faculty of Engineering, Diponegoro University, Indonesia through Strategic Research Grant 2020 No: 2496/ST!!)/UN7.5.3.2/PP/2020".

References

[1] Center for Disease Control and Prevention. 2019. *Arthritis Basics: Osteoarthritis (OA)*. Retrieved from Center for Disease Control and Prevention: https://www.cdc.gov

[2] Pratiwi AI. Diagnosis and Treatment Osteoarthritis. *Journal Majority.* 2015 Jan 25;**4**(4): 10-17.

[3] Holland K. & William M. 2018. *Stages of Osteoarthritis of the Knee*. Retrieved from Healthline: https://www.healthline.com/health/osteoarthritis-stages-of-oa-of-the-knee

[4] Marlina TT. Effectiveness of Knee Exercises Against Pain Reduction in Knee Osteoarthritis Patients in Yogyakarta. *Sriwijaya Nursing Journal*. 2015;**2**(1):44-56.

[5] Goh S-L, Persson MSM, Stocks J, Hou Y, Welton NJ, Lin J, et al. Relative Efficacy of Different Exercises for Pain, Function, Performance and Quality of Life in Knee and Hip Osteoarthritis: Systematic Review and Network Meta-Analysis. *Sports Med.* 2019 Mar 4;**49**(5):743–61. doi.org/10.1007/s40279-019-01082-0

[6] Kusumawati K. 2003. *Effects of Isotonic Exercise with EN-Tree on Pain Reduction and Functional Capacity Improvement in Knee Osteoarthritis*. Semarang: Diponegoro University, Medical Rehabilitation Study Program FK.

[7] Karman WOSC. 2016. *Effects of Sandbag Exercise and Static Bike Exercise on Functional Activities of Patients with Knee Osteoarthritis*. Muhammadiyah Surakarta University.

[8] Ulrich, K. T., & Eppinger, S. D. (2012). *Product Design and Development* (5th Edition ed.). McGraw-Hill.

[9] Otto KN, Wood KL. 2001 *Product Design: Techniques in Reverse Engineering and New Product*

Development. Retrieved from https://books.google.com/books?hl=id&lr=&id=0X54fSKq7bkC&oi=fnd&pg=PR13&ots=53vaeBMtWY&sig=IgWWbZtzEsTg7wsC71g56-7LWPA
[10] Cohen L. 1995. *Quality Function Deployment, How to Make QFD Work for you.* New York: Addison-Wesley Publishing Company

University organizational culture mapping using Organizational Culture Assessment Instrument

A Rahman[1]*, S G Partiwi[1] and D S Theopilus[1]

[1]Department of Industrial and System Engineering, Sepuluh Nopember Institue of Technology, Surabaya, 60111, Indonesia

Email: *arief@ie.its.ac.id

Abstract. Culture is the core element of an organization. Organizational culture must reflect the shared values and beliefs that show people meaning and provide them the guidance to working behaviour. The organizational culture should be constructed by following the expected future values and considering the common value of people in the organization. Organizational Culture Assessment Instrument (OCAI) is one of the structural approaches to evaluate the implementation of organizational culture. Nowadays, many of the public university in Indonesia has been transforming into a more productive and agile organization. The university transformation also needs redefine or evaluate their organization's values as a part of strategic plan. Applying OCAI framework is necessary to assess the excellence of the transformation process in university. OCAI has applied in our university, i.e.: Sepuluh Nopember Institute of Technology (ITS), to define the organizational culture for the higher agility of the organization in the future. More than 400 university employees, including lecturers and administration staffs, have involved in organization culture's survey. The results of OCAI assessment showed that ITS tend to perform as the clan and the hierarchy type of organizational culture. While the management capabilities of ITS leaders showed that, the leadership characteristic is related to the clan culture.

1. Introduction

An organization must improve their productivity by optimizing their resources. The higher the productivity of an organization, the bigger achievement level of output and the lower the cost of input or resources [1]. Human resources are the most important resources who determine the level of productivity in every business [1] [2] [3]. The work performance of employee is important to run the high productivity organization, but the commitment of employee might not be disregard [1]. The employee's commitment presents the intangible closeness of employees to the organization [4]. Personal commitment will lead to the higher performance directly, but higher performance of work will not always have linear relation with level of commitment of employee [5]. High productivity should be influenced by both the commitment and the ability of employee in a balanced relation of work [6]. The ideal relationship between workability and commitment is described in Figure 1.

Figure 1. Balance of work ability and commitment.

Working ability and commitment need a balance relation on each side. Organizations with better learning towards the ability of work would have employees with expertise and good skills. Instead, low commitment of work may lead the employee's performance decrease. On the other condition, several organizations have been growth more with the high commitment of employees and the high loyalty to the organization but still stuck with unachievable performance because of less ability of skill and competencies. Both conditions will not result the highest productivity until the ability and commitment of employee growth in the balanced [7]. The commitment factor can determine the success of an organization as the competitive advantage [2].

Corporate culture will portray the behavior and thoughts of people in an organization. A good corporate culture will make the people belong the same values and beliefs which present where they work, the business, and their view as a part of the organization [2] [5]. An organization with the employee who has same values and standards of behavior usually will more successful rather than the organization with a lack of corporate culture [5]. Organizational culture is used to address the problems of external and internal adaptation as the proper way to understand, think, and feel associated with problems on the organization [8]. An organization with good culture will achieve the goals by all employee with less entropy of behavior. On the contrary, several problem or conflict related to people behavior or dysfunctional behavior will appeared and become the interaction problem while achieving the goals of the organization [1] [2] [8]. The illustration of organization with good culture and without culture implementation is depicted in Figure 2.

Figure 2. Comparison of the organizational with and without culture.

The organizational culture will change the viewpoints and behavior of employee. The culture change may lead the employee who had a different purpose and direction of work and will be moved in the same direction and purpose. These changes may affect the form of the movement that is efficient and effective in achieving the objectives of the organization [7]. ITS is one of the top sciences and technology in Indonesia. Being more productive and agile university is the future challenge for this university. ITS must able to increase the productivity of people by managing the performance achievement for each people and the unit of organization. ITS need to reform their process business and procedure and must eliminate the bureaucracy style of work. ITS need to identify the organization weakness related to corporate culture to transform organization's with excellence human resources.

Competition of global business in education sector are now threatening university in Indonesia. ITS must transform to be the lean organization and must apply the right business strategy to cope with quick environmental changes. Strategy in accordance with the dynamic environment, full of change and the demands for the adaptive nature are several characteristic of adhocracy cultures [2]. Adhocracy culture will put the flexibility and the discretion rather than the stability and the regularity [9]. In addition to this focus on culture change in external factors with the minimize attention to internal factors organizations [2].

Therefore, this research will be conducted the measurement of culture characteristic and mapping using OCAI. The result of the ITS organizational culture mapping and profiling will be compared with the expected culture condition. Some improvement to reach the ideal characteristic of organization will be proposed based on the gap of culture's strength and weakness.

2. Methodology

OCAI is one of familiar and useful frameworks for assessing organizational culture profile [10]. OCAI may provide the evaluation of current culture characteristic and the potential actions should be done. The culture assessment through a survey will determine the tendency of organization follow 4 types of organization, i.e.: clan, hierarchy, market, or adhocracy. By obtaining the current cultural tendency, the expected gap will show several lacks aspect to reach the future model of organization. OCAI is useful to redefine the true identity and extended rework of organizational cultures [2]. The basic theory of the OCAI is a model of Competing Value Framework (CVF). CVF has two-dimensional approaches where each dimension consists of two aspects of adversarial [10]. The first dimension of CVF will identify the aspect of flexibility, discretion, discretion and dynamic, with an emphasis on the aspects of stability, order, and control. The second distinguishing orientation on the internal environment of the company and integration, with orientation on the external environment of the company. The difference between these two dimensions would create four alternatives to the organizational culture [5].

According to [2], the organization culture can be classified into four types of cultures. Firstly, the Hierarchy culture has the main characteristics in the form of the organizational structure composed properly and efficiently. All work procedures are set by default and systematic. A good coordinator is the primary parameter of the effectiveness of the leader. The organization will emphasize on the stability and work's efficiency. The management model used is centered on a strict system of control. Secondly, the Market culture that has a major characteristic of the result's orientation. Effective leaders who able to see the opportunities, a hard worker but also flexible, will dominate the competition inside the organization. The leader will encourage the employee to compete for each other. Thirdly, the Clan culture may associate with a comfortable working environment and cozy welcoming room. The effective leader performs with a mentor role rather than as a boss role. The loyalty of employee will be the effective adhesive showed inside the organization. Fourthly, the Adhocracy culture will show the dominant character is the dynamic of the system and the entrepreneurial spirit. The leader has a great visionary, innovative, always pushing do increase and dare to take risks. Adhocracy is not generally bound by the formal organizational structures. The structure of work can be easy to change and quick to adapt to the dynamic market. The adhesive of adhocracy is the high commitment to the chance and to be agile in any condition.

Meanwhile, there are six dimensions of organizational culture should be considered while applying OCAI's framework [5]. First, the dominant characteristic which indicates what is easily seen and most prominent in a neighborhood organization. Second, the leadership of the organization which observes the leadership style. Third, the management of employees which exploring the human resources management in the organization, both individual and groups. Fourth, the adhesive of organization which indicates the value must be used to glue all resources of the organization. Fifth, the strategic emphasis indicates how the organization's focus on the achievement and the objectives of the organization. Sixth, the success criteria which exploring how the company set a parameter in the achievement of the objectives.

In addition, management capabilities should be described using Management Skills Assessment Instrument (MSAI). Management capabilities present more specific of culture characteristic by following four types of culture [5]. MSAI is the complement tool of OCAI to describe the capabilities of each organization's culture. By applying the values model assessment chart as depicted in Figure 3, it can be shown the tendency of management capabilities based on four culture's quadrants of OCAI. The clan quadrant will consist of several aspects to be considered, those are team management, relationship management with team members, and management of the employee development. The adhocracy quadrant also has three aspects should be evaluated, i.e. management of innovation, management of future, and management of ongoing improvement. For market quadrant, there will be three aspects to consider, those are the management of competitiveness, the motivation of members, and customer service. The hierarchy quadrant will explore the management of acculturation, the management control system, and the management coordination [2].

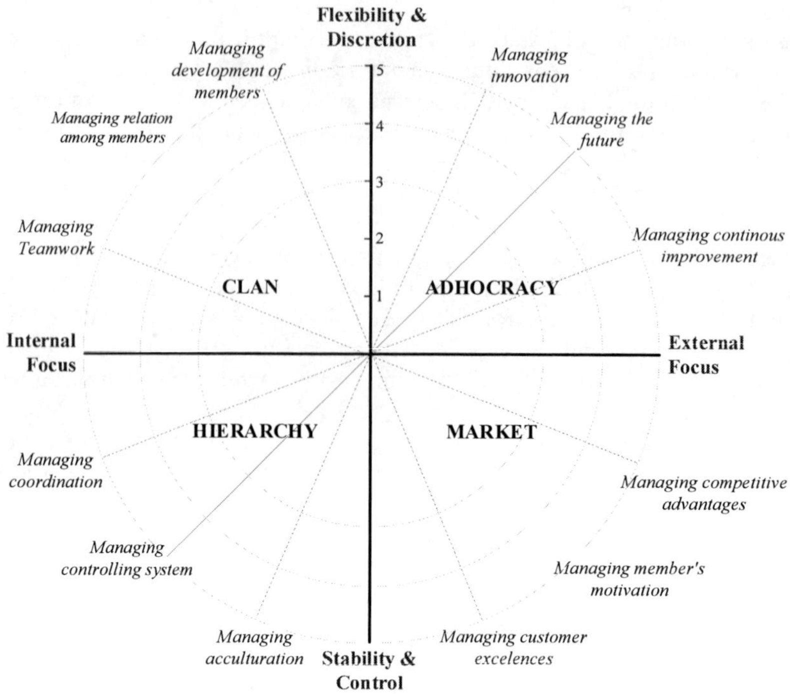

Figure 3. Values model assessment of management capabilities.

3. Results and Discussion

Corporate culture survey has done by involving 269 employees (including lecturer and operational staff) and 164 management team to obtain their responses and assessment related to culture implementation in ITS. There are three levels of management in ITS namely top management, middle management, and supervisor. Each level has a specific class of jobs, those are top management with job class equal to 16, middle management with job class equal to 14-15, and supervisor level with job class equal to 12-13. Most of the management team has submitted their responses and more than 50% of each level of management given their responses. 14 leader or 60% of top management has submitted their responses, 73 middle managers or 54% of totals middle management has submitted their response, and 77 supervisors or 50% of them has submitted their responses.

Based on the organizational culture survey which obtains several responses of ITS's employee, it showed that the clan characteristic reaches the highest 3.8 scores (maximum = 5). The scores mean that clan culture becomes the dominant characteristic of culture in organization. Dynamic working environment and informal communication among the employee will be preferred in organization. Besides, sub-coordinate unit will be performed more exclusive, the business process tends to be run autonomously, and some silos will appear as the barrier of integration. The hierarchy characteristic will also present significantly since the score reaches 3.5 points. This score of hierarchy type of culture will indicate that formal bureaucracy of process still appeared during the services. Stability of processes and systematic procedure of work will be the major character of organization. Table 1 present the employee culture survey. The indication of the market type of culture is presented, with a score equal to 3.0. Which mean that customer service orientation already had by the organization even the number is not dominant. According to this employee survey, the organization still need to transform the behavior of culture and redefine the mindset of organizational to reach the adhocracy type of culture. Since the agile organization near to adhocracy type of culture, ITS need the detail gap requirement from their baseline condition of culture toward adhocracy's culture. MSAI analysis are required to identify some detail gap of culture.

Table 1. OCAI scores based on employee survey.

Organizational Culture	Scores	Level
Clan (A)	3.8	1
Adhocracy (B)	2.5	4
Market (C)	3.0	3
Hierarchy (D)	3.5	2

The second step of the survey was done by applying the MSAI questionnaire to get the detail aspect of university culture. Every level of organization manager in the university has involved with various percentage of participation. The survey result of this stage is presented in Table 2.

Overall, the organization manager survey has a similar tendency of culture type compare with all employee survey. Clan's type of organization also took the highest score (4.22), like the all-employee survey's result. Thereby, the dominant culture characteristic of the university is the Clan organization. Figure 4 depicts the tendency of clan culture in the university. Besides, according to this manager survey, ITS also have a strong character of hierarchy and market type of organization. Moreover, the manager survey showed that some of the adhocracy's characteristic has performed inside the organization with the confident number of a score (4.10). The university management already have good potential resources to achieve the agility of organization. A change management strategy might be applied to lead that potential resources to have one direction of the goal.

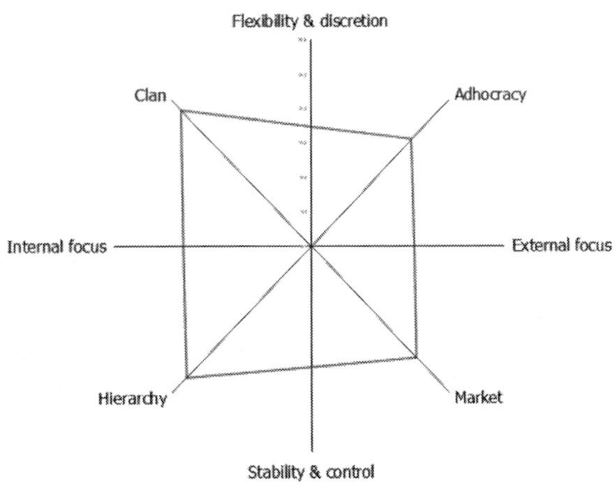

Figure 4. Overall culture characteristic of ITS.

Each level of manager in ITS has indicated some different dominant culture aspects. The top manager has already headed towards to the adhocracy organization where innovation and future-oriented become the priority to be achieved. Moreover, the middle manager has a different tendency, they tend to follow the clan organization. The priority aspect develops by this type of culture are build the teamwork and manage their member. The internal capacity building still becomes the focus of clan culture. Otherwise, most of the supervisor level presents a different tendency of culture, they have a strong aspect of the hierarchy type of organization. The strong character of the hierarchy organization usually close to the bureaucracy and have a good controlling system.

Table 2. Results score MSAI for several level of job in ITS.

Descriptor	Senior Manager	Manager	Supervisor	ALL
Clan Quadrant	4.3	4.17	4.20	4.22
Managing Teamwork	4.6	4.4	4.2	4.4
Managing relation among members	4	4.1	4.2	4.1
Managing development of members	4.3	4	4.2	4.17
Adhocracy Quadrant	4.4	4.07	3.83	4.1
Managing Innovation	4.2	4	3.9	4.03
Managing the future	4.6	4.2	3.8	4.2
Managing continuous improvement	4.4	4	3.8	4.07
Market Quadrant	4.13	3.77	3.87	3.92
Managing Competitive Advantages	3.8	4	4.1	3.97
Managing Member's Motivation	4	3.5	3.6	3.7
Managing Customer Excellences	4.6	3.8	3.9	4.1
Hierarchy Quadrant	3.73	4.03	4.27	4.01
Managing Acculturation	3.4	4	4.3	3.9
Managing Controlling System	4	4.1	4.4	4.17
Managing Coordination	3.8	4	4.1	3.97

A detailed mapping of exiting culture's profile of ITS is depicted on Figure 5 where the strength and the weakness of organization can be evaluated. The adhocracy quadrant becomes the expected position of the organization. The flexibility of resources to anticipate the very quick changes of market or the dynamic customer requirement must be the priority aspect to achieve by the university. The adhocracy organization will focus more on external excellences rather than the internal arrangement and will proposes the flexibility rather than the stability. Every resource including the organizational unit must be ready to change and response to adapt quickly. The leader of organization should have the capability as a risk-taker's leader and work with innovation to get the loop of business opportunity. The university should run a systematic action plan to encourage every employee to perform their innovation. When innovation management reaches all the part of organization, significant benefit or higher productivity can be achieved.

Figure 5. Organization culture map based on MSAI survey.

The university has several opportunities to improve the current culture characteristic toward the expected quadrant. Since the supervisor level showed the lowest score on the flexibility and external focus, continuous improvement program customer-oriented services might be chosen by the organization as the priority of action for supervisor. A knowledge transformation program also becomes an important action for the supervisor to re-orientate their basic perception about the importance of external focus and flexibility. Meanwhile, the middle manager which already have the flexibility character but still focus on internal orientation can be re-orientate by preparing the strategic future program such as business innovation program and organizational transformation program. Change the mindset of middle manager toward external focus orientation will improve cultural development.

The measurement of organizational culture at universities, especially in Indonesia, is rarely implemented as a part of strategic planning. Some universities have already stated their values of the organization but still lack structural programs or actions to implement and disseminate to their employee [9]. Assessing the performance of employees or the unit of the organization has been established periodically in most universities. Unfortunately, most leaders in the university might not realize that measuring the behaviour and values of the organization is the strategic policy. Developing a practical framework for culture-based university development can be future research development to provide the benefit of culture.

Innovation and flexibility are the two main capabilities that should be a strong identity of the university which produce academic and intellectual property. So, the adhocracy should be the most suitable design of an organization for the university. Meanwhile, the bureaucracy's characteristic seems to become the dominant culture set presented by the most university in Indonesia. Our finding showed that ITS, as one of the best universities in Indonesia, provide a strong hierarchy type of organization. The need for cultural transformation in ITS with conscious policy and program became the top priority to be implemented toward the adhocracy organization. Stimulating the individual resources in university who have already had an innovative mindset should be the high potential opportunity in the implementation stage. A strong culture-based oriented leader can be the primary key to conduct the university with less entropy.

4. Conclusions

Two stage of organizational culture using OCAI has applied in ITS as a university in Indonesia. Both all employee culture survey and the organization leader survey has confirmed than ITS tend to have a strong character as a clan organization. Based on top manager's perspective, ITS present some potential culture toward the flexibility and external focus orientation or following the adhocracy type of organization. However, the supervisor needs some cultural development program to re-orientate their mindset towards the adhocracy quadrant. Strategic change management program might be arranged as the priority action by ITS to accelerate the people transformation.

References

[1] Guiso L, Sapienza P, Zingales L. The value of corporate culture. *J financ econ*. 2015 Jul;**117**(1):60–76. doi.org/10.1016/j.jfineco.2014.05.010
[2] Cameron KS, Quinn RE. 2006. *Diagnosing And Changing Organizational Culture. Third Edition*. San Francisco: John Wiley & Sons, Inc.
[3] Galpin T, Whitttington JL, Bell G. Is your sustainability strategy sustainable? Creating a culture of sustainability. *Int. J. Bus. Soc.*, 2015 Feb 2;**15**(1):1–17. doi.org/10.1108/cg-01-2013-0004
[4] Cui X, Hu J. A Literature Review on Organization Culture and Corporate Performance. *Int. J. Bus. Adm.*, 2012 Mar 14;**3**(2). doi.org/10.5430/ijba.v3n2p28
[5] Hitka M, Vetráková M, Balážová Ž, Danihelová Z. Corporate Culture as a Tool for Competitiveness Improvement. *Procedia Econ.*, 2015;**34**:27–34. doi.org/10.1016/s2212-5671(15)01597-x
[6] Luthans F. 2011. *Organizational Behavior: An Evidence-Based Approach Published. 12th ed.* New York: McGraw-Hill/Irwin.
[7] O'Reilly C. Corporations, Culture, and Commitment: Motivation and Social Control in Organizations. *Calif Manage Rev*. 1989 Jul;**31**(4):9–25. doi.org/10.2307/41166580
[8] Schein EH. *Organizational Culture and Leadership, 4th Edition*. San Francisco: Jossey-Bass; 2010. p 436.
[9] Adeinat IM, Abdulfatah FH. Organizational culture and knowledge management processes: case study in a public university. *VINE J. Inf. Knowl. Manag. Syst*. 2019 Mar 6;**49**(1):35–53. doi.org/10.1108/vjikms-05-2018-0041
[10] Devi Ramachandran S, Choy Chong S, Ismail H. Organisational culture. *Int. J. Educ. Manag.*, 2011 Aug 16;**25**(6):615–34. doi.org/10.1108/09513541111159086

Quantifying energy consumption and greenhouse gas emissions of construction projects: A case study of Semarang - Bawen road project

J U D Hatmoko[1,*] and L Lendra[1,2]

[1]Department of Civil Engineering, Diponegoro University, INDONESIA
[2]Department of Civil Engineering, Palangka Raya University, INDONESIA

E-mail: *jati.hatmoko@ft.undip.ac.id

Abstract. Roads have significant impacts on the environment, economy and society, and the choice of pavement thus has broad implications. All phases of road development, from construction to operations, consume a vast amount of resources and produce waste and emissions. This study aims to examine energy consumption and greenhouse gas (GHG) emissions of a road construction project using Life Cycle Assessment (LCA) approach. Semarang-Bawen section road project was used as a case study, which uses Asphalt Concrete-Wearing Course (AC-WC) and Asphalt Concrete-Binder Course (AC-BC) layers for flexible pavement, and concrete slabs for rigid pavement. The methods used to estimate energy consumption and GHG emissions refers to the Intergovernmental Panel on Climate Change (IPCC) procedure, i.e. the Energy Use and GHG Emissions for Pavement Construction, and Fuel Conversion Method. The results show energy consumption of the project for the flexible and rigid pavements are 1193.79 GJ and 952.056 GJ, respectively. This indicates that the flexible pavement contributes to CO_2 emissions greater than the rigid one. The flexible pavement has a total emission of 89257.96 $kgCO_2/km$, while the rigid pavement is of 70464.53 $kgCO_2/km$. Based on these results it can be concluded that rigid pavement is considered more environmentally friendly than flexible pavement.

1. Introduction

Roads are one of important components of the infrastructure development of a country, including Indonesia. Roads are one of the transport infrastructures which plays a major role in people's lives and promotes public welfare [1]. Road construction and regular maintenance work require materials which can be produced through a process that is very carbon intensive and consume lots of energy [2]. Road construction is heavily dependent on natural materials, such as soil, stone, limestone, cement and asphalt. A significant number of studies have found that the conventional installation of asphalt pavements can contribute to vast amount of energy consumption and carbon emissions that are not favourable to environmental conservation and sustainable construction [3]. The use of non-renewable materials and natural resources can be a problem for future generations. Development programs that pay attention to social, economic, and environmental aspects have been launched by many countries and have become a joint agreement as sustainable development. Efforts leading to sustainable road development in Indonesia are based on the principles of sustainable road development used by other countries [4].

Roads have a major effect on the climate, economy and society, and pavement options also have wide consequences. All phases of road production, from construction to operations, consume vast amounts of environment products and resources and generate waste and emissions [5]. During the paving construction process significant volumes of energy consumption and particulate emissions are generated. Roads are constructed in multiple layers, consisting of subgrade, subbase, base and surface layers. Together these layers make up the pavement. The paving may be constructed from a wide range of materials and mixtures of gravels, cement, asphalt, concrete or strengthened soils. The types of materials and thickness of the pavement layers shall be determined by the projected traffic level. There are three primary types of pavements, i.e. flexible pavement (asphalt), rigid pavement (portland cement concrete) and hybrid pavement (both flexible and rigid layers exist in the same pavement) [6].

Despite growing exposure to sustainable development, the feasibility study is gradually emphasizing environmental considerations and future impacts of prospective projects. Greenhouse gas (GHG) emissions have become essential environmental standards after the Kyoto Protocol initiated a more coordinated international initiative to limit GHG emissions in 1997. As one of the main sources of GHG emissions, the construction industry has started to implement GHG pollution evaluations to evaluate project viability or sustainable development goals. Most studies have, however, centred on measuring GHG pollution from buildings. There has been relatively little attention paid to estimating GHG emissions from civil engineering systems, such as pavement. A potential explanation for this is the absence of sufficient knowledge in the early stages of the civil engineering project [7].

This study aims to examine energy consumption and GHG (CO_2) emissions of a road construction project with flexible and rigid pavements using Life Cycle Assessment (LCA) approach. Semarang-Bawen road project section Km. 11 + 500 - Km. 34 + 000 was used as a case study for this purpose, allowing a comparison to be made between the flexible and rigid pavements in terms of energy consumption and GHG (CO_2) emissions.

2. Literature Review

Table 1 shows several studies on consumption of energy also GHG emissions on road works which have been carried out. In Indonesia a research was done by collecting data on fuel consumption used at various stages of road flexible pavement, which refers to Intergovernmental Panel on Climate Change (IPCC) procedures. This research observed three stages of the process cycle, i.e. hot asphalt mixture production stage, material transportation stage, and asphalting work stage. The results show that the aggregate drying process is the most dominant process consumed energy, and emitted emissions of all stages [8]. Another research was carried out analysing the energy consumption and GHG emissions during the construction phase of a rigid pavement of the Cisumdawu toll road, with an observed road length of 800 m. This research found that the production and construction phases are the most energy consuming stages and produce the highest GHG [9]. An effort has been done to track and measure the carbon footprint of flexible pavement, from the production and distribution of materials raw material, and factory operations in road construction projects. The highest CO_2e emissions were produced by the plant used for laying asphalt concrete and material transportation [10]. Research on quantitative estimates of GHG emissions from road construction projects, complete assessments and life cycle measurements in the construction process for rigid and flexible pavements. The findings showed that production and transportation of materials contributed to the highest GHG emissions [11].

In other countries, research has been carried out to compare the implications on environment of asphalt pavement and reinforced concrete pavement using the LCA method from the material extraction stage, the production phase, the use phase, and the final disposal phase. Asphalt tends to have higher energy demand, lower production demands for materials and fertilizers and lower environmental emissions for the original development of similar paving designs [12]. Short life cycle inventory evaluation of continuous reinforced concrete pavement (CRCP) and asphalt pavement in relation to energy consumed by each type of road construction pavement. For CRCP, energy is used mainly in the production of cement and reinforcing steels, while energy use exists for asphalting during asphalt mixing, aggregate drying and bitumen production [13]. Based on project case studies in China, the research findings suggested a new inventory analysis approach for the evaluation of GHG pollution from portland cement pavement. The construction process of concrete pavement is carried out in three phases, i.e. the production of raw materials, the manufacture of concrete, and the construction at the site. The results show that the highest GHG emissions occur in the production phase, followed by the manufacturing phase, and the construction phase [14]. Another work is to use a life cycle assessment method to determine the environmental effects of maintenance of pavements. The categories of environmental impacts considered in the LCA can include global warming, human health, quality of ecosystem, acidification, land use, and etc. In this research, assessment of the impact is limited to carbon dioxide (CO_2) emissions for global warming potential (GWP). The results show that the preservation of the pavement brings significant environmental benefits to the reduction of CO_2 emissions due to the improved surface condition of the pavement, although the emissions generated at the construction stage [15].

The research findings of energy consumption of road projects may vary due to differences in construction technology, construction equipment and raw materials, and relatively large variability in sources of data base which leads to a lack of consistency of the results [16]. Although many studies have been carried out on this subject, each of the energy and emission studies has limitations depending on the nature and quality of the data used. This is also due to differences in system boundaries including energy inputs, so that the results cannot easily be compared [17], making this area of research is therefore still an interesting subject to study.

Table 1. Research on LCA of road pavement.

Research Scope	System Boundary	Output	Reference
Applying EIO-LCA model and analysis uncertainty in input and environmental outcomes of CRCP and hot mix asphalt (HMA) pavements	Materials production, construction, end-of-life phases	Energy, Toxics Release Inventory (TRI) chemical emissions (H_2SO_4), hazardous waste, pollutant emissions to air (SO_2, NO_2, particulars)	[12]
Using SETAC-EPA model for assessing the energy consumption of CRCP and HMA pavement	Process materials, construction	Energy consumption	[13]
Estimating energy consumptions and CO_2 emissions of HMA calculated based on IPCC guidelines.	Process of mixing, transportation and application at project sites	Consumption of energy and emissions of GHG	[8]
Analysis of life cycle inventory impact of concrete pavement construction on environment	Production of raw material, concrete production and pavement construction phase	GHG emissions	[14]
Consumption of energy and GHG emissions of a rigid pavement at the construction stage	Material, transportation and construction	Consumption of energy and emissions of GHG	[9]
Assessment of carbon footprint of flexible pavement	Off site and on-site construction project activities	GHG emissions in term of CO_2e	[10]
Estimates of GHG emissions from local road construction projects, rigid and flexible pavements	Production of raw materials, transportation, and construction	GHG emissions	[11]
Environmental effect of asphalt pavement preservation at stage of construction and operation	Production of materials, construction, use phase and end-of-life phases	CO_2 emissions	[15]

2.1. Sustainable Road Construction

Indonesia has proven its concern on environmental issues through Law no. 23/1997 on environmental management, with the main goal is to realize sustainable development. Indonesia also ratified the Kyoto Protocol through Law no. 17/2004. At The Conference of Parties (COP) in Copenhagen, Denmark in 2009 Indonesia is committed to reducing emissions by 26% by 2020 [9]. The Ministry of Public Works and Public Housing Regulation No. 5/2015 on General Guidelines for the Implementation of Sustainable Construction in the Implementation of Infrastructure in the Field of Public Works and Settlements, states that road operators must implement principles of sustainable construction by meeting the technical reliability requirements and sustainable principles, i.e. the social, economic, and environmental aspects must be considered in the use of resources, keeping the natural resources for future generations [18].

2.2. *Life Cycle Assessment (LCA)*

Since the 1980s, the use of LCA has gained consideration in environmental evaluation of specific goods and processes. Unlike processes and services in other industries, however, the LCA continues to be in the early stages of infrastructure implementation. Despite the growing number of LCA literature on highways, there is no agreement about what kind of pavement has environmental efficiency advantages [19].

The LCA studies generally consist of four phases, i.e. the purpose and scope of the definition, life cycle inventory (LCI), impact assessment and interpretation of results [20]. The LCI involves the collection of data and calculations to measure the input and output of materials and energy from a system, and the impact assessment evaluates the significance of potential environmental impacts based on LCI [21]. In road construction, both flexible pavement and rigid pavement stages of the LCA consist of:

- *Material Production.* The stage in which the process of making a product from raw materials. Generally, it uses asphalt mixing plant to produce hot asphalt mixes, and batching plant to produce concrete mixes, wheel loaders, dump trucks and mixer trucks.
- *Construction.* The process of assembling an infrastructure to support human life.
- *Maintenance.* The process to maintain road pavement conditions to serve traffic loads and the environment. The maintenance is in the form of routine, periodic, upgrading, and reconstruction.
- *End-of-Life.* There are two possibilities in this last stage of the life cycle. The first is that there is no waste and emissions generated at this stage, because the previous pavement becomes a new sub-pavement. The second possibilityis waste and pollution production as the old pavement is demolished and dumped into landfills.

2.3. *Flexible Pavement*

The flexible pavement refers to road with an asphalt surface. It consists of a road structure which includes asphalt-bound and unbound materials. In general, this pavement is made from various layers, such as hot mixed asphalt layer, unbound road base, unbound sub base and subgrade [6]. Figure 1 shows a typical cross section of a road with a flexible pavement structure consisting of three pavement layers [22].

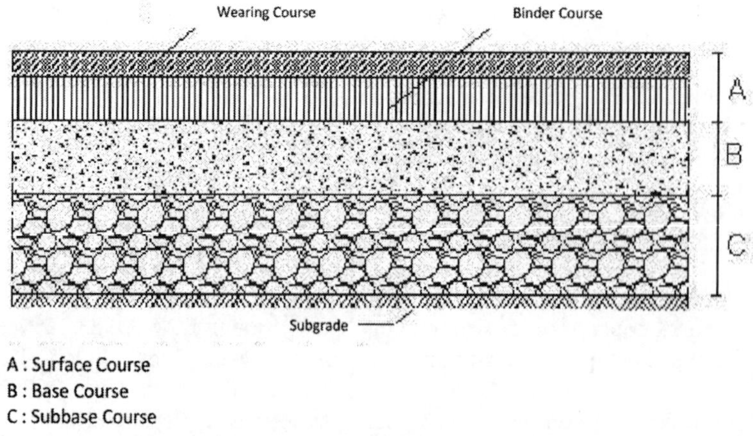

Figure 1. A typical flexible pavement structure.

2.4. Rigid Pavement

The rigid pavement, also called cement concrete pavement, is a pavement that uses cement as a binder so that the level of stiffness is relatively high, compared to asphalt pavement. Its structure consists of hydraulic cement concrete surface layers, foundation layer (base), and if needed, the sub base course layer [6]. Figure 2 shows a typical cross section of a road with a rigid pavement structure (reinforced concrete) consisting of three layers, i.e. sub grade, sub base course, and concrete slab [23].

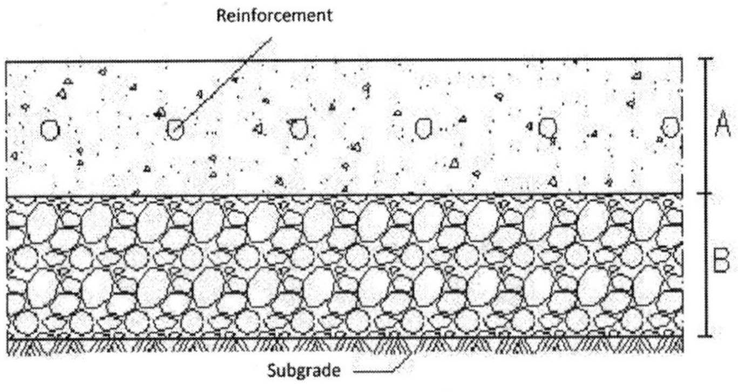

Figure 2. A typical rigid pavement structure.

3. Research Method

This research used the Semarang - Bawen road project as a case study for flexible and rigid pavements. The observed flexible pavement is located at Km. 16 + 650 - Km 17 + 375, and the rigid pavement is located Km. 11 + 990 - Km. 12 + 980. The analysis was carried out on the surface layers of the pavements, i.e. the AC-WC and AC-BC layers for the flexible pavement, and the concrete slab layer for the rigid pavement. The calculation of energy usage uses energy conversion method based on fuel consumption, while the GHG emissions analysed in this study are only Carbon Dioxide (CO_2). The formula refers to the Guidelines for National Greenhouse Gas Inventories Volume 2 guide [24] [25]. The analytical method used for estimating energy consumption is the conversion of fuel usage to a standard energy unit (Joule). To get the amount of energy consumption in each production of 1 Mg (tons) of pavement material, the calculation uses equation 1.

$$\text{Energy Consumption}\left(\frac{\text{MJ}}{\text{ton}}\right) = \frac{\text{Fuel Consumption (liter)} \times \text{Calorofic Value}\left(\frac{\text{MJ}}{\text{liter}}\right)}{\text{Total Production (ton)}} \tag{1}$$

The estimated amount of CO_2 emissions per tonne of pavement material production, referring to the equation in the IPCC guidelines, as described in equation 2.

$$\text{GHG Emissions}\left(\frac{\text{kgCO2}}{\text{ton}}\right) = \frac{\text{Energy Consumption (MJ)} \times \text{Emission Factor}\left(\frac{\text{kgCO2}}{\text{MJ}}\right)}{\text{Total Production (ton)}} \tag{2}$$

To calculate the estimate energy consumption using equation 3.

$$E = K_B \times Cv \tag{3}$$

where: E = energy consumption (MJ); K_B = fuel consumption (liter); Cv = calorific value (MJ/liter)

Calculation of CO_2 emissions per ton of material production using equation 4.

$$GHG = K_B \times F_E \tag{4}$$

where: GHG = greenhouse gas ($kgCO_2$); K_B = fuel consumption (liter); F_E = emission factor ($kgCO_2$/liter)

Data was collected in the construction site through observation of the construction process and interviews with engineers on site, and through literature research. The data collected is used to assess the life cycle of the materials following the stages, below:

1. Estimating the need for asphalt mixture in flexible pavement and concrete mixture in rigid pavement.
2. Calculating the fuel and energy consumption of equipment used at each stage of the life cycle on the pavement, starting from the stage material production to the end-of-life stage. Estimation of fuel consumption is done by converting units of fuel demand by multiplying fuel consumption by the conversion rate from Table 2 of fuel conversion, which refers to the guidelines of the IPCC.
3. Calculating emissions at each stage of the life cycle of road pavement, starting from the material production stage to the end-of-life stage. This is done by multiplying the fuel demand by the emission factor from the fuel conversion table, which refers to the guidance of the IPCC.
4. Comparing the results of the calculation of fuel consumption, energy consumption and emissions between the flexible pavement and the rigid pavement used in the Semarang - Bawen Road improvement project.

Table 2. Energy conversion factor [24].

Type of Fuel	Density (kg/ltr)	Calorific Value (MJ/ltr)	Emission Factor ($kgCO_2$/ltr)
Crude Oil	0.847	35.83	2.63
Diesel	0.837	35.99	2.67

4. Results

The stages of the analysis in this study are described as follows: (1) Collecting data in the form of flexible and rigid pavement technical data, such as: job mix formula, quantity of asphalt mixed material, quantity of concrete mix material, heavy equipment data, heavy equipment fuel consumption data, distance of asphalt mixing plant (AMP) and concrete batching plant to the location project, as well as the distance of the location of the extraction of aggregate and bitumen material to the AMP and concrete batching plant; (2) Analysis of estimated fuel consumption; (3) Analysis of estimated energy consumption; and (4) Analysis of CO_2 GHG emissions.

4.1. Consumption of Energy and Emissions in Flexible Pavement

The technical data of the flexible pavement includes 1000 m length and 14 m width of the road, 0.05 m and 0.09 m thickness of the AC-WC and AC-BC layers. Table 3 shows the material data for the structural components of the flexible pavement per 1 km. The results of the calculation of fuel consumption at the production, construction and maintenance stages, can be seen in Tables 4, 5 and 6 below.

Table 3. Material data for the structural components of 1 km flexible pavement.

Pavement Layer	Aggregate (ton)	Filler (ton)	Asphalt (ton)	Total (ton)
AC-WC	1518.91 (93.65 %)	15.41 (0.95 %)	87.58 (5.40 %)	1621.9 (100 %)
AC-BC	2749.34 (93.85 %)	27.83 (0.95 %)	152.33 (5.20 %)	2929.5 (100 %)
Total	4268.25 (93.78 %)	43.24 (0.95 %)	239.91 (5.27 %)	4551.4 (100 %)

Table 4. Fuel consumption of flexible pavement in the production stage.

Activity	Heavy Equipment	Heavy Equipment Fuel Consumption	Material Quantity 1 Km of Flexible Pavement	Total Fuel Consumption (litre/km)
Transportation:				
Aggregate	Dump Truck	0.00196 ltr/kg	4268250 kg Aggregate	8365.77
Asphalt	Dump Truck	0.0045 ltr/kg	239910 kg Asphalt	1079.59
Asphalt Mixture Production	Wheel Loader	0.677 ltr/m^3	2371.25 m^3 Aggregate	1605.34
	Generator Set	2.03 ltr/m^3	1960.96 m^3 Asphalt Mixture	3980.75
Asphalt Mixture Transportation	Dump Truck	4.752 ltr/m^3	1960.96 m^3 Asphalt Mixture	9318.48

Table 5. Fuel consumption of flexible pavement in the construction stage.

Process	Heavy Equipment	Fuel Consumption (litre/km)
Overlay	Asphalt Paver	455.14
Compacting	Tandem Roller	455.14
	Tire Roller	455.14

Table 6. Material and fuel consumption of flexible pavement in the maintenance stage.

Material Consumption (ton)		Fuel Consumption (litre)	
Asphalt	71.97	Production	7304.98
Aggregate	1280.47	Construction	409.62

For this research, the data needed to calculate the energy consumption and emissions using the energy conversion method is data of fuel consumption at each stage of the flexible pavement life cycle which then is converted to energy. The total amount of fuel consumption, energy consumption and emissions of the flexible pavement can be seen in Table 7. It shows that the estimated fuel consumption for the maintenance phase for the design life is worth 30% of each stage of production and construction. The assumption of end of life in this study is to reuse the old pavement as a subbase layer of the new pavement, so that there is no waste of material and energy at the end of life stage of this flexible pavement.

Table 7. The calculation of fuel, consumption of energy and emissions of flexible pavement.

LCA Stage	Total Fuel Consumption (litre/km)	Total Energy Consumption (GJ/km)	Total Emissions ($kgCO_2$/km)
Production	24349.93	867.00	65014.31
Construction	1365.42	49.14	3645.67
Maintenance	7714.60	277.65	20597.98
End of life	0.00	0.00	0.00
Total LCA	33429.95	1193.79	89257.96

4.2. Consumption of Energy and Emissions of The Rigid Pavement

The technical data of the rigid pavement includes a road length of 1000 m, the width of 14 m, and the thickness of the concrete slab layer of 0.27 m. Tables 8 and 9 show the results of the calculation of fuel consumption at the production and the construction stages for rigid pavement.

Table 8. Fuel consumption of the rigid pavement production stage.

Activity	Heavy Equipment	Heavy Equipment Fuel Consumption	Material Quantity for 1 km Rigid Pavement	Total Fuel Consumption (litre/km)
Transportation:				
Sand	Dump Truck	0.00208 ltr/kg	3179000 kg	6612.32
Gravel	Dump Truck	0.00137 ltr/kg	3825000 kg	5420.25
Cement	Bulk Truck	0.00436 ltr/kg	1512000 kg	6592.32
Concrete Mix Production	Wheel Loader	0.677 ltr/m^3	4395.71 m^3	2975.90
	Generator Set	0.875 ltr/m^3	3780 m^3	3307.50
Concrete Mix Transportation	Mixer Truck	0.377 ltr/m^3	3780 m^3	1425.06

Table 9. Rigid pavement fuel consumption in the construction stage.

Process	Equipment	Fuel Consumption (litre/km)
Moving bars	Pick Up	78.52
Pouring	Mixer Truck	0.80
Vibrating	Concrete Vibrator	171.60
Cutting	Concrete Cutter	54.60

Similar to the flexible pavement, the calculation of estimated consumption of energy and emissions in this study was carried out using a fuel conversion table. The results of the calculations can be seen in Table 10, assuming there is no maintenance during the design life of the rigid pavement, as one of the advantages of rigid pavement is the very minimal routine maintenance. It is assumed that at the end of life of the rigid pavement, the concrete slab that has lost its structural function and will be replaced by other new concrete slab by overlaying or reconstruction. The old concrete slab serves as a new concrete subbase.

Table 10. The calculation of fuel, consumption of energy and emissions of rigid pavement.

LCA Stage	Total Fuel Consumption (litre/km)	Total Energy Consumption (GJ/km)	Total Emissions ($kgCO_2$/km)
Production	26333.35	941.00	69829.49
Construction	305.52	11.06	635.04
Maintenance	0.00	0.00	0.00
End of life	0.00	0.00	0.00
Total LCA	26638.87	952.06	70464.53

5. Discussion

Tables 7 and 10 show a comparison of the results of the fuel and energy consumption, also the emissions for the flexible and the rigid pavements. It can be seen that at the production stage, apparently the flexible pavement consumes less fuel and energy of 24349.93 litres/km and 867.00 GJ and produces less emissions of 65014.31 $kgCO_2$/km compared to the rigid one. For flexible pavement, it is known that the asphalt production of AMP consumes the most dominant energy and generate greenhouse gas emissions [8]. The highest CO_2e emissions are typically generated by the use of plant for on-site activities, i.e. for asphalt overlaying activities accounted for 34,827 tons of CO_2e (49,130%), and material transportation accounted for 24,921 (35,155%) [10].

At the construction stage, the rigid pavement consumes less fuel and energy than the flexible one, of 305.52 litres/km and11.06 GJ, and produces less emissions of 635.04 $kgCO_2$/km. Whereas in the maintenance and end of life stages, the flexible pavement consumes more fuel and energy of 7714.60 litres/km and 277.65 GJ, respectively and produces fewer emissions of 20597.98 $kgCO_2$/km. Typically, the consumption of energy and GHG emissions of rigid pavement mostly in the stages of production and construction indeed, while the processing of materials requires the most energy and emits emissions [9].

For all the LCA stages, the total energy consumption of the flexible pavement is 1193.79 GJ, while the rigid pavement is 952.056 GJ, showing that the flexible pavement contributes greater CO_2 emissions than the rigid pavement. As for the total emissions, the flexible pavement has a total emission of 89257.96 $kgCO_2$/km, greater than the rigid pavement of 70464.53 $kgCO_2$/km.

This research found that there is a significant difference of the results of energy consumption and emissions per stage between the flexible pavement and the rigid pavement. This study assumes that the rigid pavement is without maintenance, although in fact there could be maintenance work that is carried out even at the minimum level. Assuming an equivalent pavement design, the flexible pavement seemed to have a higher energy input, and lower emissions than the rigid one. For the construction of 1 km long road with two typical lanes requires energy of 7.0×10^6 MJ for flexible pavement, and 5.0×10^6 MJ for rigid pavement [12]. The Swedish Environmental Research Institute (IVL) reports that concrete pavement requires 37% more energy in comparison to asphalt pavement [26], this dissimilar results occurred because of significant differences in system boundaries between the two studies [13].

6. Conclusion

This research aims to examine the consumption of energy and GHG emissions of a road construction project using Life Cycle Assessment (LCA) approach using a road project of Semarang - Bawen section as a case study. The research found that the energy consumption for the flexible pavement is greater than the rigid pavement, which is 1193.79 GJ and 952.056 GJ, respectively. For emissions, the flexible pavement contributes to greater CO_2 emissions than the rigid pavement. Flexible pavement has a total emission of 89257.96 $kgCO_2$/km while rigid pavement is 70464.53 $kgCO_2$/km.

Based on the results of the case study of all LCA stages, it can be concluded that the rigid pavement can be considered more environmentally friendly and hence is more sustainable than flexible pavement, because it consumes less fuel, less energy and produce less CO_2 emissions. These results are true for the project of the case study, and hopefully may represent other similar projects in general. It should be noted, however, as every project is actually unique with all various factors which might impact on the project, various results may be expected. Based on the limitation of this study, further research can be done by calculating the energy consumption in the mining process and the processing of raw materials in the quarry, as well as the calculation of consumption of energy and emissions produced in the maintenance and end of life stages of the rigid pavement.

Acknowledgement

The authors wish to thank Drajad Bagus Adhiprasetyo and Yudi Hidayat as research assistants for the field survey and data collection of this research. This research was financially supported by The Faculty of Engineering, Diponegoro University, Indonesia through Strategic Research Grant 2020.

References

[1] Isya M, Rani HA, Utama FP. Effect of Green Road Concept on Waste Management on Road Construction in The Banda Aceh City. *IOP Conf. Ser. Mater. Sci. Eng.*, 2019 Jan 16;**469**:012061. doi.org/10.1088/1757-899x/469/1/012061

[2] Santos J, Ferreira A, Flintsch G. A life cycle assessment model for pavement management: road pavement construction and management in Portugal. *Int. J. Pavement Eng.*, 2014 Aug 13;**16**(4):315–36. doi.org/10.1080/10298436.2014.942862

[3] Zhang Z, Gao X, Wang J, Ji X. Prediction model for energy consumption and carbon emission of asphalt surface construction. *IOP Conf. Ser.: Earth Environ. Sci.*, 2019 Nov 9;**330**:022052. doi.org/10.1088/1755-1315/330/2/022052

[4] Lawalata GM. Sustainable Road Development Principles. *Jurnal Transportasi.* 2013;**13**:115–24

[5] Kucukvar M, Tatari O. Ecologically based hybrid life cycle analysis of continuously reinforced concrete and hot-mix asphalt pavements. *Transp Res D Transp Environ.* 2012 Jan;**17**(1):86–90. doi.org/10.1016/j.trd.2011.05.006

[6] Thives LP, Ghisi E. Asphalt mixtures emission and energy consumption: A review. *Renew. Sust. Energ. Rev.*, 2017 May;**72**:473–84. doi.org/10.1016/j.rser.2017.01.087

[7] Kim B, Lee H, Park H, Kim H. Framework for Estimating Greenhouse Gas Emissions Due to Asphalt Pavement Construction. *J. Constr. Eng. Manag.*, 2012 Nov;**138**(11):1312–21. doi.org/10.1061/(asce)co.1943-7862.0000549

[8] Wirahadikusumah RD, Sahana HP. Estimasi Konsumsi Energi dan Emisi Gas Rumah Kaca pada Pekerjaan Pengaspalan Jalan. *Jurnal Teknik Sipil.* 2012 Apr 1;**19**(1):25. doi.org/10.5614/jts.2012.19.1.3

[9] Mulyana A, Wirahadikusumah R. Analysis of Energy Consumption and Greenhouse Gas Emissions in the Construction Phase Case Study: Construction of Cisumdawu Road. *Jurnal Teknik Sipil ITB.* 2017;**24**:269–80

[10] Utomo Dwi Hatmoko J, Hidayat A, Setiawati A, Catur Adi Prasetyo S. Measuring Carbon Footprint of Flexible Pavement Construction Project in Indonesia. Hadiyanto, Sudarno, Maryono, editors. E3S Web Conf., 2018;31:07001. doi.org/10.1051/e3sconf/20183107001

[11] Handayani FS, Pramesti FP, Wibowo MA, Setyawan A. Estimating and Reducing the Release of Greenhouse Gases in Local Road Pavement Constructions. Int J Adv Sci Eng Inf Technol. 2019 Oct 31;9(5):1709. doi.org/10.18517/ijaseit.9.5.9705

[12] Horvath A, Hendrickson C. Comparison of Environmental Implications of Asphalt and Steel Reinforced Concrete Pavements. Transp. Res. Rec., 1998 Jan;1626(1):105–13. doi.org/10.3141/1626-13

[13] Zapata P, Gambatese JA. Energy Consumption of Asphalt and Reinforced Concrete Pavement Materials and Construction. J. Infrastruct. Syst., 2005 Mar;11(1):9–20. doi.org/10.1061/(asce)1076-0342(2005)11:1(9)

[14] Ma F, Sha A, Yang P, Huang Y. The Greenhouse Gas Emission from Portland Cement Concrete

Pavement Construction in China. Int. J. Environ. Res. Public Health. 2016 Jun 24;13(7):632. doi.org/10.3390/ijerph13070632

[15] Wang H, Al-Saadi I, Lu P, Jasim A. Quantifying greenhouse gas emission of asphalt pavement preservation at construction and use stages using life-cycle assessment. Int J Sustain Transp. 2019 Jan 11;14(1):25–34. doi.org/10.1080/15568318.2018.1519086

[16] Fei L, Zhang Q, Xie Y. Study on energy consumption evaluation of mountainous highway based on LCA. IOP Conf. Ser.: Earth Environ. Sci., 2017 Jun;69:012036. doi.org/10.1088/1755-1315/69/1/012036

[17] Dixit MK, Culp CH, Fernandez-Solis JL. Embodied Energy of Construction Materials: Integrating Human and Capital Energy into an IO-Based Hybrid Model. Environ. Sci. Technol., 2015 Jan 15;49(3):1936–45. doi.org/10.1021/es503896v

[18] Lawalata GM. Pemeringkatan Jalan Hijau untuk Mendukung Implementasi Program Konstruksi Jalan Berkelanjutan. *Jurnal HPJI.* 2019;**5**:21–30

[19] Inyim P, Pereyra J, Bienvenu M, Mostafavi A. Environmental assessment of pavement infrastructure: A systematic review. J. Environ. Manage., 2016 Jul;176:128–38. doi.org/10.1016/j.jenvman.2016.03.042

[20] Environmental management. Life cycle assessment. Principles and framework. BSI British Standards; doi.org/10.3403/01139131u

[21] Ramesh T, Prakash R, Shukla KK. Life cycle energy analysis of buildings: An overview. Energy Build., 2010 Oct;42(10):1592–600. doi.org/10.1016/j.enbuild.2010.05.007

[22] Wang Y, Chong D. Determination of Optimum Pavement Construction Alternatives to Minimize Life-cycle Costs and Greenhouse Gas Emissions. Construction Research Congress 2014. 2014 May 13. doi.org/10.1061/9780784413517.070

[23] AASHTO. 1993. *AASHTO Guide for Design of Pavement Structures* (Washington, DC: American Association of State Highway and Transportation Officials)

[24] Intergovernal Panel on Climate Change. 2006. *2006 IPCC Guidlines for National Greehouse Gas Inventories* ed S Eggleston, L Buendia, K Miwa, T Ngara and K Tanabe (Hayama, Kanagawa: Institute for Global Environmental Strategies (IGES))

[25] Kim B, Lee H, Park H, Kim H. Framework for Estimating Greenhouse Gas Emissions Due to Asphalt Pavement Construction. *J. Constr. Eng. Manag.*, 2012 Nov;**138**(11):1312–21. doi.org/10.1061/(asce)co.1943-7862.0000549

[26] Stripple H. 2001. *Life Cycle Assessment of Road A Pilot Study for Inventory Analysis* (Gothenburg, Sweden: IVL)

IOP Conf. Series: Materials Science and Engineering 1072 (2021) 012071 doi:10.1088/1757-899X/1072/1/012071

How sustainable are flexible and rigid pavement? A Life Cycle Impact Assessment (LCIA) approach

J U D Hatmoko[1]* and L Lendra[1,2]

[1]Department of Civil Engineering, Diponegoro University, Indonesia
[2]Department of Civil Engineering, Palangka Raya University, Indonesia

Email: *jati.hatmoko@ft.undip.ac.id

Abstract. The number of road construction projects continues to grow, potentially causing more impacts on the environment. This research analyses the impacts of road construction projects of flexible and rigid pavement by adopting the Life Cycle Impact Assessment (LCIA) approach, and by means of the Eco-Indicator 99 (EI99). The impact values at the production, construction and maintenance stages for the flexible pavement are 86.60 kPt (49.34%), 65.40 (37.26%) and 23.50 kPt (13.39%), while for the rigid pavement equals 47.70 kPt (60.46%), 31.20 (39.54%) and 0 kPt (0%), respectively. It is smaller than the flexible pavement, i.e. 175.50 kPt and 78.9 kPt, respectively. The results of the EI99 show the impact values of the flexible pavement from the aspects of human health, ecosystem quality, and resources are 11.99 kPt (6.83%), 2.17 kPt (1.24%), and 161.40 kPt (91.93%), while for the rigid pavement of 29.49 kPt (37.43%), 37.80 kPt (4.79%), and 45.54 kPt (57.78%), respectively. The total impact value of the three aspects of the rigid pavement is 78.9 kPt, which is smaller than, and worth 45% of the flexible pavement of 175.50 kPt. It can be concluded that the rigid pavement can be considered more sustainable than flexible pavement.

1. Introduction

The construction industry plays a very important role in the national infrastructure growth. A variety of construction activities, while benefiting the national economy, have the potential to cause adverse environmental impacts. In the construction sector, road construction is one of the major emitters of greenhouse gasses and continues to grow along with the road developments by the government [1]. The construction of concrete road (rigid pavement) continues to grow and is progressively becoming the main type of roadways for heavy traffic loads, due to its long service life and less frequent maintenance requirements compared to the asphalt road (flexible pavement) [2]. The asphalt road is slowly being replaced by the concrete road as it breaks down faster [3]. In terms of structural strength and maintenance costs, the rigid pavement is superior to the asphalt roads [4]. There has been a question, however, on the environmental impacts from the construction process of these two types of pavement. The used of different materials in a project may impact on the energy consumption of the construction process, and ultimately can have impacts on the environment, including global warming.

The concept of sustainable construction exists and becomes a necessity in the midst of the phenomenon of global warming and the issue of environmental damage that is afflicting humanity. Based on Global Status Report 2017, buildings and construction combined compensate for 36% of the final electricity usage and 39% of energy-related emissions of carbon dioxide (CO_2) with the inclusion of upstream power production [5]. The idea of sustainable construction is seen as one of the strategies for mitigating environmental degradation and lowering carbon emissions, which is the key source of global warming in the construction industry. This research aims to analyse the environmental impacts of road construction projects of flexible and rigid pavement by adopting the Life Cycle Impact Assessment (LCIA) approach, and by means of the Eco-Indicator 99 (EI99).

2. Literature Review

2.1. Sustainable Development

Sustainable development is an attempt to fulfil all basic needs and to create resources to satisfy the human expectations for a better life. There are three pillars that support the nature of sustainability, i.e. social aspects, environmental aspects and economic aspects, which interact with one another [6] [7], as shown in Figure 1.

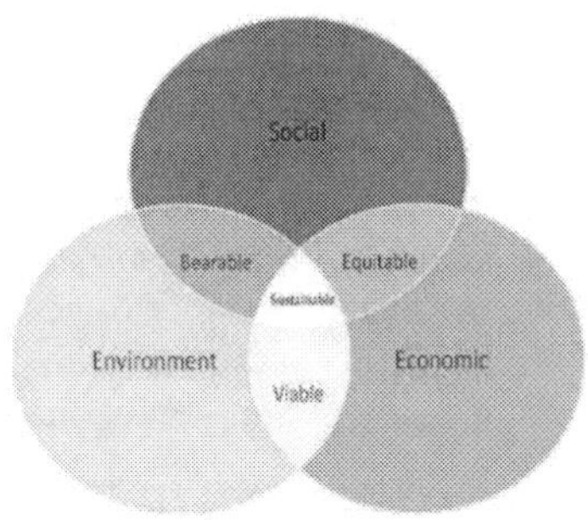

Figure 1. Supporting pillars for sustainability.

The Indonesian government's contribution to rising global change is accomplished by reducing emissions by 26 percent by 2020. This was then updated separately in 2015, with an emission reduction goal of 29% by 2030. One of the follow-up duties in the construction industry is the issuing of Ministry of Public Works regulation no. 11/2012 concerning the National Action Plan for Mitigation and Adaptation to Climate Change in 2012-2020, and the regulation no. 02/ PRT/M/2015 concerning Green Buildings for reducing greenhouse gas (GHG) emissions sourced from buildings. Emissions reducing strategies must be implemented in all sectors and regions [8]. Road operators must also implement a sustainable construction approach by meeting the requirements of technical reliability and sustainable principles in accordance with Ministry of Public Works and Public Housing (PUPR) regulation no. 5/2015, concerning General Guidelines for the Implementation of Sustainable Construction in the Implementation of Infrastructure in the Field of Public Works and Settlements [9].

2.2. Life Cycle Impact Assessment (LCIA)

Life Cycle Impact Assessment (LCA) is a method used in the life cycle stage from the material up to the product stage is used by consumers [10]. The ISO 14040 Standard divides the process of LCA in four stages, i.e.: (1) The goal and scope definition; (2) Life Cycle Inventory (LCI) analysis; (3) Life Cycle Impact Assessment (LCIA); and (4) interpretation of results [10] [11].

The LCIA has been adopted in previous studies, e.g. evaluation on the environmental impacts of mining and mineral processing, material and equipment transportation, construction, maintenance, and reconstruction of a 35-km long dual road in Abu Dhabi [12]; estimation of the asphalt and concrete impact on from the production process to the disposal(cradle to grave boundary system) by using the ReCiPe and SimaPro software [13]; calculation the effect of using polyurethane bounded pavement to determine whether it satisfies the energy efficiency, environmental protection and sustainable development requirements [14]; assessment of energy and environmental life cycles for renewable materials and urban road surface technologies focused on the use of bitumen mixtures (hot mix asphalt and wet asphalt mixtures) with recycled materials (reclaimed asphalt pavement, crumb rubber and used plastics) [15];valuation of the environmental impact of recycled aggregates by process separation into raw matter, transport and recycling aggregates, and identification of materials used at each point and amount of energy consumption for environmental impact evaluation [16]; assessment of the energy use and environmental effects of asphalt pavement from used tires during the life cycle and end of material production, construction and maintenance [17].

In addition, the LCA approach was implemented for building a Triple-Bottom-Line (TBL) model for sustainable assessment of the environmental and socio-economic impacts of pavements mixtures constructed with Warm Asphalt mix (WMA) to compare with traditional Hot Mix Asphalt (HMA) [18]; impact assessment of substances emitted from the concrete manufacturing cycle in six areas of environmental effects, i.e. global warming, acidification, eutrophication, abiotic depletion, ozone depletion, and photochemical oxidizing manufacturing [19], and examining the environmental impact of the increasing use of recycled materials in asphalt pavement construction and maintenance [20].

2.3. Eco-Indicator 99

Eco-Indicator 99 (EI99) denotes the impacts on the environment of materials or process based on data from the LCA. The higher value of the indicator indicates the greater impact on the environment. EI99 helps manufacturers or project managers to evaluate a company's environmental effects over its life cycle [21]. Three types of impacts are defined in EI99, as follows:

1. Impact on human health: articulated as the number of years lost, also the number of years living with disabilities. The index is Disability Adjusted Life Years (DALYs). The impact of these aspects includes climate change, depletion of ozone layer, effects of carcinogenic and respiratory, and ionizing radiation.
2. Impact on ecosystem quality: the end-point indicator used in this aspect is the Potentially Disappeared Fraction of species (PDF), which is the loss of a species in a certain area over a certain period of time (PDF.m2.years). Effects in this category include: ecotoxicity, acidification, eutrophication, and land-use.
3. Impact on resources: future excess energy requirement for extraction of minerals and fossil fuels, with a surplus MJ unit. The impact of this aspect is minerals and fossil fuels.

2.4. Types of Pavement of Road Infrastructure

Typically, the main types of pavement of road infrastructure are flexible pavement, rigid pavement and composite (both flexible and rigid layers in the same pavement). The designer must be able to choose the appropriate pavement type, and ensure the road can serve users to travel on the pavement that is safe, durable, quiet, smooth, economical, and using sustainable materials [22]. Road network in Indonesia is characterized by flexible pavement, which is assumed to contain high GHG emissions due to shifts in land settings and functions as well as resource use and utilization [1].

3. Research Method

This research adopted a sustainable framework of the LCA method to find out how much environmental impact arising from flexible pavement and rigid pavement. Data was collected from a road construction project of Semarang - Bawen section Km. 11 + 500 - Km. 34 + 000, via site observation, observing project documents, followed by interviews with the consultant and the contractor to understand the construction process more thoroughly. This research applies the four basic stages of the LCA, i.e.: (1) The goal and scope definition; (2) Life Cycle Inventory (LCI); (3) Life Cycle Impact Assessment (LCIA); and (4) Interpretation of Results.

3.1. The Goal and Scope Definition

The goal is to analyse the environmental impact of flexible pavement and rigid pavement through the life cycle and make a comparison of the two to learn which type of pavement is friendlier to the environment. The scope consists of two parts, i.e.: (1) Functional Unit: road pavement with a length of

1 kilometre and width of 14 m (4 lanes 2 directions) for each type of pavement being compared; and (2) Boundary System: material production, construction, and maintenance.

3.2. Life Cycle Inventory (LCI)

The purpose of the LCI study is to identify and determine the extent and value of the system's environmental impacts by measuring the product's inputs (i.e. resources and energy) and outputs (i.e. emissions and waste) over its life cycle [23]. It is a quantitative analysis of energy consumption and emissions of products, processes or activities with the process of collecting and calculating input and output data in the entire system life cycle [14]. For this purpose, the data collected include: job mix formula, quantity of asphalt mix material, quantity of concrete mix material, heavy equipment data, fuel consumption data, the distance of AMP (Asphalt Mixing Plant) and concrete batching plant to the project location, and the distance of quarries of aggregate and bitumen to AMP and concrete Batching Plant.

3.3. Life Cycle Impact Assessment (LCIA)

The LCIA analyse the type and value of each impact produced by means of the EI99 using SimaPro software. The LCIA calculation phase is divided into four steps, i.e.: (1) classification : the appropriate impact aspect is assigned to the inventory items taken out from the inventory analysis; (2) characterization: the effect of each object shall be categorized into the effect group for each quantified dimension; (3) normalization: the environmental impacts given to the environmental aspect are divided into environmental impacts, locally or globally; and (4) weighting: the comparative importance among the impact aspects is determined [19]. Table 1 shows the characterization-normalization-weighting indicator with EI99, which will be used for calculation in the following stages:

- the normalization, i.e. the characterization value is multiplied by the normalization factor
- the weighting, i.e. the normalization value is multiplied by the weighting factor
- single score, i.e. classification the values of the impact category based on the activity or process

Table 1. Indicator characterization-normalization-weighting with eco-indicator 99.

Impact Category	Unit	Normalization	Weighting
Carcinogens	DALY	65.1	300
Resp. Organics	DALY	65.1	300
Resp. Inorganics	DALY	65.1	300
Climate Change	DALY	65.1	300
Radiation	DALY	65.1	300
Ozone Layer	DALY	65.1	300
Ecotoxicity	PDF.m2.yr	1.95E-5	400
Acidification / Eutrophication	PDF.m2.yr	1.95E-4	400
Land Use	PDF.m2.yr	1.95E-4	400
Minerals	MJ surplus	1.95E-4	200
Fossil Fuels	MJ surplus	1.95E-4	200

The environmental impacts are analysed based on the damage caused to three aspects, i.e. human health, ecosystem quality and resources. Each of these aspects has their respective impacts categories, as follows:

- Human Health: carcinogens, respiratory organics, inorganic respiration, climate change, radiation, and ozone layer.
- Ecosystem Quality: ecotoxicity, acidification, eutrophication and land use.
- Resources: minerals and fossil fuel.

3.4. Interpretation of Results
This phase is a combination of the results of the LCI and LCIA, which are then used to interpret, draw conclusions and recommendations in accordance with the goals and scopes identified earlier. The findings are recorded in the most detailed manner and incentives are regularly evaluated to reduce the effect of goods or services on the ecosystem [10].

4. Results
This section describes the results of the environmental impacts on the flexible pavement, rigid pavement, and the comparison of them.

4.1. Environmental Impacts of The Flexible Pavement
The flexible pavement being studied consists of two layers of asphalt pavement, i.e. asphalt concrete-wearing course (AC-WC) and asphalt concrete-binder course (AC-BC), each with a length of 1000 m and the width of 14 m. The AC-WC layer is 0.05 m thick, while the AC-BC layer is 0.09 m thick. Table 2 shows the results of the flexible pavement with the EI99 method in accordance with the LCIA characterization factor values. The total impact value of the three aspects of damage category, i.e. human health, ecosystem quality, and resources of the flexible pavement is 175.50 kPt, with the largest is related to resources (91.93%). For the LCIA approach, among the three stages, the largest impact value is at the production stage of 86.60 kPt (49.34%), followed by the construction, and the maintenance stages of 65.40 (37.26%) and 23.50 kPt (13.39%), respectively.

Table 2. The results of EI99 and LCIA of the flexible pavement.

Damage Category	Impact Category	Total (kPt)	Production	Construction	Maintenance
Human Health	Carcinogens	0.27	0.16	0.07	0.04
	Resp. Organics	0.06	0.03	0.02	0.01
	Resp. Inorganics	9.20	6.27	1.86	1.06
	Climate Change	2.41	1.48	0.64	0.29
	Radiation	0.05	0.02	0.02	0.01
	Ozone Layer	0.01	0.00	0.00	0.00
	Sub total	11.99 (6.83%)	7.96 (66.3%)	2.61 (21.8%)	1.41 (11.8%)
Ecosystem Quality	Ecotoxicity	0.80	0.53	0.17	0.10
	Acidification / Eutrophication	1.47	1.01	0.28	0.17
	Land Use	-0.09	0.10	-0.17	-0.03
	Sub total	2.17 (1.24%)	1.64 (75.2%)	0.28 (12.8%)	0.24 (11.0%)
Resources	Minerals	0.40	0.23	0.12	0.05
	Fossil Fuels	161.00	76.80	62.40	21.80
	Sub total	161.40 (91.93%)	77.03 (47.7%)	62.52 (38.7%)	21.85 (13.5%)
	Total	175.50 (100%)	86.60 (49.34%)	65.40 (37.26%)	23.50 (13.39%)

Table 3 shows the values of materials or processes which cause the environment impacts for the three stages of LCA, i.e. production, construction, and maintenance. It can be seen that the largest impact for the three stages is caused by material and process of asphalt, with impact values of 64.4 kPt, 65.10 kPt, and 19.30 kPt, respectively, and with percentages of more than 74%.

Table 3. Contributors to the environmental impact of the flexible pavement.

Production			Construction			Maintenance		
Material / Process	**Impact (kPt)**	**(%)**	**Material / Process**	**Impact (kPt)**	**(%)**	**Material / Process**	**Impact (kPt)**	**(%)**
Asphalt	64.40	74.00	Asphalt mixture	65.10	99.00	Production:		
Aggregate	8.27	9.50				- Asphalt	19.30	82.00
Heavy fuel equipment	0.35	0.40	BBM	0.30	1.00	- Aggregate	2.48	11.00
Generator set fuel	1.75	2.00				- Fuel	1.60	6.80
Transportation	11.80	14.00				Construction:		
						- Fuel	0.09	0.40
Total	86.60	100		65.40	100		23.50	100

4.2. Environmental Impacts of The Rigid Pavement

The rigid pavement section observed for this study is of 1000 m length, the width of the road on the concrete slab is 14 m, and the thickness of the concrete slab layer is 0.27 m. Table 4 shows the EI99 results of the rigid pavement in accordance with the LCIA characterization factor values. It is assumed that there is no maintenance during the design life of the rigid pavement, as one of the benefits of the rigid pavement is the routine maintenance is minimal and there could be no maintenance at all if the materials and construction process conform to specifications. At the end of life, the concrete that has lost its structural function will be replaced overlaying with new concrete layer or reconstruction. The total impact value of the three aspects of damage category, i.e. human health, ecosystem quality, and resources of the flexible pavement is 78.9 kPt, with the largest is related to resources (57.72%). For the LCIA approach, among the three stages, the largest impact value is at the production stage of 47.70 kPt (60.46%), followed by the construction stage of 31.20 (39.54%), respectively.

Table 4. The results of EI99 and LCIA of the rigid pavement.

Damage Category	**Impact Category**	**Total (kPt)**	**Production**	**Construct**	**Maintenance**
Human Health	Carcinogens	0.55	0.25	0.30	0
	Resp. Organics	0.02	0.01	0.01	0
	Resp. Inorganics	17.40	10.10	7.26	0
	Climate Change	11.40	5.98	5.44	0
	Radiation	0.12	0.06	0.06	0
	Ozone Layer	0.00	0.00	0.00	0
	Sub total	29.49 (37.38%)	16.40 (55.6%)	13.07 (44.3%)	0 (0%)
Ecosystem Quality	Ecotoxicity	1.18	0.66	0.52	0
	Acidification / Eutrophication	2.66	1.64	1.02	0
	Land Use	-0.05	0.05	-0.10	0
	Sub total	3.79 (4.80%)	2.35 (62.0%)	1.44 (38.0%)	0 (0%)
Resources	Minerals	1.04	0.47	0.57	0
	Fossil Fuels	44.50	28.40	16.10	0
	Sub total	45.54 (57.72%)	28.87 (63.4%)	16.67 (36.6%)	0 (0%)
	Total	78.90 (100%)	47.70 (60.46%)	31.20 (39.54%)	0 (0%)

Table 5 shows that the largest impact value at the production stage is caused by cement with an impact value of 31.4 kPt (66% of the impact value of the production stage). Whereas at the construction stage, concrete mixture has the largest impact with the value of 28.50 kPt, which is equal to 91% of the impact values of this stage.

Table 5. Contributors to the environmental impact of the rigid pavement.

Production			Construction		
Material / Process	**Impact (kPt)**	**Percentage (%)**	**Material / Process**	**Impact (kPt)**	**Percentage (%)**
Sand	6.27	13.00	Concrete Mixture	28.50	91.00
Gravel	5.35	11.00	Fuel Equipment	0.05	0.16
Cement	31.40	66.00	Dowel bars and tie bars	2.67	8.56
Water	0.01	0.01			
Electricity (pump)	0.00	0.002			
Heavy fuel equipment	0.65	0.014			
Generator set fuel	1.46	3.06			
Transportation	2.60	5.50			
Total	47.70	100	Total	31.2	100

4.3. Comparison of Environmental Impacts between the Flexible and Rigid Pavement

Table 6 and Figure 2 show the comparison of the total environment impact by damage categories between the flexible and rigid pavement. It can be seen that the total impact value of the three damage categories of the rigid pavement is 78.9 kPt, which is smaller than, and worth 45% of the flexible pavement of 175.50 kPt. However, among the three-damage category, the flexible pavement has less impact value for human health and quality of ecosystem than the rigid pavement.

Table 6. Total environmental impact by damage category.

Damage Category	**Flexible Pavement (kPt)**	**Rigid Pavement (kPt)**
Human Health	11.99 (6.83%)	29.49 (37.43%)
Ecosystem Quality	2.17 (1.24%)	3.78 (4.79%)
Resources	161.40 (91.93%)	45.54 (57.78%)
Total	175.50 (100%)	78.90 (100%)

Table 7 compares the environmental impact between the flexible pavement and the rigid pavement for the LCA stages of production, construction, and maintenance. It can be seen that the total impact value of all the LCA stages of the rigid pavement is smaller than the flexible pavement, i.e. 175.50 kPt and 78.9 kPt, respectively. This total result is also consistent with the result of each individual stage of production, construction, and maintenance, where the impact value of the rigid pavement is also smaller than the flexible one.

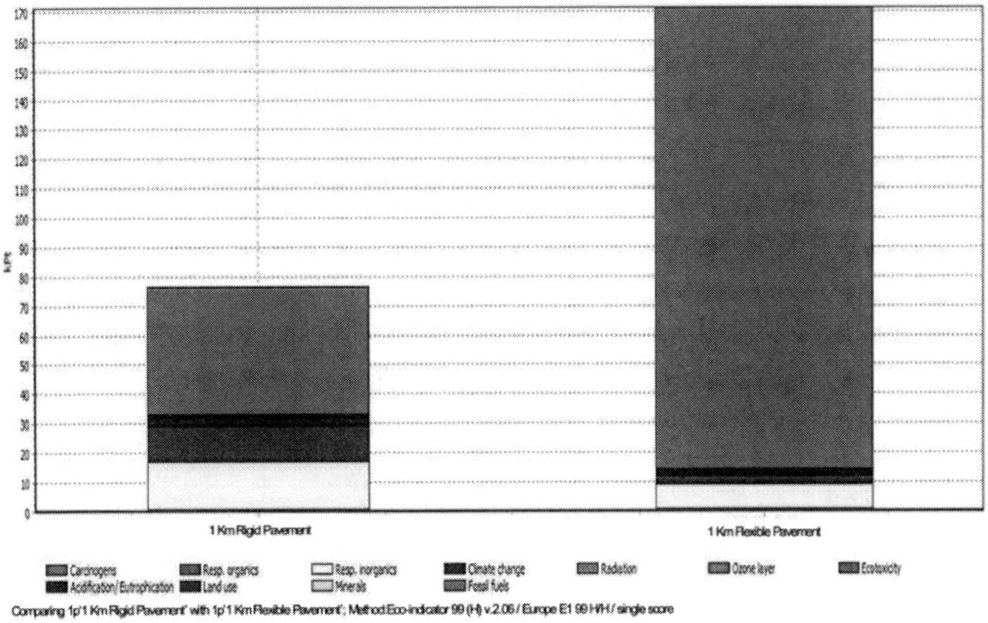

Figure 2. Comparison of the single score value of environmental impacts between flexible pavement and rigid pavement.

Table 7. Comparison of the environmental impact based on the LCA stages.

LCA Stage	Flexible Pavement (kPt)	Percentage (%)	Rigid Pavement (kPt)	Percentage (%)	Smaller Environmental Impact Value
Production	86.60	49.34	47.70	60.46	Rigid Pavement
Construction	65.40	37.26	31.20	39.54	Rigid Pavement
Maintenance	23.50	13.39	0	0	Rigid Pavement
Total	175.50	100	78.90	100	

5. Discussion

This research found that for the flexible pavement, the production stage is the stage with the largest environmental impact value of 86.60 kPt (49.34%), followed by the construction, and the maintenance stages of 65.40 (37.26%) and 23.50 kPt (13.39%), respectively. This largest impact is mainly caused by fossil fuels and asphalt material and the related processes. These findings are supported by a research which stated that the main energy consumption at the production stage occurs during the mixing and drying of asphalt aggregates [24]. This means that the production stage is the most critical stage when it comes to managing the environmental impact of the flexible pavement.

This research also found that for the rigid pavement, the production phase has the greatest environmental impact of 47.7 kPt (60.46%), followed by the construction phase of 31.2 kPt (39.54%). The biggest impact is caused by the fossil fuels of 44.50 kPt, the cement material at the production stage of 31.40 kPt (66% of the impact value of the production stage), and the concrete mixing process at the construction stage of 28.50 kPt (91% of the impact values of this stage). These research findings are consistent with a study which found that for the rigid pavement, the material contributed 92.80% of the total amount of environmental impact [4], while another research states that the three biggest impacts of the rigid pavement is caused by material, fuel and transportation [25].

The effect on human health and the environment is very similar between the two forms of pavement, but in both instances, asphalt production is still higher. The effect of asphalt production on the resources is approximately 100 mPt greater than the influence of concrete production [13].

Both the flexible and rigid pavement have the biggest impact in the production stage, followed by construction and maintenance. It is almost similar with an empirical results in the life cycle of the fast lane transportation project, demonstrating that the construction process has the biggest environmental effect (62.7%), followed by the demolition (35.8%) and the maintenance phase (1.7%) [26].

This research found that the total environmental impact on the rigid pavement is 78.9 kPt, which is smaller than, and equals 45% of the total impact of the flexible pavement of 175.50 kPt. This indicate that the rigid pavement can be considered more environmentally friendly, hence also more sustainable. These findings explain why more countries are now replacing the flexible pavement with the rigid pavement [2]. Not only because the concrete pavement is structurally superior to the asphalt pavement, has long service life and very minimal maintenance [3] [4], but also due to better environmental performance as suggested by the findings of this research.

6. Conclusion

The aim of this research is to analyse the environmental impacts of road construction projects of the flexible pavement and rigid pavement based on LCIA approach, by means of the Eco-Indicator 99 (EI99). For the LCIA approach, the impact values at the production, construction and maintenance stages for rigid pavement are 86.60 kPt (49.34%), 65.40 (37.26%) and 23.50 kPt (13.39%), while for rigid pavement respectively 47.70 kPt (60.46%), 31.20 (39.54%) and 0.00 kPt (0%), respectively. The results of the EI99 indicate the impact value of the aspects human health, ecosystem quality, and resources for flexible pavement are 11.99 kPt (6.83%), 2.17 kPt (1.24%) and 161.4 kPt (91.93%), respectively, while for the rigid pavement 29.49 kPt (37.43%), 37.8 kPt (4.79%), and 45.54 kPt (57.78%), respectively. The total impact value of the three aspects of the rigid pavement is 78.9 kPt, which is smaller than, and worth 45% of the flexible pavement of 175.50 kPt. Based on the results of the LCIA approach and the EI99, it can be concluded that for the project under review, the rigid pavement can be considered more sustainable than the flexible pavement. As every project is unique, however, various factors may influence the results of even similar research, including the design life of the project. Further research could investigate how the design life of a project may impact on the environmental performance.

Acknowledgment

The authors appreciate Yudi Hidayat and Drajad Adhiprasetyo for supporting site survey and collecting data for this research. This research was financially supported by The Faculty of Engineering, Diponegoro University, Indonesia through Strategic Research Grant 2020.

References

[1] Utomo DHJ, Hidayat A, Setiawati A, Catur APS. Measuring Carbon Footprint of Flexible Pavement Construction Project in Indonesia. Hadiyanto, Sudarno, Maryono, editors. *E3S Web of Conference*. EDP Sciences; 2018;**31**:07001 .doi.org/10.1051/e3sconf/20183107001

[2] Shaban AM, Alsabbagh A, Wtaife S, Suksawang N. Effect of Pavement Foundation Materials on Rigid Pavement Response. *IOP Conf. Ser. Mater. Sci. Eng*. IOP Publishing; 2020 Jan 17;**671**:012085. doi.org/10.1088/1757-899x/671/1/012085

[3] Handayani FS, Pramesti FP, Wibowo MA, Setyawan A. Estimating and Reducing the Release of Greenhouse Gases in Local Road Pavement Constructions. *International Journal on Advanced Science, Engineering and Information Technology*. Insight Society; 2019 Oct 31;**9**(5):1709. doi.org/10.18517/ijaseit.9.5.9705

[4] Mulyana A, Wirahadikusumah R. Analisis Konsumsi Energi dan Emisi Gas Rumah Kaca pada

Tahap Konstruksi Studi Kasus : Konstruksi Jalan Cisumdawu (Analysis of Energy Consumption and Greenhouse Gas Emissions during the Construction Stage Case Study: Cisumdawu Road Construction). *Jurnal Teknik Sipil ITB*. 2017;**24:**269–80.

[5] Abergel T, Dean B, Dulac J. *Towards a zero-emission, efficient, and resilient buildings and construction sector*. United Nations Environment Programme; 2017.

[6] Lawalata GM. Prinsip-Prinsip Pembangunan Jalan Berkelanjutan (Principles of Sustainable Road Development). *Jurnal Transportasi*. 2013;**13**:115–24

[7] Beatley T. *The Sustainable Urban Development Reader*. Routledge; 2004 Jan 22. doi.org/10.4324/9780203501627

[8] Agung WM, Uda SAKA, Zhabrinna. Reducing carbon emission in construction base on project life cycle (PLC). Hajek P, Han AL, Kristiawan S, Chan WT, Ismail M b., Gan BS, et al., editors. *MATEC Web of Conferences*. EDP Sciences; 2018;**195**:06002. doi.org/10.1051/matecconf/201819506002

[9] Lawalata GM. Pemeringkatan Jalan Hijau untuk Mendukung Implementasi Program Konstruksi Jalan Berkelanjutan (Green Road Ratings to Support the Implementation of the Sustainable Road Construction Program). *Jurnal HPJI* 2019;**5**:21–30.

[10] ISO. *Environmental Management - Life Cycle Assessment - Principles and Framework (ISO 14040:2006)* vol 44. British Standard; 2006.

[11] Araújo JPC, Oliveira JRM, Silva HMRD. The importance of the use phase on the LCA of environmentally friendly solutions for asphalt road pavements. *Transp Res D Transp Environ*. Elsevier BV; 2014 Oct;**32**:97–110. doi.org/10.1016/j.trd.2014.07.006

[12] Hasan U, Whyte A, Al Jassmi H. Life cycle assessment of roadworks in United Arab Emirates: Recycled construction waste, reclaimed asphalt pavement, warm-mix asphalt and blast furnace slag use against traditional approach. *J. Clean. Prod.* Elsevier BV; 2020 Jun;**257**:120531. doi.org/10.1016/j.jclepro.2020.120531

[13] Ivel J, Watson R, Abbassi B, Abu-Hamatteh ZS. Life cycle analysis of concrete and asphalt used in road pavements. *Environ. Eng. Res.* Korean Society of Environmental Engineering; 2019 Mar 12;**25**(1):52–61. doi.org/10.4491/eer.2018.399

[14] Cong L, Guo G, Yu M, Yang F, Tan L. The energy consumption and emission of polyurethane pavement construction based on life cycle assessment. *J. Clean. Prod.* Elsevier BV; 2020 May;**256**:120395. doi.org/10.1016/j.jclepro.2020.120395

[15] Praticò FG, Giunta M, Mistretta M, Gulotta TM. Energy and Environmental Life Cycle Assessment of Sustainable Pavement Materials and Technologies for Urban Roads. *Sustainability*. MDPI AG; 2020 Jan 18;**12**(2):704. doi.org/10.3390/su12020704

[16] Park W-J, Kim T, Roh S, Kim R. Analysis of Life Cycle Environmental Impact of Recycled Aggregate. *Appl. Sci.* MDPI AG; 2019 Mar 12;**9**(5):1021. doi.org/10.3390/app9051021

[17] Wang T, Xiao F, Zhu X, Huang B, Wang J, Amirkhanian S. Energy consumption and environmental impact of rubberized asphalt pavement. *J. Clean. Prod.* Elsevier BV; 2018 Apr;**180**:139–58. doi.org/10.1016/j.jclepro.2018.01.086

[18] Kucukvar M, Noori M, Egilmez G, Tatari O. Stochastic decision modeling for sustainable pavement designs. *Int. J. Life. Cycle. Assess*. Springer Science and Business Media LLC; 2014 Feb 25;**19**(6):1185–99. doi.org/10.1007/s11367-014-0723-4

[19] Kim T, Tae S. Proposal of Environmental Impact Assessment Method for Concrete in South Korea: An Application in LCA (Life Cycle Assessment). *Int. J. Environ. Res. Public Health.*

MDPI AG; 2016 Nov 2;**13**(11):1074. doi.org/10.3390/ijerph13111074

[20] Huang Y, Bird R, Heidrich O. Development of a life cycle assessment tool for construction and maintenance of asphalt pavements. *J. Clean. Prod.* Elsevier BV; 2009 Jan;**17**(2):283–96. doi.org/10.1016/j.jclepro.2008.06.005

[21] Louzguine DV, Inoue A, Saito M, Waseda Y. *Eco-indicator 99 Manual for Designers*. The Hague: Ministry of Housing, Spatial Planning and the Environment Communications Directorate; 2000.

[22] Thives LP, Ghisi E. Asphalt mixtures emission and energy consumption: A review. Renewable and Sustainable Energy Reviews. Elsevier BV; 2017 May;**72**:473–84. doi.org/10.1016/j.rser.2017.01.087

[23] Chang Y, Ries RJ, Lei S. The embodied energy and emissions of a high-rise education building: A quantification using process-based hybrid life cycle inventory model. *Energy and Buildings*. Elsevier BV; 2012 Dec;**55**:790–8. doi.org/10.1016/j.enbuild.2012.10.019

[24] Z Zapata P, Gambatese JA. Energy Consumption of Asphalt and Reinforced Concrete Pavement Materials and Construction. *J. Infrastruct. Syst.* American Society of Civil Engineers (ASCE); 2005 Mar;**11**(1):9–20. doi.org/10.1061/(asce)1076-0342(2005)11:1(9)

[25] Cass D, Mukherjee A. Calculation of Greenhouse Gas Emissions for Highway Construction Operations by Using a Hybrid Life-Cycle Assessment Approach: Case Study for Pavement Operations. *J. Constr. Eng. M. Asce*. American Society of Civil Engineers (ASCE); 2011 Nov;**137**(11):1015–25. doi.org/10.1061/(asce)co.1943-7862.0000349

[26] Li H, Deng Q, Zhang J, Olubunmi Olanipekun A, Lyu S. Environmental Impact Assessment of Transportation Infrastructure in the Life Cycle: Case Study of a Fast Track Transportation Project in China. *Energies*. MDPI AG; 2019 Mar 15;**12**(6):1015. doi.org/10.3390/en12061015

Designing integrated testing for product development of prosthetic hand

N Susanto[1,2,*], H Prastawa[1,2] and A Meilaningrum[1,2]

[1]Industrial Engineering Department, Faculty of Engineering, Diponegoro University
[2]Center for Bio Mechanics, Bio Material, Bio Mechatronics and Bio Signal Processing (CBIOM3S), Diponegoro University

E-mail: *nophie.susanto@gmail.com

Abstract. In order to produce a standardized product, a series of product testing is needed from the beginning of the product design to the final product. The development of medical devices that is very rapid and demands excellent product quality needs to be accompanied by a very rigorous testing process because the use of the product is related to the user's physical and mental health condition. This study aims to summarize the ongoing testing carried out during product design to the use of the product by the user. The product in this case study is prosthetic hand made by CBIOM3S Diponegoro University. The research method used is based on research and direct surveys of designers, users and experts as well as literature review confirmation. The results showed a variety of stages and standardized starting from design testing, material testing, failure analysis and simulation, prototype testing for both the bionic and mechanical hands, usability and psychological testing by respondents and ended with medical confirmation.

1. Introduction

Human health conditions become an important part of the quality of human life. Parts of the body that are widely used in daily activities are the hands. If there is interference on the hand, then the patient will have difficulty to move. For patients with hand disorders who are forced to amputate their hands due to accidents or other causes, the health sector seeks in terms of mobility to aid in the form of assistive devices [1]. These aids can be in the form of prosthetic hands or robot hands. Before being used by people with disabilities, tools are needed to test the feasibility of an artificial hand, especially robots, both as a learning tool and as an aid to the adaptation process during using artificial hands.

With limitations in carrying out human daily activities, rehabilitation tools are needed to recover hand movements quickly and assistive devices in the form of prosthetic hands or robotic rehabilitation media as a substitute for previous hands. Before use, it will be very important to test the feasibility of a prosthetic hand or robot hand that will be used by patients or people with disabilities on the hands.

Recently, test of this product development is performed separately to encourage the speed of product design and testing. However, in terms of documentation and alignment of perception and mindset, it is necessary to consider integrated testing of products made. Although there are a variety of products made, basically the thinking framework refers to and is based on the similar pattern.

Several studies conducted related to prosthetic hand testing are based on experimental evaluations on 3D-printed prosthetic hand prototype [2], tests on appearance and functionality [3], mechanical testing methods [4], Anthropomorphic Hand Assessment Protocol (AHAP) [5] and various designs for artificial hands [6-20]. The aim of this study is integrating testing concept of the prosthetic hand to produce a standardized product based on literature and actual standard.

2. Methods

The method used in the design of this integrated test is a survey and interviews with users. This was accompanied by a literature review. In accordance with the initial stages in the product life cycle, product design begins with user need requirements [21], but to accommodate the speed of development of existing medical devices, the concept of product development is shifting from user needs based on functionality and prototype development. The proposed product improvements refer to the ergonomic concept and product design.

The research method was carried out starting from mapping the stages traversed by prosthetic hand as products form Center for Bio Mechanics, Bio Material, Bio Mechatronics and Bio Signal Processing (CBIOM3S) Diponegoro University. At each stage, a defined testing is carried out to achieve product standards in accordance with technical and ergonomic standards. The integrated test results are intended for the construction of an integrated testing center.

Interviews were conducted with three product designer respondents from CBIOM3S, two persons with disabilities respondents and two product experts. Prototype trials and comments from respondents provide improvements that are used to build the next version of the product. The tested product is divided into two types namely bionic hands and mechanical hands.

3. Result and Discussion

The study results show testing is divided into several parts, namely testing in product design, material testing, running simulation testing, prototype testing and usability testing.

3.1. Design Testing

In design testing, there are several stages of testing that are accompanied by physical parameters. The stages in design testing can be seen in Figure 1.

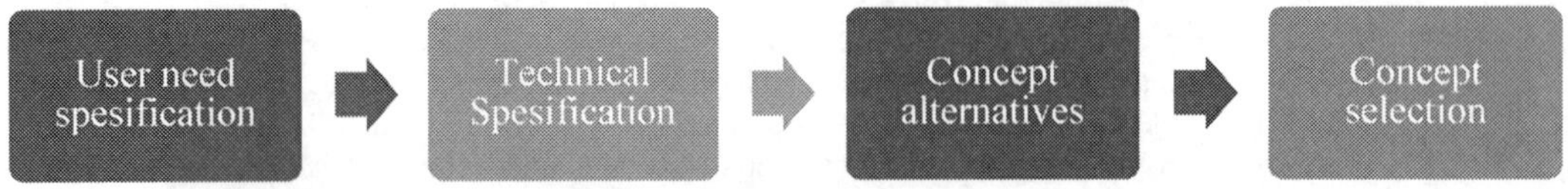

Figure 1. Phase of design testing.

User need specifications are obtained from interviews result with disabilities respondents. The initial specifications obtained are functional hands with relatively cheap prices, easy and comfortable, ergonomics size according to the user, strong durability, a human hand-like shape; easy and fast to use, high safety during adaptation process, simple, convenient, lightweight, intuitive actuation, can be produced with 3D printing and there are sensors at each finger. In addition, several previous studies also provide initial discussions related to user needs [21-27].

The technical specification phase provides the initial technical assumptions set by the product designer. Examples of assumptions are that the product is safe to use when carrying a 1.5 kg load, the weight of a light product (defined as having a weight of 300-500 grams), the use of sufficient and inexpensive quality materials such as PLA and resin [25] [26].

At the stage of making alternative concepts, the product is differentiated first for bionic and mechanical hand products. Definition of the alternatives is distinguished by sub-functions up to three levels. Sub-functions for bionic hands can be seen in Table 1, while sub-functions for mechanical hands can be seen in Table 2.

Alternative of the concepts are arranged in three concepts for example for alternative bionic hand concepts are as follows [26]:

Concept 1: A1+B1+C1+D1+E1+F1+G1
Concept 2: A1+B2+C2+D2+E2+F1+G1
Concept 3: A1+B2+C2+D2+E2+F2+G2

Alternative of the concepts for mechanical hands are arranged as follows [25]:

Concept 1: A1+B2+C2+D2+E1
Concept 2: A1+B1+C3+D1+E1
Concept 3: A1+B1+C1+D2+E1

Table 1. Sub function of bionic hand.

No	Sub Function	Sub-sub Function	Sub-sub-sub Function
1	Material selection	Polymer	A1 PLA
2	Index finger	Mechanism	B1 Gear
			B2 Linkage
		Motor	C1 Servo MG995
			C2 Micro servo MG90
3	Middle finger	Mechanism	D1 Couple little finger
			D2 Linkage
4	Thumb	Mechanism	E11 Degree of Freedom
			E2 2 Degree of Freedom
		Motor	F1 Servo Mini Micro MG90
			F2 Micro Gear motor DC N20
5	Linkage	Material	G1 PLA
			G2 Acrylic

Table 2. Sub function of mechanic hand.

No	Sub Function	Sub-sub Function	Sub-sub-sub Function
1	Material selection	Polymer	A1 PLA
2	Fingers	Material	B1 PLA
			B2 PLA and Flexible PLA
3	Linkage	Material	C1 PLA
			C2 Flexible PLA
			C3 Acrylic
4	Spring	Material	D1 Metal
			D2 Flexible PLA
5	Rope	Material	E1 Fishing line

The basis of concept selection is determined based on the results of weighting and the highest criterion value of all alternative concepts. For example, designers of bionic and mechanical hand products decide to make products with Concept 3 which has the highest criterion value. Some assessment criteria that need to be considered are price, product weight, degree of freedom, grasping ability, ease of manufacture, ease of assembly, flexibility, ergonomic factors, ease of care, product shape, and ease of lifting [25] [26].

3.2. Material Testing

There are three tests on the material carried out namely tensile test, impact test and bending test. Tensile Test gives consideration of unit factors related to proportionality limit, elastic limit, yield point and yield strength, maximum tensile strength, breaking strength, ductility, modulus of elasticity, modulus of resilience, modulus of toughness, modulus of stress and strain engineering and actually. The test equipment used was Universal Testing Machine with ASTM D638-02 standard. Existing standards indicate the testing process by pulling a material until it breaks. The basic analysis used in the product is that the ductile material is preferred because it is generally more resilient and gives advance warning before damage occurs.

Impact testing considers unit factors related to temperature. Transition temperature is the temperature which indicates the change in fracture type of a material when tested at different temperatures. In testing with different temperatures, it will be seen that at high temperatures, the material will be ductile while at low temperatures the material will be brittle. Physical parameters show that a tough material can absorb more energy and lower the position h'. A material is said as a tough material if it has the ability to absorb large shock loads without cracking or deforming easily. The test equipment used was an impact testing machine with ASTM 32 standard, British Standard 4360 which states that the test of material strength is determined in the form of absorption of potential energy from a load that swings from a certain height and mashing the test material so that deformation occurs.

The final stage in material testing is the bending test. Physical parameters used are a maximum crack of 3 mm measured from all directions on the surface, a maximum crack of 10 mm from the sum of all the biggest cracks between 1 - 3 mm, a maximum crack angle of 6 mm unless cracks originate from several types of cracks, a maximum crack of 3 mm. Considerations of the most difficult factors include tensile strength, chemical composition and microstructure, yield stress (yield). Test equipment that can be used is AU-1016A stiffness tester (Taber Type) with ASTM D790-2 standard, ASME. In the standard described testing does not require a testing machine to determine the quality of a material visually, to measure the strength of the material due to loading and suppleness of the results of the connection.

3.3. Simulation Running

The simulation is used to analyze important parts of the components, especially the most potential failures and critical pressures. There are two stages of testing, namely displacement and stress analysis (Von Mises Stress).

Displacement analysis shows areas of stress concentration and deformation points. In addition, by using maximum load, test can estimate almost tangible results to determine weak points in the design. For simulation parameters, the force is assumed to be the main force broken into x, y and z components. The unit factor considerations are unit displacement (URES) [28] with ANSYS/SolidWork software test kits and finite element analysis standards. By using software-based testing, the impact of design changes through simulation can be analyzed objectively. This analysis forms the basis and generates broad assumptions, so that their values and displacements can be used for design purposes.

Another analysis is the stress analysis using Von Mises Stress. Material is said to begin to melt when the Von Mises voltage reaches a critical value known as yield strength. Material is also determined as in good level when the Von Mises Stress value is below the yield strength of the material. Unit factor considerations related to the Von Mises Stress are switch, material yield stress and safety factor [28]. Test equipment used was ANSYS / SolidWork software and finite element analysis standard. Examples of displacement analysis can be seen in Figure 2 (a), while the Von Mises Stress analysis can be seen in Figure 2 (b).

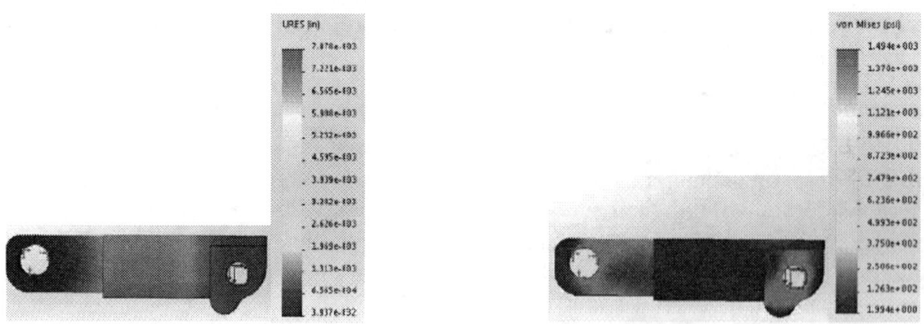

(a) Displacement analysis (b) Von Mises stress analysis

Figure 2. Simulation running test.

3.4. Prototype Testing

Prototype testing is distinguished by the type of prosthetic hand, namely the bionic hand or mechanical hand. Bionic hand testing refers to various types of grips such as precision grip, power grip/cylindrical grasp, key grip/lateral pinch, ball grasp, open palm grasp and index point as shown in Figure 3 [28]. Testing is performed based on functionality aspect only by the designer.

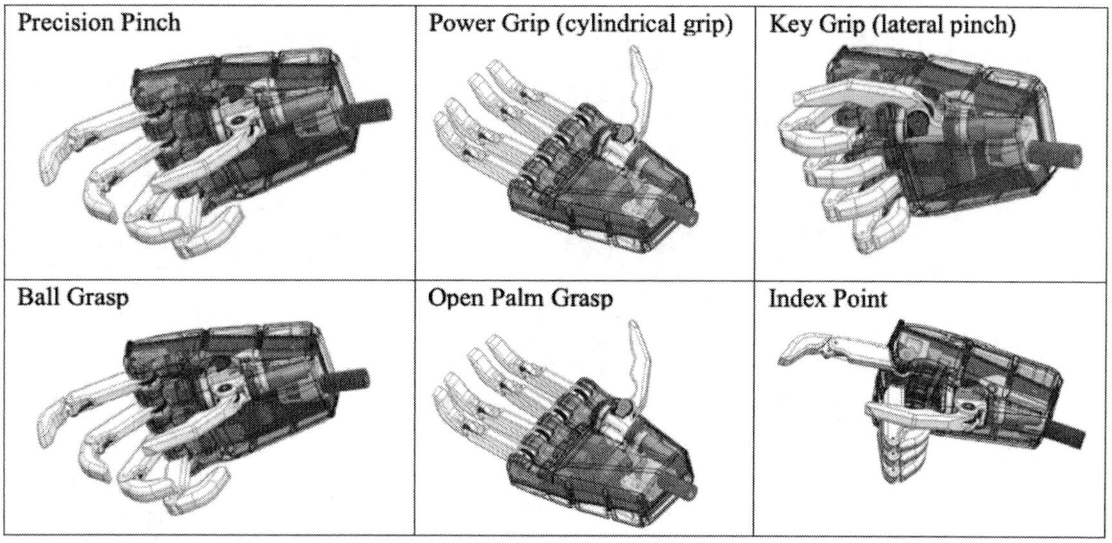

Figure 3. Prototype testing for bionic hand [28].

Mechanical hand testing is only carried out using a hand tester and a suspended system. In the use of a testing hand, if the bag load can be maintained in the air for about 10 seconds without falling, a higher weight needs to be tested on the prosthetic hand until the weight can no longer be maintained. Consideration of related factors is the power generated on the dynamometer and the weight of the load that can be lifted. Tests carried out can vary with weightlifting tests on bag loads containing loads 2, 3, 5, and 10 lb. or 0.91 kg, 1.36, 2, 27 and 4.53 kg. When the maximum load generated in the final configuration setting on the prosthetic hand is the same as the power output on the dynamometer, the hand must then be assembled into a new configuration and the test needs to be repeated. Examples of tests can be seen in Figure 4.

Suspended system testing is almost similar to previous tests. The difference lies in the utilization of the PVC pipe segment to provide force on the pipe surface. The examiner also must determine whether the hand can support the weight for 10 seconds or the weighted iron handle will fall. If it fails to maintain grip for 10 seconds, then the test will be repeated to ensure that the weight is above the holding capacity. Test equipment used in mechanical hand testing is a dynamometer as shown in Figure 5 [29].

By utilizing the strength of the grip from outside the prosthetic hand, it is expected that the larger muscle groups in the arms can be used and it is possible to load the hands with higher weights. This test emphasizes the optimal angle bending that can be done to support the highest possible weight. The examiner also must determine whether the hand can support the weight for 10 seconds or the weighted iron handle will fall. If it fails to maintain grip for 10 seconds, then the test will be repeated to ensure that the weight is above the holding capacity.

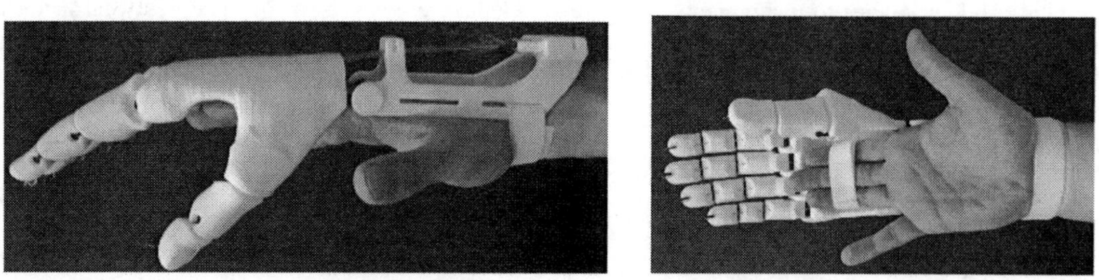

Figure 4. Prototype testing for mechanic hand based on tester hand [29].

Figure 5. Prototype testing for mechanic hand based on suspended system testing [29].

3.5. Usability Testing

Usability testing has been investigated for respondents [22] [23] [24] [29 [30]. In general, the testing phase includes measuring the time to complete the activity and filling out the Usefulness, Satisfaction, and Ease of Use (USE) questionnaire. The respondents are given time for adaptation and learning to complete the given activities. Time measurement and questionnaire filling were conducted after completing each activity to obtain ISO 9241-11 criteria data, namely effectiveness, efficiency and satisfaction. The time calculation starts with instructions from the moderator as the stopwatch holder and ends when the respondent's hand is lifted off the object as a sign that the activity is said to have been completed. Completion of the activity was carried out several times to get at least 5 times the activity was successfully carried out [22] after repeating the type of activity given, respondents were given a USE questionnaire to find out the level of satisfaction of using prosthetic hands in completing that type of activity [31].

During the, moderators are encouraged to write small notes related to things that happened during the respondent trying to complete the activity. This can be in the form of obstacles experienced by respondents, complaints that are expressed related to the use of prosthetic hands, restrictions on the function of prosthetic hands that are found when unsuccessful in completing an activity and things that are considered important to be incident diaries when respondents respond.

Test results are documented for each usability criteria. The testing equipment used is a test kit that is continuously improved to accommodate daily activities as well as hand activity therapy [22].

3.6. Emotional Testing

Product evaluation was also carried out by means of product emotional assessment by respondents consisting of persons with disabilities and the community at large. The test tool used in this study was Premo Tool. The advantage of Premo Tool is the respondent giving emotional product does not have to see the physical/direct product to be assessed so as to facilitate filling out the questionnaire by focusing on product design. Stages in taking product emotional data is done by distributing questionnaires and showing product images to respondents. Then respondents were asked to fill in scores for 14 types of emotions that were felt after seeing the product [32]. The final stage is then a brief interview between the moderator and the respondent to find out the reason for choosing the most dominant emotional product. This product emotional assessment focuses on the achievement of research objectives so that the number of respondents is not a primary consideration in retrieving product emotional data. An example of the form of analysis for one prosthetic hand is shown in Figure 6.

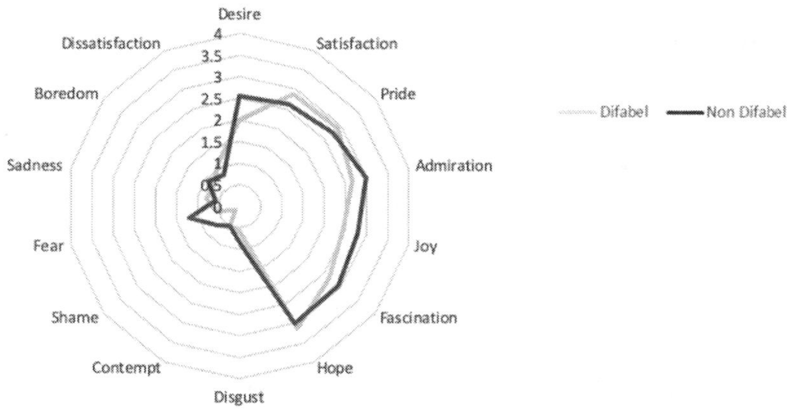

Figure 6. Premo analysis example for prosthetic hand.

4. Conclusion

The results showed several stages of integrated testing for prosthetic hand products ranging from product design, material testing, running simulation, prototype testing and testing by respondents. Differentiation of products based on characteristics and functions needs to be done in each test to produce products in accordance with user needs and standards set. The final stage that needs to be done in subsequent research there is medical confirmation related to the results of this study. In addition, the development of the test can be done for other medical devices with similar models such as prosthetic foot or stroke therapy aid.

Acknowledgments

This research was financially supported by Diponegoro University in RPP (Riset Pengembangan dan Penerapan) 2020.

References

[1] Brey. Descriptive of Prosthetic Hand Appliance. United States: Naval Medical Research Institute; 2005

[2] Llop-Harillo I and Perez-Gonzales A. International Biomechanics; 2017

[3] Schweitzer W. Improving Prosthetic Arms through Better Testing. *The O&P Edge*; Aug 2014

[4] Mio R, Sanchez M, Valverde Q. Mechanical Testing Methods for Body-Powered Upper-Limb Prostheses: A Review. *2018 IEEE 18th Int. Conf. Bioinform. Bioeng.* (BIBE). IEEE; 2018 Oct; doi.org/10.1109/bibe.2018.00040

[5] Llo-Harillo I, Perez-Gonzales A, Starke J, Asfour T. Robotics and Autonomous Systems; 2019

[6] Controzzi M, Cipriani C, Carozza MC. The Human Hand as an Inspiration for Robot Hand Development, part of Springer Tracts in Advanced Robotics book series. *STAR vol 95*; 2014

[7] Belter T, Segil JL, Dollar AM, Weir RF. *J. Rehabil. Res.* Dev. 50; 2014

[8] Bostelman R, Messina E, Foufou S. *Front. Inform.Technol. Electron. Eng*; 2013

[9] Kate J, Smit G, Breedveld P. *Disabil. Rehabil. Assist. Technol.* 12 300–314

[10] Falco J, Van Wyk K, Liu S, Carpin S. IEEE *Robot. Autom. Mag*; 2015

[11] Baiyang Sun, Caihua Xiong, Wenrui Chen, Qiaofei Zhang, Liu Mao, Qin Zhang. A novel design method of anthropomorphic prosthetic hands for reproducing human hand grasping. *36th Annu. Int. Conf. IEEE Eng. Med. Biol. Soc.* IEEE; 2014 Aug; doi.org/10.1109/embc.2014.6945049

[12] Liarokapis MV, Artemiadis PK, Kyriakopoulos KJ. Functional Anthropomorphism for human to robot motion mapping. 2012 IEEE RO-MAN: *21st IEEE Int. Symp. Rob. Hum. Interact. Commun.* IEEE; 2012 Sep; doi.org/10.1109/roman.2012.6343727

[13] Krumlinde-Sundholm L, Holmefur M, Kottorp A, Eliasson AC. *Dev. Med. Child Neurol*; 2007
[14] Duong T, Wagner B, Abraham T, Davidson M, Bains G, Daher N, Friedrich A. *J. Prosthet. Orthot*; 2017
[15] Belter JT, Leddy MT, Gemmell KD, Dollar AM. Comparative clinical evaluation of the Yale Multigrasp Hand. *6th IEEE Int. Conf. Biomed. Rob. Biomech.* IEEE; 2016 Jun; doi.org/10.1109/biorob.2016.7523680
[16] Fukaya N, Toyama S, Asfour T, Dillmann R. Design of the TUAT/Karlsruhe humanoid hand. *Proc. 2000 IEEE/RSJ Int. Conf. Intell. Rob. Sys.* (IROS 2000) (Cat No00CH37113). IEEE; doi.org/10.1109/iros.2000.895225
[17] Vergara M, Agost MJ, Gracia-Ibáñez V. *Hum. Factors Ergon. Manuf. Serv. Ind*; 2018
[18] Weiner P, Starke J, Hundhausen F, Beil J, Asfour T .The KIT prosthetichand: design and control. *Int. Conf. Intell. Robot. Syst*; 2018
[19] Bullock IM, Zheng JZ, De La Rosa S, Guertler C, Dollar AM. Grasp Frequency and Usage in Daily Household and Machine Shop Tasks. *IEEE Trans. Haptics*. Institute of Electrical and Electronics Engineers (IEEE); 2013 Jul;6(3):296–308. doi.org/10.1109/toh.2013.6
[20] Cipriani C, Zaccone F, Micera S, Carrozza MC. On the Shared Control of an EMG-Controlled Prosthetic Hand: Analysis of User–Prosthesis Interaction. *IEEE Trans. Rob.* Institute of Electrical and Electronics Engineers (IEEE); 2008 Feb;24(1):170–84. doi.org/10.1109/tro.2007.910708
[21] Ulrich KT, Eppinger SD. *Product Design and Development.* Mc Graw-Hill New York; 2012
[22] Susanto N, Prastawa H, Mahachandra M, Rakhmawati DA. JITI; 2019
[23] Susanto N, Mahachandra M, Purwaningsih R, Budiawan W, Afhami AS. Defining quality function deployment to improve the usability of prosthetic hand. *5th Int. Conf. Ind. Mech. Electr. Chem. Eng.* AIP Publishing; 2020; doi.org/10.1063/5.0000984
[24] Susanto N, Budiawan W, Ismail R, Ariyanto M, Sahal A. Usability Evaluation of Anthropomorphic Prosthetic Hand Based on ISO 9241-11. Setiadji BH, Han AL, Widodo A, Setiawan JD, Kurdi O, Hatmoko JUD, editors. *MATEC Web Conf.* EDP Sciences; 2018;159:02069. doi.org/10.1051/matecconf/201815902069
[25] Perdana YK. Rancang Bangun dan Pengujian Mechanical Prosthetic Hand (Metic Hand) untuk Pasien Amputasi Tangan Bawah Siku. Mechanical Engineering, Diponegoro University; 2009. (unpublished)
[26] Oktavianto AI. Perancangan & Pengembangan Sistem Mekanik 5 Degree of Freedom pada Low Cost Bionic Anthrophomorphic. Mechanical Engineering, Diponegoro University; 2017. (unpublished)
[27] Llop-Harillo I, Pérez-González A. System for the experimental evaluation of anthropomorphic hands. Application to a new 3D-printed prosthetic hand prototype. International Biomechanics. *Informa UK Limited*; 2017 Nov 3;4(2):50–9. doi.org/10.1080/23335432.2017.1364666
[28] Harris JW. Optimization of Prosthetic Hands : Utilizing Modularity to Improve Grip Force, Grasp, and Versatility. Las Vegas. University Libraries; 2017
[29] Rubin J, Chisnell J. Handbook of Usability Testing. United States of America: Wiley Publishing; 2008
[30] Kyberd PJ. *Prosthetics and Orthotics International*; 2011
[31] Gusti MF. Evaluasi Usability pada Fortrero Spoon dan Fortrero Cup untuk Pasien Penyandang Tremor pada Tangan. Industrial Engineering Department, Diponegoro University; 2018 (unpublished)
[32] Desmet P. Measuring Emotion: Development and Application of an Instrument to Measure Emotional Responses to Products. *Funology 2.* Springer International Publishing; 2018;391–404. doi.org/10.1007/978-3-319-68213-6_25

Automated fish feeding system for an offshore aquaculture unit

P D Karningsih[1,4], R Kusumawardani[1], N Syahroni[2], Y Mulyadi[2] and M S B M Saad[3]

[1]Department of Industrial and System Engineering, Sepuluh Nopember Institute of Technology, Indonesia
[2]Department of Ocean Engineering, Sepuluh Nopember Institute of Technology, Indonesia
[3]Department of Marine Electric and Electronics, Universiti Kuala Lumpur, Malaysia

Email: [4]dana@ie.its.ac.id

Abstract. One of the major challenges in aquaculture is fish feeding management. Inappropriate feeding practices could threat the fish's growth and health, may lead to negative environment impact due to excess food contamination, and more importantly high operations cost. This study is aiming at proposing a fish feeding management system for OceanFarmITS, an offshore aquaculture unit designed by Marine Research Centre of Institut Teknologi Sepuluh Nopember Surabaya. One of the most commercially popular fish, Tiger grouper is selected to be cultivated in OceanFarmITS pilot project. The fish feeding management systems include fish feeding arrangement and fish feeder system that is customized specifically to OceanFarmITS operations nature. Automatic fish feeder is an alternative to solve existing feeding problems such us inefficiency of conventional or manual feeding. The proposed system consists of a control system using microcontroller Arduino board to receive data from sensors then send the instruction to the motor driver. Moreover, it has alarm system that sends notification to Aquaculture operator when food storage in silo tank has reached a limit level. The proposed design of feeding system would support operations of OceanFarmITS to conduct feeding activity efficiently by minimizing operations cost and environmental impact while keep fish optimal growth.

1. Introduction

Aquaculture is one of the fastest-growing food sectors worldwide [1] [2] and it provides almost half of the world's fish for human consumption [3]. Aquaculture is considered one of the most profitable businesses because there are many varieties of fish that is not only taste good and constitute healthy food [4]. As per global aquaculture production statistics, the contribution of aquaculture to the world total fish production reached 171 million tonnes in 2016, about 88% (over 151 million tonnes) was utilized for direct human consumption [5]. Asia is the highest seafood-consuming region in the world, accounting for two-thirds of the world's food fish supply [2]. The development of aquaculture needs to be compatible with the environment and it is essential that the industry is ecologically sustainable [6].

Aquaculture can be applied in fresh, brackish and marine waters. Application of aquaculture method in Indonesia has been well known for more than a couple hundreds of years. The oldest practice is the brackish water pond which suitable for shrimp and milkfish. Marine aquaculture can be applied on coastal/inshore (i.e. less than three-mile territorial limit of the coast) or offshore/open ocean (i.e. from the three-mile territorial limit of the coast to two hundred miles offshore). Since 1990, seaweed is started to produce using marine aquaculture, then is followed by pearl culture. Today, one of popular marine culture practice is groupers which mostly is exported with a very high value [7]. While offshore aquaculture is a common practice in several countries (e.g. Turkey, Norway, China), it is not the case for Indonesia. Indonesia has a lot of potential on practicing Offshore aquaculture as around 3.5 million square kilometres of its area is ocean and it has suitable condition for growing fish and other type of seafood. This condition provides a perfect environment for offshore aquaculture, however up to now there is only a few offshore aquacultures that is operated in Indonesia.

Some important aspects of any types of aquaculture operations are the fish's growth, health and ability to reproduce. They depend on adequate (quality and quantity) feed supply for the fish. The feeding time and amount must be appropriate. Feed adjustment to meet the fish requirement is very important for income/ benefit maximization. Feeding frequency is thus an essential consideration that effect on growth performance, feed utilization, and economic viability [8]. Inaccurate feeding practices in aquaculture may lead to feed being wasted. As a result, it leads to higher production cost and could contaminate the aquatic environment. On the other hand, when insufficient feed is being provided, the growth and health of the fish will be jeopardized [9]. Since, cost for fish feeding is typically the greatest operating expense for aquaculture operations, therefore it is a challenge to provide a proper management of the feeding.

Furthermore, as offshore aquaculture is located far away from its support facilities in the shore, it adds more challenge and complexity from pond installation to all aquaculture operations including feeding [10]. Traditional feeding techniques rely on the aquaculture owner/operator to manually feed the fish at a particular time and at certain amount of food. Thus, there is a possibility that aquaculture owners/operator do not feed the fish as it is scheduled or may inaccurately give the proper food quantity and quality [11].

The efficiency and profitability of aquaculture practice could be enhanced with improved technology. This has necessitated the design, development, and construction of the automatic feeding device to meet feeding needs and to reduce labour requirements, thereby reducing the cost of fish production [1]. The automatic fish feeder can be an alternative to solve these problems that are effectively implemented for the cage culture system in Indonesia ocean territory, such us inefficiency of conventional feeding, inaccurate amount of fish pallets, and lack of user control interface and warning system in fish feeder design. There are many different designs and brands on the automatic fish feeder on the market, but it needs a specific fish feeder system for offshore aquaculture due to its particular operations environment. Therefore, the aim of this study is to propose a fish feeding system to be installed on one of offshore aquaculture unit. The systems include design and development of an "Automatic Fish Feeder Machine" and the system communicates a specific amount of fish feed at various times intervene that conform to open ocean environment.

2. Literature Review

According to [12] and [13], definition of Open Ocean or Offshore Aquaculture is "rearing of marine organisms under controlled conditions in the Exclusive Economic Zone—from the three miles territorial limit of the coast to two hundred miles offshore. Facilities may be floating, submerged, or attached to fixed structures". Offshore aquaculture is becoming more popular due to several reasons. Nearshore aquaculture could lead to potential conflict with other sectors due to limitation of nearshore areas. Even though a country has a long coastal line, several sectors such as tourism, hotels, real estates, and shipping also depend on nearshore area for their operations. Condition of open ocean is considered to be more favourable for better aquaculture operations due to low pollution, current, oxygen availability and many more [14].

Norway has started their offshore aquaculture for growing Salmon for more than decades ago. Today, Norway produces 1.2 million tons of Salmon yearly and it is the largest producer of farmed salmon in the world. Turkey is the second largest fish producer in Europe after Norway. It has 425 marine farms that mostly is offshore aquaculture and its main product are sea bass and sea bream. China has a plan to achieve fish production by 15.4 million metric tons by 2025 by using a large-scale aquaculture structure. It is a large deep-water aquaculture farms (each has diameter of 139 meters and height of 12 meters) which is designed and built by De Maas and located off the coast of China [15]. While in 2018 Indonesian Ministry of Marine Affairs and Fisheries launched and installed the first pilot Offshore aquaculture project in Pangandaran (West Java), Jepara (Central Java) and Sabang (Aceh). The target of these farms is to produce 946 tonnes of Barramundi and Seabass when it is fully operated [16]. In conjunction with Indonesian Ministry of Marine Affairs and Fisheries strategy, since 2017 Marine Research Centre of Institut Teknologi Sepuluh Nopember (ITS) has developed an offshore aquaculture design, which is called Ocean FarmITS (Figure 1). The design also considers potential to integrate fish production with hotel rooms (accommodation for visitors or tourists) on top of the fish farm structure. Figure 1 describes the offshore aquaculture preliminary design by ITS Marine Research Centre.

Figure 1. Offshore aquaculture preliminary design by ITS Marine Research Centre.

One of important support facility for offshore aquaculture is fish feeding management systems that includes equipment and also feeding activity management. As offshore aquaculture unit will be installed distantly from coastal area therefore automated fish feeding equipment is a perfect option. An automatic fish feeder is a device that could provide food for the fish at a predetermined time. It controls the fish feeding activity by combining mechanical and electrical systems to form a device instead of manually feeding fish by human [17]. The system consists of parts that work automatically in order to save time and labour.

There have been several studies on fish feeder design. The first fish feeder design is proposed by David C Smeltzer which is patented on 4th April 1985. It capable of dispensing various sizes of food grains with a wide range of dispensing volumes. The device has disadvantages of less accuracy, lack of efficiency, and its large in size [17].

Automatic feeder with adjustable parameters has been developed by [3]. It was developed to fulfil certain objectives and requirements with added advantages such as detachable hopper to accommodate various sizes of hopper according to user's requirement. It was also designed to have adjustable height, speed, and opening angle to accommodate different sizes of fish tanks and ponds, as well as provide a mobile fish feeder with a pneumatic system for safety reasons. The system to be specifically used for rearing Tilapia and Catfish. The dimension of this equipment is 0.8 m (length) x 0.5 m (width) x 1.6 m (height). It consists of few components such us hopper, lift, compressor, propeller, dosing valve, and process control panel. The machine also controlled by a digital timer with a feeding rate is 250 gr/ min. This automatic fish feeder can be implemented in any aquaculture by adjusting to the available technology and required equipment size, but it is not equipped with an alarm system to monitor the feed supply [3].

The application of technology for automatic fish feeder using PLC microcontroller is reported by [18] and it has successfully been implemented. The system is simple in construction and operations, while works based on a timer. The storage that is attached at the top. Some number of pellets (food) is decreased during the time as the motor start running. Pellets in the sphere former is dispensed at target area based on the motor speed. The shape of the former was designed corresponding to sphere shape ensuring the spreading process more efficient compare to the other shape, especially on the number of pellets stored. It was proven by mathematical calculation that given the same surface area; the sphere outperformed other shapes in terms of their volume. This study's is able to control motor speed rotation and voltage drop at DC motor output to determine suitable speed that should be used. The disadvantages of this system that it is not feasible to be applied in offshore aquaculture due to its material vulnerability.

In the more recent study of [19], two latest fish feeder machines are compared and set as a benchmark as well as reference to assist in design improvement. Both fish feeder machines had been located at the real marine agriculture site to testing their functionality and suitability. Machine A is utilized solar energy to open the gate while fish feeder schedule is set in the system. Machine B uses Programmable Logic Controller (PLC). It has a setting for 24 hours timer with a direct electricity supply. Result of this study shows that both of machines are made with heavy structure with material that has low resistance to water and temperatures. Thus, [19] proposed new design of an automatic feeder machine with a maximum storage of 5 kg and is operated with two DC motors. It able to distribute 500 grams of pellets within 90 seconds with the longest distance of 4.7 meters.

An automatic fish feeder design using IoT (Internet of Things) is proposed by [20]. It consists of Wi-Fi to connect the system and user command, GUI (Graphical User Interface) to run on Android devices, and Arduino to control the system. This system has a storage container and a dispatch container while owners can monitor the feeding process from Android devices and set many different time variations based on species of fish [20]. It is a flexible system and designed for small aquatic area. However, it is not always feasible to be applied on offshore aquaculture in Indonesia due to limitations of internet connection in several places especially in the remote area. Table 1 represents comparison of this study with previous studies according to the system used, dimension, and the integrated system.

Table 1. Comparison of related studies.

Author(s)	**System**	**Dimensions Length x width x height/ Capacity**	**Controller**	**HMI**	**Alarm System**	**IoT**
Smeltzer, et al. (1986)	Vibration System	-	-	-	-	-
Yeoh, et al. (2010)	Pneumatic System	0.8 m x 0.5 m x 1.6 m / 10 kg	v	-	-	-
Noor, et al. (2012)	Timer System (PWM for rotation motor speed control)	-	v	v	-	-
Ogunlela, et al. (2016)	Timer System	0.3 m x 0.26 m x 0.79 m	v	-	-	-
Salleh, et al. (2017)	Timer System (Solar Recharge)	5 kg	v	v	-	-
Meshram, et al. (2019)	Customize feeding time using controlled mobile phone	-	v	v	-	v
This study	Feeding management System	2 m x 2.4 m x 4.3 m/ 1000 kg	v	v	v	v

Aquaculture operations could be divided into two main activities, they are: (a) Cultivating the fish, from stocking (fish seed/fingerling supply), feeding, nursing/monitoring, and harvesting; and (b) Maintaining the floating net cage.

Feeding management is also important issue in any aquaculture operations. In order to achieve maximum business profit or minimizing cost of food, is by ensuring fish size can quickly reach to market size by optimizing feed consumption and minimizing food waste/excess. In general, the rule of thumb for fish feeding are: (1) slowly serve the food, (2) make sure accessibility for all fish in the pond to get enough food, (3) stop the feeding when fish loses their interest to food, (4) try to avoid feeding times at noon and early afternoon due to strong sunlight, (5) serve the food relatively at the same place time every time [21] [22]. There are two types of fish food they are: fresh trash fish and commercially pellets. Pellet feed has an advantage that it is always available in any season, while it is not necessary applicable for fresh trash fish as its availability heavily depends on seasons. In addition, fish also needs vitamin A guideline for fish feeding amount and frequency specifically for groupers [22] can be seen in Table 2.

Table 2. Feeding dose for groupers.

Fresh Trash fish			Pellet		
Fish weight (gr)	**Daily feed (% fish biomass)**	**Frequency (times)**	**Fish weight (gr)**	**Daily feed (% fish biomass)**	**Frequency (times)**
5-10	15-20	4-6	5-20	2-4	2-3
10-50	10-15	2-3	20-100	1.5-2	2
50-100	8-10	1-2	100-200	1.2-1.5	1-2
150-300	6-8	1	200-300	1-2	1
300-600	4-6	1	>300	0.8-1	1

3. Methodology

As a part of Ocean FarmITS development project, this study proposes development of automated fish feeder design that could accommodates two types of food (pellets and fresh trash fish). Research methodology of this study adopts product development stages according to [23], which are including three main stages they are: (1) Pre Design/Development (i.e. idea generation), (2) Design and Development Process (i.e. concept, design, preproduction validation) (3) Post Design Development (i.e. production, post-company). While, this study covers Pre-Design and Design and Development Process stages that focuses in developing automated fish feeder for Ocean FarmITS (Figure 2).

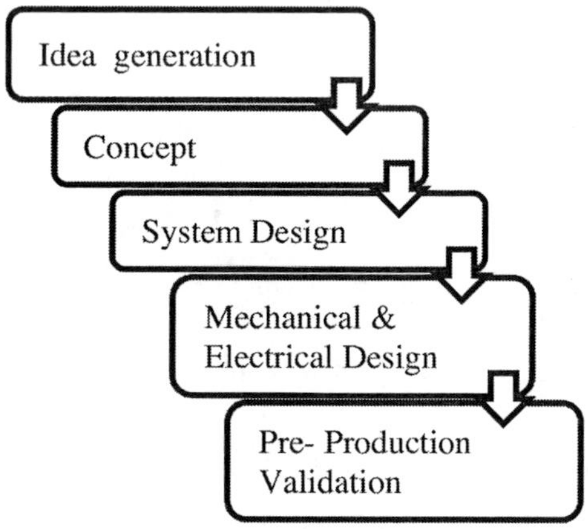

Figure 2. Automated fish feeder development process.

The first stage, Pre-Design/Development, is to generate idea by reviewing existing studies on automated feeding systems as well as focus group discussion with several experts from Indonesian Ministry of Marine Affairs and Fisheries (MMAF) Department and academics, and interviewing aquaculture business owners in East Java and Bali.

The next stage is Design and Development Process which is started by formulating concept of automated fish feeder according to Ocean FarmITS operations characteristics. Ocean FarmITS is an offshore aquaculture unit that has a fish farm that is 15 meters in diameter and 3 meters in height. As a pilot project, the structure is going to be installed at Eastern Indian Ocean, southern part of Malang. Based on discussion with local fishermen and also Head of Fishing Port, Brown-Marbled or Tiger Grouper (*Epinephelus fuscoguttatus*) is selected to be cultivated in this floating net. It is projected to grow around 14.500 fingerlings (a young/small fish) with average fingerling weight is 50 grams. With approximately 25% mortality rate, this farm is expected to harvest up to around 9.000 Tiger groupers in 7 months. Each fish should reach about 500 grams in weight which is required mostly by market (local and overseas). To ensure the fingerlings can grow optimal to the targeted weight, therefore the daily feeding arrangement from the first day up to the next 7 months is developed, which can be seen in the Table 3. This table could be updated every month, as mortality rate and growth rate sometimes could be higher or lower than the estimation. Fresh trash fish is decided to be the main food as most of the time it is quite easy to get in the area of farm location with low cost. However, between November to February is usually a difficult time to get fish due to weather, thus pellet could be used to replace a certain amount of fresh trash fish.

Table 3. Feeding arrangement for OceanFarmITS.

Day	Fresh Trash Fish			Commercially Pellets		
	Amount (kg)	Frequency/day	Total (kg)	Amount (kg)	Frequency/day	Total (kg)
1	74	1	74	11	2	22
2	75	1	75	11	2	22
3	76	1	76	11.5	2	23
...	...	...	...	...	...	...
30	104	1	104	15.5	2	31
31	105	1	105	16	2	32
....	...	...	...	...	...	...
60	135	1	135	16.5	2	33
61	135	1	135	16.5	2	33
....	...	...	...	...	...	...
90	135	1	135	17	2	34
91	135	1	135	17	2	34
....	...	...	...	...	...	...
120	175	1	175	44	1	44
121	175	1	175	44	1	44
....	...	...	...	...	...	...
150	251	1	251	63	1	63
151	254	1	254	63	1	63
....	...	...	...	...	...	...
180	278	1	278	85	1	85
181	278	1	278	86	1	86
...	...	...	...	...	...	...
210	323	1	323	108	1	108
211	326	1	326	109	1	109
212	328	1	328	109	1	109

As a result, there are several requirements for Automated fish feeder of Ocean FarmITS that should take into account into its design. First, it should be able to monitor the total amount of pellets/ fresh trash fish and also can send a notification through the mobile phone application to the aquaculture owner/operator. Moreover, it should be able to synchronize with the ultrasonic sensor to calculate the level of pallets/ raw fish in the silo tank and update the user by wireless communications.

3.1. System Design

The main system design of this automatic fish feeder machine is a control system using a microcontroller Arduino board, it received the data from sensors in the aquaculture farm and sent the instruction to the motor driver. The block diagram of the automatic fish feeder as shown in Figure 3, where the electronics components are includes:

a) Arduino Microcontroller

Arduino is an open-source computing microcontroller board based on ATmega32, and it is used for programming electronic devices (sensor and actuator). This device has 14 digital input/ output pins that some pins can be used as PWM outputs and simply connect it to a computer with a USB cable or power with AC-to-DC adapter to run. Arduino is low cost and easy to reprogram when the system needs to upgrade with the control algorithm like Proportional-Integral (PI) loop, so it is the right choice for this system. Arduino receives data from the level sensor and sends the instruction to the motor driver for running the servo motor. Arduino also programmed the alarm system to send an alert using IoT system to the user when the ultrasonic sensor detects limit level in the Silo Tank.

b) User Interface

The user interface for automatic feeding system is using Human Machine Interface (HMI), and it installed in the system that could be controlled and set by the operator. The owner also could monitor the system using a smartphone by installing the application HMI Controller for Arduino in App Store to receive the alarm notification also change the feeding schedule.

c) Ultrasonic Level Sensor

The level food in Silo Tank is monitoring by Ultrasonic level sensor. It has transmitter and receiver that installed in the top of Silo Tank. The Ultrasonic sensor measuring the distance from the transmitter to the surface of a food fish in Silo Tank by calculate the time for a sound wave to reflected by the object and echo returns to the receiver. The ultrasonic level measurement principle is also used to find out of fish position in ocean, locate submarines below water level. Therefore, ultrasonic level sensor is suitable to be implemented in offshore aquaculture area.

d) Buzzer Alarm System

The buzzer alarm system is used to send the alert to the operator/ owner when the level sensor reaches the limit level in Silo Tank and its reminders the operator to fill the food fish. The buzzer alarm system integrated with a level sensor to create a sound alarm when reached the limit level in offshore aquaculture area, and Arduino also sends the notification alarm to the owner. The buzzer alarm system using a Piezo Buzzers type that has a piezoelectric element, with the characteristic is utilized to make the ceramic plate vibrate rapidly to generate sound waves.

e) *Servo Motor Driver*

The function of servo motor drive is received the voltage command signal from Arduino controller and send the necessary signal voltage and current to the servo motor to achieve the required motion to spread the food from two pipelines. The servo motor drive is also could control the torque, velocity, or position of the servo motor.

f) Servo Motor

The servo motor is used for dispensing the food into the farm through two pipelines. The servo motor can generate high torque over a wide speed range on demand also wider torque ranges. Although servo motor designed to run at high speeds, it could accurately run at extremely low speeds under precise control and repeatable because it runs closed-loop system. That is the reason why the servo motor is suitable for automated fish feeding system, so the wide area could be covered when food is being dispensed.

g) Power Supply

A power supply is a device used to provide the electric energy to operate the devices running by electric power [24]. The function of power supply in this system is to convert Alternative Current (AC) voltage to regulated Direct Current (DC) voltage to supply electronic devices in the automated fish feeding management system.

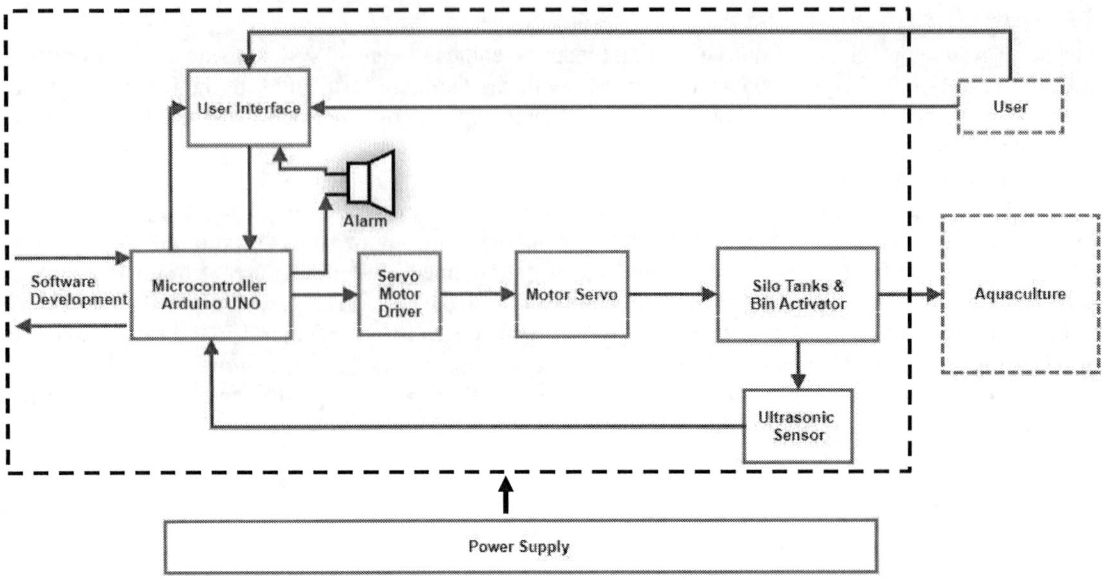

Figure 3. Block diagram of the system.

Figure 3 describes the block diagram of the system. The design for one silo tank should hold up to 1000 kg of fresh trash fish. The bin activator could be activated using timer control from user interface input and forwarded to the control system to be data input, it generates the instructions for bin activator to release or close. This system can be also controlled by the user for setting the food timing schedule when there is no one in the aquaculture unit. The amount and frequency of feeding, which is shown in Table 3, is then converted into Arduino programs. The program could be altered when the feeding arrangement is updated.

To ensure proper amount of food and its frequency being dispensed to the farm, a timer is used to open the circuit and dispense the food or to cut off the circuit to stop feeding. The food is dispensed into the farm through two pipelines that are attached to the servo motor. This action is to ensure a wide area covered when food is being dispensed and ensuring fair share for all fish in the farm. When the trash fish or pellet reaches limit level, it will trigger the buzzer alarm and send the alarm signal to the user (farm operator/owner).

After distribution of food has finished, the pipe is flushed by water that is supplied from a small tank on the top of silo in order to keep the pipe clean from remaining food that may blocking it or contaminated food on the next feeding schedule. The ultrasonic sensor is constantly checking the amount of food level in the silo tank and trigger the alarm to notify farm operator/owner for refilling. This system is important to estimate and arrange the supply vessel to bring new inventory of pellet/ fresh trash fish.

3.2. Mechanical and Electrical Design

The overall design of automatic fish feeder is shown at Figure 4. This device is mostly constructed using stainless steel and aluminium to ensure high resistance from saltwater and to protect electronics component, where Table 4 shows the detail of construction material. Figure 5 shows the detail of the position of each part of the automatic fish feeder, which consists of a level sensor, silo tank, bin activator, pipe, and servo motor. The detail size of every part can be seen at Figure 6. The size of the silo tank is 2 m in diameter with 2.4 m height and the length of each pipe 2.5 m with total height of the fish feeder machine unit is 4.3 m.

The schematic diagram for electronics designed is shown in Figure 7. Arduino controller receives an input signal from the sensor and user interface then processes the data and sends the program instruction to the motor. The buzzer alarm is triggered if the food reaches the limit levels, and the ultrasonic sensor sends the data of food level continuously.

Table 4. Construction materials.

Components	Materials Used	Dimensions	Remarks
1. Lid	Aluminium	1 m (diameter)	Aluminium is lightweight
2. Silo Tank	Aluminium	2,4 m x 2 m	Aluminium is high resistance from saltwater due to protective oxide layer, lightweight, and low cost
3. Bin Activator	Stainless Steel	2 m (diameter)	Stainless Steel is corrosion resistant and high strength
4. Base	Aluminium	2,5 m x 1 m	Aluminium is high resistance from saltwater due to protective oxide layer and low cost
5. Pipe	PVC	2,5 m	PVS is lightweight, good in water and temperature resistance
6. Panel Box	Fiberglass	50 cm x 40 cm x 20 cm	Fibreglass used for panel box casing to protect electronics components

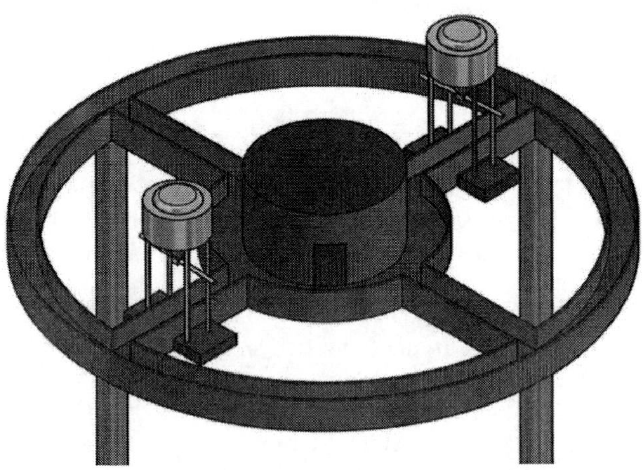

Figure 4. Overall design of the fish feeder on the fish farm.

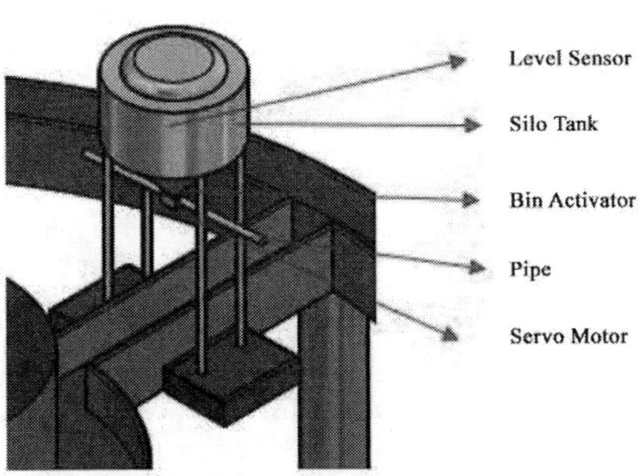

Figure 5. The main components of the fish feeder.

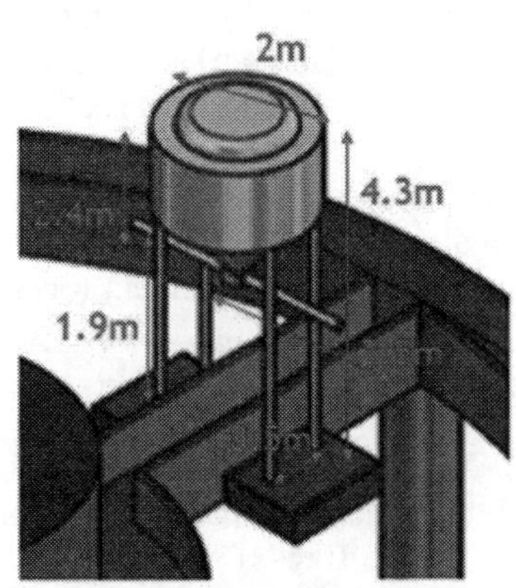

Figure 6. Prototype size.

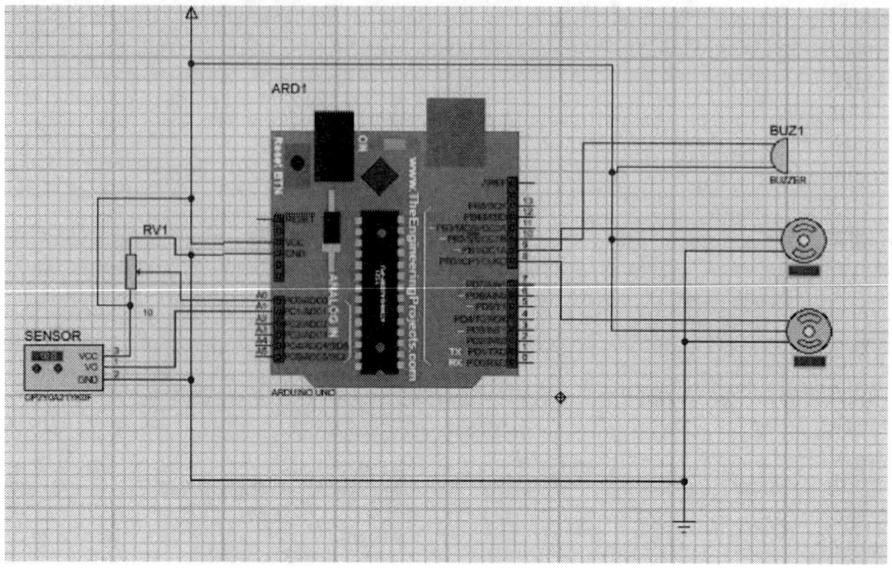

Figure 7. Schematic diagram control system.

4. Conclusion

This paper proposes fish feeding system for Ocean FarmITS pilot project which located in Eastern Indian Ocean. OceanFarmITS is an offshore aquaculture that has a fish farm of 15 meters in diameter and three meter in height. It consists of fish feeding arrangement and fish feeder system that is customized for growing Tiger grouper in the farm for around seven months. The automatic fish feeder is designed to be able to provide two types of food (i.e. fresh trash fish and pellet), while trash fish is preferable as the main dietary. It serves food according to the daily food arrangement that is formulated based on estimation of number of fish in the farm, fish size (average weight), as well as availability of trash fish. The feeding arrangement should be changed and updated following real condition in the farm. To ensure availability of food in the tank, this system is equipped with alarm system that send notification to Aquaculture operator when food storage in silo tank has reached its limit level. The prototype of this fish feeding system for Ocean FarmITS is going to be built and tested by installing it in one of near or onshore fish farm in one of Malang beach for two months. Based on the testing result, some changing, or improvement is made and then then final design of the fish feeding system is then build and installed in OceanFarmITS.

By providing an automatic fish feeder, it would assist or replaces offshore aquaculture operator/owner to feed the fish at the proper volume and frequency (avoid human error) or during hostile weather that is commonly happened. Thus, operator could focus on performing other tasks that could not replace by automated system, such taking sampling of water or fish for monitoring. As a result, OceanFarmITS could operate its farm efficiently by reducing excessive food and preventing from contaminating the environment (ocean). At the same time, the growth and health of fish is guaranteed with minimum operations cost.

References

[1] Ogunlela AO, Adebayo AA. Development and Performance Evaluation of an Automatic Fish Feeder. *J. Aquac. Res. Dev.* OMICS Publishing Group; 2016;**07**(02). doi.org/10.4172/2155-9546.1000407

[2] Liao IC. Trends and prespectives in Asian aquaculture. *Report of The APO Seminar*; 2001.

[3] Yeoh SJ, Taip FS, Endan J, Talib RA, Mazline MS. Development of automatic feeding machine for aquaculture industry. *Pertanika Journal of Secience & Technology*. 2010;**18**(1):105-10.

[4] Lehmkoster J, Schroder T, Ladischensky O. 2013. The future of fish - the fisheries of the future. *World Ocean Review*; 2013.

[5] FAO. 2018. *The State of Word Fisheries and Aquaculture: Meeting the Sustainable Development Goals*. Rome: Food and Agriculture Organization of the United Nations; 2019:2-6.

[6] Prem R, Tewari VK. Development of human-powered fish feeding machine for freshwater aquaculture farms of developing countries. *Aquacultural Engineering*. Elsevier BV; 2020 Feb;**88**:102028. doi.org/10.1016/j.aquaeng.2019.102028

[7] National Aquaculture Sector Overview Indonesia. *Food and Agriculture Organization of the United Nations*. Retrieved from: https://www.fao.org/fishery/countrysector/naso_Indonesia/en.

[8] Aderolu AZ, Seriki BM, Apatira AL, Ajaegbo U. Effect of feeding frequency on growth, feed efficiency and economic viability of rearing African catfish (clarias gariepinus, burchel 1822) fingerlings and juveniles. *African Journal of Food Sciencee* 2009;**4**(5):286-90.

[9] Azzaydi M, Madrid JA, Zamora S, Sanchez VFJ, Martinez FJ. Effect of three feeding strategies (automatic *adlibitum* demand-feeding and time-restricted demand-feeding) on feeding rhythms and growth in European sea bass (*dicentrarchus labrax L*). *Aquaculture* 1998;**163**:285-296.

[10] Knapp G. The development of offshore aquaculture: an economic perspective. *In A. Lovatelli, J. Aguilar-Manjarrez & D. Soto, eds. Expanding mariculture farther offshore: technical, environmental, spatial and governance challenges. FAO Technical Workshop*. FAO Fisheries and aquaculture; 2013.

[11] Dada EG, Theiphine NC, Adekunle AL. Arduino UNO microcontroller based automatic fish feeder. *The Pasific Journal of Science and Technology*. 2018;**19**:1.

[12] Upton HF, Buck EH. 2010. Open Ocean Aquaculture. *Congressional Research Service* 7–5700 (RL32694). Retrieved from: www.crs.gov.

[13] Edwards P. Aquaculture environment interactions: Past, present and likely future trends. *Aquaculture*. Elsevier BV; 2015 Oct;**447**:2–14. doi.org/10.1016/j.aquaculture.2015.02.001

[14] Gentry RR, Froehlich HE, Grimm D, Kareiva P, Parke M, Rust M, et al. Mapping the global potential for marine aquaculture. *Nat. Ecol. Evol.* Springer Science and Business Media LLC; 2017 Aug 14;**1**(9):1317–24. doi.org/10.1038/s41559-017-0257-9

[15] Offshore Finfish Aquaculture Global Review and U.S. Prospects 2018. *California Environmental Associates (CEA) Report 2018*. Retrieved from: https://www.packard.org/insights/resource/offshore-finfish-aquaculture-global-review-and-u-s-prospects/

[16] Soebjakto S, Pregiwati LA. Mengenal Sistem Budidaya KJA Offshore Pertama di Indonesia. *KKP Kementerian Kelautan dan Perikanan*. 2018. Retrieved from: http://kkp.go.id/an-component/media/upload-gambar-pendukung/kkp/SIARAN PERS/April/SP73 MENGENAL SISTEM BUDIDAYA KJA OFFSHORE PERTAMA DI INDONESIA.pdf.

[17] Dhavale SD, Sagar AC, Chavan MS, Jagtap SC. A literature survey on automatic fish feeder *International Journal of Research in Engineering, Science and Management*. 2019;**2**(2).

[18] Noor MZH, Hussian AK, Saaid MF, Ali MSAM, Zolkapli M. The design and development of automatic fish feeder system using PIC microcontroller. *2012 IEEE Control and System Graduate Research Colloquium*. IEEE; 2012 Jul. doi.org/10.1109/icsgrc.2012.6287189

[19] Wei HC, Salleh S., Mohd Ezree A, Zaman I, Hatta M., Md Zain BA, et al. Improvement of automatic fish feeder machine design. *Journal of Physics: Conference Series*. IOP Publishing; 2017 Oct;**914**:012041. doi.org/10.1088/1742-6596/914/1/012041

[20] Mesharam S, Meshram G, Rokde B, Kapse R, Hedaoo O, Mandhata C. Fish feeder using internet of things. *International Research Journal of Engineering and Technology*. 2016;**6**(02).

[21] Sim S Y, Rimmer M A, Toledo J D, Sugama K, Rumengan I, Williams K C, Phillips M J. *A Practical Guide to Feeds and Feed Management for Cultured Groupers. NACA*. Bangkok: Thailand; 2005.

[22] WWF. *Better Management Practices Small Scale Fisheries Guideline Series Tiger Grouper Culture–Floating Cage System* Vol 2. Jakarta: WWF-Indonesia; 2015.
[23] Peters AJ, Rooney EM, Rogerson JH, McQuater RE, Spring M, Dale BG. New product design and development: a generic model. *The TQM Magazine*. Emerald; 1999 Jun;**11**(3):172–9. doi.org/10.1108/09544789910262743
[24] Brown M. *Power Supply Cook Book*. MA U.S.A, Butterworth- Heinemann; 2001.

IOP Conf. Series: Materials Science and Engineering 1072 (2021) 012074 doi:10.1088/1757-899X/1072/1/012074

Determination of route for estimating number of truck needed in periodic distribution: A case study

D Hartanto[1]* and R Wahyuningtyas[1]

[1] Industrial and Systems Engineering, Faculty of Industrial Technology and Systems Engineering, Institut Teknologi Sepuluh Nopember (ITS)

Email: *dodyhartanto@ie.its.ac.id

Abstract. Transportation is an important supply chain element. Lowering logistics costs can be effectively carried out by lowering transportation costs. PT. IAP is a distributor company that supplies fast-moving consumer goods products to more than 500 Modern Trade. There was 77% of delivery have been carried out by PT. IAP has achieved the targeted service level. The root cause of this problem is delivery which is triggered by purchase order (PO) from retailers. IAP plans to change the delivery procedure that has been triggered by the PO to periodic delivery based on historical data of each retailer's demand. Therefore, this research aims to determine the route to estimate the number of trucks needed is periodic delivery. The distribution route will be developed using a saving matrix algorithm for clustering retailers in one route while the order of delivery is obtained by the Nearest Neighbor method. Based on the results of the saving matrix developed in visual basic for application (VBA), it is found that the number of vehicles needed by Colt Diesel Double (CDD) trucks is 23 units and Colt Diesel Engkel (CDE) trucks are 11 units. The reduction in the number of trucks needed is significant, before the study, the company needs 40 trucks CDD and 15 trucks CDE.

1. Introduction

Transportation is responsible for the physical movement of materials between points in the supply chain [1]. Transportation is an important supply chain element because products are rarely produced and consumed in the same location [2]. Transportation costs are costs that dominate the total logistics costs. The amount of these costs can reach 60% of the total logistics costs of the company. Therefore, lowering logistics costs can be effectively carried out by lowering transportation costs. Transportation costs can be reduced by increasing the utility of transport vehicles, balancing capacity between stages of distribution, determining efficient routes of distribution and good coordination between companies and consumers. In many cases, distribution is carried out by several different parties. Therefore, good coordination is also needed between the parties involved at all stages of distribution.

The Case Study in this research is conducted at IAP, a distributor company. The company distributes products produced by 24 companies to hundreds of retailers located throughout Surabaya. These retailers send POs to IAP and IAP make deliveries to fulfill these POs tomorrow. This causes the number of trucks needed every day to keep changing. IAP classifies retailers into two categories of retailers namely general trade and modern trade. In general trade, product shipments are carried out in two stages: the product is sent to the stocking point, the product is sorted and then sent to retailers. IAP has a target service level for delivering goods to retailer not more than 24 hours after receiving a

Purchase Order (PO). Based on historical data in October 2018 to September 2019, there was 116,023 PO and only 77% of the POs which can be delivered not more than 24 hours. Figure 1 explains flow of good in general and modern trade.

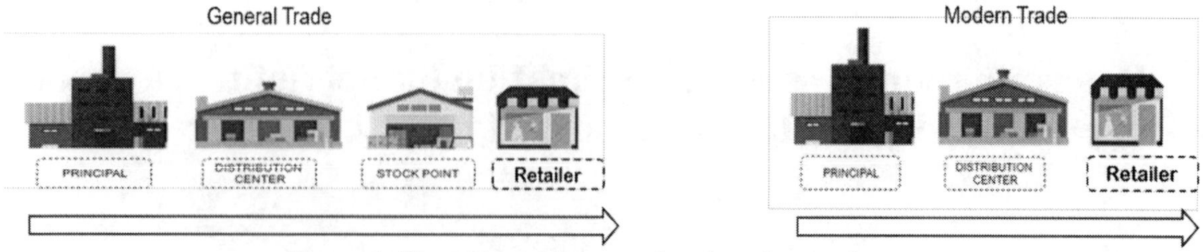

Figure 1. Flow of good in general and modern trade

Currently, IAP has a total of 55 trucks with details of 40 Colt Diesel Double (CDD) trucks with a capacity of 4 tons and 11 m^3 and 15 Colt Diesel Engkel (CDE) trucks with a capacity of 2.5 tons and 7 m^3. Currently, the IAP has not been able to achieve the expected target service level so it has to rent 8 vehicles when there is a surge in demand or POs that do not follow the RPS. The addition of this vehicle is quite helpful to cover delivery when the IAP gets a surge in demand from the retailer, but on the other hand the rented truck will be idle when demand is not more than average demand. This causes the truck utility to be low.

Moreover, IAP delivery team determines the distribution route manually or according to the knowledge and experience of drivers. Therefore, the routes are not optimal. The root cause of this problem is delivery which is triggered by PO from retailers. IAP plans to change the delivery procedure that has been triggered by PO to routine delivery based on historical data of each retailer's demand. So the formulation of the problem in this research is if routine delivery is applied, how to determine the delivery route and how many trucks are needed.

2. Methods

This study aims to determine the number of trucks needed by determining a good (near optimal) route. The determination of the route is done using a saving matrix algorithm. This study was carried out in three main stages:

1. Estimating the distance between each retailer and between each retailer with distribution center which is required in the saving matrix algorithm.
   Distance is estimated using google maps and haversine formula. Google maps are used for retailers that are located with quite a high circularity factor.
2. Converting demand that is stated in units into units of volume.
   Historical demand data contains the number of products that need to be sent to each retailer. Truck capacity is not measured in units but in weight and volume so conversion is needed. Conversion becomes even more complicated because there are many types of products and packaging that need to be distributed. Based on experience, the volume capacity is met before the weight capacity. Therefore, the capacity parameter that needs to be considered is only the volume capacity.
3. Determine the route with a saving matrix algorithm using visual basic for application (VBA) software.
   At this stage, the node is grouped points into one cluster using the Saving Method and then sorted using the Nearest Neighbor with the help of Visual Basic for Application (VBA) tools. The algorithm which can be seen in Figure 2 becomes a reference for making algorithms in VBA to simplify and precipitate the calculation of all modern trades served by PT. IAP.

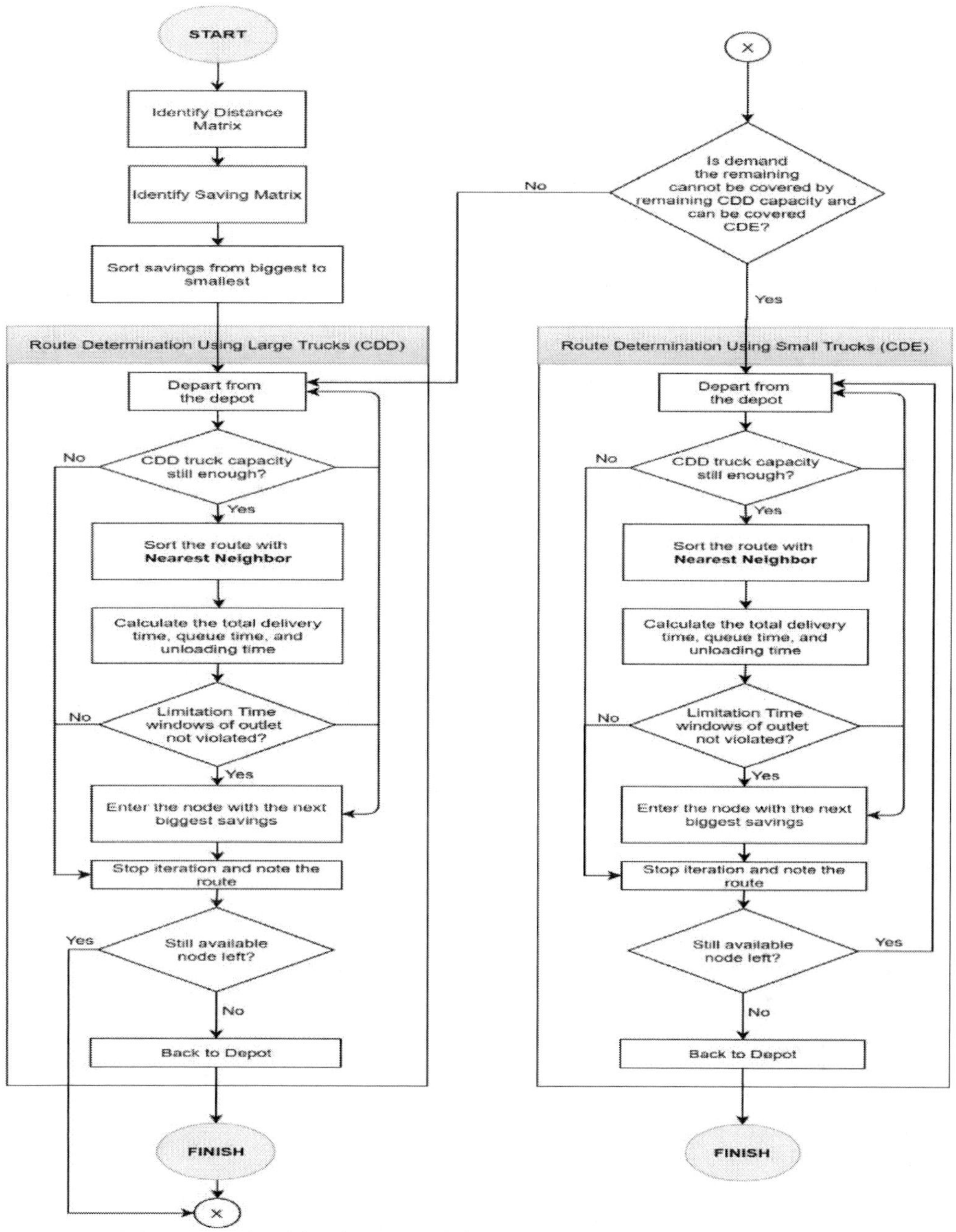

Figure 2. Algorithm in Determining Route of Truck (CDD and CDE)

Prior to use the developed VBA was verified and validated to ensure that the VBA was developed correctly. Validation is done by testing VBA with a group of destination points with certain demands. VBA solutions will be compared with manual calculations, if the results are the same then VBA can be said to be valid so that it can be used to determine the overall route. The samples used were retailers located in Semampir District, Surabaya. The results of manual calculations and VBA software can be seen in Table 1and Table 2.

Table 1. The result of VBA

Route	Demand (m^2)	Load (m^2)	Delivery Time (min)	Arrived Time	Finish Service
			Truck CDD 1 Day 1		
0					
2	0.1473	0.1473	1.0000	08:00	08:25
6	0.6147	0.7620	0.8583	08:26	14:26
3	4.8174	5.5794	1.0000	14:27	14:57
8	2.1557	7.735	0.9142	14:58	17:58
0			2.1920	18:00	
			Truck CDD 1 Day 2		
0					
4	5.71	5.71	2.07679	08:00	13:00
1	2.6368	8.3468	2.17193	13:02	14:02
5	2.5992	10.946	2.13776	14:04	14:34
0			2.11096	14:36	
			Truck CDD 2 Day 1		
0					
9	1.9027	1.9027	1.28101	08:00	08:30
4	9.1	11.0027	0.79576	08:31	13:31
0			2.07679	13:33	
			Truck CDE 1 Day 1		
0					
4	7	7	0.79576	08:31	13:31
0			2.07679	13:33	
			Truck CDE 1 Day 1		
0					
7	2.5804	2.5804	2.45754	08:00	13:00
0			2.45754	13:02	

Table 2. The route as a result of manual calculation

Vehicle ID	Type of Vehicle	Route
1	CDD	0(1) -> 2(1|08:25|0.15) -> 6(1|14:25|0.61) -> 3(1|14:55|4.82) -> 8(1|17:55|2.16) -> 0
		0(2) -> 4(2|13:00|5.71) -> 1(2|14:00|2.64) -> 5(2|14:30|2.6) -> 0
2	CDD	0(1) -> 9(1|08:30|1.9) -> 4(1|13:30|9.1) -> 0
3	CDE	0(1) -> 4(1|13:30|7) -> 0
		0(2) -> 7(2|13:00|2.58) -> 0

Based on the verification process, it can be concluded that the route algorithm built on the VBA tools is valid.

3. Result and Analysis

The results of the route are formed using a routine delivery scenario for one month. algorithm is run once for all modern trades served. The number of trucks needed under the routine delivery scenario every month is 23 CDD trucks and 11 CDE trucks. The scenario of routine delivery every month can reduce the need for vehicles up to 42.5%. This is because of the application of VMI, where delivery to the retailer is fully regulated by PT. IAP as a supplier of FMCG in retailers.

Sensitivity analysis is performed to determine changes in the solution if there are changes in the model parameters [2]. In this case, the parameter that is most likely to change in demand.

- Demand Increased by 20%.
  The need for trucks increased to 28 CDD trucks and still needed 11 CDE trucks. Retailers that are visited are reduced because trucks load more demand for the same retailer, so the need for trucks increases. The route formed changes from normal conditions. However, there are 20% of the unchanged routes. Utilities have increased from 96.07% to 97.64% due to the greater amount of demand carried.
- Demand is increased 100%.
  The number of trucks needed increased to 43 CDD trucks and 19 CDE trucks. This is because the load carried by the vehicle is greater than usual so that the vehicle volume capacity limit is easily reached even though the number of retailers sent is less. Utilities increased to 98.06%.
- Demand Decreased by 15%
  The need for trucks has decreased from what was originally needed to be 20 CDD trucks and 10 CDE trucks. As many as 28% of the routes have not changed due to the decrease in demand which is not too significant.
- Demand Decreases 50%
  The need for trucks decreased to 12 CDD trucks and 10 CDE trucks. This can occur because of the increase in retailers visited in one route. The route formed changes from normal conditions. However, from the sampling of 100 routes from CDD trucks and CDE trucks with ID Vehicle 1, 2 there is 1 unchanged route, namely CDD 1 truck with routes going to retailers 2, 6, 9, 3.

This study has not yet implemented the levelling criteria or the equalization of vehicle loads to the retailers visited has not been done so it cannot prevent imbalances. The clustering process is only used as a tool to simplify the algorithm to find out which retailers are the closest to visit afterwards, but in the reality the closer retailers may be different retailers in the sub-district. Therefore, determining the route cannot be done separately between one cluster with another cluster. The distance used in this study does not use the distance calculated by Google Maps (using estimates with the Haversine Formula) so that it is less representative of the actual distance. In addition, the amount of demand entered is the average demand every month, so the demand processed has not accommodated the surge in demand during the season. Vehicle assignments do not consider which locations are only able to be passed by CDE trucks. The algorithm that was formed also did not consider shipping for the second, third and so on. Because of these limitations, easing time windows in the sensitivity analysis does not significantly influence the route formed and the number of vehicles needed.

4. Conclusion

The conclusions of this study are the schedule and route obtained are better than existing conditions because routine delivery can reduce the possibility of trucks to experience idle without having to reduce the level of demand fulfillment. From the 40 CDD trucks provided by the company, only 23 were needed and only needed 11 CDE trucks out of 15. With routine deliveries, higher trucking utility is generated. Based on a comparison of existing routes and running results, utility utilization of trucks has increased from 80% to 90%.

The sensitivity analysis that has been carried out proves that the modified parameters are able to influence the route decisions made. The demand parameter that was changed to double shows the need for 43 CDD trucks and 19 CDE trucks. The total demand for these trucks has increased by 80% of the needs of normal trucks using the routine shipping business process. On the other hand, the demand for trucks decreased by 47% when the demand dropped to half. This shows that the demand parameters are very sensitive to the decision of the route and the number of trucks needed.

5. References

[1] Widjaja, A. B. (2019). Developing Model of Flexible Periodic Vehicle Routing Problem (FPVRP) Considering Color Combination Requirements for A Paint Distribution System. Surabaya: ITS.
[2] Chopra, & Meindl, P. (2001). Supply Chain Management Strategy, Planning, And Operation.
[3] Agung Hadhiatma, A. P. (2017). Vehicle Routing Problem untuk Distribusi Barang Menggunakan Algoritma Semut. Prosiding SNATIF, 139.
[4] Ahmad Naihadi, S. (2017). Minimasi Biaya Dalam Penentuan Rute Distibusi Produk Minuman Menggunakan Metode Saving Matrix. 2-3.
[5] Ballou, R. H. (2004). Business Logistics Management 5th. ed. Pretince Hall.
[6] Billy, A. (2019). Developing Model of Flexible Periodic Vehicle Routing Problem (FPVRP) Considering Color Combination Requirement for A Paint Distribution System. Surabaya.
[7] Bureau of Transportation Statistic. (2019, September 12). July 2019 Freight Transportation Services Index (TSI). Retrieved from News and Statistical Releases: https://www.bts.gov/newsroom/july-2019-freight-transportation-services-index-tsi
[8] Chelsea Sabo, Kelly Cohen. (2011). SMART Heuristic for Pickup and Delivery Problem (PDP) with Cooperative UAVs. 2-3.
[9] Ferdinand, F. N. (2013). A Genetic Algorithm Based Heuristic for the Design of Pick-up and Delivery Routes for Strategic Alliance in Express Delivery Services. In Conference on Manufacturing Modelling, Management, and Control (pp. 1938 - 1943).
[10] Firmandani, R. A. (2017). Penjadwalan Bus Pada Transportasi Karyawan PT. Kaltim Prima Coal dengan Model Split Delivery Vehicle Routing Scheduling Problem. Surabaya: ITS.
[11] Ghiani, Laporte, Musmanno. (2004). Introducing to logistic system planning and control. Halsted Press.
[12] Indrawati, E. L. (2016). Penentuan Rute Optimal pada Pengangkutan Sampah di Kota Palembang dengan Menggunkan Metode Saving Matrix. 105-110.
[13] Kenton, W. (2019, Juni 25). Visual Basic for Applications (VBA). Retrieved from Investopedia: https://www.investopedia.com/terms/v/visual-basic-for-applications-vba.asp
[14] Kleywegt, Nori, & Savelsbergh,. (2004). Dynamic programming approximations for a stochastic inventory routing problem. In Transportation Science (pp. 42-70).
[15] Nugroho, A. T. (2014). Perancangan Algoritma Heuristik untuk Penyelesaian Permasalahan Swap-Body Vehicle Routing Problem. Surabaya: ITS.
[16] Pemerintah Kota, S. (2019). Kecamatan di Surabaya. Retrieved from Web Resmi Surabaya: surabaya.go.id
[17] Pujawan, N. (2017). Supply Chain Management.
[18] Tempo. (2015). Biaya Distribusi Indonesia Mahal di ASEAN, Ini Penyebabnya. Jakarta: Tempo.co.id.
[19] Terapan, U. L. (2017, Agustus 25). Pengenalan VBA (Visual Basic for Applications) pada Macro Excel. Retrieved from SAS Laboratory: https://fit.labs.telkomuniversity.ac.id/pengenalan-vba-visual-basic-applications-pada-macro-excel/
[20] Thangiah, S. R. (1995). Vehicle Routing Problems with Time Windows Using Genetic Algorithms. Working Paper.
[21] Toth & Vigo. (2014). Vehicle Routing: Problems, Methods, and Applications. Philadelpphia: Society for Industrial and Applied Mathematics.
[22] Waller, Johnson, & Davis. (1999). Vendor managed inventory in the retail chain. Journal of Business Logistics, 20(1).
[23] Wuryono, T. (2015). GINSI: Biaya Distribusi Barang di Indonesia Tertinggi Se-ASEAN. Semarang: Metrosemarang.com.

[24] Yosefa, Carles Sitompul, Alfian. (2015). Perancangan Model VMI (Vendor Managed Inventory) dengan Satu Pemasok dan Banyak Retailer yang Meminimasi Ongkos Total Rantai Pasok. 88-90.
[25] Yulianto, R. (2018). Penerapan Formula Haversine pada Sistem Informasi Geografis Pencarian Jarak Terdekat Lokasi Lapangan Futsal. Jurnal Ilmiah Ilmu Komputer, 15-16.

Development of virtual laboratory applications based on virtual reality using the requirements engineering method

S Saptadi[1,2], M A Arliawan[1,3], D Nurkertamanda[1,4], Y Widharto[1,5], P A Wicaksono[1,6]

[1]Diponegoro University, Semarang, Indonesia

E-mail: [2]singgihs@ft.undip.ac.id, [3]aswinarliawna29@gmail.com, [4]nurkerta@gmail.com, [5]yududito@gmail.com, [6]nawanadi@gmail.com

Abstract. The virtual laboratory application that has been designed in Industrial Engineering Department of Diponegoro University can simulate several work operations on turning machines as well as the introduction of tools to the work process bench. In addition to these aspects, applications that have been designed also do not provide a virtual environment that has complete equipment such as in a production system laboratory. Another drawback is that there is no usage guide in the form of a video tutorial for an initial understanding of users when operating a work operation on a turning machine. In practice, mistakes still occur frequently in working on turning machines. According to students, errors often occur due to a lack of understanding from students themselves even though virtual reality tools have been provided in the training of turning machine operations. Based on that problem this research tries to develop virtual laboratory using the Requirement Engineering method and HTC VIVE Pro Full Kit hardware that can correct the shortcomings of the previous development. The output of this research is a virtual reality-based virtual laboratory application that has been tested and evaluated by considering several aspects such as interaction, visualization, information, reusability, performance, and reliability.

1. Introduction

Education is important for the progress of a nation, therefore, to become a great nation, good education technique is certainly being considered as a way to achieve the goal for every country in the world. Along with the times, the world of education must innovate quickly and integrated [1]. Indonesia plans to overhaul the education curriculum with more emphasis on STEAM (Science, Technology, Engineering, the Arts, and Mathematics), aligning the national education curriculum with future industrial needs [2]. The industrial revolution 4.0 is a concept of automation carried out by machines without the need for human labour in its application. Diponegoro University (Undip) as one of educational institutions in Indonesia try to implement it. The technology applied in each faculty is different, depending on the needs of the lecture activities. Industrial Engineering Department has conducted learning using technology, as an example preliminary practices that held by production system laboratories which intended as training media before conducting practicum manufacturing processes using virtual laboratory applications based on virtual reality. The virtual laboratory application that has been designed can simulate several work operations on turning machines which include zero, facing, conventional, and cut off settings, as well as the introduction of tools to the bench work process. The application has been designed to meet several aspects such as visualization and information, animation, reliability, and reusability. In addition to these aspects, applications that have been designed also do not provide a virtual environment that has complete equipment such as in a production system laboratory. Another drawback is that there is no usage guide in the form of a video tutorial for an initial understanding of users when operating a work operation on a turning machine [3]. In carrying out practical activities, errors still often occur in working on turning machines. There are 57.9% of students making mistakes on turning machines with the total number of respondents being 75 students. According to students, mistakes often occur due to a lack of understanding from students themselves in operating the turning machine. Although there are virtual reality tools in the training of turning machines, there are still many mistakes in turning machines. Based on the problems and conditions that have been explained, an improvement in the development of a virtual laboratory will be carried out using methods and hardware that can correct the shortcomings of the previous development. The method used in developing virtual laboratory applications in this study is the requirements engineering method. Requirements engineering is a systematic and disciplined approach to the specification and management of requirements to identify relevant needs, reaching an agreement between stakeholders regarding these requirements, documenting them according to given standards, and managing them systematically and also to understand and document their wants and needs. stakeholders, determine and manage requirements to minimize the risk of delivery of the system that does not meet the wants and needs of the stakeholders [4]. The output of this research is a virtual reality-based virtual laboratory application that has been tested and evaluated by considering several aspects such as interaction, visualization, information, reusability, performance, and reliability.

2. Literature Review

This section explains the literature used in designing a virtual laboratory based on virtual reality. Some important literature that will be used will be briefly explained as follows:

2.1. Human-Computer Interaction Framework

Human-computer interaction (HCI) is a discipline that examines communication or interaction between users and systems. Users are individuals or groups that work together, the order of users in an organization, each of which deals with some part of the task or process [5].

2.2. Virtual Laboratory Based on Virtual Reality

A virtual laboratory is an interactive environment for creating and conducting simulation experiments consisting of dependent domains of simulation programs, experimental units that include data files, tools that operate on objects, and reference books [6]. The virtual laboratory simulates the device, including all necessary parameters, so students can work in a complete and realistic environment. The level of reality that is simulated and the software used are different. Some show temporary two-dimensional images that use virtual reality to simulate real [7].

2.3. Requirement Engineering

Requirements engineering is a systematic and disciplined approach to the specification and management of requirements to identify relevant needs, reaching an agreement between stakeholders regarding these requirements, documenting them according to given standards, and managing them systematically and also to understand and document their wants and needs. stakeholders, determine and manage requirements to minimize the risk of delivery of the system that does not meet the wants and needs of the stakeholders [4]. The Requirements Engineering method consists of four phases as follows Requirements Elicitation, Requirements Documentation, Requirements Validation and Negotiation, and Requirement Management [4].

2.4. Unity 3D Game Engine

Unity 3D is a cross-platform based game engine. Unity can be used to make a game that can be used on computer devices, Android smartphones, iPhone, PS3, and even X-BOX. Unity 3D is a software designed to create or develop applications with 2D and 3D views. Unity 3D has the function of making 3D objects like real objects that are affected by gravity, colliding, sound, script, animation, artificially intelligent, network, threading, and animated graphics [3].

2.5. HTC Vive Pro
HTC VIVE Pro Full Kit is a VR headset developed by an American company, HTC Corporation. This company is seen as Oculus's main competitor in the VR entertainment field (especially in PC games). The goal of HTC is to attract users of users who are interested in users in virtual collaboration and design of products and other applications. The HTC VIVE Pro Full Kit combines sensors and translates them into motion in the VR system in Real-time. HTC VIVE Pro Full Kit has a precise tracking feature to use two controllers that are used to facilitate user interaction and two base station 2.0 which are used to monitor room 5m x 5m where the user is located. HTC VIVE Pro Full Kit has input specifications such as a multifunction trackpad, grip buttons, dual-stage trigger, system button, and menu button. HTC VIVE Pro Full Kit also has sensors such as SteamVR Tracking, G-sensor, gyroscope, proximity, and IPD sensor. Also, the HTC VIVE Pro Full Kit has a connection using Bluetooth, and a USB type C port to connect to the computer.

2.6. Evaluation Method
Evaluation is the final stage of the design process that used to validate the results of the design needs, desires, and expectations of users. In designing virtual reality applications, evaluation is included in the design phase that must be done. The main purpose of the evaluation is to measure the suitability of the product. In digital products such as virtual reality, there have been a variety of previous studies using various aspects of the evaluation conducted. In previous studies, we have used several aspects such as visualization and information, animation, reliability, and usability [3]. Evaluation methods in the form of filling out questionnaires to several questionnaires using the scale of Strongly Agree (SS), Agree (S), Disagree (TS), and Highly Disagree (STS). The results of the questionnaire are then processed and generalized to get the conclusions of the evaluation results [3].

3. Research Design
3.1. Research Objective
This study aims to: (1) identify and develop user persona, user goals, and user requirements in the virtual laboratory application to be designed; (2) creating a virtual laboratory application based on user persona, user goals, and user requirements compiled using Unity 3D; (3) explain the process of integrating virtual laboratory applications in virtual reality using the HTC Vive Pro Full Kit Headset; and (4) evaluate the developed virtual laboratory applications.

3.2. Research Instrument
This research was conducted in the Industrial Engineering Department at production system laboratory of Diponegoro University. This study involved industrial engineering students and production systems laboratory assistants as users, Lecturers of the Department of Industrial Engineering, ergonomists, and software experts as an expert. This design includes depicting a production system laboratory situation and providing simulations of turning machine operations. This research was carried out by three other industrial engineering students who were distinguished by the type of turning machine operation. In this study, turning machine operations consist of boring operations, drilling operations, reaming operations, and tapering operations.

3.3. Research Method
This research uses the Requirements Engineering method which is used to design the needs and design of the virtual laboratory that will be made and consists of four phases, namely beginning with requirements elicitation, requirements documentation, requirements validation and negotiation, and requirements management. The following research schemes will be conducted as shown in Figure 1.

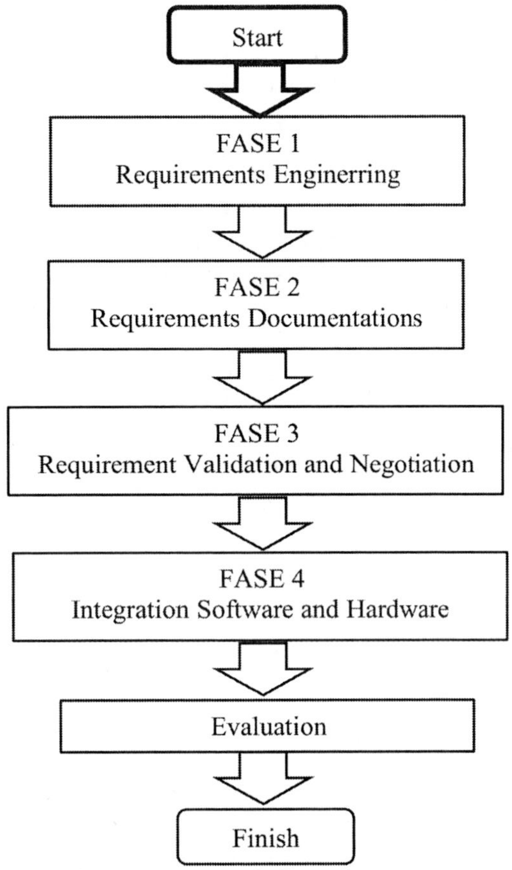

Figure 1. Research flow diagram.

3.3.1. Elicitation requirements. In the requirements elicitation phase, there are two activities namely observation and interview. At the observation, stage to see the user's behavior and the actual situation of the laboratory directly in carrying out work processes that occur and identifying and seeing potential hazards that might occur, as well as identifying and looking at the preferences, errors, and frequency of users in carrying out a process. At the interview stage or interviews conducted on experts and users. Interview the user to identify the goals and motivations of the user, identify the needs, desires, and expectations of users, get the user's aspirations, identify the level of problems and frustration of users. User interviews were conducted with ten students and five laboratory assistants. Interviews with experts were conducted to obtain detailed information related to the field of expertise needed in designing a virtual laboratory. Interviews of experts were carried out on ergonomists, software experts, virtual reality experts.

3.3.2. Requirement documentation. In the requirements documentation phase in this study using the results of the requirements elicitation phase. In this phase, the needs and information that must be produced by the application both functional and non-functional needs have been obtained. These needs are obtained based on observations that have been made in the previous stage. In this phase, the user model and domain user are created. The user model is a model of the user that will be the basis for designing a Virtual laboratory. In this phase also a confession of needs has been determined which is translated into a design that is following the principles of interaction. At this stage, it produces an interaction framework in the form of sequential and more detailed steps. Use cases are also used to model learning that must be done by the practitioner. Activity diagrams are used to frame the application activation process. Sequence diagrams are used to create information exchanging interactions between users and systems. After that, the design of the framework is made with visuals that are determined by several things related to the color, type, widget treatment, and overall dimensions, as well as the material properties of the interface used to achieve the goals of the user following predetermined needs. Then the existing design is made into 3 dimensions related to the work operation of the turning machine and a virtual environment will be made according to the actual conditions of the laboratory.

3.3.3. Requirements validation and negotiation. In the requirements validation and negotiation phase, the validation of the requirements that have been determined, and the refinement of the design concept that has been done are carried out. The framework that has been created will be evaluated and converted into the final design form used in the virtual laboratory. If there is a part that is not following the output of the previous stage, it will be revised from the part that is not appropriate.

3.3.4. Requirements management. This phase is the last phase of the requirements engineering method. In this phase, a learning application is made on the working of a lathe in the form of a virtual laboratory based on virtual reality. The software used in designing virtual laboratories is to use Blender and Unity 3D. making applications based on the final design that has been generated from the previous phase.

3.3.5. Application and hardware integration. The application design that has been done previously will produce virtual laboratory software. Integration needs to be done between software and hardware devices HTC Vive Pro Full Kit in the world of virtual reality. The results of the installation will then be used for the trial process as an initial step in the evaluation process of the virtual laboratory product.

3.3.6. Evaluation. The evaluation phase carried out with aspects that see the aspects used in the design. The aspects of evaluation consist of functional and non-functional aspects. Functional aspects include visualization, interaction, and information. While the non-functional aspects consist of aspects of performance, reliability, and reusability.

Evaluation is carried out by respondents consisting of users and experts. Users consist of ten industrial engineering students and five production system laboratory assistants. The experts consist of lecturers from the undip industrial engineering department, ergonomics experts, and software experts. The evaluation begins with conducting a virtual laboratory trial. Then the respondent is asked to fill out a questionnaire that refers to the evaluation aspects that have been determined. The questionnaire was filled out using a Likert scale consisting of Strongly Agree (SA), Agree (A), Neutral (N), Strongly Disagree (SD), and Disagree (D). Data processing is done by calculating the number of answer choices from respondents consisting of Strongly Agree (SA), Agree (A), Neutral (N), Strongly Disagree (SD), and Disagree (D) on each aspect of the assessment. The calculation results are then processed into a proportion of answers to each aspect into diagrammatic form to facilitate the reading of the information contained. Analysis of the evaluation results is based on the results of data processing. The analysis is done by comparing the proportion of answers from each aspect of the assessment. The assessment aspect is considered to have fulfilled the user goals and requirements if it has a higher proportion of the responses of the Agree (SA) and Agree (A) compared to the total proportion of the Neutral (N), Strongly Disagree (SD), and Disagree (D) answers.

4 Conclusion

This research is intended to develop a virtual reality-based virtual laboratory using the requirements engineering method in the production system laboratory. This development is based on the lack of interaction aspects in previous research. The development of virtual laboratory applications is oriented towards user goals in defining functional and non-functional requirements. Applications that have been installed in the HTC Vive Pro are used to conduct virtual laboratory evaluations by users and experts.

References

[1] Nurdyansyah N, Rais P, Aini Q. The Role of Education Technology in Mathematic of Third Grade Students in MI Ma'arif Pademonegoro Sukodono. *Madrosatuna: Journal of Islamic Elementary School.* Muhammadiyah University Sidoarjo; 2017 Dec 7;**1**(1):37.

doi.org/10.21070/madrosatuna.v1i1.923
[2] Prianto Y, Subaidah, Rohmah Z, Firdaus F, Analisis Pendidikan Indonesia di Era Revolusi Industri 4.0. Universitas Muhammadiyah Sidoarjo; 2019
[3] Nurkertamanda D, Saptadi S, Widharto Y, Maliansari AN. Feasibility Evaluation: Virtual Laboratory Application Based on Virtual Reality for Lathe Engine Training Simulation. *2019 6th Int. Conf. Front. Ind. Eng*. IEEE; 2019 Sep. doi.org/10.1109/icfie.2019.8907772
[4] Pohl K, Rupp C. *Requirements Engineering Fundamentals: A Study Guide for the Certified Professional for Requirements Engineering Exam*, 2nd ed. IREB Rocky Nook Inc; 2015
[5] Fix A, Finlay J, Abowd GD, Beale R. *Human-Computer Interaction (3rd ed.)*. Pearson Education; 2004
[6] Gall MD, Gall JP, Borg WR. *Educational Research: An Introduction* (7th Edition). Allyn & Bacon; 2003
[7] Budhu M. Virtual Laboratories for Engineering Education. *Int. Conf. Eng. Ed*. 2002:1–6

The virtual laboratory for turning machine operations using the goal-directed design method in the production system laboratory as simulation devices

Y Widharto[1,2], M R Tahqiqi[1,3], D Nurkertamanda[1,4], S Saptadi[1,5], P A Wicaksono[1,6]

[1]Industrial Engineering Department, Engineering Faculty, Diponegoro University

E-mail: [2]yudidito@gmail.com, [3]mohrizalt@gmail.com, [4]nurkerta@gmail.com, [5]singgihs@ft.undip.ac.id, [6]nawanadi@gmail.com

Abstract. Virtual laboratory is an interactive learning media that is more effective than conventional media. The implementation and development of the virtual laboratory technology was carried out in the production system laboratory of the Industrial Engineering Department. This is an improvement effort from previous research that still has some shortcomings, especially in aspects of user interaction with the virtual environment. This development is increasingly important because of the frequency of human error when practical activities are running. The development of this virtual laboratory follows the ADDIE model which includes the process of analyzing, designing, developing, implementation, and evaluation. The method for developing virtual laboratories is carried out using the principle of Goal-Directed Design which focuses on user goals. GDD consists of several phases, namely research, modeling, requirements definition, framework definition, refinement, and development support. The output of this method is a virtual laboratory application that has been created with a unity 3D game engine and will be installed on HTC Vive Pro hardware. Evaluation is carried out using a Likert scale by considering aspects of evaluating functional and non-functional requirements. Functional aspects consist of aspects of interaction, visualization, and information. Non-functional aspects consist of aspects of usability, performance, and reliability.

1. Introduction

The development of the world nowadays is in the era of the Industrial revolution 4.0 with a massive role of technology. The technology development has positive impacts in education [1]. The latest technology which often used in educational activities is virtual reality. Diponegoro University as an instrument of Higher Education seeks to adapt to technological developments. One of the efforts taken is to develop a machine training simulation media using virtual laboratory based on virtual reality in the Industrial Engineering Department. The design of the Virtual Laboratory application is carried out in order to improve the quality of learning in the field of the manufacturing process. Based on preliminary research that has been done by observing and filling out the questionnaire, it shows that the frequency of human error in students when operating machines is high. This is due to the lack of students' understanding of how to operate the machine correctly. The analysis shows that conventional machine operation training is not effective. This is because there are various constraints on the resources such as time, machines and operators. Based on the drawbacks of using conventional methods, it is necessary to develop training methods that are more modern and effective. Virtual laboratory based on virtual reality is a technology that can be used to create a modern training media.

The development of a virtual laboratory in the production system laboratory of the Industrial Engineering department was carried out in 2019, but it was lacked visualization and interaction. Aspects of functional requirements number four and five relating to aspects of user interaction and virtual environment are not designed according to actual conditions. This is due to limitations on the method and VR Headset used. The development of the virtual laboratory was carried out to improve and enhance the capabilities of the virtual laboratory that had been previously designed. Using the Goal-Directed Design (GDD) method as the design method and the HTC Vive Pro Full kit as the hardware to run the program. The GDD method design based on user needs on the defined persona. The GDD method consists of several stages such as research, modeling, requirements definition, framework definition, and design support [2]. The output of this research is a virtual laboratory based on virtual reality application that has been tested and evaluated by considering several aspects such as interaction, visualization, information, usability, performance, and reliability.

2. Literature Review

This section explains the literature used in designing a virtual laboratory based on virtual reality. Some substantial literature that will be used will be briefly explained as follows:

2.1 Human Computer Interaction Framework

Human-computer interaction (HCI) as a discipline that concentrates on the design, implementation, and evaluation of interactive computer systems for human use with various particular fields of study [3]. There are three key aspects of HCI, which are users, systems and interactions.

2.2 Virtual Laboratory Based on Virtual Reality

A virtual laboratory is a media that supports traditional practice activities. The presence of a virtual laboratory is expected to provide an effective, efficient, and enjoyable training facility so that students can learn independently and flexibly [4]. The virtual laboratory obtains several advantages in its use such as more complete experience, detailed understanding, low cost, and assisting in real-time [5]. The design of the virtual laboratory is carried out by taking into account design strategies such as Goal Oriented, Learning by Doing, Interactive Environment and Flexibility System.

2.3 Goal Directed Design

Goal-Directed Design (GDD) is an interaction design method [6]. The GDD method consists of six phases: research, modeling, requirements definition, framework definition, refinement, and development support [2]. The GDD method is oriented towards the right target users with the persona model identified at the research stage. gaps between the research and design phases frequently, the GDD method bridges this gap by defining user models, building user needs, and translating them into a framework [2]. The Goal-Directed Design method uses a persona in the scenario and develops a design sketch by following the flow through the interface and applying design principles and patterns to construct a solution [6].

2.4 Unity 3D Game Engine

Unity 3D is a software that was developed to produce an object or application using 2D and 3D views but is more focused on 3D. Unity 3D can be used to provide tangible effects such as the effects of gravity, collisions, sound, scripts, animations, networks, streaming, memory management, threading, artificial intelligence, and motion animation [7].

2.5 HTC Vive Pro

HTC Vive Pro is a VR Headset that was developed by an American technology company, HTC. HTC Vive Pro Full Kit combines sensors and calculates user movements and translates them into VR system movements in Real-Time [8]. This VR headset is supported by using two controllers used to facilitate user interaction and two base station 2.0 which are used to monitor the room where the user is placed. The controller has input specifications such as multifunction trackpad, grip button, dual-stage trigger, system button, and menu button. This hardware contains sensors like Steam VR Tracking, G-sensor, Gyroscope, Proximity and IPD Sensor.

2.6 Evaluation Method

Evaluation is a stage in the design process used to validate the design results with the needs, desires, and expectations of users. Evaluation is critical parameter as the feasibility of a design result. In digital products such as virtual reality, there have been a variety of preliminary studies using various aspects in the evaluation phase. Some aspects often used are visualization, user friendly, learnability, usability, interaction, and effectiveness. Nurkertamanda et al used an evaluation method in the form of filling out questionnaires with several respondents consisting of experts and users. Respondents were asked to fill out a questionnaire using the scale of Strongly Agree (SS), Agree (S), Disagree (TS), and Strongly Disagree (STS). The results of the questionnaire are then processed and generalized to approach the conclusions of the evaluation results.

3. Research Design

3.1 Research Objective

This study aims to: (1) identify the user's persona, user's goal, and user's requirements; (2) create a virtual laboratory application using Unity 3D; (3) integrate applications with the VR Headset HTC Vive Pro; and (4) evaluating virtual laboratory applications.

3.2 Research Instrument

This research was conducted in Production Systems Laboratory of the Industrial Engineering Department of Diponegoro University. Involve several stakeholders to build the basic framework of design, application development, and evaluation. Stakeholders consist of two categories, users and experts. Users consist of industrial engineering students and production systems laboratory assistants. The experts consisted of three lecturers from the Industrial Engineering department, ergonomics expert, software expert, and virtual reality software development expert. The design of a virtual laboratory involves depicting a complete production system laboratory and providing operating simulations for several turning machine operations. This research is a project carried out by three industrial engineering students which are distinguished by the type of turning machine operations performed. In this research, a turning machine operation consisting of facing operation, conventional operation, contour operation, and threading operation will be carried out.

3.3 Research Method

This study combines the principles of Goal-Directed Design and the principles of ADDIE (Analyse, Design, Develop, Implementation, and evaluation) models [9]. Therefore, this research will be divided into 5 stages of work steps. Each stage will contain design principles which are part of the Goal-Directed Design. This is done to produce virtual laboratory applications based on virtual reality that are in accordance with the needs and rules of product development. The following research schemes will be conducted:

3.3.1 Stage 1: Analyse

The analyse stage consists of the research phase, persona modelling, and requirements definition. The research phase uses ethnographic studies by carrying out several steps aimed at gathering user and system information. The first step undertaken is a review of the development of virtual laboratories that have been carried out previously and several similar studies to gather information about the features, functions, designs, and interactions of virtual reality. Then the user was observed, and the focus group discussion consisted of ten students and five laboratory assistants. This step is carried out to identify user behaviour, user character, goals and motivation, and identify user requirement. The final step in the research phase is conducting interviews with experts to get detailed information in the perspective of virtual reality systems. The Output Research Phase is the pattern of behaviour and workflow of the user. The results of the Research Phase then become input into the Modelling Phase, modelling virtual laboratory users who are designed based on the information obtained. Output Modelling Phase is the User Model and Domain model that explains the workflow diagram of the persona that has been designed. The end of the analyse phase is defining the functional and non-functional requirements of the person. This phase will compile functional requirements, non-functional requirements, and scenario of the virtual laboratory based on information gathered in the research phase.

3.3.2 Stage 2: Design.
The Design Stage consists of the Framework Definition and Refinement phases. The Framework phase designs the information obtained at the analyse stage into several design frameworks, Interaction Framework (IF), Visual Design Frame (VDF), and Industrial Design Framework (IDF). Interaction framework will model the interaction between the user and the system by using several diagramming tools such as use case diagrams, activity diagrams, and sequence diagrams. The design of the framework is based on scenarios and work patterns of users when interacting with work systems in the production system laboratory. VDF defines design guidelines related to visualization of real environments in a virtual environment. Some things to consider include the colour, dimensions, widget treatment, and material properties of various elements of the virtual environment. IDF is carried out to model the equipment related to turning machine operations into a 3D form. The output refinement Framework phase is a design framework that matches the behaviour patterns and actual environmental conditions. The output of this phase will be refined in the refinement phase. The refinement phase will produce a final framework design from the virtual laboratory as input to Software Development at the develop stage.

3.3.3 Stage 3: Develop
This stage contains the final phase of the principle of Goal-Directed Design, Development Support. At this stage a virtual laboratory application will be made based on the final framework design. Making an application is done using the Unity 3D game engine. The output from this stage is the application of a virtual laboratory turning machine based on virtual reality.

3.3.4 Stage 4: Implementation
This stage is a step to integrate virtual laboratory software with HTC Vive Pro hardware. This process is carried out by performing an installation which is followed by a trial of the tool by users and experts. Each respondent will try to use a virtual laboratory to conduct an evaluation at a next stage.

3.3.5 Stage 5: Evaluation
The evaluation model is carried out by considering the aspects design. Evaluation aspect for developing virtual laboratory consists of two parts, such as functional needs aspects and non-functional needs aspects. The following are aspects that will be evaluated:

Evaluation aspects of functional needs consist of interaction aspects, visualization, and information. Interaction, an aspect that evaluates the suitability of interactions in a virtual laboratory application with actual interactions in the real world. Visualization, aspects that appraise the level of visualization from various content, assets, and graphics in a virtual laboratory application. Information, aspects that appraise the level of information availability required by users in using virtual laboratory applications. The evaluation aspects of non-functional needs consist of usability aspects, performance, and reliability. Usability, aspects that appraise the level of user's convenience to learn, understand, operate, and interact with virtual laboratory applications [10]. Performance, an aspect that evaluates the pace of response for virtual laboratory applications. Reliability, the aspect for evaluating the level of system completeness in providing services when it's needed by users and the reliability of virtual reality applications in running tasks in the system that mainly related to errors that occurs.

Evaluation is carried out by respondents consists of users and experts. users consist of ten industrial engineering students and five assistant of production system laboratory. Expert respondents consisted of three Industrial Engineering department lecturers, ergonomics expert, software expert, and virtual reality development expert. The evaluation steps begin with conducting a virtual laboratory trial. Then Respondents are asked for completing questionnaires with a total of about 30 questions. The questionnaires questions refer to evaluation aspects that have been determined before. The questionnaire was filled out using a Likert scale that consist of Strongly Agree (SA), Agree (A) , Neutral (N), Strongly Disagree (SD) , and Disagree (D) . Data processing is done by calculating the number of respondents' choices that consists of Strongly Agree (SA), Agree (A), Neutral (N), Strongly Disagree (SD), and Disagree (D) in each aspect of assessment. The results of the calculation then processed to an aspect answers proportion into diagrammatic form to make contain information reading easier. Analysis of the evaluation results is based on the results of data processing. The analysis is done by comparing the proportion of answers from each the assessment aspects. The assessment aspect that fulfil users goals and requirements are an aspect that has a higher response proportion of the Agree (SA) and Agree (A) than the total proportion of answers Neutral (N), Strongly Disagree (SD), and Disagree (D).

4. Conclusion

This research is aimed for developing a virtual laboratory based on virtual reality by using the goal-directed design method in the production system laboratory. This development is based on the lack of interaction aspects in previous research. The development of virtual laboratory applications is oriented towards user goals in defining functional and non-functional requirements. Applications that have been installed in the HTC Vive Pro are used to conduct virtual laboratory evaluations by users and experts.

REFERENCES

[1] Hanson, K. B. E. S. 2008 Design and Development of Virtual Reality: Analysis of Challenges Faced by Educator *Journal of Educational Technology & Society*, 11(educational technology), 15.

[2] Cooper, A., Reimann, R., and Cronin, D. 2007 *About Face 3: The essentials of interaction design. In Information Visualization* (Vol. 3). https://doi.org/10.1057/palgrave.ivs.9500066

[3] Keraminiyage, K., Amaratunga, D., and Haigh, R. 2009 A Human-Computer Interaction Principles Based Framework To Assess the User Perception of Web Based Virtual Research Environments *International Journal of Strategic Property Management*, Vol. 13, pp. 129–142. https://doi.org/10.3846/1648-715x.2009.13.129-142

[4] Puspita, R., and Yamin, M. 2008 Sistem Informasi Aplikasi Virtual Lab *Proceeding, Seminar Ilmiah Nasional Komputer Dan Sistem Intelijen* (KOMMIT 2008), (Kommit), 190–198.

[5] Mosterman, P. J., Dorlandt, M. A. M., Campbell, J. O., Burow, C., Bouw, R., Brodersen, A. J., and Bourne, J. R. 1994 Virtual Engineering Laboratories: Design and Experiments *Journal of Engineering Education*, 83(3), 279–285. https://doi.org/10.1002/j.2168-9830.1994.tb01116.x

[6] Halley, L. 2019 *Six Sigma and goal-directed design*. Retrieved from www.cooper.com website: https://www.cooper.com/journal/2008/05/six_sigma_and_goaldirected_des/

[7] Nurkertamanda, D., Saptadi, S., Widharto, Y., and Maliansari, A. N. 2019 Feasibility evaluation: Virtual laboratory application based on virtual reality for lathe engine training simulation. *6th International Conference on Frontiers of Industrial Engineering*, ICFIE 2019, 6–10. https://doi.org/10.1109/ICFIE.2019.8907772

[8] Lynch, G. 2019 *Oculus Rift S review*. Retrieved from Techradar.com website: https://www.techradar.com/reviews/oculus-rift-s-review

[9] Moradmand, N., Datta, A., and Oakley, G., 2014 The Design and Implementation of an Educational Multimedia Mathematics Software: Using ADDIE to Guide Instructional System Design, *The Journal of Applied Instructional Design*. Volume 4 · Issue 1 https://www.jaid.pub/vol-4-issue-1-2014

[10] Silva, F., Lucena, M., and Lucena, L., 2013 STREAM-AP: A process to systematize architectural patterns choice based on NFR, *3rd International Workshop on the Twin Peaks of Requirements and Architecture (TwinPeaks)*, DOI: 10.1109/TwinPeaks-2.2013.6617356

Virtual laboratory application based on virtual reality simulation as training tool of turning machine using goal-directed design method

D Nurkertamanda[1], F Frendiansyah[1]*, S Saptadi[1], Y Widharto[1], P A Wicaksono[1]

[1]Diponegoro University, Semarang, Indonesia

Email: *ffrendiansyah@gmail.com

Abstract This study aims to develop a virtual laboratory application based on virtual reality simulation as a training tool for turning machines. The application is built as a solution for manufacturing engineering education problems at the Production System Laboratory (PSL) in the Industrial Engineering Department, Diponegoro University. Problem based on work accident data shows human error from the student when using turning machines. The data shows that lack of student understanding due to insufficient of preliminary training at the beginning of manufacturing engineering education especially physical skills training. Virtual Reality Simulation is chosen because conventional machining training has some restrictions in execution. Furthermore, virtual reality simulation can reduce cost, improve safety, and add flexibility along with advances in computer technology. For this Research the Goal-Directed Design (GDD) method was chosen to develop VR simulation. This method is used as a procedure to accomplish the good design of application which including research, modelling, requirements, framework, refinement, and support. The application is built using Unity 3D software and run on Oculus Quest hardware. At the evaluation phase for assessing the needs, researchers use functional and non-functional requirements. Functional requirements consist of interaction, visualization, and information. Non-functional requirements consist of usability, performance, and reliability.

1. Introduction

Physical skills training for manufacture engineering education is crucial [1] [2] [3]. It also happened at Industrial Engineering Department, Diponegoro University that one the educational purpose activity is to improve the physical skills and the understanding of students. One of the activities for improving physical skill was held in Production System Laboratory (PSL) which consist of preliminary training, practicum implementation, and evaluation [4]. At practicum implementation in 2019, there are 55.7% of students that making an error at the turning machines, where 60.5% of the error was caused by lack of understanding. This lack of understanding problem happened by lack of preparation. Nowadays, the preparation is using conventional method that have several restrictions as lack of facilities, limited time, less of funds, and safety factors [1] [2]. Along with the development of human-computer interaction technology, there is a possible solution that can be chosen to develop a tool based on virtual reality training simulation as a virtual laboratory application [5]. Moreover, study about the implementation of virtual reality technology for higher education especially manufacturing education is still rare, even though it has a lot of potentials that can be developed [6].

This study is research development from [4]. In previous, the development of a virtual laboratory application is using the User-Centered Design (UCD) method, built at Unity software, and run at Oculus Go hardware. The result of the study show limitation such as limited interaction, and less of immersive. This problem happened because the hardware has some limitations and the visual is not too immersive. Thus, this study is focused to improve the shortages of earlier study by redesign process using the Goal-Directed Design (GDD) method as a procedure to accomplish a good design of the application. The application is developed at Unity 3D software and run at Oculus Quest hardware. Then, evaluation phase is based on functional and non-functional requirement aspects. Functional requirement aspects consist of interaction, visualization, and information. Non-functional requirement aspects consist of usability, performance, and reliability. Figure 1 explains the goal-directed design (GDD) method flowcharts.

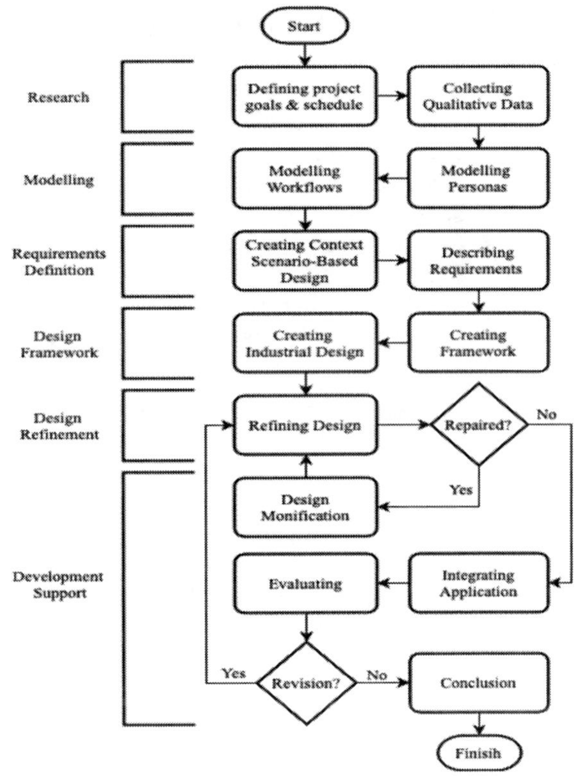

Figure 1. The Goal-Directed Design (GDD) method flowcharts [7].

2. Design Method

The Goal-Directed Design (GDD) method is used in this research to accomplish the good design that meets the need and the goal of users. The method has six phases shown in Figure 1: research, modelling, requirements definition, framework definition, refinement, and development support [7].

2.1. Research

In this phase, qualitative data is collected by researcher from users, stakeholders, and subject-matter expert (SME). It aims to find the scope of research include goal and schedule, explore the concept of design, workflow, also understand the user needs and behaviour.

2.2. Modeling

In this phase, researchers generate user models through a persona, that is representative of the type of users based on qualitative data in the research phase. It is useful for knowing goal and behaviour of user to define the design requirements [7]. The persona sorted to three types: high user persona, middle user persona, and low user persona, that can be seen in Table 1. The high user persona is the target of design and the others is the complement.

IOP Conf. Series: Materials Science and Engineering 1072 (2021) 012077 doi:10.1088/1757-899X/1072/1/012077

Table 1. The user personas.

High User Persona	Middle User Persona	Low User Persona
	Goal	
Understand the workflow in the laboratory and can operate the turning machine in more detail.	Understand the workflow and environment in the laboratory and can operate the turning machine in more detail.	Understand the workflow in the laboratory, can operate the turning machine in more duration, and obtain safety induction.
	Behaviour	
Feel comfortable alone, experience and serenity are affecting self-confidence, feel safe when obeying the rules, less duration has an impact on feeling pressured, and silent for thinking if make an error.	Feel comfortable with some of others, experience is most affecting self-confidence, keep feel safe when disobedient the rules, the assistant has an impact on feeling pressured, and just silent if make an error.	Feel comfortable with many of the people, the experience and friends are most affecting self-confidence, keep feel safe when disobedient the rules, the environment has an impact on feeling pressured, and panic if make an error.
	Attitude	
Has motivated by watching videos, always obey the rules, and feel having difficulties due to gap of knowledge.	Has motivated by learning with others, obey the rules, and feel having difficulties due to nervousness.	Has motivated by reading the books, not to obey the rules, and feels having difficulties due to nervousness.
	Aptitudes	
Requires instructors to give detailed explanations, easy to understand with audio-visual, skilled enough in motoric.	Requires instructors to avoid mistakes, easy to understand visually, skilled in motoric.	Requires instructors to give examples, easy to understand visually, less skilled in motoric, have self-control with self-suggestion.
	Mental Models	
Obtain in more detail information, can operate machines longer, need an aid media to understand like videos, and a conducive work environment.	Obtain in more detail information, can operate machines longer,	Need an aid media to understand like videos, a conducive work environment, and a good implementation of safety rules.

Other model that used by researchers is workflow model. Shown in Figure 2.

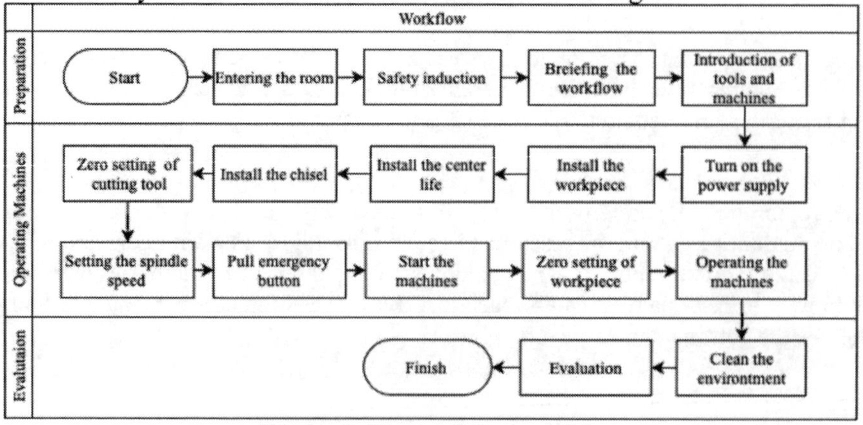

Figure 2. The workflow model.

2.3. Requirement Definition

In the requirement definition phase, the persona that created in the modelling phase positioned as an actor in the context scenario-based design. The scenario aims to explore and describe how users can accomplish their tasks according to goals and needs [7]. After creating the context scenario-based design that use the persona as an actor, researchers define the requirements. The requirements explained as a description of various services and restrictions that can be given by the application system [8]. In this study, researchers use functional requirements and non-functional requirements. Functional requirements consist of visualization, information, and interaction. Then, the non-functional requirements consist of usability, performance, and reliability. The requirement is used to develop the system design and the evaluation of the design.

2.4. Framework Definition

In the framework definition phase, the requirements are transformed to framework design. There are 3 framework design: interaction framework, visual design framework, and industrial design framework [7]. Interaction framework is created using the Unified Model Language (UML). UML is notation based on object as standard to describe design object-oriented software [9]. In this study, that helps to build the design framework in more detail according to requirement definition. Also, interaction is described using a diagram, consist of use case diagram for represents the system and the user behaviour, activity diagram for represents the system and the user workflows, and sequence diagram for represents the system interaction base on a time sequence.

After designing the interaction framework, the unified model language is generated to the visual design framework. Consist of menu design, scene, and information dialogue. Figure 3 show that the user is interacting with the menu of briefing the workflow as a scene. Also, Figure 4 show that there are error messages in the system.

The entire visual design framework is united as an industrial design framework shown in Figure 5.

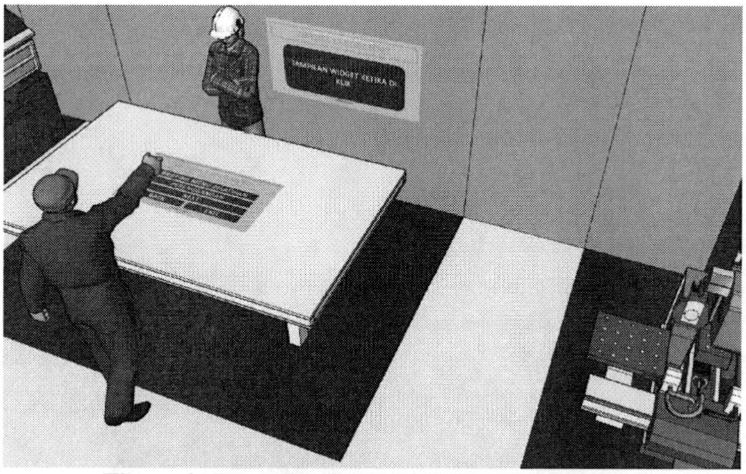

Figure 3. Example of scene briefing the workflow.

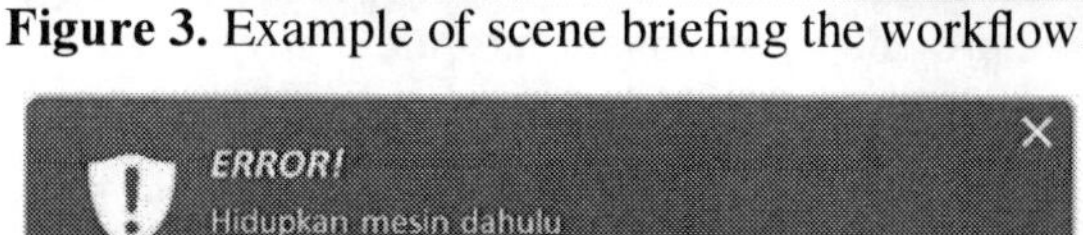

Figure 4. Example of error information dialogue design.

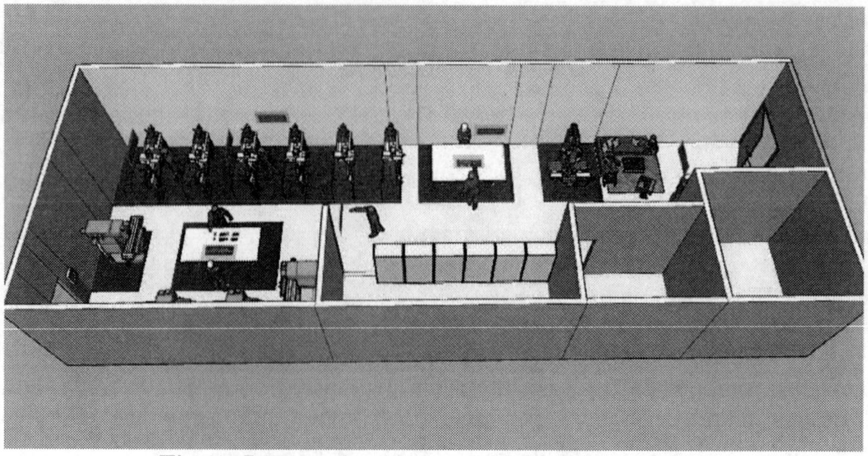

Figure 5. Example of industrial design framework.

Table 2. The result of the validation shown.

Attribute	Value	Category
The design provides a good impression of the production system laboratory environment and the equipment. (Visualization)	82.5	Strongly Agree
The design provides a detailed explanation of the workflow, safety, turning machines operation and equipment through an assistant. (Information)	81.25	Strongly Agree
The design provides feature room-scaling and/or teleporting to reach a section of the laboratory room. (Interaction)	81.25	Strongly Agree
The design provides a choice of various turning operations processes for training. (Interaction)	80	Strongly Agree
The design provides animation and sound effect according to real conditions for feedback. (Visual)	76.25	Agree
The design provides a menu hierarchy for selecting various explanations and operating processes. (Interaction)	77.5	Agree
The design of the user interface allows being used of various users with a level of ability. (Usability)	80	Strongly Agree
The design has feedback such as error messages on the page and dialogs with status. (Usability)	76.25	Agree
The design supports various components or training sessions to meet the user experience. (Usability)	77.5	Agree
The design has tooltips that provide an explanation of the information on the designated object. (Usability)	80	Strongly Agree
The design has guidance for the user and procedures for operating and controlling the device. (Usability)	81.25	Strongly Agree

The questionnaire uses a Likert 5 scales that consist of Strongly Disagree, Disagree, Neutral, Agree, and Strongly Agree.

2.5. Design Refinement
The design refinement phase is carried out about validating and refining the design framework. There are two ways to validate the framework design, it can be formal or informal [7]. A formal way can use usability testing for users. While informal way can use a feed-back session to users. In this study, researchers use an informal way. It is because of some restrictions at the time and resources in this research. Thus, the informal way is chosen. Furthermore, validation is based on aspects including visualization, interaction, information, and also usability. These aspects are according to the requirements except performance and reliability. That because the performance and reliability aspects are related to the running application system in hardware. The result of the validation shown as Table 2.

2.6. Development Support
In the development support phase, the final framework design is generated to the application. The application of a virtual laboratory is built at Unity Game Engine. Also, the application is built as an android application to support the hardware used. The result of this phase is the application of a virtual laboratory based on virtual reality simulation.

3. Integration and Evaluation
3.1. Integration
Application integration is merging applications that have been made with Unity to the Oculus Quest hardware. Oculus Quest is chosen because the feature supports the design. Consist of two hand controller, head-mounted display, gyroscope, proximity, speaker, and others [10].

3.2. Evaluation
In the evaluation, applications that have been integrated with hardware will be tested by testing based on the requirement in requirement definition phase. The testing addressed to users consisting of ten students and five production system laboratory assistants. Also, the experts and stakeholders. In the beginning, user use the application of virtual laboratory. Then, user given a questionnaire according to the requirements. Similar in refinement phase, evaluation use Likert scale too. The questionnaire uses a Likert 5 scales that consist of Strongly Disagree (SD), Disagree (DA), Neutral (N), Agree (A), and Strongly Agree (SA). If there is revision after evaluating, then it is repaired. Else, the project is done.

4. Conclusion
In this study, the design process is using the Goal-Directed Design method where the fundamental is the personas. Step by step of this method generate the detail of the requirement and need from the user. That allows researchers to make a good design for users to accomplish their goals. Thus, the virtual laboratory application based on virtual reality simulation can be a solution to the problems that occur in manufacturing engineering education, especially physical training.

References
[1] Dela Cruz DR, Mendoza DMM. Design and Development of Virtual Laboratory: A Solution to the Problem of Laboratory Setup and Management of Pneumatic Courses in Bulacan State University College of Engineering. *2018 IEEE Games, Entertainment, Media Conference (GEM). IEEE*; 2018 Aug; doi.org/10.1109/gem.2018.8516467.
[2] Fang, X. D., Luo, S., Lee, N. J., & Jin, F. 1998 Virtual Machining Lab for Knowledge Learning and Skills Training. *Comput. Appl. Eng. Educ.*, ***6***(2), 89–97. doi.org/10.1002/(sici)1099-0542(1998)6:2<89::aid-cae4>3.0.co;2-w.
[3] Numfu M, Riel A, Noel F. Virtual Reality Based Digital Chain for Maintenance Training. *Procedia CIRP*. Elsevier BV; 2019;**84**:1069–74. doi.org/10.1016/j.procir.2019.04.268.
[4] Nurkertamanda D, Saptadi S, Widharto Y, Maliansari AN. Feasibility Evaluation: Virtual Laboratory Application Based on Virtual Reality for Lathe Engine Training Simulation. *2019 6th International Conference on Frontiers of Industrial Engineering* (ICFIE). IEEE; 2019 Sep;

doi.org/10.1109/icfie.2019.8907772.

[5] Ivanović M, Milićević AK, Aleksić V, Bratić B, Mandić M. Experiences and perspectives of Technology-enhanced learning and teaching in higher education – Serbian case. *Procedia Computer Science*. Elsevier BV; 2018;**126**:1351–9. doi.org/10.1016/j.procs.2018.08.086.

[6] Radianti J, Majchrzak TA, Fromm J, Wohlgenannt I. A systematic review of immersive virtual reality applications for higher education: Design elements, lessons learned, and research agenda. *Comput. Educ*. Elsevier BV; 2020 Apr;147:103778. doi.org/10.1016/j.compedu.2019.103778.

[7] Cooper A, Reimann R, Cronin D. *About Face: The Essentials of Interaction Design* (3 ed.). Indianapolis, IN 4652: Wiley Publishing Inc. 2007.

[8] Ganney, P. S., Pisharody, S., & Claridge, E. In Clinical Engineering: A Handbook for Clinical and Biomedical Engineers. *Softw. Eng.* 2013. doi.org/10.1016/B978-0-12-396961-3.00009-3.

[9] Pooley, R., & King, P. The unified modelling approach and performance engineering. *IEE Proc.-Sofw*. 1999;*146*(**1**), 2–10.

[10] Getting Started with your Oculus Quest. *Facebook Technologies.* 2019.

AUTHOR INDEX

Institute of Physics
Dirac House, Temple Back
Bristol BS1 6BE UK

ISSN: 1757-8981
ISBN 978-1-7138-4579-9

CURRAN ASSOCIATES INC.
proceedings.com